Architectural Drafting
for Interior Design

Architectural Drafting for Interior Design

Third Edition

Lydia Sloan Cline

FAIRCHILD BOOKS
NEW YORK · LONDON · OXFORD · NEW DELHI · SYDNEY

FAIRCHILD BOOKS
Bloomsbury Publishing Inc
1385 Broadway, New York, NY 10018, USA
50 Bedford Square, London, WC1B 3DP, UK
29 Earlsfort Terrace, Dublin 2, Ireland

BLOOMSBURY, FAIRCHILD BOOKS and the Fairchild Books
logo are trademarks of Bloomsbury Publishing Plc

First edition published 2008
Second edition published 2014
This edition published 2022

Copyright © Bloomsbury Publishing Inc, 2022

For legal purposes the Acknowledgments on p. xiii constitute an extension of this copyright page.

Cover design by Louise Dugdale

All rights reserved. No part of this publication may be reproduced or transmitted in any form or by any means, electronic or mechanical, including photocopying, recording, or any information storage or retrieval system, without prior permission in writing from the publishers.

Bloomsbury Publishing Inc does not have any control over, or responsibility for, any third-party websites referred to or in this book. All internet addresses given in this book were correct at the time of going to press. The author and publisher regret any inconvenience caused if addresses have changed or sites have ceased to exist, but can accept no responsibility for any such changes.

Library of Congress Cataloging-in-Publication Data
Names: Cline, Lydia Sloan, author.
Title: Architectural drafting for interior design / Lydia Sloan Cline.
Other titles: Architectural drafting for interior designers
Identifiers: LCCN 2020057255 | ISBN 9781501361197 (paperback) | ISBN 9781501361142 (pdf)
Subjects: LCSH: Architectural drawing—Technique. | Interior decoration—Design.
Classification: LCC NA2708 .C58 2022 | DDC 720.28/4—dc23
LC record available at https://lccn.loc.gov/2020057255

ISBN: 978-1-5013-6119-7

Typeset by Lachina Creative, Inc.
Printed and bound in India

To find out more about our authors and books visit
www.fairchildbooks.com and sign up for our newsletter.

Extended Contents

PREFACE	xi
ACKNOWLEDGMENTS	xiii

CHAPTER 1

Drafting and the Design Process — 1

What Is Drafting?	2
History of Manual Drafting	3
History of Computer Drafting	4
Professional Groups and Certifications	5
Design Thinking	6
The Design Process	7
Programming	7
Schematic Design	7
Bubble Diagrams	7
Mood Boards	13
Design Development	14
Construction Documentation	16
Contract Administration	16
Summary	17
Classroom Activities	17
Questions	17
Further Resources	17
Keywords	18

CHAPTER 2

Tools, Scales, and Media — 19

Drafting Board and Accessories	20
Drafting Board	20
Drafting Chair	20
Lamp	20
Parallel Bar	21
T-Square	21
Triangle	21
Protractor	24
Circle Template	24
Architect's Scale	24
Subdivided Scale	25
Proper Architectural Notation (Imperial)	28
Scale Conversions	29
Metric Scale	30
Enlarging and Reducing	30
Proper Architectural Notation (Metric)	30
Scales of Different Drawings	31
Engineer's Scale	31
Proportional Scale	31
Pencils, Leads, and Erasers	32
Mechanical Pencil	32
Lead Holder	33
Leads	33
Eraser	33
Eraser Shield	34
Other Tools	34
Drafting Brush	34
Drafting Tape	34
Divider	34
Compass	34
Ames Lettering Guide	35
French Curve	36
Ink Pens	36
Furniture Template	38
Cutting Mat	38
Utility and Art Knives	38
Drafting Media	39
Tracing Paper	40
Vellum	40
Plastic Film	40
Care and Storage of Tools	40
The Copying Process	40
Paper Copies	40
Digital Scans	41
Phone Apps	42
Summary	42
Classroom Activities	42
Questions	42
Further Resources	43
Keywords	43

CHAPTER 3

2D and 3D Sketching — 45

Three Physical Dimensions	46
Why Are 2D Drawings Needed?	47
Orthographic Projection Drawing Technique	47
Sketching Orthographically	49
Square Table	49
Round Table	50
Drafting Orthographically	51
File Cabinet	51
Chest of Drawers	52
Mission Style Chest	54
Why Are 3D Drawings Needed?	55

Paraline Drawing	56
Types of Paraline Drawings	56
Perspective Drawing	59
Perspective Grid	60
One-Point Perspective Grid	60
Two-Point Perspective Grid	61
Sketch Over a Photograph	62
Sketching Tips	63
Hone Observational Skills	63
Estimate Proportions	63
Scale the Picture to the Paper	64
Take Photos of Existing Spaces	64
Summary	65
Classroom Activities	65
Questions	65
Further Resources	65
Keywords	66

CHAPTER 4

Drafting Conventions — 67

Standards	68
The Line: Weight and Type	68
Line Quality	68
Line Weight	68
Line Type	69
Visible Object	69
Hidden Object	69
Long Break	70
Center Line	70
Short Break	70
Cutting Plane	70
Identifiers	72
Elevation Callout	72
ID Label	72
How to Draw an Elevation Callout	72
How to Draw an ID Label	72
Hatch Lines and Poché Symbols	73
Hatch Lines	73
Poché	73
Dimensioning Symbols	74
Dimension and Extension Lines	74
Dimension Notes	74
Tick Marks	74
Leader Line	74
Other Line Types	75
Match	75
Border	75
Construction	75
Enlargement Box	75
Structural Grid	75

Keynotes	76
Discipline Designator	76
Sheet Composition and Organization	77
Sheet Sequence	77
Title Block	77
Title Sheet	77
Sheet Size	77
Drawing Orientation	77
North Arrow	77
Sheet Layout for Multiple Drawings	78
Sheet Layout	78
Lettering	80
Lettering Tips	80
Notes	81
Summary	81
Classroom Activities	82
Questions	82
Further Resources	82
Keywords	82

CHAPTER 5

The Architectural Floor Plan — 83

What Is an Architectural Floor Plan?	84
What the Architectural Floor Plan Shows	84
Drawing Scale	84
How to Sketch a Plan of an Existing Space	85
The Visual Inventory	88
Visual Inventory Tools	88
How to Measure a Room	90
Measuring Tips	91
Measuring Heights	91
Sketching Tips	91
Utilize Good Line Quality	91
Sketch to Scale	91
Layer Sheets of Tracing Paper	91
Rendering	92
Photogrammetry	93
Wall Thicknesses and Pochés	94
Floor Pochés	96
Symbols	97
Line Quality	104
Line Hierarchy	104
Drafting a Floor Plan	106
Second-Floor Plan	110
Inking	110
Inking Tips	110
Drafting a Presentation Plan	111
CAD Standards	113
Line Weights and Colors	113

Text and Dimension Heights	113
Font Styles	113
Space Planning	113
Space Planning Tips	114
Room Sizes	114
Layouts	115
Kitchen Space Planning	115
U-Shape	117
G-Shape (Peninsula)	117
L-Shape	117
Galley	118
Single Wall	118
Islands	118
Kitchen Appliance Sizes and Clearances	119
General Kitchen Clearances	120
Dining Room Furniture	123
Bathroom Space Planning	124
Bathroom Fixture Sizes	124
Vanity	125
Placement of Fixtures	125
Bathroom Fixture Clearances	126
Accessible and Universal Design Accessibility	129
Accessible Kitchen Design	129
Accessible Bathroom Design	130
Closets	131
Laundry Rooms	133
Bedrooms	133
Bedroom Furniture	133
Living Room	134
Living Room Furniture	134
Egress and Exits	135
Means of Egress	135
Half-Diagonal Rule	136
Egress Windows	136
Egress Doors	136
Hallways	137
Habitable Space	137
Summary	137
Classroom Activities	137
Questions	137
Further Resources	138
Keywords	138

CHAPTER 6

Interior Elevations and Sections 139

What Is an Elevation View?	140
The Elevation Callout Symbol	141
How to Draw an Elevation View	142
How to Draw a Sloped Ceiling in an Elevation View	144
How to Draw Caddy-Corner Items in an Elevation View	145
What Is a Section Drawing?	146
Section Scale	146
Section Location Symbol	148
Section Detail	148
How to Draw a Longitudinal Section	148
Poché Symbols in Section and Elevation	150
Hatch Lines	152
Entourage	152
Sheet Layout	153
Summary	158
Classroom Activities	158
Questions	158
Further Resources	158
Keywords	159

CHAPTER 7

Dimensioning Floor Plans and Elevations 161

Why Dimension Drawings?	162
Dimension Notes	162
Dimension Lines	162
American National Standards Institute (ANSI) Conventions	162
Wood Frame vs. Masonry Dimensions	164
Wood Frame	164
Masonry	164
Masonry Veneer on Wood Frame	166
Dimensioning an Interior Elevation	166
National Kitchen and Bath Association (NKBA) Drawings	166
NKBA Conventions for Floor Plans	166
NKBA Conventions for Elevations	173
Section and Elevation Views of Cabinets	175
Summary	177
Classroom Activities	177
Questions	177
Keywords	177

CHAPTER 8

Door and Window Symbols 179

What Do Door and Window Symbols Show?	180
Door Types	180
Swing	180
Flush and Panel Doors	184
Sliding	185

Accordion	187
Pivot	188
Revolving	188
Full Height	189
Overhead Sectional	189
Cased Opening	190
Door and Rough Opening Sizes	190
Egress Requirements and Egress Doors	190
Egress Door	191
Half-Diagonal Rule	191
Door Hardware	192
Hinges	193
Handle and Knob	193
Lock	193
Bolt	193
Exit Device	193
Closer	193
Holder	193
Stop	193
What Does a Window Symbol Show?	193
Window Types	194
Double-Hung	194
Fixed	195
Sliding	196
Casement	196
Awning	196
Hopper	197
Jalousie	198
Pivot	199
Bay	200
Box Bay	201
Bow	201
Specialty	202
Clerestory and Skylights	203
Window Placement in Elevation	203
Window Definitions	204
Window Sizes	206
Summary	*215*
Classroom Activities	*215*
Questions	*215*
Further Resources	*215*
Keywords	*215*

CHAPTER 9

Building Construction 217

Benefit of Building Construction Knowledge	218
The Foundation	218
Structural vs. Partition Wall	218

Structural System	218
Foundation Types	219
Slab-on-Grade	219
Spread Footing	220
Post-on-Pad	222
Other	223
Masonry	224
Concrete Block Types	224
Masonry Sizes	224
Masonry Wall Construction	225
Masonry Bonds	227
Tilt-Up Construction	228
Wood Framing Components	228
Wood Framing Definitions	231
Wood Frame Types	234
Other Framing Items	236
Other Construction Plans	237
Modular Construction	238
3D Printing	239
Steel Frame Components	239
Steel Framing	242
Finish Materials	244
Detail Drawings	244
Door and Window Details	248
Fireplace	251
Traditional Fireplace	252
How to Draw a Traditional Fireplace	252
Firebox Size	252
Non-Traditional Fireplace	254
Millwork	255
Cabinets	256
Range Hoods	258
MasterFormat	259
Site Plan	260
North Arrow	260
Engineer Scale	260
Footprint	261
Property Lines	261
Hard Surfaces	261
Utility Lines	261
Vegetation	261
Setbacks and Easements	261
Contour Lines	261
Dimensions and Bearing Angle	261
Summary	*263*
Classroom Activities	*263*
Questions	*263*
Further Resources	*263*
Keywords	*264*

CHAPTER 10

Utility Systems — 265

Benefit of Utilities Systems Knowledge	266
Electrical Drawings	266
Electrical Components	266
Service Panel	266
Outlets	266
Switches	267
Permanently Attached Light Fixtures	268
Electrical Plans	268
Electrical Plan Design	270
Drafting an Electrical Plan	274
Power Plans	275
Reflected Ceiling Plan	278
Drafting a Drop-In Ceiling	280
Heating, Ventilation, and Air-Conditioning (HVAC) Systems	282
Forced Air	282
Mini-Split	284
Radiant Floor Heating	284
Solar Heating	285
Evaporative Cooler	285
HVAC Drawings	286
Plumbing System	288
Plumbing Plan	291
Summary	293
Classroom Activities	293
Questions	293
Further Resources	293
Keywords	293

CHAPTER 11

Stairs — 295

Staircase Parts and Sizes	296
Treads and Risers	296
Baluster, Guard, and Handrail	297
Wall Rail	299
Headroom	300
Landing	300
How to Draft Stairs	300
Stair Types	301
Straight Run	301
L	302
Turning	303
U	304
Spiral	305
Circular	306
Other Configurations	307
Exit Stairs	308
Exterior Stairs	308
How to Calculate Riser Height, Tread Width, and Total Run	309
Use a Grid to Draft a Staircase's Elevation View	310
Ramp	311
Elevator	312
Copying Images into Plans	313
Summary	313
Classroom Activities	314
Questions	314
Further Resources	314
Keywords	314

CHAPTER 12

Legends and Schedules — 315

What Are Legends and Schedules Used For?	316
Callouts	316
Legends	317
Schedules	318
Types of Schedules	318
Creating a Schedule	320
Summary	325
Classroom Activities	325
Questions	325
Keywords	325

CHAPTER 13

Hand Drafting Isometric and Perspective Pictorials — 327

Why Construct 3D Drawings by Hand?	328
What Is an Isometric Pictorial?	328
What Is a Perspective Pictorial?	328
Draft an Isometric Pictorial of a Gable House	328
Isometric Circle	330
Draft an Isometric Circle	331
Cutaway Isometric	333
Draft a Cutaway Isometric Room	333
Draft a Cutaway Isometric of a Floor Plan	334
Perspective Drawing Manipulation	338
Draft a Two-Point Interior Perspective Using the Rotated Plan Method	341
Construct a Scale Figure	353

Draft a Two-Point Perspective of a Circle	353
Draft a One-Point Perspective	354
Construct a Sloped Ceiling	359
Troubleshooting	359
Summary	360
Classroom Activities	360
Questions	360
Further Resources	360
Keywords	361

CHAPTER 14

Digital Drafting 363

2D Drawing Programs	364
3D Drawing Programs	364
Which Program Is Best?	365
Other Programs	365
Hardware	366
Mouse	367
Plotter	367
Workflow Techniques	368
Virtual Reality Headset	370
Augmented Reality	370
3D Printing	370
Generative Design	372
Photogrammetry	373
Summary	375
Classroom Activities	375
Questions	375
Further Resources	376
Keywords	376

APPENDIX I

Floor Plans for General Use 377

APPENDIX II

Worksheets 395

GLOSSARY 478

INDEX 490

PREFACE

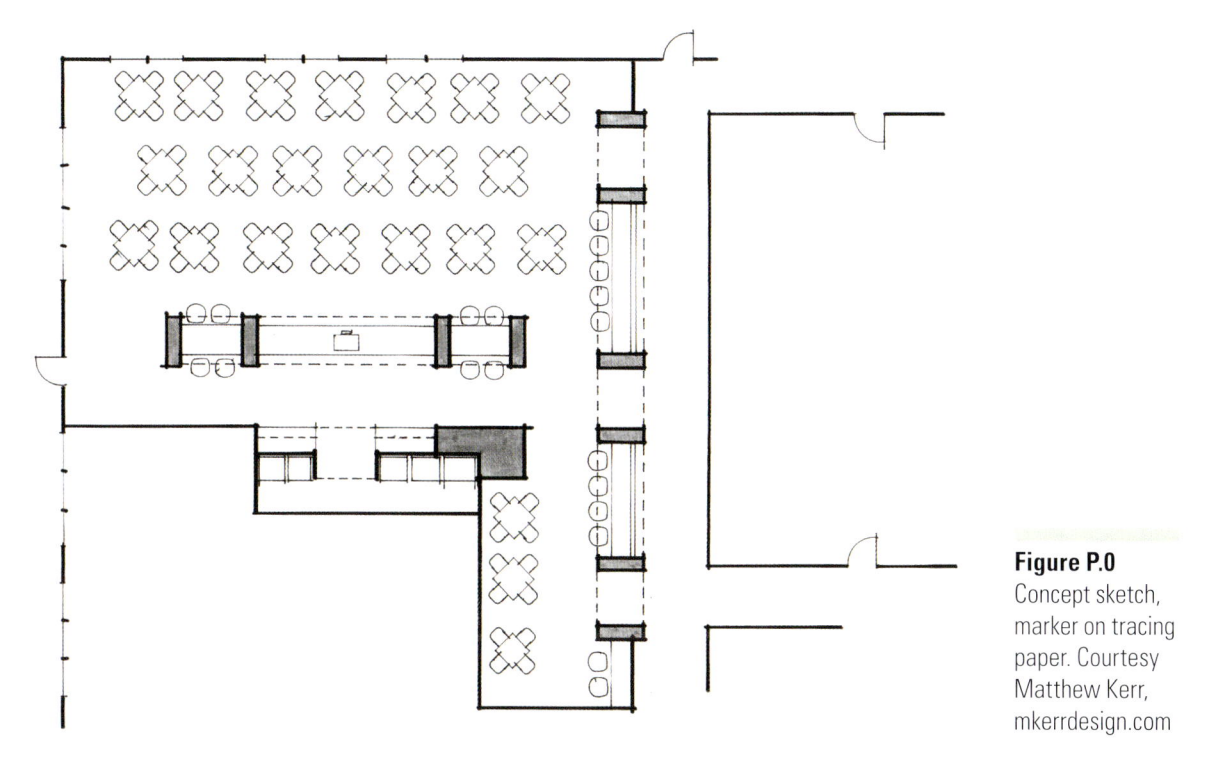

Figure P.0
Concept sketch, marker on tracing paper. Courtesy Matthew Kerr, mkerrdesign.com

Interior design is a field that enjoys great popularity. The number of accredited programs continues to grow, as do TV shows and websites about interior design. Practitioners of this profession, who make our homes and offices attractive, functional, and satisfying to the eye and soul, enjoy a robust market for their services.

It is a profession that requires a broad range of knowledge, critical to which is the ability to communicate ideas. There are two ways to do this: **verbally and visually**. Visual communication is done via **drafting**, which is the art and science of creating graphics specific to the architecture field (Figure P.0). *Architectural Drafting for Interior Design* was written specifically for the student learning how to create those graphics. It includes the following:

- Examples relative to interior design.
- Content applicable to beginning drafting students.
- Worksheets that reinforce chapter concepts.
- Scale floor plans.
- Suggested exercises in each chapter.

It was written to comply with **National Council for Interior Design Qualification (NCIDQ), Council for Interior Design Accreditation (CIDA), National Kitchen and Bath Association (NKBA)** standards, and **IIDA (International Interior Design Association)**, the governing authorities in interior design education and practice.

Although computer software is used for technical drawings, the critical thought process and information needed to create those drawings is a skill set independent of computer drafting. That thought process is what this text emphasizes, rather than any specific software. Many designers continue to manually sketch and draft as a means of thinking through ideas. Also, manual drafting solutions to space planning and other design problems are required for the **NCIDQ test**, hence it continues to be taught in interior design programs.

Coverage and Organization

No prior background or experience in drafting is assumed. The reader starts with an introduction to manual drafting tools and progresses through hand sketching, floor plans, dimensioning, elevations, sections, utility plans, and construction details. The relationship between 2D and 3D views is discussed throughout. The text ends with a discussion of digital drafting and how it can be combined with manually created work. This transitions the student to the computer drafting and presentation visuals classes that typically follow an introductory drafting class.

Features

Each chapter begins with objectives and keywords and ends with questions, classroom activities, and further resources. Step-by-step, illustrated instructions are given for difficult concepts. The text is arranged so that material builds from one chapter to the next, but an extensive glossary is provided to assist those who wish to skip chapters. You can find drafting tutorial videos at the author's YouTube channel: youtube.com/profdrafting.

New to This Edition

This edition contains enhanced and new worksheets, updated visuals and symbols, new design and drafting information, discussion of current software, emerging technologies such as photogrammetry and 3D printing, and suggested exercises for each chapter.

Instructor and Student Resources

Architectural Drafting for Interior Design STUDIO

The book is accompanied by an online multimedia resource, *Architectural Drafting for Interior Design STUDIO*. The online *STUDIO* is specially developed to complement this book with rich media ancillaries that students can adapt to their visual learning styles to better master concepts and improve grades. Within the *STUDIO*, students will be able to:

- Study smarter with self-quizzes featuring scored results and personalized study tips
- Review concepts with flashcards of essential vocabulary
- Download floor plan templates and worksheets to practice your drafting skills

STUDIO access cards are offered free with new book purchases and also sold separately through Bloomsbury Fashion Central (www.BloomsburyFashionCentral.com).

Instructor Resources

- The Instructor's Guide provides suggestions for planning the course and using the text in the classroom, supplemental assignments, grading rubrics, and a CIDA Professional Standards Matrix mapped to the chapters in the book.
- The Test Bank includes sample test questions for each chapter.
- PowerPoint® presentations include images from the book and provide a framework for lecture and discussion.

Instructor's Resources may be accessed through Bloomsbury Fashion Central (www.BloomsburyFashionCentral.com).

It is hoped that the student will discover that drafting is not only a necessary skill to the design field but an enjoyable part of the practice of design as well. The author welcomes your comments at lcline@jccc.edu.

ACKNOWLEDGMENTS

Many thanks to Wolfgang Trost, Matthew Kerr, and Jay Colestock for their beautiful drawings that so enhance this text. Thanks, too, to Sonia Levin for contributing her drafting skills, and to all the companies and organizations that provided their great product photos.

The publisher wishes to gratefully acknowledge and thank the editorial team involved in the publication of this book:

- Acquisitions Editor: Emily Samulski
- Senior Development Editor: Corey Kahn
- Editorial Assistant: Jenna Lefkowitz
- Art Development Editor: Edie Weinberg
- In-House Designer: Lachina Creative
- Production Manager: Ken Bruce
- Project Manager: Molly Montanaro, Lachina Creative

CHAPTER 1

Drafting and the Design Process

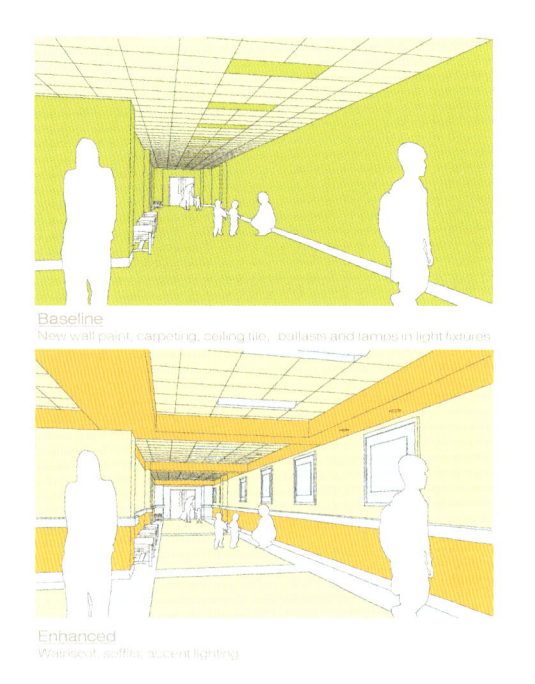

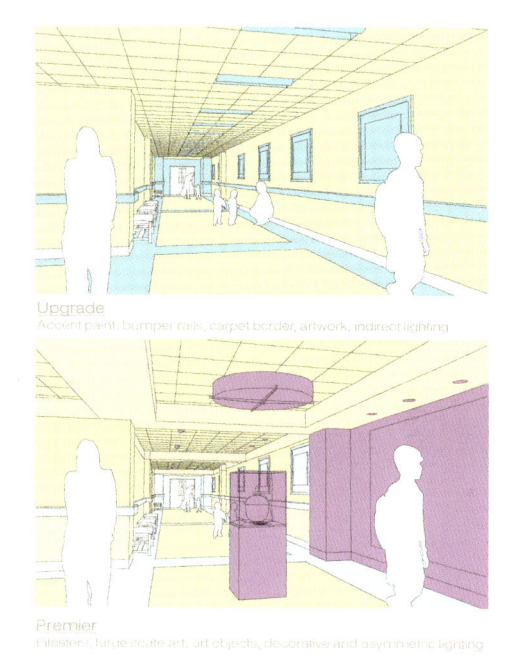

Figure 1.0
Design development drawings created in SketchUp and enhanced in Photoshop. Courtesy of mkerrdesign.com

OBJECTIVES

Upon completion of this chapter you will be able to:

- Explain what drafting is and why it is done
- Identify relevant industry groups
- List the components and documents of the design process

What Is Drafting?

Drafting, also known as **graphic** or **visual communication**, is the art of putting ideas to paper in picture form to explain ideas and create instructions. Pictures convey ideas in a way that written descriptions cannot. All manufactured items—whether they are clothing (Figure 1.1), cars, toys, furniture, or buildings (Figure 1.2)—started with drafted pictures. We see these pictures everywhere. For instance, the pictures in product assembly instructions are drafted drawings. When we sketch a map giving directions to our home, we are drafting a simple picture. We imagine our surroundings before we create them. Drafting enables us to communicate those ideas to others. Architectural drafters use universally recognized symbols and protocols to enable everyone involved in a project to interpret the drawings in a consistent manner, no matter in what country the drawings are made or read.

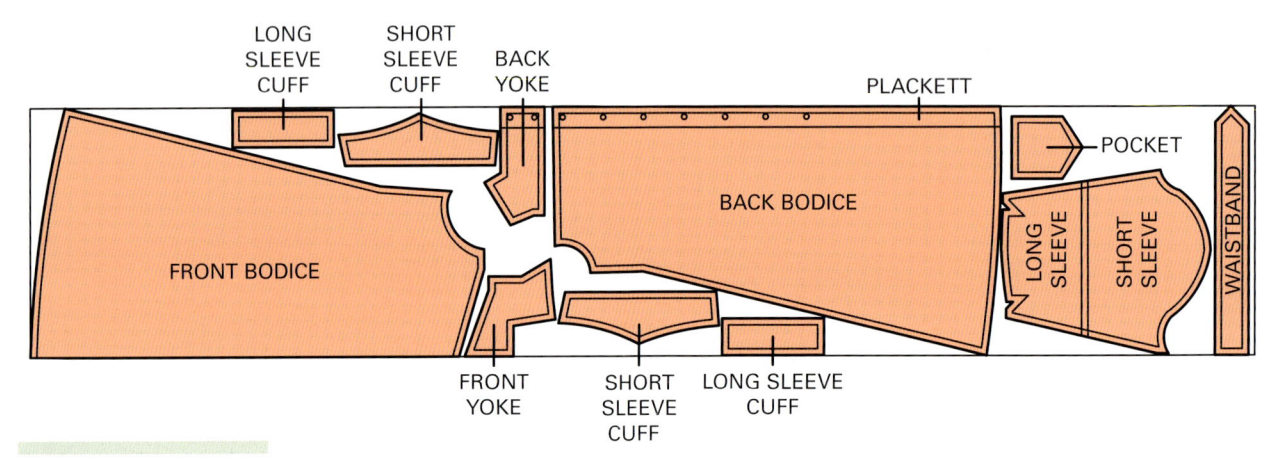

Figure 1.1
A clothing pattern is a drafted set of instructions.

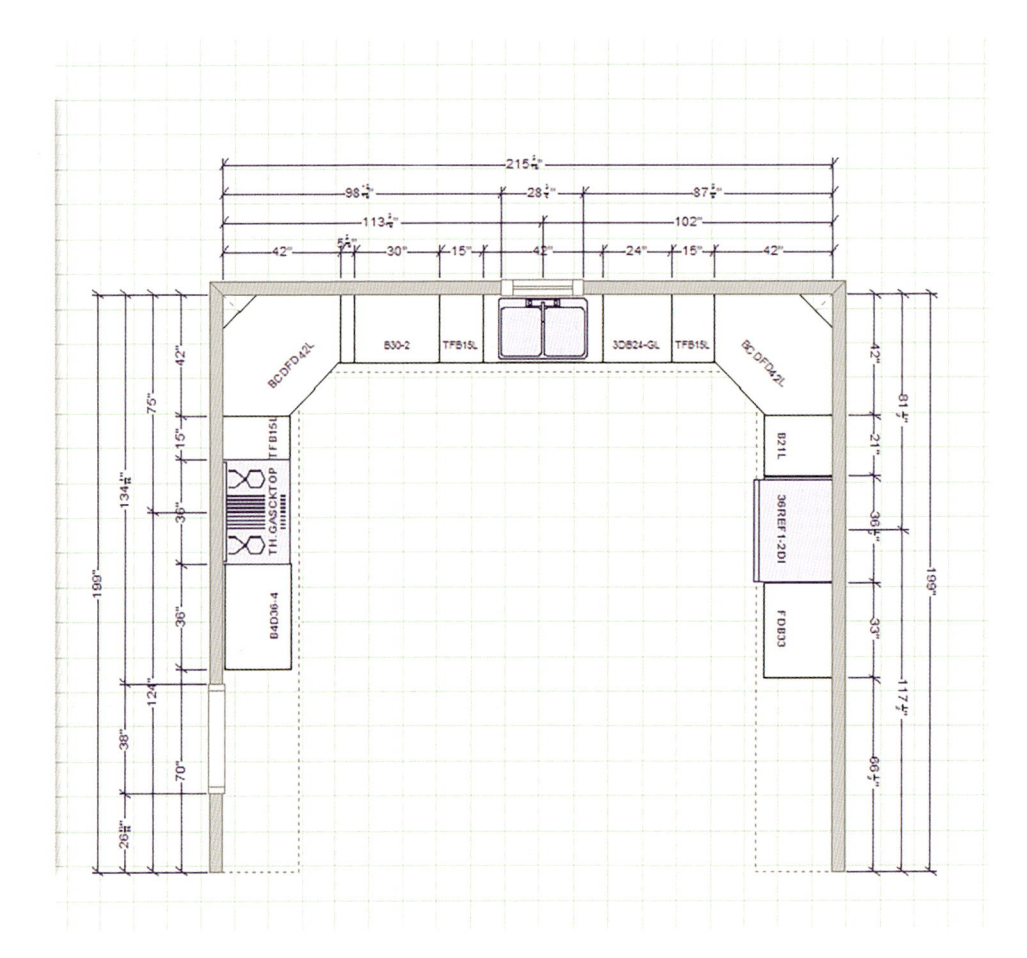

Figure 1.2
A floor plan is part of instructions for a building. This plan was created in 20/20.

2 CHAPTER 1 DRAFTING AND THE DESIGN PROCESS

History of Manual Drafting   Architectural drafting originated when the first designer needed to give instructions. Ancient civilizations sketched elaborate structures onto papyrus (paper made from reed pulp) using reeds dipped in ink. Drawings were also carved into flat stone panels that were later installed into the base of a building. Since ancient structures could take centuries to build, this served the purpose of showing future workers the designer's intent. Examples of these "blueprints" are found in Egyptian, Greek, and Roman buildings, as well as medieval castles (construction plans are etched into the walls and floors of the cathedrals at York, Chartres, and Rheims). Most of the drawings are full size. The temple of Apollo at Didyma has straight lines, circles, quarter circles, and more complicated shapes so precisely drawn that they were clearly made by experienced draftsmen. It is even evident where some of the designs had been changed and where small deliberate mistakes had been made, possibly as a show of humility by the designer.

For thousands of years there was little change in the instruments that craft guild members used to draw their pictures; in fact, the T square and compass were carefully guarded secrets. The tools of modern board drafting have been largely unchanged for the past 100 years (Figure 1.3). Although some tools have been modernized (e.g., technical pens have replaced inkwell-dipped quill pens), others such as triangles, scales, and dividers have remained the same.

Figure 1.3
Drafting classroom circa 1906.
Courtesy of Colorado College,
Colorado Springs, CO.

History of Computer Drafting

In 1950 Dr. Paul Hanratty developed a **Computer Aided Drafting (CAD)** program that let users create simple line drawings. In 1957 researchers at MIT developed **Pronto**, a program that enabled more complicated drawings. However, computers were physically enormous and very costly, so these programs were not widely adopted. During the next 20 years more **Computer Aided Drafting and Design (CADD)** programs were developed but limited to large manufacturing companies for their own use.

In 1982 the Autodesk Company introduced its **AutoCAD** architectural drafting program (Figure 1.4). This program replicated the manual drawing process where drafters used the mouse like a pencil to create 2D drawings. For the next decade AutoCAD was primarily used by large companies as it, and the computers needed to run it, were too expensive for small businesses and personal users. But in the 1990s both software and hardware prices dropped and computers became smaller and cheaper, which resulted in AutoCAD becoming the dominant drafting program for personal users and companies of all sizes.

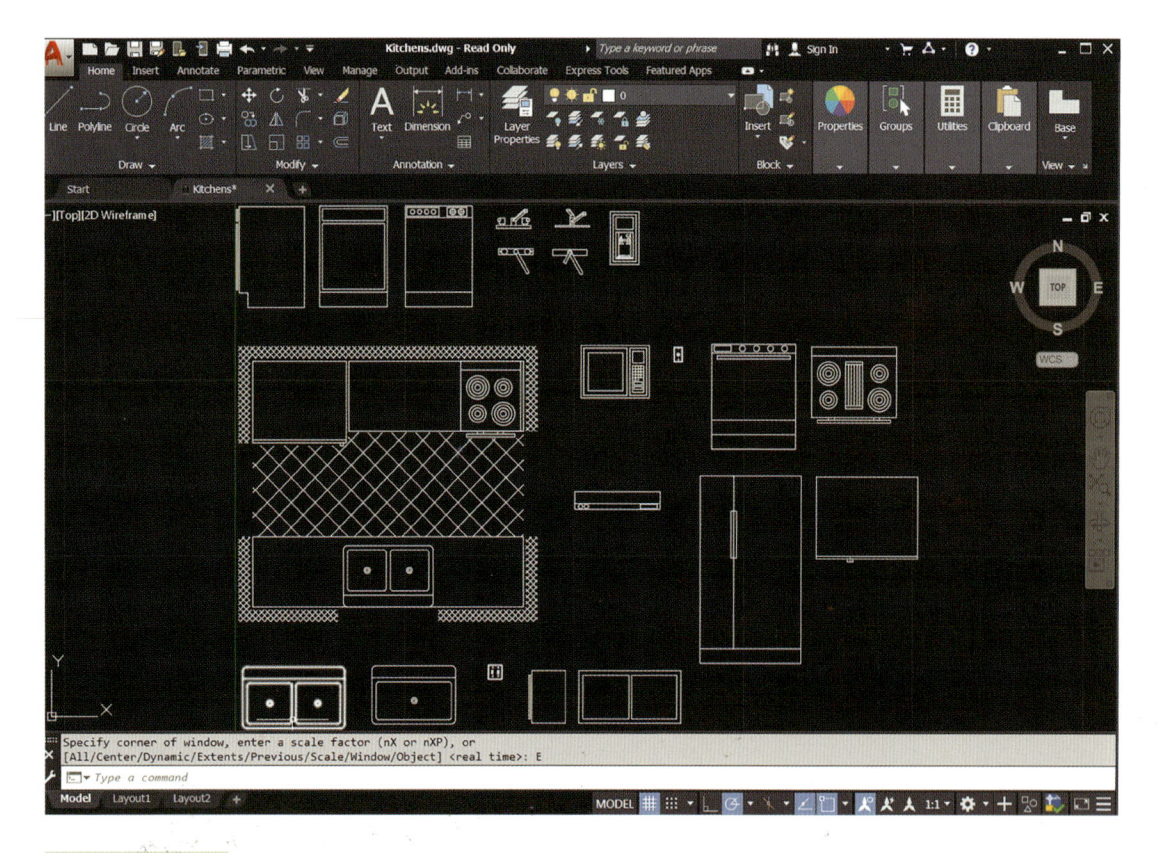

Figure 1.4
AutoCAD uses the mouse like a pencil to create 2D pictures.

CADD evolved to **modeling programs**. This is software in which everything is drawn in 3D; the drawing is called a **model**, and 2D views are generated from it (Figure 1.5).

Some modeling software has **Building Information Modeling/Management (BIM)** capabilities. With BIM software, the model is hosted on a server and everyone in the project works collaboratively on it. A BIM model contains more than just construction information; it also has document management, coordination, simulation and life cycle operation and maintenance information. Currently, the dominant architectural BIM program is **Revit**.

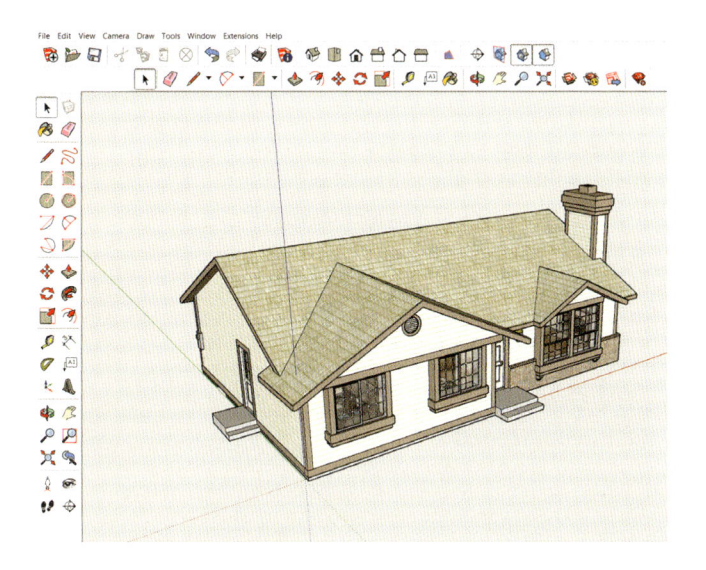

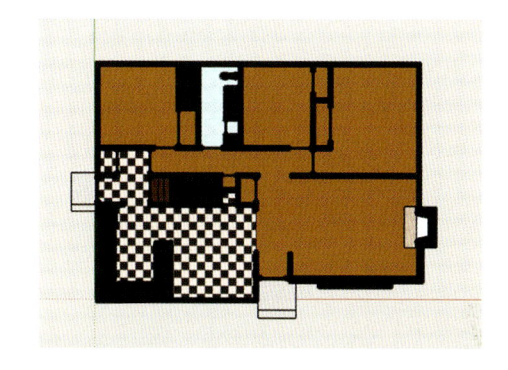

Figure 1.5
A SketchUp model of a house and 2D views generated from it.

Professional Groups and Certifications

Architectural drafting is done by those who work in the interior design and architecture fields. Professional groups and certifications include:

- **American Society for Interior Designers (ASID).** This national organization represents interior designers and interior design students. It advocates for the profession, provides continuing education and advice on running a practice, and hosts awards and competitions.
- **National Council for Interior Design Qualification (NCIDQ).** This nonprofit organization oversees the eligibility, examination, and certification of interior designers.
- **American Institute of Architects (AIA).** This national organization represents licensed architects and architectural interns. Its activities include influencing legislation for professional liability, licensing requirements, building codes, preservation, and environmental concerns. The AIA also provides professional development opportunities and hosts a yearly convention. There is an AIA for each state.
- **National Council of Architectural Registration Boards (NCARB).** This organization recommends model laws, regulations, and other guidelines. NCARB also develops, administers, and maintains the **Architect Experience Program (AXP)** and the **Architect Registration Examination (ARE).** Each state also has its own registration requirements for licensing architects.

- **Council for Interior Design Accreditation (CIDA)**. This organization accredits interior design programs to assure the public that an education in any program prepares students to be responsible, well-informed, skilled professionals.
- **Council for Interior Design Qualification (CIDQ)**. This is the certifying body that develops and administers the NCIDQ.
- **International Interior Design Association (IIDA)**. This global organization supports commercial interior designers, industry affiliates, educators, students, companies, and their clients.
- **National Kitchen and Bath Association (NKBA)**. This national organization provides education and support to the kitchen and bath industry. Professional certifications that can be obtained through it include **Associate Kitchen & Bath Designer (AKBD), Certified Kitchen Designer (CKD), Certified Bath Designer (CBD),** and **Certified Master Kitchen & Bath Designer (CMKBD)**. It has many local chapters.
- **US Green Building Council (USGBC)**. This organization provides a framework for identifying and implementing sustainable building design, construction, operations, and maintenance. Its internationally recognized green building certifications include **Leadership in Energy and Environmental Design (LEED)** and **Regreen**.

> **Exercise**
> Visit the USGBC website and write short descriptions about each LEED rating system found there.

There are currently over 90 building certifications in the United States. The three largest are LEED, **National Association of Home Builders (NAHB)**, and **EnergyStar**.

Design Thinking This is the process of creative problem solving. It's utilized in every field, not just the obviously creative ones; technology and manufacturing benefits from **design thinking**, too. An interior design project calls for problem-solving skills at every stage of the process—from thinking what it will look like, to drafting it, to fixing unexpected issues during construction. Design thinking requires thoroughly understanding the needs and wants of the client, being agile and adaptive, brainstorming ideas fast and prolifically, being a critical thinker, and being bold, innovative, and creative.

Design thinking is iterative and non-linear. Intuition, empathy, and pattern recognition must be employed to create meaningful, functional solutions. Developing **prototypes**, which are preliminary models of the solution that test both the design and user reaction to it, are done during this thinking process. Prototypes can be anything from a sketch to a **pop-up space** (short-term, temporary retail store) to a **3D print** (physical object made from a digital model) (Figure 1.6).

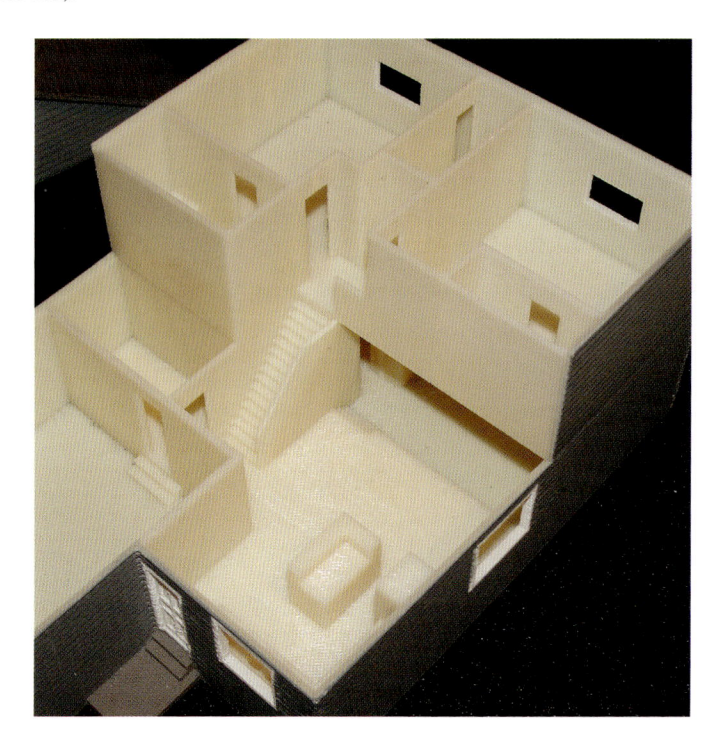

Figure 1.6
3D print of a house.

The Design Process   Design thinking is interwoven with the design process. The AIA, ASID, and NKBA list five stages in this process: **programming**, **schematic design**, **design development**, **construction documentation**, and **contract administration**. Drafted drawings and other visuals are used at all of them. Following are descriptions of those stages, the activities done in them, and the documents produced.

Programming   This is the research stage. Information that will affect the design must be gathered, analyzed, and synthesized. Such information includes spaces required; their functions and square footage; how many people will use them and how; which spaces should be adjacent, separated, public, private, or secured; the traffic flow of goods, services, and people; and the furnishings and equipment needed. Regional limitations, context, cost, energy usage, accessibility, and aesthetic and functional requirements (mechanical, electrical, and plumbing systems) are evaluated. Life cycle cost, organizational and image goals, and environmental concerns may also be considered. Some projects go through the site selection process at this stage.

Other issues include physical conditions that affect occupant health and safety (air quality and circulation, temperature control, ergonomic layout, and the physical circulation plan); maintenance concerns (ability of the products and materials to be kept in good condition, as well as the work required to keep that condition over the material's life span); and sustainability (using resources in a manner that does not deplete them and will have the least long-term effect on the environment).

An **architectural program**, also called a **project statement**, is written. This is a description of client needs and wants. It describes the problem, scope, goals, requirements, and constraints (Figure 1.7).

Depending on complexity, a program can be one or one hundred pages long. **Adjacency matrixes** (Figure 1.8) and **criteria matrixes** may also be drawn, which show space relationships and requirements in a visual manner.

The **occupancy** and **occupant load** are also determined. *Occupancy* is a category that describes the building's use, and occupant load is the number of people the building is expected to hold. There are ten occupancy categories: assembly, business, educational, factory, high-hazard, institutional, mercantile, residential, storage, and utility/misc. There is also **mixed occupancy**, meaning more than one type of occupant.

Schematic Design   In this phase ideas are developed. **Concept** and **parti sketches** are drawn. Concept sketches sum up the "big idea," that is, the design's main function or appearance (Figure 1.9).

Parti sketches show the organizing thoughts of a design and the relationship of parts to the whole. For instance, in Figure 1.10 we see the beginnings of a hospital design. The sketch shows the main areas, their spatial relationships, and where/how the user starts experiencing it all.

Bubble Diagrams   A **bubble diagram** is a concept sketch that organizes and articulates ideas and helps the designer visualize how everything will work together. It contains labeled circles that represent spaces and functions occurring inside. Different size circles reflect different space needs. Rooms, orientations, traffic patterns, views, and physical and visual access are shown, along with arrows that describe traffic patterns. Other design considerations, such as landscaping, direction of breezes, sun, and buffer zones or sound, may also be shown.

Program for a Custom-Built Vacation Home

Goal: Design a family vacation home in Rio Rancho, NM.

Client Description: The Eve family owns a corner lot with a mountain view in a town outside Albuquerque. They want a ranch style home that maximizes the views, has a Southwest style, and accommodates their family of two adults and two minor children. Short term renters will also use the house.

Required Rooms:
1. Kitchen that accommodates gourmet cooking and family/friend gatherings
2. Walk-in pantry
3. Family room that accommodates the Eves plus up to eight guests.
4. Exercise room that accommodates five machines and storage for weights
5. Theater with a large screen that accommodates 8 viewers.
6. Game room that accommodates 5 vintage gaming machines
7. Laundry room with sink and space for ironing
8. Master bedroom with private sitting area and office
9. 2 children's bedrooms with shared bathroom
10. 1 guest bedroom with bathroom
11. Outside living spaces with outdoor kitchen
12. Two-car garage

Required Adjacencies:
1. Kitchen must have physical access to family room and convenient access to a large, walk-in pantry.
2. All private spaces must have direct access to private bathrooms.
3. Public spaces must have convenient access to outside living spaces and views.
4. Master bedroom and office must have direct access to outdoor space.
5. All bedrooms must have direct access to large closets.
6. Garage must be directly attached to house.

Other:
Exterior finish material to be stucco. House should be adobe style with exposed beams, tile flooring and wrought iron gates. Natural light and breezes should be incorporated. House should project an upscale Southwest look. Security, intercom, wireless Internet and streaming must be incorporated in all rooms. Front and back lawn to be rock and indigenous plants for easy maintenance.

Figure 1.7
A program is a written description of client needs and wants.

ADJACENCY MATRIX FOR THE EVE FAMILY VACATION HOME	KITCHEN	FAMILY ROOM	EXERCISE ROOM	THEATER	GAME ROOM	LAUNDRY ROOM	MASTER BEDROOM	CHILDREN'S BEDROOMS	GUEST BEDROOM
KITCHEN		○	✕			✕		○	○
FAMILY ROOM	●			●	○				○
EXERCISE ROOM	✕			✕			●		
THEATER	○	○			●		○		○
GAME ROOM				●			○		
LAUNDRY ROOM	●		✕						
MASTER BEDROOM	✕		●	○	○	○			○
CHILDREN'S BEDROOMS	○	○	✕						
GUEST BEDROOM	○	○		○			○		

● DIRECTLY ADJACENT ☐ NO ADJACENCY NEEDED
○ CONVENIENTLY ADJACENT ✕ DO NOT ADJOIN

Figure 1.8
Adjacency Matrix for the program in Figure 1.6.

Tip
Habitable spaces are those that building codes consider capable of being fit to live in. Other words for habitable spaces are occupiable, dwelling, sleeping, and living.

Figure 1.9
A concept sketch for office cubicle walls.

PROFESSIONAL GROUPS AND CERTIFICATIONS

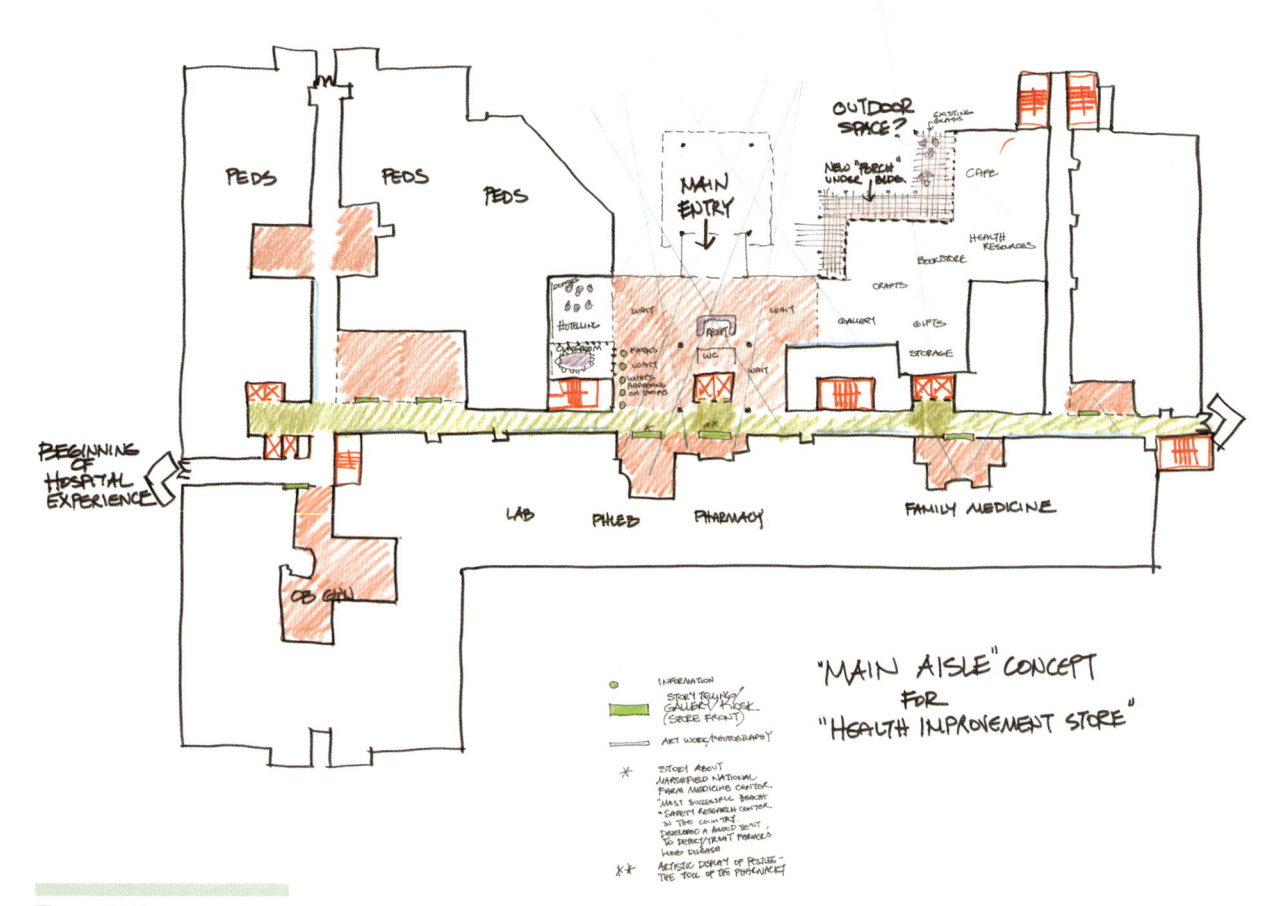

Figure 1.10
Parti sketch for a hospital layout.
Courtesy of mkerrdesign.com

As the bubble diagram evolves from rough to refined, it becomes a **block diagram**, a graphic that resembles a floor plan. Once scale is introduced, the block diagram transitions to **floor plan**, a scaled diagram of room arrangement. Figures 1.11 through 1.16, done by Jay Colestock, AIA, of Colestock and Muir (www.cmarchitects.com), show the diagram development of a floor plan from bubble to refined block.

Figure 1.11
Start the architectural design process with property characteristics.

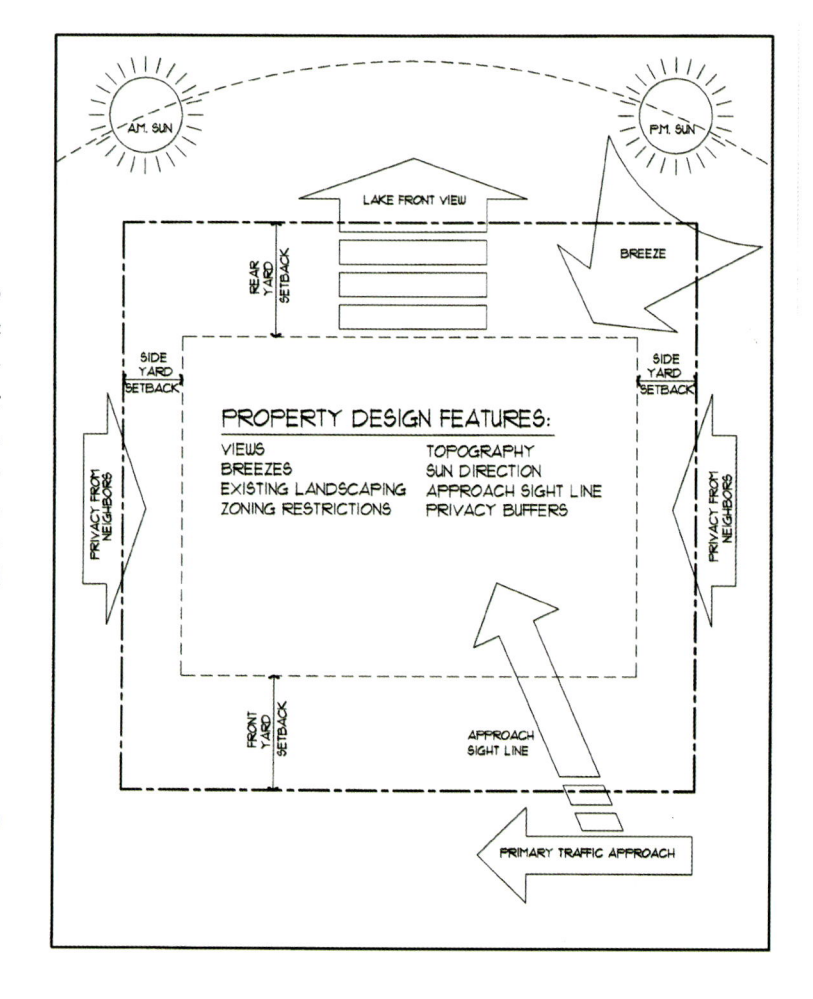

10 CHAPTER 1 DRAFTING AND THE DESIGN PROCESS

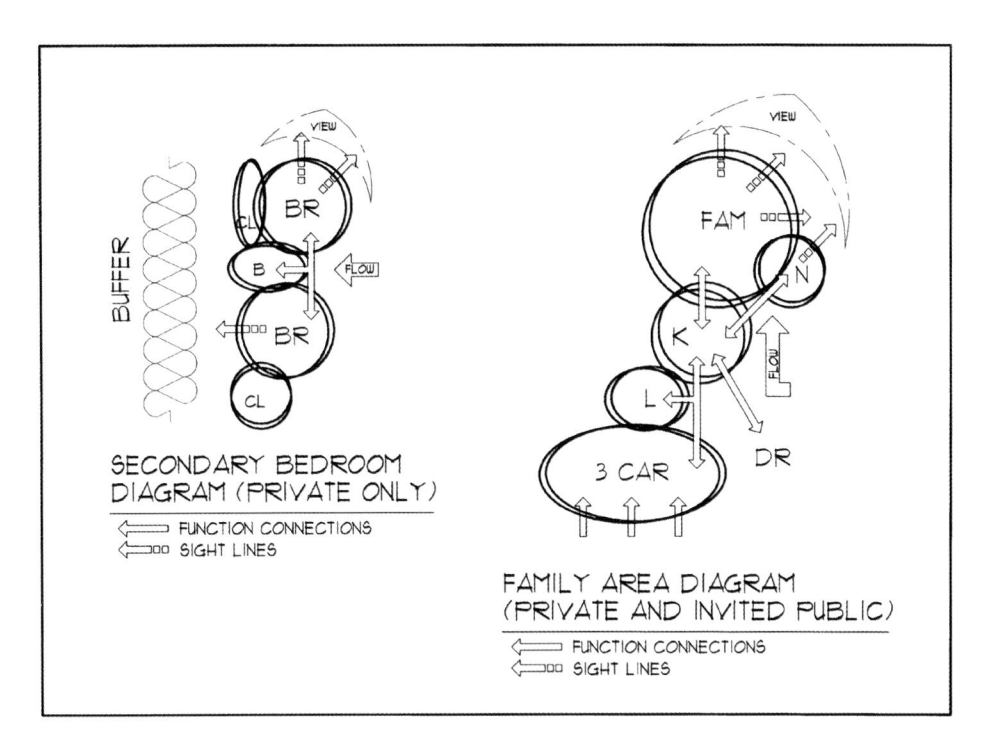

Figure 1.12
Diagramming the bedroom and family areas.

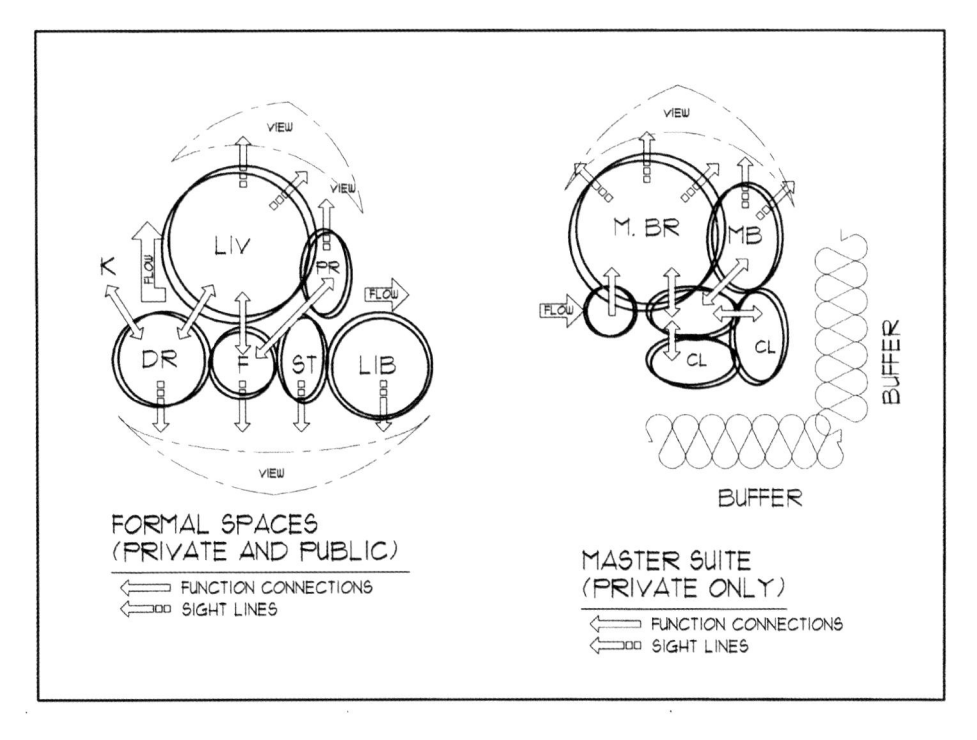

Figure 1.13
Diagramming the formal spaces and master suite.

PROFESSIONAL GROUPS AND CERTIFICATIONS 11

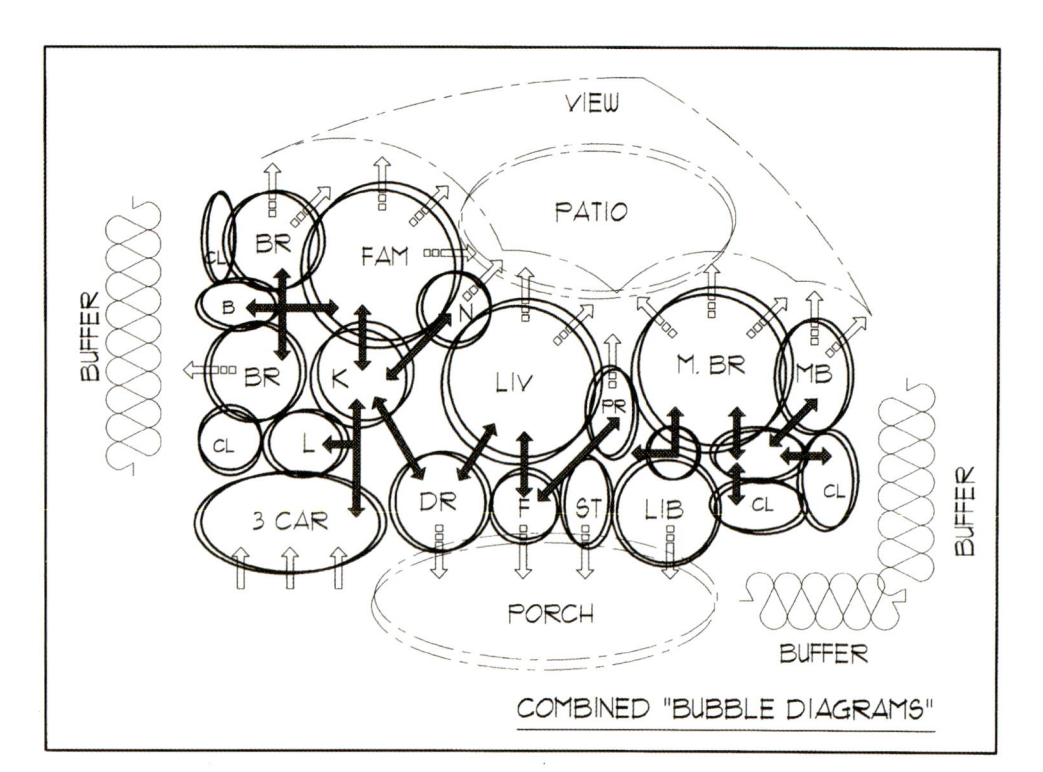

Figure 1.14
Pulling all the bubble diagrams together.

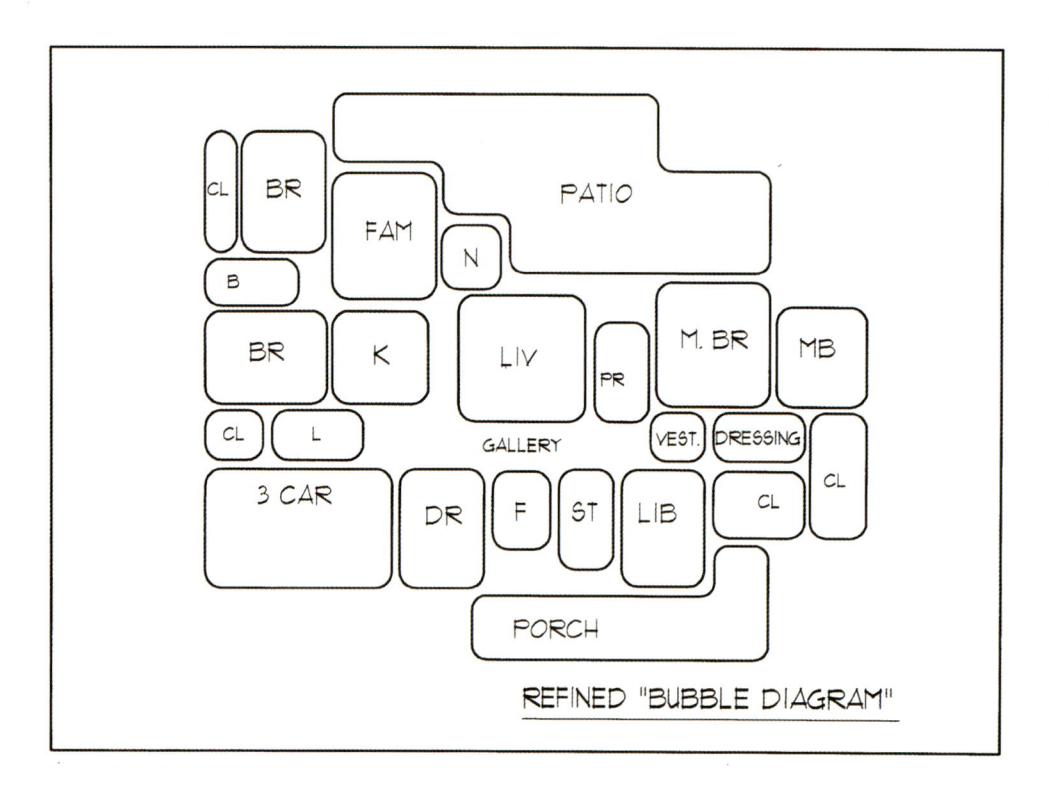

Figure 1.15
Refining the bubble diagram into a block diagram.

12 CHAPTER 1 DRAFTING AND THE DESIGN PROCESS

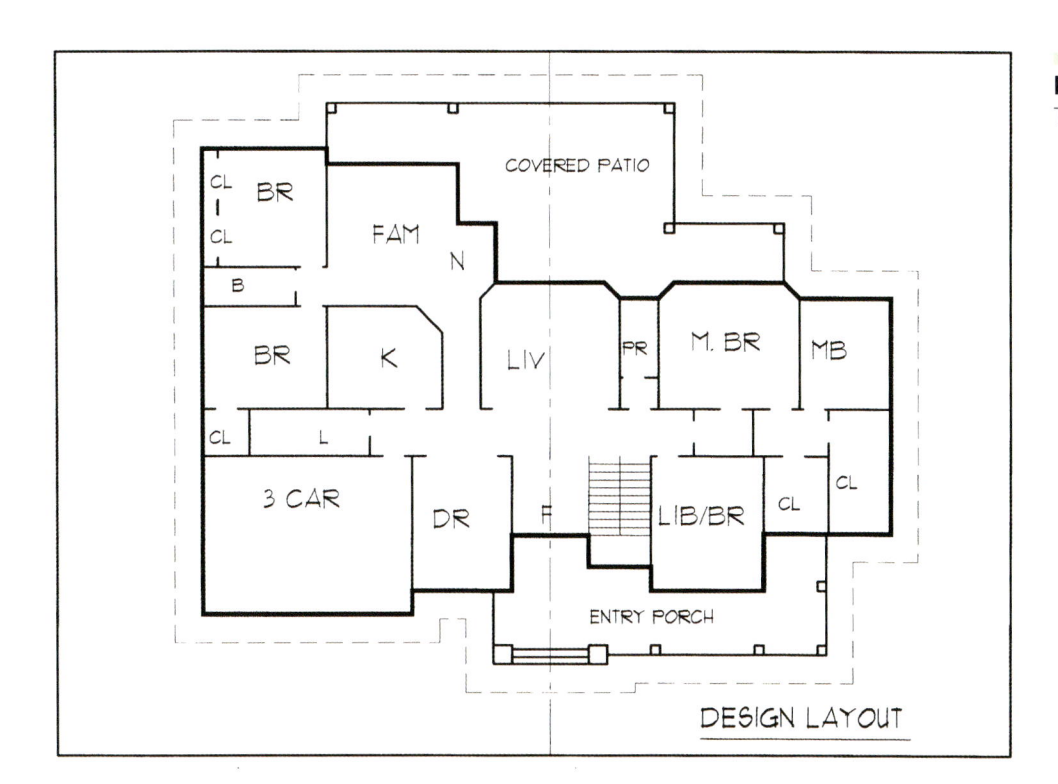

Figure 1.16
The design layout.

Mood Boards **Mood boards**, which are visuals that describe the design essence without being specific, may also be presented (Figure 1.17). A mood board's goal is to evoke emotion and convey the design direction in an unspecific way. Any photos, symbols, materials, and textiles that capture the designer's thoughts for how the space will affect the user or hint at its appearance can be used. A mood board is not the same as a **color board**, which shows the actual materials selected.

Figure 1.17
Mood board for a vacation home in the Southwest.

PROFESSIONAL GROUPS AND CERTIFICATIONS

Design Development The project now goes from concept to workable design. Visuals may include color sketches (Figure 1.18), scale models (Figure 1.19), color boards (Figure 1.20), or anything else that communicates the idea. The visuals selected should address the specific audience. For instance, the floor plan shown in Figure 1.21 may be appropriate for a client presentation, but not for a builder.

Figure 1.18
Two-point interior perspective sketch created with SketchUp and hand-colored with marker. Courtesy of mkerrdesign.com.

Figure 1.19
Scale model of a custom-designed residence. Paper elevations are glued to foam core board. Courtesy of wolfgangtrost.com.

Figure 1.20
Color board for an office building lobby. Courtesy of Matthew Kerr, mkerrdesign.com.

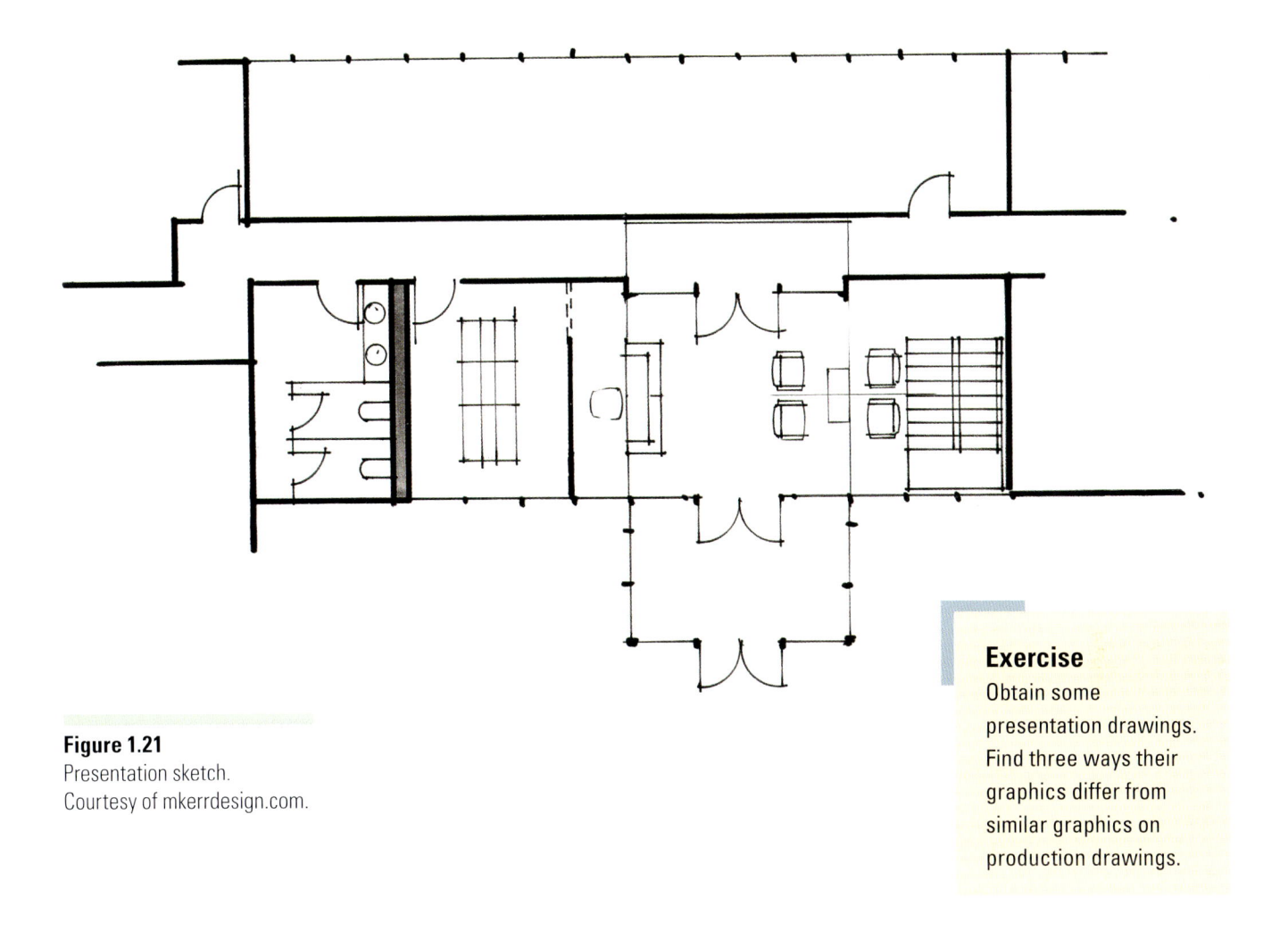

Figure 1.21
Presentation sketch.
Courtesy of mkerrdesign.com.

> **Exercise**
> Obtain some presentation drawings. Find three ways their graphics differ from similar graphics on production drawings.

Local **codes**, **ordinances**, and **zoning**, which are regulations for aesthetics, traffic, and activities inside a building are researched, as are **building codes**, and a detailed code review is done. Building codes are rules that govern design and construction and address risk factors that characterize the people and activities in the space. For instance, a nightclub has dim light and loud music; offices have upholstered furniture, paper, and other flammables; nursing homes have occupants with limited mobility; prisons have occupants with restricted mobility; and theaters have many occupants in one space. Codes help ensure that all types of buildings provide their users with an appropriate level of safety to live and work. Room size, number of exits, lighting, hallway length, and interior finish selection are just a few examples of code-dictated features.

The **International Code Council (ICC)** (Figure 1.22) is a standards organization that writes 15 separate **model codes**. These are building codes written by a standards organization independent of the governing authority (e.g., state government, fire district, or municipality) that adopts and enforces them. Model codes do not become law until a governing authority adopts them. The two ICC codes most applicable to interior designers are the **International Residential Code (IRC)**, which applies to one- and two-family homes of three stories or less, and the **International Building Code (IBC)**, which applies to commercial construction.

Other relevant codes are the **Life Safety Code (LSC)** and **National Electric Code (NEC)**, written by the **National Fire Protection Association (NFPA)**, which is another codes standards organization. The LSC covers hazards to human life in buildings and the NEC covers the safe installation of electrical wiring and equipment. The **Americans with Disability Act (ADA)** is a civil rights law that prevents discrimination to the disabled, and applies to public and commercial buildings; **ANSI-A117.1** is the Accessible & Usable Buildings & Facilities code; and the **Occupational Safety and Health Administration (OSHA)** writes workplace safety and health codes.

Figure 1.22
The International Codes Council (iccsafe.org) writes model codes for the design and construction industries.

Figure 1.23
Set of construction drawings for apartment villas.

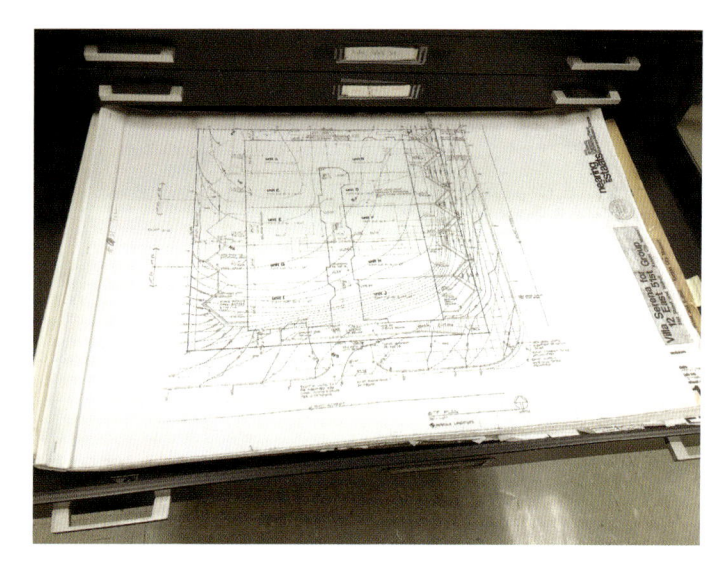

Construction Documentation   The design is finalized in this stage, and documents that serve as legal and binding instructions for building are drafted (Figure 1.23). These instructions include architectural, structural, electrical, mechanical, plumbing, and any specialty drawings.

Specifications are written descriptions of quality of materials and workmanship standards. For instance, the drawings show where carpet is to be laid; the specifications discuss the carpet's material, pile height, backing, and glue. **Schedules**, which are charts of information, are done for the fixtures, furniture, and equipment to be purchased. Producing a complete set of construction documents is a team effort to which many professionals contribute.

Contract Administration   In this phase the job is awarded to a contractor (assuming the contractor wasn't selected at the outset; in such cases the contractor may even provide the drawings). **Time schedules**, which describe workflow, are created by construction managers. **Shop drawings** are also created at this phase. These are highly detailed production drawings of items shown in the design drawings and are created by contractors, subcontractors, manufacturers, or fabricators. For example, where the design drawing might show a simple drawing of a steel beam, the shop drawing shows how many bolt holes it has, the exact size of its flanges, and pictures of the bolts themselves. They are a communication from the fabricator to the designer that says, "As I understand it, this is what you want, and this is how I plan to produce it."

> **Exercise**
> Obtain some production drawings and their accompanying specifications. Find three materials and compare/contrast how they are described in the drawings and how they're described in the specifications.

Summary

Drafting is graphic communication and is used to explain ideas and create instructions. It has a recorded history that spans thousands of years. Standardized ways of presenting those ideas and instructions have evolved to make drawings readable wherever they are created or read. Professionals who draft include interior designers and architects and various industry groups. Certifications are available to advocate quality and standards. Organizations also exist to regulate design and construction quality in the form of building codes. The design process requires design thinking and visuals for different stages of the documentation process. Different graphics and visuals are used, with all having the same purpose of communicating intent.

Classroom Activities

1. Draw a bubble diagram for a residence.
2. Create a mood board for a residential kitchen or small retailer.
3. Create an adjacency matrix for the part of the building your classroom is in.
4. Visit the ICC website and list the model codes it writes.
5. Visit your local city website and research what codes it has adopted.
6. Obtain a copy of the International Residential Code for One- and Two-Family Dwellings and research some specific topics, such as exit requirements or hallway lengths.

Questions

1. What is drafting?
2. Why is drafting used?
3. What does "LEED" mean?
4. What are the five stages of design, as recognized by the ASID, AIA, and NKBA?
5. What is an architectural program?
6. Why are building codes used?
7. What is a model code?
8. What is an adjacency matrix?
9. What is a bubble diagram?
10. What is a mood board?

Further Resources

American Institute of Architects (AIA). www.aia.org

American Society for Interior Designers (ASID). www.asid.org

Council for Interior Design Accreditation (CIDA). www.accredit-id.org

Free online access to an abbreviated version of ICC codes. https://codes.iccsafe.org/content/IRC2018

International Code Council (ICC). www.iccsafe.org

National Council for Interior Design Qualification (NCIDQ). www.ncidq.org

National Kitchen and Bath Association (NKBA). www.nkba.org

U.S. Green Building Council (USGBC). www.usgbc.org/LEED

Keywords

- 3D print
- adjacency matrix
- American Institute of Architects (AIA)
- American Society for Interior Designers (ASID)
- Americans with Disability Act
- ANSI-A117.1
- Architect Registration Examination (ARE)
- architectural program
- Associate Kitchen and Bath Designer (AKBD)
- AutoCAD
- Architecture Experience Program (AXP)
- block diagram
- bubble diagram
- building code
- Building Information Modeling/Management (BIM)
- Certified Master Kitchen and Bath Designer (CMKBD)
- Certified Kitchen Designer (CKD)
- Certified Bath Designer (CBD)
- color board
- Computer Aided Drafting (CAD)
- Computer Aided Drafting and Design (CADD)
- concept sketch
- construction documentation
- contract administration
- Council for Interior Design Accreditation (CIDA)
- Council for Interior Design Qualification (CIDQ)
- criteria matrix
- design development
- design thinking
- drafting
- dwelling
- EnergyStar
- floor plan
- graphic communication
- habitable
- International Building Code (IBC)
- International Code Council (ICC)
- International Interior Design Association (IIDA)
- International Residential Code (IRC)
- Leadership in Energy and Environmental Design (LEED)
- Life Safety Code (LSC)
- mixed occupancy
- model
- model code
- modeling programs
- mood board
- National Association of Home Builders (NAHB)
- National Council for Interior Design Qualification (NCIDQ)
- National Council of Architectural Registration Boards (NCARB)
- National Electric Code (NEC)
- National Fire Protection Association (NFPA)
- National Kitchen and Bath Association (NKBA)
- occupancy
- occupant load
- Occupational Safety and Health Administration (OSHA)
- ordinance
- parti sketch
- pop-up space
- programming
- project statement
- Pronto
- prototype
- Regreen
- schedule
- schematic design
- shop drawings
- specifications
- time schedules
- US Green Building Council (USGBC)
- visual communication
- zoning

CHAPTER 2

Tools, Scales, and Media

Figure 2.0
Manual drafting tools.

OBJECTIVES

Upon completion of this chapter you will be able to:

- Identify manual drafting tools
- Select the appropriate tools for different manual drafting tasks
- Use the tools correctly
- Explain historic and modern copying processes

19

Like all jobs, drafting projects require the proper equipment. Although the prevalence of computer software has made drafting as fine art irrelevant, any manual drafting done for school or work still must be neat, accurate, and presentable. Let's discuss the tools and supplies needed and how to use them.

Drafting Board and Accessories

Drafting Board A **drafting board** is a large, smooth surface, which may be mounted on a base (Figure 2.1) or **portable**, without a base (Figure 2.2). Those are useful for carrying between home and class. A minimum size of 2' × 3' with tilt and height adjustment will probably serve all your manual drafting needs.

It must be topped with a **vinyl cover**, which is a semi-soft surface placed on top of the board and needed for quality drafting. A hard surface does not facilitate good drafting. Sheet vinyl cut to standard drafting board sizes and rolls of vinyl (Figure 2.3) are available. In a pinch, a smooth-surface cutting mat taped to the board can work.

Drafting Chair Pair the board with a cushioned chair that has rolling castors.

Lamp This light-providing appliance consists of an electric bulb, holder, and cover. Choices include fluorescent, incandescent, halogen, or a combination, with the optimal output being 40 foot-candles. Some have built-in magnifying glasses, helpful for small details. A lamp that clamps to the board, is adjustable in length, and swivels works best.

Tip
Don't draft on a kitchen table or other makeshift surface. A practiced eye can discern work that was done on a proper drafting surface and work that wasn't.

Figure 2.1
A drafting board, parallel bar, chair, and lamp with built-in magnifying glass.

Figure 2.2
A portable drafting board and attached parallel bar. Courtesy Alvin & Co.

Figure 2.3
A roll of vinyl for the top of a drafting board. Courtesy Alvin & Co.

CHAPTER 2 TOOLS, SCALES, AND MEDIA

Parallel Bar   The **parallel bar** is a long, straight tool used for drawing horizontal lines (Figure 2.4).

Some bars can be held in place on a tilted board with a built-in thumbwheel lock. The best quality bars have ball bearings and pulleys and require assembly, which includes screwing hardware directly on the board to enable the bar to glide up and down on cables. Cheaper bars are attached at the board's ends and are often the type preassembled and preinstalled on portable boards.

The bar's purpose is to draw horizontal lines, so use it for this. Don't draw horizontal lines by placing any straight-edge tool, such as a triangle or scale, on the board and "eyeballing" it straight (Figure 2.5). This is poor drafting practice. Whether you're drawing a line for the first time or tracing it, use the bar to ensure consistently horizontal lines.

Figure 2.4
Use a parallel bar to draw horizontal lines. Glide the pencil directly on it and hold it perpendicular to the board.

Figure 2.5
Don't draw horizontal lines by "eyeballing" with a straight-edge tool.

T-Square   The **T-square** is an alternative to the parallel bar. It consists of a head and blade and is not installed on the board (Figure 2.6). T-squares are available in different lengths and materials (plastic, metal, wood/plastic combination). A T-square's advantages are portability, the ability to position it on a board to draw long angled lines, and a cheaper price than the parallel bar. When using a T-square, you must constantly hold its head firmly against the board to keep the bar straight.

Triangle   The **triangle** (Figures 2.7, 2.8) is used for drawing vertical lines. There are three types:

- **45°** draws lines that angle 45° and 90° to the horizontal
- **30°–60°** draws lines that angle 30°, 60°, and 90° to the horizontal
- **Adjustable** can be set to draw all angles between 0° and 90° It's an alternative to a protractor.

DRAFTING BOARD AND ACCESSORIES

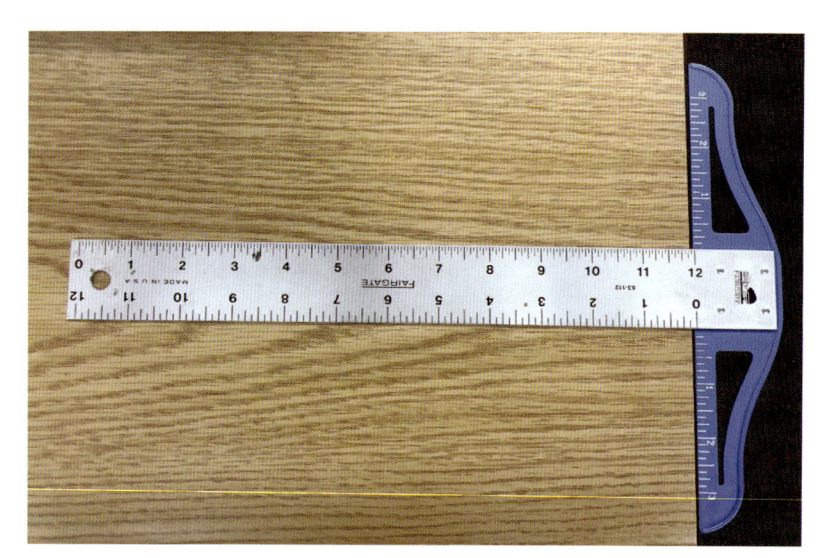

Figure 2.6
The T-square is an alternative to the parallel bar for drawing horizontal lines.

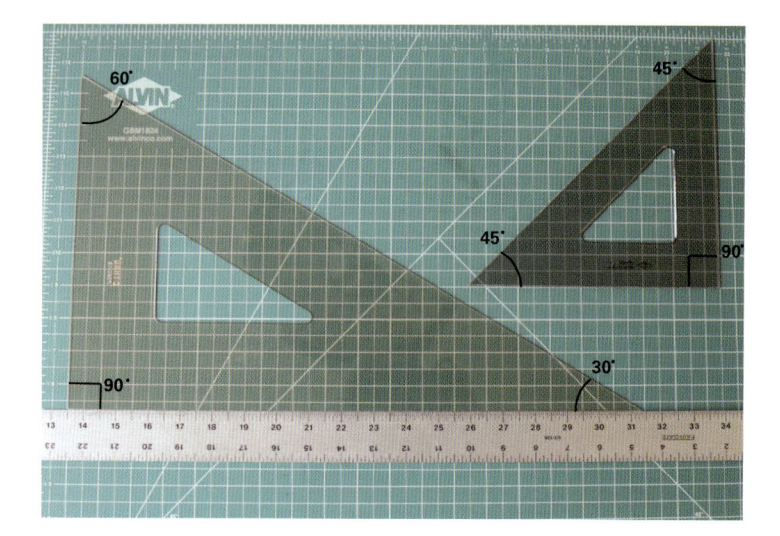

Figure 2.7
30°–60° and 45° triangles.

Figure 2.8
An adjustable triangle can be set to any angle between 0° and 90°.

22 CHAPTER 2 TOOLS, SCALES, AND MEDIA

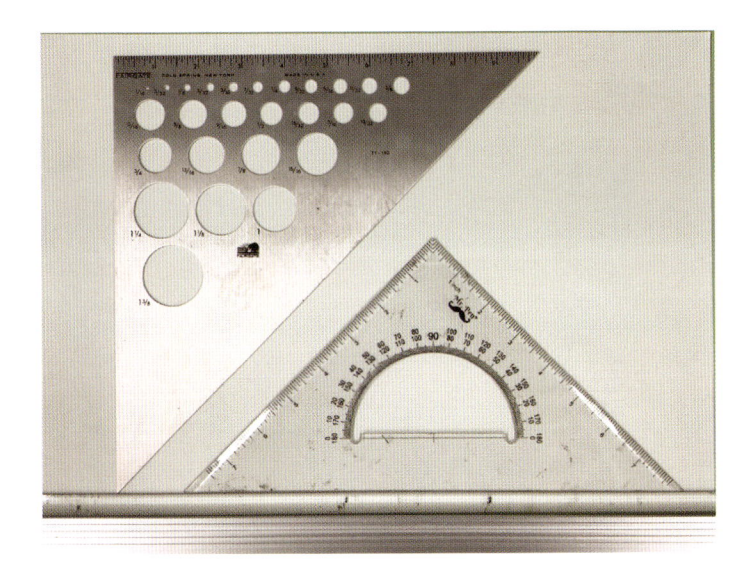

Figure 2.9
Specialty triangles

Figure 2.10
Proper way to draw a vertical line. The triangle rests on the parallel bar and slides left and right.

Figure 2.11
Improper way to draw a vertical line. The parallel bar is ignored, and the triangle is "eyeballed" straight.

Triangles come in different sizes suitable for different drafting chores. For instance, tall ones (36") are useful for perspective drawing, and small ones (6") work well for lettering. Medium height triangles (8–12") work for most general drafting purposes. There are also specialty triangles, such as ones with built-in protractors and circle templates (Figure 2.9).

The proper way to use the triangle is to seat it on the parallel bar or T-square and slide it back and forth (Figure 2.10). Don't hold a triangle or any other straightedge on your paper and "eyeball" it vertically (Figure 2.11). You cannot make consistently vertical lines like this.

DRAFTING BOARD AND ACCESSORIES

Figure 2.12a, b
Protractor showing 40° and 220° angles.

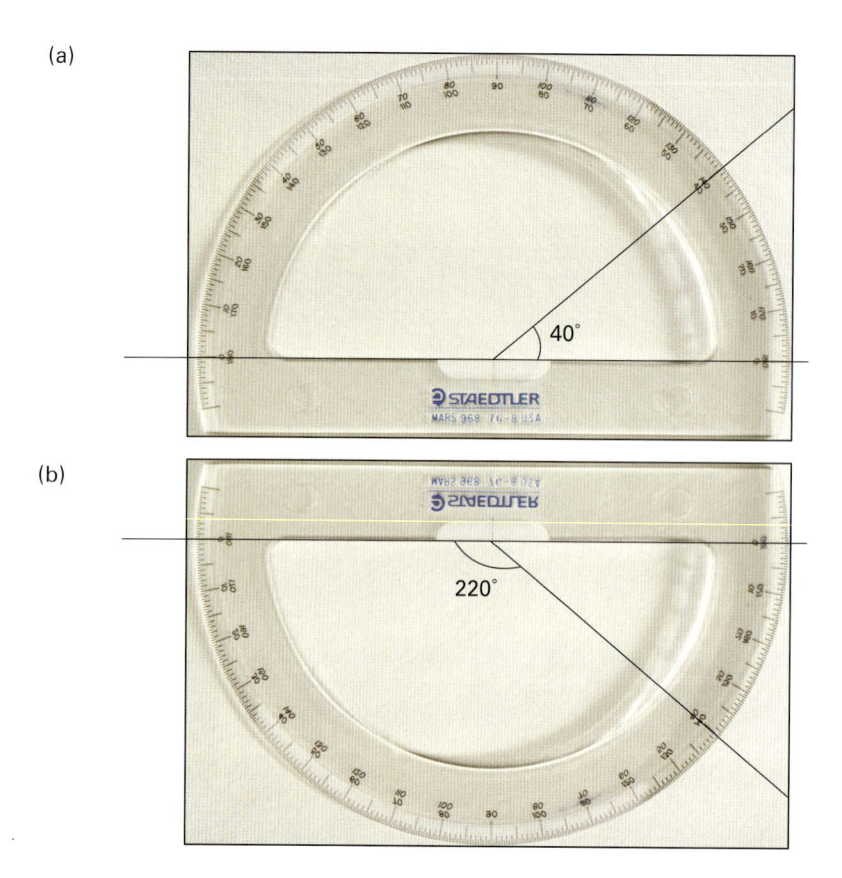

Protractor A **protractor** is used for drawing angles (Figures 2.12a, b). Numbers run from both left to right and right to left to enable an angle to be drawn on either side.

To draw an angle, align the protractor horizontally with the parallel bar. Find the crosshair in the center. Draw a line from it to a number on the curved part. That number is the angle. Figure 2.12a shows 40°. To measure an existing angle, align the protractor's flat side with one line, placing the crosshair at the vertex (point where the two lines intersect). Note where the other line intersects the protractor's curve; this number is the angle. To decide whether to read the upper or lower row of numbers, use the number that is smaller than 90° for an acute angle and the number that is larger than 90° for an obtuse angle.

To draw angles larger than 180°, draw a horizontal line, flip the protractor upside down, and mark off the number that, when added to 180, is the angle wanted. Then draw a line from the crosshair through it. Figure 2.12b shows 220° (180° + 40°).

Circle Template A **circle template** is a flat piece of plastic with cutouts of different size circles (Figure 2.13). The diameter (a line through the center to opposite points on the perimeter) is printed next to each circle. Each circle also has lines or tiny holes at each quadrant called quadrant markers. These are for placing on drafted lines to accurately position the circle's center point (Figure 2.14)

Architect's Scale The **architect's scale** enables a drafter to draw large objects on a small sheet of paper while remaining proportionally accurate. It measures in **imperial** (also called **English** or **U.S. Customary**) units of feet and inches. Triangular models have six edges, with a total of 11 scales. All edges except the imperial rule side have two scales printed on them. One scale reads left-to-right, the other right-to-left. The left-to-right scale is twice the size of the right-to-left. Select a scale based on the size and detail of what is being drawn. Large items, such as floor plans, are drawn with a small scale (e.g., 1/4" = 1'-0" or 1/8" = 1'-0") to fit on the paper. Small items, such as window details, are drawn with a large scale (e.g., 3/4" = 1'-0") to adequately show their features.

24 CHAPTER 2 TOOLS, SCALES, AND MEDIA

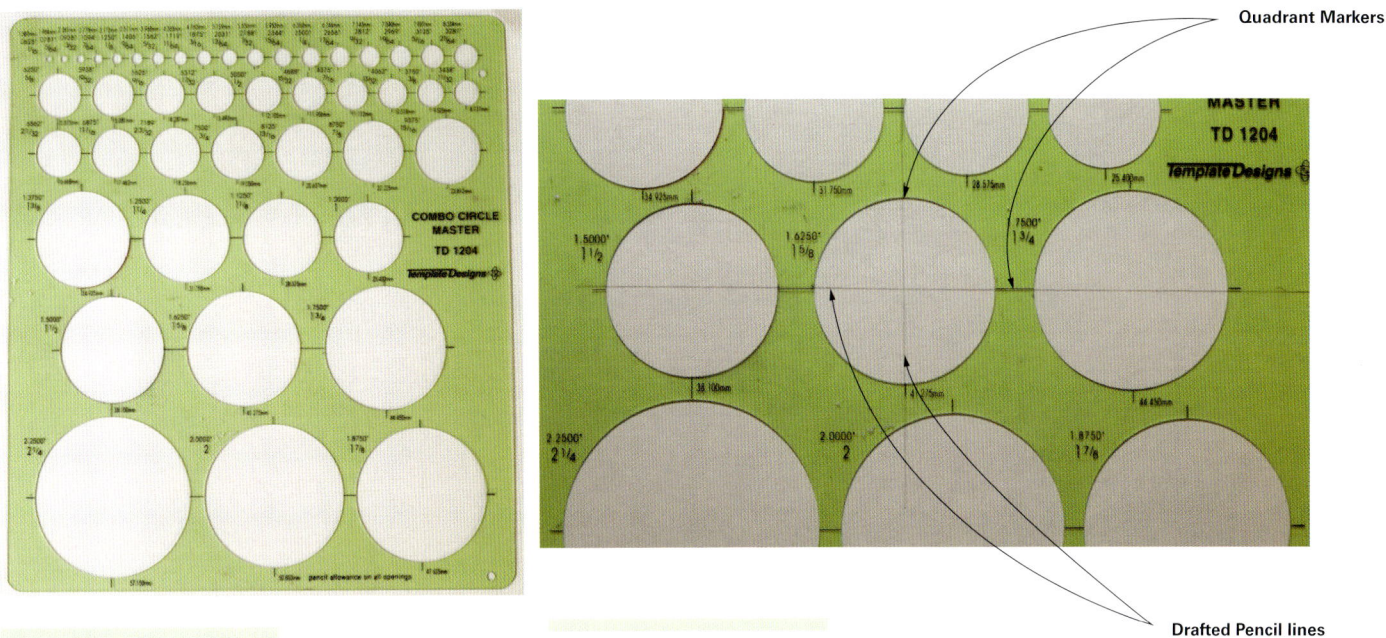

Figure 2.13
Circle template.

Figure 2.14
Line up the circle's quadrant markers with drafted lines to position the center.

Look at the 1/4" = 1"-0" scale on the right side of Figure 2.15. Each mark represents one real-life, or true size, foot. A line drawn from 0 to 4 represents a line 4' long. A line drawn from 0 to 14 represents a line 14' long. Always start from zero when measuring or drawing a line. A way to avoid confusing the 1/4" = 1'-0" scale marks with the 1/8" = 1'-0" scale marks on the opposite side is to note that numbers corresponding to a scale will ascend from its zero.

Subdivided Scale The architect's scale is open-divided, meaning whole feet are marked off. To measure inches, look at the marks to the right of the zero (Figure 2.16). These marks divide a foot into one-inch increments.

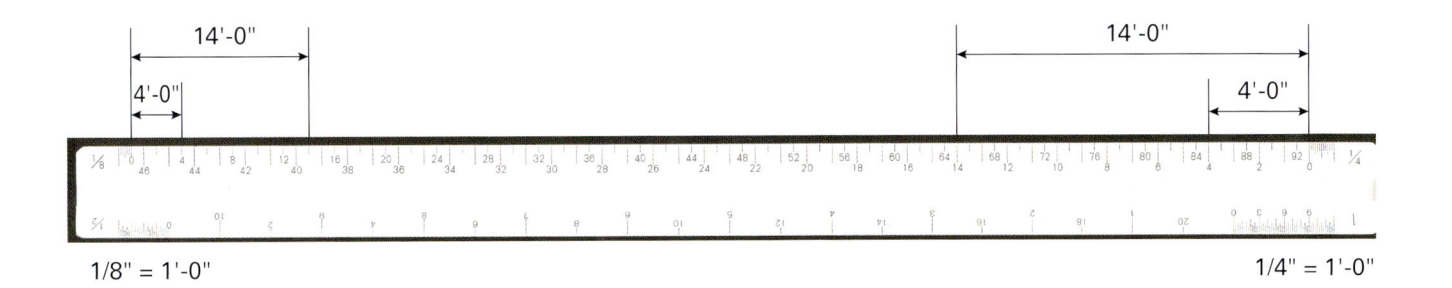

1/8" = 1'-0"   1/4" = 1'-0"

Figure 2.15
Identical measurements shown on the 1/8" = 1'-0" and 1/4" = 1'-0" scales.

DRAFTING BOARD AND ACCESSORIES   **25**

Figure 2.16
The subdivided 1/4" = 1'-0" scale.

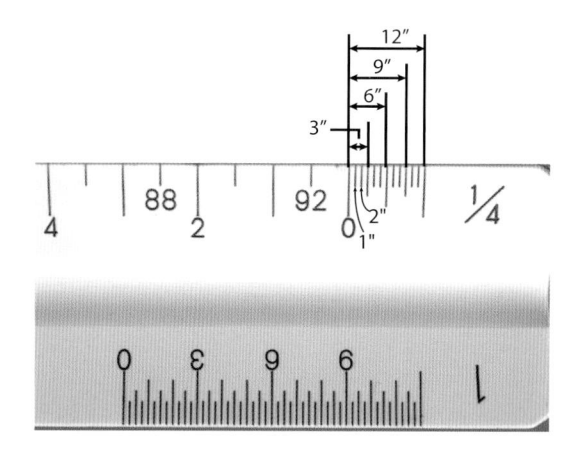

Figure 2.17
The subdivided 3/4" = 1'-0" scale.

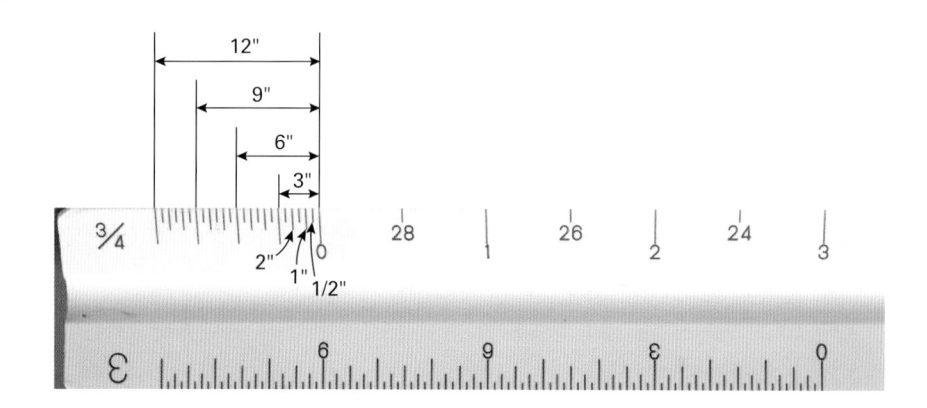

Every scale's zero has a fully divided foot to its left or right. One-quarter of the total distance is 3"; 6" is in the middle, and three-quarters of the way is 9".

On larger scales we can easily see the 3", 6", and 9" marks, as well as 1" marks between them. When confused, remember that the distance between zero and the last mark is always 12", so the 6" mark will always be in the middle. The 3" mark is exactly between the 0 and 6" marks, and the 9" mark is exactly between the 6" and 12" marks. Note how much larger the subdivided 3/4" scale (Figure 2.17) is than the subdivided 1/4" scale.

The subdivided 3/4" = 1'-0" scale has half--inch increments as well as inch increments. Find the 6" mark, then the 3" mark. Notice that there are even half-inch increments. These are suitable for drawing objects that must be dimensioned to a fraction of an inch.

Look at the 1 1/2" = 1'-0", scale in Figure 2.18. It is easy to confuse the lines delineating a large scale like this with the lines delineating the scale opposite it. A tip for telling them apart is to put your thumb and forefinger on the subdivided inch scale as shown in Figure 2.18. Then holding that position, move them to the open-divided scale as shown in Figure 2.19 to find 1'. The distance between your thumb and forefinger represents 1'.

Although single inches are delineated on the larger scales, the tiny 1/8" and 3/32" scales lack room for this. Therefore, each mark on their subdivided scales represents 2 inches (Figure 2.20). The 1/8" = 1'-0" scale is suitable for drawing large items such as site plans or large commercial plans. Obviously if detail and fractional inch measurements are needed, this is not a good choice.

26 CHAPTER 2 TOOLS, SCALES, AND MEDIA

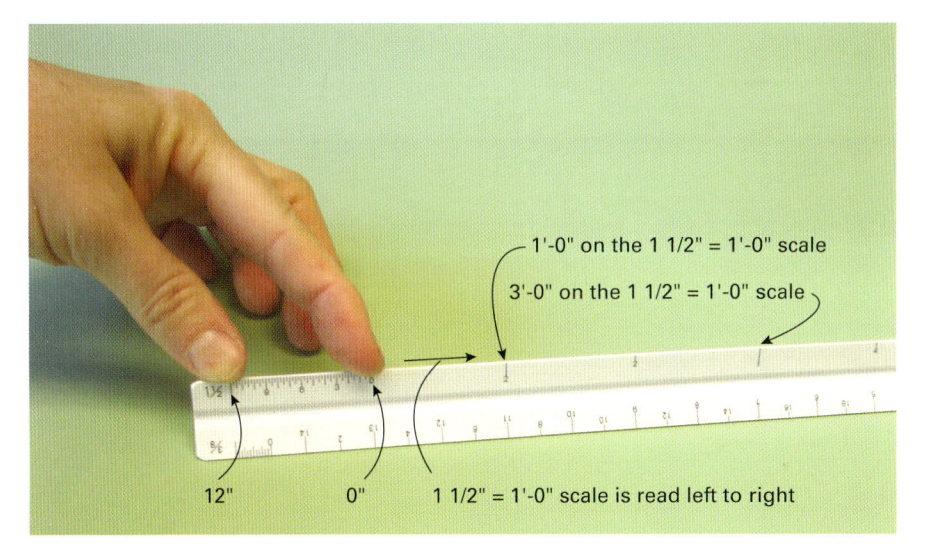

Figure 2.18
To verify which numbers go with the 1 1/2" =1'-0" scale, put your forefinger on the 0 and your thumb on the 12 of the subdivided scale. The distance between these two fingers is 12".

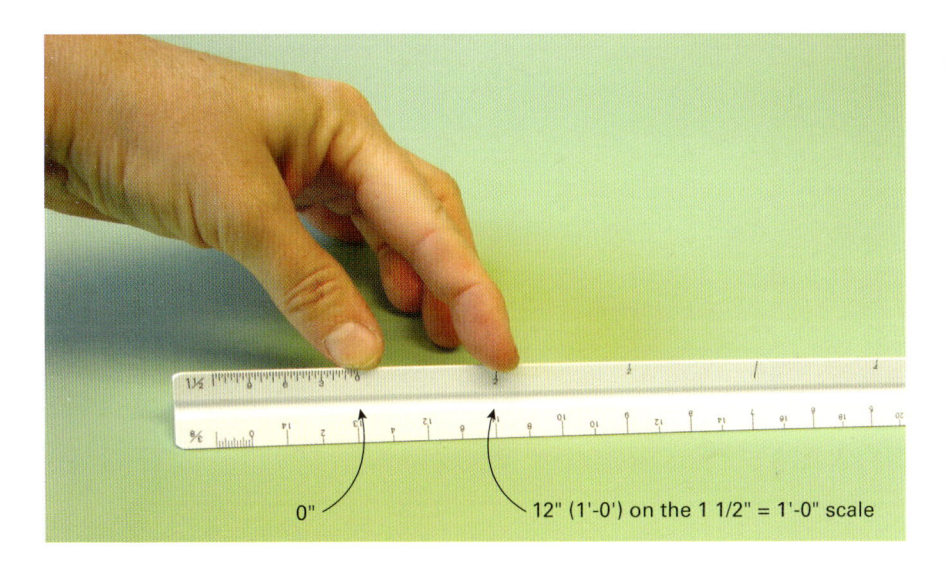

1'-0" on the 1 1/2" = 1'-0" scale
3'-0" on the 1 1/2" = 1'-0" scale
12"   0"   1 1/2" = 1'-0" scale is read left to right

Figure 2.19
Holding that position, move your fingers so your thumb lands on the 0. Your forefinger will land on the 1' increment on the open scale.

0"   12" (1'-0') on the 1 1/2" = 1'-0" scale

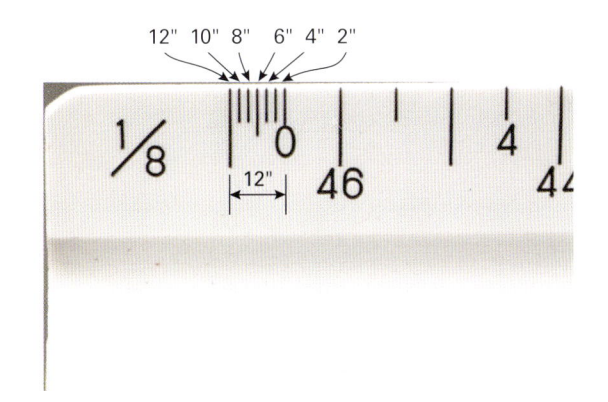

12" 10" 8" 6" 4" 2"

Figure 2.20
The subdivided 1/8" = 1'-0" scale.

DRAFTING BOARD AND ACCESSORIES   **27**

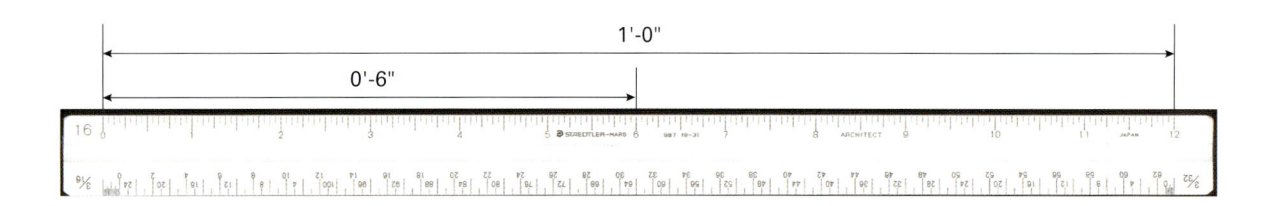

Figure 2.21
Imperial rule, also called the full (1" = 1") or true scale, is for drawing objects life-size.

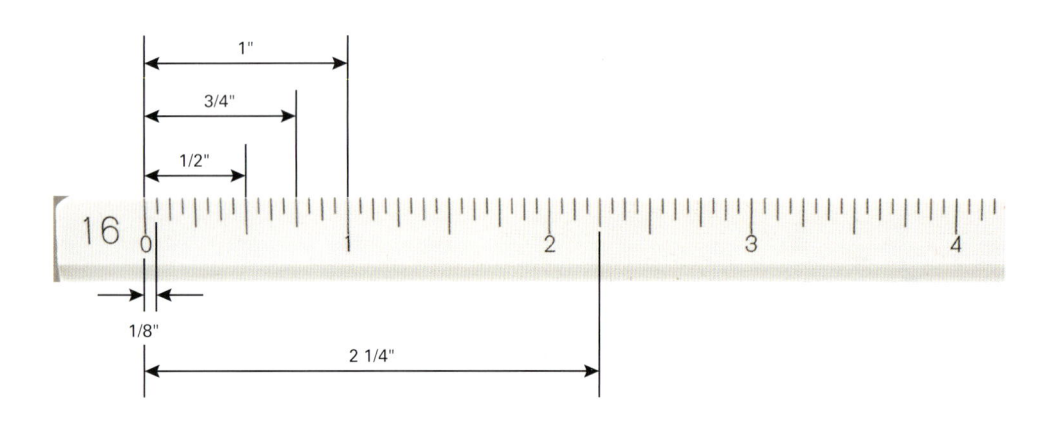

Figure 2.22
Partial and whole inch increments on the imperial rule.

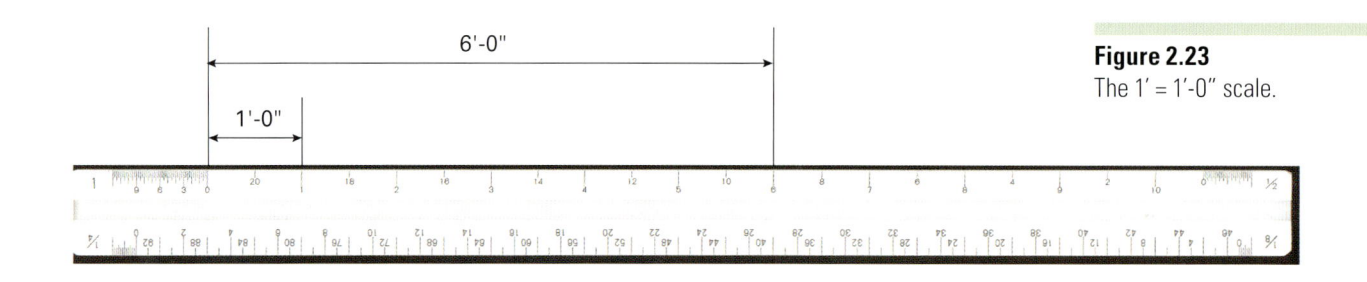

Figure 2.23
The 1' = 1'-0" scale.

Now look at the side marked 16 (Figure 2.21). This is the **true scale**, or **imperial rule**: 1" = 1". It enables drafting at life size. It is labeled 16 because each inch is subdivided into 16 segments. Each segment is 1/16th of an inch; thus, two segments are 1/8th of an inch (Figure 2.22).

Don't confuse this scale with the 1" = 1'-0" scale (Figure 2.23), where each inch represents a foot.

Finally, it is incorrect to call the scale tool a "ruler", as you can see that only one of its sides has an imperial rule.

Proper Architectural Notation (Imperial)
Look at the dimension notes (numbers) on Figure 2.23. They are written in proper architectural format.

Five feet is written *5'-0"*. It is incorrect to write *5'* or *5 1/2"* or *5.5'*. The first way is wrong because a hyphen and inch symbol should be included for clarity. The second way is wrong because it could be mistaken as 5 feet, half an inch (which would be written *5'-0 1/2"*). The third way is wrong because decimals aren't used on the architect's scale. Decimals are based on units of 10 and the architect's scale is based on units of 12. The following are examples of dimension notes written in proper architectural format:

0'-3"   0'-4 1/2"   6'-7 3/4"   9'-0"   100'-7"

When writing the scale itself, the entire equation should be written. For example:

1/4" = 1'-0"   1/8" = 1'-0"   1" = 1'-0"

Table 2.1 shows reductions achieved by different architect's scales:

> **Exercise**
> Draw horizontal lines of random length with a parallel bar or T-square. Then place a triangle on the bar and draw some vertical lines. Measure those lines with different architect's scales.

Table 2.1

1" = 1'-0"	makes a drawing	1/12 true scale
1 1/2" = 1'-0"	makes a drawing	1/8 true scale
3" = 1'-0"	makes a drawing	1/4 true scale
3/4" = 1'-0"	makes a drawing	1/16 true scale
1/2" = 1'-0"	makes a drawing	1/24 true scale
1/4" = 1'-0"	makes a drawing	1/48 true scale
1/8" = 1'-0"	makes a drawing	1/96 true scale

Scale Conversions   To convert a number from architectural to decimal units, divide the inch portion by 12. For instance, 7'-3" on the architect's scale is 7.25' in decimal units, since 3 divided by 12 is 0.25. 8'-9" is 8.75' (9 divided by 12 is 0.75). To convert from decimal to architectural units, set up an equation and cross-multiply to solve (see Table 2.2).

Table 2.2

Example 1: What is 0.5' in architectural units?	Example 2: What is .9' in architectural units?
5/10 = x/12	9/10 = x/12
10x = 60	10x = 108
x = 60/10	x = 108/10
x = 6"	x = 10.8
	x = 10 2/3" (divide 8 by 12 to get 2/3)
Example 3: What is .8' in architectural units?	Example 4: What is .2' in architectural units?
8/10 = x/12	2/10 = x/12
10x = 96	10x = 24
x = 96/10	x = 24/10
x = 9.6	x = 2.4
x = 9 1/2" (divide 6 by 12 to get 1/2)	x = 2 1/3" (divide 4 by 12 to get 1/3)

DRAFTING BOARD AND ACCESSORIES

> **FYI**
> When this specialty scale is rolled over a drawing, dimensions appear on a screen. It finds linear distances, rectangular areas, and volumes. It also converts between scales and dimensions. (Figure 2.24)
>
> **Figure 2.24**

Metric Scale

Like an architect's scale, a **metric scale** is used for drawing items proportionally accurate; however, it measures **System International (SI)** or metric units. Designers should be familiar with metric scales to specify the many products that come from other countries. Meters and millimeters are the most common metric units for architectural dimensioning. Look at the 1:100 scale in Figure 2.25.

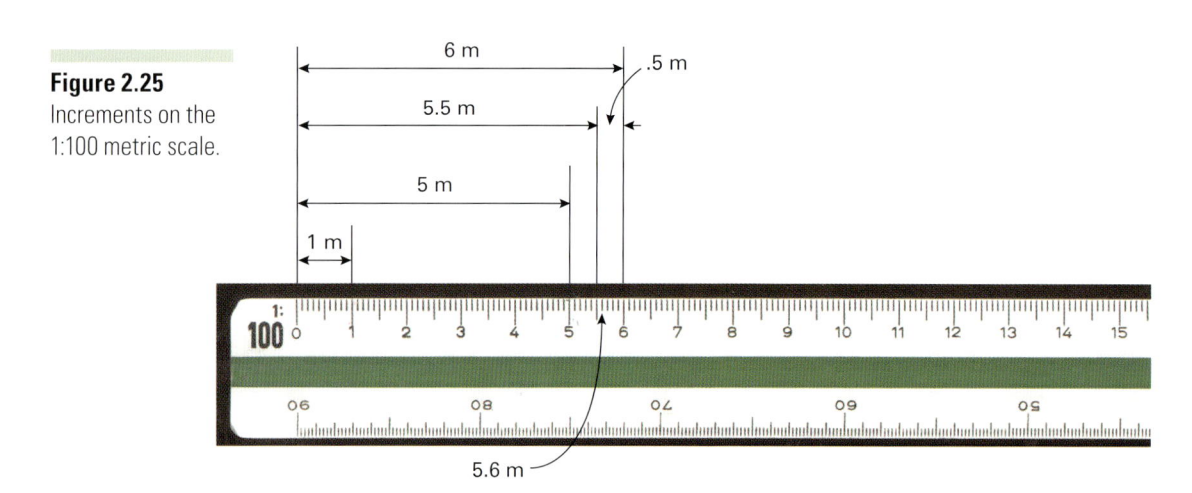

Figure 2.25
Increments on the 1:100 metric scale.

Its parts can represent any metric unit; we'll use meters for this discussion. The distance between 0 and 1 represents 1 meter. There are ten marks between 0 and 1, so each of those marks represents 1 (1/10) of a meter. There are ten spaces between 0 and 1, so each of those spaces represents 0.1 (1/10) of a meter. A line from 0 to 5 represents 5 meters (5 m). Look at the marks between 5 and 6. The mark exactly between them represents 5.5 meters. The line next to it represents 5.6 m (5 6/10 meters). All the metric scales are read similarly.

Enlarging and Reducing   Decimal places can be added to or subtracted from each scale to enlarge or reduce it. Removing a zero (moving the decimal one place to the left) from the 1:100 scale makes it a 1:10 scale, and the distance between 0 and 10 becomes 1 meter, with each of the 10 marks in between them becoming 0.01 m (1/100 of a meter). Adding a zero (moving the decimal to the right) makes the scale 1:1000, which turns the distance between 0 and 1 into 10. The 1:1 ratio is the metric true scale. At this ratio, the distance between 0 and 1 not only represents 1/100th of a meter; it is a full scale 0.01 meter (1/100 meter), or 1 centimeter.

Proper Architectural Notation (Metric)   Metric units are written lowercase with no s or period after them. Leave a space between the number and abbreviation. Place a zero in front of any number less than one unit and don't use a prefix alone to indicate a metric unit. The following are examples:

   10 dm   0.89 m   7.65 m   1,000 mm

Scales of Different Drawings

Table 2.3 shows drawing types and the scales commonly used for them:

Table 2.3		
Drawing	**Architect's Scale**	**Metric Scale**
Site Plans	1/8" = 1'-0"	1:100 m or 1:150 m
Floor Plans	1/4" = 1'-0"	1:40 m, 1:5 m, 1:7 m
Reflected Ceiling Plans	1/4" = 1'-0"	1:50 m
Electrical Plans	1/4" = 1'-0"	1:50 m
Sections, Details	3/4" = 1'-0" to 3" = 1'-0"	1:5 m to 1:10 m
Kitchen/Bath Cabinetry	1/2" = 1'-0"	1:5 m, 1:10 m, 1:20 m
Interior Elevations	1/4" = 1'-0", 1/2" = 1'-0"	1:20 m

Engineer's Scale   The **engineer's scale** is used for site plans, which are top-down views of a property. It measures in decimal units and is for large items like roads and topographical features. Its scales run from left to right and are 1:10, 1:20, 1:30, 1:40, 1:50, and 1:60. Meaning one inch equals 10', one inch equals 20', etc. Figure 2.26 shows distances on the engineer's 1:20 scale.

Figure 2.26
Distances on the 1:20 engineer's scale, often used for site plans.

Proportional Scale   A **proportional scale** (Figure 2.27) enables you to calculate what percentage to enlarge or reduce a picture on a copy machine to make it a desired size. Unlike the architect's scale, the proportional scale doesn't measure sizes; it just does the math to enlarge or reduce a picture. Common uses are to reduce photos to fit behind presentation board mats or reduce drawings to make them traceable onto a floor plan.

How to Use the Proportional Scale   You want to trace a photo of a plan view of a spiral staircase into your scaled floor plan instead of drawing that plan from scratch. The photo needs reducing to match that scale. So:

1. Measure the diameter of the photo with the true scale. Say it's 2".
2. Figure out the diameter of the circle you want to draw into your floor plan. Assume your stairs will have a 6'-6" diameter. Draw a circle with that 6'-6" diameter at the scale you're using for the floor plan. Then measure that circle with the true scale. Let's say it's 1 1/2".
3. Spin the inner wheel of the proportional scale so that the 2 (what you have) lines up with the outer wheel's 1 1/2" (what you want). Then look in the center window for the number needed (75%). If you don't have a physical scale, there are websites that will do this calculation; search for "proportional scale calculation."
4. Set the zoom function on a copy machine to that percentage. This will provide a correctly sized copy to trace.

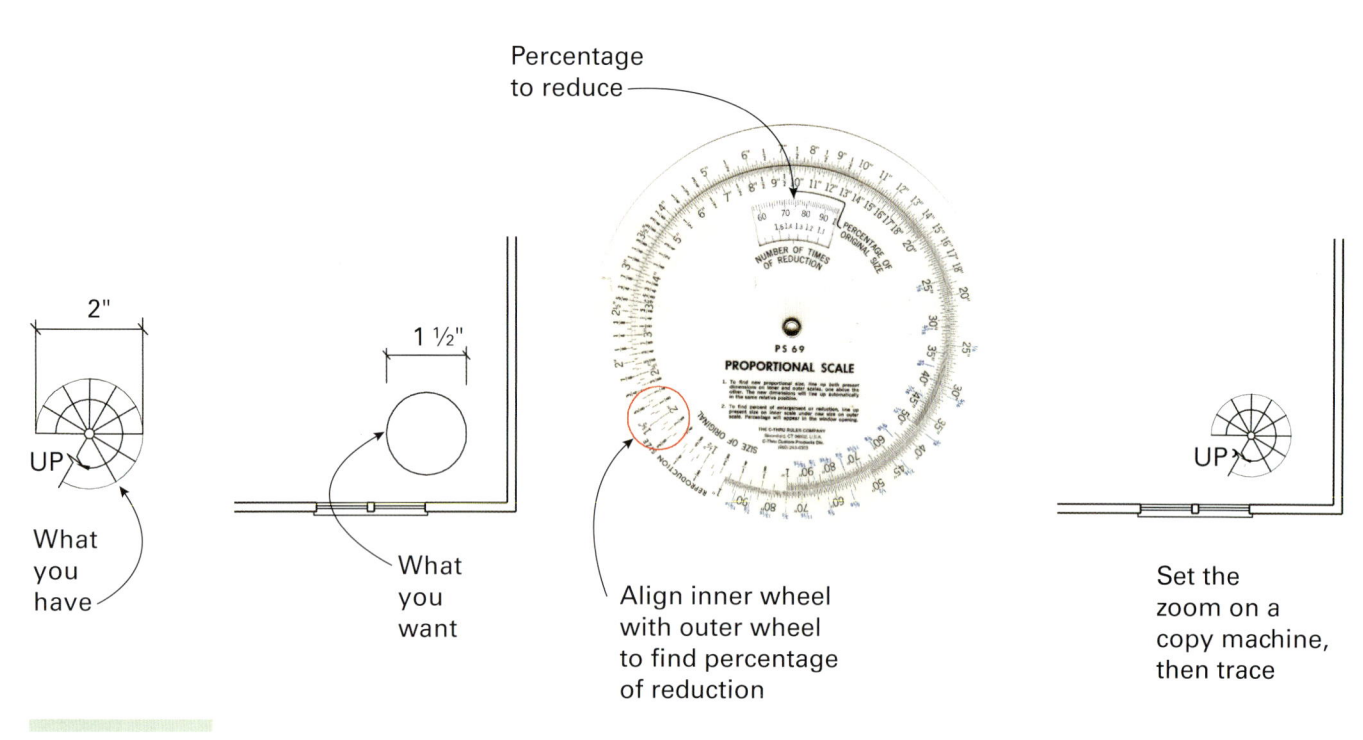

Figure 2.27
Use the proportional scale to reduce and trace a drawing.

Labels in figure: 2" / What you have / UP · 1 ½" / What you want · Percentage to reduce / Align inner wheel with outer wheel to find percentage of reduction · Set the zoom on a copy machine, then trace / UP

Pencils, Leads, and Erasers

Mechanical Pencil A **mechanical pencil** is a multi-piece, refillable tool. It is available in 0.3 mm, 0.5 mm, 0.7 mm, and 0.9 mm sizes, which refer to the tip opening. A quality model holds the lead sturdily and has a conical tip specifically angled to glide over a parallel bar, and an eraser. When drawing with a mechanical pencil, hold the pencil perpendicular to the board and rotate it as you pull it across the paper to obtain a consistent line width.

To fill a mechanical pencil, remove the cap and eraser, exposing a hollow area called the lead reservoir (Figure 2.28). Put a few leads in it (not a whole package, as they won't all fit and will jam the pencil) and then replace the eraser and cap. Click the cap until a lead protrudes through the opposite end. To change leads, empty the reservoir, then remove

Figure 2.28
Remove the cap and eraser to fill a mechanical pencil. Remove old lead in the tip by clicking the cap so the lead protrudes enough to grasp and remove.

Labels: Tip · Eraser · Cap · Reservoir

Exercise
Obtain two mechanical drafting pencils. Fill one with soft (dark) lead and one with hard (light) lead. (Know that when you buy one, it is prefilled with a soft lead). On vellum, draw three rows of guidelines 1/2" tall separated by 1/4" spaces; three rows of guidelines 1/4" tall separated by 1/8" spaces; and three rows of guidelines 1/8" tall separated by 1/6" spaces.

any lead already threaded through the tip by clicking the cap so enough lead protrudes that you can grasp and remove it (keep the cap depressed while removing the lead).

When leads get short they will push back into the pencil when you write. When this happens, remove the short piece and click a new piece through. If a piece of lead becomes jammed inside, remove it by unscrewing the tip from the pencil and pulling the lead out.

Lead Holder Also called a **clutch pencil**, a **lead holder** is an alternative to the mechanical pencil. It holds a 2 mm lead (Figure 2.29). This larger lead doesn't break as easily and can create multiple **line weights** or thicknesses by rubbing one side on a sandpaper block. It requires frequent sharpening with a **lead pointer** (rotary blade sharpener) or **sandpaper block** (sheets of light grit sandpaper).

Leads Drafting leads are actually sticks of powdered **graphite** (carbon), fillers, and clay binder. More graphite makes a dark line; more filler makes a light line. Leads come in multiple grades, meaning levels of hard/lightness and soft/darkness (Figure 2.30).

Figure 2.29
Clutch pencil, 2 mm leads, sharpener and sand block.

Figure 2.30
Leads range from hard (light) to soft (dark).

The **lead grades** most relevant to drafting are B, F, HB, H, 2H, 3H, and 4H. Soft and medium grade leads such as B, F, and HB draw dark, easily smudged lines. Hard grade leads such as H, 2H, 3H, and 4H draw light, hard to smudge lines. Pressure applied to the pencil also affects the line's lightness and darkness. A heavy-handed drafter may need to use an H for a dark lead and a 4H for a light lead; a light-handed drafter may need to use an F for a dark lead and an H for a light lead. Most drafters only need a 4H and HB for pencil work, as these two weights provide sufficient contrast. Leads also come in diameters to match the pencil tip size.

Soft leads are used for **object lines**. Hard leads are used for **construction lines**. An object line defines the idea being communicated, hence must be dark. For instance, the walls on floor plans are object lines. They are intended to reproduce on copies. A construction line simply helps the drafter create object lines. It is not meant to reproduce on a copy, hence is light. Since pencil drafting utilizes both light and dark lines, it is convenient to keep two mechanical pencils on hand, one filled with each. A hard lead will appear thinner than the same size soft lead.

Eraser An **eraser** is a piece of soft rubber or plastic used to remove pencil lines. A white plastic eraser is formulated for drafting on vellum and works better on that medium than a general-purpose eraser, which is too hard on it. A yellow **imbibed eraser**

Tip
A soft lead drawn with light pressure does not create the same quality light line that a hard lead does and should not be substituted for one.

PENCILS, LEADS, AND ERASERS

Figure 2.31
Cut ragged eraser ends off with an art knife.

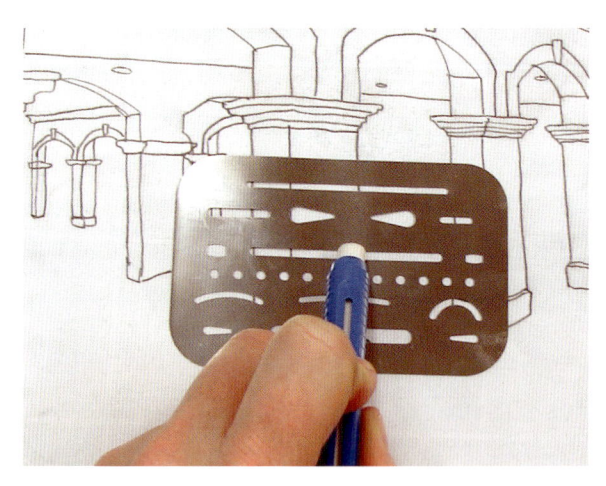

Figure 2.32
An eraser shield protects surrounding areas from being erased.

is formulated for drafting on plastic film. When an eraser end gets ragged, cut it smooth with an **art knife**, a precision cutter with a sharp, carbon blade (Figure 2.31). Only use fresh erasers, because hard, stale ones leave indelible marks. There are also electric erasers, handheld tools that erase more powerfully, and may eliminate line traces that handheld ones don't.

Eraser Shield An **eraser shield** is a metal template that has openings of different sizes and shapes (Figure 2.32). It allows the drafter to erase small areas without removing surrounding lines.

Other Tools

Drafting Brush A **drafting brush** is used to remove eraser crumbs from the paper. Don't sweep the paper by hand, as that rubs oil and dirt into it. A wide, soft paintbrush is an alternative.

Drafting Tape **Drafting tape** is a sticky-backed paper that resembles masking tape but has less glue, making it less likely to damage paper when removed. It's available in rolls and dots. Roll tape is better for connecting long sheets of paper. **Painter's tape** is an alternative to drafting tape. Cut tape off the roll instead of tearing it because ragged edges curl up under the parallel bar.

Divider The **divider** has points at the end of both legs and is used for marking off distances where numerical measurement isn't needed (Figure 2.33).

To use it, place its points at the endpoints of the line whose length you want to measure. Then, holding the dividers in that position, move it to where you wish to mark that distance off (Figure 2.34). To double that distance, swing one point around the other.

Compass A **compass** is used for drawing circles and arcs (Figure 2.35). It has a point at one end and a piece of graphite on the other. Rub one side of the graphite on a sandpaper block to **bevel** (angle) it, which creates a

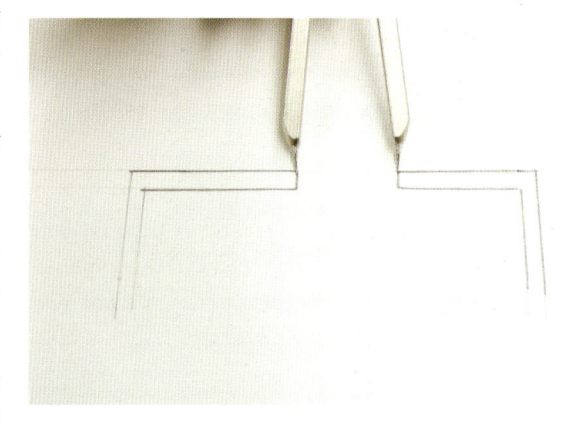

Figure 2.33
The divider marks off distances.

34   CHAPTER 2   TOOLS, SCALES, AND MEDIA

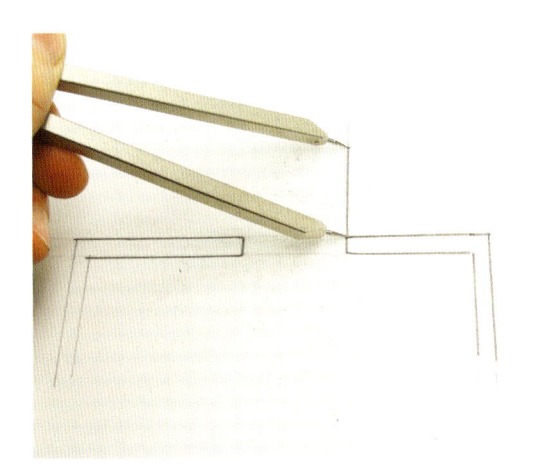

Figure 2.34
A common use for a divider is to mark the width of a door opening and then transfer that distance to the line representing the door.

Tip
Frequently tighten the thumbwheels that hold the divider points in, because when they loosen, the points fall out. Reinserting the points requires lining up small holes in the bars that hold the points in.

sharp point. Place it in the compass with the bevel facing out. Some compasses are **universal**, meaning you can insert a whole pencil or pen into it, which is more convenient to work with. **Technical compasses** have a thumbwheel in the middle to adjust the diameter. This makes more accurate circles.

To use a compass, place the point where the center of the circle or arc is to be, adjust the distance between the point and the lead to the desired radius, and swing the lead (Figure 2.36). Placing tape at an often-used compass point location will help keep the paper from tearing at that spot.

Ames Lettering Guide Lettering is the art of drawing notes inside guidelines and is discussed in Chapter 4. Here we'll discuss guidelines, which are construction lines that control letter height. The **Ames lettering guide** is used for drawing multiple rows of guidelines spaced consistently apart (Figure 2.37).

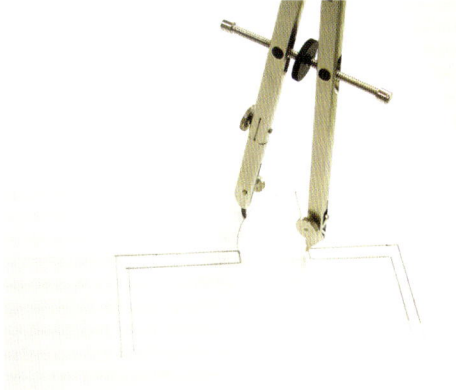

Figure 2.36
Using a compass to draw a door arc.

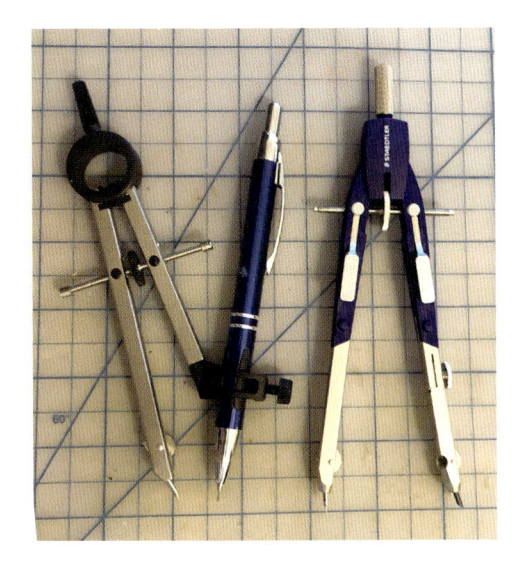

Figure 2.35
A universal compass (left) and a traditional compass (right). Both have thumbwheels, which enable more accurate drawing.

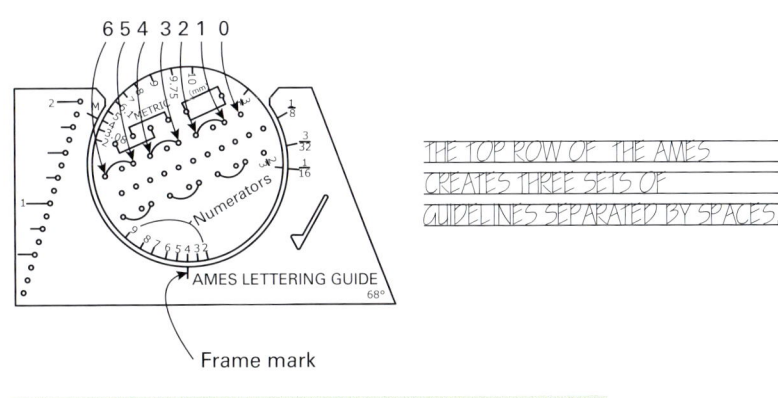

Figure 2.37
Use the guide's top row of holes to draw lines. Here, the 4 is lined up with the frame mark to draw 1/8" guidelines.

OTHER TOOLS 35

The guide consists of a stationary part and a wheel that rotates. The numbers at the bottom of the wheel are numerators, and all have a denominator of 32. That is, the #2 is 2/32"(1/16"), the #3 is 3/32", the #4 is 4/32"(1/8"). Metric units are at the top of the wheel. Choose the height you want the letters to be, and then set that height by rotating the wheel so the mark under the numerator number lines up with the frame mark. For example, to draw letters 1/8" tall, choose the 4, since 4/32 is 1/8".

Place the lettering guide on the parallel bar, put the pencil in the holes shown in Figure 2.37 starting with hole (1), and slide the lettering guide back and forth across the paper. After drawing a line at hole (6), you'll have three rows of guidelines spaced 1/16" apart. To draw another set of three guidelines spaced consistently from the first set, line up the (0) hole with the last line drawn. Then put the pencil back in hole (1) and start over.

When only one set of guidelines is needed, it's easier to use a scale to make two marks denoting the top and bottom of the guideline, and then draw lines through those marks with a parallel bar (Figure 2.38).

> **Tip**
> Make notes and title block information 1/8" tall. Make drawing titles 1/4" tall.

French Curve   French curves, also called **irregular curves**, are for drafting **arcs** (parts of circles or other curves). Use the French curve by finding a place on it that matches the arc to be drafted, and trace as much of the arc possible with that portion of the curve (Figure 2.39). Then find a different section on the curve that matches more of the arc being drawn. Overlap the connecting arcs a bit to ensure continuity.

Ink Pens   Ink pens are used for fine drafting and make it easy to apply different line weights. See Chapter 5 for a chart that describes which pen weights to use for different lines. They are either **disposable** or **technical** (Figure 2.40), and both can be bought singly or in sets. Disposable pens are felt-tips. When they're empty, throw them out. Most have a line quality good enough for manual drafting and are low cost and low maintenance. Disadvantages include running out of ink quickly and nibs (tips) deforming under pressure. Nib sizes vary with the manufacturer, but generally are 01 mm, 02 mm, 03 mm, and 04 mm, with 01 mm being the smallest.

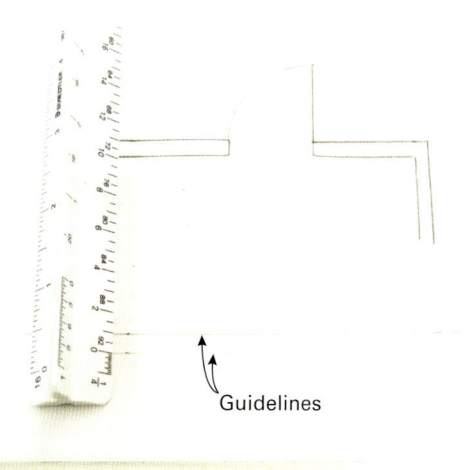

Figure 2.38
When just one row of guidelines is needed, use a scale to mark its height, and then draw lines through those marks.

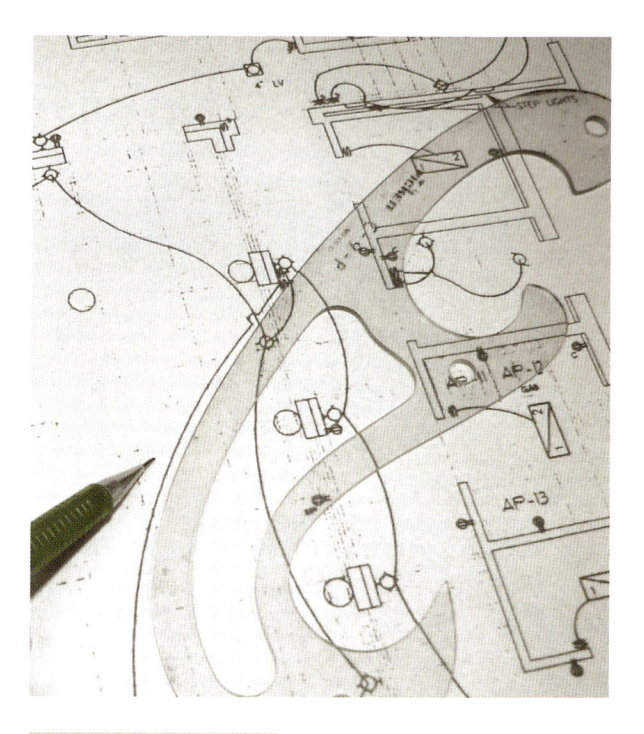

Figure 2.39
Draw arcs with a French curve.

36   CHAPTER 2   TOOLS, SCALES, AND MEDIA

Figure 2.40
Technical and disposable pens.

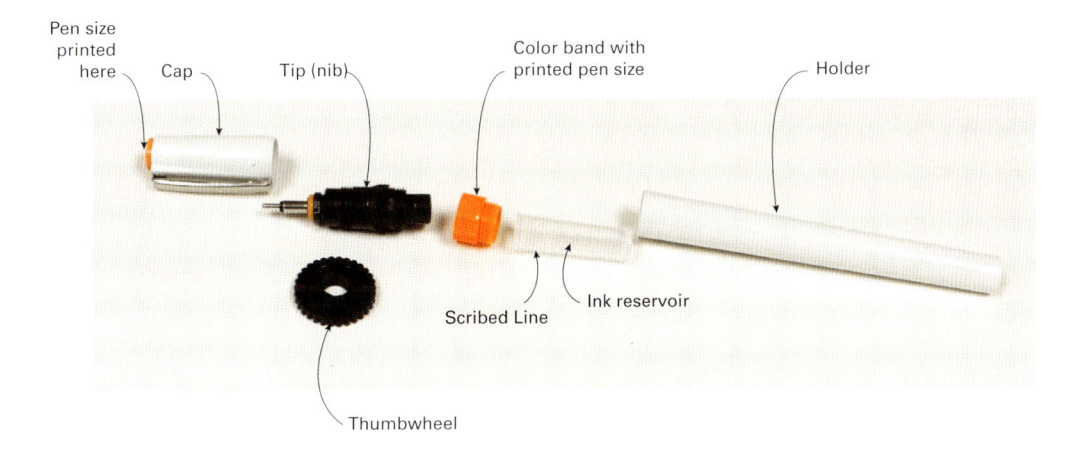

Figure 2.41
Parts of a technical pen.

Technical pens have multiple parts (Figure 2.41). They offer more sizes than disposable pens, draw nicer lines, and can be refilled when empty. Disadvantages include expense and maintenance; they require cleaning after each use or they will become clogged. Their tip sizes are notated as 000, 00, 0, 1, 2, 3, 4, 5, 6, 7, and 8, with 000 being the smallest.

A functional pen tip makes a rattling noise when shaken. Take the pen apart to fill the ink reservoir and don't fill it past the scribed line near its top. Reassemble and shake gently to start the ink flowing or tap it lightly on the board with the cap on. To clean, unscrew the tip from the rest of the pen, rinse upside down under water (Figure 2.42), and blot dry.

If it is difficult to unscrew the tip, use the included thumbwheel to apply extra torque. Every few weeks pen tips should be soaked in an ultrasonic cleaner (like a jewelry cleaner) with pen cleaning solution. If there is no sound when the pen tip is shaken or if the pen leaks around the tip, an ultrasonic cleaning will usually fix the problem.

Figure 2.42
Rinse a technical pen tip after each use.

OTHER TOOLS

Figure 2.43
Trace the body of the sofa bed and chairs along a lightly drafted line so they're straight. Then move the edge on the template up to touch the body and trace it with a single line.

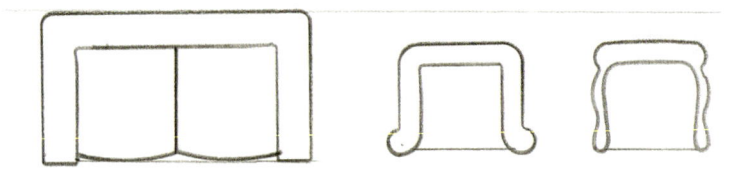

Figure 2.44
Trace the tub and shower and then draft their supporting structures as per the template's printed lines.

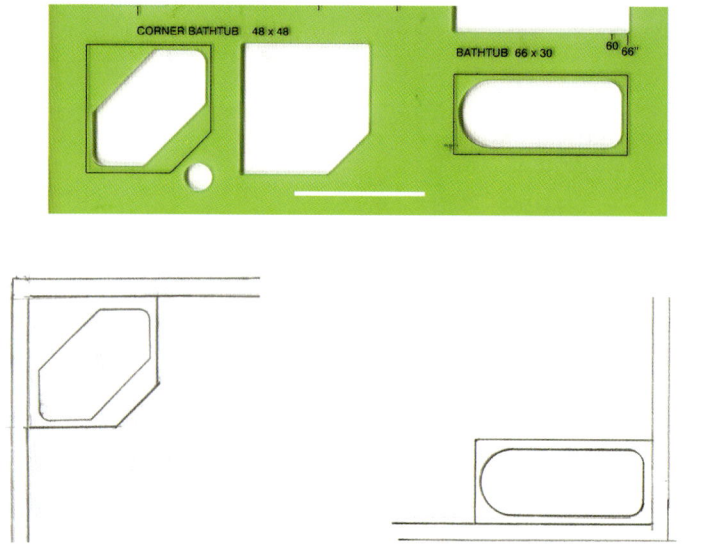

Furniture Template   A **furniture template** is a flat piece of plastic with plan symbols for tracing. While most symbols can be traced by simply running a pencil around the outline, some need more work. For example, look at the loveseat and chair symbols in Figure 2.43. Trace the body first and then move the edge so that it touches the body. Trace that edge with a single line and then add a vertical line to represent cushions. Shapes for tubs need to be enclosed in a supporting structure because the template just has the basin. The printed lines around the basin show how to add that structure (Figure 2.44).

All template shapes should be **rendered**, or completed with lines that describe what the piece of furniture is. For example, add pillows and a turndown to a bed template instead of leaving it as a rectangle (Figure 2.45).

Cutting Mat   A **cutting mat** is a vinyl surface used to cut paper on (Figure 2.46). Good quality ones are smooth, self-healing, and have a printed grid to facilitate straight cutting. They're available in different sizes.

Utility and Art Knives   The **utility knife**, also called a **box cutter**, has a thick blade that cuts **vinyl substrate** and heavyweight board. The art knife is a pencil-like tool that cuts plastic tape, paper, and thin board. Buy their replaceable blades in packs for the greatest economy, and change them often, as dull blades make sloppy cuts.

38   CHAPTER 2   TOOLS, SCALES, AND MEDIA

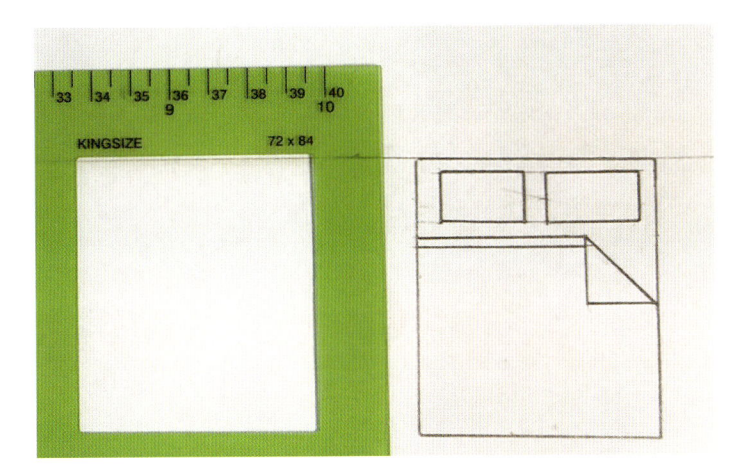

Figure 2.45
Add pillows and a turndown to a bed symbol.

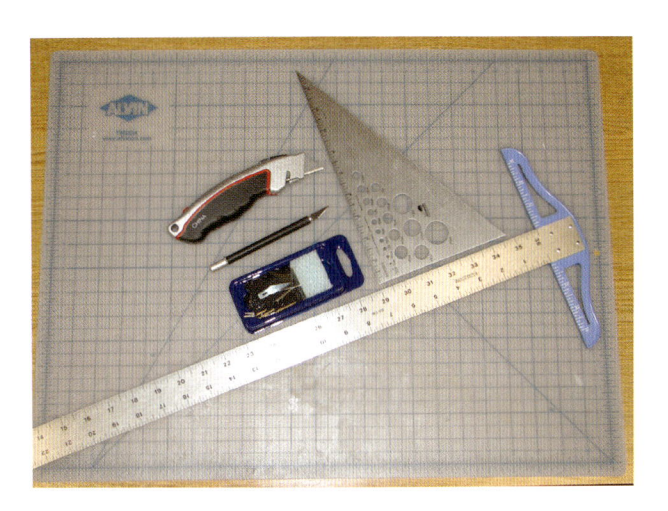

Figure 2.46
A cutting mat, metal straightedges, utility and art knives.

Tip
Don't cut on surfaces that you draft on or with edges of tools you draft with. A sliced surface will catch the pencil, creating a jagged part in the drawing.

Drafting Media

This is what you draw on (Figure 2.47). There are three types: **tracing paper**, **vellum**, and **plastic film**. Vellum and film come in standard size sheets, rolls, and tablets (Figure 2.47). A fourth media, graph paper, is used for sketching and is discussed in Chapter 3.

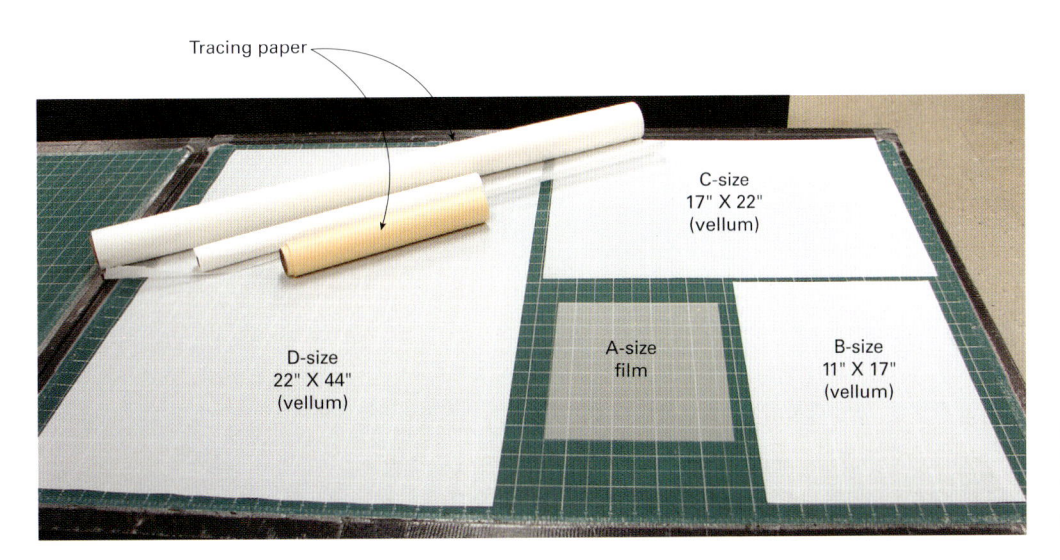

Figure 2.47
Vellum, film, and tracing paper.

DRAFTING MEDIA 39

Tracing Paper Tracing paper, also called **trash paper or bumwad**, is thin, cheap, and semi-transparent. Its semi-transparency enables layering, which lets you trace parts of your design while refiguring others. Use it for brainstorming, sketching, and making a final drawing to trace. It's available in yellow or white.

Vellum This is a semi-opaque, quality paper used for presentation ink or pencil work. Its fine, white background makes good copies. Some vellum sheets are smooth on one side and have a rough tooth on the other. Vellum is not ink-erasable.

Plastic Film Plastic film, such as the brand Mylar, is polyester sheets used for ink work. It's ink-erasable and stronger than vellum. Film is unsuitable as a final presentation material due to its high translucency, but this characteristic makes for great, sharp copies. Film can be single sided or double sided, the latter meaning either side can be drawn on. On single-sided film, the **matte** (dull) side is drawn on. Graphite pencils can't be used on it due to smearing.

Care and Storage of Tools

Drafting tools are expensive and fragile, so take care of them and keep them clean. Also clean the underside of the parallel bar and wipe down the drafting board after each use, as dirt from them will embed into the paper. "Green" cleaners work better than regular household ones because the latter remove painted lines from the scale and other instruments and may also leave a film. Use tools for their intended purpose only and don't leave plastic ones in a hot car, because heat will warp them.

Tools can be kept in anything that keeps them compartmentalized and from jumbling during transport. Art boxes are the most common solution (Figure 2.48). A portfolio envelope or tube (Figure 2.49) keeps large papers from creasing.

Figure 2.48
An art box holds tools well.

Figure 2.49
Portfolio envelopes and tubes keep design work from creasing.

The Copying Process

Most designers don't permit their original drawings to leave the office. Instead, they make paper or electronic copies.

Paper Copies Sets of architectural drawings are often called **blueprints**, but this is a misnomer. *Blueprinting* is an old, ammonia-based copying method. A transparent original was needed through which a lamp could shine. This method produced copies that had white lines on a dark blue background (Figure 2.50).

Digital Scans Today, print shops have **wide-format photocopiers** (Figure 2.51), printers that support paper roll widths between 18" and 100". They digitally scan the original and can produce hard copies (Figure 2.52) or PDF files (Figure 2.53). PDFs are used in electronic design portfolios and construction management software. These photocopiers can also read files on inserted thumb drives. So, while some still refer to the sets of drawings as blueprints, a more accurate term would simply be *the drawings*. Calling them *the plans* is also inaccurate because a plan is just one drawing; a set contains many kinds of drawings.

Figure 2.50
Blueprinting is an old copying process that produced white lines on a dark blue background. These are elevation drawings of a Joy Gas service station in Toronto.
Source: Wikimedia Commons.

Tip
Don't use a scale to measure photocopied drawings. Each time a photocopy of a drawing is made the lines become slightly stretched; therefore, taking measurements directly from a copy may not be accurate. Read the dimensions on a drawing instead.

Figure 2.51
Wide-format photocopier.

Figure 2.52
Set of architectural drawings.

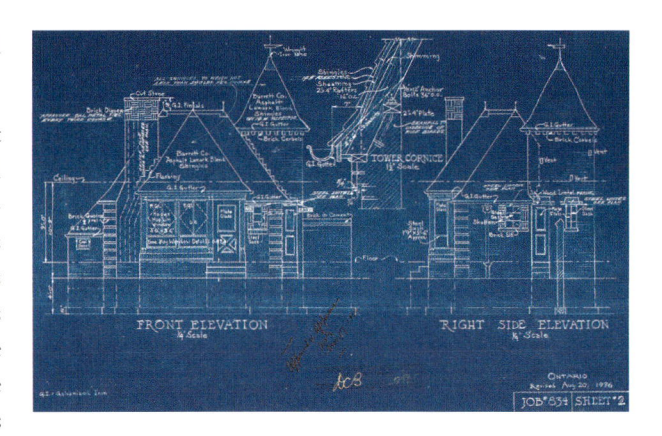

Figure 2.53
Digital copiers can generate PDF files.

THE COPYING PROCESS   **41**

Copies are printed on rolls or standard size sheets of bond paper. They can make copies from other copies, so a transparent original isn't needed. But vellum or film originals produce the sharpest copies. Don't apply correction fluid, plastic tape, rub-on letters, or any other adhesive-backed object on an original, because they may come off or melt inside the copier, causing damage. Print shops will typically reject such originals.

Phone Apps   Apps that scan and save as PDF files exist, but they are best suited for text documents, not architectural drawings. The PDFs they create are **low-resolution** (few pixels spread throughout the image), which precludes zooming in for a closer look or printing full-size paper copies. Results may also include background (such as the table the sheet is on), dim lighting, and distortion from the phone not held parallel to the sheet. These apps also create **raster** (pixel) not **vector** files, which can scale up or down without losing resolution. Vector files are needed for import into construction software (Figure 2.54).

Figure 2.54
Construction software reads vector PDFs. A touchscreen enables zooming, searching, and placing drawings side-by-side.

Summary

Manual drafting requires specialized tools, the most basic being a drafting board, parallel bar, triangles, pencils, hard and soft leads, and the architect's scale. Once you know what these tools are and how to use them, you are on your way to creating architectural drawings.

Classroom Activities

1. On a sheet of vellum, draw lines with different hard and soft leads and observe their differences.
2. On a sheet of drafting film, draw lines with different ink pen tip sizes and observe their thickness differences.
3. Measure this textbook with the 1:1 metric scale and with the imperial true scale.

Questions

1. What is the purpose of the architect's scale?
2. What is the "true scale"?
3. Describe the correct way to draft a vertical line.
4. What number should the Ames lettering guide be set to for 1/8" tall guidelines?
5. What is the difference between vellum, film, and tracing paper?

6. Describe the purpose of an object line and what it should look like.
7. Describe the purpose of a construction line and what it should look like.
8. Which draws a darker line, an HB or a 4H lead?
9. Why is "blueprint" an incorrect term to describe today's copied architectural drawings?
10. Describe how today's architectural drawings are mechanically reproduced.

Further Resources

History of the drafting pencil. https://blog.pencils.com/brief-history-mechanical-pencil/

https://www.metric-conversions.org/converter.htm

Imperial to Metric Conversions

Online Proportional Scale www.gieson.com/Library/projects/utilities/proportion_wheel/

Sources of competitively priced drafting supplies: amazon.com, draftingsteals.com, and utrechtart.com.

Keywords

- Ames lettering guide
- arc
- architect's scale
- art knife
- bevel
- blueprint
- box cutter
- bumwad
- circle template
- clutch pencil
- compass
- construction lines
- cutting mat
- disposable pen
- divider
- drafting board
- drafting brush
- drafting chair
- drafting tape
- engineer's scale
- English units
- eraser
- eraser shield
- French curve
- furniture template
- grades
- graphite
- imbibed eraser
- imperial rule
- imperial units
- ink pen
- irregular curve
- lamp
- lead grade
- lead holder
- lead pointer
- line weights
- low-resolution
- matte
- mechanical pencil
- metric scale
- object lines
- painter's tape
- parallel bar
- portable drafting board
- proportional scale
- protractor
- raster file
- render
- sandpaper block
- subdivided scale
- System International (SI)
- technical compass
- technical pen
- tracing paper
- trash paper
- triangle
- true scale
- t-square
- universal compass
- U.S. Customary units
- utility knife
- vector file
- vellum
- vinyl cover
- vinyl substrate
- wide-format photocopier

CHAPTER 3

2D and 3D Sketching

Figure 3.0
3D sketch used for a presentation. Courtesy of wolfgangtrost.com.

OBJECTIVES

Upon completion of this chapter you will be able to:

- Understand the difference between perspective and paraline views
- Identify the different kinds of perspective and paraline views
- Visualize 3D space from 2D views
- Explain why 2D views are needed for architectural drafting.
- Create 2D and 3D sketches

All built items start out as an idea. That idea is brainstormed, problem-solved, refined, and communicated with **sketches**. These are quick, freehand drawings created to problem solve and assist a more finished picture. **Sketching**, the process of creating sketches, is an important skill for designers as it enables them to quickly create visuals. It precedes the more labor-intensive work of drafting. Sketches can also serve as guides to the finished product while drafting.

Three Physical Dimensions

All objects have three physical dimensions: height, depth (also called width), and length. A **two-dimensional sketch** means only two of their three physical dimensions are shown; a **three-dimensional sketch** means all three of their dimensions are shown (Figures 3.1, 3.2).

Figure 3.1
Plan and elevation views of a living room (2D).
Courtesy of mkerrdesign.com.

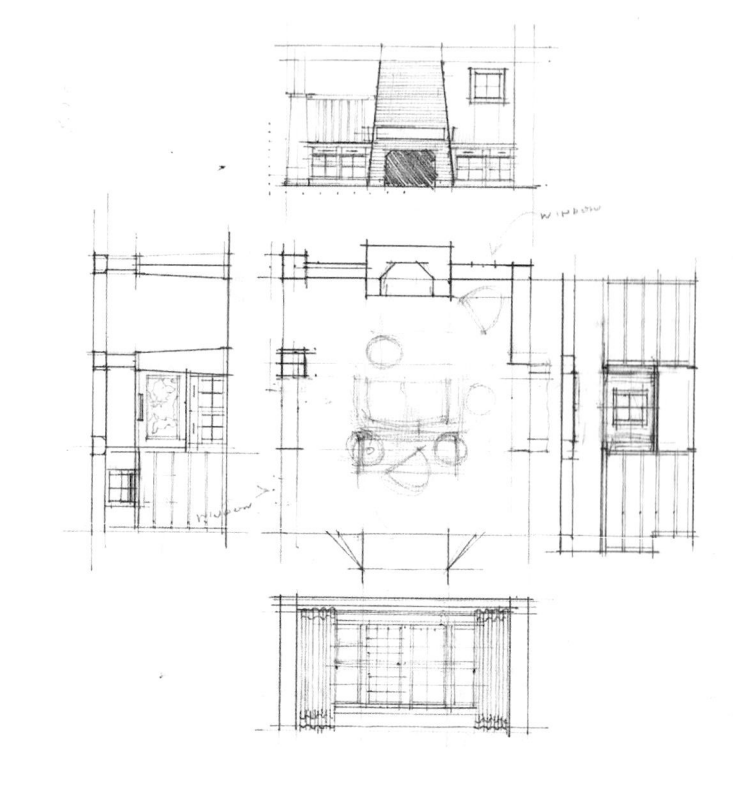

Figure 3.2
Two-point perspective view of the living room (3D).
Courtesy of mkerrdesign.com.

Why Are 2D Drawings Needed?

A 2D drawing, also called an **orthographic drawing**, is one that shows two of an object's three physical dimensions. It's used because it is measurable and depicts spaces and objects as they look, not as we see them. Because of the way our brains perceive depth, we see things in **perspective**, meaning objects appear smaller as they get farther away and sets of parallel lines appear to converge to a common point. Each time we change our location relative to the object, we see a new shape and size. This is not helpful to a builder. Therefore, orthographic drawings, created with a drawing technique called **orthographic projection**, are used. This technique deconstructs a 3D object into multiple 2D views. The result is a set of related drawings that documents a space or object in a measurable, buildable way.

Orthographic Projection Drawing Technique

Visualize an object suspended in a glass cube (Figure 3.3). Each side of the cube is a **picture plane**, a flat surface onto which you **project**, or bring forward, the suspended object. Now visualize the object exploding apart, with each side projected onto the plane facing it. The resultant views are flat and depthless, because only two of the object's three dimensions appear.

If this cube lay flat on a table with its sides unglued, each side, or plane, would contain a view of the object. And those views are located within each plane so that their corners and edges align with each other.

Lines that are parallel to the plane they're projected on are **true length**. Lines that are skewed to the plane they're projected on are **foreshortened**, meaning they appear shorter than they really are, hence are not measurable. When all the lines projected onto a plane are true length, the entire view appears as its true shape and size.

Look at the chair in Figure 3.4. Except for the back cushion, all planes on the chair are parallel to their respective orthographic views, hence they are true shape and size and measurable. The back cushion is parallel to the side plane, so it is true shape and size in that view, but appears foreshortened in the top and front views.

FYI
Ortho is a prefix with Greek origins. It means straight, perpendicular, or vertical. A non-drafting example of use is "orthodontist." In drafting, to project orthographically means to project perpendicular (at a 90° angle) to something.

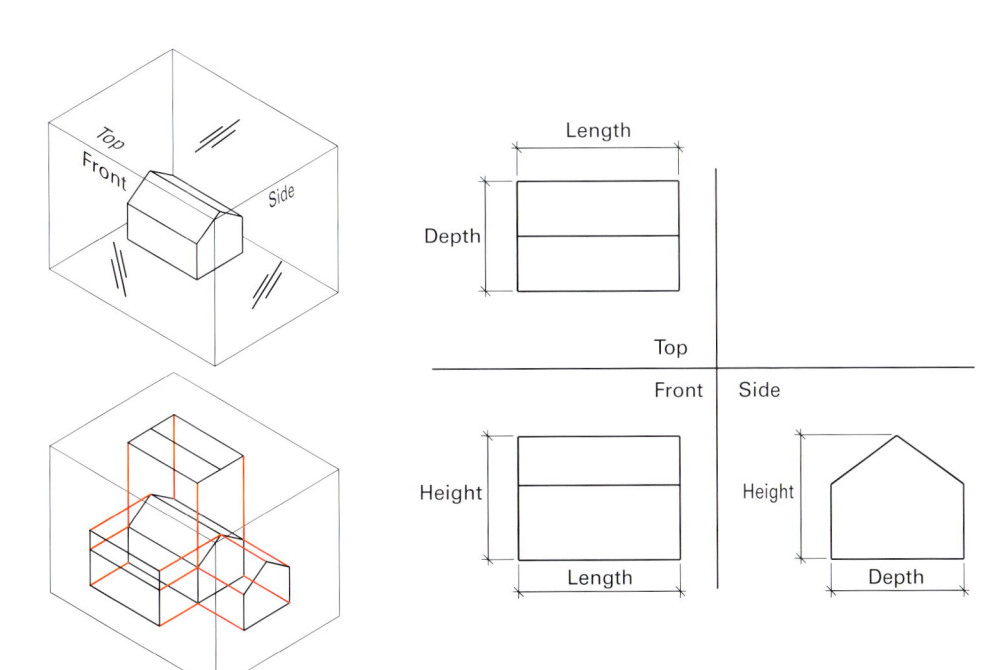

Figure 3.3
Visualize an item inside a glass box and project its endpoints perpendicular to the box's sides.

Now look at the 3D (oblique) drawing of the corner cabinet in Figure 3.5. The middle of the cabinet is parallel to the front picture plane; thus it appears true shape and size. But the sides of the cabinet slope away from the front picture plane, making them foreshortened and unmeasurable in that view.

Views of objects are called **top**, **front**, and **side**. Views of architectural spaces (Figure 3.6), are called **plans** and **elevations**.

Figure 3.4
3D and 2D views of a chair. Lines parallel to their respective picture planes appear as true length.

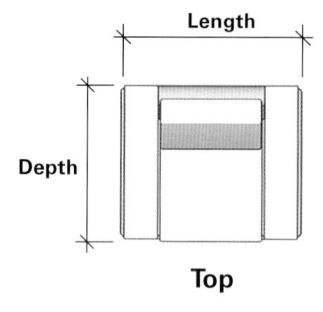

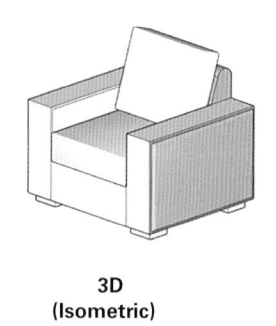

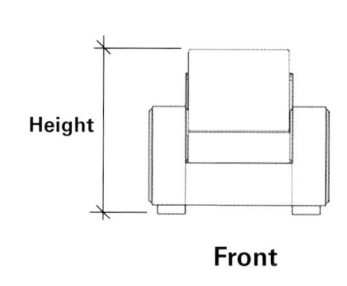

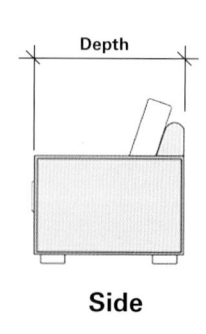

Figure 3.5
3D (oblique) and 2D views of a corner cabinet. In the front view, the cabinet's middle is true shape and size, and the sides are foreshortened.

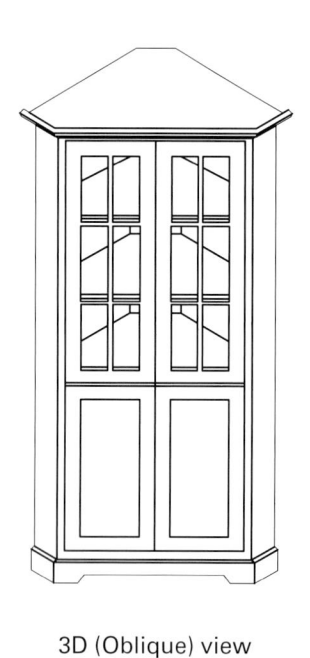

48 CHAPTER 3 2D AND 3D SKETCHING

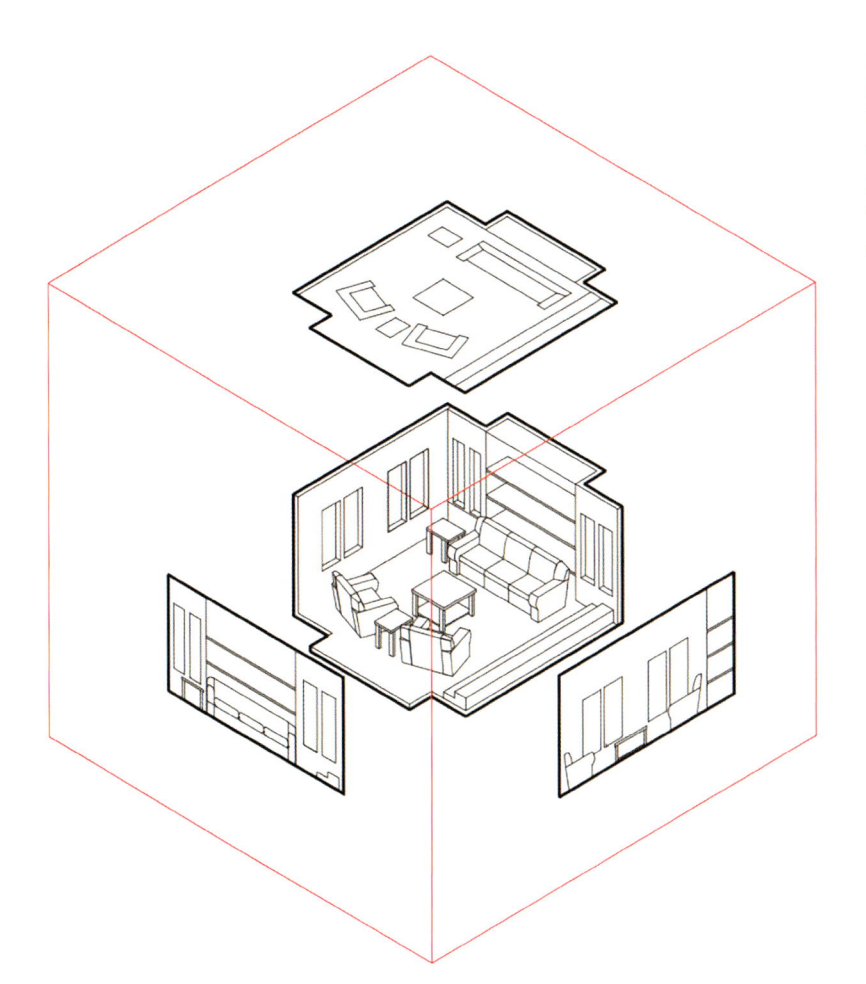

Figure 3.6
The top and side views of this living room are called the plan and elevations.

Sketching Orthographically

Let's sketch orthographic views of some furniture pieces. We'll use **grid**, also called **graph paper**, which is paper printed with four or eight squares per inch. It can be used similarly to the 1/4" = 1'-0" or 1/8" = 1'-0" scales, facilitating quick, proportionally accurate work. It also facilitates sketching straight lines. Grids are available on both bond paper and vellum. The type with faint lines allows pencilled lines to show up better. Draw on it directly or use as an underlay for tracing paper.

Square Table Figure 3.7 shows a perspective drawing of a square table. Figure 3.8 shows its top and front views.

Tip
When sketching, use a dark lead (HB, B) and draw each line continuously from point to point. Don't make small, short overlapping lines. Lines may overlap a bit at corners.

Figure 3.7
3D drawing of a square table.

SKETCHING ORTHOGRAPHICALLY **49**

Figure 3.8
Top and front views of the square table.

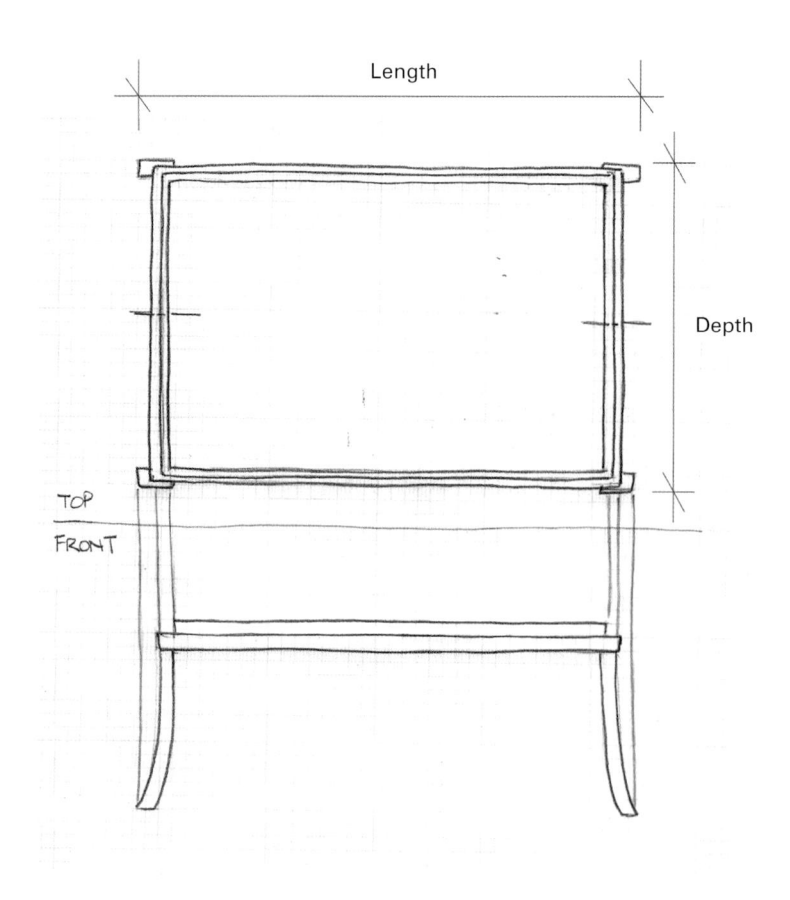

The top view is sketched as though the viewer is looking straight down on it (birds-eye view). Mark its overall dimensions on the grid, sketch the outline, and then sketch details. All edges are parallel to the top picture plane, so all lines in that view are true length, making it true shape and size. Note that the top view is oriented so lines can be projected straight down to create the front view. The front view is drawn with the viewer standing on the floor, looking straight at the table. The projection lines from the top view define the table's width. All edges are parallel to the front picture plane, so this view is also true shape and size.

Round Table   Figure 3.9 shows a 3D drawing of a round shelf-table. Figure 3.10 shows its top and front views.

Sketch the top view first. Because the table top is parallel to the top picture plane, it appears true shape and size in that view—we see a perfect circle. The circular **hidden line** represents the round platform under the table's base. Hidden lines represent features that can't be seen in a view but must still be acknowledged. The platform is drawn hidden because it cannot be seen from the top view.

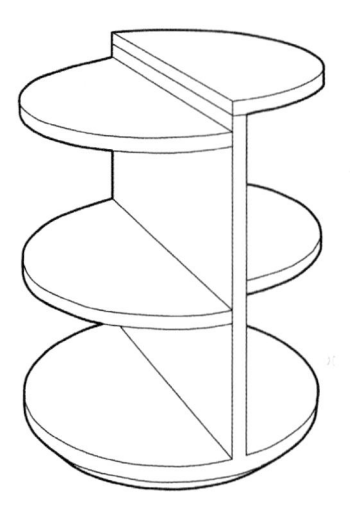

Figure 3.9
3D (perspective) drawing of a round shelf-table.

Project lines from the front view down to define the table's width. The shelves curve away from the front picture plane, thus they appear foreshortened in the front view. That is, we can't tell that they're circular.

50   CHAPTER 3   2D AND 3D SKETCHING

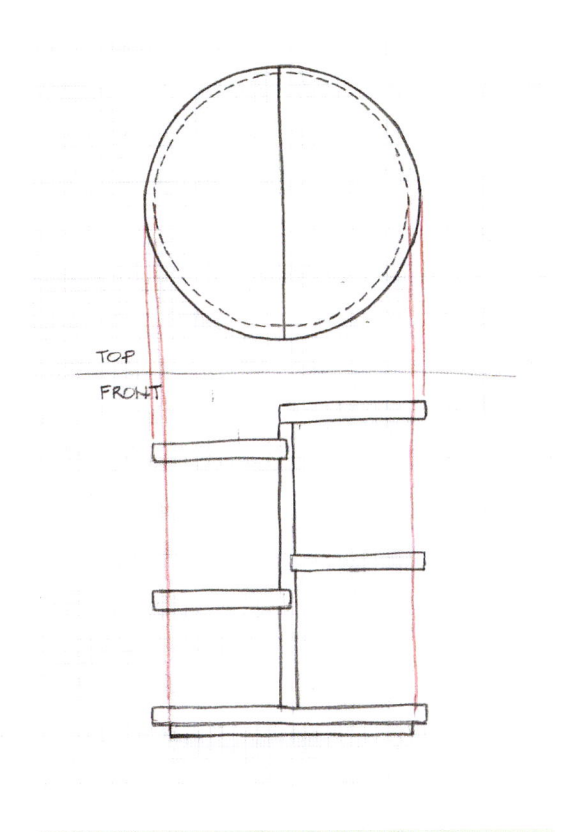

Figure 3.10
Top and side views of the round shelf-table. The shelves appear foreshortened in the front view.

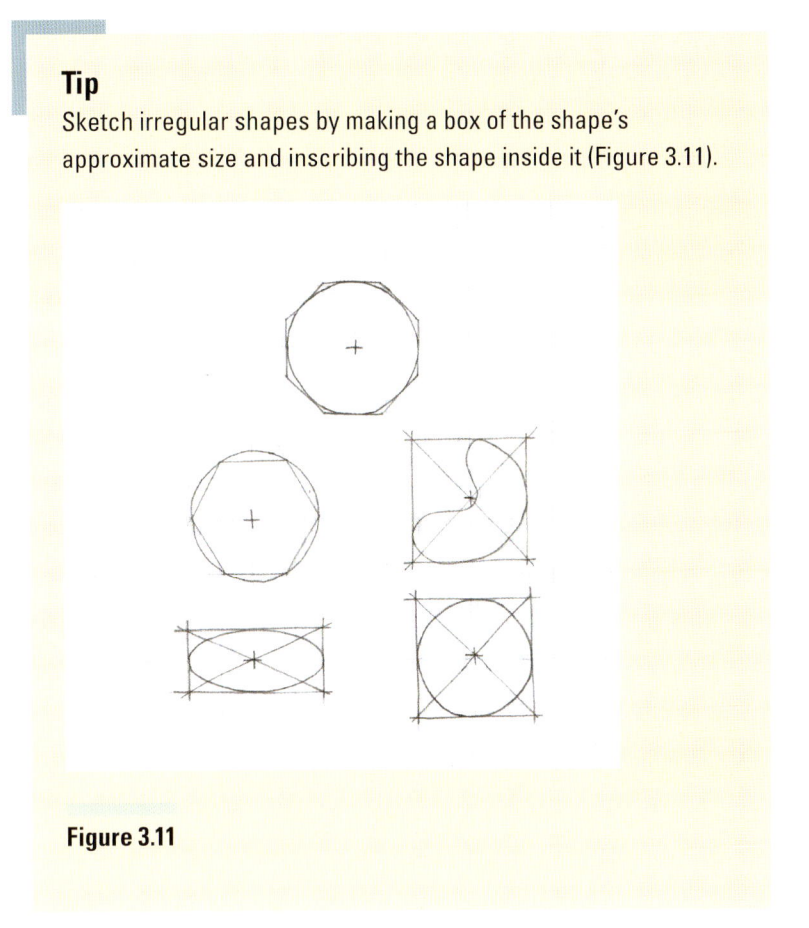

Tip
Sketch irregular shapes by making a box of the shape's approximate size and inscribing the shape inside it (Figure 3.11).

Figure 3.11

Drafting Orthographically

Let's now use drafting tools—a parallel bar, triangle, and divider—to create orthographic views. We'll use a divider instead of a scale because we only need line lengths, not numeric dimensions.

File Cabinet   Figure 3.12 shows a 3D drawing of a file cabinet. In Figure 3.13 we see its top, front, and side views.

Top   Measure the cabinet's length and depth with the dividers and draw a rectangle with those lengths. Orient the rectangle so that lines can be projected from its edges down to the front plane to create the front view. Then add a drawer and knob.

Front   Project lines from the top view down to define the cabinet's length. Mark the cabinet's height, then add drawers and knobs.

Side   Project lines from the front view to bring the cabinet's height into the side view. Mark its depth, which is found in the top view. Then project the drawers and knobs over from the front view.

DRAFTING ORTHOGRAPHICALLY   **51**

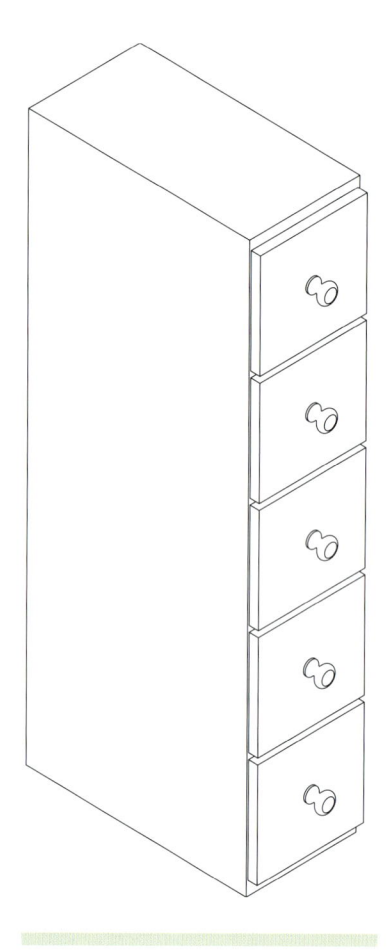

Figure 3.12
3D (isometric) drawing of a file cabinet.

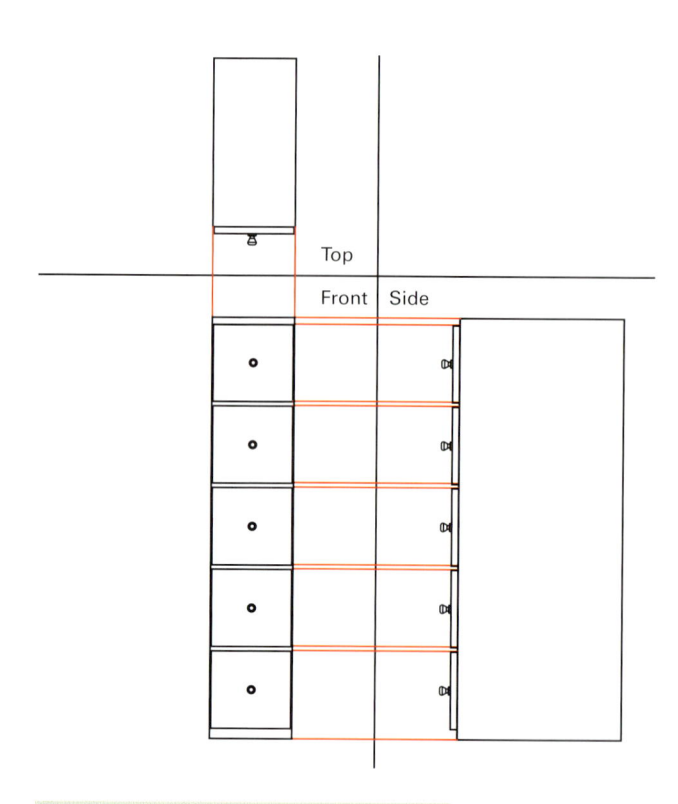

Figure 3.13
Top, front and side views of the file cabinet.

Chest of Drawers Figure 3.14 shows a 3D view of a chest of drawers. In Figures 3.15, 3.16, and 3.17 we'll draw its orthographic views.

Figure 3.14
3D (perspective) drawing of a chest of drawers.

52 CHAPTER 3 2D AND 3D SKETCHING

Top   Measure and mark the chest's length and depth, the lid's overhang, and the plate rail (Figure 3.15). Draw the lid first, since it's the biggest, and then add the other details. Note the chest and legs underneath are drawn with hidden lines. Although there would also be hidden lines for the drawers, they are left out for clarity. Hidden lines are drawn at drafter discretion. If all hidden features were included the drawing would become cluttered. Drafters must decide which hidden features are best left for the larger scale detail drawings.

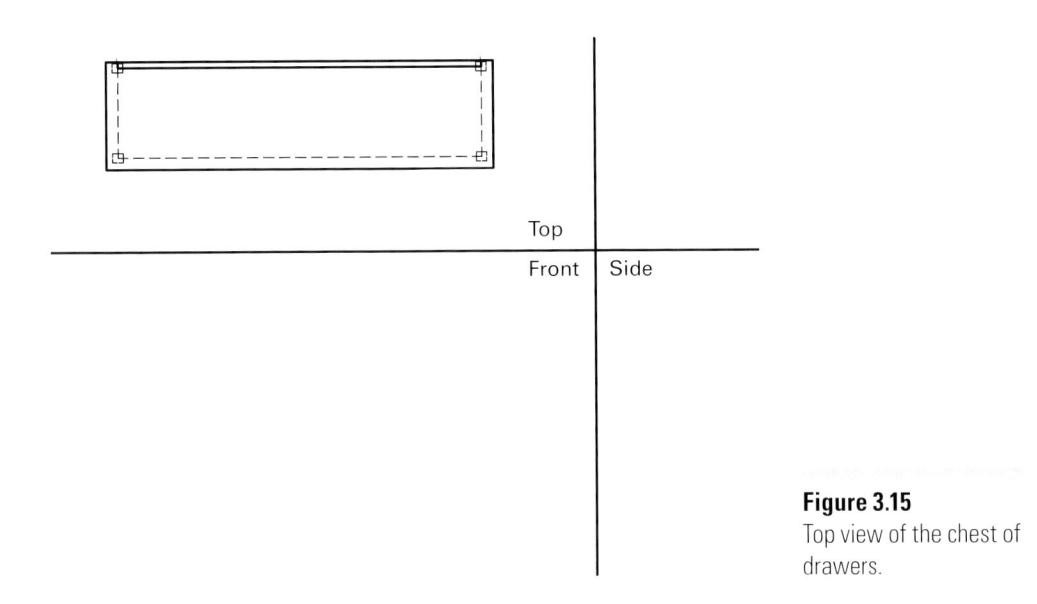

Figure 3.15
Top view of the chest of drawers.

Side   Measure and mark the chest's height and depth. Add details (Figure 3.16). Hidden lines for the drawers are appropriate in this view.

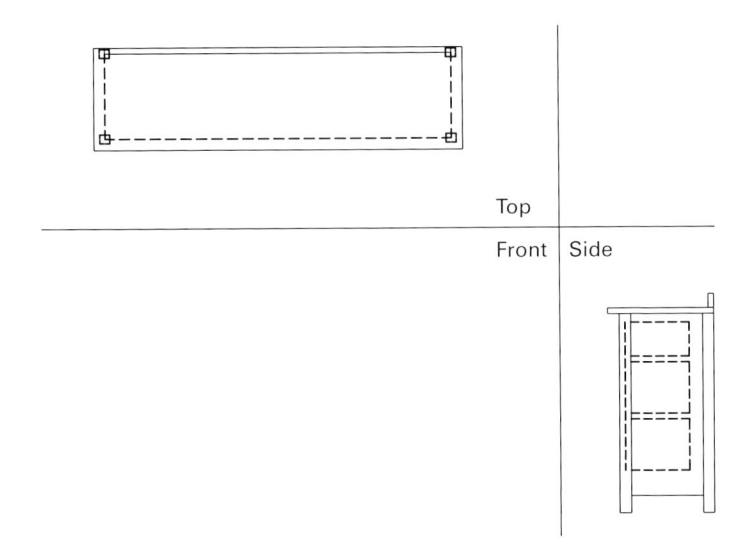

Figure 3.16
Top and side views of the chest of drawers.

DRAFTING ORTHOGRAPHICALLY

Front   Project the width of the chest and lid from the top view down to the front view (Figure 3.17). Project the chest's height over from the front view, and then draw the drawer details. There are no hidden features in the front view; hence there are no hidden lines.

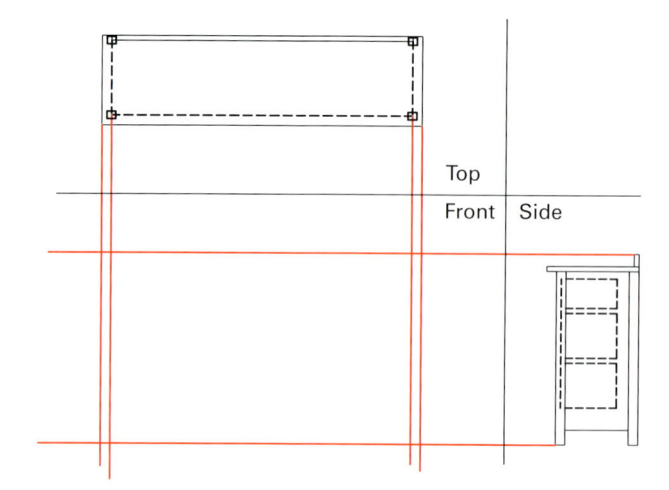

Figure 3.17
Constructing the front view of the chest of drawers.

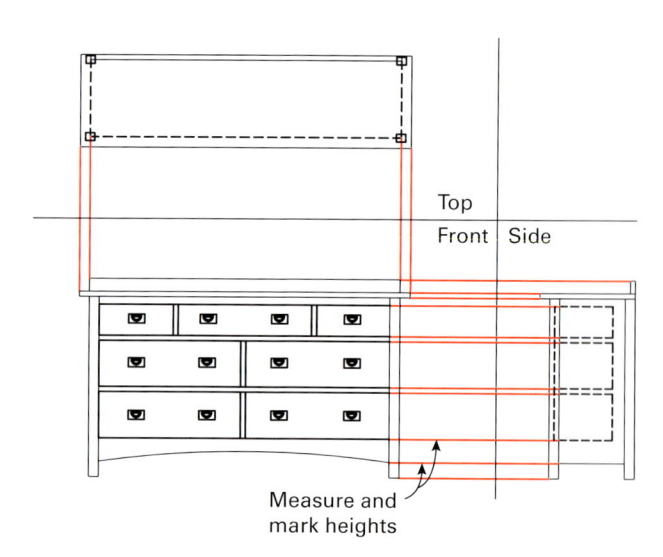

A divider was used to measure line length in these examples, because transferring distances with it is faster. But when an object is more detailed, better accuracy is achieved by using the graphics technique shown in the next example.

Mission Style Chest   Figure 3.18 shows front and side views of a Mission style chest. We want to construct a top view. Transfer the chest's length directly from the front to the top plane. But width information from the side view must be transferred via a 45° angle in the quadrant above the side view. (A 45° line is used because when a 90° angle is bisected the result is two 45° angles.) Project lines from the side view up until they hit the 45° angle, and from where they hit, project them horizontally over to the top view. Then darken in the perimeter line and turn the construction line into a hidden line (Figure 3.19).

54   CHAPTER 3   2D AND 3D SKETCHING

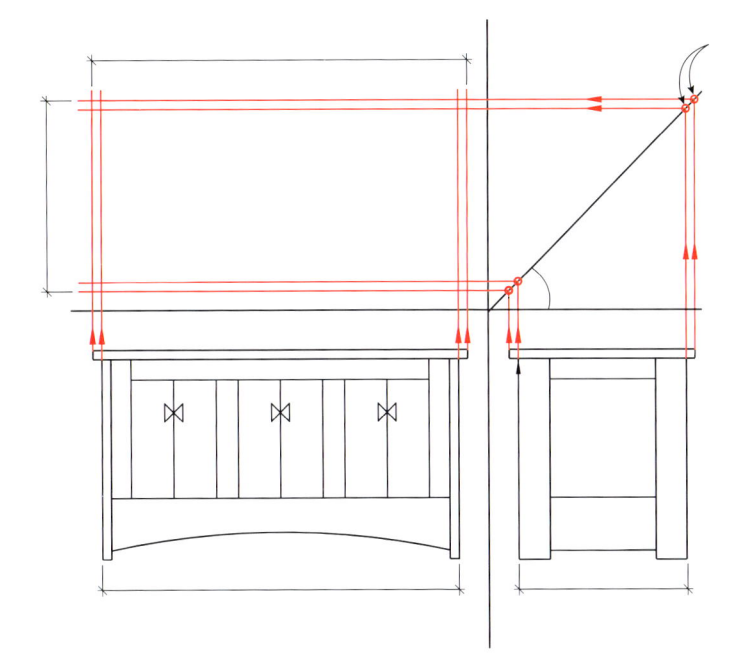

Figure 3.18
Use a 45° angle to transfer measurements from the side to the top view.

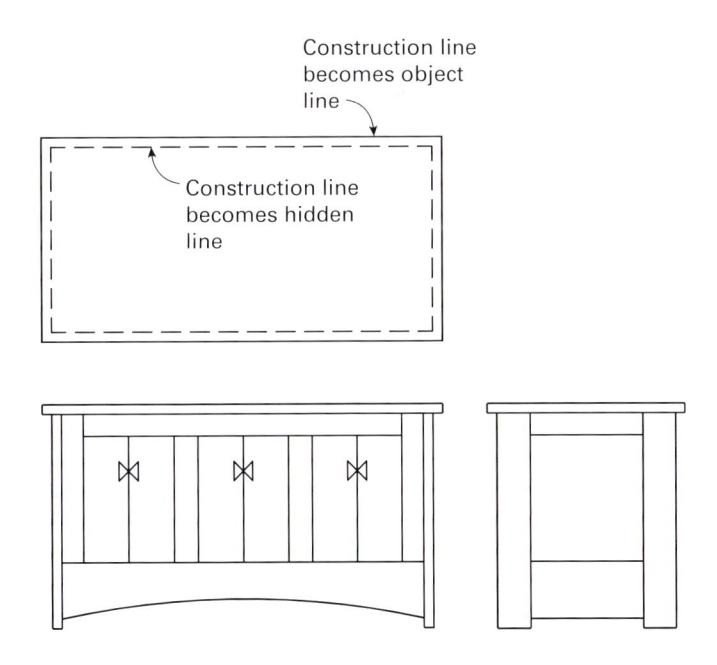

Figure 3.19
Turn the construction lines into object and hidden lines.

Why Are 3D Drawings Needed?

3D drawings, also called **pictorials**, show an object's three physical dimensions in one view. They communicate ideas to clients better than orthographic drawings do, as most people not trained in drafting cannot visualize an item by looking at orthographic views. They benefit from a 3D picture.

There are two umbrella categories of 3D drawings: paraline and perspective. Perspective drawings have line convergence; paraline drawings do not. (Figure 3.20).

Figure 3.20
The perspective drawing shows line convergence. The paraline drawing's horizontal lines remain parallel to each other.

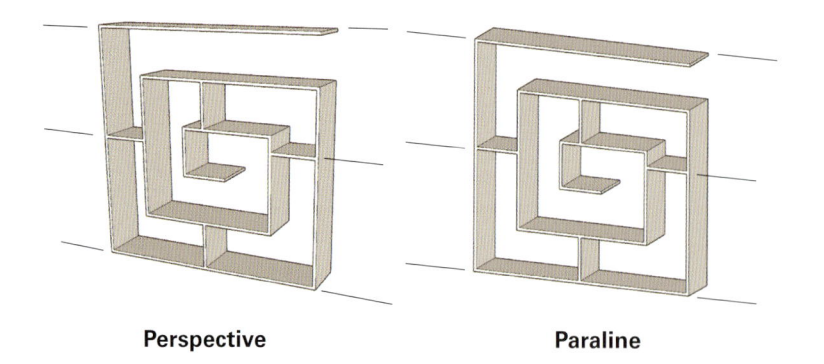

Perspective Paraline

Paraline Drawing

A **paraline drawing** shows an object skewed along two axes, making it look three-dimensional. Sets of parallel lines remain parallel and distances and angles are preserved, so they are scaled and measurable. This technique evolved from technical drafters' need to be able to describe an object's shape simply and accurately without artistic talent. Paraline drawings do contain some distortion or deviation from a true perception of 3D space. Different types evolved to minimize this distortion and achieve the best picture possible.

Types of Paraline Drawings The two types of paraline drawings are **axonometric** and **oblique** (Figure 3.21). Axonometric drawings are skewed along both horizontal axes. Oblique drawings are skewed along one horizontal axis and have an orthographic front face. Both those types have subtypes.

Figure 3.21
Axonometric drawings are skewed along two axes. Oblique drawings are front-facing and skewed along one axis.

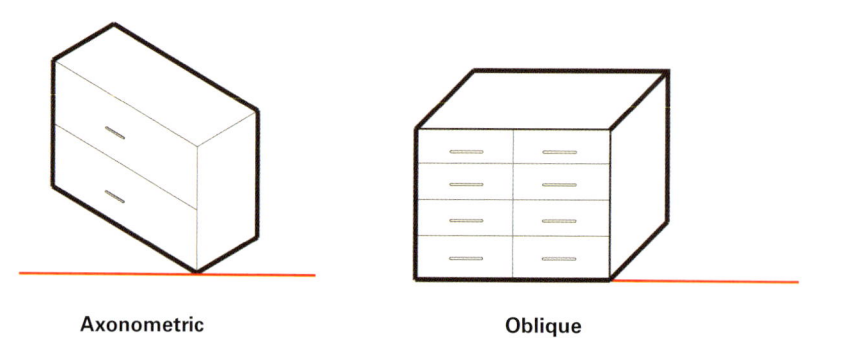

Axonometric Oblique

Axonometric Drawing There are three types of axonometric drawings: **isometric**, **diametric**, and **trimetric** (Figure 3.22). Isometric drawings have two horizontal axes skewed at a 30° angle. Measurements along them, and along the vertical axis, are done at the same scale. Circles appear as ellipses. Dimetric drawings have horizontal axes skewed at any angle between 0° and 90°, both angles may be different, and the horizontal axes have a different scale than the vertical axis (Figure 3.23). Trimetric drawings have horizontal axes skewed at any angle between 0° and 90°. Both axes can be skewed at different angles and all three axes have different scales.

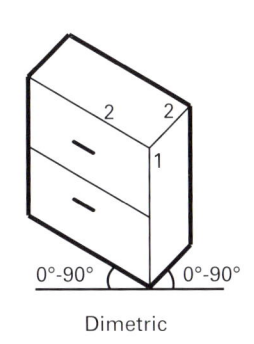

2D (Orthographic) views

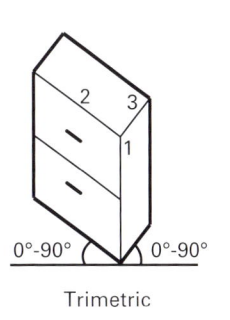

Isometric 30° 30°

Dimetric 0°–90° 0°–90°

Trimetric 0°–90° 0°–90°

Figure 3.22
The three types of axonometric drawings.

Figure 3.23
Orthographic and diametric sketches. Courtesy of wolfgangtrost.com.

PARALINE DRAWING 57

Isometric and diametric drawings are most relevant to architects and interior designers. Grid paper and ellipse templates are available to help make them. Figure 3.24 shows the steps in an isometric sketch. Figure 3.25 shows a 35° ellipse template used on an isometric sketch.

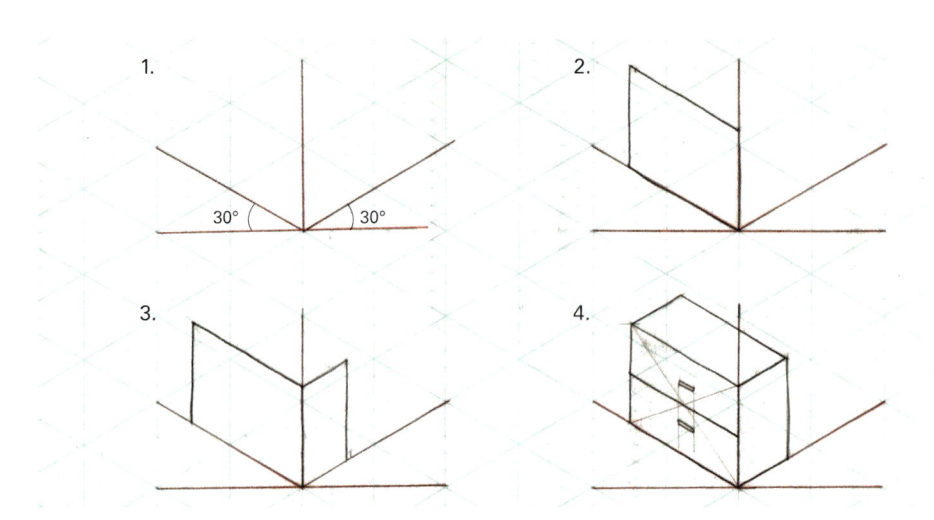

Figure 3.24
Steps in an isometric sketch.

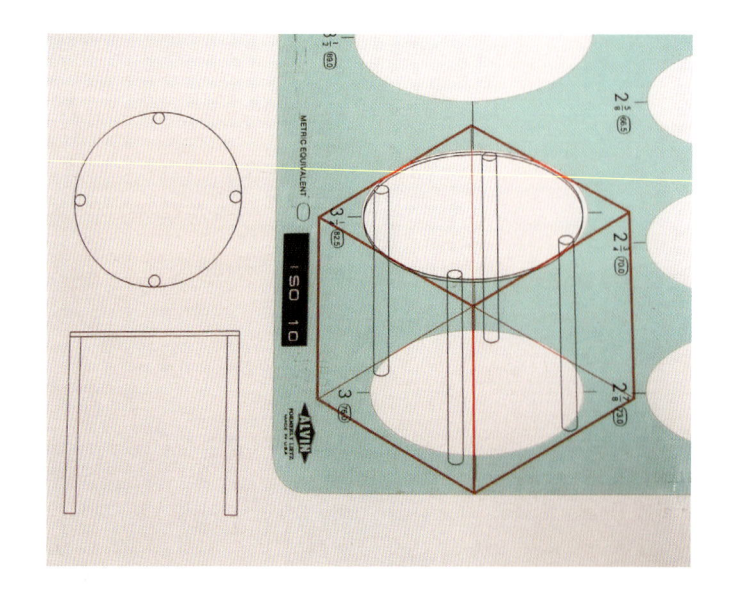

Figure 3.25
A 35° ellipse template used for a round isometric sketch.

Oblique Drawing Also called an **elevation oblique**, this is a 3D drawing with only one receding axis. The receding axis may be the same scale as the front view, or it may be a smaller scale. Circles appear as ellipses. Obliques are used for technical drawings such as kitchen and bath construction documents.

There are three types of oblique drawings: **cavalier**, **cabinet**, and **general** (Figure 3.26).

Cavalier has a receding axis that may be any angle but is usually 30°, 45°, or 60°. Measurements on it are marked at the same scale as on the front view. Cabinet has a receding axis that may be any angle but is usually 30°, 45°, or 60° and measurements on it are marked at half the scale of the front view (Figure 3.27). It is the most realistic looking of all obliques and is often used to draw cabinetry and furniture, hence its name. General has a receding axis angle that may be any angle but is usually 30°, 45°, or 60°. Measurements on it may be marked at any scale between full and half of the front view's scale.

A **plan oblique**, also called a **planometric**, skews the horizontal axes 45°. This results in a top view that is true shape and size (Figure 3.27). Plan oblique drawings give the clearest, most unobstructed views into interior spaces.

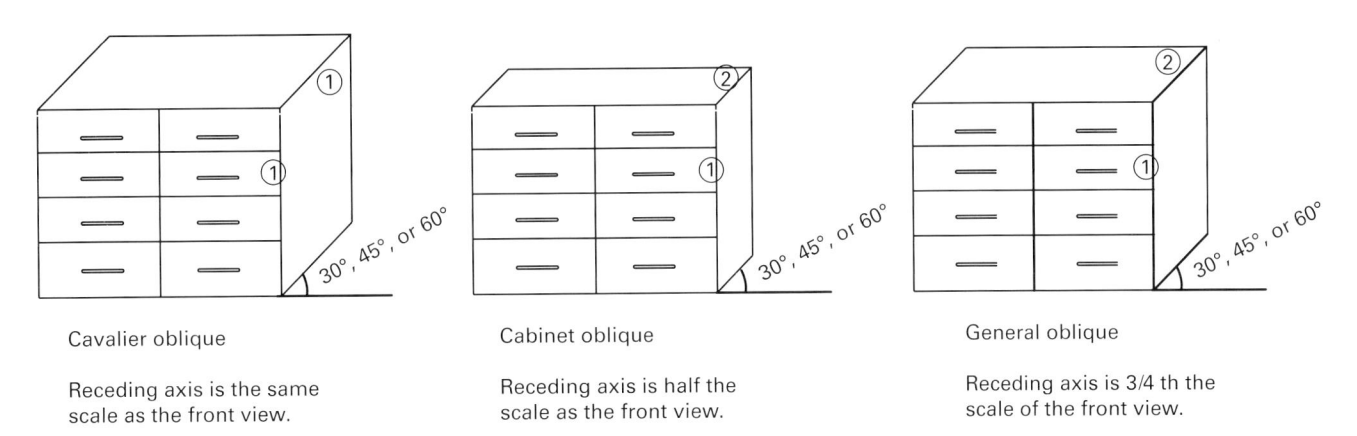

Cavalier oblique

Receding axis is the same scale as the front view.

Cabinet oblique

Receding axis is half the scale as the front view.

General oblique

Receding axis is 3/4 th the scale of the front view.

Figure 3.26
Oblique views show the front in elevation and the side at varying angles and depths.

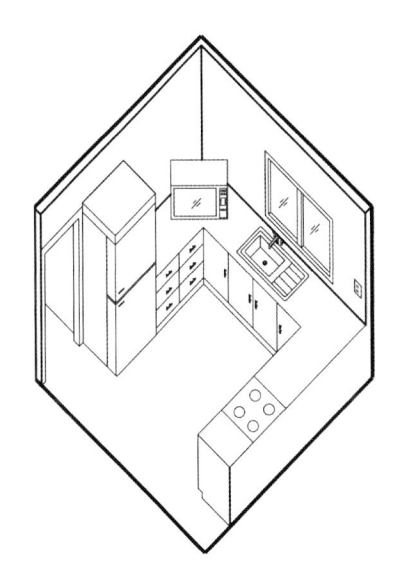

Figure 3.27
A plan oblique skews both horizontal axes at 45°, resulting in a top view that's true shape and size.

Perspective Drawing

This technique shows spaces and objects the way the eye sees them. These drawings are not measurable because parallel lines converge to a point on the horizon, resulting in objects appearing smaller the farther away they are. However, measurability is not their purpose; their value is in their realism. A **one-point perspective** is a drawing where all parallel lines converge to one vanishing point. That vanishing point may be in the middle to create a symmetrical drawing, or it may be located off-center to emphasize one wall (Figure 3.28). A **two-point perspective** has two vanishing points, one for each set of parallel lines.

Figure 3.28
One-point perspective sketch. The vanishing point is off-center, emphasizing the right wall.
Courtesy of mkerrdesign.com.

Perspective Grid

A **perspective grid** is a 3D chart that enables quick perspective sketching. It works similarly to a 2D grid in that each square represents a unit of measurement. You can download premade perspective grids from various websites, but it's useful to be able to create your own.

One-Point Perspective Grid   Figures 3.29 and 3.30 show how to draw a one-point perspective grid and items inside it.

Figure 3.29
Steps for drawing a one-point perspective grid.

1. Draw the back wall, choose a VP location, and draw grid squares to represent distance increments.

2. Place the front wall far from the back wall to obtain depth. Use diagonal lines to ensure that the walls' corners align.

3. Draw diagonal lines at the ceiling and floor.

Radiate lines from the vanishing point

4. Draw horizontal lines where the radiating lines in Step 3 intersect the diagonal lines. Then draw vertical lines.

Move the vanishing point left or right to emphasize the left or right wall.

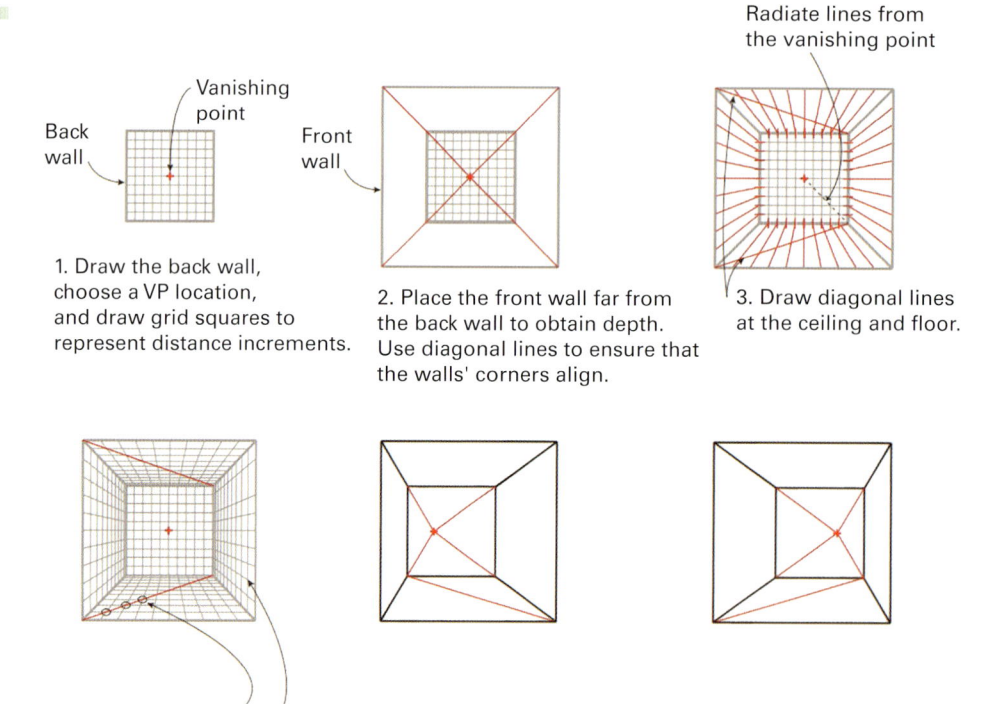

Figure 3.30
Steps for drawing objects inside the perspective grid.

1. Draw "footprints" of the objects in plan.

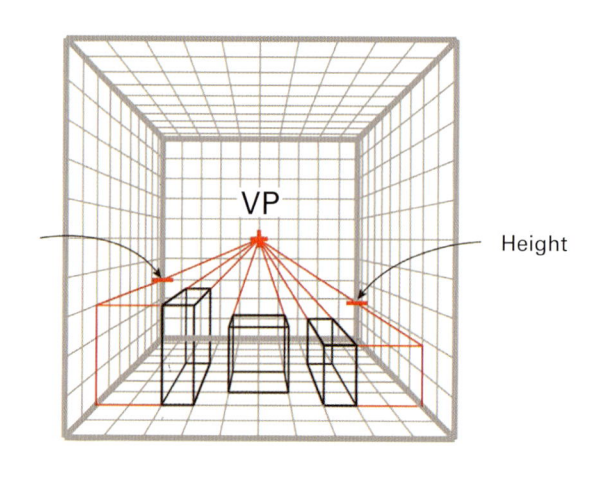

2. Mark the objects' heights. Radiate those heights through the VP, and project them to the footprints.

CHAPTER 3   2D AND 3D SKETCHING

Two-Point Perspective Grid Figure 3.31 shows the steps for making a two-point perspective grid. Draw the true height of the room on a *height line*. This height line has been arbitrarily divided into four parts for a four-part grid. A *horizon line* is drawn at the viewer's eye level, left and right *vanishing points* are found, and the vanishing points are connected to the top and bottom of the height line to create the floor and ceiling lines. Figure 3.32 shows a sketch made by overlaying tracing paper onto a two-point perspective grid.

Grids can be manipulated to show any view wanted. Figure 3.33 shows a top-down view created on a grid. This gives a nice interior view of a space.

Figure 3.31
Steps for drawing a two-point perspective grid.

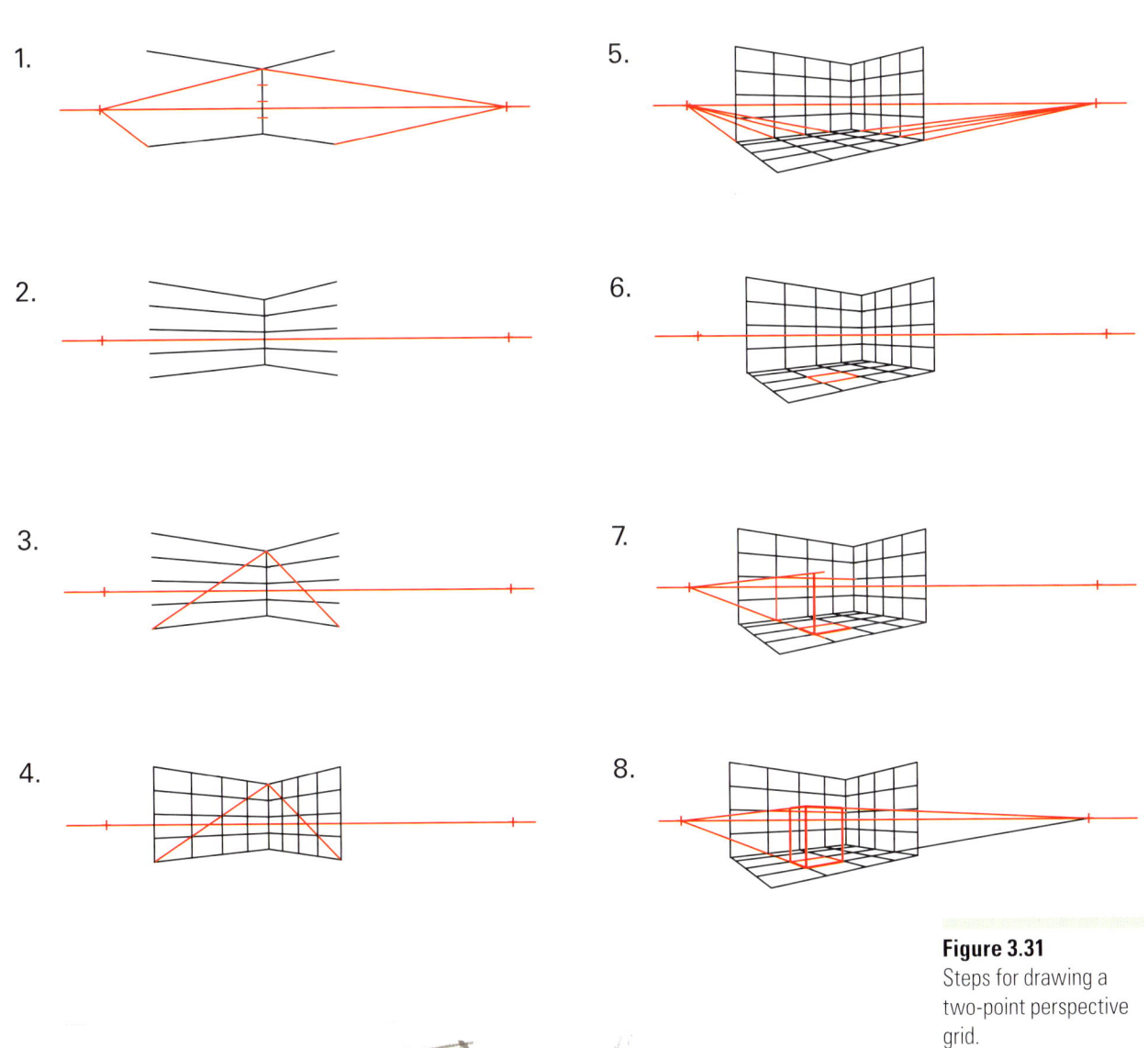

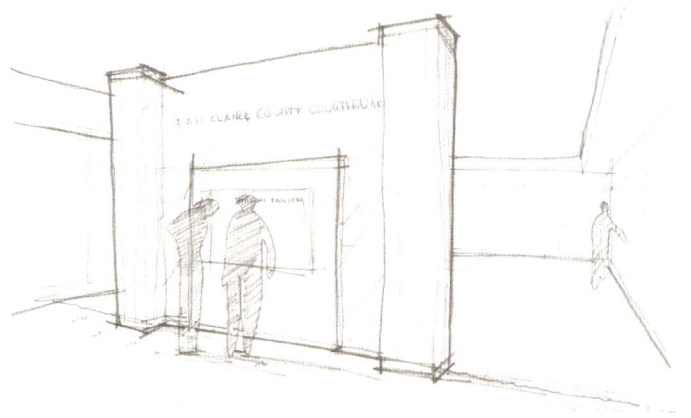

Figure 3.32
This scene was sketched over a perspective grid.

PERSPECTIVE GRID **61**

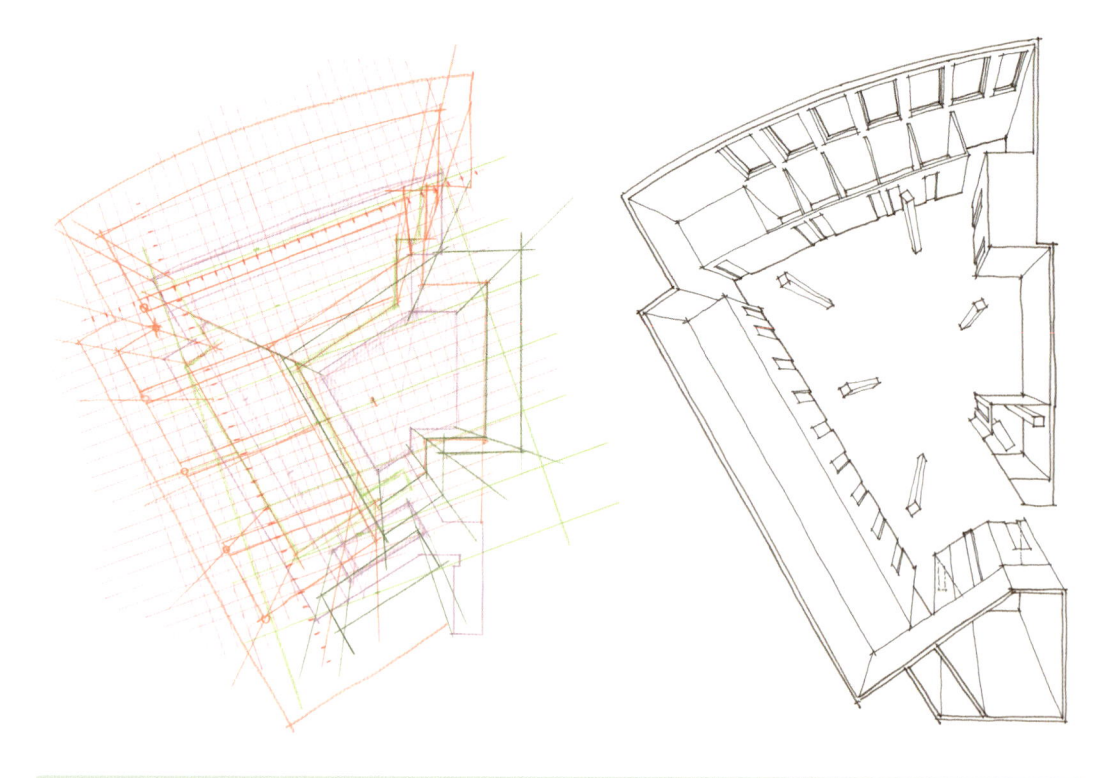

Figure 3.33
Birds-eye view of an office space. The left is a pencil underlay for the ink sketch on the right. Walls, cubicles, and ceilings were then overlaid.
Courtesy mkerrdesign.com.

Sketch Over a Photograph

A technique for making a quick perspective drawing is to sketch over a photograph (Figure 3.34). This allows you to use existing vanishing points and features. Use digital photos as an underlay for sketching design ideas. Enlarge or reduce the photo to the size of the desired sketch.

Figure 3.34
Trace a photograph for a quick perspective drawing and then sketch new ideas over it.
Courtesy wolfgangtrost.com

CHAPTER 3 2D AND 3D SKETCHING

Sketching Tips

The study and practice of creating orthographic and perspective drawings will help students become more proficient at sketching. Following are some techniques to help achieve that.

Hone Observational Skills Seeing and observing are two different things. Seeing is giving a cursory glance to identify. We might see a wallpaper pattern and remember its general appearance. However, if we are asked what the specific shape in the pattern was or if it was more denim than dark blue our recall is less precise because we didn't *observe,* or look closely. Observation is a disciplined, learned skill. Learning to look at the built environment with a critical eye will improve your sketching ability and help you put on paper exactly what you see.

> **Exercise**
> Practice observation by studying and sketching your surroundings. Sketch the room first, then add its furnishings. Start with simple rooms and then sketch larger, more detailed rooms.

Estimate Proportions Knowing an object's actual measurements isn't necessary to sketch it proportionally, but estimating relative heights and widths is. Here are some techniques.

Eyeball Apparent-Size Relationships. To accurately draw images smaller than actual size, study the relative sizes of the parts of your subject and their relationship to the whole. For example, if you are sketching a dresser, estimate how wide it is relative to its height. The dresser in Figure 3.35 appears as wide as it is tall, and its depth appears half the width. So, sketch it that way.

Figure 3.35
Sketch a dresser by estimating its width, height and depth.

Pencil Trick Stand parallel to the dresser. To estimate its width, hold your pencil at arm's length. Turn the pencil sideways, align one end with the right or left side of the dresser, and move your thumb to mark the end of the dresser's other side on the pencil's length. Mark that length on your paper. Then turn the pencil vertically with the bottom aligned with the bottom of the dresser and slide your thumb until it aligns with the top. Mark that length on the paper.

To sketch a foreshortened line, align your pencil with one and, keeping that angle, bring your arm to your paper. Draw a line parallel to it in the place desired. Accuracy depends on maintaining the pencil at a constant distance from your eyes (full arm's length is best) and keeping the pencil parallel to the line being sketched.

Point-to-Point This method involves drawing and connecting points on an object to form its shape, analogous to drawing pictures of star constellations. Constellations are actually three-dimensional; the stars making up a constellation are at light-year-different distances from Earth. But because we cannot perceive this depth from where we stand, we just see a flat group of stars with measurable proportions. When we look at the Big Dipper, we see the 2D shape of a 3D cluster of points. So, if drawing an object's proportions is difficult, try ignoring its 3D form and focus on the 2D shape (Figure 3.36).

Figure 3.36
Trace around the 3D form as a guide to drawing the 2D shape.

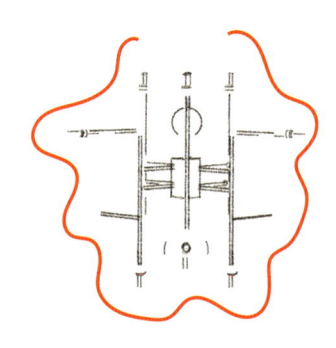

Count Units in Existing Finishes Count the tiles in a floor or in an acoustic ceiling grid. You can also pace off a room with long strides, each equalling about one yard. This has the additional benefit of developing a feel for the size of spaces.

Scale the Picture to the Paper Part of layout and proportion is figuring how large to sketch the subject relative to the paper size. Too-large sketches fall off the paper and too-small sketches are difficult to detail. Establish a scale on the grid paper and then mark the four corners of the object you're drawing. Put those four corners close to the corners of the paper for the largest sketch. If the subject is square or rectangular, position your clipboard accordingly. Binder clips keep the paper in place when using the clipboard horizontally.

Take Photos of Existing Spaces Photograph close-ups of beams that intersect walls, vaulted ceilings, and other difficult-to-sketch architectural features. Print the photos and trace them onto vellum. This will help you learn how objects are drawn in perspective.

> **Exercise**
> Sketch orthographic drawings of common objects. This will help you understand the relationship between 2D and 3D drawings. Keep them in a journal as a record of progress. Figure 3.37 shows student work from an assignment to orthographically sketch two items a week.

Figure 3.37
Student journal drawings. Courtesy Allison Noyes.

64 CHAPTER 3 2D AND 3D SKETCHING

Summary

Designers visualize and communicate with 2D and 3D drawings. 2D drawings are orthographic; 3D drawings are paraline or perspective. The ability to estimate proportions and sizes accurately is important to good sketching. Grid paper and sketching techniques can help you hone your sketching skills. After the space or object has been visualized, designed, and refined, drafting with hard-line instruments commences.

Classroom Activities

1. Draft your name with isometric drawing technique.
2. Sketch orthographic and isometric views of objects in your classroom.
3. Draw a piece of furniture in your classroom as a cabinet oblique.
4. Draw a one-point perspective of your classroom and its furniture on the grid provided in this text's worksheets section.
5. Overlay vellum on a photograph of a room and trace. Make design changes.

Questions

1. Why do designers sketch?
2. What is orthographic projection?
3. Why is orthographic drawing technique used?
4. What does "foreshortened" mean?
5. When does an element appear foreshortened?
6. What is paraline?
7. Name the two types of paraline drawings.
8. What is the difference between axonometric and oblique drawings?
9. What angle is used to create an isometric drawing?
10. What is the advantage of a perspective drawing?

Further Resources

Ching, F., and S. Juroszek. 2019. *Design Drawing*. Hoboken, NJ: Wiley. 3rd ed.

Sloan Cline, L. 2011. *Drafting and Visual Presentation for Interior Designers*. Englewood Cliffs, NJ: Prentice Hall.

Keywords

- axonometric
- cabinet oblique
- cavalier oblique
- diametric
- elevation
- elevation oblique
- foreshortened
- front view
- general oblique
- graph paper
- grid paper
- hidden line
- isometric
- oblique
- one-point perspective
- ortho
- orthographic drawing
- orthographic projection
- paraline
- perspective
- perspective grid
- pictorial drawing
- picture plane
- plan
- plan oblique
- planometric
- project
- side view
- sketch
- sketching
- three-dimensional sketch
- top view
- trimetric
- true length
- two-dimensional sketch
- two-point perspective

CHAPTER 4

Drafting Conventions

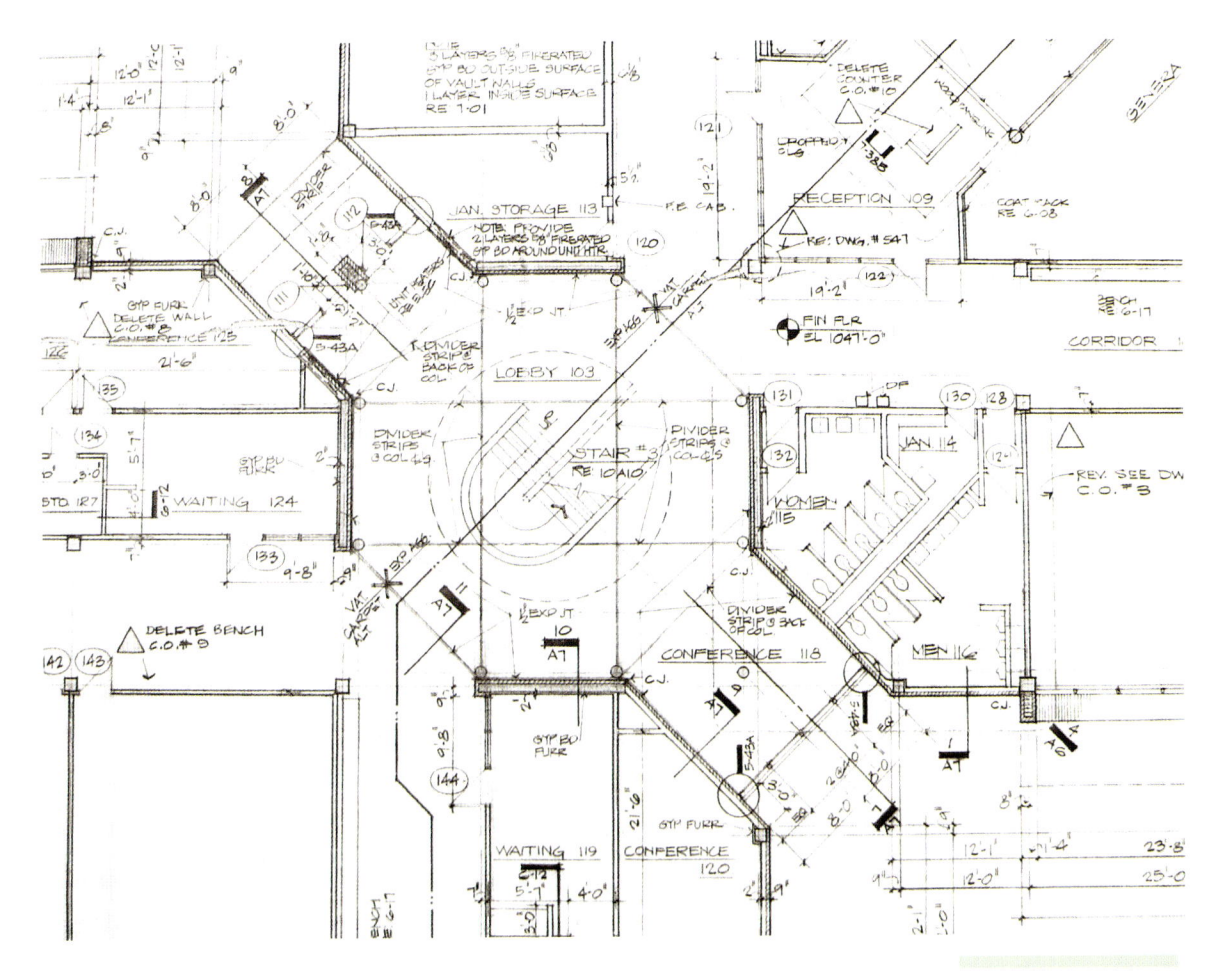

Figure 4.0
Hand-drafted floor plan.
Courtesy Edwin Korff,
Prairie Village, KS

OBJECTIVES

Upon completion of this chapter you will be able to:

- Identify architectural symbols
- Choose appropriate line weights and line types
- Apply conventional sheet layout standards

Once a sketch represents a design, it is turned into a construction-ready document. This is where the drafting starts.

Standards

A drafter breaks complex features down into simple, representational lines. Universally recognized **standards**, or conformities, are followed to help everyone interpret the drawing similarly. A drawing must be read the same way in any language or country it's in and by everyone who reads it. Those readers will have different backgrounds, such as designer, contractor, fabricator, owner, vendor, and banker. While companies may have variants of those standards, they are similar enough to be recognizable.

Sheets and drawings must be drafted, labeled and laid out in a manner that makes their content and referencing clear. The **National CAD Standard (NCS)** is widely used for this. It combines the American Institute of Architect's CAD Layer Guidelines, the Construction Specification Institute's Uniform Drawing System, and the National Institute of Building Sciences BIM Implementation and Plotting Guidelines. The NCS discusses sheet layout and order, line weights, types, widths and colors, code conventions, symbols, terms and abbreviations, layer names, fonts, scale factors, dimensions, keynotes, and more for manual and computer drafting. Another standard is the CSI's **MasterFormat**, a coding system for organizing construction documents and design specifications. Website links for these standards are at the end of the chapter. We'll discuss the standards most relevant to beginners in interior design drafting.

> **FYI**
> A standard is a recognized conformity, necessary for a common understanding of what to expect. If mass-produced goods like queen-sized sheets, three-hole paper, and lightbulbs weren't standardized, we couldn't buy them with any certainty that they'd fit the intended bed, binder, or socket. Standards on architectural drawings include methods of dimensioning, labeling, symbols, and line types.

The Line: Weight and Type

The **line** is the fundamental tool of graphic communication. It can be long, short, straight, curved, light, dark, solid, patterned, freehand, horizontal, vertical, diagonal, thick, or thin, with each type communicating a different idea. Every line on a drawing means something, so never draw random lines. Know what each line you draw represents. Construction drawings are meant to be abstract, so use as few lines as possible to describe an object.

Line Quality

This is appearance. A well-drawn line has consistent width along its length and is crisp, clean, with no erasure marks or smears. All **object lines** are equally dark; it's their width and type that varies. Press down firmly on the pencil when drawing lines; sometimes rotating the pencil as you draw helps keep line thickness consistent. Avoid drawing over a line a second time because that creates double lines or lines thicker than similar surrounding ones. Connect all lines at corners; don't leave gaps between them.

> **Tip**
> Before turning in a photocopy of original work, make a test copy to check **line quality**. Weak, light lines and smudges are more apparent on a copy than on the original.

Line Weight

This is thickness. Apply different **line weights** depending on the lines' importance in the drawing. Complicated drawings utilize multiple line weights. This makes them easier to read than drawings with just one weight. Looking at Figure 4.1, the left drawing is easier to read because the thick outline leads the eye to the form first and then to interior details. On the right drawing, the eye doesn't know where to focus first. Thicker lines engage the reader's eye first, so draw features that should be read first with thicker lines.

Figure 4.1
A thick outline directs the eye to interpret the object's form first, and then to interior details.

On floor plans, the walls are drawn thickest, furniture and door swings thinner, window glazing, floor poché thinnest. You may also adjust line weight for distance; for example, draw a rug on a plan view thinner than a table on top of it because the rug is farther from the viewer's eye. Keep a consistent pressure and angle on your pencil or pen to keep the line weight consistent.

Line Type

This is pattern (Figure 4.2). For example, a line may be drawn as a series of long dashes, short dashes, or a combination of both. Each pattern represents a different concept. Following is a description of different architectural **line types**.

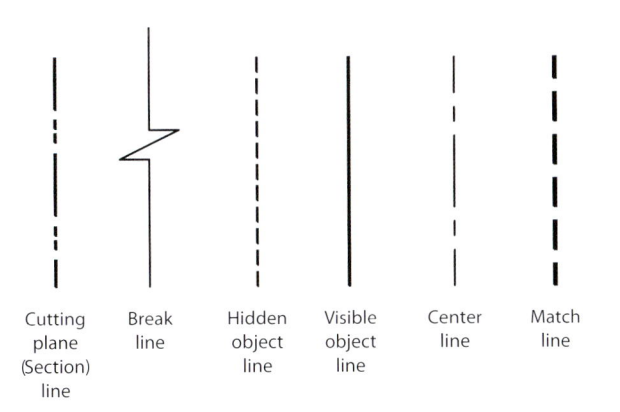

Figure 4.2
Line types.

Cutting plane (Section) line | Break line | Hidden object line | Visible object line | Center line | Match line

Visible Object Visible object lines define a physical item's outline. Walls, doors, cabinet edges, and floor tiles are examples of items drawn with visible object lines. These lines are continuous and non-patterned.

Hidden Object Hidden object lines are made with short, evenly spaced and sized dashes that define an item not visible in the reader's current view. For instance, the wall cabinets in Figure 4.3 are shown with a **hidden line** because they are above the 5'-0" **cutting plane** with which floor plans are made. Hidden lines may also show items destined for removal, or lines below the reader's view. This drawing shows hidden lines representing a dishwasher under the countertop. Draw the dashes 3/8" long, space them evenly 1/8" apart.

Exercise
Draw different line types and line weights on a sheet of vellum. Use hard and soft leads.

LINE TYPE

Figure 4.3
The hidden lines in this drawing represent wall cabinets and a dishwasher.

Long Break A long **break line** ends a feature when drawing it in its entirety is unnecessary and the feature's defining characteristics have already been shown. The stair in Figure 4.4 is a common example of long break line usage. This line can be drawn vertically or diagonally and has a small jag in the center.

Center Line The **center line** is placed through a feature's center. It's used for dimensioning and to denote curved items that appear foreshortened in orthographic views (Figure 4.5).

Short Break Also called a *cylindrical* break, this is a curved line placed through a cylindrical object, such as a pipe, to end that feature when drawing it in its entirety isn't needed (Figure 4.5).

Cutting Plane Also called a section plane line, a **cutting plane** is drawn on the floor plan to show where a building is sliced to create a section drawing (Figures 4.6, 4.7). The arrows point in the direction of view and the section drawing shows everything in front of the arrows. Numbers at both endpoints tell the reader where to find the section drawing. In Figure 4.7, they say to go to drawing 1 on sheet 2.

Figure 4.4
A long break line means the feature continues.

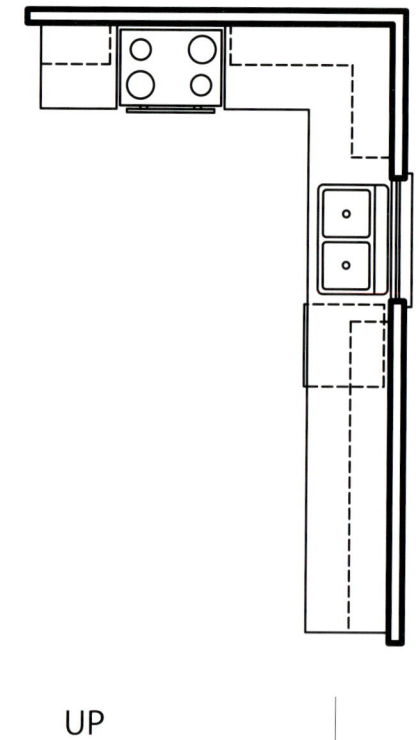

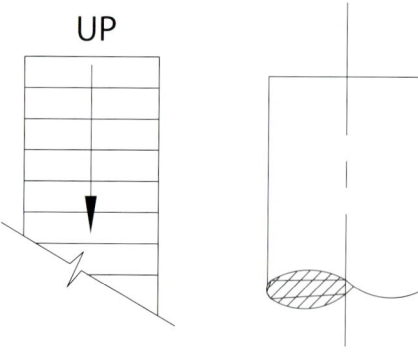

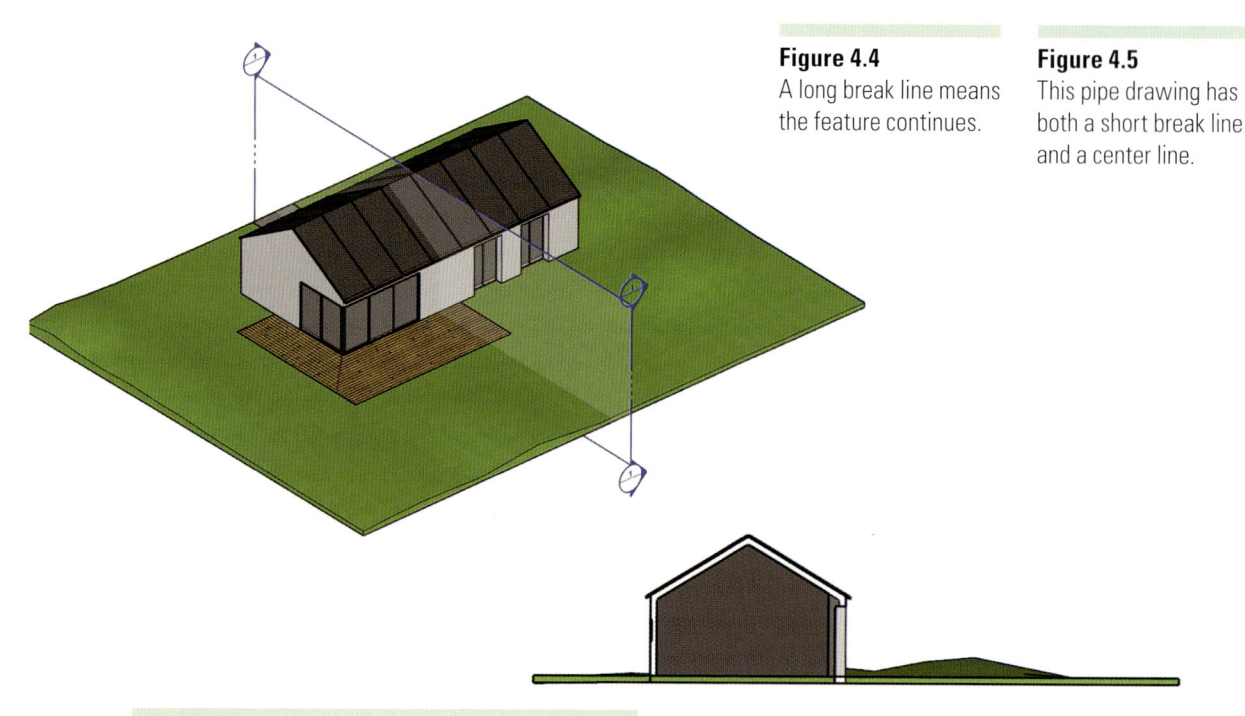

Figure 4.5
This pipe drawing has both a short break line and a center line.

Figure 4.6
A section drawing is a slice through the whole building.

70 CHAPTER 4 DRAFTING CONVENTIONS

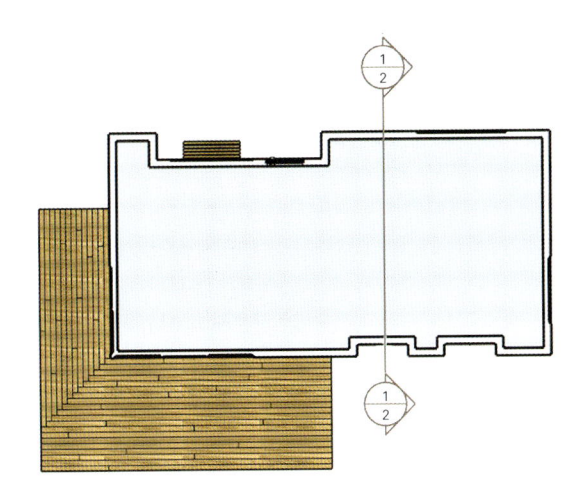

Figure 4.7
A cutting plane line is placed where the section is taken. The arrows point in the direction of the section and the ends direct you to drawing 1 on sheet 2 to find it.

The endpoints may include numbers telling the reader where to look or they may be simple arrows with letters such as A–A or BB–BB (Figure 4.8). The cutting plane line itself may be drawn solid, dashed, or with just the two ends to avoid obscuring other lines on the plan (Figure 4.9).

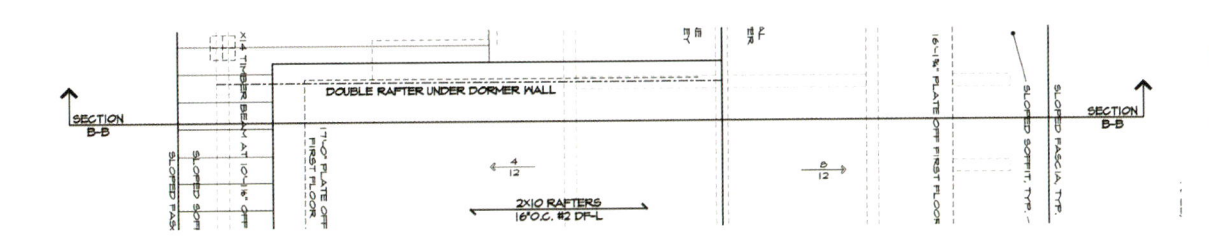

Figure 4.8
This cutting plane line is named BB–BB.

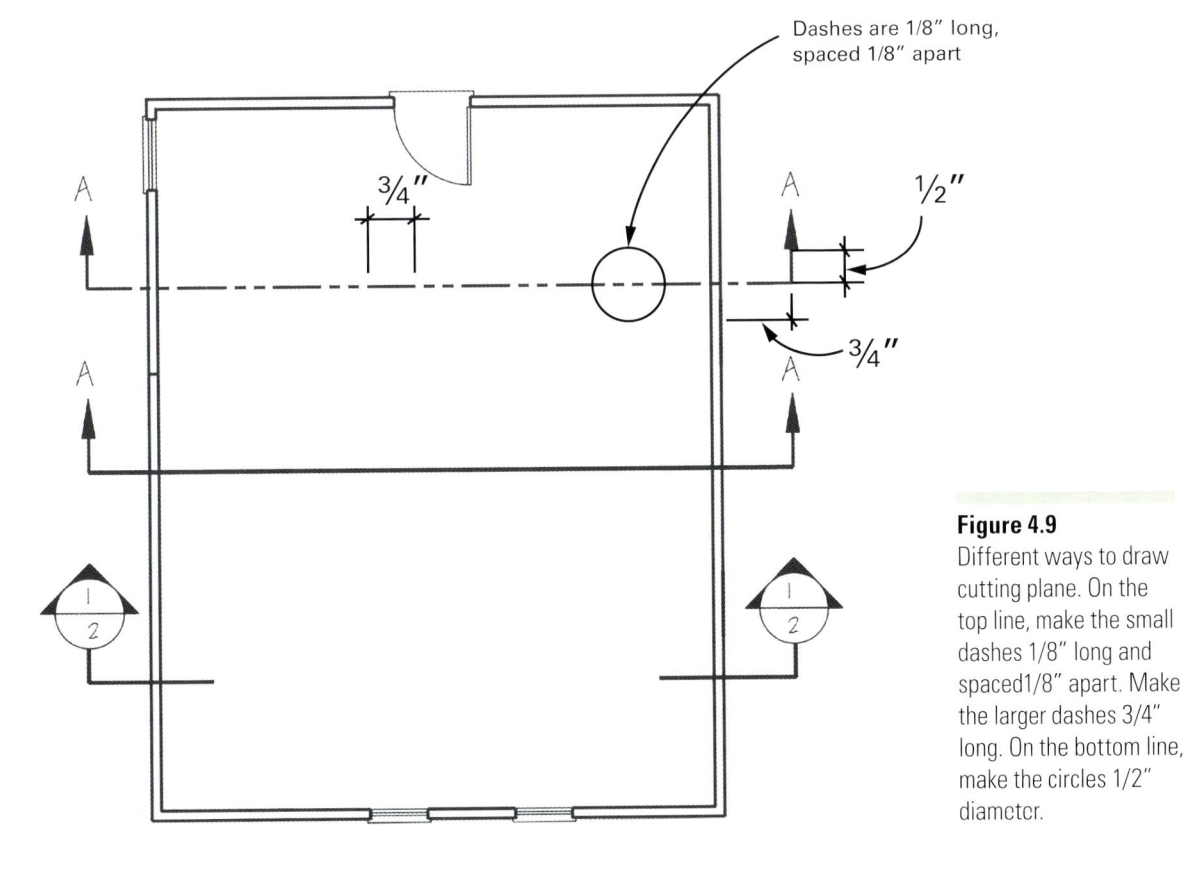

Figure 4.9
Different ways to draw cutting plane. On the top line, make the small dashes 1/8" long and spaced 1/8" apart. Make the larger dashes 3/4" long. On the bottom line, make the circles 1/2" diameter.

LINE TYPE

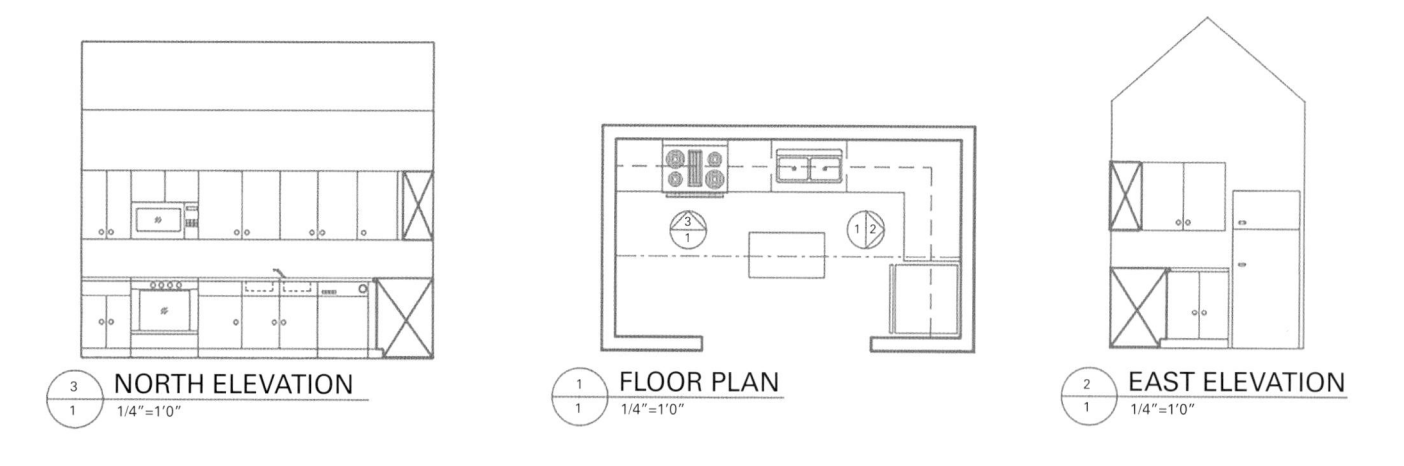

Figure 4.10
Floor plan with elevation callouts and the elevation drawings referenced.

Identifiers

These **annotation** (explanatory note or comment) symbols identify drawings and reference the reader to other, related drawings. Many of these symbols are collectively called **callouts**, **marks**, or **keys**.

Elevation Callout   This symbol goes on the floor plan. It points to a wall and tells the reader where to find an elevation drawing of that wall. The floor plan in Figure 4.10 has two **elevation callouts**. One points to the north wall and tells the reader to go to drawing 3 on sheet 1. The other points to the east wall and tells the reader to go to drawing 2 on sheet 1.

ID Label   This identifier symbol goes underneath each drawing, typically on the left side or center. Some offices place it on the left for floor plans and on the right for all other drawings.

Look at the **ID labels** in Figure 4.10. The top number in the circle is the drawing number; the bottom is the sheet number. So, the floor plan is drawing 1 on sheet 1; the east elevation is drawing 2 on sheet 1; and the north elevation is drawing 3 on sheet 1.

How to Draw an Elevation Callout   Make a 1/2" diameter circle. Place the arrow's base in the middle. All points should be **tangent** (touch) to the circle. Letter the numbers 1/8" tall and they must read straight up, no matter which way the arrow points. Don't letter them sideways or upside down. (Figure 4.11).

Figure 4.12 shows alternative ways to draw an elevation callout. The arrow can be **inscribed** (drawn inside the circle) or **circumscribed** (drawn outside the circle). If

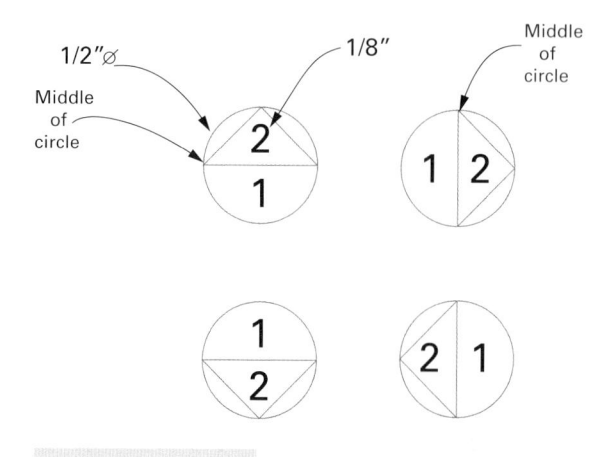

Figure 4.11
Elevation callouts.

space is tight, combine the bottom (sheet) number of multiple callouts. Figure 4.13 shows steps for creating a callout using a 45° triangle. Shading in the arrow on the last step is optional.

How to Draw an ID Label   Draw a 3/4" diameter circle. Draw a horizontal line through its middle and continue it until it's long enough to underline the title. Put the drawing and sheet numbers inside the circle; the title above the line; and the drawing's scale below the line. Leave space between the horizontal line and the title and scale; that is, don't letter the title or scale directly on the horizontal line. (Figure 4.14)

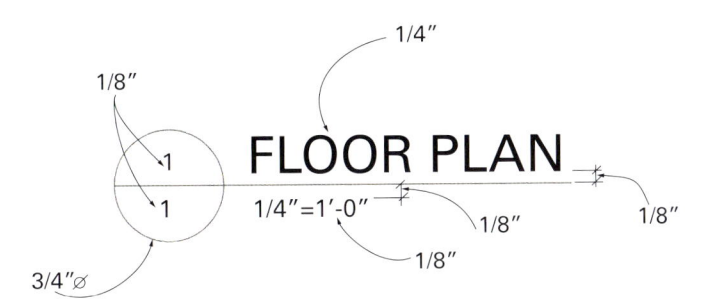

Figure 4.12
Alternative ways to draw elevation callouts.

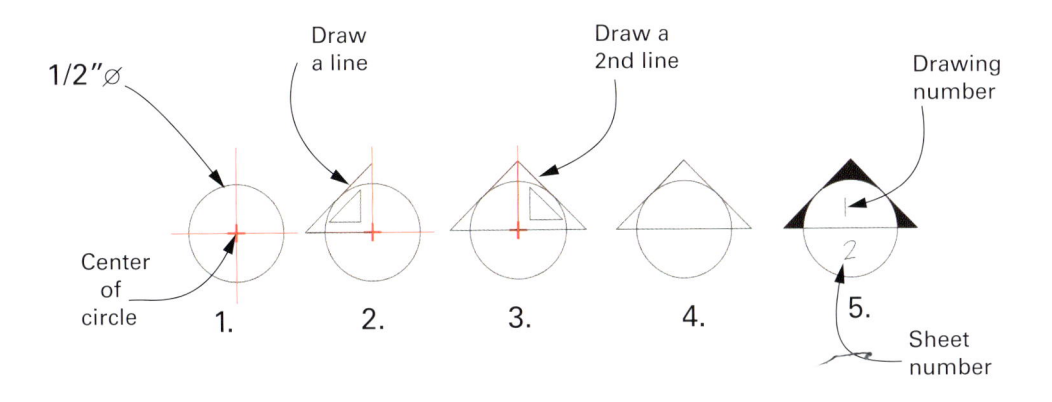

Figure 4.13
Steps for drawing an elevation callout.

Figure 4.14
ID label.

Hatch Lines and Poché Symbols

Hatch lines are generic angled lines that tell the reader a surface has been sliced. They don't give information on material. **Poché**, pronounced "po´-shay," is a textural pattern that represents a specific material. Here's how to draw these lines. and patterns.

Hatch Lines Draw 45° angled lines through the whole object. Space them evenly; see Figure 4.15 as a guide.

Poché Apply this to materials in elevation, plan, and section. Figure 4.16 shows plan view pochés for carpet and paver stones.

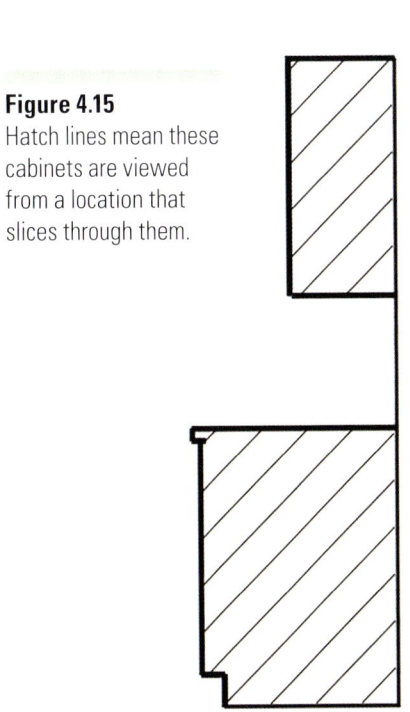

Figure 4.15
Hatch lines mean these cabinets are viewed from a location that slices through them.

HATCH LINES AND POCHÉ SYMBOLS   73

Figure 4.16
Plan view pochés for carpet and paver stones. Courtesy wolfgangtrost.com.

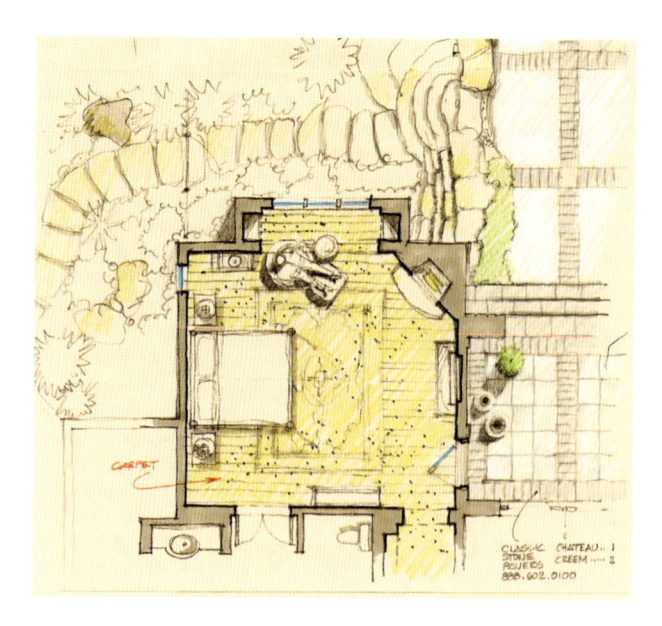

Dimensioning Symbols

Dimensions symbols (Figure 4.17) include **extension** and **dimension lines**, **dimension notes**, **tick marks**, and **leader lines**.

Dimension and Extension Lines   These describe lengths of building components such as walls and the distances between them. Extension lines emanate from wall endpoints and centers of doors and windows. Dimension lines run perpendicular to the extension lines.

Dimension Notes   These are numbers parallel to, or inside, the dimension lines. They describe size. Letter these notes 1/8" tall.

Tick Marks   These are short, angled lines placed at dimension and extension line intersections to indicate the specific length of the dimension note. Draw tick marks with a 45° triangle.

Leader Line   This is an annotation line that has an arrow at one end and a **local note** at the other. A local note describes the feature the leader line points to, as opposed to a **general note**, which is placed elsewhere on the sheet and applies to everything on it. Draw leader lines angled or curved because straight can be confused with the drawing's straight lines. Use an arrow template to keep the arrows consistent, and shade them opaque. A standard proportion is three times as long as they are wide. 3/16" x 1/6" is a good size.

Figure 4.17
A dimensioned wall.

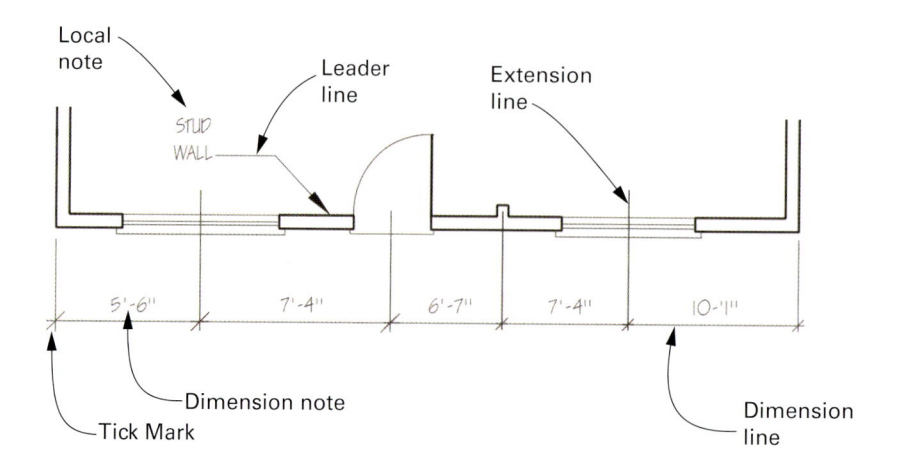

74   CHAPTER 4   DRAFTING CONVENTIONS

Other Line Types

Depending on the complexity of the project, these lines may appear on a set of drawings.

Match This shows where to align a large drawing that spans two or more sheets of paper. Like hidden lines, **match lines** are dashed, but are thicker and longer.

Border These are thick lines around the perimeter of a drawing sheet.

Construction Also called **layout** lines, these are thin, light lines that help create object lines but are not part of the object itself. Typical uses include positioning a drawing, making guidelines for letters, and projecting features from one orthographic view to another. Draw them light enough so it's obvious they're not object lines and so they don't appear on a photocopy. If it's clear they're not object lines, you don't need to erase them when the drawing is finished. Use a hard pencil, such as 4H. Never ink construction lines. In Figure 4.18 the bed's dimensions were marked along the light vertical and horizontal construction lines, and the pillows were placed with help from the light intersecting construction lines.

Enlargement Box An **enlargement box** is a rounded rectangle of thick dashed lines. It encircles an area that is drawn to a larger scale elsewhere. An attached ID label tells the reader exactly where to find that enlarged drawing (Figure 4.19).

Structural Grid Drawings are laid out on a grid that shows the location of **structural** (load-bearing) components such as columns (Figure 4.19). These lines are used mostly for dimensioning and schedule reference.

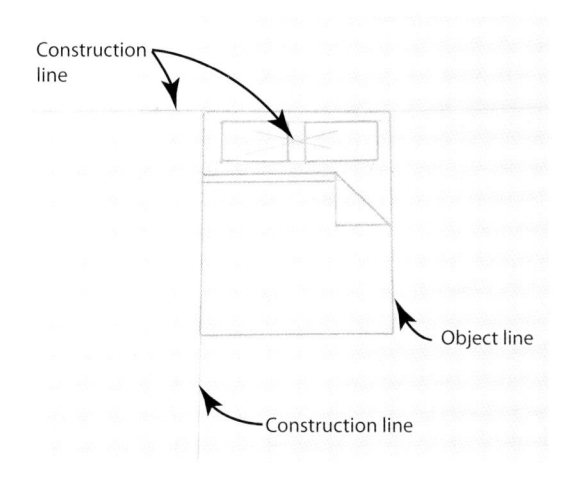

Figure 4.18
The bed's object lines are dark and the construction lines are light.

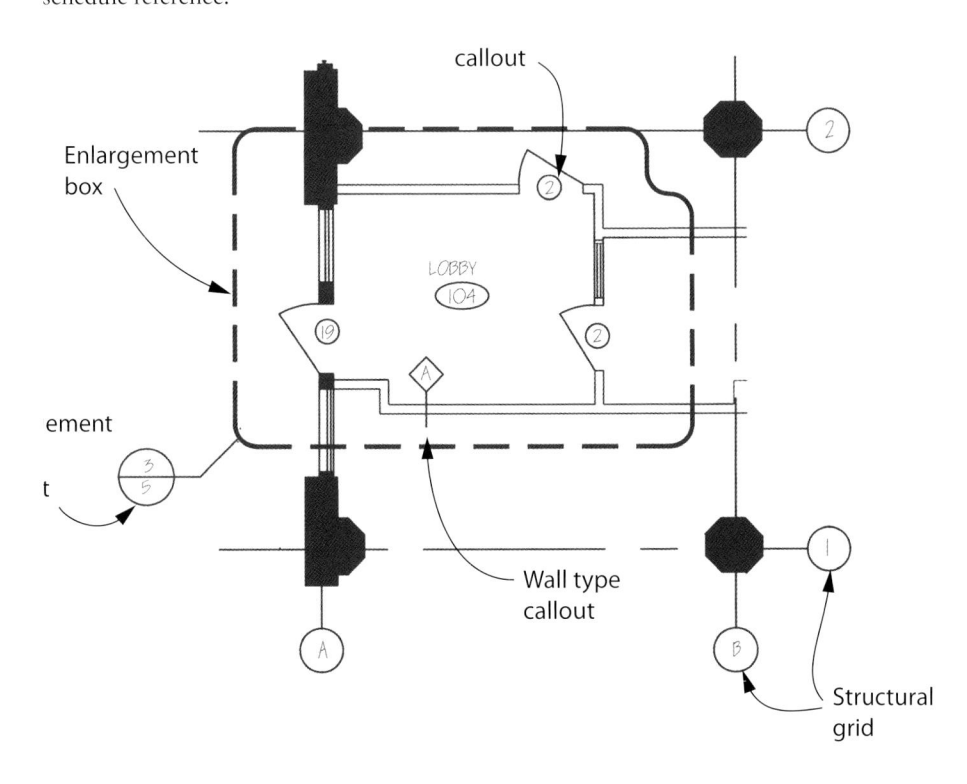

Figure 4.19
Enlargement box, structural grid, wall type, and door callouts.

Keynotes

Keynotes are a geometric shape with a number or letter inside. Ovals, rectangles, diamonds, triangles, and circles are the most common shapes used. Place them on or near doors, windows, finishes, wall types, and any other repeating components (Figure 4.19) and reference them to a **legend** or **schedule**, which are charts of information.

Figure 4.20 shows a commercial floor plan with many of the line types and symbols just discussed.

Figure 4.20
Symbols and line weights are shown in this floor plan.
Courtesy PGAV Architects, Westwood, KS

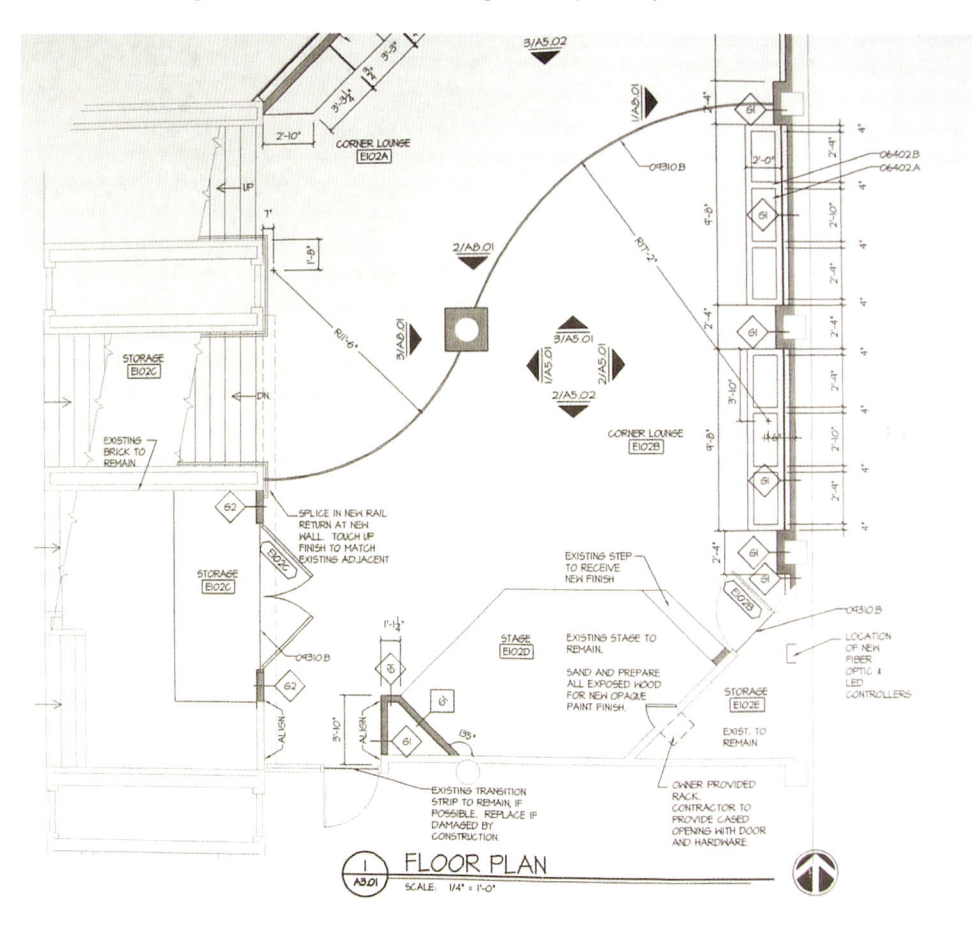

Discipline Designator

Each sheet in a drawing set has its own page number or **designator**, which is a code. The designator is in the **title block** and looks like **A-XX** or **A-X-XX**. The letter is the discipline of the sheet's drawings. The numbers are the sheet type and/or sheet sequence number. Tables 4.1 and 4.2 show discipline designators and sheet type designators.

Table 4.1 DISCIPLINE DESIGNATORS

G	General	P	Plumbing
H	Hazardous Materials	D	Process
V	Survey/Mapping	M	Mechanical
B	Geotechnical	E	Electrical
C	Civil	W	Distributed Energy
L	Landscape	T	Telecommunications
S	Structural	R	Resource
A	Architectural	X	Other Disciplines
I	Interiors	Z	Contractor/Shop drawings
Q	Equipment	O	Operations
F	Fire Protection		

CHAPTER 4 DRAFTING CONVENTIONS

Table 4.2
SHEET TYPE DESIGNATORS

0	General (symbols, legend, notes)
1	Plans
2	Elevations
3	Sections
4	Large-scale views
5	Details
6	Schedules and Diagrams
7	User defined (anything that doesn't fit in other categories)
8	User defined (anything that doesn't fit in other categories)
9	3D Representations (isometrics, perspectives, photographs)

Tip
When preparing a multi-sheet or multi-board presentation, create a consistent organization system between them. Use a repeating graphic symbol or line.

Sheet Composition and Organization

Each office has its own standards, but here are some common conventions.

Sheet Sequence   This is a two-digit number that identifies each sheet in a series of the same discipline and sheet type. Numbering starts with **01**. So, **A-01** means first sheet in the architectural discipline. **A-1-01** means first plans sheet in the architectural discipline.

A complete set of drawings has all the discipline sheets stapled together. The architect or designer coordinates the entire project and creates the A drawings. Consultants to the project are professionals such as mechanical and electrical engineers, and they contribute the rest of the drawing set.

Title Block   A title block is a square or rectangular box with general information about the project (Figure 4.21). It goes on each sheet. The amount and type of information varies, but will always include the project name, client, designer, professional seal, date, drawing types on the sheet, designator, and scale. Rectangular-shape title blocks are placed vertically along the right-hand side of the sheet set, opposite the stapled end. Square-shape blocks are placed in the lower-right corner.

Title Sheet   Also called a cover sheet, this is the first sheet in a set and typically contains the sheet index, project name and address, designer name, and sometimes a location map and graphic.

Sheet Size   This is selected early in the design phase and is based on client or designer standards. 22" × 44" (also called "D" size) are the most common. Student work is usually done on 17" × 22" ("C" size), 11" × 17" ("B" size), or 8 1/2" × 11" ("A" size). When choosing the paper size, account for the drawings' dimensions, labels, and notes. All sheets in a set must be the same size and oriented the same way.

Drawing Orientation   All drawings on each sheet must read straight up. Never place drawings sideways. Orient all drawings consistently throughout the sheet set.

North Arrow   This graphic tells the reader which direction is north and should be prominently placed on the plan. Draft plans parallel to the sheet edges, with the north arrow at the top. If true (**compass**) north isn't straight up, then an arrow called **plan north** is added to the graphic that shows how far the plan deviates from true north (Figure 4.22). This enables the drafter to draw the plans parallel to the sheet edges and give simple names to the elevation drawings.

Figure 4.21
Title blocks vary in information and placement, but one goes on each sheet.

Figure 4.22
True north (left) and plan north (right).

Sheet Layout for Multiple Drawings

Here are some tips on laying out multiple drawings on one sheet.

Consider design. There's **contrast** (different-appearing items in a composition); **spatial tension** (perceived line or link through space); **balance** (distributing text and graphics evenly); **movement** (the path the viewer's eye takes); proportion (relative size and scale of a sheet's items), **scale** (size of one item in relation to another); and **repetition** (using the same items throughout a sheet). Use grids or modules to place each graphic in and line up whatever you can.

Only place related drawings on a sheet. This enhances reader comprehension. The most common combinations are plan/plan, elevation/section, elevation/plan, elevation/detail, and detail/detail. Organization can also be based on the construction trade: for instance, architectural details on one sheet and casework details on another.

Don't fill a partially completed sheet with drawings that don't reference or relate to the others. Leave extra space blank.

You may grid sheets into modules and place one drawing placed in each. This is so the structural grid or dimension lines on separate drawings don't overlap. On commercial sets, the graphics on a sheet are often numbered from bottom right to top left, with any unused space on the left (stapled) side.

Orient large-scale drawings that are keyed from enlargement boxes on the plan the same as the plan.

Sheet Layout

When there is just one drawing on a sheet, center it. On sheets with several drawings, create a balanced layout. It can be **symmetrical** (graphics are arranged the same on both sides of an axis), **asymmetrical** (graphics are arranged differently on both sides of an axis), or **radial** (graphics are arranged like rays around a center point) (Figure 4.23). Figure 4.24 shows a balance of plans and graphics.

Symmetrical　　　Asymmetrical　　　Radial

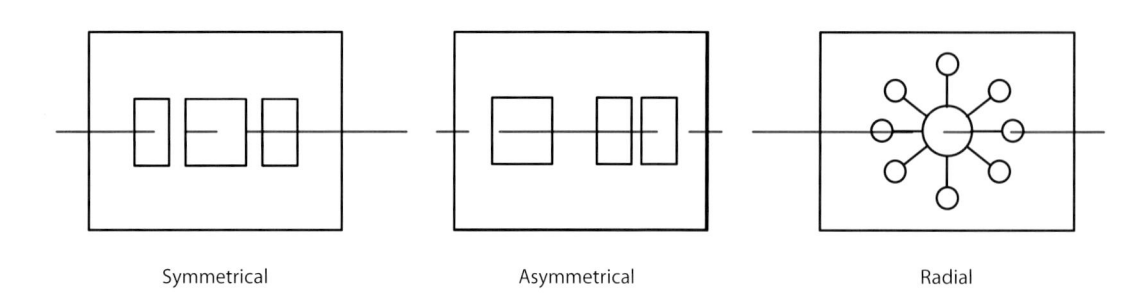

Figure 4.23
Symmetrical, asymmetrical, and radial.

CHAPTER 4 DRAFTING CONVENTIONS

Figure 4.24
The asymmetrical arrangement of plans and graphics counter-weigh each other.
Courtesy mkerrdesign.com

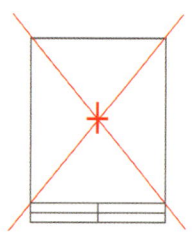

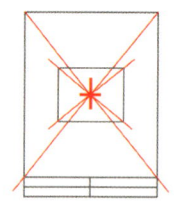

Figure 4.25
Draw the title block first. Find both its center and the paper's center by connecting their corners. Use only the space *above* the title block to center the drawing.

A simple student assignment might have one centered drawing, an ID label, and a title block. Figure 4.25 shows how to lay that out. If the drawing is asymmetrical, enclose it in a rectangle and center the rectangle.

Lettering

Lettering is note making. Its purpose is to keep notes legible no matter how small the drawing set is shrunk to when copied. Legibility is important because builders need to understand what you've written, and drawings are legal documents. Lettering is not the same as printing in capital letters; the latter doesn't provide the level of legibility a set of documents needs. Look critically at the letters in Figure 4.22 to see what differentiates lettering from merely printing in capitals. Namely:

Height is consistent, which is achieved by drawing guidelines. Letters should be the entire height of the guidelines.

Spacing is consistent between words and within words.

Appearance is consistent; all same-type letters should look identical

Vertical strokes are vertical, horizontal strokes are horizontal, and all angled strokes angle to the same degree. There are no gaps between a letter's individual strokes.

Each number of a fraction is drawn the entire height of the guideline.

Notes are 1/8" tall and drawing titles are 1/4" tall. One "point" in lettering size is 1/72". So, 1/2" high letters are 36 points. Separate multiple rows of guidelines with spaces half the height of the guidelines.

Use one lettering style throughout a project.

Lettering Tips Good lettering takes a lot of practice, but these suggestions will help achieve that.

Use a simple font style, such as Ariel (sans serif) or Times Roman. Don't include serifs (the crossbars on I's and J's). Numbers should match the letters' font.

Use a straightedge for the vertical strokes. Slide a small triangle along the parallel bar and use its vertical edge (Figure 4.26) as the straightedge. Freehand all the other strokes; slightly angled horizontal strokes look good and are easy to draw.

Make the strokes quickly; slow, labored-over strokes look shaky.

Rotate the pencil while drawing to obtain slightly different line weights.

> **Tip**
> Complete all pencil drawings in a project before inking any of them. Don't ink drawings individually as you complete them, as this will likely result in a poorly composed sheet. Lay out all pencil drawings on top of a sheet to ensure there's enough space for them plus their ID labels, dimensions, title block, and border. Then start inking.

Figure 4.26
Guidelines are light, letters are dark, and the letters are the full height of the guidelines.

Exercise
Fill a sheet of size "A" vellum with letters. Draw words and sentences instead of the alphabet, because words and phrases incorporate spacing. You can letter your favorite poem, philosophy, or a biography of yourself.

Reduce smudging by choosing a harder lead or keeping a clean sheet of paper under your hand to protect the letters already drawn.

All letters should touch the upper and lower guidelines; some strokes may go slightly above or below the guidelines, but none should be shorter than the guidelines.

Notes

These are lettered comments on the drawings. (Figure 4.27). *General notes* apply to the whole drawing set. *Local notes* discuss a specific feature and have a leader line and arrowhead pointing to that feature.

Guidelines for local note placement are:

Letter them horizontally. In the rare case they must be vertical, rotate them 90° degrees to read from the right side of the sheet.

Group the notes around the features they refer to. For instance, on a desk, place notes that refer to its feet close to the bottom (Figure 4.27).

Place notes close to the items they point to, to avoid crossed leader lines.

Align notes at the left. An even margin improves readability.

Use a French (irregular) curve to draw curved leader lines.

Start leader lines at the beginning or end of the note. Don't start them somewhere in the middle of the note.

Arrowheads must touch the object they refer to. Use a template to keep them a consistent size.

Make notes short and to the point. Keep terms consistent throughout the drawing set.

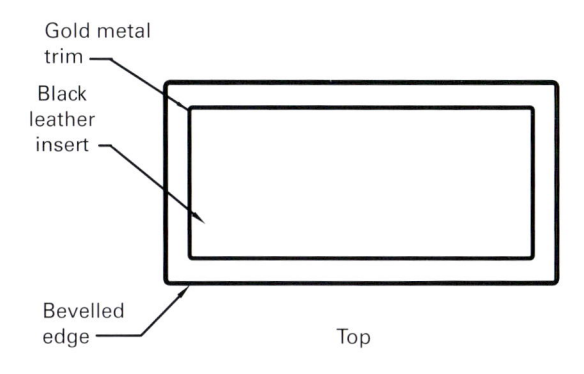

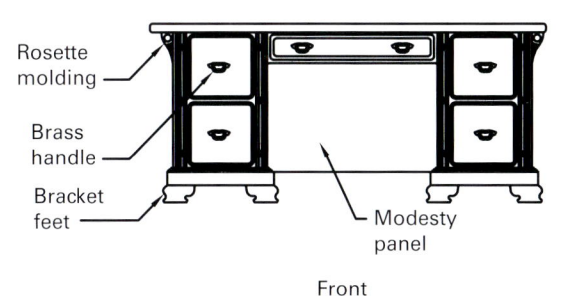

Figure 4.27
Notes, leader lines, and arrowheads.

Summary

The complexity of architectural drawings is managed by representing their features with abstract line types and symbols. Using recognized standards for drawing those lines and symbols ensures that everyone involved in the project will interpret them the same way.

Exercise
Visit a website in the Further Resources section, choose an area of it that interests you, and write a short essay about what you learned.

Classroom Activities

Obtain two sets of construction drawings from different sources. Then:

1. Find and identify different line types and symbols on each set.
2. Compare how the same symbols are drawn on different sets.
3. Find five objects (e.g., doors, cabinets) on each set and describe the line types used to draw them.
4. Find and read the general notes. Discuss their relationship to the drawings.

Questions

1. What is the difference between a visible object line and a hidden object line?
2. Why are different line weights used on a drawing?
3. What is the purpose of a local note?
4. What are three types of balance for laying out a sheet?
5. What information is found in a title block?
6. Explain how an elevation callout and a drawing ID label are related to each other.

Further Resources

CSI https://www.csiresources.org/home

NCS https://www.nationalcadstandard.org/ncs6/

NCS-UDS https://www.scribd.com/document/331308165/NCS-Uniform-Drawing-System

NIBS https://www.nibs.org/default.aspx

Keywords

- balance
- break line
- callouts
- center line
- circumscribe
- compass north
- contrast
- cutting plane
- dimension line
- dimension note
- elevation callout
- enlargement box
- extension line
- general note
- hidden line
- identification (ID) label
- inscribe
- key
- keynotes
- layout line
- leader line
- legend
- lettering
- line
- line quality
- line type
- line weight
- local note
- mark
- MasterFormat
- match line
- movement
- National CAD Standard
- object line
- plan north
- poché
- radial
- repetition
- scale
- schedule
- spatial tension
- standard
- structural
- structural grid lines
- symmetrical
- tangent
- tick mark
- title block

CHAPTER 5

The Architectural Floor Plan

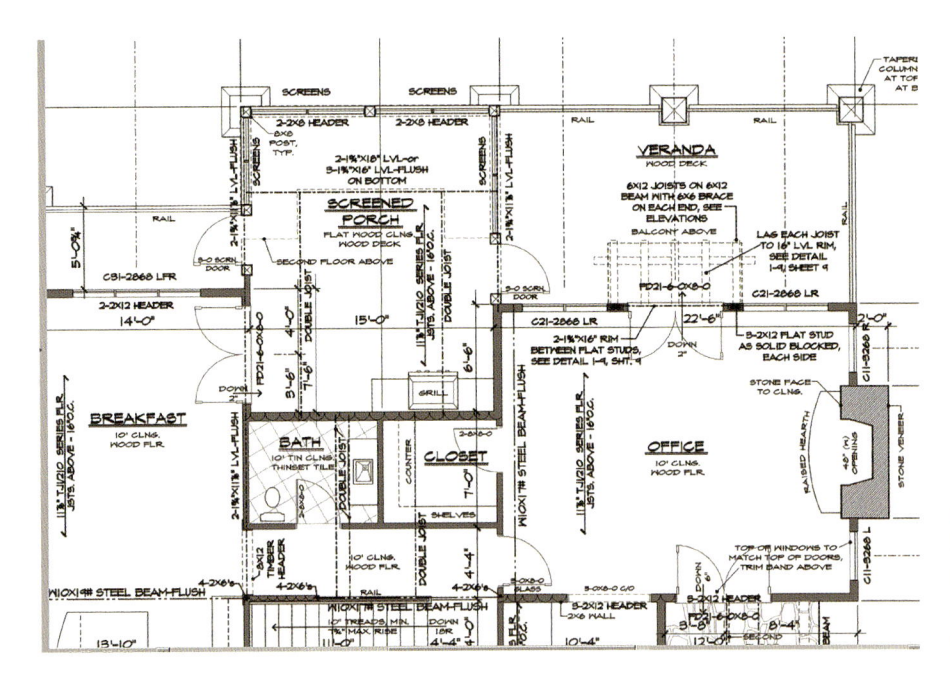

Figure 5.0
A CAD floor plan.
Courtesy of Bickford
and Company, Overland
Park, KS.

OBJECTIVES

Upon completion of this chapter you will be able to:

- Explain what an architectural floor plan is
- Design, measure, sketch and draft an architectural floor plan
- Select appropriate wall thicknesses
- Apply material pochés
- Use ink pens
- Recognize CAD standards.

The **architectural floor plan** is the heart of a set of drawings. Most of the other drawings are referenced from it. It is used for both client presentation and construction documentation. While a plan for presentation may be sketched, a plan for construction is always drafted.

What Is an Architectural Floor Plan?

An architectural floor plan is a 2D, top-down view made by inserting a horizontal cutting plane through a room or whole building 4' or 5' above the ground (Figure 5.1). The roof and walls above the cutting plane are removed, and the reader looks straight down into the rooms. Sometimes the cutting plane is offset at different levels to show non-aligning features, such as the staggered levels of a split-level house.

What the Architectural Floor Plan Shows This drawing shows **fixed** (non-movable) features: walls, windows, doors, skylights, exposed beams, columns, soffits, cabinetry, built-ins, appliances, plumbing fixtures, stairs, and fireplaces. Both finished and unfinished space is shown. Balconies, attached garages, decks, patios, pools, and detached garages may also be included. Plans done for construction contain dimensions, room names, and symbols that link it to schedules, sections, and details.

Only built-in furniture is shown. Moveable furniture goes on a **space plan**, which is a floor plan that shows furniture arrangements. Or it may go on a **commercial furniture plan**, which is a non-residential floor plan and is done to **keynote** or reference the plan to a **key** (legend). Floor plans may also contain **entourage**, which is people, furniture, décor, and accessories that provide visual scale. Figures 5.2 and 5.3 show floor plans done for client presentation.

Floor plans have standard symbols for doors, windows, draperies, furniture, appliances, drains, materials, and identifiers. Electrical and mechanical symbols usually have their own plans so as not to clutter the architectural one.

Each level of the building has its own plan. This applies even if the second floor is just a partial plan, such as a loft that overlooks the first floor. The loft must have its own plan. Plans show length and width. Ceiling height may be noted, but a ceiling with multiple heights is better noted on a **reflected ceiling plan**, which is a plan that shows lighting fixtures and everything else attached to the ceiling.

Drawing Scale

This determines the amount of detail and information on it. Wall attachments like marker boards, mirrors, and wood trim aren't

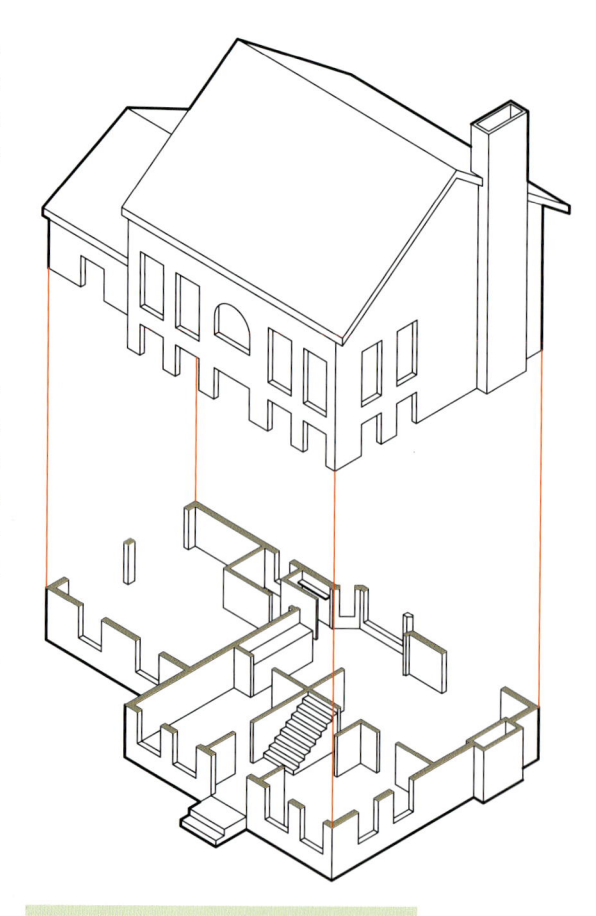

Figure 5.1
A floor plan is a view made by inserting a horizontal cutting plane 4' or 5' above the ground.

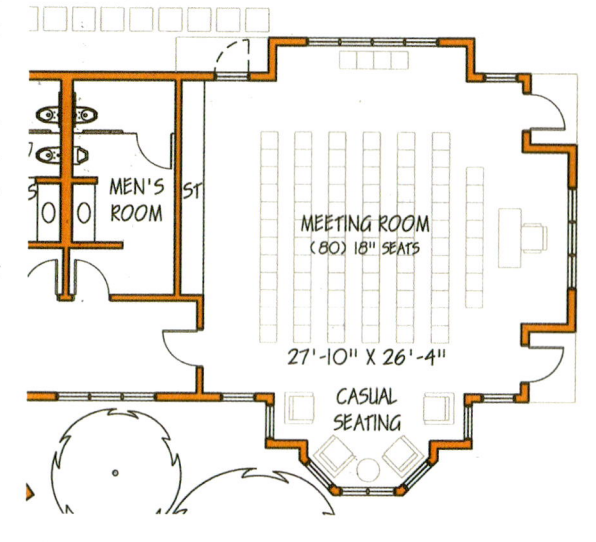

Figure 5.2
Space plan. Courtesy of mkerrdesign.com.

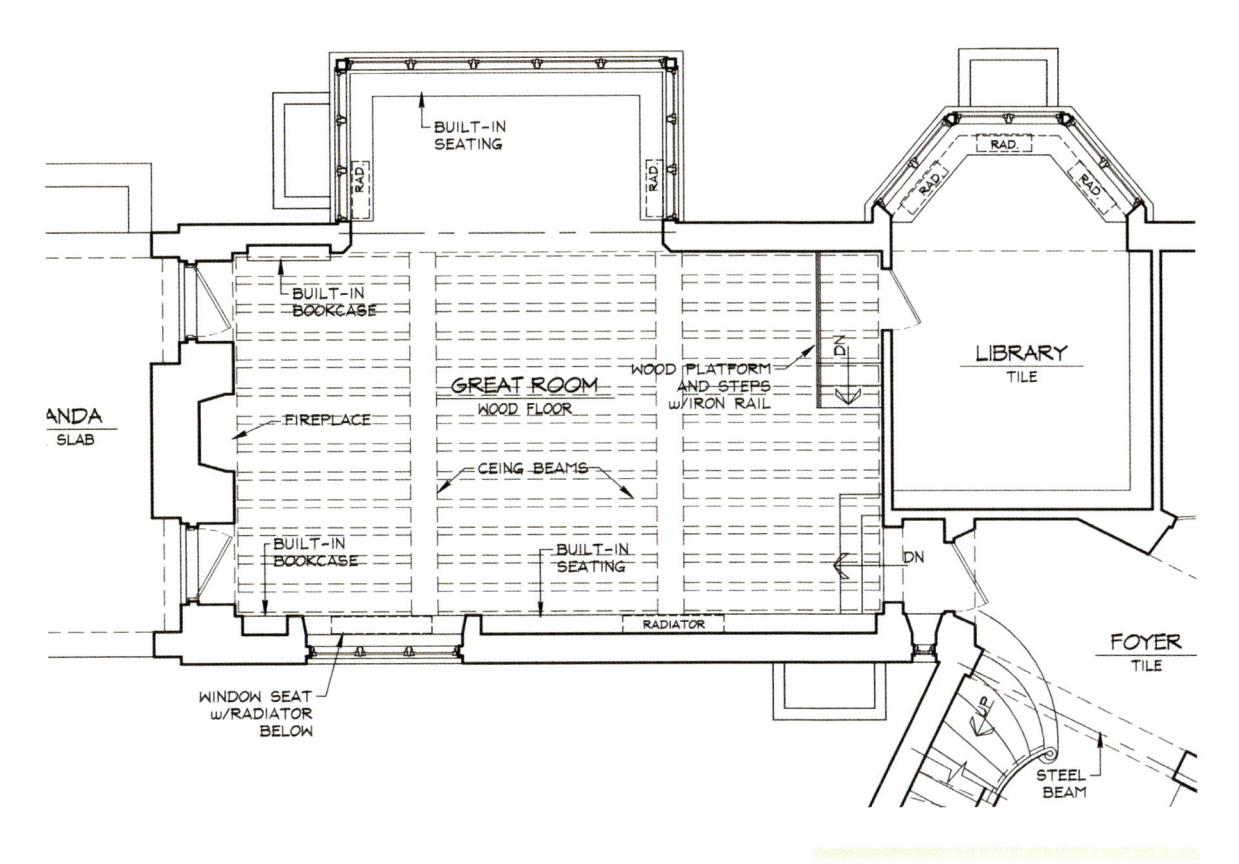

Figure 5.3
Floor plan of a great room. The dashed lines are overhead beams. Courtesy of mkerrdesign.com.

drawn unless the scale is large (at least 1/2" = 1'-0"). Smaller scale drawings show less detail; for example, you would draw sinks but not faucets and controls. Tracing **CAD blocks** (digital templates) into your sketch is convenient, but adjust the block's detail to match the detail of your sketch. Most CAD blocks are too detailed to be fully traced in a 1/4" =1'-0" plan.

How to Sketch a Plan of an Existing Space

Interior designers often need to sketch an existing space to have a canvas on which to brainstorm, problem-solve and make changes. If you are sketching an existing space, you can either measure it first, or eyeball the proportions, sketch the plan, and then measure and add those measurements to the plan.

Let's sketch a plan of the room in Figure 5.4.

Choose a scale. Base it on the amount of detail needed, which depends on the plan's purpose and the amount of accuracy needed. Let's sketch the room at 1/4" = 1'-0". Our grid paper has 8 grid squares per inch, meaning that two grids together are 1/4" wide, and represent one foot. One inch represents four feet.

Starting at the lower-right corner, sketch the plan's outline (Figure 5.5).

Add a second line for the walls' thickness (Figure 5.6). Don't draw walls as single lines, because that will make adding details later difficult. For sketching purposes, a thickness that easily matches the grid lines is fine. Apply exact material thickness when drafting the plan.

Note which side the door is hinged on and the direction it swings, and sketch accordingly. Record door and window header heights and windowsill height.

Figure 5.4
A simple room with some furniture.

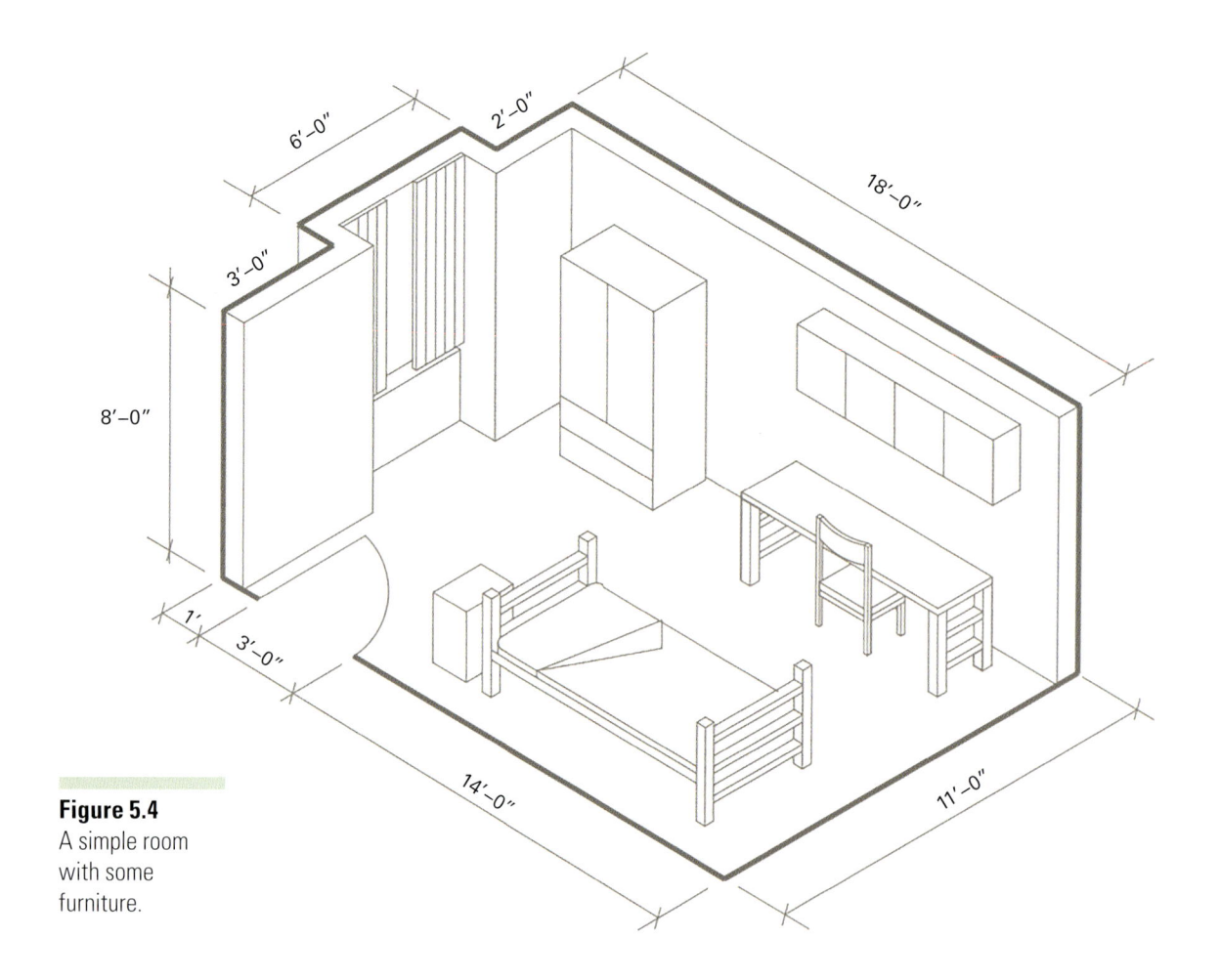

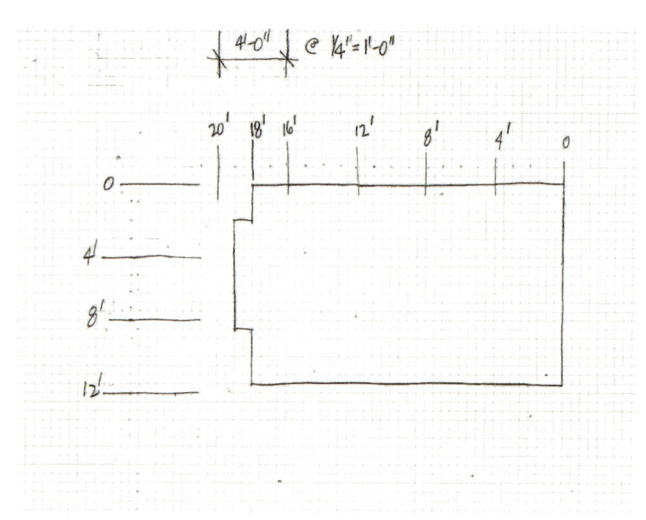

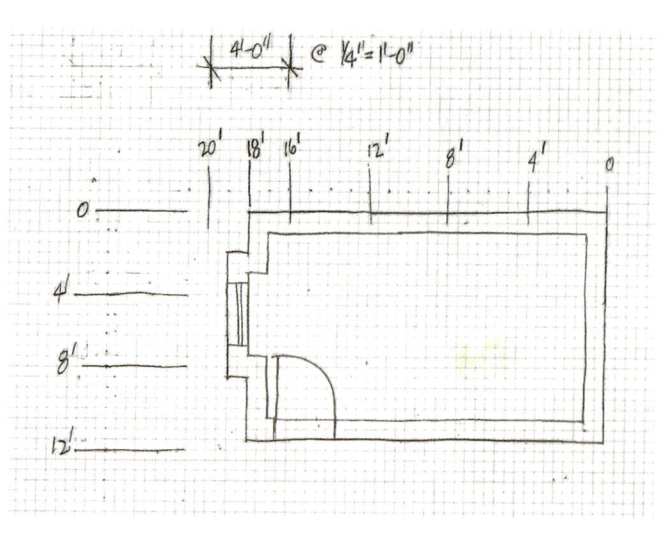

Figure 5.5
Sketch the plan's outline.

Figure 5.6
Add wall thickness, doors and windows.

Add built-in cabinetry. Add furniture if needed.

You may want to make your sketch a **fit sketch**, which is a plan that shows how all furniture, fixtures, and appliances will fit in a room (Figure 5.8). **Furniture templates**, which are plastic sheets with cutouts, are useful for this. Trace the furniture in the arrangement you want in a room, add clearance space around the tracings, and then enclose it all with walls. Make sure the templates are the same scale as the sketch. Know

86   CHAPTER 5   THE ARCHITECTURAL FLOOR PLAN

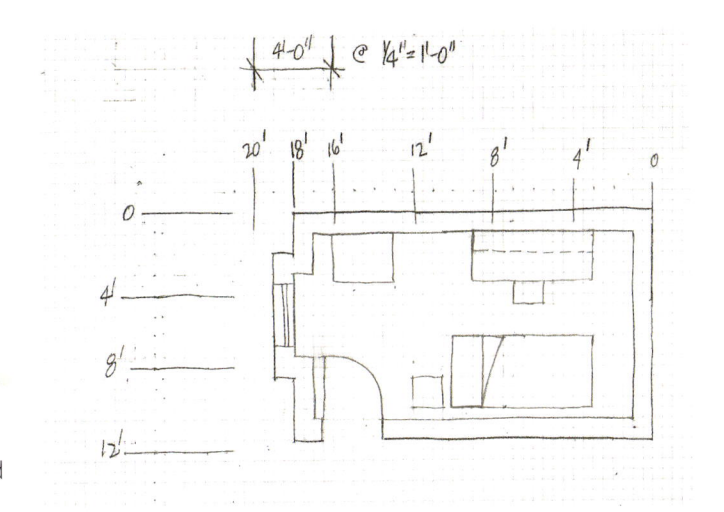

Figure 5.7
Add built-in cabinetry and furniture.

Exercise
Using furniture and fixture templates, make a fit sketch of a room. Then trace over the template drawings to "loosen them up" and add your own stylistic touches.

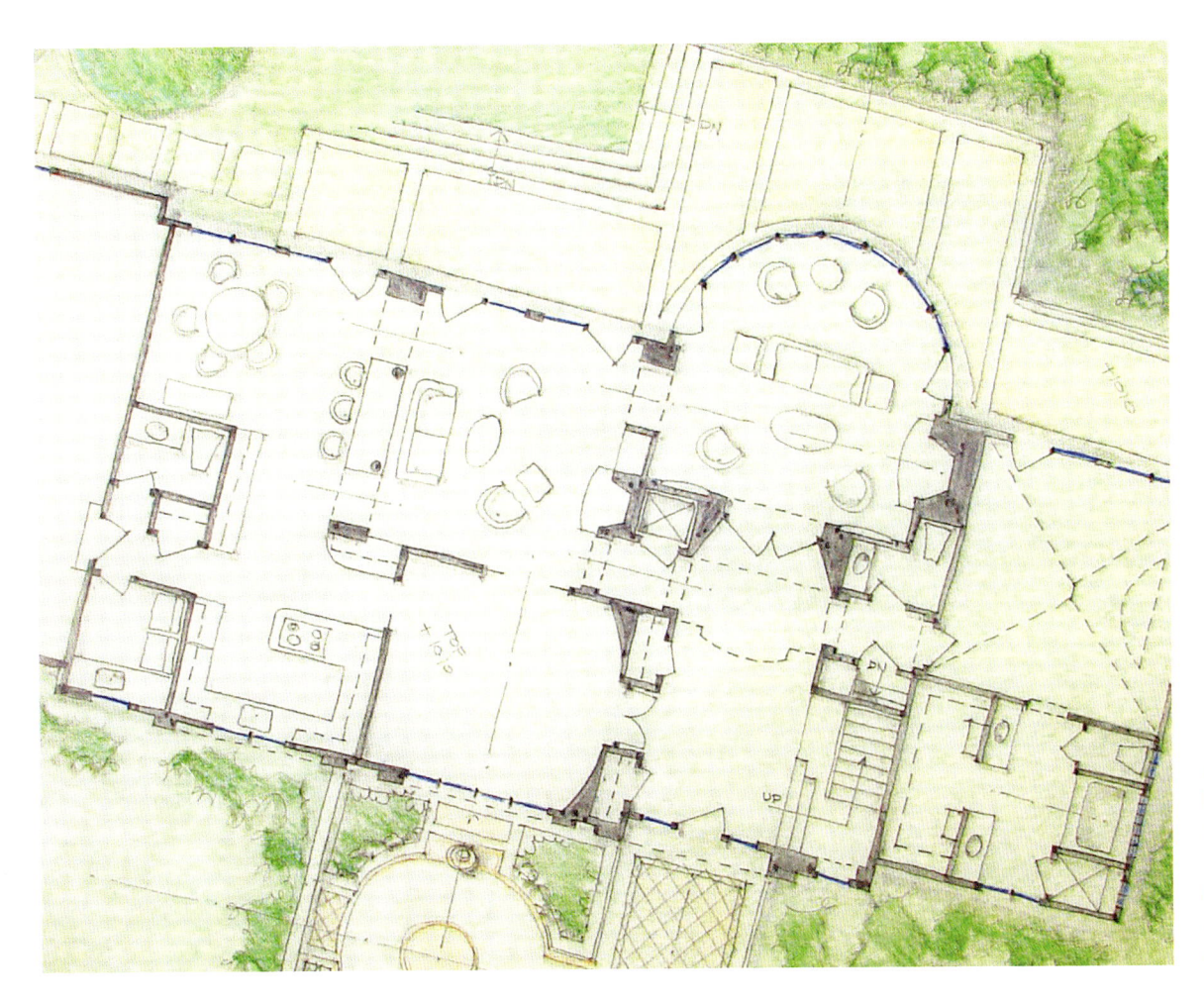

Figure 5.8
A fit sketch shows how furniture and equipment fit in a room.

that the template cutout may or may not be the annotated size. Even if it is, lead thickness and pencil tilt may cause the drawn symbol to be smaller. Check the measurements of any template cutout instead of trusting the annotation if precision is required.

Layer plans on multiple sheets with different information on each sheet if a room has lots of items in it. For example, put walls on one and furniture on another.

If you include an adjacent room, show the wall between it and the first room as shared; don't draw two separate floor plans or double the width of the shared wall.

HOW TO SKETCH A PLAN OF AN EXISTING SPACE

Add room names, the name/address of the building, the name/contact information of anyone who helped measure it, and the date. Some offices leave off the north arrow if north is up, but if north is to the left (the only other option for architectural plans), a north arrow is added.

The Visual Inventory

Visual inventory is documentation of the space or spaces to be designed. When on a **site visit**, which is a trip to the project location, gather more information than just a floor plan sketch. Include measurements, photos, and interviews with the project users to learn their reactions, likes, and dislikes about the space. The designer must find out:

- What needs to be done
- The current and desired focal points
- Individual components, overall arrangement, predominant shapes, light, and view
- If the space is open to other spaces, requiring visual coordination
- Finishes and colors, traffic patterns, and what needs to be moved, removed, or replaced
- What will be a major cost to remove
- What needs to be there, what doesn't
- If there's enough storage
- Where carpet will be joined
- Where air vents, outlets, light switches, and thermostats are
- Which side of the window a drapery pull should be placed

Note all this on the sketches. You need enough detail, accuracy, and legibility to reconstruct the drawing later or to allow someone else to draft it from your notes. (Figure 5.9).

Visual Inventory Tools Time is money, so arrive properly equipped. Keep the following tools (Figure 5.10) in your car:

Metal Tape Measure One that is 25', rigid, retractable, and lockable, with a hook at the end for attaching it to a wall (useful when working solo) is best. Tapes that show increments in feet and inches or metric are available.

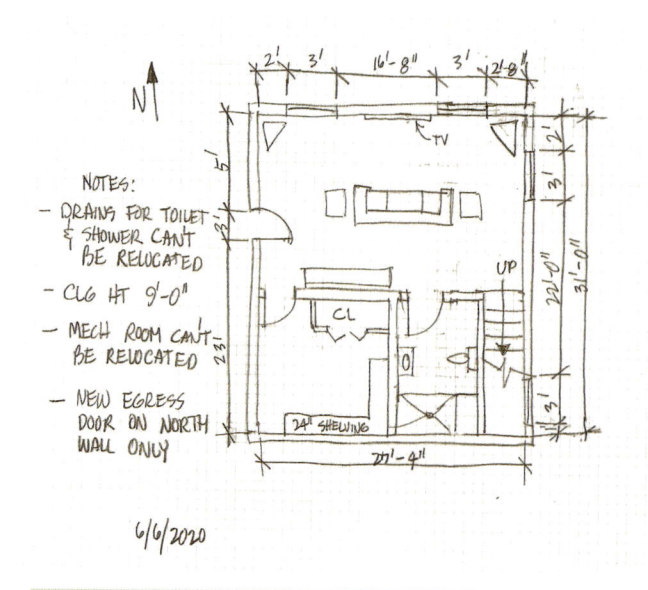

Figure 5.9
A clipboard sketch done at a job site.

> **Tip**
> Take lots of photos on a site visit. When photographing walls, hold the camera lens parallel to them to minimize distortion.

Figure 5.10
Carpenter's square, bubble level, metal tape measure, flashlight, pad of grid paper, pencil, smartphone with measuring apps.

Laser Tape Measure This device records distance by pointing a laser across a space and displaying the distance on a screen. It can calculate square footage, too.

Measuring App This phone program can measure, calculates square footage, and do other tasks (Figure 5.11). One comes with the Apple iOS on the iPhone. The App and Google Play (for Android) stores have others, both free and for pay.

Camera A smartphone that takes good, clear photos is enough for most documentation. Attachment lenses for zoom and wide angle are available.

Pad of Grid Paper and Dark Lead Pencil A pad provides a portable surface for sketching; you can also use a clipboard. Grid paper is available in 8 1/2" × 11" and 11" × 14" sizes.

Construction Calculator Besides the basic four functions, this electronic device converts between imperial and metric units; linear, square, and cubic formats; translates inches, feet, yards, and meters; and figures out board feet and stair angles. All this helps when estimating materials and costs. Figure 5.12 shows two calculator apps. There are also construction calculator phone apps.

Other A **flashlight** (this can be a portable light or a phone app) and a step ladder or **step chair** (portable stairs) always come in handy. A **carpenter's bubble level** is a tool that establishes a horizontal reference line and checks other lines against it. A **carpenter's square** is a triangular tool placed inside a corner to see if the walls align with it. There are phone apps for the level and square (Figure 5.13).

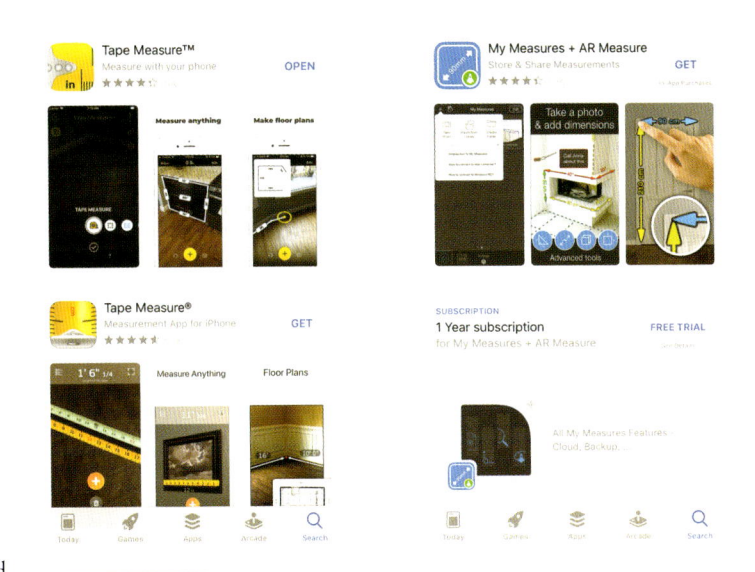

Figure 5.11
4 measuring apps.

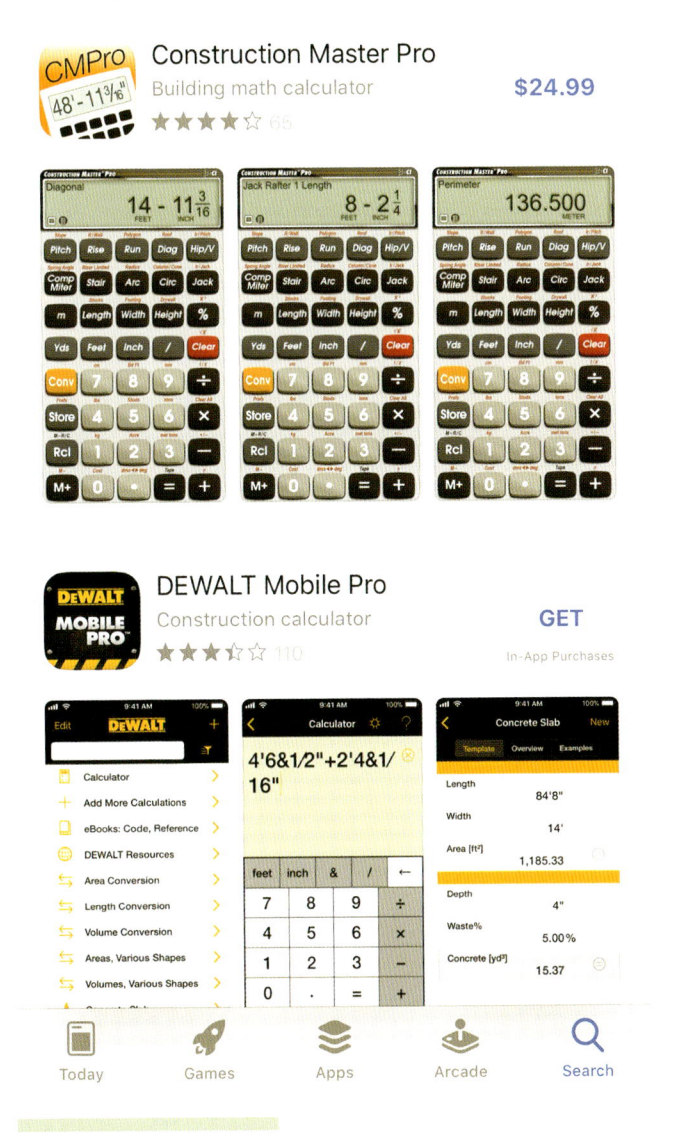

Figure 5.12
Construction calculator apps.

Exercise
Keep a tape measure and pocket journal in your bag and record the sizes of common objects and clearances. This will help you internalize the size of the built environment. Note doorknob heights, countertop widths, room lengths, and the distance between restaurant tables. Add sketches and ideas to the journal, too.

THE VISUAL INVENTORY

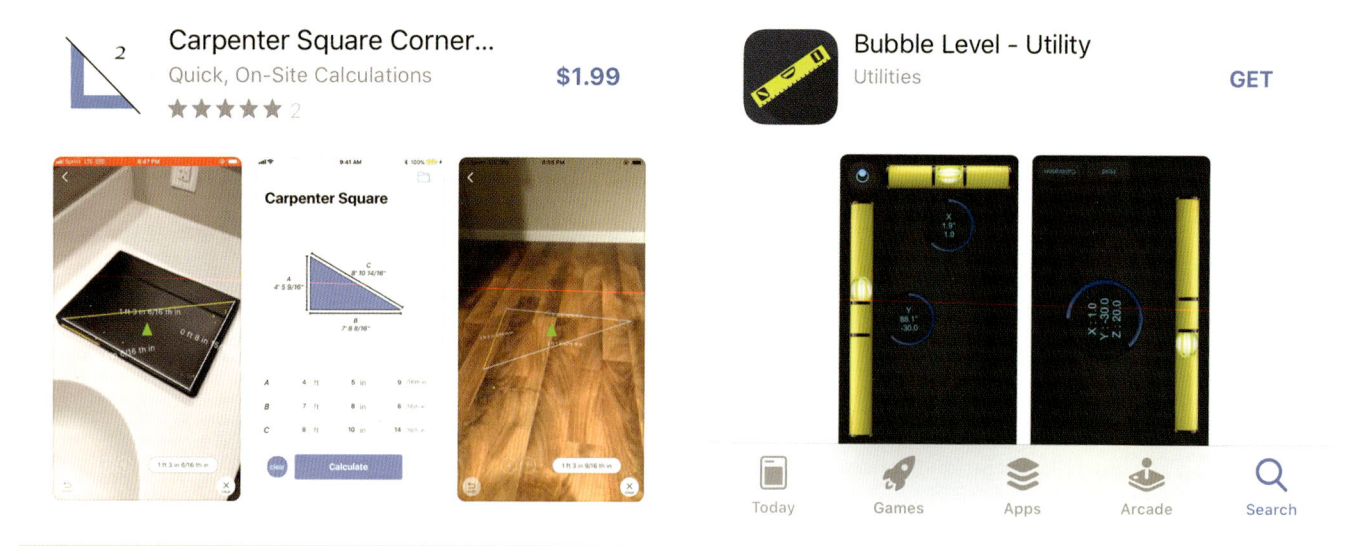

Figure 5.13
Bubble level and carpenter's square apps

How to Measure a Room

Document *size* and *location:* the size of the space and the relative locations of everything in it. The accuracy needed depends on the project. If a living room is incorrectly measured by a few inches, the new sofa will probably still fit. If a cabinet or kitchen wall is incorrectly measured by a few inches, custom work may not fit. Even if dimensioned drawings are available, careful measuring is still needed, because the "as-built" dimensions may differ from the paper ones. Here's a suggested sequence.

1. *Measure the room's overall size*. Choose a corner as a datum point to reference everything from and commence measuring from it. Work your way around the plan. Lay the tape along the floor's baseboards for greatest accuracy and measure from the walls. If furniture or obstructions prevent this, place the tape on the wall, but ensure that it is parallel to the floor. Measure each feature's distance from a wall or corner and take running, or cumulative, measurements. Hold the measuring tape at one corner and read all points along that line without moving the tape instead of continually moving the tape and taking each measurement from the last reading. This prevents small errors from accumulating and makes measuring errors apparent.

2. *Measure all the interior, permanent features* and their length, width, and distance from one another. Locate and measure doors, windows (note trim and framed opening sizes), chimneys, closets, columns, base cabinets, countertop overhangs, floor drains, air vents, outlets, switches, thermostats, other utility connections, any obstructions, and special characteristics of the space.

3. *Measure any furniture* that is part of your drawing. Record its size and location. Note odd and oversized pieces, and any existing furniture you'll be working with.

4. *Draw any radiators or air ducts in the walls*, the location of water supply drainage pipes and vents, gas pipes, and electrical outlets. Note door swing direction. Indicate wall thickness and, when possible, indicate whether the walls are **load-bearing** (structural) or **non–load-bearing**, also called **partition wall**.

5. *Measure stairs*. Count the number of steps. Measure the **riser** (vertical board) height and any landings. Measure the length and width of one **tread** (the horizontal board). When measuring for carpet and stair runners, the width of carpet bolts and the placement of joins will add to the square footage of material needed.

90 CHAPTER 5 THE ARCHITECTURAL FLOOR PLAN

Measuring Tips For greatest accuracy:

6. *Hold the tape level and taut.* Sagging tape gives incorrect measurements. Know where the tape's zero point is, since it's not always at the end. Hold the tape at a height that lets you get measurements for key features; for instance, don't start so low on the wall that the tape runs below the bottom of the windows. Measure from the wall surface, not the baseboards.

7. *Don't assume that rooms are square*, walls are plumb, or floors are level. Take diagonal measurements and check walls and floors to determine distortion early in the documentation process.

8. *Check corners for squareness.* If corners are not square, adjustments must be made during cabinet installation. There are two ways to check for squareness. The first is with a carpenter's square, The second is the 3/4/5 method. Choose a corner and measure 3' along the wall in one direction and 4' in the other direction. Connect the points with a straightedge. If the distance is 5', the corner is square.

9. *Compare the wall's overall measurements* with the sum of its subsections. Measure within 1/4".

Measure walls with windows:

- From wall to wall
- From corner to outside trim of window
- From outside window trim to outside window trim
- From outside trim of window to corner
- Find wall thickness by measuring the width of door jambs (correcting for the size of the frame).

Measuring Heights Collect height data on a site visit, too:

- Measure ceiling height, baseboard, chair rail, and other trim width.
- Measure the distance from the top of the window trim to the ceiling and from the bottom of the window trim to the floor. Measure any **soffit** dimensions (enclosed area below the ceiling and above the wall cabinets). Measure the distance from the soffit to the floor.
- Measure flooring and wall material thickness where possible.

Sketching Tips

Sketches are best done freehand, as it frees up the brainstorming and problem-solving process. Straight-edged tools serve little purpose during this part of the design process and slow it down. Drawings done with straight-edged tools also imply that the design stage is over, which may make clients hesitate to suggest changes. Sketches should look like what they are: early stages of design that invite input. But freehand is not synonymous with sloppy. Sketches must be to scale and proportionately accurate.

Utilize Good Line Quality This is the appearance of lines. Sketch continuous lines from one endpoint to another, and make them dark, crisp, and confident. No half-erased lines. Don't draw long wall lines as a series of short, overlapping ones. When sketching with pencil, pull, don't push it across the paper. In manual work, slightly darken the ends of lines and cross (overlap) them at corners. Rotate the lead as you pull it to help keep lines a consistent thickness.

Sketch to Scale Sketch on grid paper—assign a scale to it and count grid squares—or sketch on tracing paper with a scale.

Layer Sheets of Tracing Paper Layer sheets of tracing paper to develop your design. (Figure 5.14). This enables you to keep what you want and redesign what you don't. Each sketch is a developmental sketch; design is an iterative process.

Figure 5.14
Layer pieces of tracing paper for design iterations.

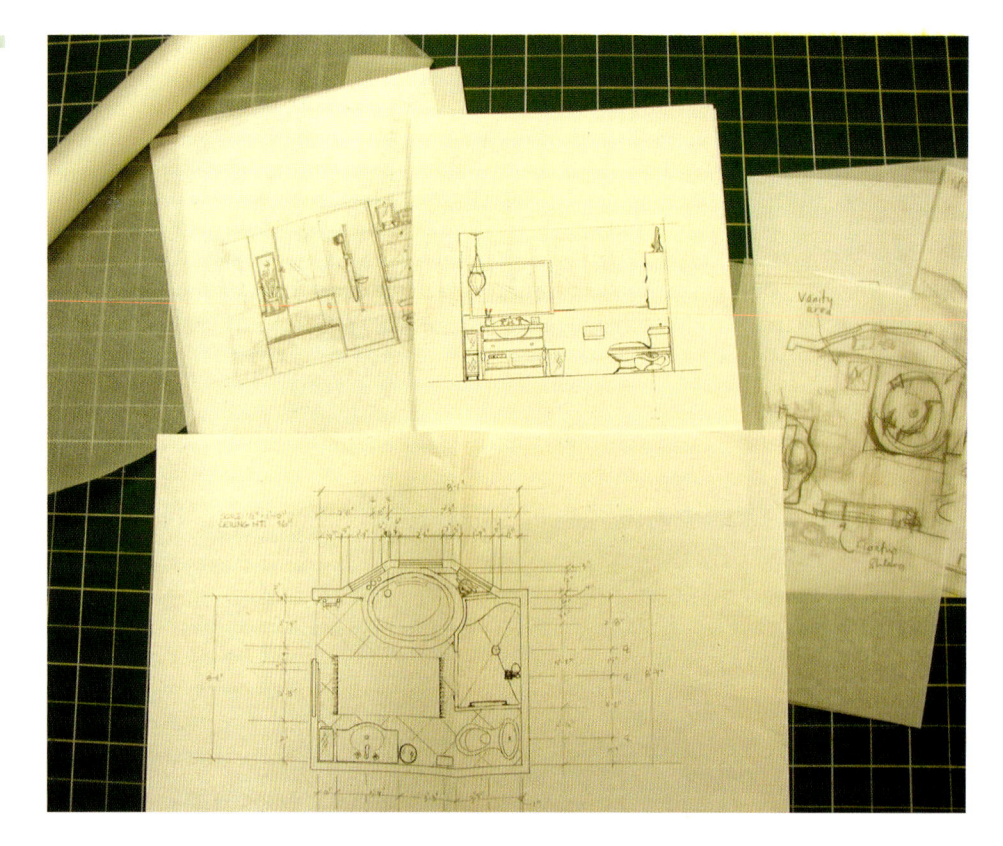

Rendering

Rendering is the addition of color, shade, shadow, and entourage. Such additions help clients understand 2D drawings. Figure 5.15 shows the same sketch in black marker and with color pencil and gray markers. It does take a lot of practice to render well, and poor rendering will ruin an otherwise great sketch. An option is using a tablet and **Autodesk Sketchbook**, which is a program that lets you import images, such as your hand-drawn sketches, and apply color with digital tools.

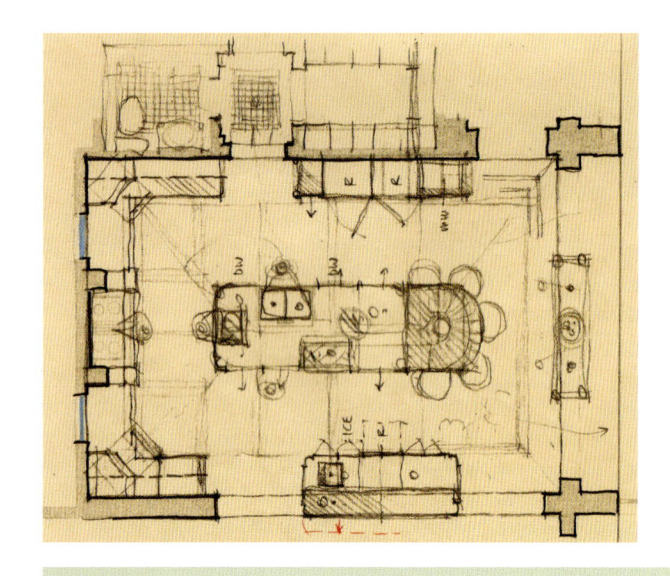

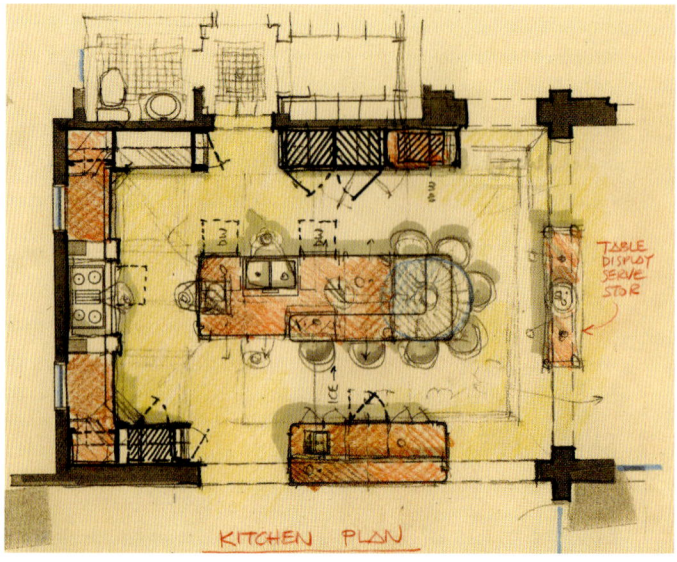

Figure 5.15
Colors, shades, and shadows make a sketch pop.
Courtesy wolfgangtrost.com

Figure 5.16
Screenshots from an Autodesk ReCap Photo tutorial on YouTube.

Photogrammetry

An alternative to manually measuring and sketching a floor plan is **photogrammetry**. This is the use of photography to create a floor plan or an entire **digital model** (3D picture) of a space. There are several ways to implement this.

1. Photograph a space with an ordinary camera and upload the photos to a photogrammetry program. The program stitches the photos together on a cloud server and returns a **photo-textured** (photo-covered) **mesh model**. Figure 5.16 shows a scanned room as a plain **mesh** (the polygons that define the model) and a photo-textured mesh. It was created with smartphone photos and **Autodesk ReCap Photo**.

2. Use a **3D camera**. This is an imaging device that creates depth in photos. (Figure 5.17). You place the camera at different

Figure 5.17
A Matterport 3D camera.

PHOTOGRAMMETRY 93

locations throughout a house, take a photo at each, and upload the photos to the manufacturer's website for stitching. You can then download a file of the model or a floor plan to import into a modelling program of your choice.

3. Hire someone with a 3D camera to photograph the site and provide you with files. This is a burgeoning business in the real estate and short-term rental industries. Figure 5.18 shows choices that are partnered with VRBO (Vacation Rental by Owner) for host use.

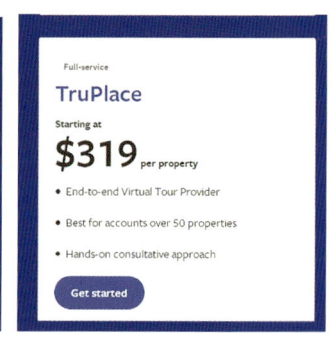

Figure 5.18
Companies that scan and/or sell scanning cameras.

Wall Thicknesses and Pochés

Floor plans for construction documentation need different treatment than sketches. Draw them with hard, not freehand, lines; with wall thicknesses that represent materials they're made of; and with appropriate symbols, pochés, and dimensions. **Poché** is a textural pattern that represents material. Figure 5.19 shows dimensions for different wall types and their material pochés.

Note that:

- A **knee wall** (4' or lower) is not pochéd. This is because it is under the cutting plane. A spindle (decorative round rod) on top of it, however, *is* cut by the plane, so it is pochéd.
- A **chase wall**, also called a stack or **wet wall**, is built behind plumbing fixtures to house the pipes. Chase walls are between 6" and 18" wide, depending on how many fixtures are hung on them and if the fixtures are floor or wall mounted.
- The **profiles** (shapes) of floor- and wall-mounted toilets are different, as is their placement. Draw floor-mounted fixtures slightly away from the wall; draw wall-mounted fixtures directly against the wall.

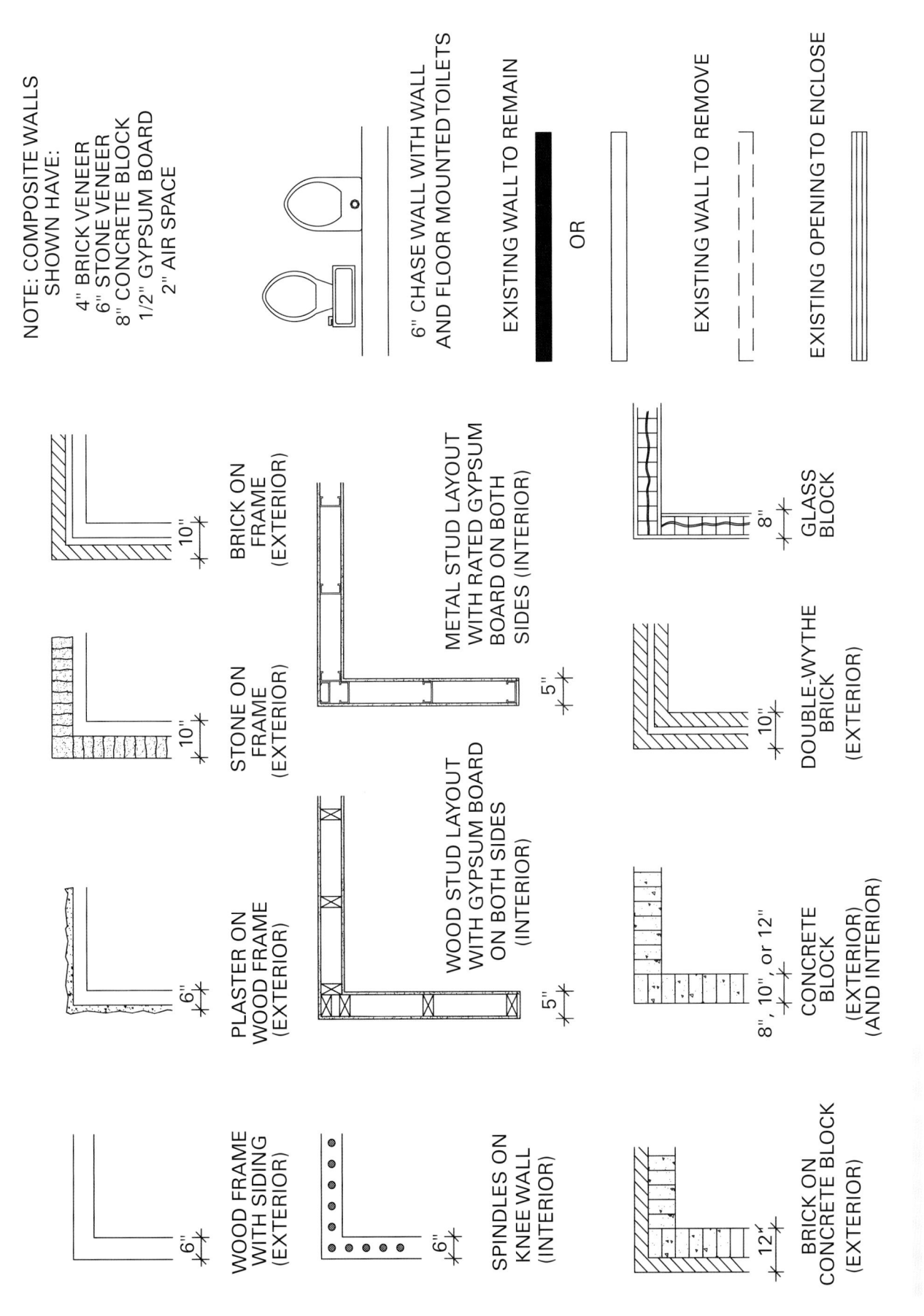

Figure 5.19
Wall types, thicknesses, and pochés.

Floor Pochés

A material viewed from a distance away—such as looking down at the floor—is pochéd differently than when it is sliced with a cutting plane. This helps the drawings maintain clarity. The walls in a floor plan are sliced, as they're touched by the cutting plane. But the floors below them are not. Hence, a material used in both a wall and a floor is pochéd differently. For example, note the poché for a stone veneer wall and a stone floor (Figure 5.20). The left shows the stone veneer wall. You are slicing through the wall, so the cutting plane is touching the stone. The right shows a stone floor. It's the same stone as in the wall, but you're looking down at it, not slicing through it. So, its poché is different.

Figure 5.21 shows pochés for floors, counters, and rugs as seen from a distance, looking down at them. Figure 5.22 shows section pochés—materials that have been sliced by the cutting plane.

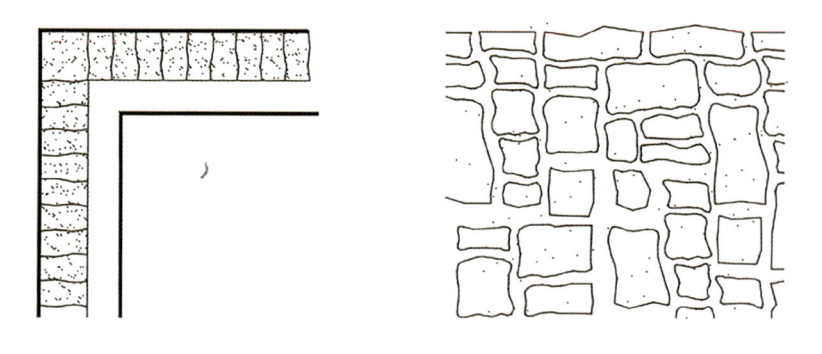

Figure 5.20
Stone in the wall (left) is sliced through. Stone in the floor (right), is seen from a distance. Same stone, different poché.

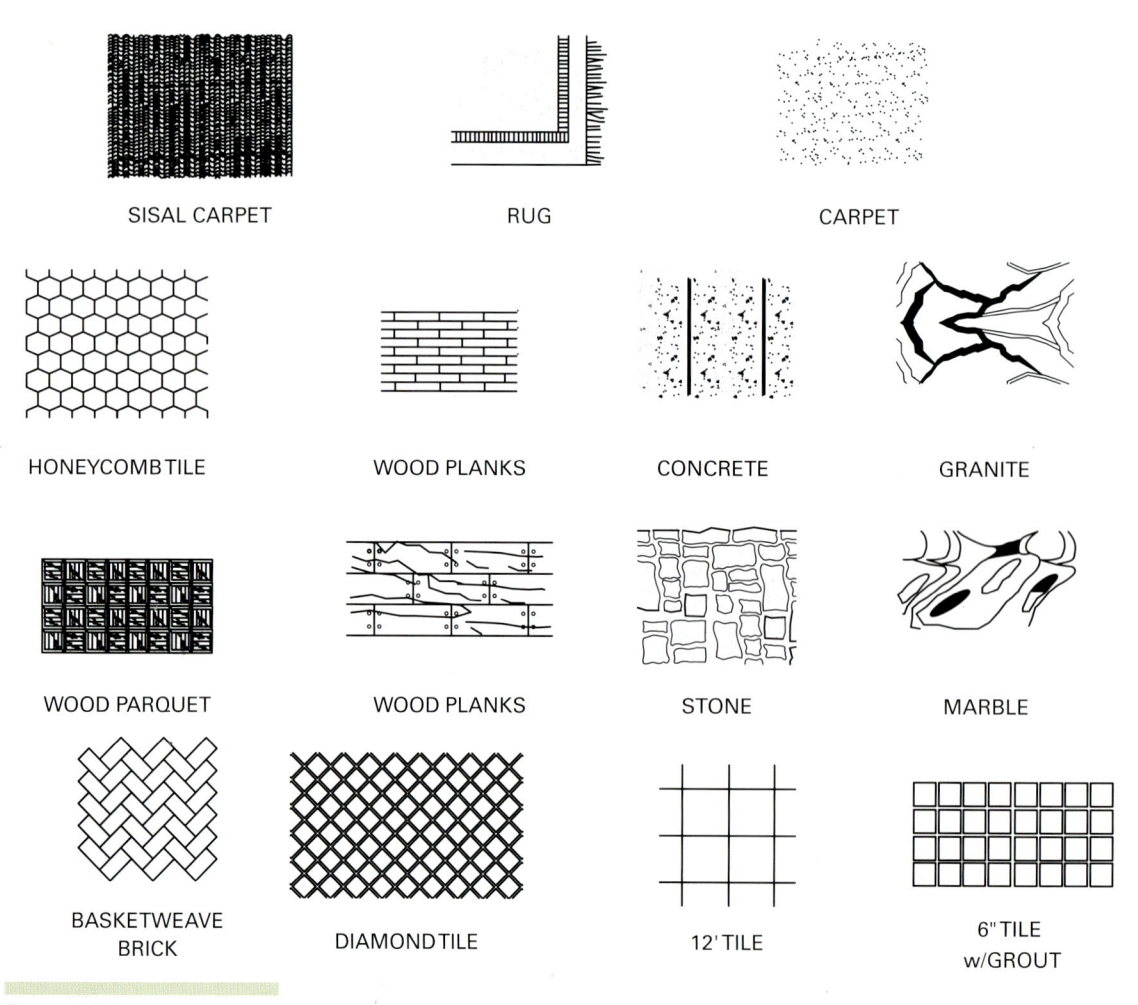

SISAL CARPET

RUG

CARPET

HONEYCOMB TILE

WOOD PLANKS

CONCRETE

GRANITE

WOOD PARQUET

WOOD PLANKS

STONE

MARBLE

BASKETWEAVE BRICK

DIAMOND TILE

12' TILE

6" TILE w/GROUT

Figure 5.21
Floor, counter and rug pochés.

96 CHAPTER 5 THE ARCHITECTURAL FLOOR PLAN

Concrete Block

Cast Concrete

Sand/Mortar

Glass Block

Sheet Glass

Frosted Glass

Face Brick

Fire Brick

Ashlar Stone

Rubble Stone

Dimension Lumber

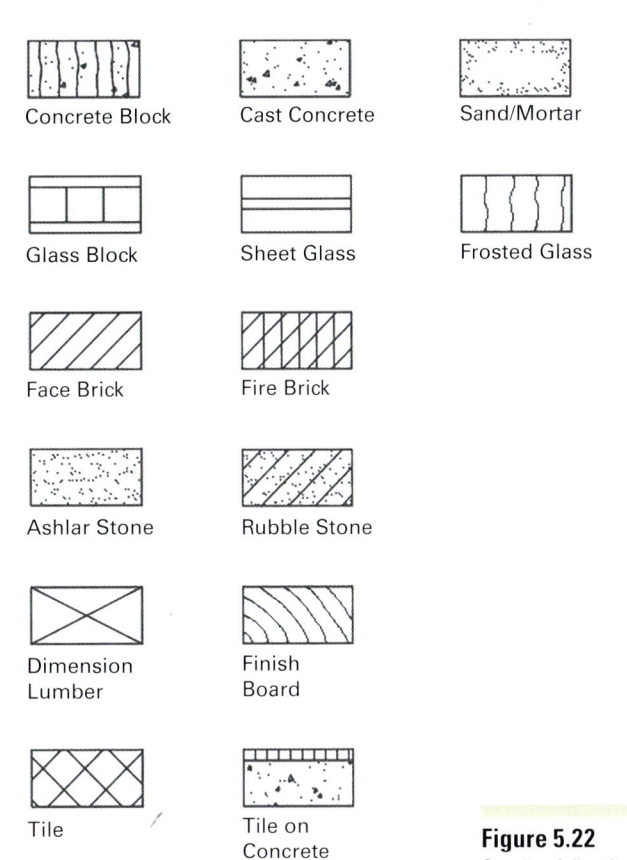

Finish Board

Tile

Tile on Concrete

Figure 5.22
Section (sliced-through) material pochés.

Symbols

Doors, windows, draperies, furniture, appliances, drains, materials, identifiers, and everything else are represented with unique symbols. Figures 5.23–5.28 show common ones. They're printed at the 1/4" =1'-0" scale, so you can freehand trace them when sketching. Place tracing paper over them, hand-trace, and then underlay your tracing under your final-project vellum to draft.

If you can't find a template for a specific symbol, use your bar, triangle, and circle template to draw it manually. These symbols are generic; each company's products have different profiles. Some companies provide manual or CAD templates of their products, downloadable from their websites or provided if asked.

Tip
If you're unsure what an item looks like in plan, search the SketchUp Warehouse (https://3dwarehouse.sketchup.com) for it and generate a top/parallel projection view. Or google an AutoCAD block of that item (search "*item* autocad block"), and download, scale and trace it. Match the symbol's detail to the plan's detail. Many AutoCAD symbols are too detailed for a typical student-created plan and must be simplified.

Exercise
Download AutoCAD plan view blocks of a toilet, sink, and cooktop. Delete their small details, such as flush handles, burner rings, and interior outlines to make them more appropriate for tracing into a manually drawn plan.

Tubs

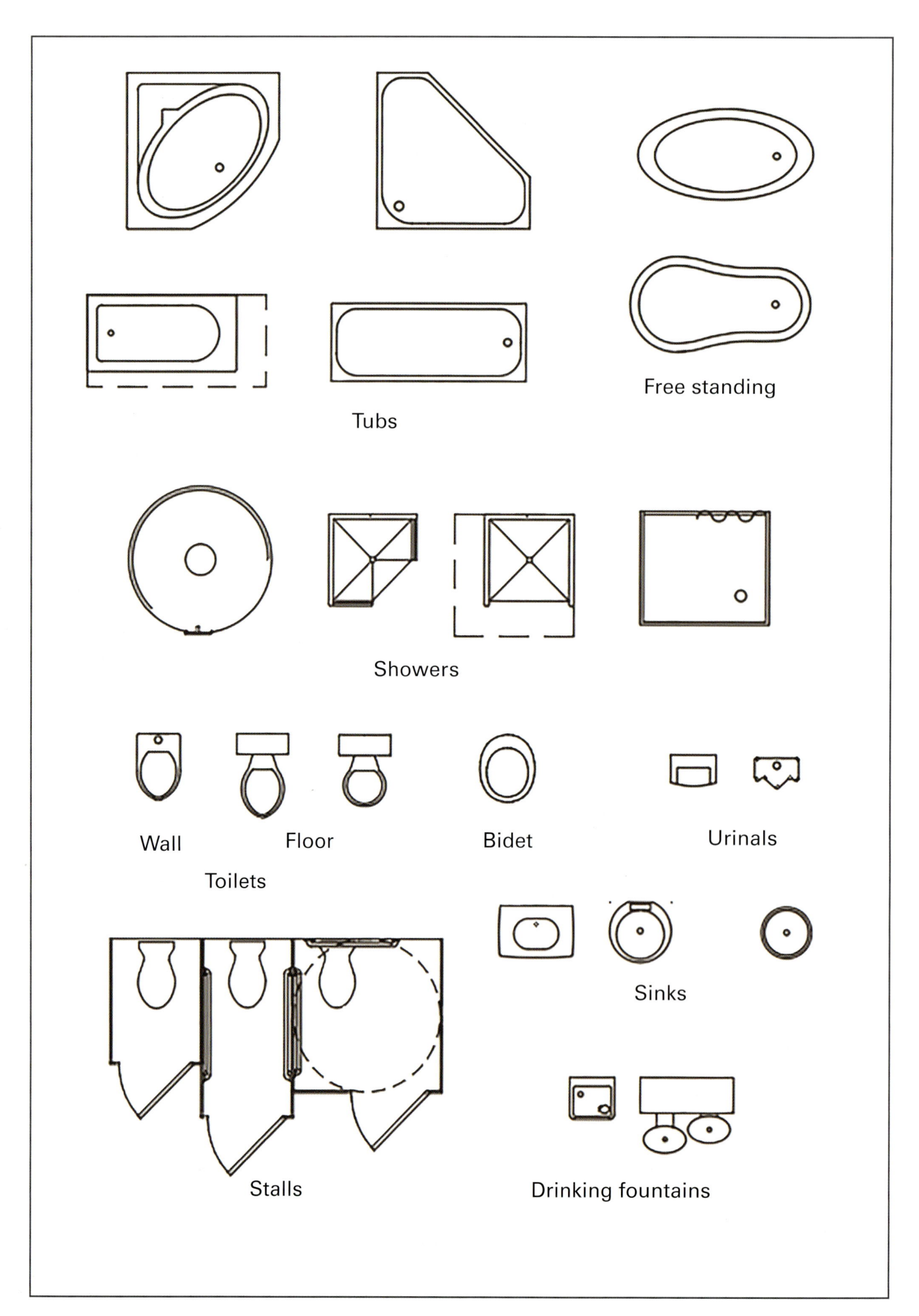

Free standing

Showers

Wall　Floor　Bidet　Urinals

Toilets

Sinks

Stalls　Drinking fountains

Figure 5.23 Plumbing fixtures. 1/4"= 1'-0"

Cooktops

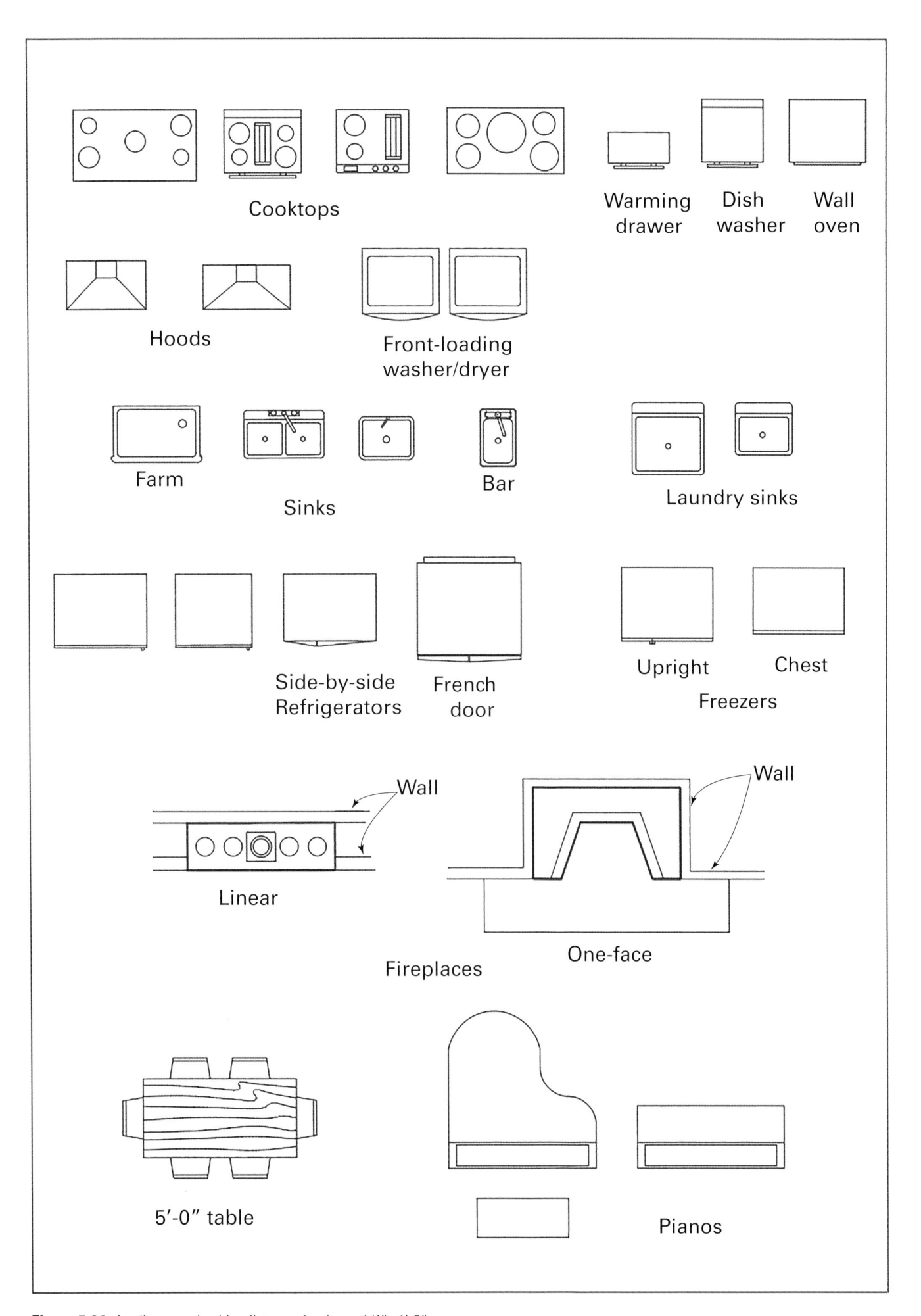

Warming drawer

Dish washer

Wall oven

Hoods

Front-loading washer/dryer

Farm

Sinks

Bar

Laundry sinks

Side-by-side Refrigerators

French door

Upright

Chest

Freezers

Wall

Wall

Linear

One-face

Fireplaces

5'-0" table

Pianos

Figure 5.24 Appliances, plumbing fixtures, furniture. 1/4"= 1'-0"

4' Flat screen on stand

4' Flat screen on wall — Wall

Exercise equipment

Seating

Twin Full Queen King

Beds

N

North Arrows

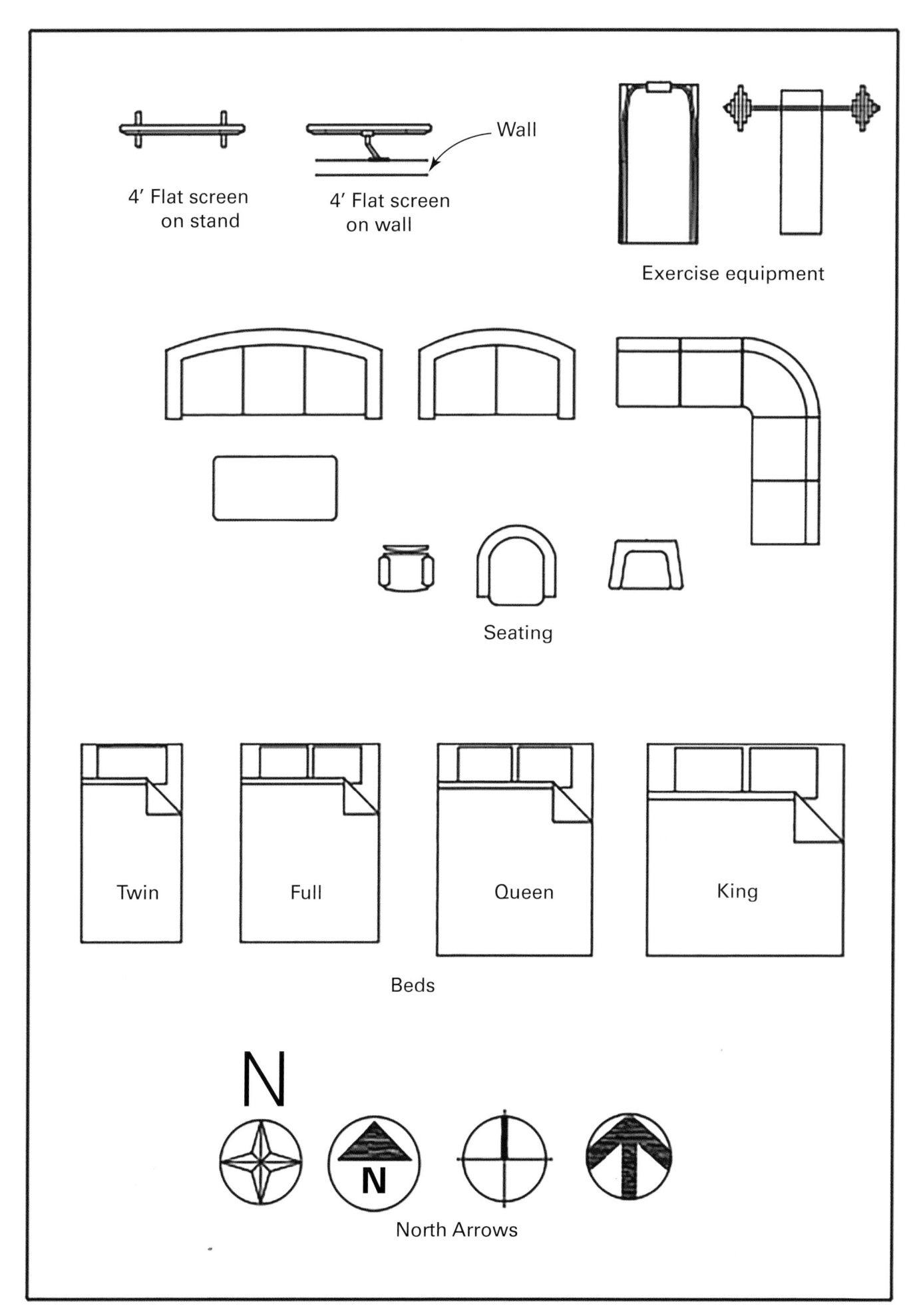

Figure 5.25 Furniture, exercise equipment, North arrows. 1/4"= 1'-0"

Tree

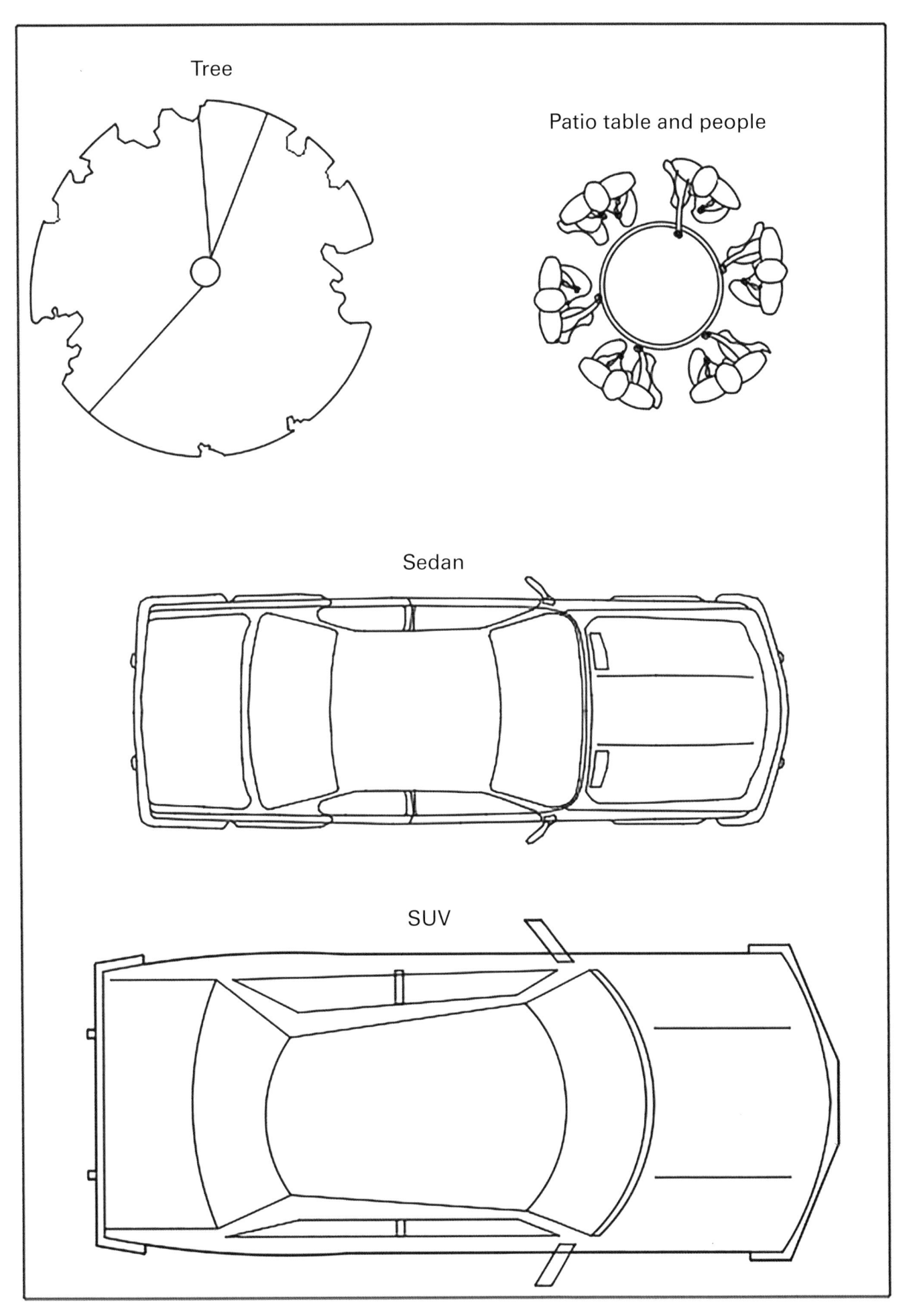

Patio table and people

Sedan

SUV

Figure 5.26 Outdoor items. 1/4" = 1'-0"

SYMBOLS 101

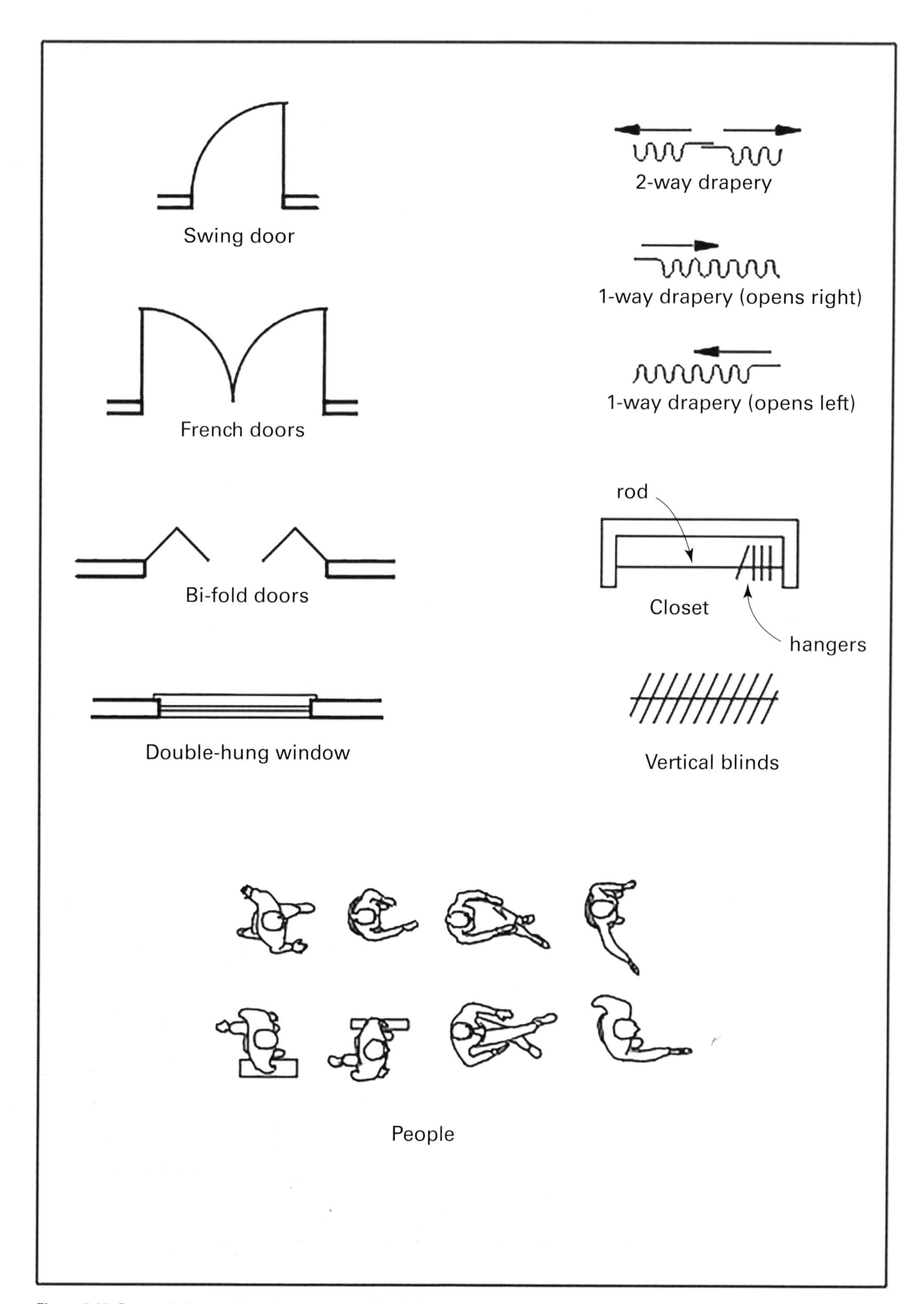

Figure 5.27 Doors, window, curtains, closet, people. 1/4" = 1'-0"

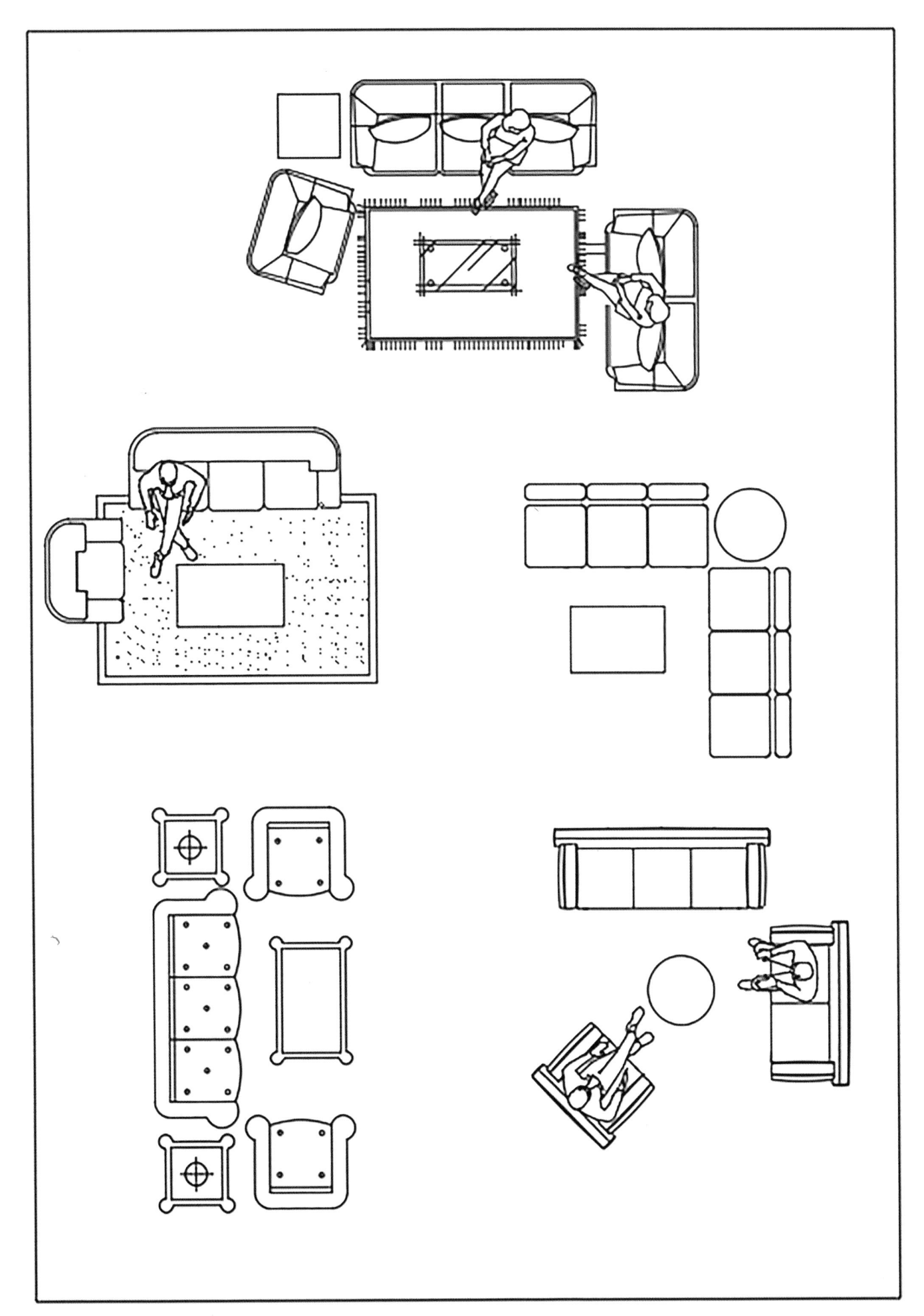

Figure 5.28 Furniture arrangements. Trace around them for approximate living room sizes. 1/4" = 1'-0"

Line Quality

A line's appearance must be excellent. Whether drawn with pencil or ink, lines must be crisp and strong with no fuzzy, smudgy, or half-erased sections. All are equally dark; their width and type are what varies. Lines should be a consistent width from end to end and the appropriate width for their place in the drawing's hierarchy. Construction lines and lettering guidelines are the only ones that should be light. Those lines are always penciled, never inked.

Connect corners and intersections (Figure 5.29); at best, gaps are poor drafting and at worst may communicate something different than what you intended. For example, a gap at an intersection could mean the walls don't physically touch. Avoid tracing over a line, as that often results in double lines. If turning in a photocopy of original work, make test copies of the assignment first, as weak lines and smudges are more apparent on a copy than on the original media.

> **Tip**
> When evaluating whether to submit your work or redo it, ask yourself what you'd think if that work was presented to you as a final product.

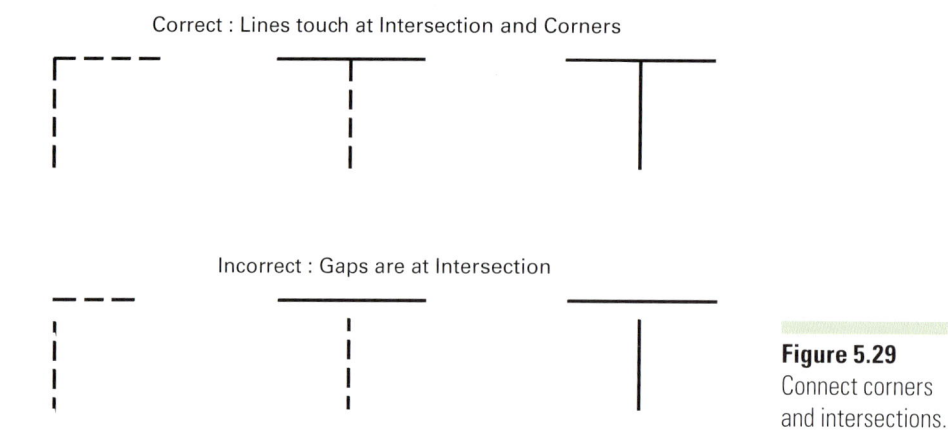

Correct : Lines touch at Intersection and Corners

Incorrect : Gaps are at Intersection

Figure 5.29
Connect corners and intersections.

When drafting with pencil, pull, don't push it across the paper, and rotate the lead as you pull it to help keep a consistent weight. If using the larger 2 mm clutch pencil, bevel one side of the lead on a sandpaper block and rotate the point as you draw to obtain different weights.

Keep your hands, drafting board, and underneath the parallel bar clean, because dirt rubs onto the paper. If you constantly smear pencil lines, put a paper under your hand to protect your work or use harder leads. You can also use an **eraser bag**, which is mesh fabric containing eraser flakes. Rub it gently over the paper to remove dirt and smudges.

Line Hierarchy

Also called **line weight**, this is thickness that varies with item importance. Walls and the profile of anything else sliced by a cutting plane are drawn thickest; furniture and doors thinner; floor pochés thinnest. Draw

thicker than their swings. Draw swings and glazing lines inside window symbols as thin, or thinner, as floor pochés. Swings, glazing lines, and floor pochés can also be drawn with a hard (4H) pencil instead of ink.

Ink pens are the best tools for achieving multiple line weights. Figure 5.30 shows suggestions for technical ink pens. Figure 5.31 shows pen and pencil suggestions for symbols.

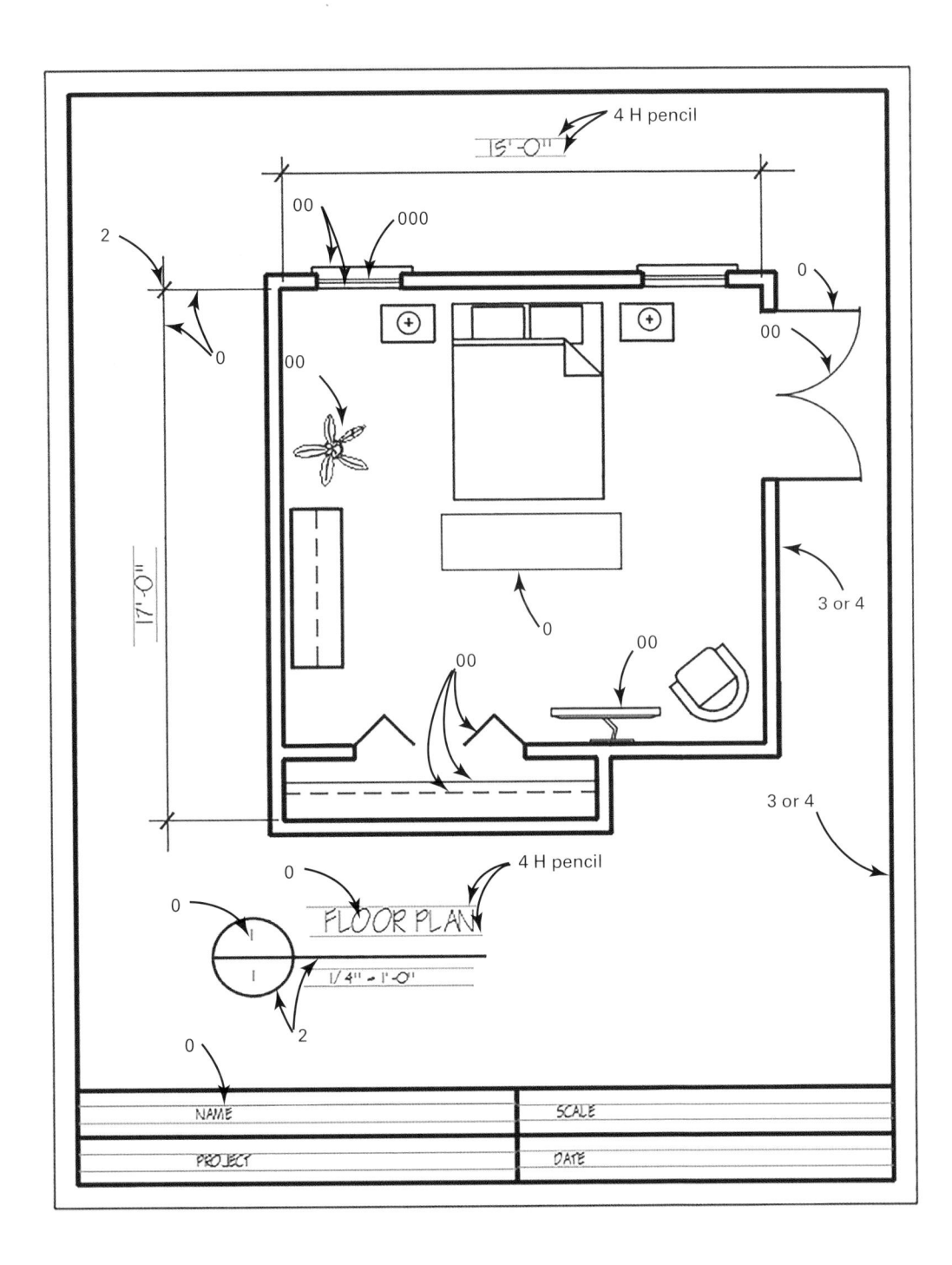

Figure 5.30
Suggested technical pen weights.

LINE HIERARCHY **105**

Line Type	Line Symbol	Pencil	Technical Pen	Disposable Pen
Center	— - —	2H	00	.2
Walls	——	B	3	.6
Hidden Object	- - - - -	HB	0	.4
Break	~~~	H	0	.3
Cutting Plane	└ - - ┘	2B	4	.8
Match	- - - - -	2B	4	.8
Border	━━━	2B	4	.8
Construction		4H	-	-
Guide Line		4H	-	-
Title Block	▭	2B	4	.8
Lettering	PROJECT 1	HB	0	.4
ID Label	⊕ FLOOR PLAN	HB	1	.5
Enlargement Box		HB	4	.8
Key Shape	⬭ ◇ ⬡ ⬠	HB	1	.5
Door	⁄	B	1	.5
Door arc	⁄	2H	00	.2
Objects cut in section view		2B	4	.8
Stairs		B	1	.5
Leader	→	HB	0	.4
Extension/Dimension	⊢——⊣	H	00	.2
Tick mark	⁄	2B	2	.4
Entourage		HB	0	.4
Poche	▨	H	00	.2
Window glass		2H	00	.2
Sketch	~~~	B	-	any weight
Callout	⊕	HB	0	.4

Figure 5.31
Suggested pencil and technical pen weights.

Drafting a Floor Plan

Let's discuss the process of using straight-edged tools to create a construction-ready drawing. We will draft the floor plan that was bubble-diagrammed in Chapter 1.

1. *Decide which scale to use.* Most residential floor plans are drawn at a 1/4" = 1'-0" scale; very large ones, and commercial plans, are drawn at 1/8" = 1'—0". Floor plans of kitchens and baths drawn to the **NKBA standard** (discussed in a later chapter) are drawn at 1/2" = 1'—0". Include space for lettering, dimensions, title block, and a border if your sheet needs one.

2. *Tape paper to the board.* Align the paper's horizontal edges with the parallel bar, and then tape the paper down at all four corners. Only taping down two corners can cause the others to bend under the bar. If drawing in multiple sessions, retape the paper by aligning a horizontal line you drew in an earlier session with the parallel bar. This ensures that all the drawing's parallel lines will remain parallel.

3. *Draw the plan's overall width and length* (Figure 5.32). This helps keep all the smaller interior measurements on track.

4. *Draw the exterior walls* (Figure 5.33). Use the parallel bar and triangle to draw horizontal and vertical lines and mark the walls' lengths on those lines. Show walls with a double line; the space in between is the wall's thickness. Due to the difficulty of manually drafting inches at a 1/4" = 1'-0" scale, you can round up. For instance, an exterior wall framed with 2" × 4" studs, with gypsum board on the inside and wood siding on the outside, is 5 1/2" wide. But you may draw it 6" wide for convenience.

5. *Draw the interior walls* (Figure 5.34). Draw interior walls thinner than exterior walls. They also need to be drawn as double lines spaced their material's thickness apart. An interior wood-framed wall with 1/2" gypsum board on each side is 4 1/2" thick but may be drawn 4" or 5" wide for convenience. Set the divider tool to the wall's thickness so you don't have to measure each time.

   When drawing walls to enclose a specific square footage, be consistent in how you measure them. For instance, if you measure one room's dimensions from interior wall to interior wall, measure all rooms that way. Periodically check corners for squareness by placing a triangle inside them (the triangle must be seated on the parallel bar). If both walls align with the triangle's edges, the corner is square (Figure 5.35).

6. *Draw door and window openings, stairs, and fireplace* (Figure 5.36). Fireplace size and type, number of stairs, and their tread widths must first be known to draw them. Draw half the full stair run and then place an arrow in the middle and letter **UP** at its foot. On the second floor, reverse the arrow and letter **DN** for down. How to draft stairs and fireplaces is discussed in later chapters.

Figure 5.32
Draw a rectangle of the plan's length and width.

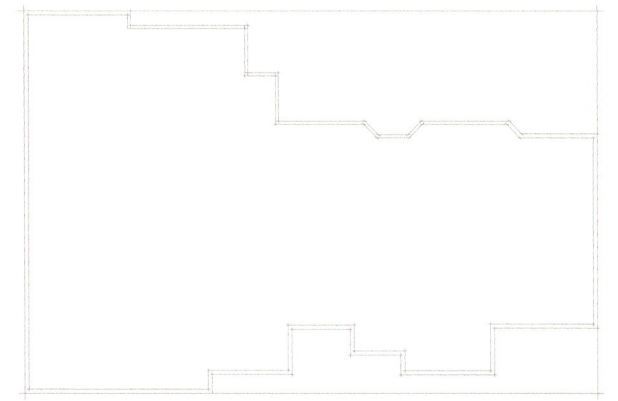

Figure 5.33
Draw the exterior walls.

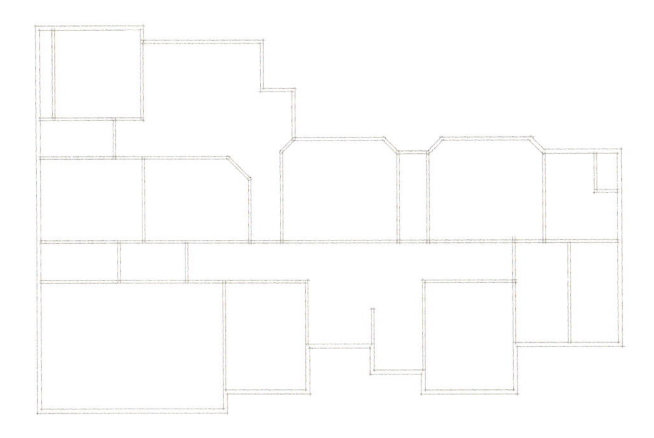

Figure 5.34
Draw the interior walls.

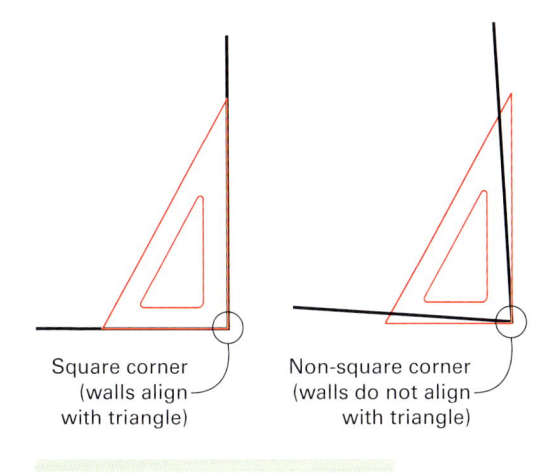

Figure 5.35
Check corners for wall perpendicularity.

DRAFTING A FLOOR PLAN

7. *Add line weight to the walls, interior details, decks* (Figure 5.37). Make the line weight for walls two very thick lines. Alternatively, you can fill the distance between them solid or hatch it with 45° angle lines.

   Draw cabinetry, appliances, plumbing fixtures, door and window symbols, and outline the front and back porches. Drafting doors and windows is discussed in a later chapter. Place floor-mounted plumbing fixtures slightly away from the wall, not directly against it. Note that the wall cabinets are drawn with a hidden line because they are above the floor plan's 5' cutting plane. Add exterior porches and decks.

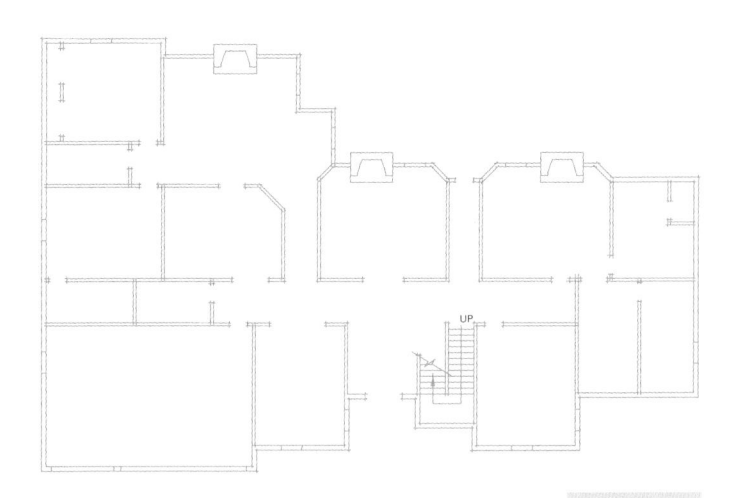

Figure 5.36 Add doors, windows, stairs and fireplace.

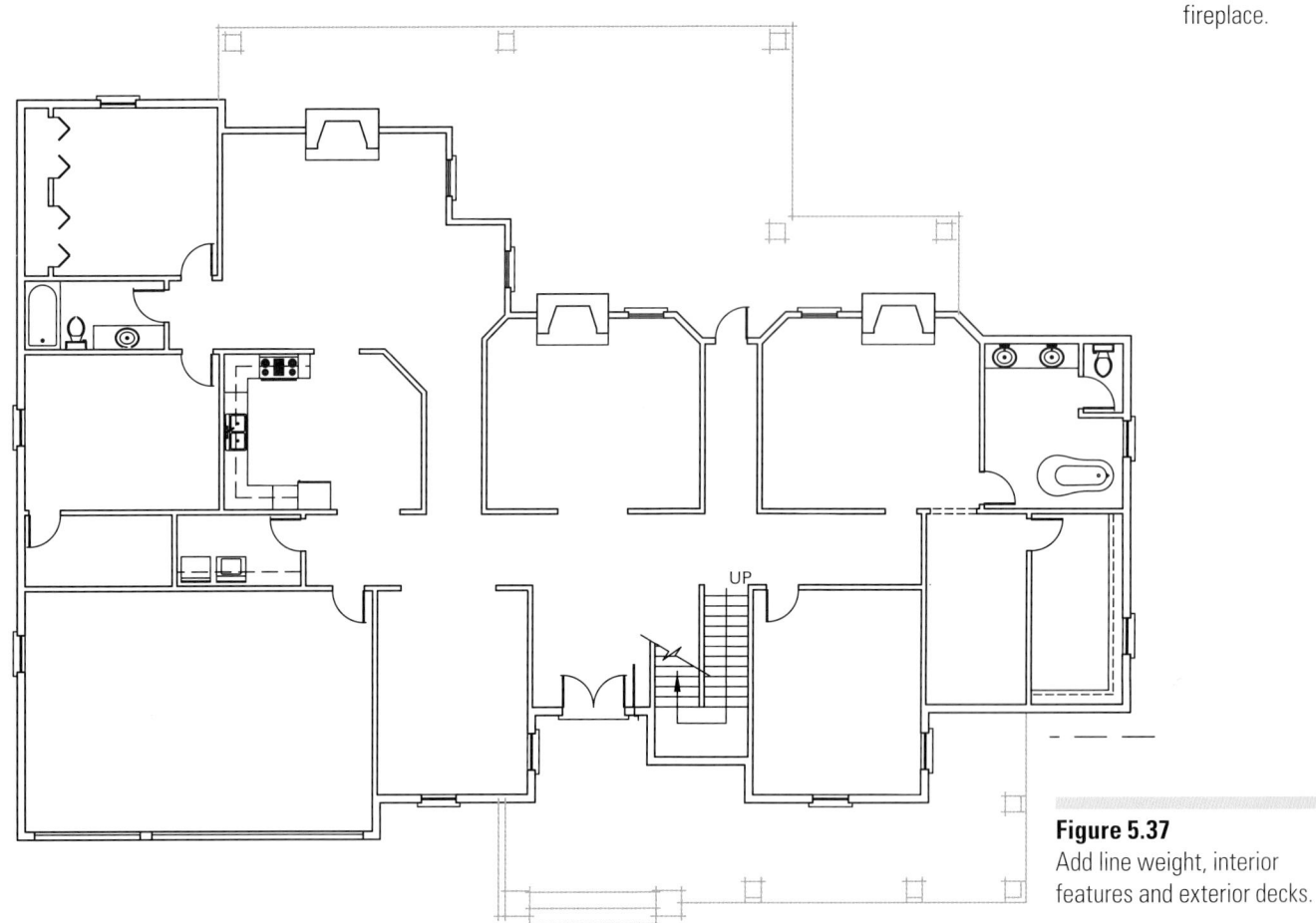

Figure 5.37 Add line weight, interior features and exterior decks.

8. *Add the finishing details* (Figure 5.38). These are ID labels, room names, notes, and flooring poché. The whole floor doesn't need to be pochéd, just patches that convey its essence. Add the roof overhang, ID label, scale, and if appropriate, electrical and mechanical symbols and annotated ceiling heights, and north arrow. Include any other details last because they have the most flexibility to be worked around the other items.

108   CHAPTER 5   THE ARCHITECTURAL FLOOR PLAN

If the plan is for construction purposes, add the symbols that reference it to other sheets and drawings. You may want to add a **bar scale** (Figure 5.39), which is a line with distance marks along its length, allowing visual estimates. A bar scale enlarges and reduces in tandem with the drawing.

> **Tip**
> Place poché near entries where flooring changes occur and make it asymmetrical for a fade-out appearance. Poché with square corners as if placed in rectangles gives the appearance of furniture.

> **Tip**
> Letter all notes 1/8" and titles 1/4". In presentation drawings, make notes large enough to be read from 6' away. Text should be horizontal; if it can't be, rotate it 90° so it's readable from the right side of the sheet. Use guidelines and a simple font. Don't use italic, script, or Gothic fonts.

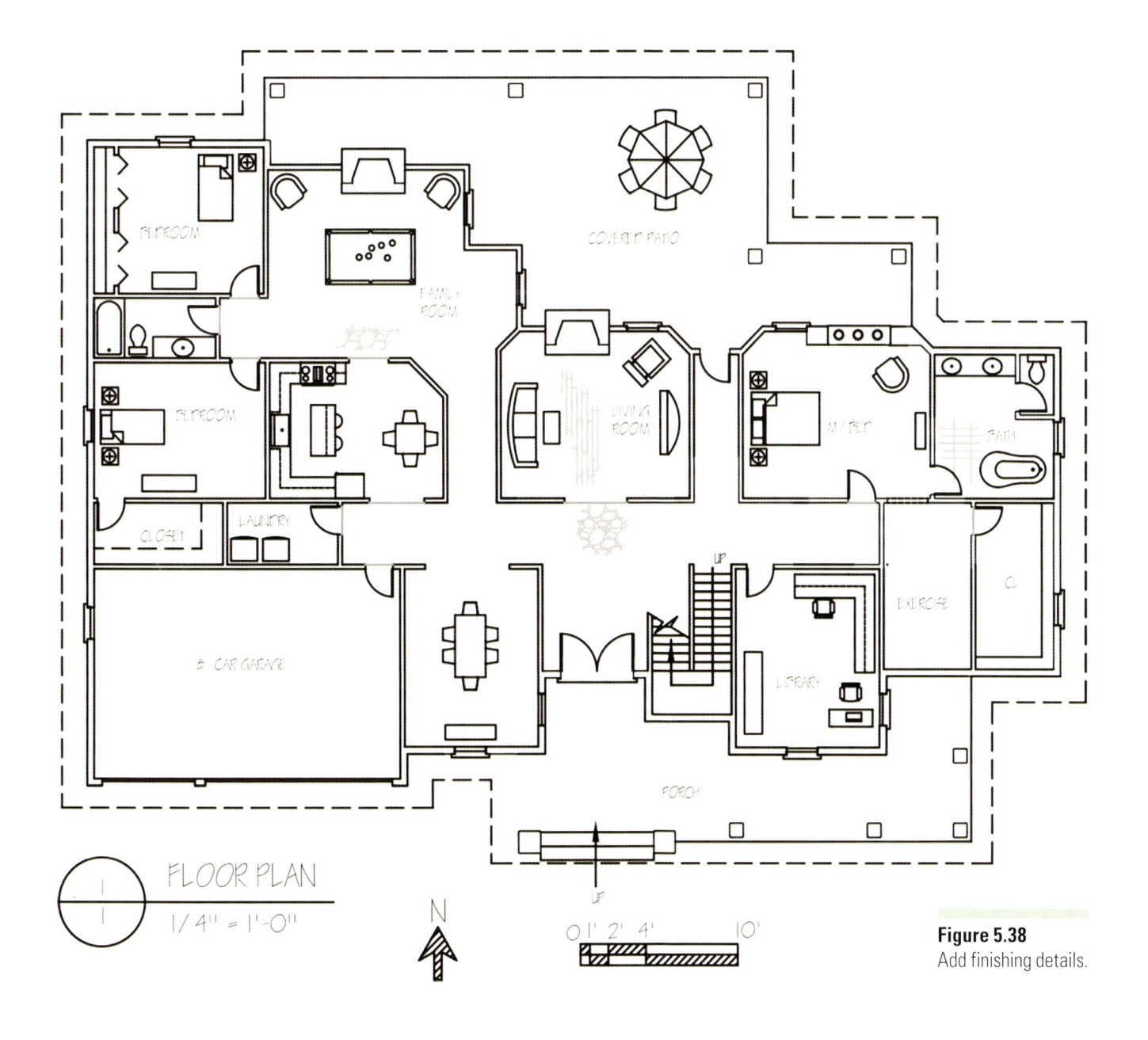

Figure 5.38
Add finishing details.

DRAFTING A FLOOR PLAN 109

Figure 5.39
A bar scale visually shows a plan's scale.

Second-Floor Plan

Each level needs its own plan. Draw a second-floor plan by overlaying a piece of tracing paper on the first-floor plan. Align stairwells, chimneys, and walls. The exterior **fenestration** (arrangement of windows and doors) may require first- and second-floor windows to align. Overlay the floor plan with all other plans, such as basement and mechanical, to check for alignment and general accuracy.

If the second-floor plan is a loft, it still needs its own plan. Add a simple, single-line outline to the loft plan that represents the first-floor below. But other than stairs, leave that first-floor plan outline blank; don't put built-ins, furniture, or anything else inside it. Orient all plans the same way, horizontally.

Inking

Manually ink a plan only after it has been completely figured out, problem-solved, annotations, symbols and dimensions added, and fully drafted on tracing paper. Vellum and film media are final-product media and don't handle multiple erasures well. Use pencil to make a traceable product, and when the plan is complete, overlay vellum or film and trace with ink.

Inking Tips   Pens require practice for good results. Here are some inking tips:

Hold the pen vertically so it is perpendicular to the board. This allows ink to flow smoothly through the tip. Angling the pen can create inconsistent line widths.

Starting at the top of a drawing, draw all horizontal lines, and then starting at the left of a drawing, draw all vertical lines. This allows ink to dry while simultaneously allowing you to continue drawing.

Avoid smears by using a triangle with an ink lip. Creating a small space between the tools and paper can also help. To do that, affix coins or strips of tape to the undersides of triangles and templates. Don't press down hard on straight-edged tools while inking, as that can cause ink to bleed underneath.

Lift triangles up and away from wet lines instead of sliding them away. This helps prevent smearing the wet ink.

Move pens across the paper at a constant speed. Slowing at the end of a line causes a small pool of ink to form.

Special erasers are formulated for drafting film, but any eraser will work if you wet it. There is such a thing as vellum-erasable ink, but the ink that comes with technical pen sets and inside disposable pens is not.

Some furniture templates have ink risers (small bumps) on one side to facilitate inking work. The risers face the paper. Don't press templates down hard; hold them lightly with one finger on a location far from the outline being traced. This helps prevent ink from bleeding underneath. Sofa and chair templates have one edge missing; add that edge with a triangle or parallel bar.

Don't use any marker or pen with a tip that's wider than 1/16". The Pentel Razor Point or Flair disposable enable extra fine points. The Pentel Sign Pen is good for lettering or thick wall lines.

Drafting a Presentation Plan

Presentation drawings are meant to sell your ideas to the client, so add furniture and entourage for visual scale and interest. When drawing furniture, draw each piece's entire footprint. This may be more than the space between the four legs; a couch may have large rolled or flared arms that lengthens its footprint. Draw furniture slightly away from the wall instead of touching it, which acknowledges the design trick of leaving a little space between the piece and the wall. (Empty space behind furniture is perceived as bigger than when furniture is pushed against the wall, so what is physically lost is visually gained).

Color makes a drawing pop. The most common media are color pencils, markers, and disposable and felt-tip ink pens. Figure 5.40 shows color pencil added to an AutoCAD drawing.

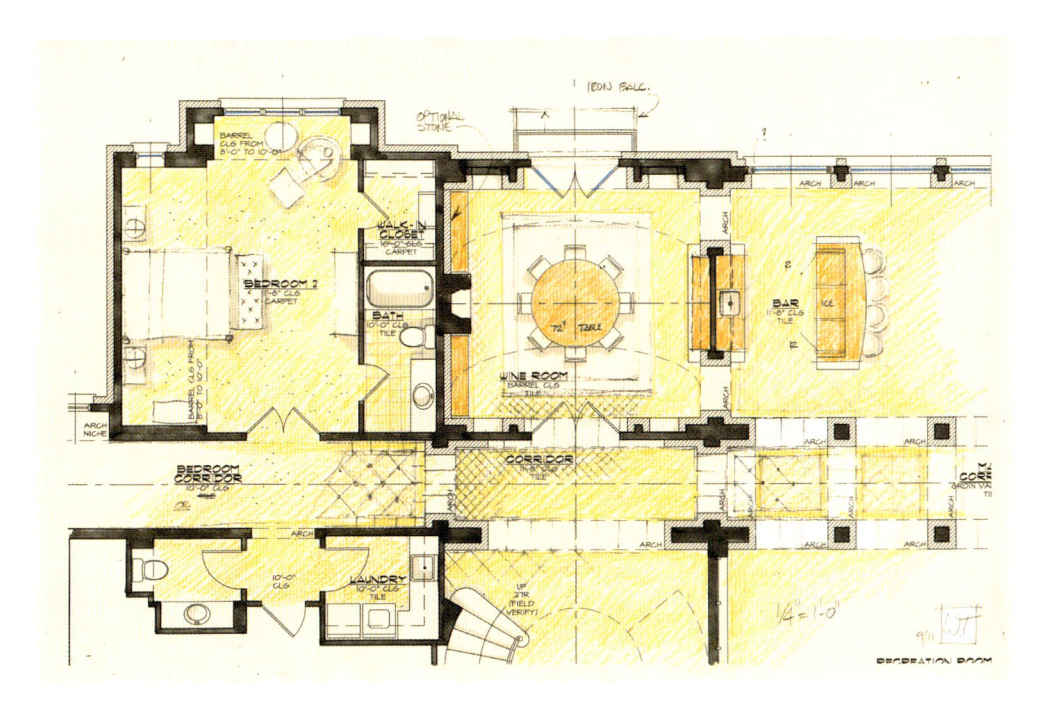

Figure 5.40
Color pencil added to an AutoCAD plan. Courtesy wolfgang trost.com

Search YouTube for tutorials on architectural rendering, as it takes a lot of study and practice to apply color well. Make lots of photocopies of hand-done original work so that poor color application doesn't require a complete redraw. Figures 5.41a–e show a process for rendering the plan we drew earlier with ink pen, marker, and color pencil. Each rendering stage was scanned so that if color media was applied poorly, a printout of the last good step made could be used.

Figure 5.41a
Walls are poché'd with marker.

Tip
The humble copier duplicates, enlarges, reduces, darkens, lightens, and alters color, often resulting in a better piece of work than the original. You can also scan artwork at each rendering stage. An electronic file provides greater detail and can be digitally manipulated.

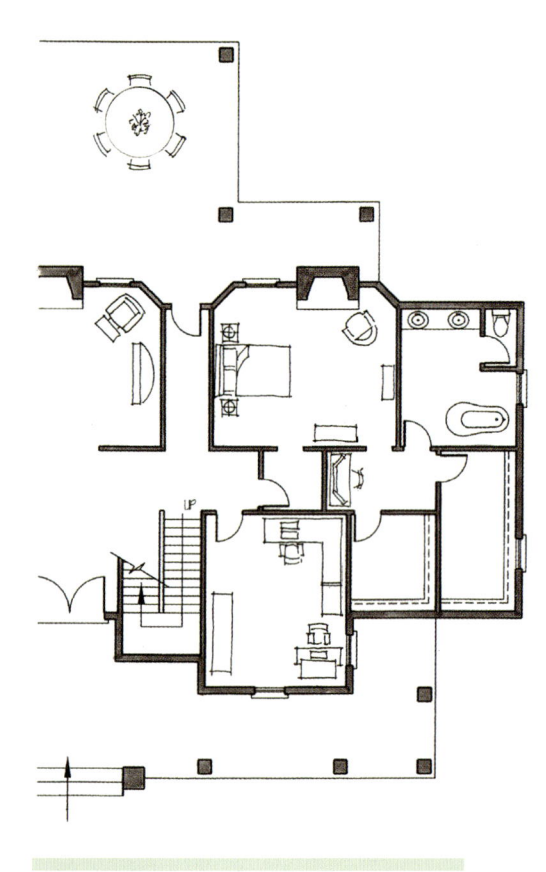

Figure 5.41b
Furniture is hand-traced for an interesting look.

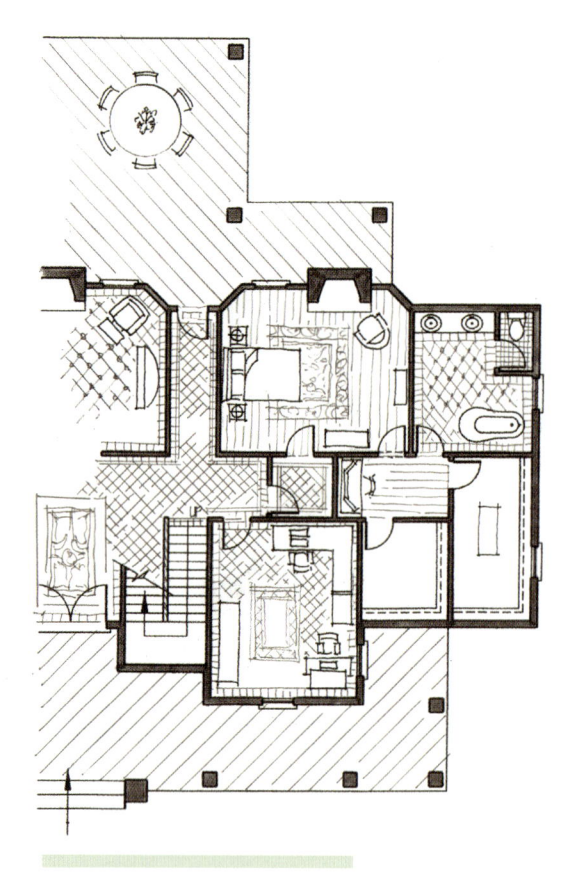

Figure 5.41c
Poché is applied with an ink pen.

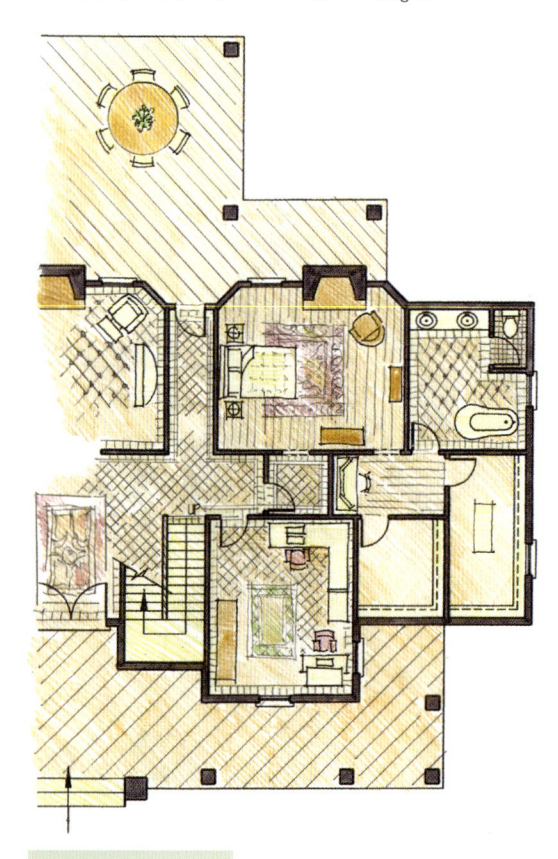

Figure 5.41d
Color pencil is applied.

Figure 5.41e
Shade and shadow is applied with a marker.

112　CHAPTER 5　THE ARCHITECTURAL FLOOR PLAN

CAD Standards

Architectural offices have requirements for line weights, colors, text fonts and heights, and other features particular to CAD drawings, such as layers. Each office has its own standards but Figure 5.42 and Table 5.1 show some generic ones.

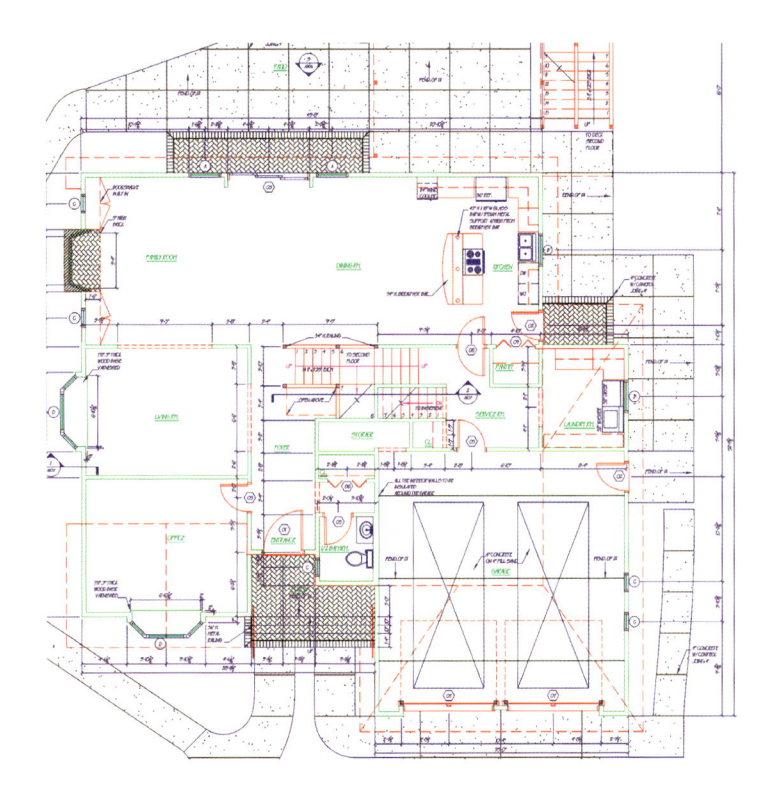

Figure 5.42
CAD plans utilize colors and other features specific to computer drafting. Courtesy Ali Mazlum: https://www.linkedin.com/in/ali-mazlum-3a5ba518/

Line Weights and Colors

Table 5.1		
Description	**Line Weight (mm)**	**Color**
Outline	.20 or .25	White, cyan, yellow, blue
Center line	.00 or .05	Blue, gray, 241
Note	.10 or .15	Green, red, blue
Thin line	.18 or .20	White, cyan, green, 41
Reference line	.00 or .05	Gray, 08, 111
Hatch line	.00	Magenta, gray
Color 9-256	.00	Magenta, green, gray, red
Dimension line/leader with arrows	.00	Gray color/9 or 8, Red
Dimension line/text	.18 or .20	Cyan, green

Text and Dimension Heights 2.5 mm, 3.5 mm, 7.0 mm. Line weight should be .1 mm of the character height.

Font Styles Romans.shx, Simplex, Arial, Times New Roman.

Space Planning

This is the process of designing an interior to make it beautiful, efficient, and functional. It goes beyond placing a bed and a wall; it is the space between the bed and the wall (Figure 5.43). Well-designed floor plans have enough room to move around the furniture, and doors that do not open into each other. While a thorough **space planning** discussion is beyond our scope, some knowledge is needed to competently draft floor plans. We'll discuss general planning themes and **clearances,** which are the distances between furniture and architectural features.

Space Planning Tips When laying out a floor plan:

Choose an **orientation** *(compass location) that maximizes the site's views*, features, and breezes. The north side is usually shaded, and the south receives heavy sun. Important spaces should face the best views and receive cooling summer breezes; closets and garages should face the worst views and block heavy winter winds.

Group areas into three zones: public, private, and dining, and plot the circulation between them. Find convenient routes among rooms or areas that have the most connecting traffic or whose functions are dependent on each other, such as kitchen and dining. This is usually done by making those rooms adjacent or near to each other. Avoid circulation paths through rooms.

Bathrooms should be convenient to bedrooms, accessible via a hall and not another room. The bathroom in a one-bedroom house should be convenient to all rooms.

Halls should be short, kept to a minimum, and not take more than 10 percent of the plan's total area.

Living rooms shouldn't have through traffic to other rooms; slightly raising or lowering the room level can set it apart. The main outside entry should not open directly into it; instead, utilize a hall or foyer as a buffer.

When locating doors, consider swing direction and furniture placement. In residential construction, exterior doors swing into the house, and into the room from a hall. Doors must completely open without being blocked by appliances or furniture.

Room Sizes

There is no such thing as an "average" or "ideal" room size. Table 5.2 shows the **Federal Housing Authority's** (government body that provides mortgage insurance on single and multi-family properties) minimum room sizes. The **International Residential Code (IRC)**, which is a set of instructions governing the design and construction on homes) requires all rooms to be a minimum of 70 square feet (ft²). Typical residential construction, however, uses 120 ft² as a standard minimum. Living, family, and great rooms may be areas instead of rooms in an open floor plan and would then need at least 220–400 square feet, with 250 square feet being average.

Know that sizes and clearances at the margins have changed over the past decade, with clearances getting tighter to accommodate the popularity of **tiny homes** (residence under 600 square feet) and plumbing fixtures getting larger to accommodate the popularity of luxury homes.

Figure 5.43
Space planning isn't just a bed and a wall; it's the space between the bed and the wall.
Courtesy Ali Mazlum: https://www.linkedin.com/in/ali-mazlum-3a5ba518/

Exercise
Find a 1/4" = 1'-0" floor plan. Lay tracing paper over it and use templates and an architect's scale to sketch furniture arrangements, incorporating proper furniture and traffic path clearances. Then draft your best furniture arrangement on the floor plan.

CHAPTER 5   THE ARCHITECTURAL FLOOR PLAN

Table 5.2

Room	Minimum	Area	Average	Area
Living Room	12' × 16'	192 ft²	15' × 21'	315
Dining Room	10' × 12'	120 ft²	13' × 14'	182
Dining Area (in Kitchen)	7' × 9'	63 ft²	10' × 10'	100
Master Bedroom	11' 6" × 12'-6"	143 ft²	13' × 15'	195
Bedroom	9' × 11'	99 ft²	11' × 13'	143
Kitchen	7' × 10'	70 ft²	9' × 13'	117
Bathroom	5' × 7'	35 ft²	6' 6" × 9'	58.5
Utility (No Basement)	7' × 8'	56 ft²	9' × 12'	108
Hall Width	3'		3'–6'	18
Closet	2' × 2'	4 ft²	3' × 5'	15
Garage (Single)	9' 6" × 19'	180 ft²	13' × 21'	273
Garage (Double)	18' × 19'	342 ft²	21' × 21'	441
Recreation Room	10' × 15'	150 ft²	12' × 20'	180
Mudroom	4' × 4'	16 ft²	4' × 9'	36

Layouts

Figure 5.44 shows examples of good and poor design layouts.

Kitchen Space Planning

Kitchens (Figure 5.45) vary from the elaborate to the simple, but the same planning basics apply to all of them. One hundred square feet of usable floor space is the minimum needed; more than 160 square feet is harder to manage well. Efficient kitchens are organized around three activity centers: cooking, cleanup, and storage. The cooking center comprises the range, cooktop, and microwave. The cleaning center comprises the sink, dishwasher, trash compactor, and garbage disposal. Storage consists of the refrigerator, pantry, and cabinets.

Figure 5.44
Good and poor design layouts.

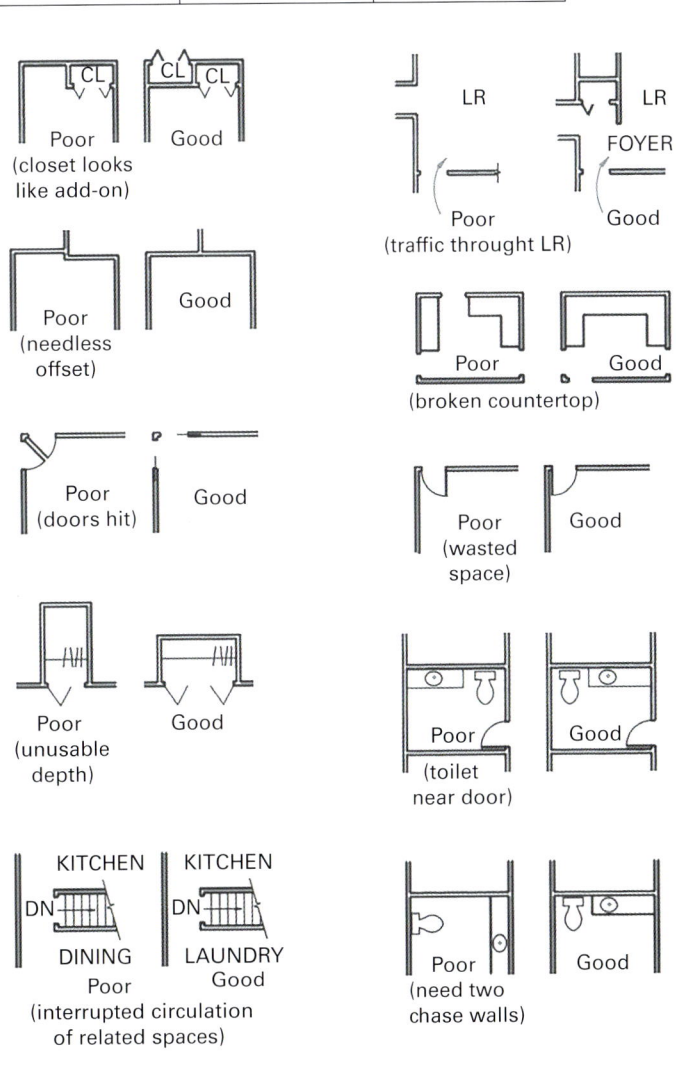

Figure 5.45
A galley kitchen.

Each center needs storage and counter space. There should be at least 50 square feet of cabinet storage in the form of 6 linear feet of base cabinets with wall cabinets above. At least 11 square feet of drawer space are needed, which can be provided by four drawers. Centers should not be subjected to pass-through traffic, and work should flow in one direction from center to center.

Kitchens have a **work triangle** (Figure 5.46), which is the area between the stove, sink, and refrigerator. Kitchen efficiency is measured by the amount of walking needed between them, so these appliances should be in proximity. The work triangle should not be intersected by an island or subject to pass-through traffic (Figure 5.47).

When the lengths of the triangle's three sides are added together, the total length should be between 12' and 26'. Each side should be between 4' and 9'. There should be 4' to 6' between the sink and range, 4' to 7' between the refrigerator and the sink, and 4' to 9' between the range and the refrigerator. The dishwasher should be within the triangle and next to the sink.

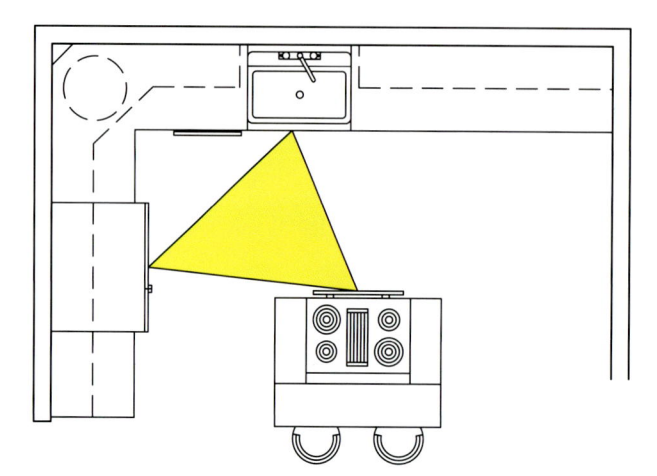

Figure 5.46
Kitchen work triangle.

116 CHAPTER 5 THE ARCHITECTURAL FLOOR PLAN

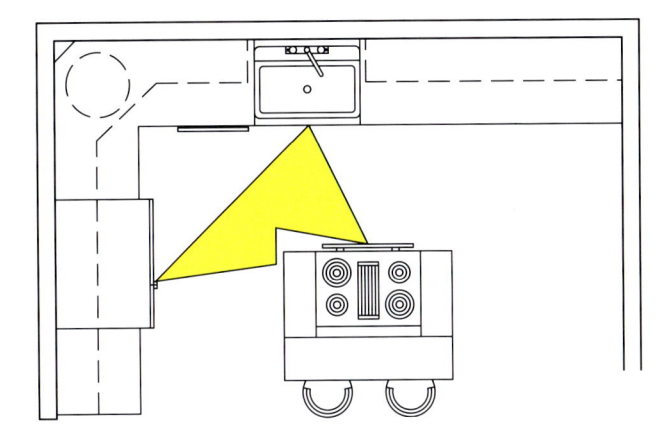

Figure 5.47
Don't interrupt the work triangle with an island.

There are five kitchen plans: **U-shape**, **G-shape (peninsula)**, **L-shape**, **galley (corridor)**, and **single wall**. The arrangement chosen depends on room dimensions and door and window location.

U-Shape   This efficient plan surrounds the cook with countertop and storage space (Figure 5.48). You can remove one wall to open up the kitchen while allowing the counter to be used for food preparation and seating. The arms of the U should be at least 5' apart.

G-Shape (Peninsula)   This (Figure 5.49) has the benefits of the U-shape plus additional storage, food preparation, and appliance areas. The peninsula is usually a "pass-through" that provides bar seating and opens up the kitchen into another space. This design typically has one opening, which means less pass-through traffic.

L-Shape   This open plan maximizes a small space's counter and storage areas and accommodates an island (Figure 5.50). It can adapt to almost any space but is often placed in a corner of a house. The open area of the L provides space for an eating area.

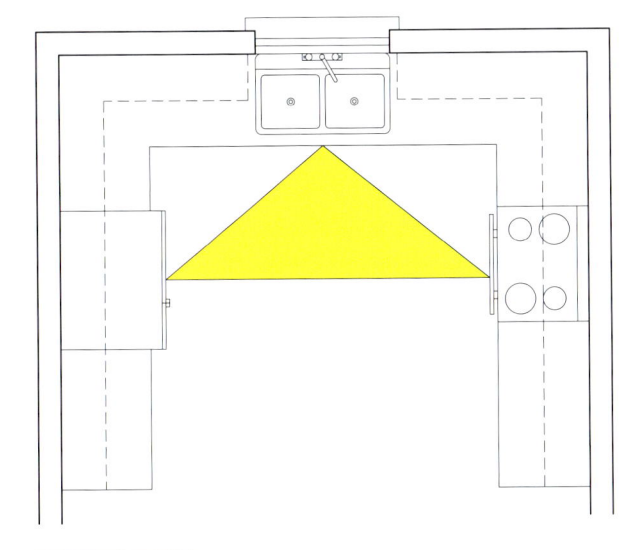

Figure 5.48
U-shaped kitchen.

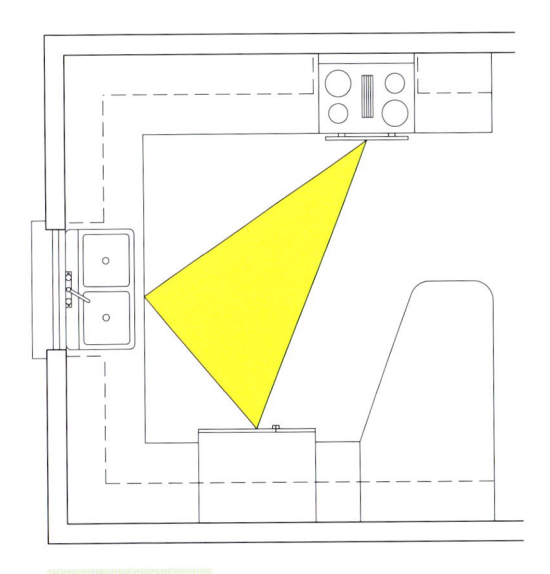

Figure 5.49
G-shaped kitchen.

KITCHEN SPACE PLANNING

Galley This maximizes a small space by aligning cabinets and appliances on two parallel walls (Figure 5.51). Long countertops uninterrupted by corners are the most efficient. A galley is best utilized when there are doorways on opposite ends of the kitchen; if there is only one doorway, traffic will be a problem. The corridor in between the counters should be between 4' and 6' wide and the whole room should be at least 8' wide.

Single Wall This arranges all appliances and cabinets along one wall. It is used in small spaces where the floor plan allows no other arrangement (Figure 5.52). Its efficiency is in its use of space, because its layout doesn't produce a work triangle.

Figure 5.50
L-shaped kitchen.

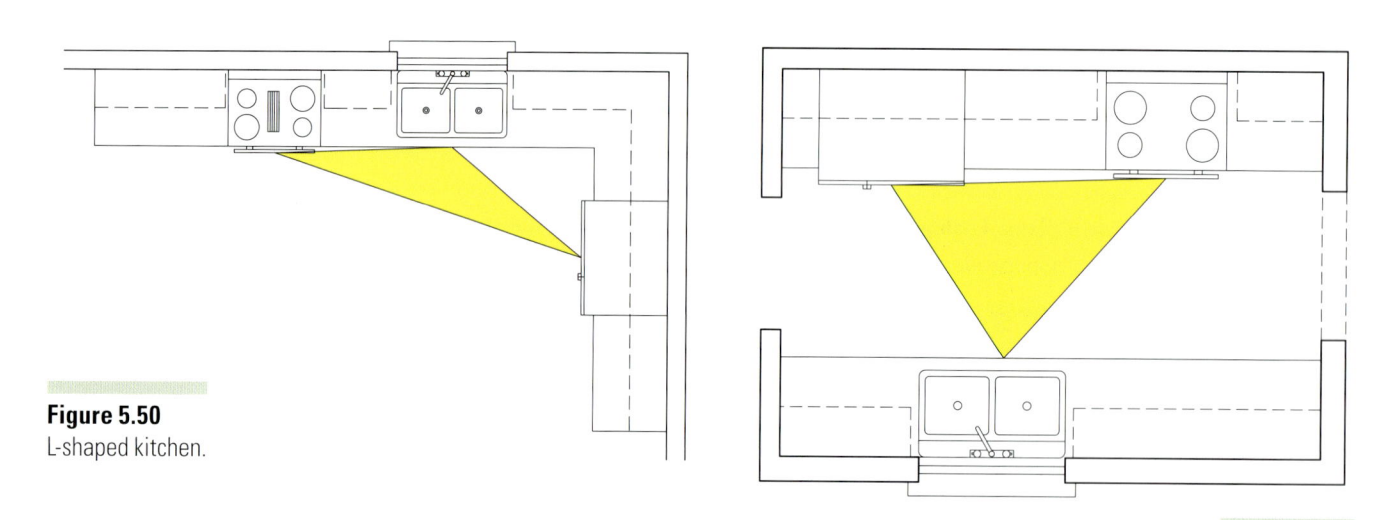

Figure 5.51
Galley kitchen.

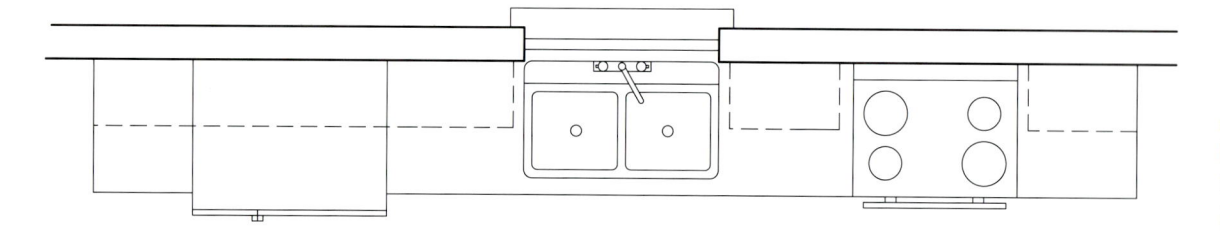

Figure 5.52
Single wall kitchen.

Islands

These are stand-alone work centers (Figure 5.53). They may have a simple countertop for workspace or multiple levels with cabinets, wine racks, appliances, a sink, and eating space (Figure 5.54). Islands work well in single-wall and L-shaped kitchens, as those usually have less space for storage or appliances. In a U-shape, islands work when the space between the U's two arms is at least 10'. An island should be at least 2"-6" × 3'-0" (larger if housing an appliance) and accessible from all sides, with at least 36" of clearance. 42" or more are needed if cabinet or appliance doors open into the clearance. A work aisle's width should be at least 42" for one cook and 48" for multiple cooks.

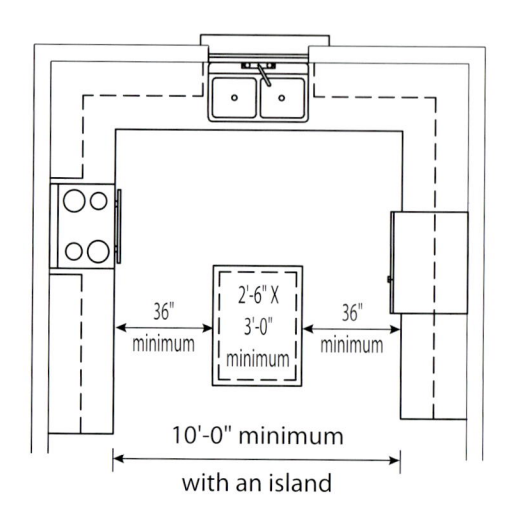

Figure 5.53
U-shaped kitchen with island.

Figure 5.54 Island layout studies done with SketchUp.

Kitchen Appliance Sizes and Clearances

These vary based on model and manufacturer, so download product information from the manufacturer website for precise dimensions. You can use the generic sizes in Tables 5.3 and 5.4 when drafting kitchen plans.

Table 5.3				
Appliance	**Width**	**Height**	**Depth**
Single-Door Refrigerator	35 ½"	69"	27"
	33 1/2"	69"	32"
French doors Refrigerator	29 1/2"–36"	68 1/2"–70"	29"–34"
Side-by-Side Refrigerator	30"–36"	67"–70"	29"–35"
Freestanding Range	20"	30"	24"
	21"	36"	25"
	30"	26"	36"
	40"	36"	27"
Double-Oven Range	30"	61"	26"
	30"	67"	27"
	30"	71"	27"
Drop-In Range	23"	23"	22"
	24"	23"	22"
	30"	24"	25"

(continued)

Table 5.3

Appliance	Width	Height	Depth
Built-In Cooktop	12"	2"	18"
	24"	3"	22"
	48"	3"	22"
Range Hood	24"	5"	12"
	30"	6"	17"
	66"	7"	26"
	72"	8"	28"

Table 5.4

Appliance	Width	Height	Depth
Single-Compartment Sink	24"		21"
	30"		20"
Double-Compartment Sink	32"		21"
	36"		20"
	42"		21"
Dishwasher	24"	34"	24"
	18"	34"	24"
Microwave	20"	17"	12"
	20"	13"	11"
	17"	12"	15"
	29"	16"	15"
Trash Compactor	14"	13"	23"
	15"	34"	24"
Warming Drawer	24"	12"	24"
	27"	12"	24"
	30"	12"	24"

General Kitchen Clearances

Countertop and appliance clearance standards (Figures 5.37, 5.38) vary depending on the source; FHA, NKBA, ANSI, and local codes all have different suggestions or requirements. Some guidelines:

- A clear floor space of at least 30" × 48" is needed at each appliance.
- A 36" × 48" table plus four chairs needs at least 8'-8" of clear space.
- A change in countertop height can separate cooking and eating areas; counter height where chairs are used should be 26" with at least 21" of knee space depth underneath; 28" to 30" of knee space is needed for **accessibility** (an overhanging or extended countertop can help with this). Knee space should be 27" to 30" high.

Figures 5.55 and 5.56 show more kitchen clearance guidelines.

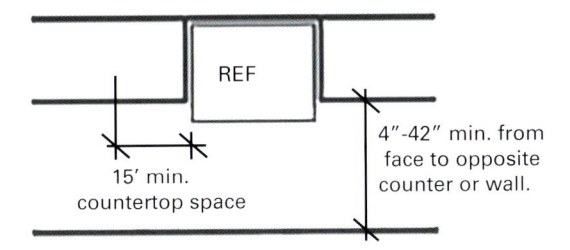

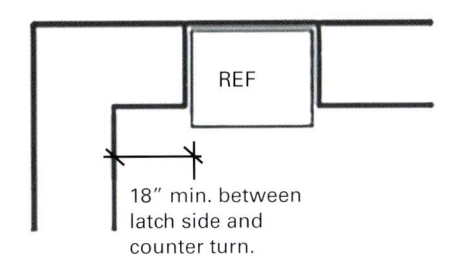

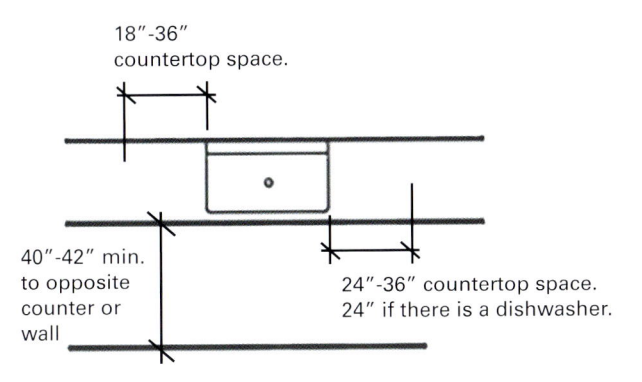

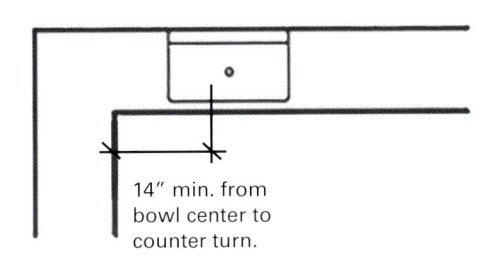

15' min. countertop space

4"-42" min. from face to opposite counter or wall.

REF

18" min. between latch side and counter turn.

REF

18"-36" countertop space.

40"-42" min. to opposite counter or wall

24"-36" countertop space. 24" if there is a dishwasher.

14" min. from bowl center to counter turn.

18"-42" countertop space on both sides of cooktop.

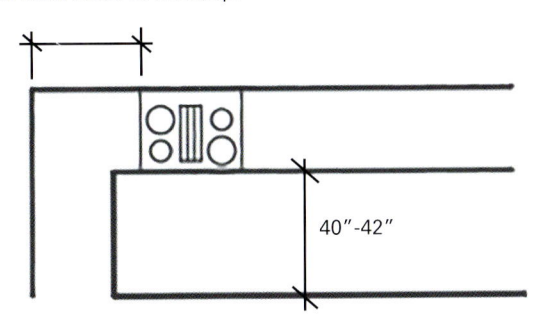

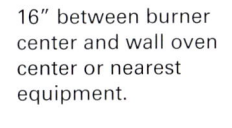

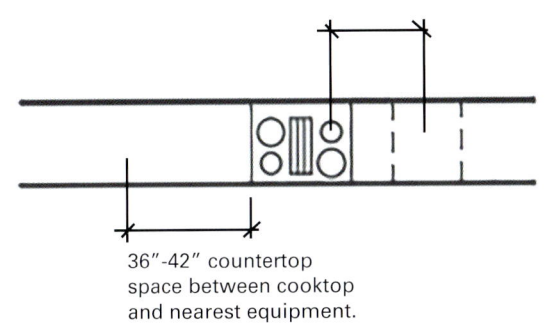

40"-42"

16" between burner center and wall oven center or nearest equipment.

36"-42" countertop space between cooktop and nearest equipment.

Figure 5.55
Kitchen clearances.

GENERAL KITCHEN CLEARANCES **121**

34"

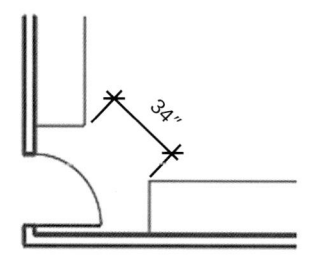

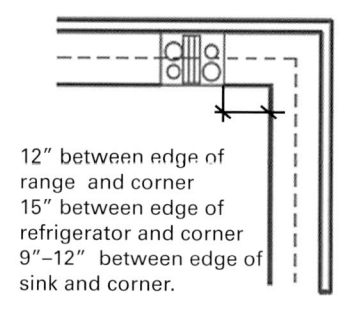

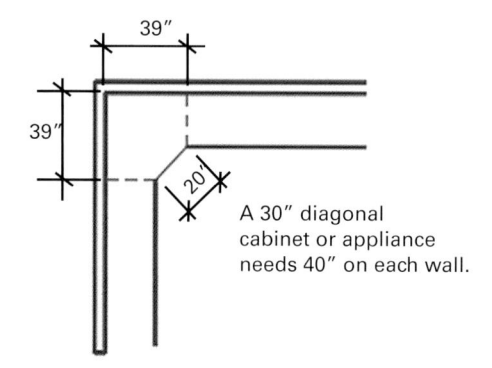

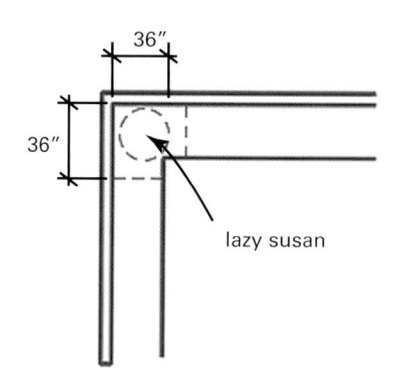

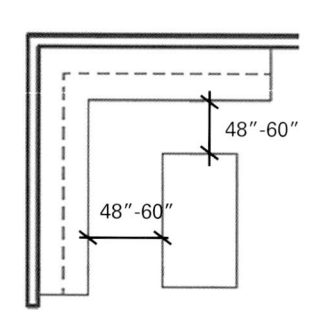

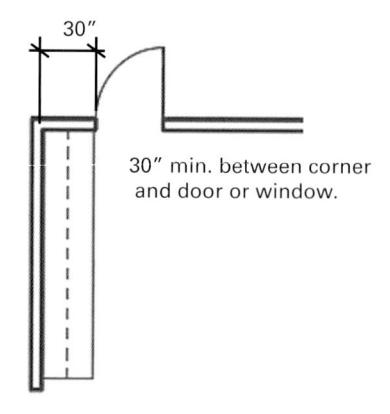

12" between edge of range and corner
15" between edge of refrigerator and corner
9"–12" between edge of sink and corner.

39"
39"
20"
A 30" diagonal cabinet or appliance needs 40" on each wall.

36"
36"
lazy susan

48"-60"
48"-60"

30"
30" min. between corner and door or window.

Figure 5.56
Kitchen clearances.

Dining Room Furniture

Table 5.5 shows some generic dining room furniture sizes.

\multicolumn{4}{c}{**Table 5.5**}			
Furniture	**Length**	**Depth**	**Height**
Rectangular Dining Table	42"	30"	29"
	48"	30"	29"
	48"	42"	29"
	60"	40"	28"
	60"	42"	29"
	72"	36"	28"
Oval Dining Table	54"	42"	28"
	60"	42"	28"
	54"	30"	36"
	72"	40"	28"
	72"	48"	28"
	84"	42"	28"
	Diameter		**Height**
Round Dining Table	32"		28"
	36"		28"
	42"		28"
	48"		28"
	Length	**Width**	**Height**
China Cabinet or Hutch	48"	16"	65"
	50"	20"	60"
Buffet	62"	16"	66"
	36"	16"	31"
	48"	16"	31"
	52"	18"	31"
	Width	**Depth**	**Height**
Corner Cabinet	36"	15"	80"
	38"	16"	80"
Dining Chairs	17"	19"	29"
	20"	17"	36"
	22"	19"	29"
	24"	21"	31"

Bathroom Space Planning

Bathrooms are categorized by their plumbing fixtures. A **half bath**, also called a **powder room**, has a toilet and lavatory. A **three-quarter bath** has a toilet, lavatory, and shower. A **full bath** has a lavatory, toilet, and tub or a tub/shower combination, and needs at least 40 square feet with minimum dimensions of 5' × 8'.

Mount **medicine cabinets** (concealed shelving for small items) so the top shelf is no more than 50 1/2" from the floor, lower if over a counter. Provide ventilation via a window, an exhaust fan, or both. To avoid drafts and contact with moisture, avoid placing windows near showers or tubs. Achieve privacy with location or semi-opaque glass.

Bathroom Fixture Sizes

The minimum interior shower size is 30" × 30" or 900 square inches; a 30" diameter circle must fit inside it regardless of shape. Recommended size is 36" × 36". Bathroom doors are typically narrower than bedroom doors, usually 2'-6". Swing them into a toilet compartment when space permits. Swing stall doors outside the stall, and into the restroom.

Use the generic fixture sizes in Table 5.6 for sketching and drafting bathroom plans. As with kitchen appliances, consult manufacturer literature for precise dimensions.

Table 5.6

Fixture	Width	Length	Height	
Standard Tub	30"	60"	12"	
	30"	60"	14"	
	30"	60"	16"	
	30"	66"	16"	
	32"	60"	18"	
	36"	72"	16"	
Whirlpool Tub	37"	42"	12"	
	42"	48"	14"	
	42"	60"	16"	
	42"	72"	16"	
	72"	42"	19"	
Freestanding Tub	30"	60"	20"	
Freestanding Japanese Tub	41"	41"	20"	
	Width	**Depth**	**Height**	
Toilet (Floor Mounted, Two-Piece)	17"	25"	19"	17 ½" Accessible
	21"	26"	18"	Seat Height
	21"	28"	18"	

Table 5.6

Fixture	Width	Length	Height
Toilet (Floor Mounted, One-Piece)	20"	27"	20"
	20"	29"	20"
Toilet (Wall Hung, Two-Piece)	22"	26"	21"
Toilet (Floor Mounted, One-Piece)	14"	24"	15"
Bidet	15"	22"	15"
Lavatory (Wall Hung)	19"	17"	
	20"	18"	
	22"	19"	
	24"	20"	
Lavatory (Floor Mounted)	27"	20"	35"
Shower	30"	30"	72"
	36"	36"	73"
	36"	48"	75"
Rectangular basin sink	19"–24"	16"–23"	
Vessel Sink	16"–20"	15"	8"

Vanity

This is a cabinet that houses a **lavatory** (bathroom) sink. It comes in three standard depths—16", 18", and 21"—and in 18", 24", 30", 36", 48", and 72" lengths. A common **vanity** size is 18"–21" deep and 24" long. Small vanities that are 16" deep and 18" wide with an integral sink and top are available for tiny bathrooms. Vanities topped with **vessel sinks** (raised above the countertop) should be 21"–26" tall when the vessel sink is 8" high (Figure 5.57).

Placement of Fixtures

When placing plumbing fixtures, think three-dimensionally and visualize the spaces above and below them. Placing fixtures on exterior walls or under windows can cause problems with **stack pipe** (vertical drainpipe) installation and cold temperatures can freeze their pipes. A wider chase wall can solve these problems but cut into usable square footage. Locate fixtures below windows only when the **windowsill** (bottom edge) is at least 48" above the floor. Utilize common chase walls by placing fixtures back to back on them. Stack chases on multi-floor homes. Arrange fixtures so that all pipes are concealed in the wall or beneath the flooring.

Figure 5.57
Vessel sinks, free-standing tub and opaque glass privacy window. Courtesy hy-lite.com

Built-in tubs must be placed in a corner or **alcove** (recess in the wall). The drain end goes against the wall. Free-standing tubs can be placed anywhere, but still must be connected to the stack pipe in the wall. They take up more space than built-ins due to the space required around the tub itself and the accompanying floor-mounted faucets. Leave at least 12"–18" clearance from the wall and 19" around a free-standing tub for access. Free-standing tubs are often placed on platforms because room is needed underneath for a P-trap and the sloping branch pipe that connects to the stack.

Bathroom Fixture Clearances

Figures 5.58 and 5.59 show NKBA-suggested sizes and clearances. They're printed at 1/4" = 1'-0" scale for tracing.

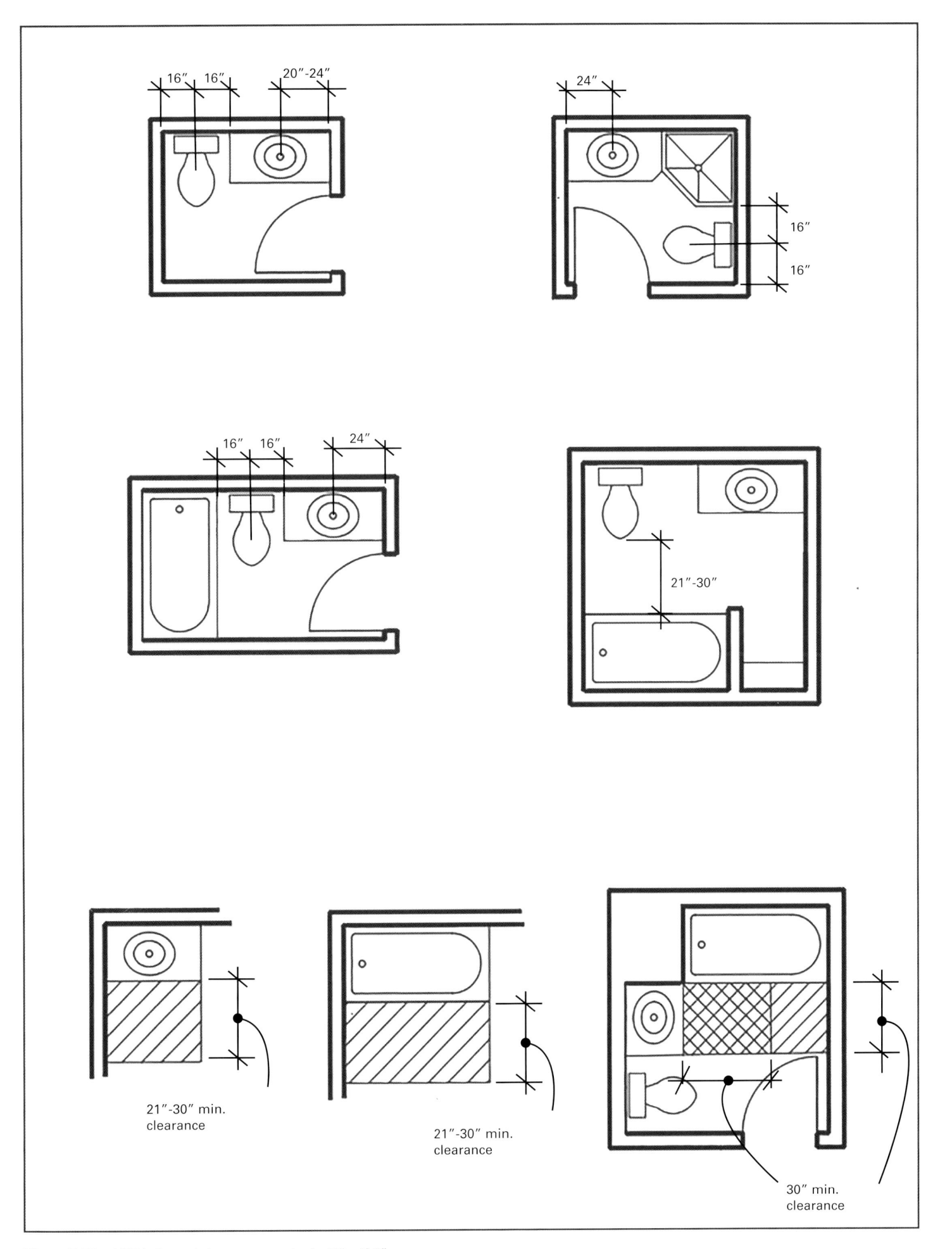

Figure 5.58 NKBA size and clearance standards. ¼" = 1'-0".

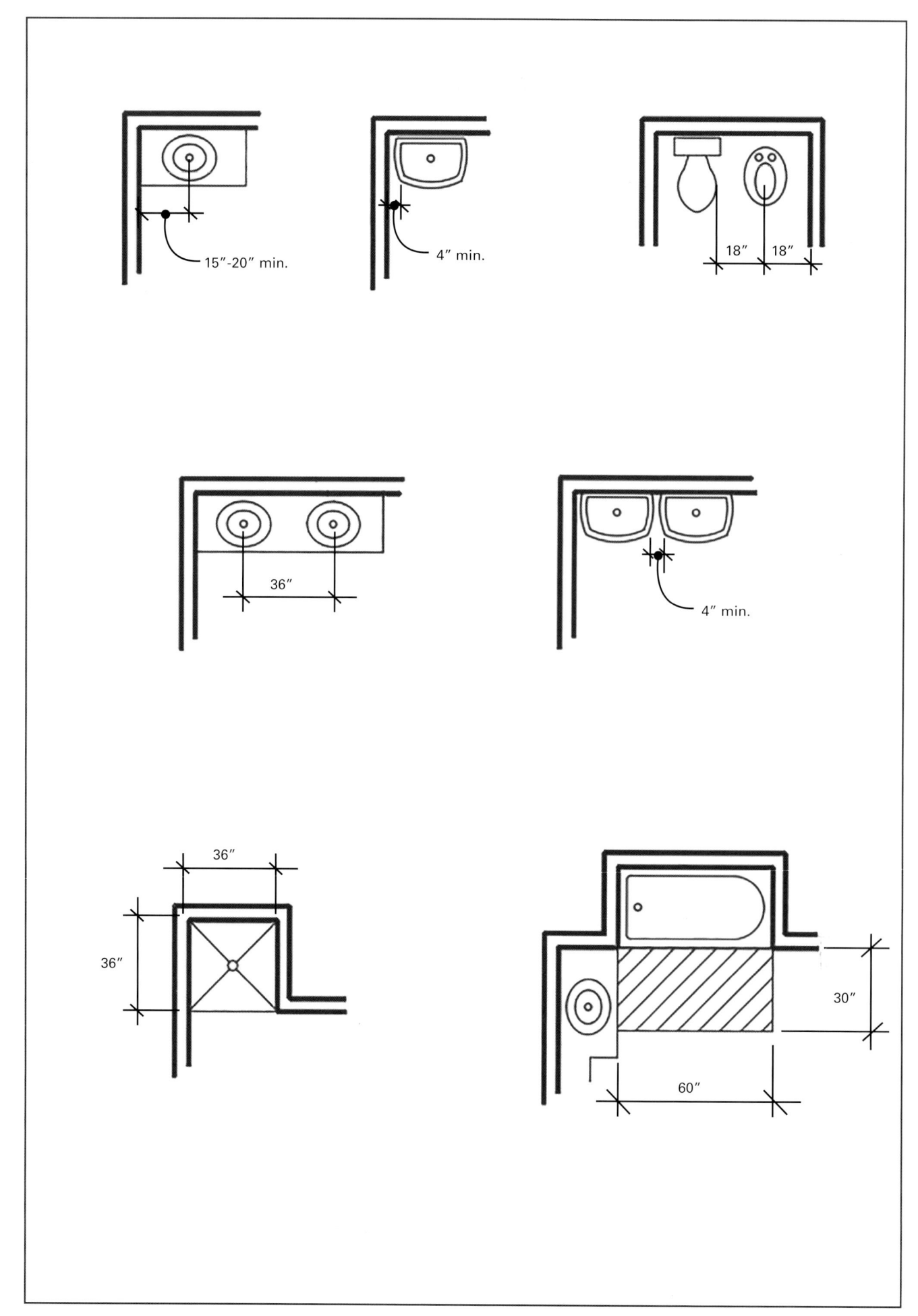

Figure 5.59 NKBA size and clearance standards. ¼" = 1'-0".

Accessible and Universal Design Accessibility

Accessible and Universal Design This is the accommodation of people with disabilities in buildings. **Disability** is a broad term that includes physical and psychological issues. The government-written **Americans with Disabilities Act Accessibility Guidelines (ADAAG)** covers design issues that affect accessibility, such as parking, entering, elevators, and communications systems.

Accessible design is not the same as **universal design**, which is the philosophy of making buildings and products usable to as many people as possible. Young, old, tall, short, abled, disabled, or temporarily injured, all should be able to use the design without adaptation, extra cost, or specialized accessibility features. Common examples of universal design are closed-captioned TV shows and curbs that slant into the street instead of dropping straight down 6 inches.

Accessible Kitchen Design
Here are some guidelines for making kitchens accessible (Figures 5.60 and 5.61).

30" to 33" cabinet and cooktop height is ideal, but the standard 36" height will work.

Provide **knee** *(open)* **space** *under the sink counter* for wheelchair users. Place a protective panel in front of the pipes to prevent leg scalding; alternatively, wrap pipes in insulation.

Cabinet **toe space** *(the indented area at the bottom to accommodate the foot) should be 6" deep and 8" to 11" high* for wheelchair footrests.

Place the bottoms of wall cabinets 17" maximum above the counter or 48" above the floor. This makes the first shelf reachable from a seated position. Make wall cabinet shelves adjustable.

Place cooktop controls at the front to prevent users from having to reach across burners and to alert the visually impaired where the burners are.

Place wall ranges so that the top of the open oven door is 2'-7" above the floor.

French door and side-by-side refrigerator/freezers are the most accessible. Refrigerators with the freezer on the bottom are also good.

Tables with pedestal legs and round tables with pedestal bases are wheelchair friendly. A 4'-0" diameter table accommodates two wheelchair users; a 4'-5" diameter accommodates four wheelchair users. 36" between any two points is enough for a wheelchair to pass, but 5' is needed to fully rotate 180°.

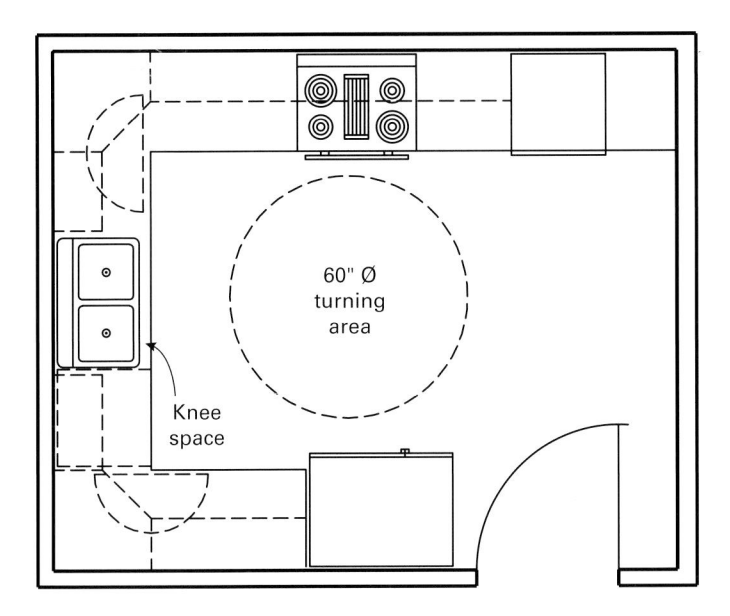

Figure 5.60
An accessible kitchen needs a 60" diameter circle for wheelchair turning and knee space under the sink.

Standard 84" high installation

Lighted interior

Special sink base conceals plumbing

Tambour door for easy access

Space for wheelchair accessibility

9" high X 6" deep toekick allows wheelchair access

Oven accessible from a seated position

32 1/2" tall cabinet

Figure 5.61
Features of an accessible kitchen.

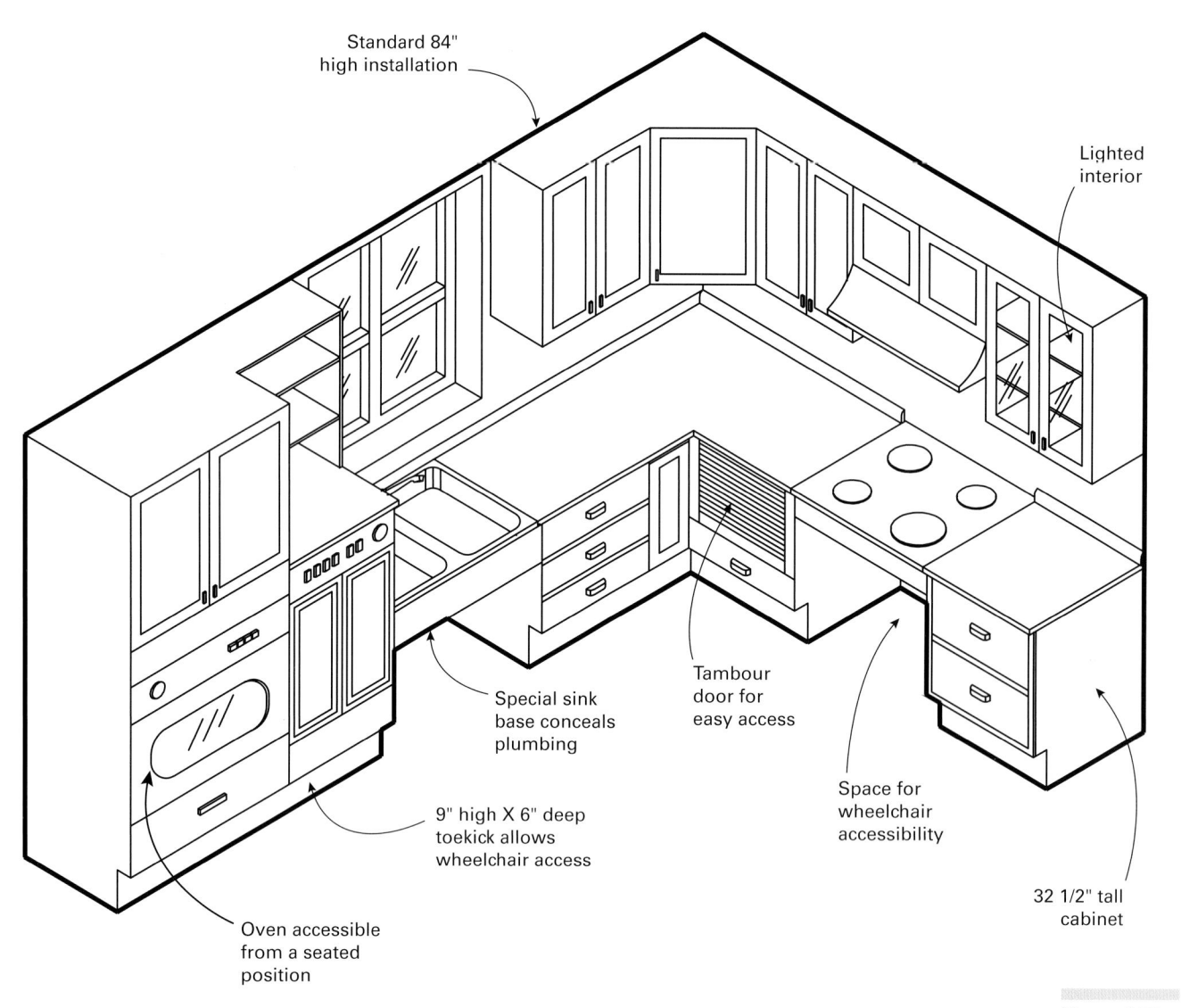

Accessible Bathroom Design   Here are some guidelines for making bathrooms accessible.

Doors must be at least 2'-8" wide; 3'-0" is better. An accessible bathroom needs at least a 60" diameter of clear rotation space between all elements to turn a wheelchair (Figures 5.62, 5.63).

There must be 60" diameter of clear rotation space to turn a wheelchair and a clear space 36" in front of the toilet. Wall-mounted toilets are more wheelchair friendly, since a 20" high toilet seat is about the same height as most wheelchair seats.

Up to 19" of a 48" floor space can extend under the sink when a knee space is provided. 26" to 30" from its underside to the floor is the minimum needed for wheelchair armrests; 30" to 34" from the rim of the sink to the floor is better. Wall-mounted and pedestal sinks provide the knee space required for wheelchair users (insulate any exposed pipes).

26" to 30" from the sink underside to the floor is the minimum needed for wheelchair armrests; 30" to 34" from the rim of the sink to the floor is better.

Place faucet handles a maximum of 18" from the front of the sink.

Install **grab bars**, which are safety devices to maintain balance, on the walls. Horizontal bars are for pushing up; vertical bars are for pulling up. They go over tubs and toilets and in showers and must be 1 1/4" in diameter and 1 1/2" away from the wall.

Figure 5.62
Features of an accessible bathroom.

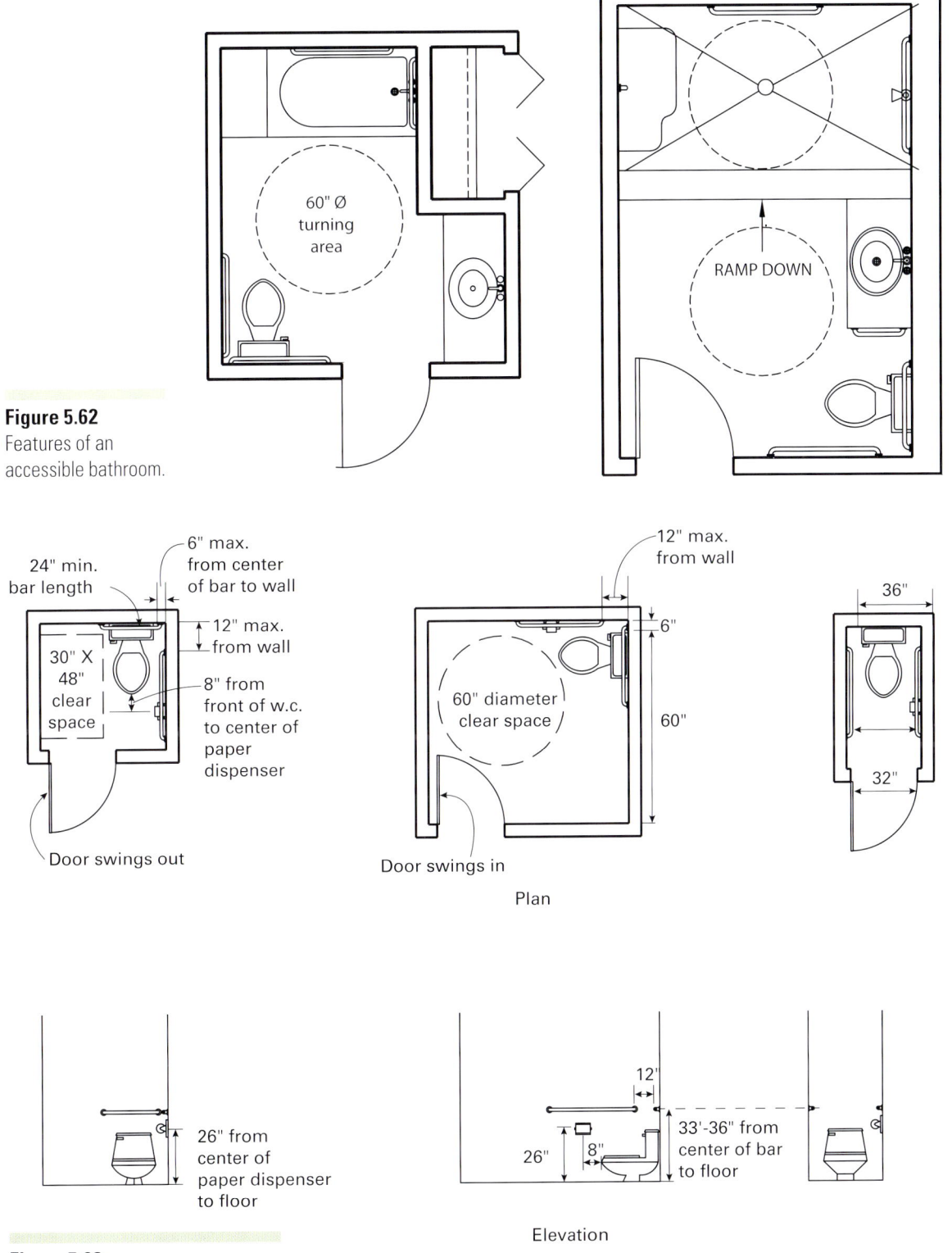

Figure 5.63
Features of an accessible bathroom.

Closets

Clothing closets should be at least 24" deep and 36" wide; 30" deep and 48" wide is preferable. Shelves should be 12"–18" deep. Place closets on interior walls to provide a noise barrier between rooms. Figure 5.64 shows some closet clearances. They're printed at 1/4" = 1'-0" for tracing.

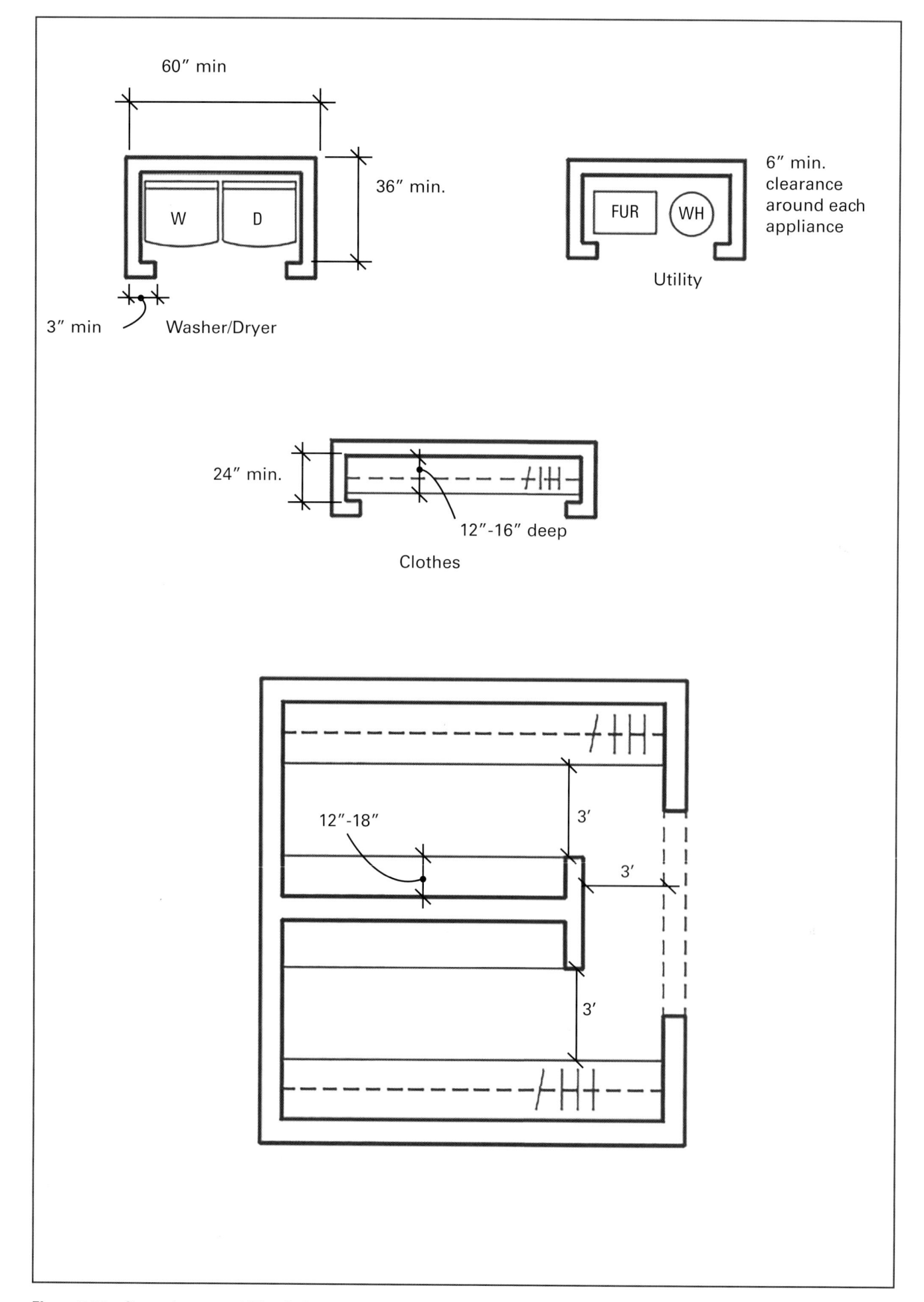

Figure 5.64 Closet clearances. 1/4" = 1'-0"

Laundry Rooms

These shouldn't be in a traffic area. Leave 48" in front of **front-loaded appliances** (doors are on the front instead of the top) to walk around open doors. If a washer/dryer has shelves above, clearance for a top-loading washer lid is 16"–18". A laundry sink is bigger than a lavatory, typically 18" wide and 9" deep. Provide a floor drain in case of flooding from a broken hose. The drain can be in a shower base with the washer positioned over it, or in the room's center with the floor slightly sloped (1/4" per foot) toward it.

Bedrooms

The IRC requires a bedroom to have at least 70 square feet of floor area, with a 7' minimum length in any horizontal dimension. 175 square feet is the minimum needed to hold a double bed, chest of drawers, dresser, and nightstand. The IRC also requires two **means of egress** (unobstructed way to exit a building), so the bedroom must be accessible from the house and have one other exit. The second exit can be a door or **egress window** (discussed later in this chapter).

Bedroom Furniture

Table 5.7 shows generic sizes for bedroom furniture. There are industry standards for beds. A **twin bed** is always 3'-3" wide and is called a 3/3. A **full bed** is always 4'-6" wide and is called a 4/6. A **queen bed** is always 5'-0" wide and a **king bed** is always 6'-6" wide (the latter is called a **6/6 bed**). A **California king bed** is not standardized and varies by manufacturer. Twin and full beds are always 75" long. They can be special ordered 80" long and those are called 3/3 XL and 4/6 XL. Queen and king beds are always 80" long.

Table 5.7			
Furniture	Width	Depth	Height
Single Bed	75"	39"	20"–36"
Double Bed	75"	54"	20"–36"
	80"	54"	20"–36'
	84"	54"	20"–36"
Queen-Size Bed	80"	60"	20"–36"
King-Size Bed	80"	72"	20"–36"
	80"	76"	20"–36"
California King-Size Bed	84"	72"	20"–36"
	84"	76"	20"–36"
Sofa Bed	87"	31"	
	91"	32"	
	79"	34"	
Nightstand	24"	15"	22"
	22"	16"	22"
	24"	18"	22"
	22"	22"	22"

Table 5.7

Furniture	Width	Depth	Height
Dresser	30"	18"	30"
Chest of Drawers	20"	16"	50"
	26"	16"	37"
	28"	15"	34"
	32"	17"	43"
	36"	18"	45"
Armoire	36"	22"	66"
	48"	22"	66"
	60"	22"	66"
Desk	33"	16"	29"
	40"	20"	30"
	43"	16"	30"
Recliner	30"	31"	
	32"	35"	
	36"	38"	
Crib	20"	50"	36"

Living Room

A 12' × 18' space fits between 6 and 10 people in a conversation area with 5 of the seats facing a screen. A 3-seat sofa will fit in that space. Following are sizes of common pieces of living room furniture.

Living Room Furniture

Table 5.8 and 5.9 show generic sizes for living room furniture.

Table 5.8

Furniture	Width	Depth	Height
Sofa	72"	36"	28"
	76"	35"	35"
	84"	36"	37"
	87"	31"	31"
	91"	32"	30"
Sofa	72"	30"	30"
	90"	30"	30"

(continued)

Table 5.8

Furniture	Width	Depth	Height
Love Seat	47"	28"	36"
	54"	30"	36"
	59"	36"	37"
Recliner Chair	31"	30"	40"
	32"	35"	41"
	36"	37"	41"
Armchair	30"	30"	25"
	36"	36"	25"
Small Armchair	18"	18"	29"
	21"	22"	32"
Coffee Table	35"	19"	17"
	50"	18"	15"
	54"	20"	15"
	57"	19"	15"
	61"	21"	17"
	66"	20"	15"

Table 5.9

Furniture	Width	Depth	Height
End Table	21"	28"	20"
	26"	20"	21"
	28"	28"	20"
Ottoman	22"	18"	13"
	24"	119"	16"
Side Table	20"	20"	15"
	21"	21"	16"
Coffee Table	48"	16"	29"
	44"	26"	30"

Egress and Exits

Building codes dictate the number of exits needed location, and width. While they are beyond this chapter's scope, some discussion is needed for competent floor plan drafting.

Means of Egress This refers to the occupants' ability to reach an exit during emergencies. Egress has three components: **exit access**, **exit**, and **exit discharge** (Figure 5.65). *Exit access* is the portion of the building that leads to an exit and includes halls and stairs. Exit is the fully enclosed, fire-protected space between the exit access and the exit discharge.

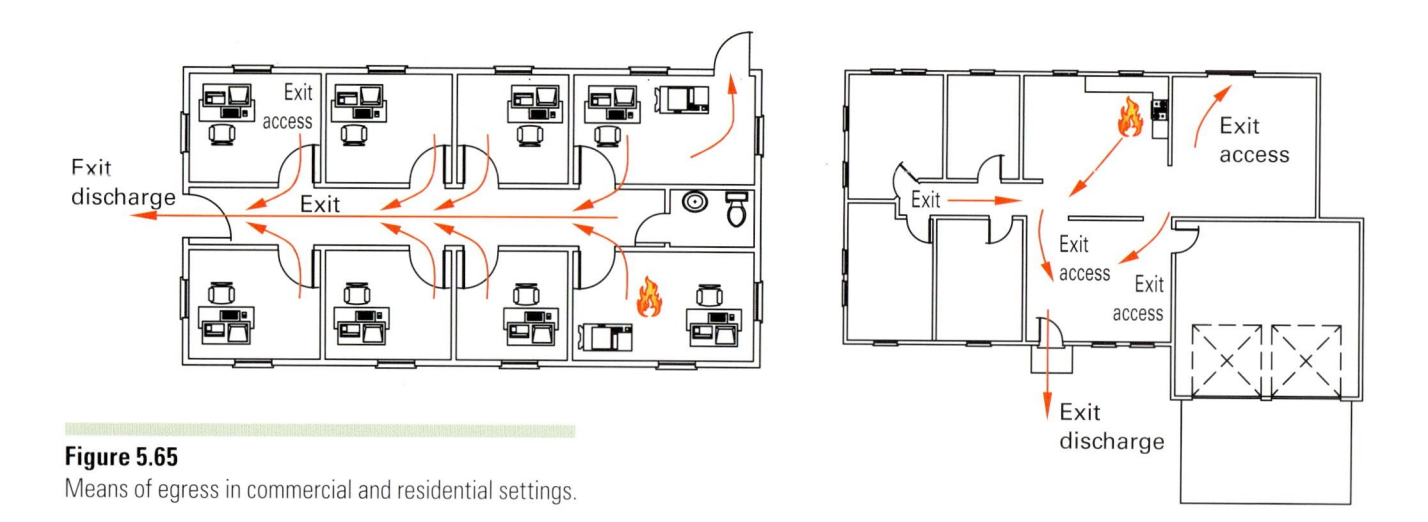

Figure 5.65
Means of egress in commercial and residential settings.

Exit discharge is the area between the termination of an exit and a **public way.** A public way is the area outside a building between the exit discharge and a public street and is usually the end destination of a means of egress. Another type of end destination is an **area of refuge,** a space protected from fire and smoke where people unable to use stairs or an elevator can wait for help.

Half-Diagonal Rule Exits must be located as far from each other as possible. When two or more exits are required, at least two of the exits must be a minimum distance apart. The **half-diagonal rule** requires that minimum distance must be at least one-half of the longest diagonal distance within the building or room that the exits serve (Figure 5.66). Some codes allow the minimum distance to be one-third instead of one-half if the building has an automatic sprinkler system.

Egress Windows Windows can be part of the exit plan if they're egress windows (Figure 5.67), which are sized to allow a fully equipped fireman to enter. Egress windows must between 24–44" from the floor, at least 24" high and 20" wide, and at least 5.7 square feet. A window well and integral ladder is needed if the window is over 44" deep. Tools must not be needed to open it.

Egress Doors These must have a minimum clear width of 32" and be 80" high. Note that a single 36" wide door yields 34" of clear width. Therefore, if 40" of exit width is required, a 44" door will be sufficient, but a 36" one will not.

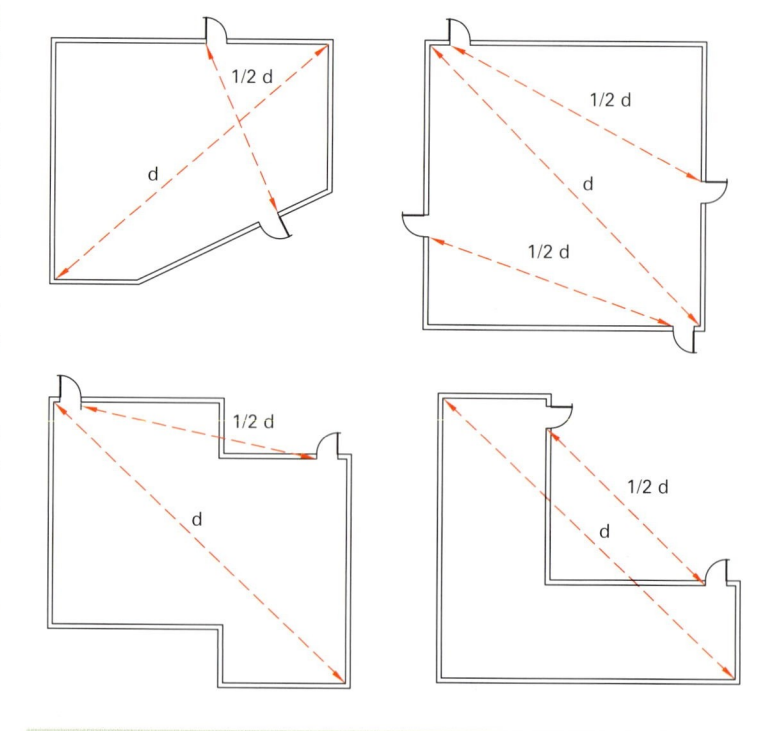

Figure 5.66
The half-diagonal rule for exit placement. D is the diagonal or maximum distance permitted. One-half D is half that distance.

Depending on their location, doors must be of a specific type, size, and swing. Usually a swing door is required, and it must swing in the direction of exit travel to prevent crowds from piling up against a door that opens inward. Sometimes revolving or power-operated sliding doors are allowed as part of the means of egress (usually in a mercantile-occupancy building). But these doors rarely meet full egress requirements; for

136 CHAPTER 5 THE ARCHITECTURAL FLOOR PLAN

instance, revolving doors are sometimes credited with meeting just 50 percent of the exit requirements, so adjacent swing doors must be included.

Hallways

These long passages with doors into rooms on both sides must be at least 36" wide in residences, 44" in commercial buildings, and 48" wide for handicapped accessibility. **Dead-end hallways** (hallways with no egress besides the one you came through) may not be longer than 20'.

Habitable Space

This is a code descriptor for any space that in which sleeping, eating, cooking, and general living, takes place. Bathrooms, toilet compartments, closets, halls, storage, and utility spaces are not considered **habitable space**. Emergency egress is required from every room and basement that contains habitable space. Egress may occur through a door or an egress window that opens directly to the outside.

Figure 5.67
An egress window in a basement bedroom.

Summary

An architectural floor plan is a 2D, top-down view of a room or building made by inserting a cutting plane 4' and 5' above the floor and drawing what is below it. Fixed architectural features and symbols that reference other drawings are shown. Existing spaces are measured and documented, and spaces are sketched and then drafted. Documenting a space can be done manually or with photogrammetry software. Drafted drawings incorporate line types and weights to make them readable. Colors and line weights are standardized on CAD drawings. Knowledge of space planning, clearances, and building codes is required for competent design and drafting.

Classroom Activities

1. Measure, sketch, and draft your classroom.
2. Document, measure, and sketch your home. Ask a classmate to draft the rooms from your notes and sketches.
3. Using a tape measure, record the length, width, and depth of two pieces of furniture in your classroom and the clearance between them. Record in a pocket notebook kept for this purpose.

Questions

1. What is a floor plan?
2. Why sketch a plan before drafting it?
3. What is a fit sketch?
4. How does a sketch's scale affect its detail?

5. What is a visual inventory?
6. What is photogrammetry?
7. What is a chase wall?
8. What is line quality?
9. What is the minimum closet shelf depth?
10. What is the minimum hall width?

Further Resources

AutoCAD blocks of Kohler products https://www.us.kohler.com/us/cad-cutout/kitchen/_/N-25am

Autodesk ReCap Photo tutorial https://www.youtube.com/watch?v=RYM7uZeiXH0

CAD libraries: https://cad-blocks.net/, https://www.firstinarchitecture.co.uk/

Educational version of Autodesk ReCap Pro https://www.autodesk.com/education/free-software/recap-pro

Educational version of Autodesk SketchBook https://sketchbook.com/education

How to Sketch a Floor Plan with Multiple Rooms https://www.youtube.com/watch?v=O5A58npxsps

Product information and downloadable files: https://sweets.construction.com/, http://www.4specs.com/

Keywords

- 3D camera
- 6/6 bed
- accessibility
- Accessible and Universal Design
- alcove
- Americans with Disabilities Act (ADA)
- Americans with Disabilities Act Accessibility Guidelines (ADAAG)
- architectural floor plan
- area of refuge
- Autodesk ReCap Photo
- Autodesk Sketchbook
- bar scale
- CAD block
- California king bed
- carpenter's bubble level
- carpenter's square
- chase wall
- clearance
- commercial furniture plan
- construction calculator
- dead-end hallway
- digital model
- disability
- egress window
- entourage
- eraser bag
- exit
- exit access
- exit discharge
- Federal Housing Authority
- fenestration
- fit sketch
- fixed features
- flashlight
- front-loaded appliance
- full bath
- furniture template
- galley (corridor) kitchen
- grab bar
- G-shape (peninsula) kitchen
- habitable space
- half bath
- half-diagonal rule
- hallway
- International Residential Code (IRC)
- key
- keynote
- king bed
- knee space
- knee wall
- laser tape measure
- lavatory sink
- line quality

CHAPTER 6

Interior Elevations and Sections

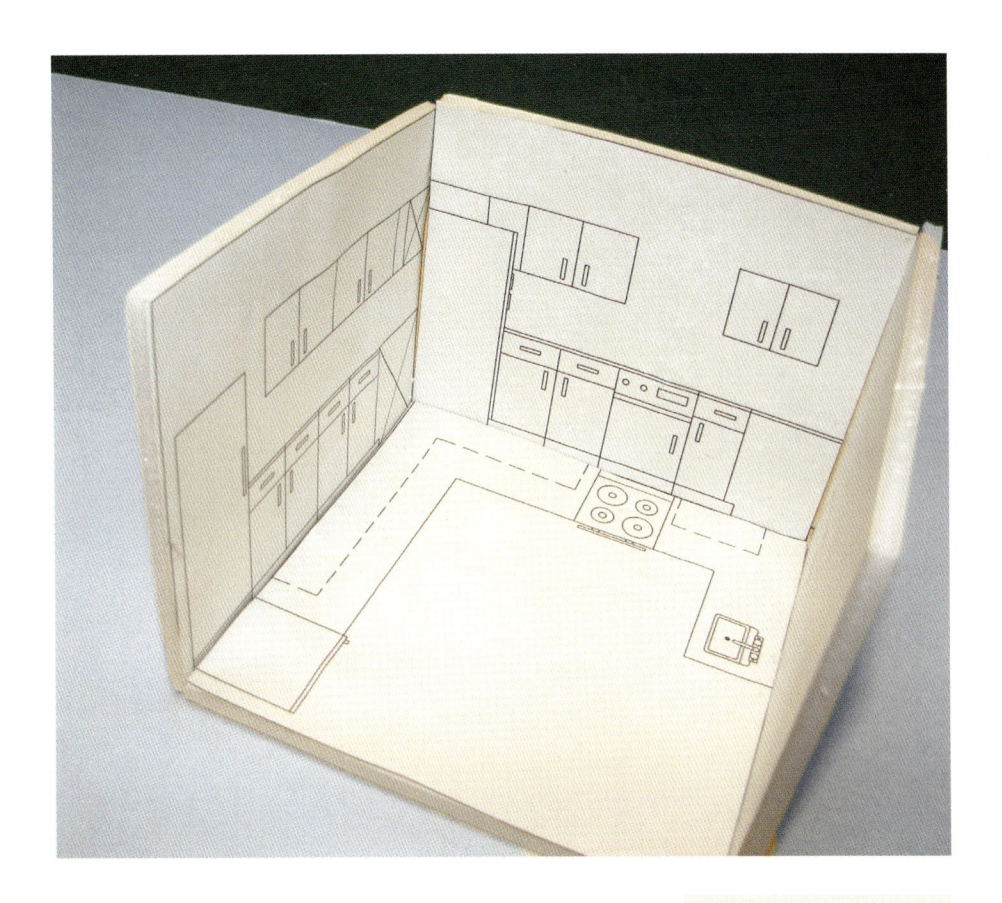

Figure 6.0
Elevation views glued to foam board for an architectural study model.

OBJECTIVES

Upon completion of this chapter you will be able to:

- Choose appropriate walls to make elevation and section views of
- Create interior elevation and section view drawings
- Apply relevant details and symbols

Height drawings of objects are called **front** and **side views**. Height drawings of walls are called **elevation** *and* **section views**. Like floor plans, it is most efficient to sketch them while problem-solving and draft them for construction documents.

What Is an Elevation View?

This is a drawing made by looking straight ahead at the wall and outlining it from corner to corner and floor to ceiling (Figure 6.1). The resultant picture shows the wall's height and width, and everything on it.

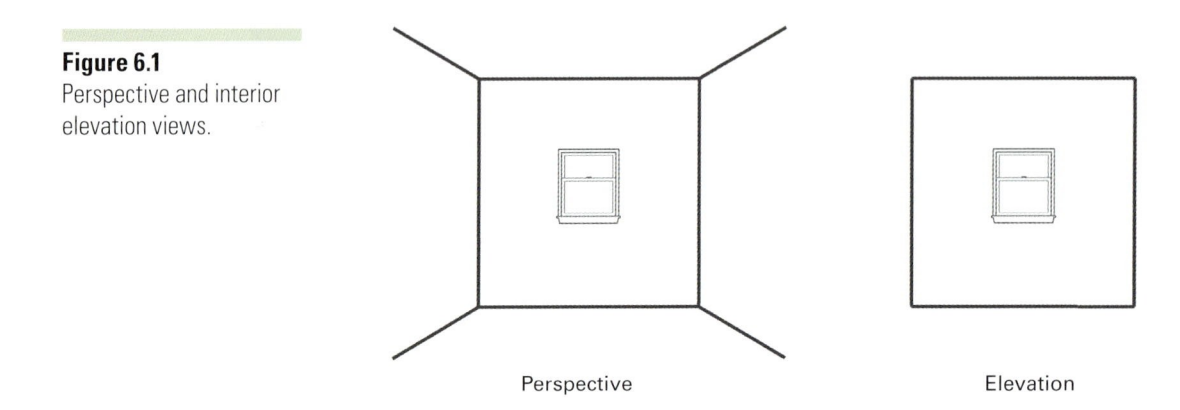

Figure 6.1
Perspective and interior elevation views.

Figure 6.2
Design ideas are explored on these interior elevations. Entourage figures provide visual scale.
Courtesy wolfgangtrost.com

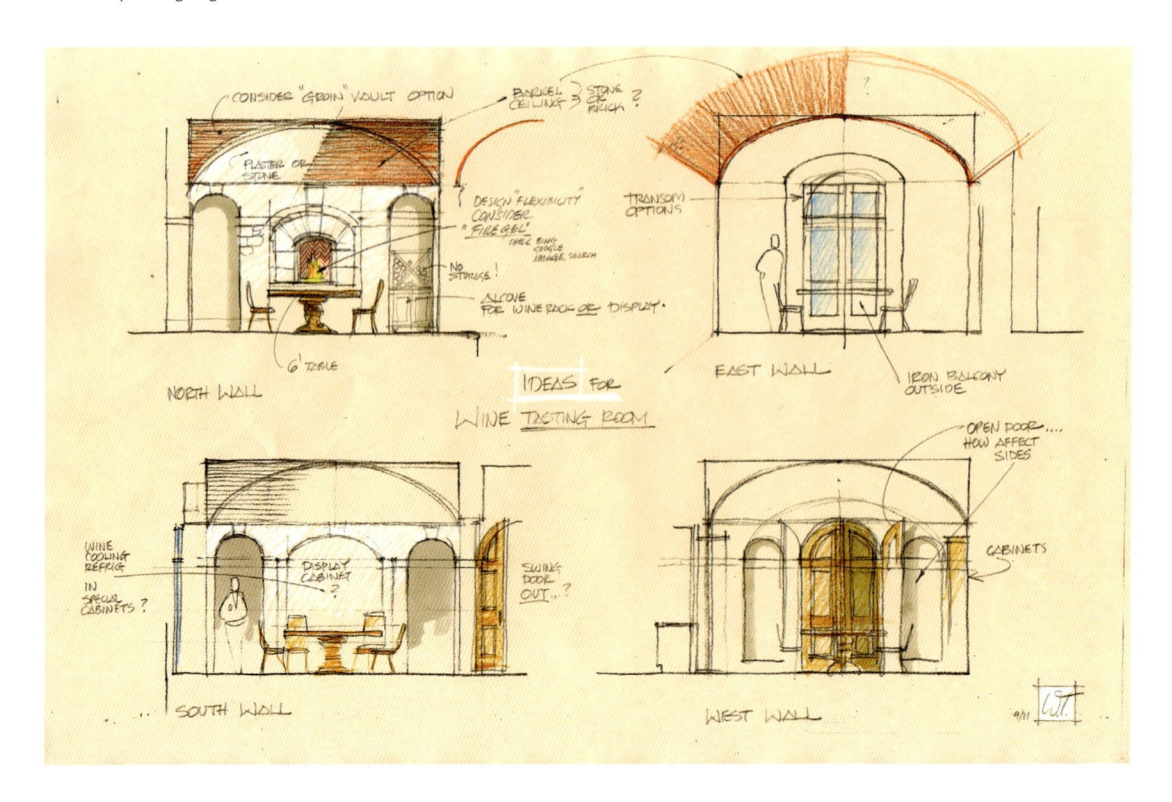

140 CHAPTER 6 INTERIOR ELEVATIONS AND SECTIONS

An **exterior elevation** is a drawing of a wall outside the building; an **interior elevation** is a drawing of a wall inside the building or of the interior side of an exterior wall. Label it by its compass direction (e.g., "North"). You can also label it more descriptively, such as "Dining Room South Wall." Its purpose is to show built-in and attached items on the walls and their dimensioned heights. Doors, windows, fireplaces, shelving, cabinetry, trim, wainscots, and built-in light fixtures are all included (Figure 6.2). Exterior elevations show windows, doors, finishes, shutters, and the **grade** (ground) **line**. Draw elevations the same scale as the floor plan.

The Elevation Callout Symbol Elevations are referenced from the plan with **callouts**, also called **marks**. This symbol points to the wall of interest. There are two numbers: the **drawing number** and the **sheet number**. The drawing number is first and tells us which drawing to look for. The sheet number is second and tells us what sheet to find the drawing on. The callout in Figure 6.3. tells us to go to drawing 2 on sheet 1 to find the elevation view of the wall the arrow points to. This elevation's ID label confirms you're indeed looking at drawing 2 on sheet 1.

On commercial drawings, callouts may contain three numbers (Figure 6.4). The third number tells you which sheet the callout symbol is on. Whichever type callout is chosen must be used throughout the drawing set.

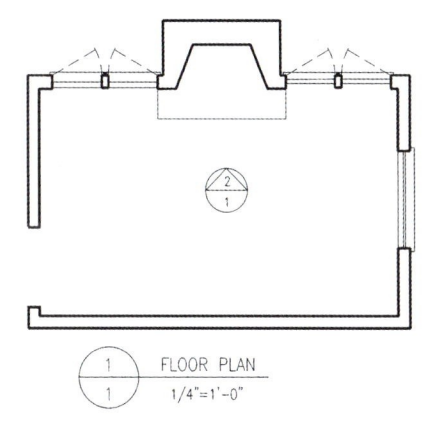

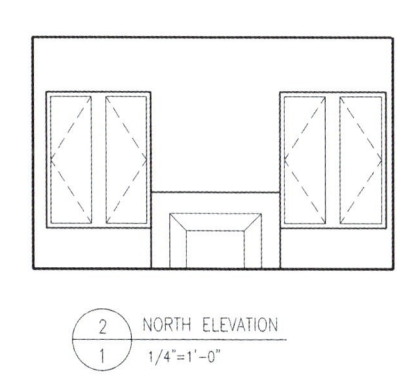

Figure 6.3
The elevation callout tells you to go to drawing 2 on sheet 1 to see the elevation view.

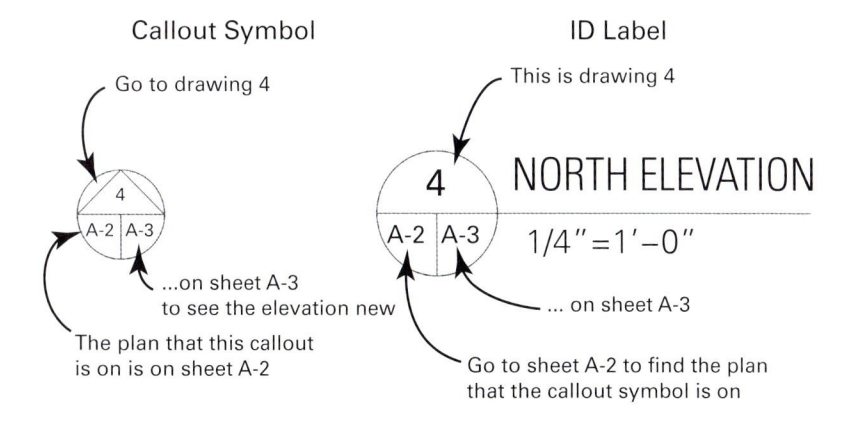

Figure 6.4
Callouts on commercial drawings may contain three numbers.

NORTH ELEVATION
1/4"=1'-0"

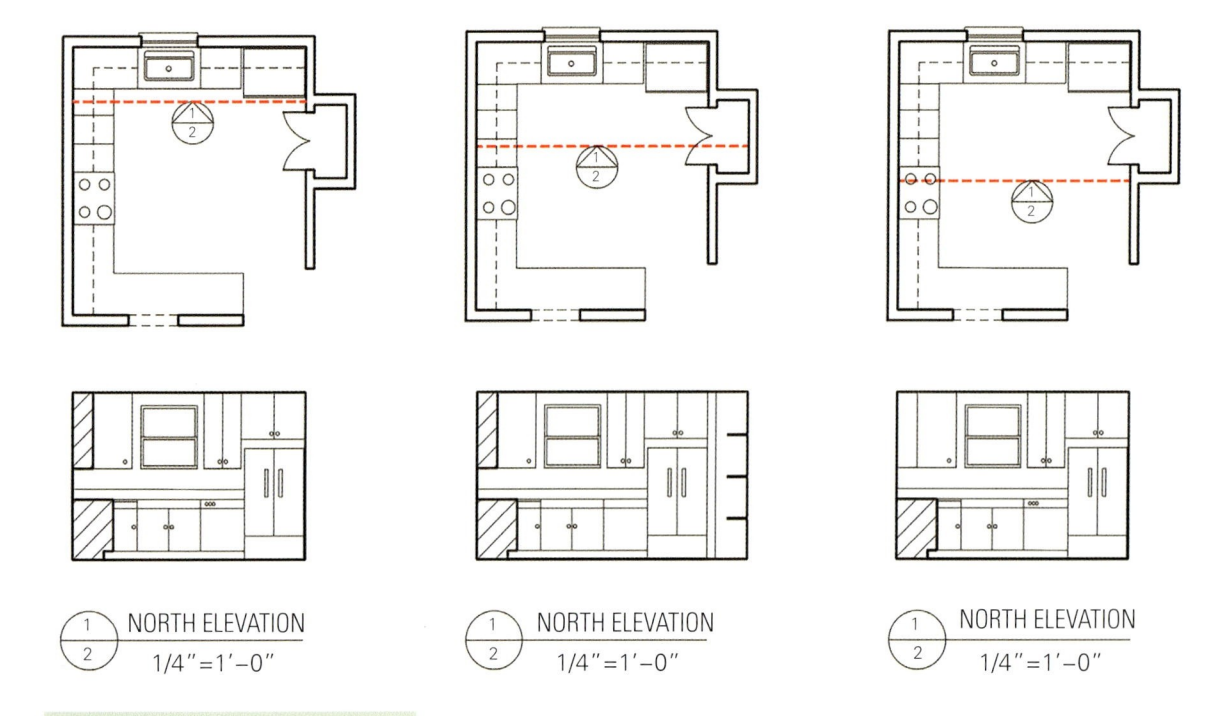

NORTH ELEVATION
1/4"=1'-0"

NORTH ELEVATION
1/4"=1'-0"

Figure 6.5
The callout's placement affects the elevation view. The red line isn't part of the symbol; it's there to illustrate what is shown in the elevation view.

Everything in front of the arrow is drawn, so it must be strategically placed. Different locations in the same room may yield different elevation views. Figure 6.5 shows three different callout locations and the resultant drawings. The first location cuts through the left side wall and base cabinets, hence those features are thickly outlined and hatched. The second location cuts through the left side wall and base cabinets and the pantry on the right. The third location cuts through the stove. The wall cabinets above aren't cut, so they aren't hatched.

How to Draw an Elevation View Following are steps for drawing an elevation view from a floor plan.

> **Exercise**
> Gather some magazine floor plans of houses and choose walls that would make informative elevation views.

1. Orient the plan so the *callout arrow points up* (Figure 6.6). This places the wall it points to on top. If the arrow points down, the wall is on bottom, and the resultant elevation drawing will be reversed; items that should be on the viewer's left will appear on the viewer's right. Align a horizontal wall with the parallel bar and tape the paper to the drafting board.

2. *Project the back wall and everything on it down.* You can draw construction lines from the floor plan's interior corners straight down or straight up. In Figure 6.7 they're drawn straight down. Draw a ground line and then measure up the desired height for the ceiling line. The resultant rectangle is the wall's outline. Alternatively, tape a detail drawing for the wall in the side view location and project its floor and ceiling lines horizontally. Wall thicknesses are not included in an interior elevation.

3. Project the base and wall cabinets down, and then measure and mark their heights. Find correct heights from website searches of specific products. If the elevation is of an existing space, measure all features for their heights yourself.

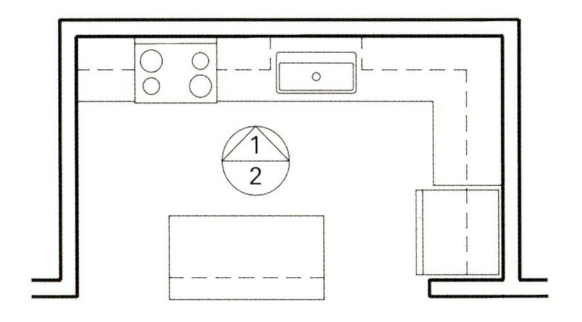

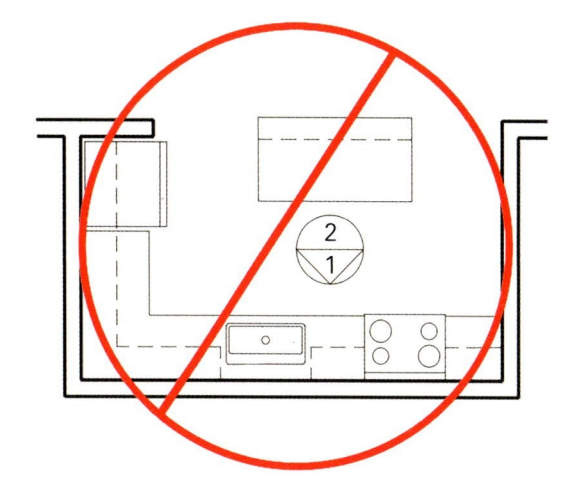

Figure 6.6
Orient the plan so the callout arrow points up. If it points down, the resultant elevation drawing will be reversed.

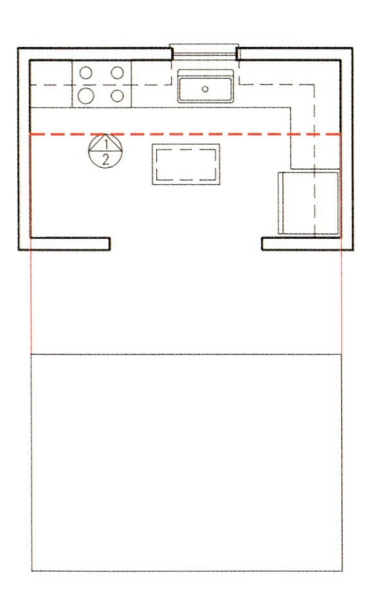

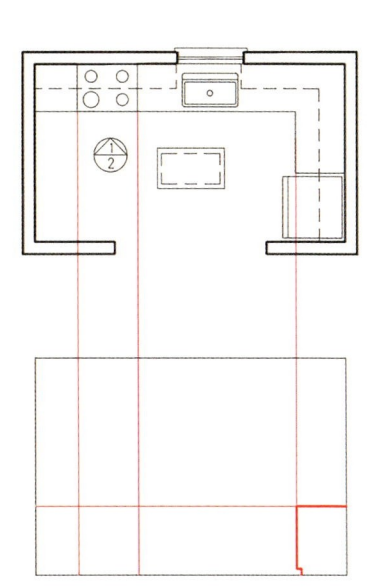

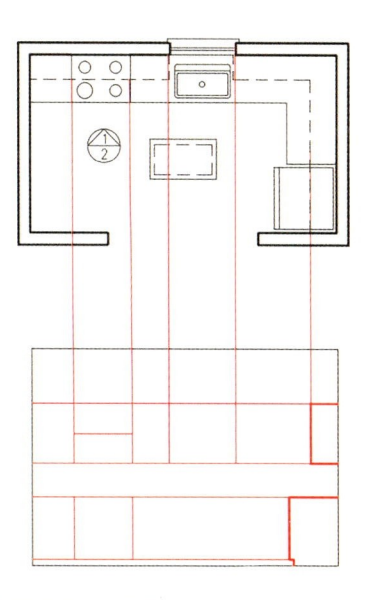

Figure 6.7
Project everything on the wall down.

4. Darken object lines and add line weights and details (Figure 6.8). Note that the base and wall cabinets on the right side have a thicker line weight and are hatched. That's because due to the callout arrow's location, this small portion of the elevation appears in section. The dashed lines on the front-facing cabinet doors are hinge lines. The doors are hinged on the point side.

An elevation for construction documentation is a simple line drawing containing the details, dimensions, symbols, and information a contractor needs to build it. A presentation elevation may include color, shade and shadow, and entourage to make it more relatable to a client (Figure 6.9).

WHAT IS AN ELEVATION VIEW?

Figure 6.8
The finished elevation view.

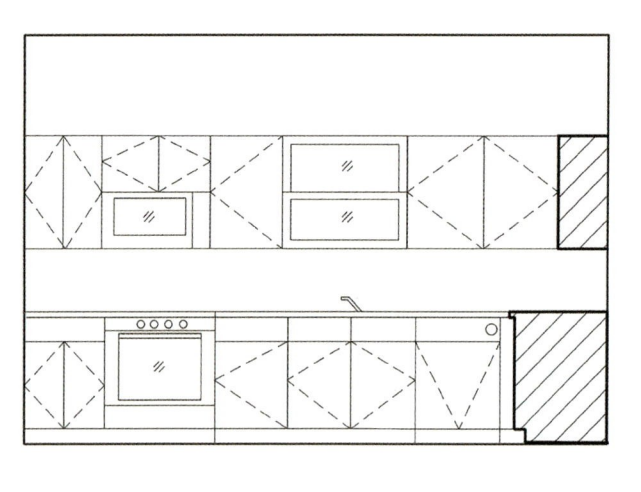

1/2 — NORTH ELEVATION 1/4″=1′-0″

Figure 6.9
Kitchen elevation.
Courtesy wolfgangtrost.com.

How to Draw a Sloped Ceiling in an Elevation View If the ceiling is sloped, its true angle appears in an elevation perpendicular to the **roof ridge** (the roof's high point; the line where two planes intersect). The ceiling's **rake,** which is a foreshortened view of an angled item, appears in a view parallel to the roof ridge. Constructing the rake view of a ceiling's height requires an elevation view that is perpendicular to the roof ridge. So, you'll need to draw that if you don't have it already. Using the divider tool, mark the distance the arrow point is from the back wall (distance X) and transfer that distance to the elevation drawing (Figure 6.10). The farther away the callout symbol is from the wall, the higher the rake height in the elevation view will be.

144 CHAPTER 6 INTERIOR ELEVATIONS AND SECTIONS

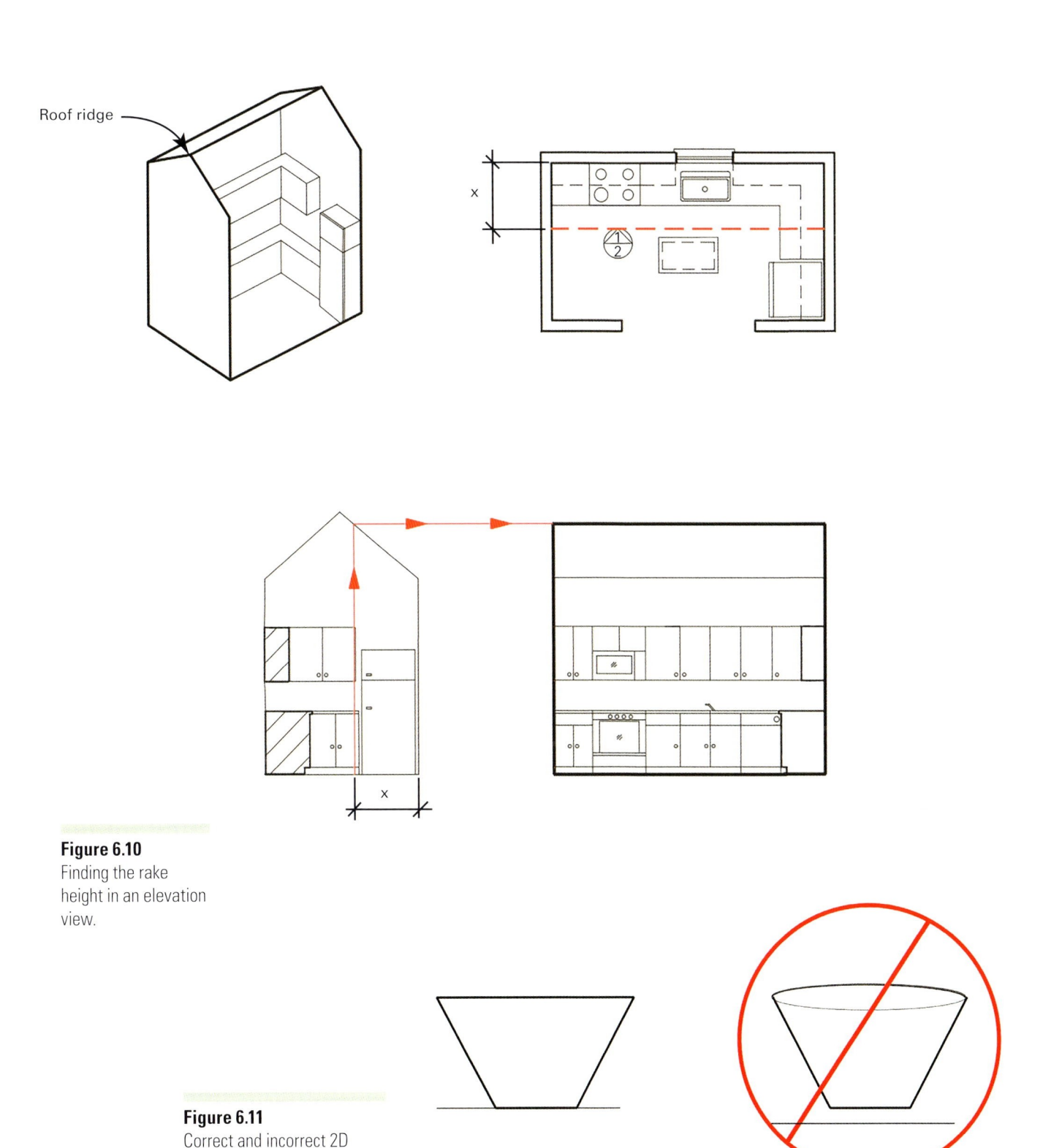

Figure 6.10
Finding the rake height in an elevation view.

Figure 6.11
Correct and incorrect 2D drawings of a flowerpot.

How to Draw Caddy-Corner Items in an Elevation View
Furniture not parallel to the front or side picture planes will appear distorted due to **foreshortening** (lines appearing shorter than true length). It's easiest if you draw a box around the whole piece of furniture in plan, project the box's corners down, and then add interior details. Correctly done, the bottom of all furniture and entourage touch the floor line and the furniture's top and bottom horizontal lines are parallel. A common beginning drafter mistake is angling the horizontal lines and not drawing them touching the floor (Figure 6.11). Figure 6.12 shows how to draft some common furniture pieces orientated caddy-corner.

WHAT IS AN ELEVATION VIEW? **145**

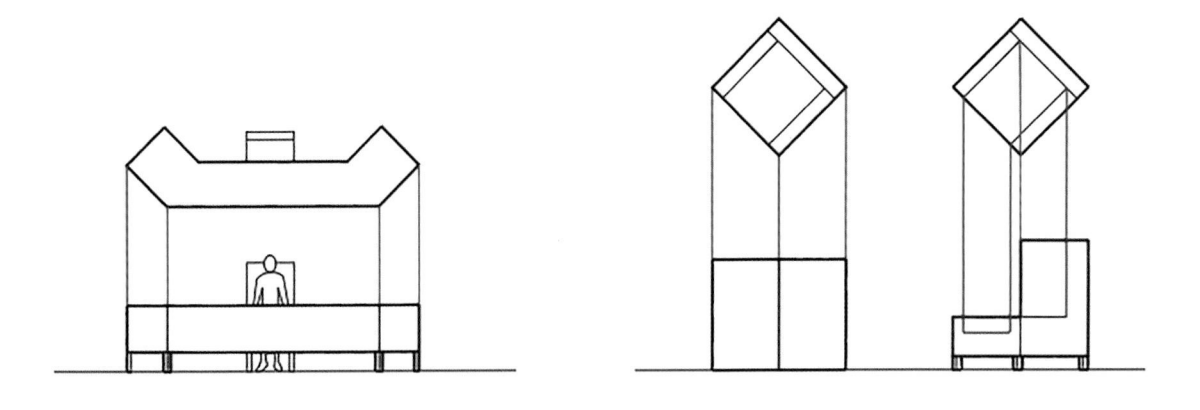

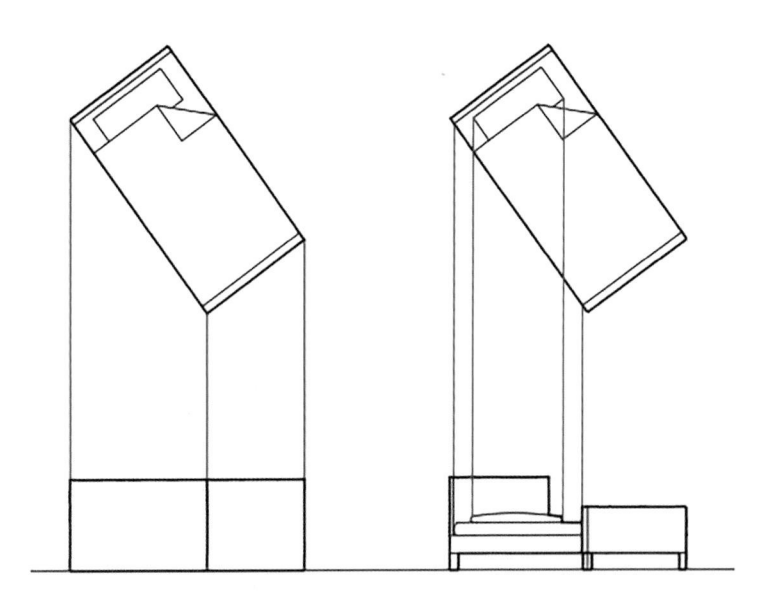

Figure 6.12
Projecting caddy-corner furniture from plan to elevation.

What Is a Section Drawing?

A **section drawing** is a cut made vertically through an entire building (Figure 6.13) from foundation to roof. A **transverse section** is a cut perpendicular to the roof ridge and a **longitudinal section** is a cut parallel to it. Its purpose is to show heights, major vertical elements (e.g., chimneys, stairs, sloped roof), spatial relationships, framing, thicknesses of floors walls and ceilings, and details of construction that are not visible on an elevation. If the building has multiple levels, all are shown on one drawing (Figure 6.14).

Section Scale This varies with its purpose. If the section is just to show shape and be a reference for enlargement details, 1/8" = 1'-0" is appropriate. If it's to show pochés, studs, insulation, and finishes, then at least 1/4" = 1'-0" is needed. A portion may be encircled with an enlargement box that directs the reader to a larger scale drawing.

Figure 6.13
Longitudinal and transverse sections.

Figure 6.14
A transverse section through a two-story house.

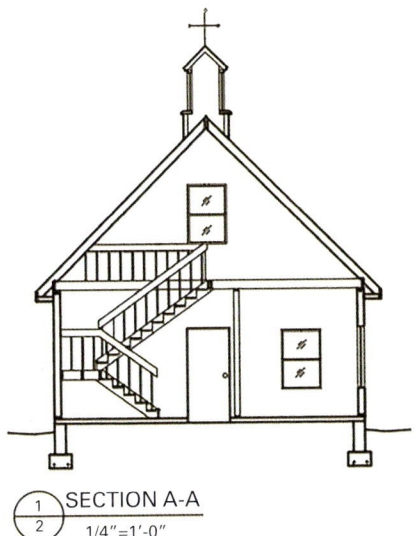

① FIRST FLOOR
① 1/4"=1'-0"

② SECOND FLOOR
① 1/4"=1'-0"

① SECTION A-A
② 1/4"=1'-0"

WHAT IS A SECTION DRAWING? **147**

Section Location Symbol   This is shown in plan with a cutting plane line; refer to Chapter 4 for how to draw it. The cutting plane line is topped with arrows that point in the direction the viewer is looking. The arrows can be labelled with a generic letter, such as **A-A** or with a 1/2" diameter circle containing numbers describing where to find the section drawing like elevation callouts. The cutting plane line location must be carefully chosen for the most informative view.

Section Detail   A **section detail** is a drawing that shows a small portion of a building, such as a wall, and can be sliced vertically or horizontally. Its symbol is only placed on one side of the plan and has one arrow. It's placed precisely where the detail drawing will be made, and its callout contains numbers that tell the reader where to find the detail. Figure 6.15 shows a cutting plane line for a wall section. Its numbers tell the reader to go to drawing 2 on sheet A-4 to see the resultant wall detail.

How to Draw a Longitudinal Section   Let's draw a longitudinal section through the house in Figure 6.16. Figure 6.17 shows the interior. We'll draw a rake view of its sloped ceiling.

Figure 6.15
This symbol tells the reader where to find a wall detail.

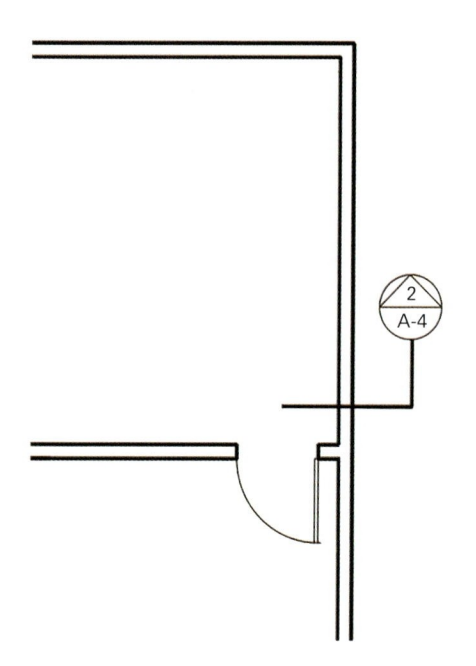

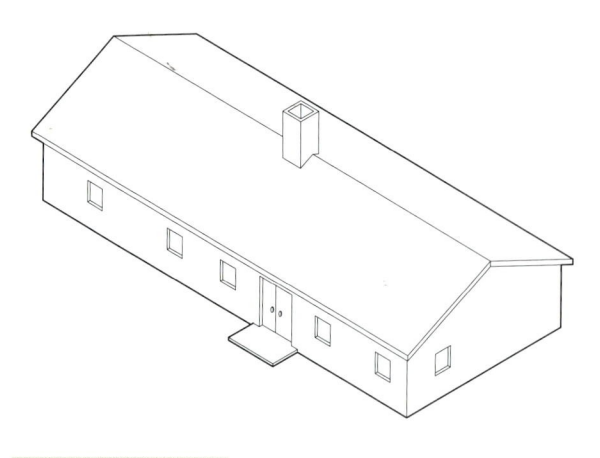

Figure 6.16
House with a gable roof.

Exercise
Find some magazine floor plans of houses and place a section cutting plane line at the most informative locations.

Figure 6.17
The house's interior.

148   CHAPTER 6   INTERIOR ELEVATIONS AND SECTIONS

1. *Choose the cutting plane location* (Figure 6.18). Find one that shows the most useful information possible in one drawing. Draw the cutting plane symbol precisely where the slice is made, with arrows pointing in the direction the viewer looks. Avoid slice locations that cut through long lengths of walls or other large items. Where possible, include exterior windows and doors. Orient the paper with the arrows pointing up, align a horizontal line on the plan with the parallel bar, and tape the plan down.
2. *Project the walls down* (Figure 6.19). Include the exterior wall thickness. Draw lines for the floor and ceiling to complete the outline of the back wall. Project windows and doors down, then measure and mark their heights. Windows and doors in walls are drawn as though they are closed.

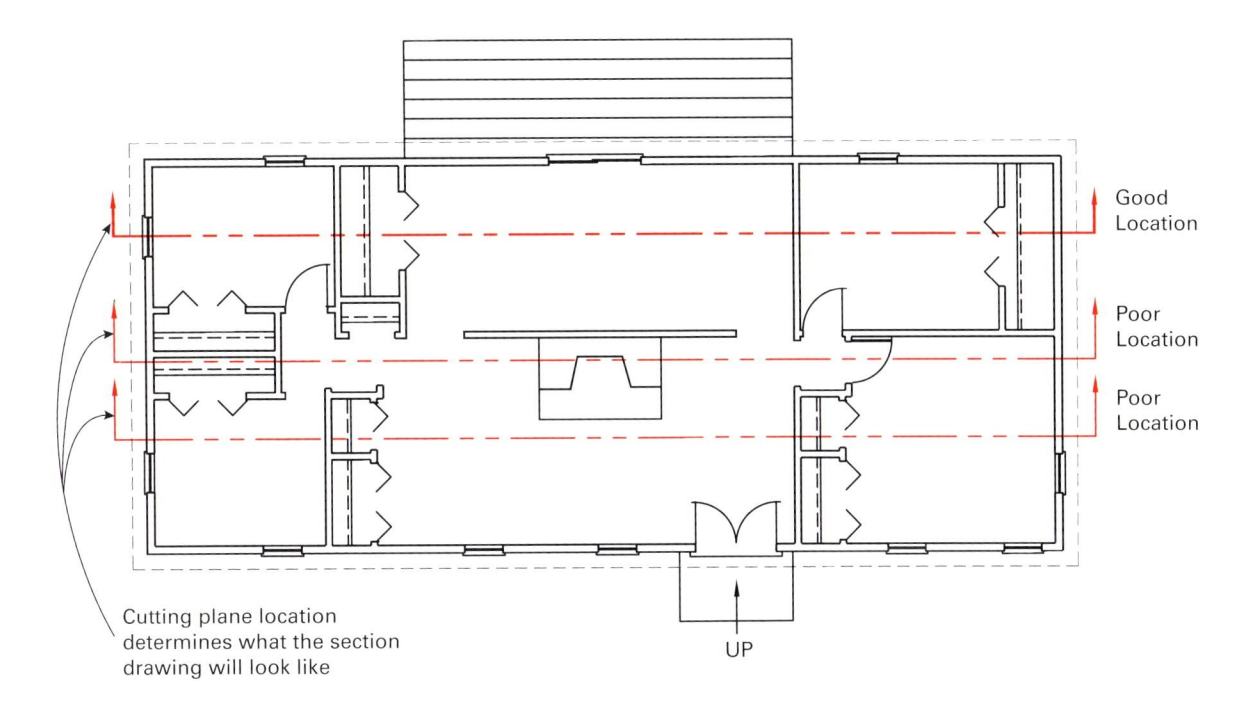

Figure 6.18
Choose an informative location.

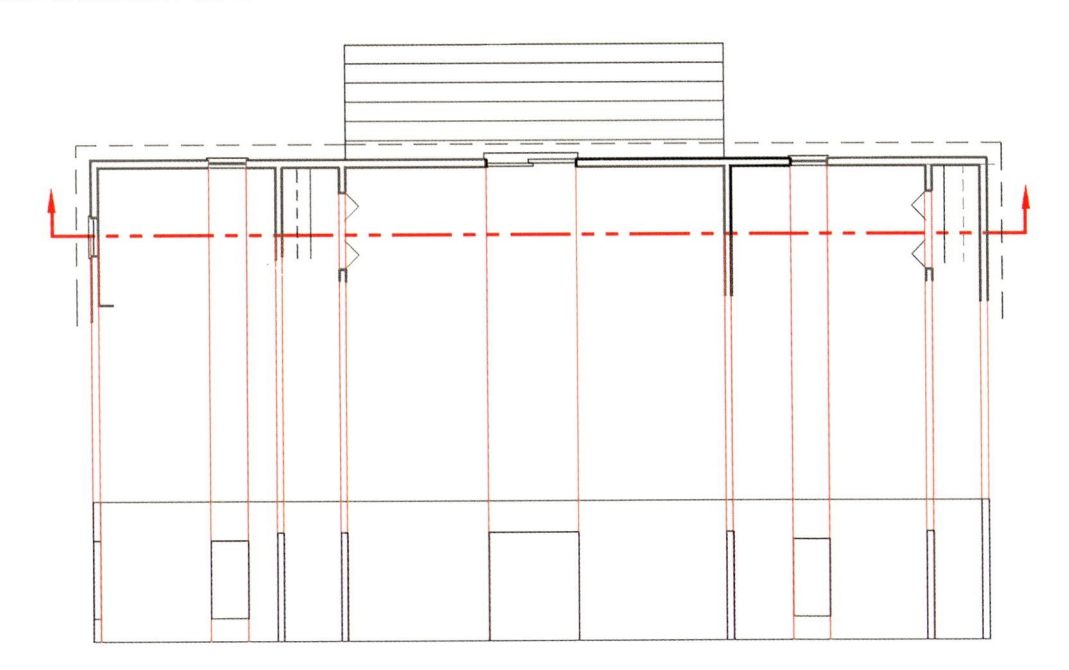

Figure 6.19
Outline the back wall and project interior doors, windows, and interior walls down.

WHAT IS A SECTION DRAWING? **149**

3. *Add line weight to the cut-through walls* (Figure 6.20). Draw whatever the cutting plane touches with a thick line. A small scale (1/4"=1'-0") section whose purpose is to show spatial relationships and reference other drawings can be drawn with thick lines or poché black. A larger section might contain specific material pochés. Project the closet shelves/rod and roof overhang down. Add window symbols, roof overhang, foundation, and a thick grade line. If the grade is unpaved, draw it freehand; if it is paved, draw it hard line.

When cutting through door openings (Figure 6.21), the **jamb** (side of the door opening) is not sliced by the cutting plane, so draw it with thinner lines than the walls that are cut through. The **head** (top of the door) is sliced by the cutting plane, so draw it with thick lines.

Figure 6.20
Add line weight and details.

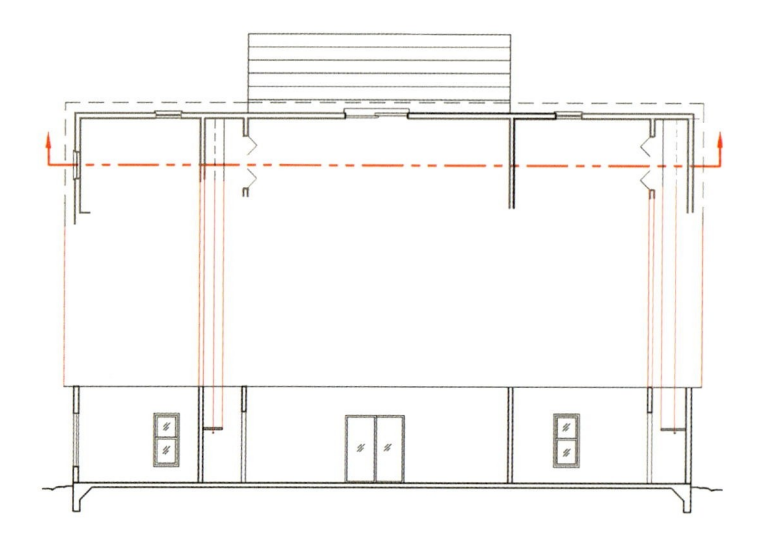

Figure 6.21
How to draw a cut-through door.

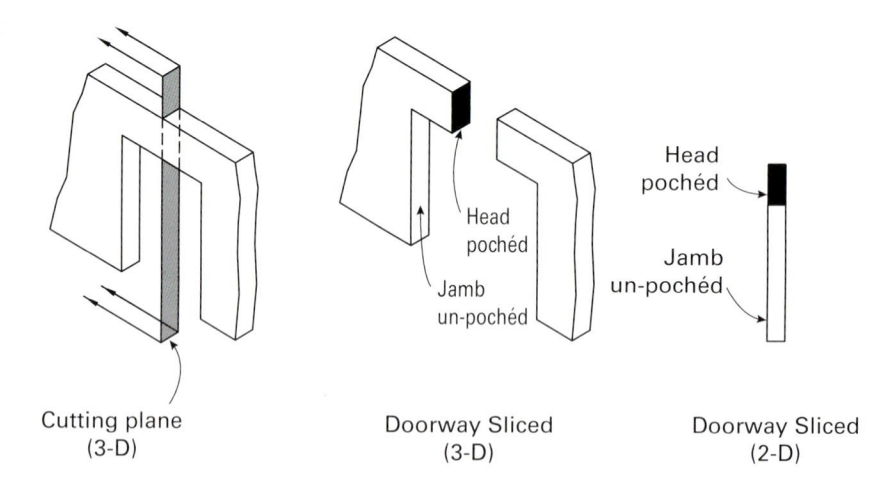

4. *Add the ceiling height* (Figure 6.22). This is a longitudinal section of a gabled roof, so we see the rake view. As with the elevation example earlier in this chapter, a side elevation is needed, and construction is done similarly.

If the ceiling's angle and true shape need to be shown, draw a transverse section (Figure 6.23). Figure 6.24 shows a **rendered** (has color and entourage) transverse section used for a presentation.

Poché Symbols in Section and Elevation   Figure 6.25 shows material pochés for section and elevation views. The same material is poched differently on elevations and sections for clarity. If a component on a complicated drawing was poched the same way in both elevation and section, the drawing would be confusing to read.

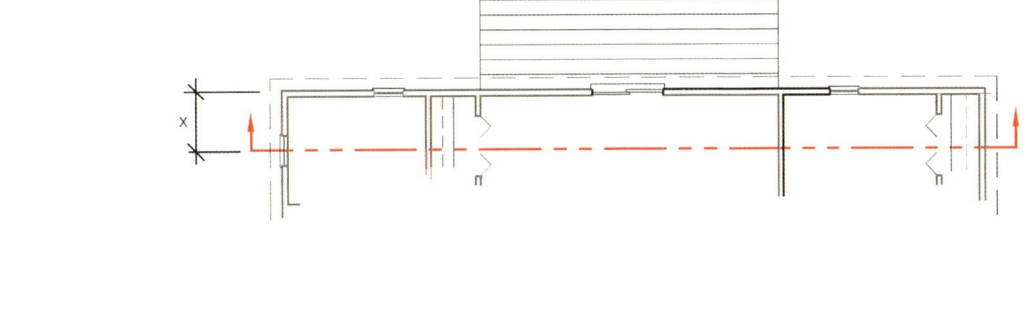

Figure 6.22
Constructing the ceiling's rake view.

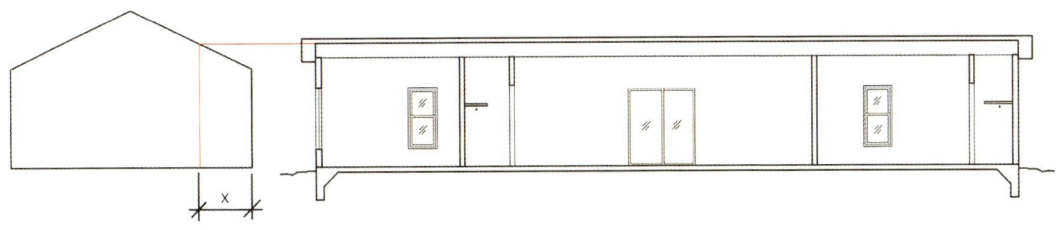

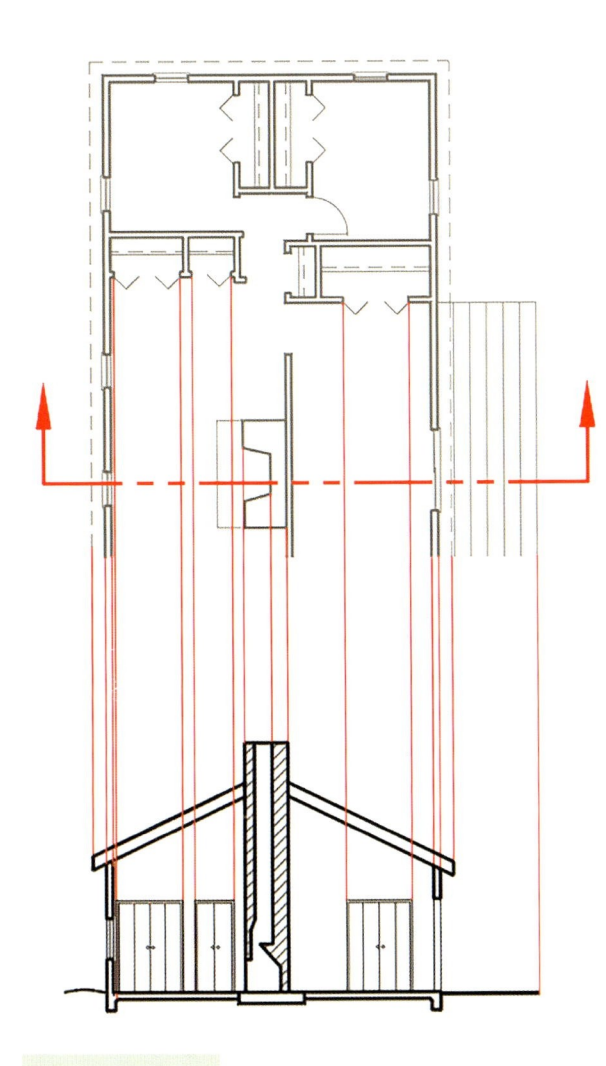

Figure 6.24
A rendered transverse section.
Courtesy wolfgangtrost.com

Figure 6.23
A transverse section shows the ceiling's angle and shape.

WHAT IS A SECTION DRAWING? 151

Figure 6.25
Material symbols in section and elevation.

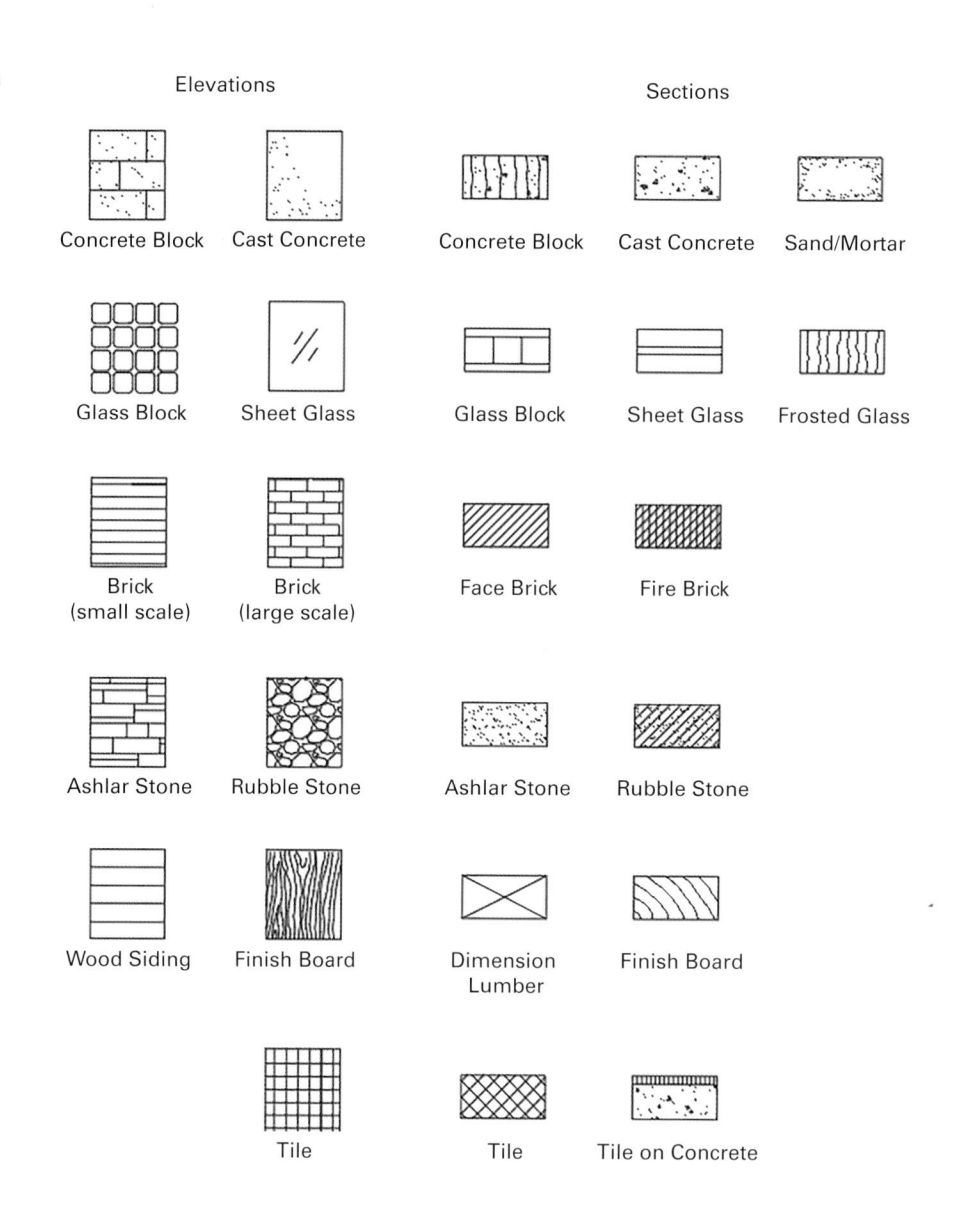

Elevations		Sections		
Concrete Block	Cast Concrete	Concrete Block	Cast Concrete	Sand/Mortar
Glass Block	Sheet Glass	Glass Block	Sheet Glass	Frosted Glass
Brick (small scale)	Brick (large scale)	Face Brick	Fire Brick	
Ashlar Stone	Rubble Stone	Ashlar Stone	Rubble Stone	
Wood Siding	Finish Board	Dimension Lumber	Finish Board	
	Tile	Tile	Tile on Concrete	

Hatch Lines Poché symbols are not the same as **hatch lines**. Hatch lines are 45° angled lines that simply indicate an object has been sliced. Space hatch lines consistently and don't draw them parallel or perpendicular to object lines. Figure 6.26 shows a technique for drafting consistently spaced lines.

Entourage This is people, décor, and accessories. Human scale figures are important in elevation and section views because they make heights immediately and visually readable. Beginning drafters often place windows, doors, and ceilings unrealistically low or high, a mistake easily avoided with the addition of a reference scale figure.

Find entourage figures by googling "architectural entourage figures." However, many you'll find are 3D so they'll need to be altered before being placed in a 2D drawing. A common alteration is to the feet. In a 3D drawing, feet are at different heights to indicate depth. But in an elevation or section drawing, both feet are level on the floor (Figure 6.27).

Entourage figures must match the drawing's style, color, and context, with detail corresponding to the detail of the space they're inserted into. Avoid details and colors that clash with or distract from the space. Simple figure outlines work well in simple line drawings but are inappropriate in highly detailed drawings.

Build an entourage library by taking photos. This enables the inclusion of items specific to a space. Such photos can be used in digital drawings or traced into hand-drafted ones. Figures 6.28 through 6.31 are

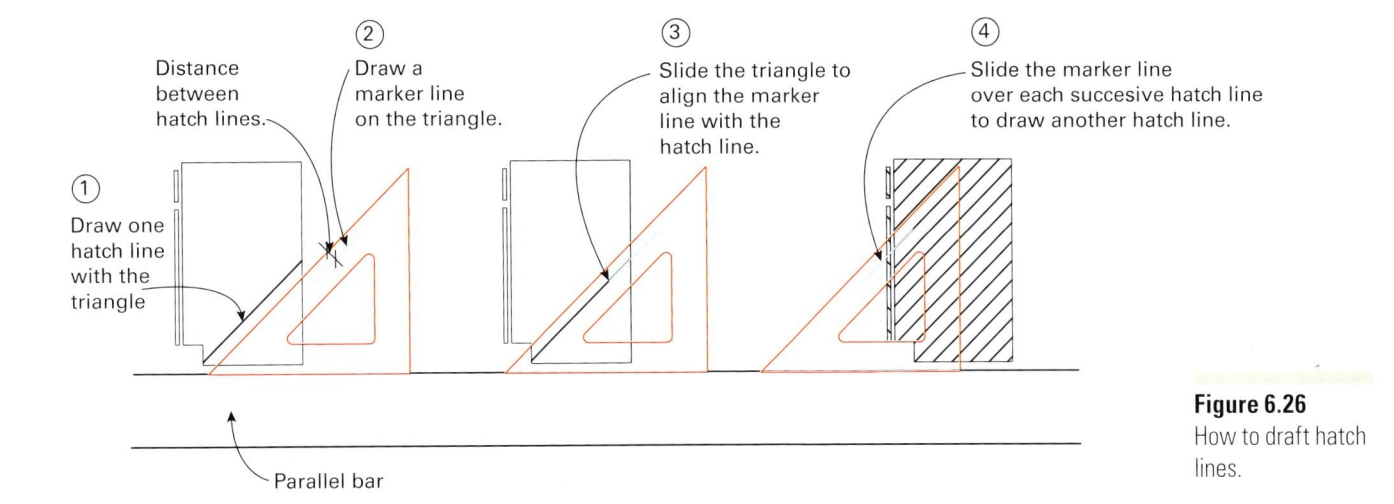

① Draw one hatch line with the triangle
② Draw a marker line on the triangle.
Distance between hatch lines.
③ Slide the triangle to align the marker line with the hatch line.
④ Slide the marker line over each successive hatch line to draw another hatch line.
Parallel bar

Figure 6.26
How to draft hatch lines.

Figure 6.27
3D entourage figures often need altering for use in a 2D drawing.

Feet are at different heights, indicating depth (3D)

Feet are level, indicating an orthographic view (2D)

Tip
Component heights are needed to draw elevation and sections. Obtain these heights from product catalogs, industry reference books, scaled entourage figures, and by measuring them yourself.

elevation entourage items scaled at 1/4" = 1'-0" for height reference and tracing. Figure 6.32 shows island and seat heights.

Sheet Layout

An elevation or section is drawn directly above or below the floor plan. However, it doesn't need to stay there in the final layout. All drawings should be arranged on a sheet in a balanced manner. Draw the elevation and plan on separate pieces of tracing paper to make arranging the final sheet easier. If multiple drawings are on the same piece of tracing paper, cut that paper into individual drawings and then arrange those drawings underneath the media they'll be traced on.

Orthographic views of small objects can be displayed on one sheet of paper with their edges aligned for easy reading, but the larger architectural drawings cannot. Typically, floor plans are drawn on one sheet, elevations on another, and sections on yet another. Always draw elevations and sections right side up, never perpendicular to other drawings on a sheet. Cross-reference all drawings with callout symbols and ID labels.

Exercise
Take photos of items for use as entourage to trace. Photograph them straight-on to make them as 2D as possible. If you can use a digital imaging program such as Photoshop, cut out the background and save for import into computer drawings.

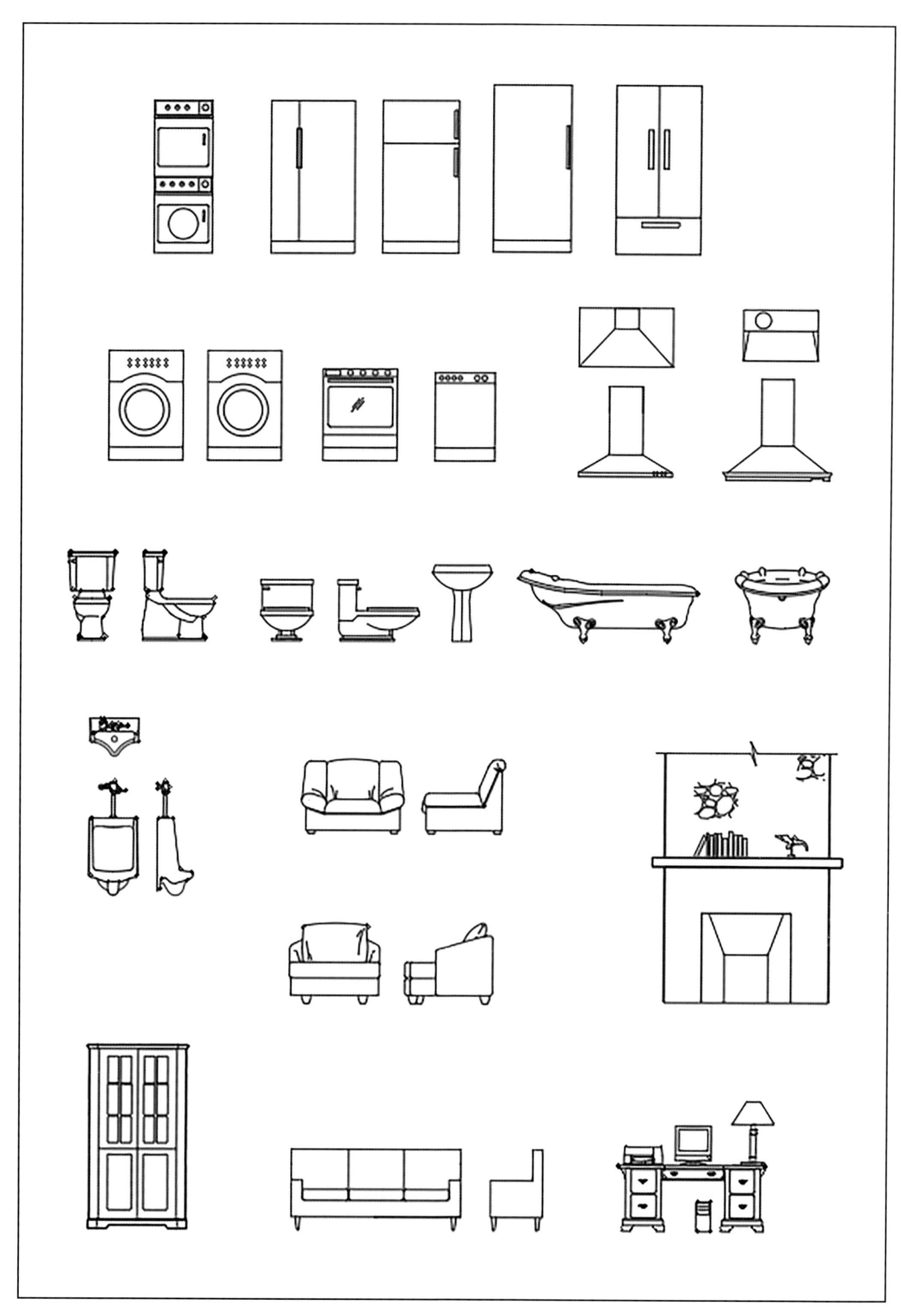

Figure 6.28 Entourage figures 1/4"=1'-0".

Back wall

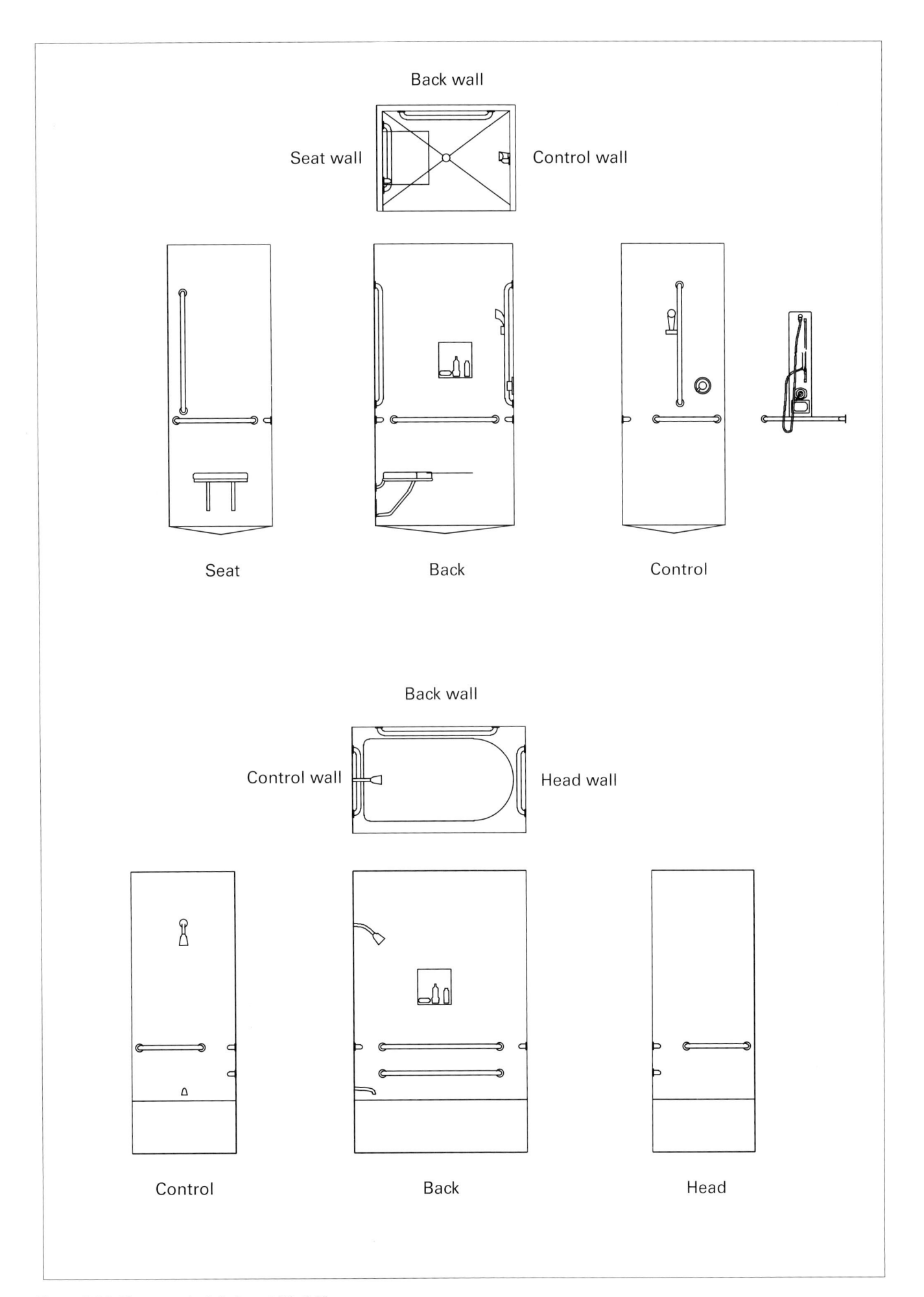

Seat wall   Control wall

Seat   Back   Control

Back wall

Control wall   Head wall

Control   Back   Head

Figure 6.29 Shower and tub fittings. 1/4"=1'-0".

Figure 6.30 Furniture. 1/4"=1'-0".

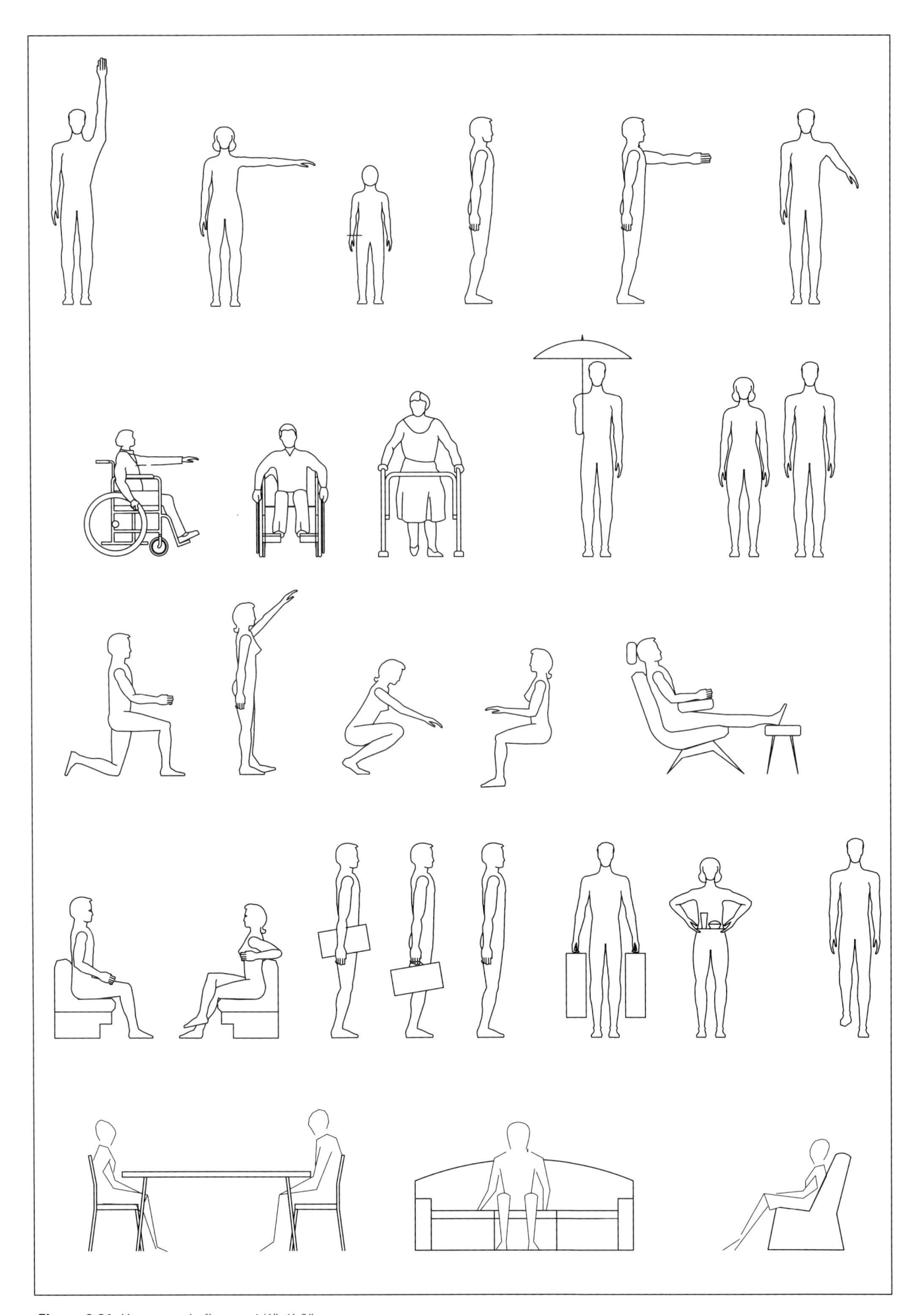

Figure 6.31 Human scale figures. 1/4"=1'-0".

SHEET LAYOUT 157

Figure 6.32
Island and seat heights. Not to scale.

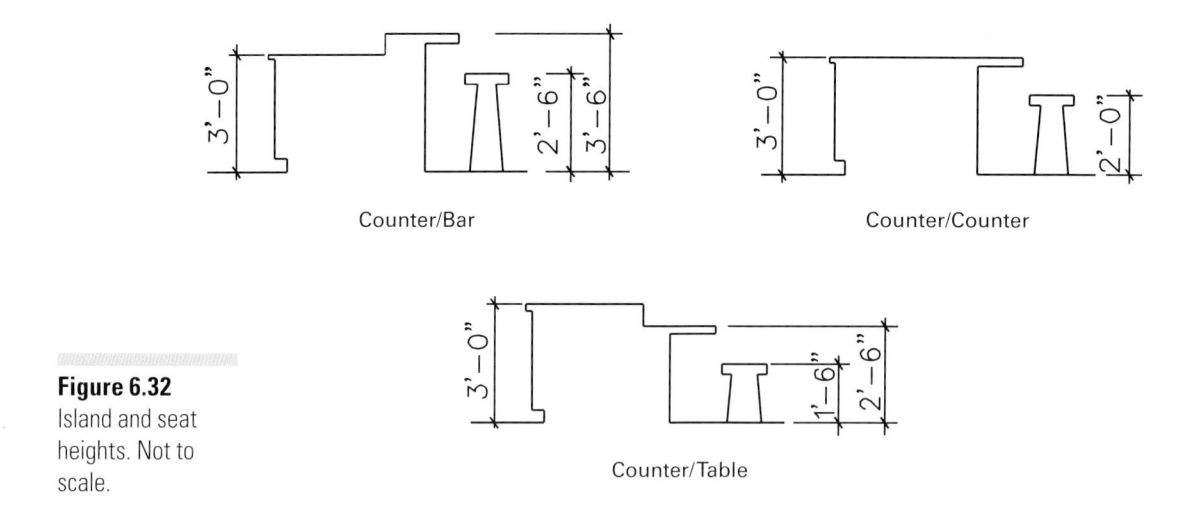

Counter/Bar

Counter/Counter

Counter/Table

Summary

An elevation view is a height drawing of one wall. A section view is a height drawing of multiple walls, made with a vertical slice through the whole building. Both drawings show architectural features and other items attached to the walls. They are referenced from the floor plan with callout and section line symbols. Materials poché is added to make the drawings more readable and entourage is added to provide visual scale.

Classroom Activities

1. Measure and draw interior elevations of walls at home and on campus. Draw elevations with both flat and non-flat ceilings.
2. Obtain a floor plan from a home décor magazine and select a cutting plane location. Overlay it with tracing paper and project lines down from it to sketch a section. Since magazine floor plans are not to scale, estimate appropriate heights or scan the drawings and scale them in SketchUp or AutoCAD.

Questions

1. What is an interior elevation?
2. What is a section view?
3. What is the difference between a transverse and longitudinal section?
4. What is a rake view?
5. How should a callout or section cut symbol be oriented on the drafting board before drafting the elevation or section?

Further Resources

Online entourage libraries: https://dwgmodels.com/, www.immediateentourage.com/, https://cad-blocks.net/people-cad-blocks.html, https://pimpmydrawing.com/

Keywords

- callout
- drawing number
- elevation view
- exterior elevation
- foreshortening
- front view
- grade line
- hatch lines
- head
- interior elevation
- jamb
- longitudinal section
- mark
- rake view
- rendered
- roof ridge
- section detail
- section view
- sheet number
- side view
- transverse section

CHAPTER 7

Dimensioning Floor Plans and Elevations

Figure 7.0
Dimensions on bathroom fixtures.
Courtesy American Standard.

OBJECTIVES

Upon completion of this chapter you will be able to:

- Identify and apply ANSI dimensioning standards
- Identify and apply NKBA drafting and dimensioning standards
- Apply dimension annotations to residential plans and elevations

Dimensioning is the process of adding measurements to a drawing. A floor plan's features—walls, columns, doors, windows, stairs, and other built-ins—must be correctly noted for construction.

Why Dimension Drawings?

A simple **annotation** (note) that gives a room's useable length and width is appropriate for a presentation drawing. But it is not enough to build from. Tradespeople need specific instructions tailored to the material they're building with. Architectural, cabinetry, mechanical, electrical, modular, and metric drawings all have different standards tailored to how they build and the materials they build with. Office practices may vary, too.

Dimension Notes Dimension notes are numbers describing the feature's size. Letter them in 1/8" tall guidelines. **Size** (how big something is) and **location** (where something is) are the intent of dimensions. Tell the reader how big the room is, how big its interior features are, and where those features are placed relative to other features. The amount of dimension notes needed depends on the plan's complexity and how much latitude the designer gives the contractor. Too many dimensions prohibit workers from making field adjustments; too few dimensions make workers guess where a feature is. When dimensioning a floor plan, analyze whether enough information has been given to allow workers to build from it.

> **Tip**
> When a drawing is copied, it becomes slightly distorted. So, don't obtain any measurements on them with a scale. Written dimensions override any measurement taken with a scale.

Dimension Lines Also called **stringers**, these are consecutive lines placed together to form a continuous string that defines a feature's length and height. **Extension lines** are lines perpendicular to the **dimension lines** that emanate from the endpoints of the feature. Both are drawn thinner than the wall lines. **Tick marks** are angular lines at the intersection of the dimension and extension lines (Figure 7.1); the dimension note describes the distance between two of them. All lines have a 1/8" overlap between them.

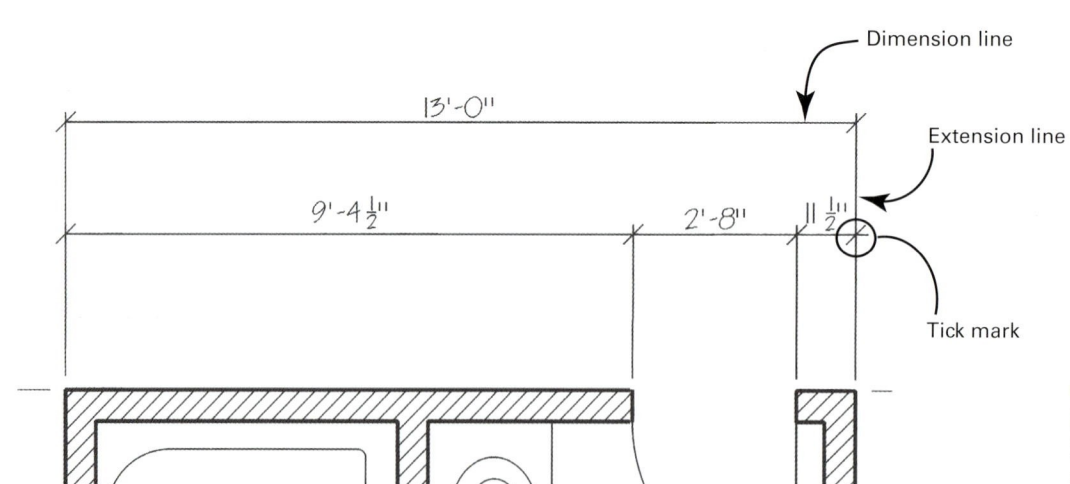

Figure 7.1 Components of a dimension line.

American National Standards Institute (ANSI) Conventions

Following are conventions for dimensioning **ANSI**-standard architectural design drawings.

Architectural dimensioning is hierarchical (Figure 7.2) Everything is dimensioned in decreasing size order. The farthest-away line is the overall dimension. Doors, windows, and interior walls go on a dimension line closer to the plan. Smaller details such as wall offsets, cabinetry, and plumbing fixtures go on a third dimension line closest to the plan.

162 CHAPTER 7 DIMENSIONING FLOOR PLANS AND ELEVATIONS

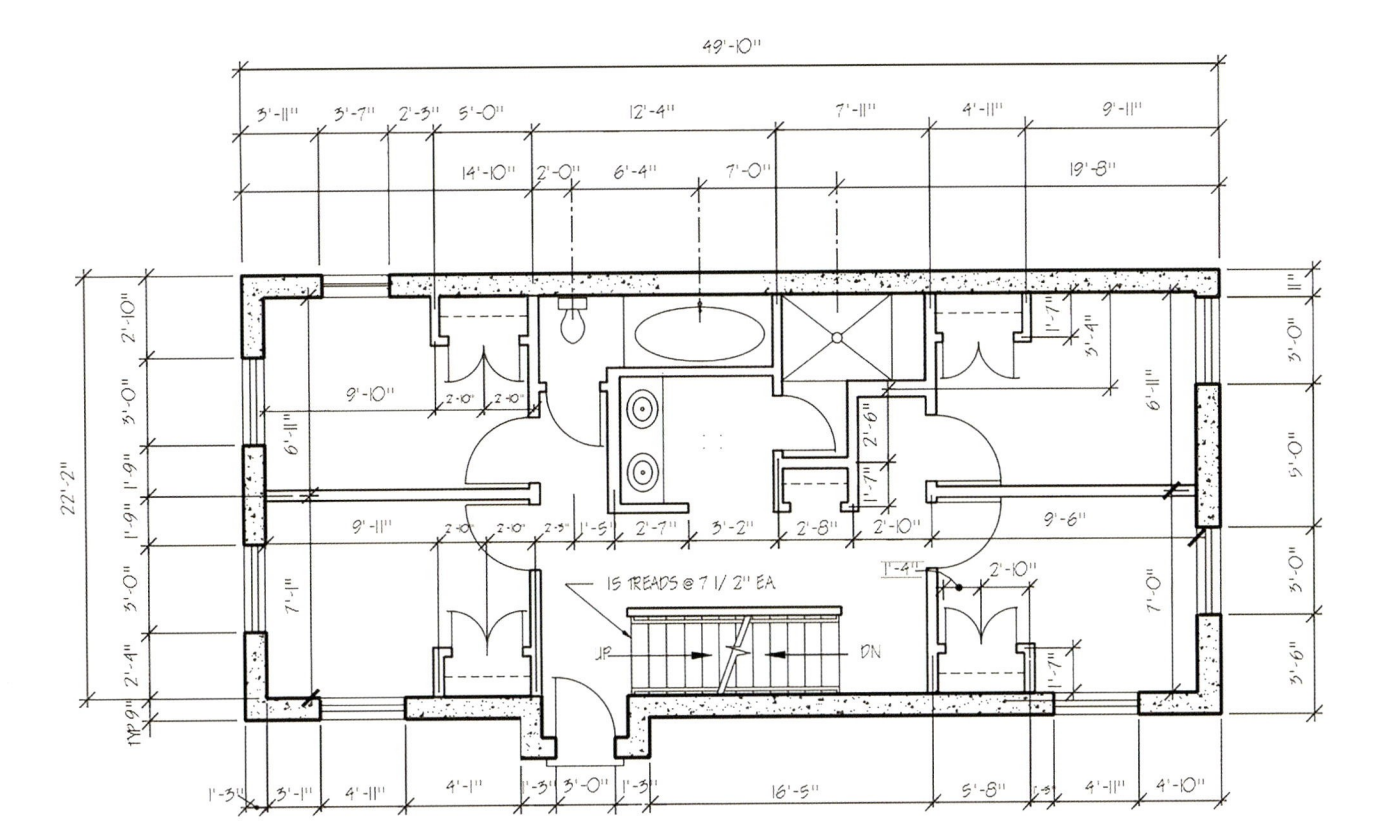

Place the first stringer 3/4" from the wall, and subsequent ones 1/2" apart. Adjust this spacing if needed. Place exterior dimension lines outside the plan and interior dimensions inside the plan. Place dimension lines close to the items they're dimensioning.

Align the dimension notes with the stringers (Figure 7.3). The notes are **aligned** with the stringers; that is, above/parallel to horizontal dimension lines, and left/parallel to vertical dimension lines (Figure 7.3). If you rotate the paper 90° to the right, the notes should read up. Center the dimension notes on the lines when possible, and letter them 1/8" above the stringers.

Angle all tick marks in the same direction. Some offices angle them left-to-right on vertical dimensions and right-to-left on horizontal dimensions.

When space is tight, pull dimension notes outside. Don't shrink notes to fit. You can place the dimension note outside the dimension line and point a leader line to it (Figure 7.4). Or abbreviate a dimension like this: 3° instead of 3'-0".

Figure 7.2
Place small features closest to the plan and the overall dimension farthest.

Tip
The overarching goal for dimensioning is accuracy, completeness, and a single, legible interpretation. Therefore, it is a drafter's discretion to place dimensions as needed to accommodate other drawing symbols.

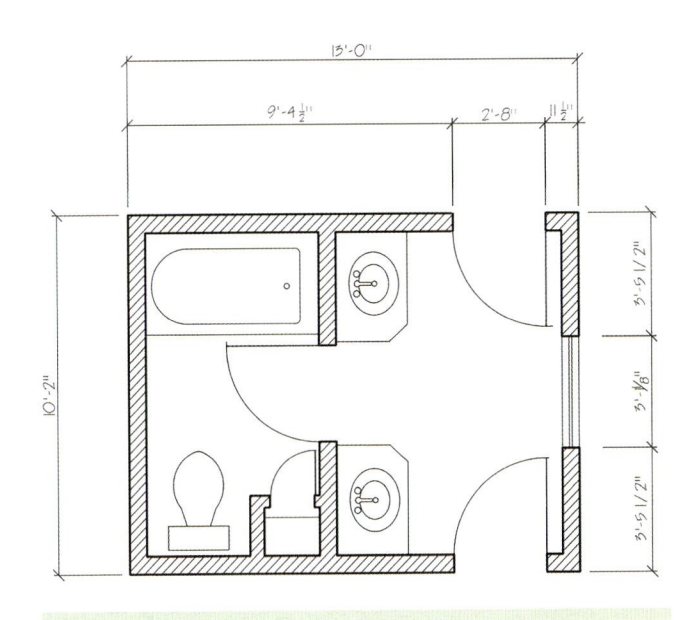

Figure 7.3
Place notes parallel to, above, or to the left of the stringers.

AMERICAN NATIONAL STANDARDS INSTITUTE (ANSI) CONVENTIONS

Write dimension notes in proper architectural format. If a number is smaller than one foot, place a 0'- in front of it (0'-3"). If the dimension is a whole foot number with no inches, place a -0" after it (3'-0").

Dimension circles and arcs from their center points (Figure 7.5). Locate the center point with two extension lines to the walls and denote the **radius** (distance from the center to one point on the circle) with R. Denote the **diameter** (distance from one point on the circle through the center to an opposite point on the circle) with D or the Ø symbol.

- *Dimension columns to their centers* This applies whether they're round or square. Place a centerline in both directions through a column. Fireplaces are dimensioned to their centers; masonry chimneys are dimensioned to their edges.
- *Don't repeat the same dimension on a plan.* However, that dimension may be shown on several different drawings.
- *Dimension individual rooms*, closets, halls, and wall thicknesses.
- *Provide some dimensions as notes.* For example, instead of dimensioning the walls around a shower, write 36" × 36" inside the shower.
- *Annotate components with their nominal, not actual, sizes.* For instance, 2" × 4" stud is a nominal dimension; the stud's actual size is 1 1/2" × 3 1/2". Annotate its nominal size.
- *Doors and windows are not dimensioned on plans.* Their sizes are on schedules.
- *Draw an enlargement box on a detailed portion* of the floor plan that refers the reader to a larger, more detailed plan.

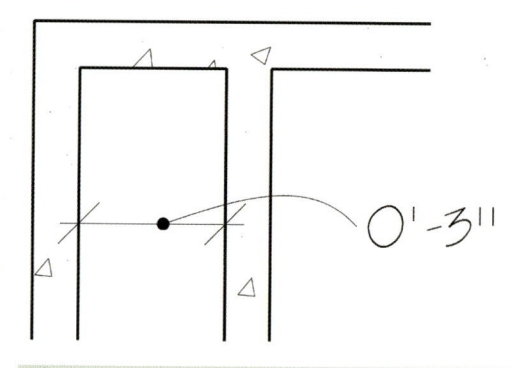

Figure 7.4
Place a dimension note outside the dimension line instead of making the note smaller.

Wood Frame vs. Masonry Dimensions

The conventions just discussed apply to all architectural design drawings. However, construction type also affects dimensioning. **Wood frame**, which is **milled lumber** (wood cut flat, square, and to specific sizes), and **masonry**, which is brick, concrete block, glass block or stone, are dimensioned differently. **Cast concrete**, which is concrete poured into forms, is dimensioned as masonry.

Wood Frame   Wood-framed buildings are dimensioned from the stud face on exterior walls and from the stud center on interior walls (Figure 7.6). This is because carpenters build the structure first and then apply the finishes. Doors and windows are also dimensioned to their centers. For ease of manual drafting, an alternative is to align the extension line with the corner and leave a small gap (Figure 7.7). However, the dimension note is interpreted the same as if the extension line were drawn to the stud face.

Masonry   Dimension masonry structures and concrete foundations to their edges. Extension lines run from corner to corner of the building, windows, and doors (Figure 7.8).

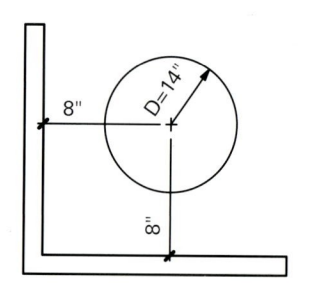

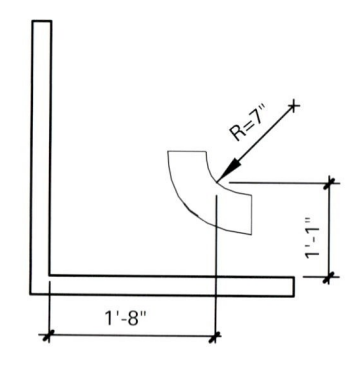

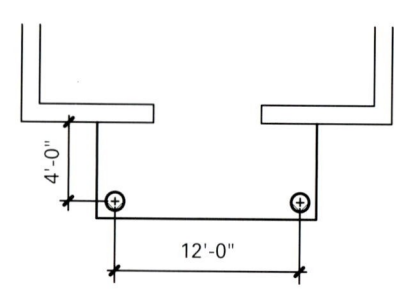

Figure 7.5
Dimension circles, arcs, and columns to their centers.

164   CHAPTER 7   DIMENSIONING FLOOR PLANS AND ELEVATIONS

Figure 7.6
Wood frame is dimensioned to the face and center of studs.

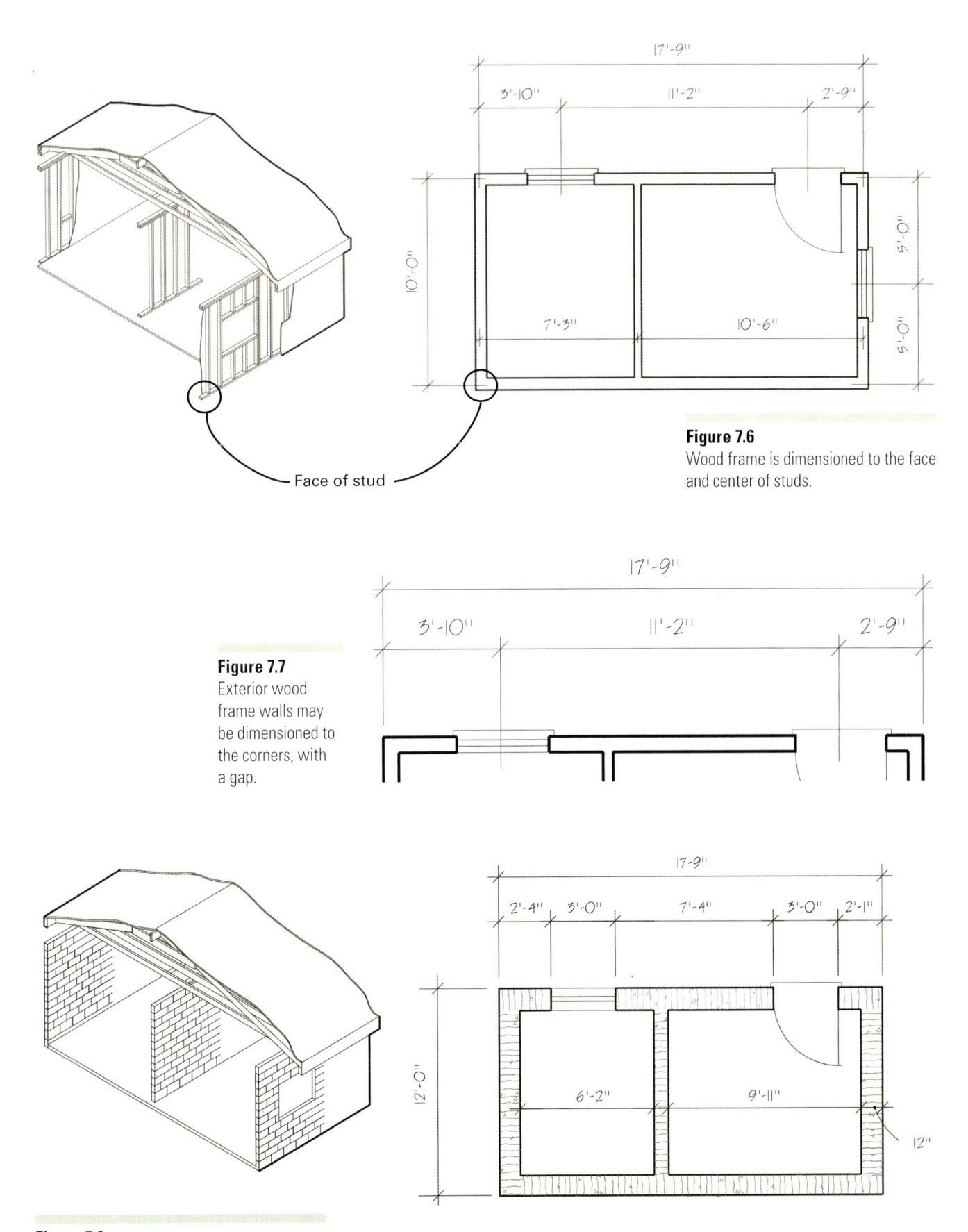

Face of stud

Figure 7.7
Exterior wood frame walls may be dimensioned to the corners, with a gap.

Figure 7.8
Dimension masonry structures to their edges.

WOOD FRAME VS. MASONRY DIMENSIONS

Masonry Veneer on Wood Frame   Dimension is the same as wood frame because the wood frame is the structural system. The veneer is simply added after it is built (Figure 7.9). Veneer thickness can be included with the dimensions or described with a note. Figure 7.10 shows a wood frame floor plan. Study it to see how exterior and interior dimensions are placed.

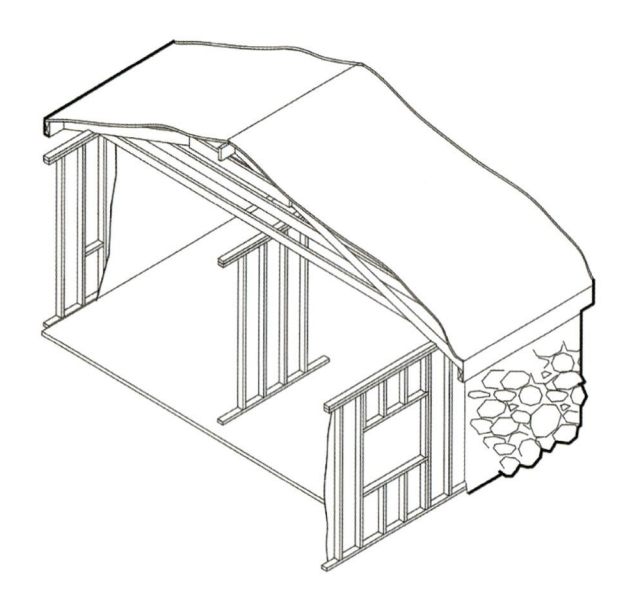

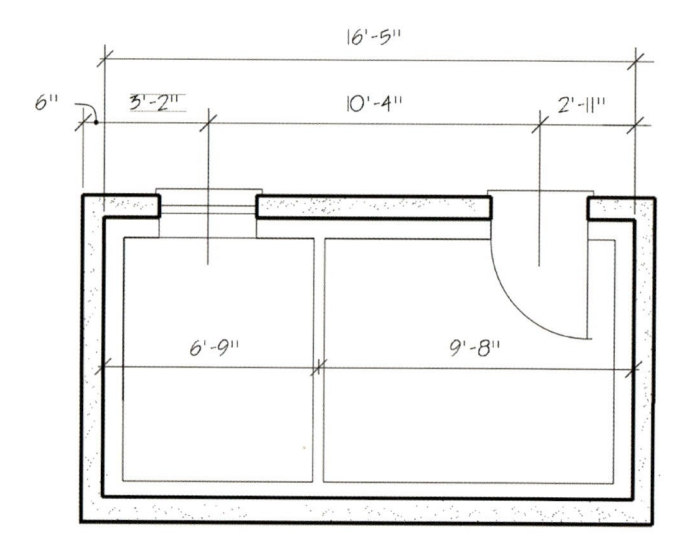

Figure 7.9
Wood frame with masonry veneer is dimensioned the same way as a wood frame.

Dimensioning an Interior Elevation

Figure 7.11 shows ANSI dimensions on a kitchen elevation. All vertical heights and distances from the floor are dimensioned. Include everything mounted on the walls: windows, cabinets, fixtures, appliances. Wall-mounted objects are dimensioned to their tops. A common abbreviation is **AFF**, which means above finished floor.

National Kitchen and Bath Association (NKBA) Drawings

This convention is used for **shop drawings**. A shop drawing is a highly detailed drawing derived from the design drawings and is both drafted and dimensioned differently from the NKBA standard. They're typically created by the contractor who will be constructing that job, such as a cabinet maker/installer.

Shop drawings show design features in more detail. For instance, design drawings show locations of cabinets and appliances, and their general appearance. Shop drawings show specifics of the cabinet construction and fixture control locations. They are drawn and dimensioned to give the cabinet and appliance installer the information needed to make everything fit. The scale is usually 1/2" = 1'-0" for plans and 1/2" = 1'-0" or 3/4" = 1'-0" for elevations.

NKBA Conventions for Floor Plans   The viewer's location is at the ceiling instead of 5' above the floor. Therefore, wall (upper) cabinets are drawn with a solid line; base (lower) cabinets are drawn with a hidden line, and the countertop edge that hides the base cabinet edge is a third, solid line. NKBA plans also include cabinet codes, a system of letters and numbers that describe **millwork** (woodwork). Figure 7.12 compares how ANSI and NKBA floor plans are drafted.

> **Exercise**
> Find and download a fully-modeled house from the SketchUp 3D Warehouse (https://3dwarehouse.sketchup.com), generate a plan view and dimension it.

> **Tip**
> If the project is of an existing space, obtain feature heights by measuring them yourself. You can also look up their sizes in product literature, on manufacturer websites and in designer reference books.

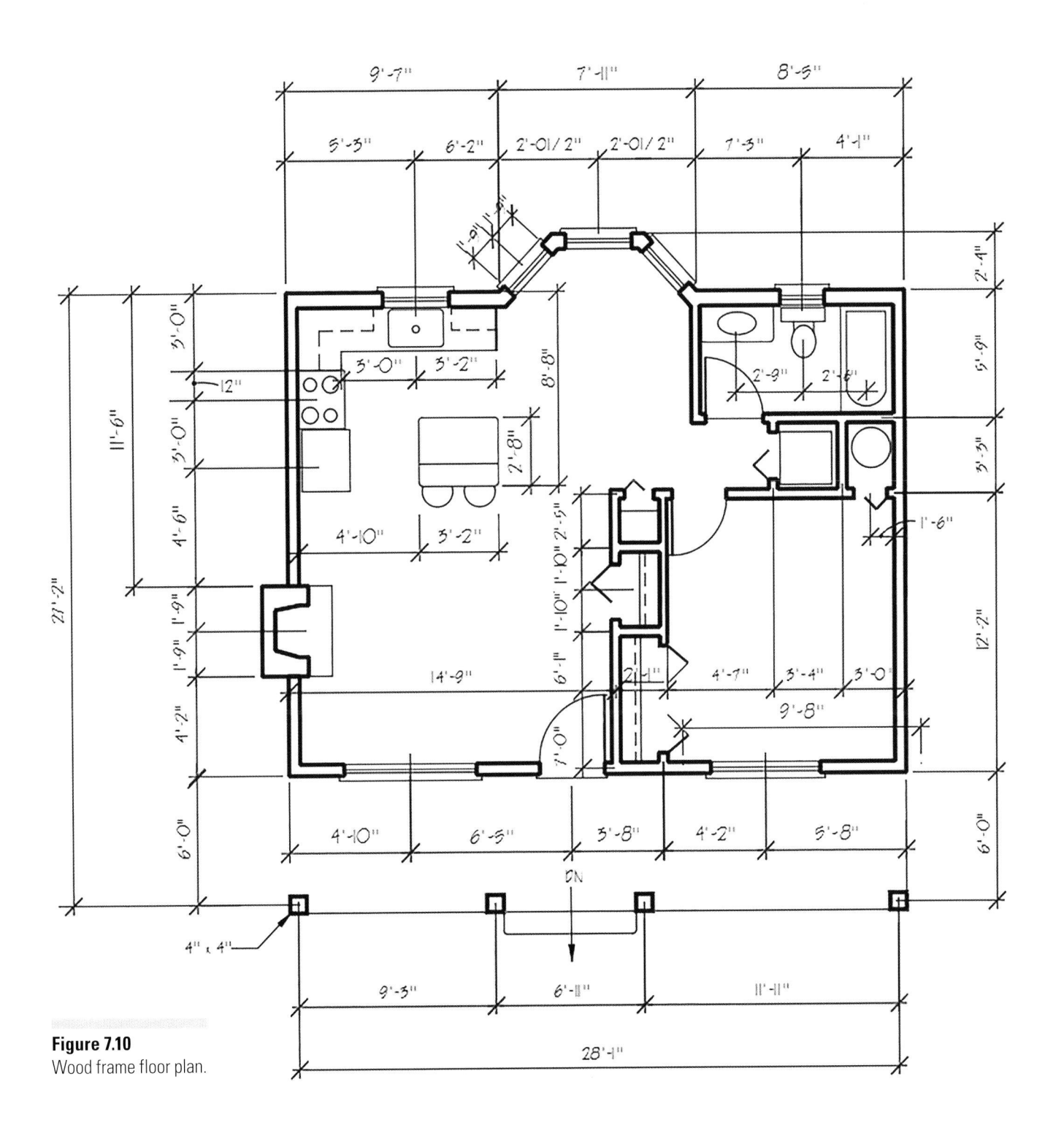

Figure 7.10
Wood frame floor plan.

NATIONAL KITCHEN AND BATH ASSOCIATION (NKBA) DRAWINGS **167**

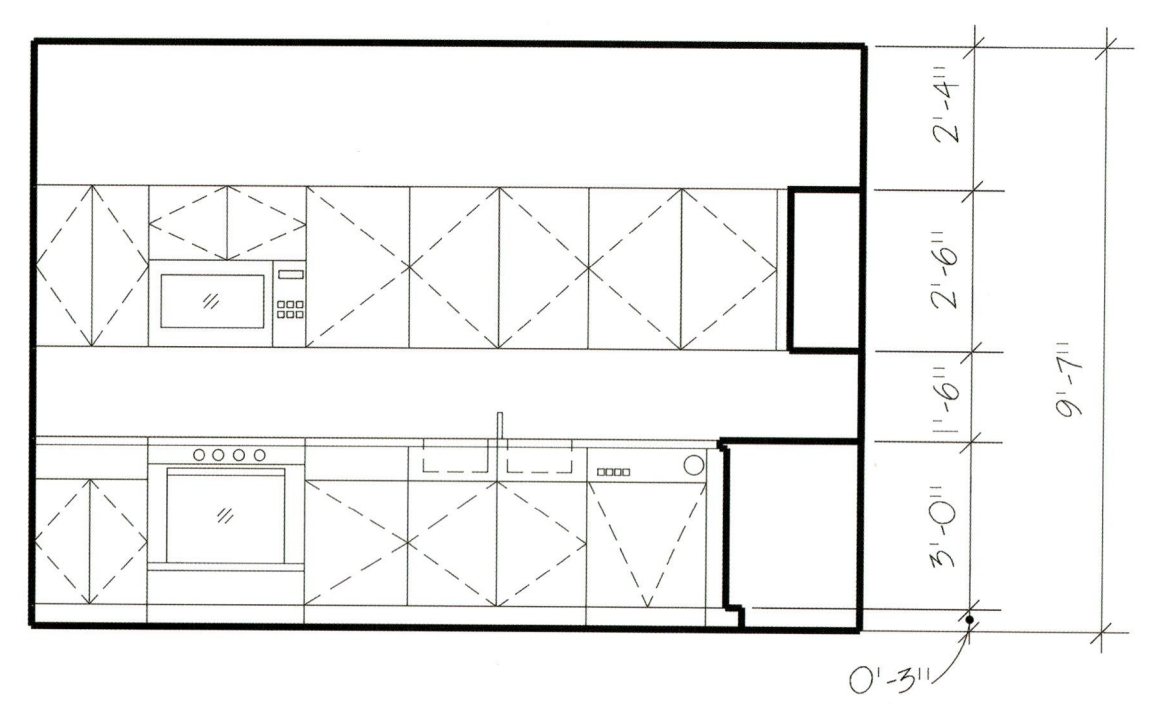

Figure 7.11 ANSI dimensioned kitchen elevation.

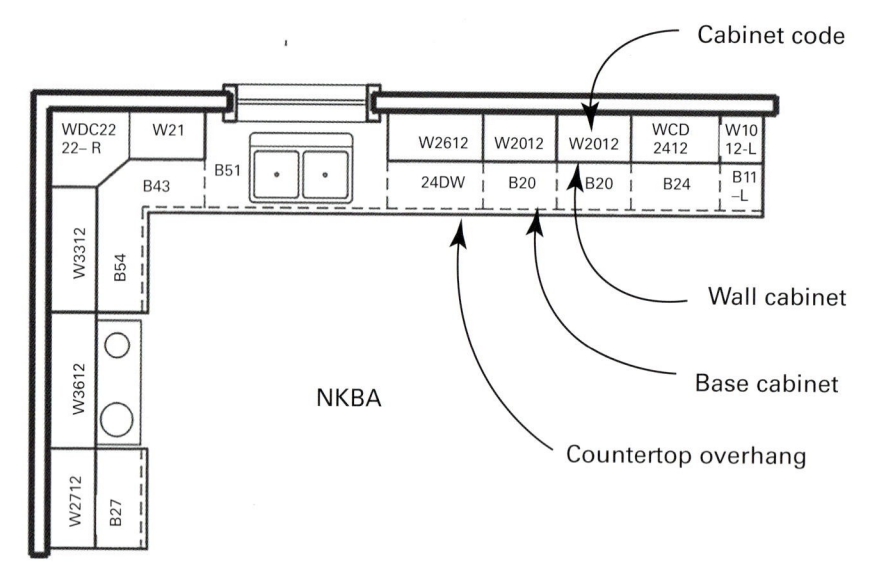

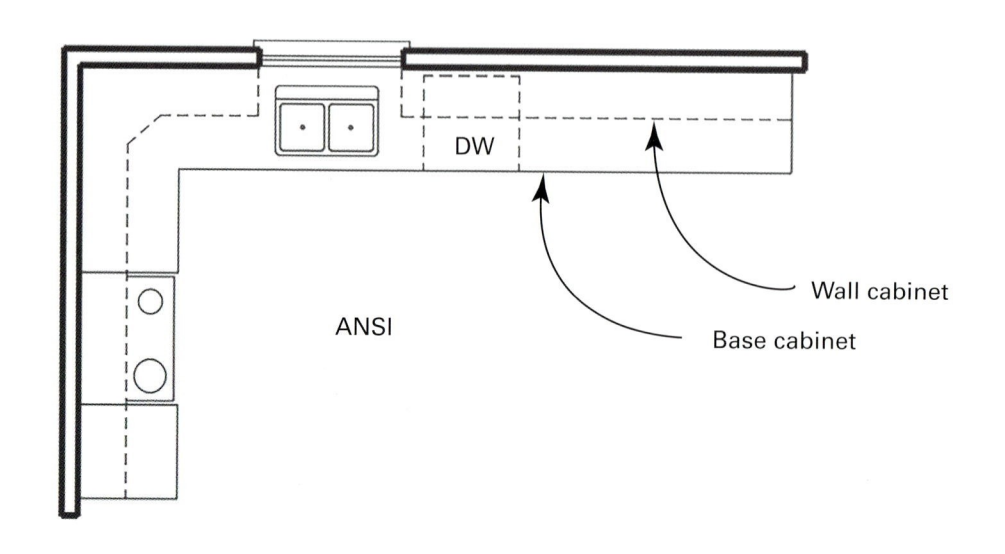

Figure 7.12 ANSI vs. NKBA drafted plans

168 CHAPTER 7 DIMENSIONING FLOOR PLANS AND ELEVATIONS

Following are conventions for dimensioning NKBA-standard plans.

There are three stringers. The first shows all opening sizes, and locates partition walls, columns, chimneys, and other fixed items. It dimensions adjacent cabinets as one "chunk," not individually. It also dimensions the **wall** (upper), not **base** (lower) **cabinets**. The second stringer locates centers of appliances. The third stringer shows the room's overall length. All are spaced 1/2" apart, with the first placed 1/2" from the wall. The extension line for the center of appliance is a center line-type with abbreviation **CL** (Figure 7.13).

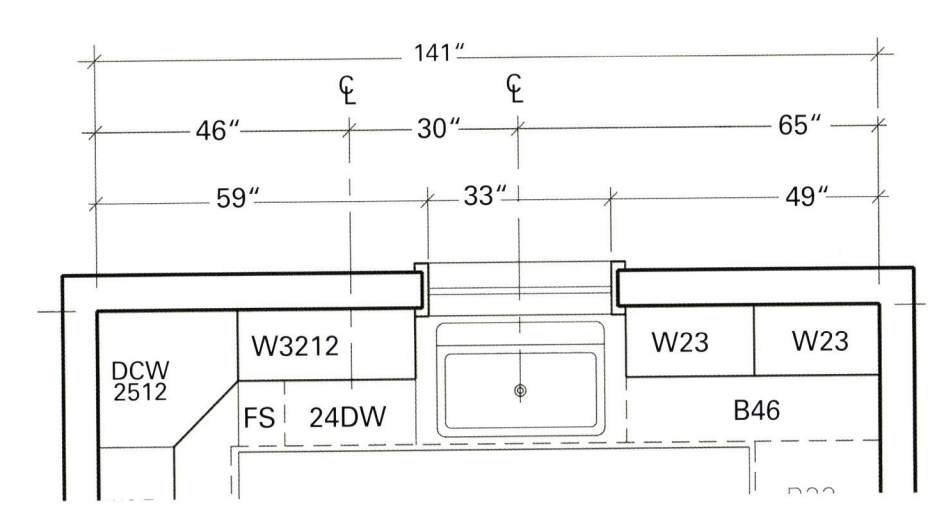

Figure 7.13
The plan has three stringers spaced 1/2" apart.

Dimension islands and peninsula cabinets to two walls and from their countertop edges. Their exact location is noted with dimension lines running in two directions to adjacent walls or to the faces of cabinets and appliances opposite them (Figure 7.14).

All numbers are in inches. In metric, use centimeters. Dimension notes are **unidirectional** (they read straight up) and placed inside the stringers. Note height from finished floor to ceiling on the plan. Dimension openings for specific cabinets or appliances with the appliance's exact height, width, and depth. Obtain that information from the manufacturer's website.

Build **clearances** *into dimensions.* These are distances between the appliance and the opening it fits into. Download installation instructions from the factory website on the appliances chosen. Not all appliances of a specified width are the same. Some are smaller than their specified width; for instance, a 30" stove may actually be 29 7/8"; the designation "30" is not the actual appliance width but the recommended opening dimension. A 36"-wide refrigerator might require a 36 1/2" opening; a 30" range might require a 30 1/4" opening. Drop-ins, slide-ins, pop-up vents, double ovens, freestanding, overlaid, inset, microwave trim kits, even different handle options for a refrigerator all require different trim panel configurations.

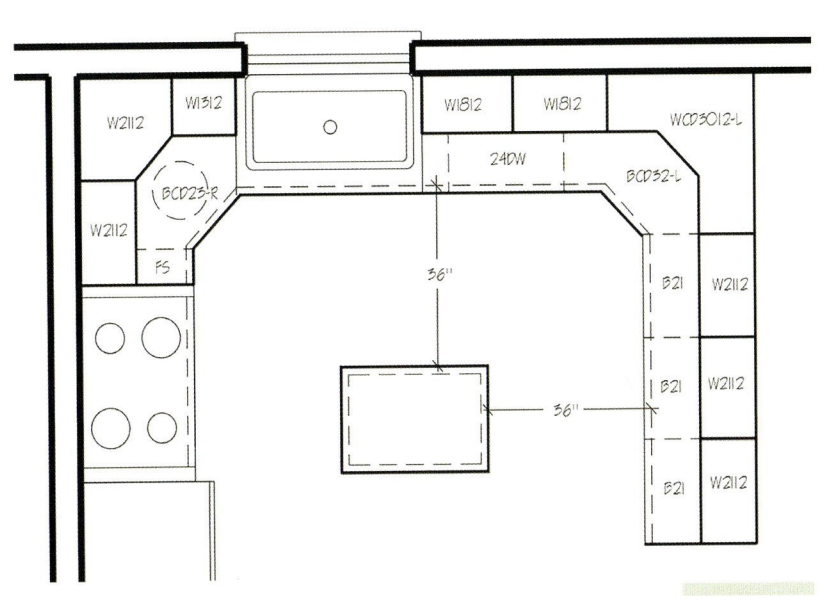

Figure 7.14
Dimension an island to two walls.

Refrigerator openings often have baseboard at the bottom and edge banding on the countertop, so a larger opening may be needed to slide the fridge in and open its doors all the way. Since many refrigerators don't have zero-clearance doors, manufacturer's specifications for the refrigerator chosen must be used. Cabinets cannot be competently built when guessing at appliance dimensions. Therefore, get all appliance specifications early and design to them.

Key special features to a legend (chart of information) Pull-out shelves, plate partitions, wine racks and such are not drawn; rather, they are keyed to a legend via a call-out placed inside that cabinet (Figure 7.15). **Toe kicks**, the indented spaces at the bottom of cabinets, are also not drawn. Window and door openings include trim, which is needed for design and installation calculations.

> **Exercise**
> Measure the width of an appliance in your own home and the width of the opening it is encased in. Calculate the clearance.

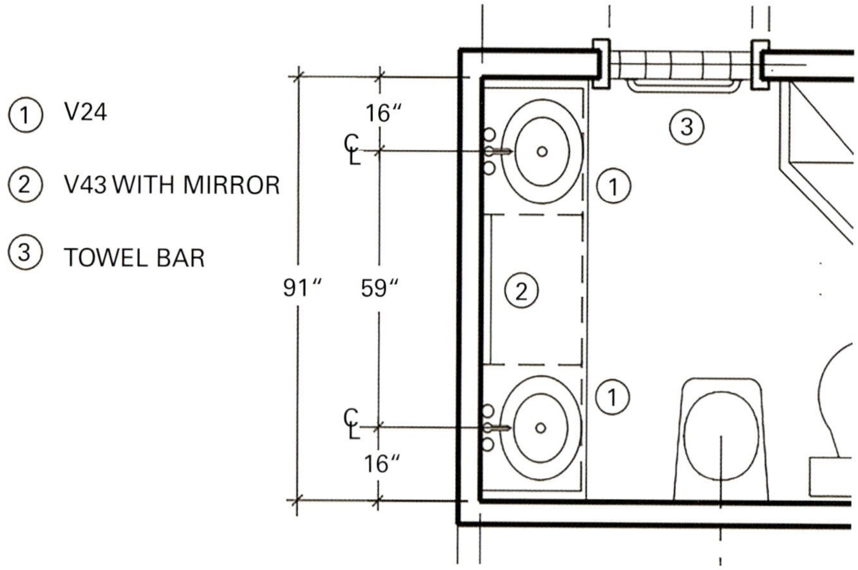

① V24

② V43 WITH MIRROR

③ TOWEL BAR

Figure 7.15
Callouts key special features to a legend.

Each cabinet has a code. These are number and letter combinations with size and detail information. For example, in W1224-L, the W means wall cabinet; it is 12" wide, 24" tall, and has one door hinged on the left (as you face it). B15-R means base cabinet; it is 15" wide and its one door is hinged on the right side (as you face it). If no L or R is noted, the cabinet has two doors that open opposite each other. Height isn't included in the base cabinet code because cabinets are standardized. A base cabinet is 34 1/2" high and the countertop is 1 1/2" thick, for a total height of 36". Depths are not included in cabinet codes either; a base cabinet is 24" deep and a wall cabinet is 12" deep. If a wall cabinet is customized to be deeper than 12", the depth is added at the end of the code; for example, W243018 means the cabinet is 18" deep.

Codes also describe millwork items such as pull-out shelves, wine racks, plate partitions, and bread boxes and such. Table 7.1 shows codes for common cabinetry items. They may differ between manufacturers. Figures 7.16 and 7.17 show kitchen and bathroom floor plans drawn and dimensioned to NKBA standards.

Table 7.1 CODES FOR COMMON CABINETRY ITEMS	
Code	**Millwork**
AB	Base End Angle Cabinet
AC or AG	Appliance Garage
B	Base Cabinet
BC	Base Corner
BEC or BEA	Base End Angle Cabinet
BEP or EP	Base End Panel

CHAPTER 7 DIMENSIONING FLOOR PLANS AND ELEVATIONS

Table 7.1
CODES FOR COMMON CABINETRY ITEMS

Code	Millwork
BES	Base End Shelf
BF	Base Filler
BFH or FHD	Base Full Height Door
BFP	Base Filler Pullout
BM	Base Molding
BP	Island Panel
BPP	Base Pantry Pullout
BWB	Base Cabinet with Waste Basket
BWR	Base Wine Rack Cabinet
CSB	Corner Sink Base Cabinet
DB	Drawer Base Cabinet
DBEP, BFD, or BDD	Base Decorative End Panel
DC, WCA, or WDC	Wall Diagonal Corner Cabinet
DD	Wall Decorative End Panel
DW, DEP, or BEP	Dishwasher Panel
FF	Fluted Filler
FS	Farm Sink
LW or WR	Lattice Wall or Wine Rack Cabinet
MCROWN or CM	Crown Molding
MLR or LR	Light Rail Molding
MOC or OC	Outside Corner Molding
MQR or QR	Quarter Round Molding
MWC, WMSC, or MC	Microwave Cabinet
NCSB	Diagonal Corner Sink
OC	Oven Cabinet
PB	Peninsula Base Cabinet
PC or WP	Wall Pantry Cabinet
PL	Plate Rack Cabinet
PW	Peninsula Wall (with doors on both sides of cabinet)
QRM	Quarter Round Molding
RP or REP	Refrigerator Panel
RW, WR	Wall Refrigerator Cabinet (double deep; example code is W362424)
SB	Sink Base Cabinet
SC or WBC	Wall Blind Corner Cabinet
SCB, BBI or BBC	Blind Base Cabinet
SCER, BRER, EZR, BLS, LS, or DCB	Lazy Susan Cabinet
SDC	Spice Drawer
SGH, GR, or GH	Stem Glass Holder
SM or SCM	Scribe Molding
TF	Tall Filler
TK	Toe Kick
VA	Vanity

(continued)

Table 7.1
CODES FOR COMMON CABINETRY ITEMS

Code	Millwork
VA60DD	Vanity 60" Double Sink
VDB	Vanity Drawer Base
W	Wall Cabinet
WCD, PL, PR	Plate Rack Cabinet or Wall China
WEA	Wall End Angle Cabinet
WEC	Wall End Cabinet
WEOS or WES	Wall End Shelf Cabinet
WEP	Wall End Panel
WER, DCR, WEZR	Wall Easy Reach or Wall Lazy Susan
WF	Wall Filler
WM	Wall Microwave Cabinet

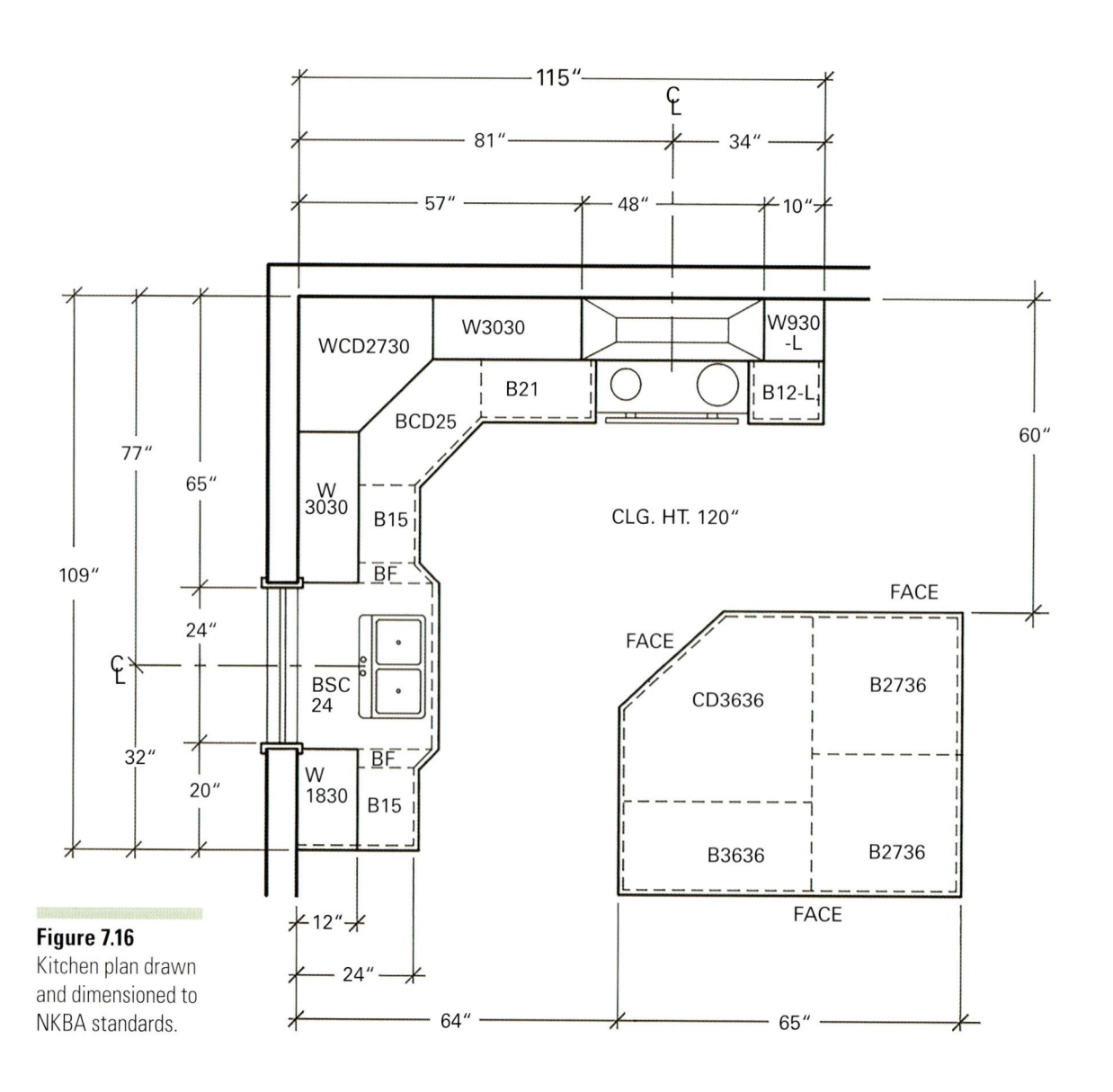

Figure 7.16
Kitchen plan drawn and dimensioned to NKBA standards.

172 CHAPTER 7 DIMENSIONING FLOOR PLANS AND ELEVATIONS

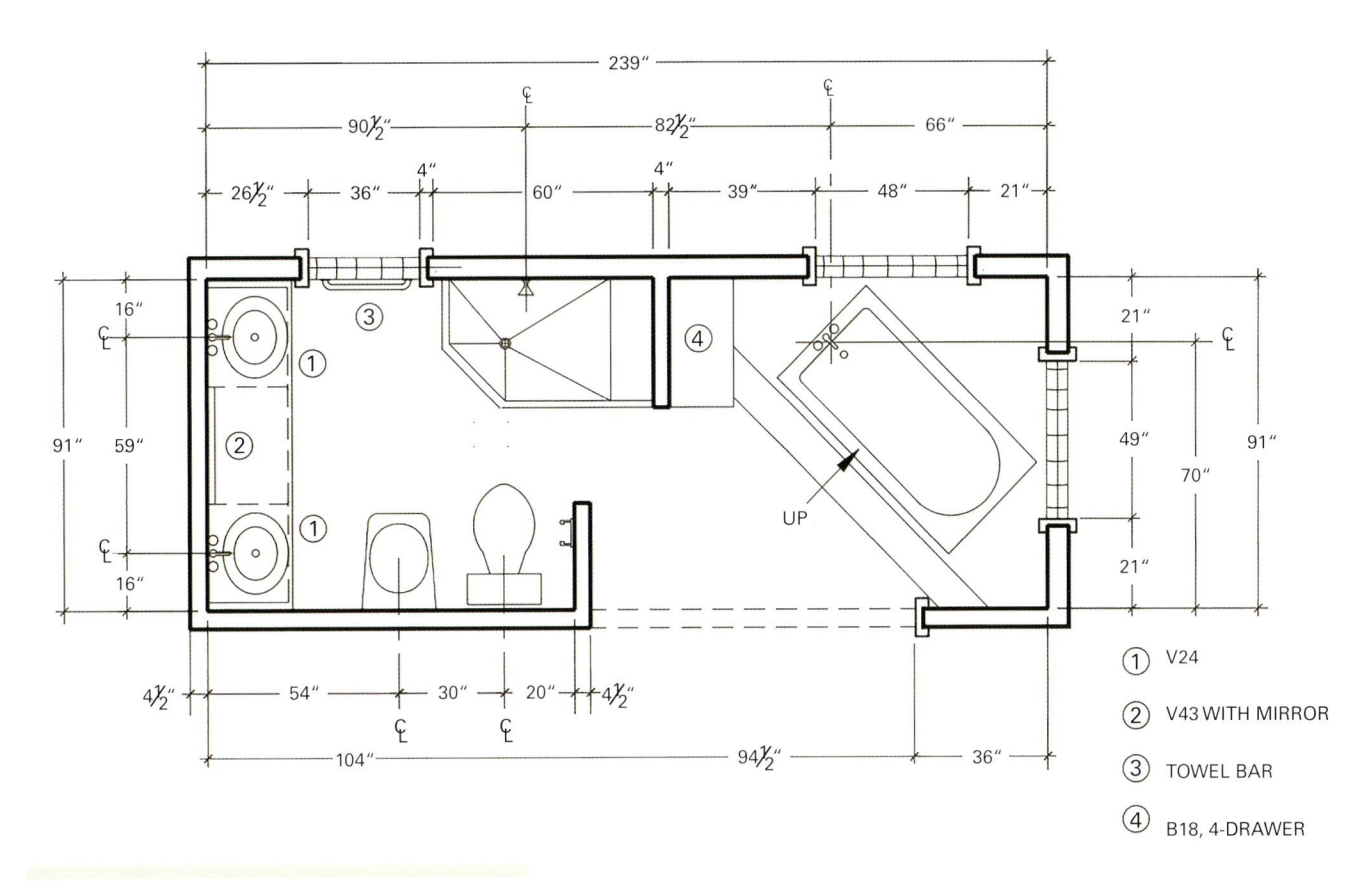

Figure 7.17
Bath plan drawn and dimensioned to NKBA standards.

NKBA Conventions for Elevations   When dimensioning elevations:

Include trim (baseboard, chair rail, casing) on doors and windows in the overall dimensions. Wall openings are measured from their outside trim.

Show a portion of doors and drawers to indicate the style and handle placement.

Dimension cabinets with the toe kick and finished height.

Show countertop thickness and backsplash height.

Figures 7.18 and 7.19 show NKBA dimension placement on elevation drawings.

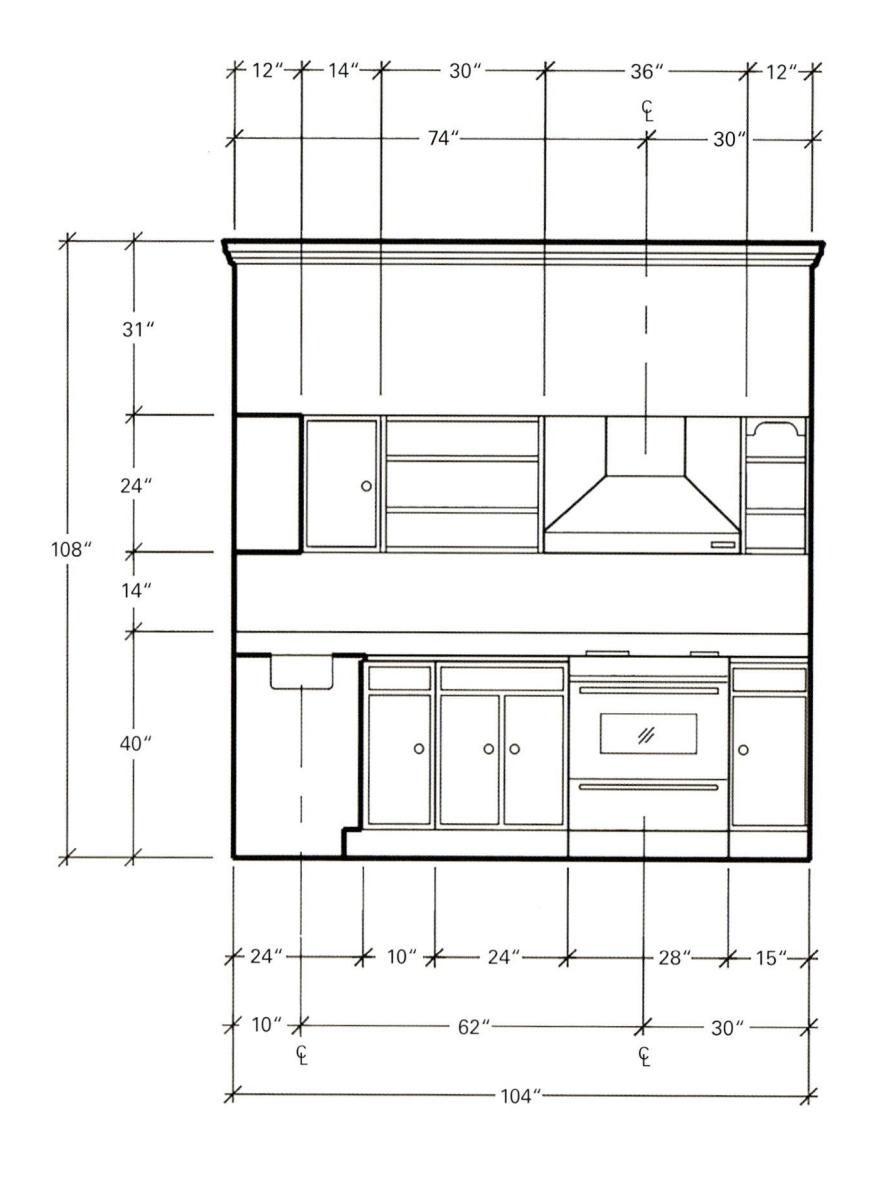

Figure 7.18
An NKBA dimensioned kitchen elevation.

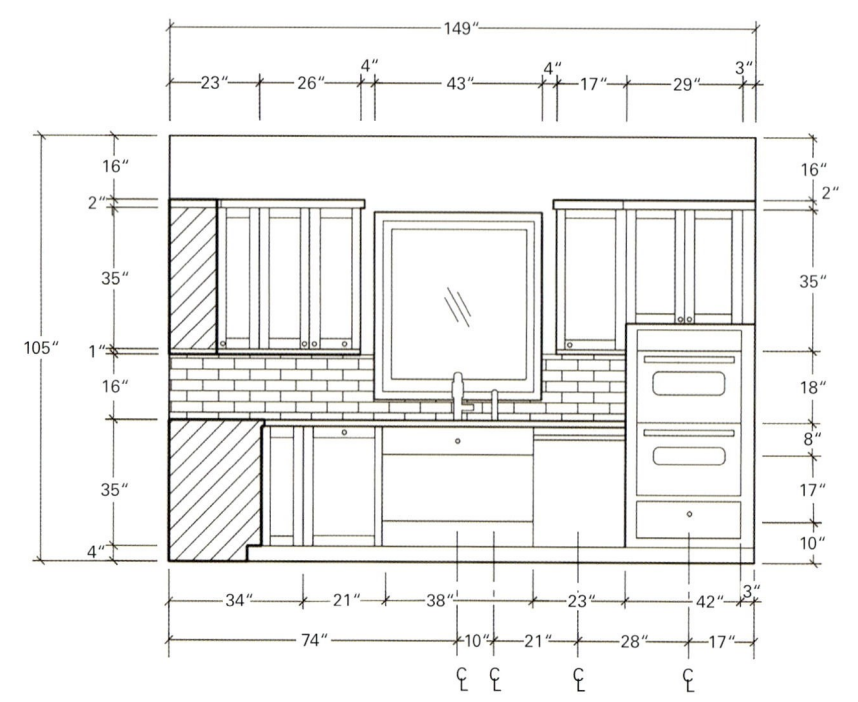

Figure 7.19
An NKBA dimensioned kitchen elevation.

174 CHAPTER 7 DIMENSIONING FLOOR PLANS AND ELEVATIONS

Section and Elevation Views of Cabinets

Figure 7.20 shows a section view through wall and base cabinets and a pictorial to help visualize what the view is describing. Everything that the cutting plane touches is hatched with 45° angle lines. The elevation views in Figure 7.21 of wall and base cabinets include standard dimensions.

Calculating Area Once you have a set of dimensioned drawings, you can calculate **area** (square footage). This is needed for cost estimates and calculating quantities of materials. Figure 7.22 shows basic shapes and formulae to calculate their areas. If a shape is irregular, split it into recognizable shapes, calculate each area separately, then add the products together. If you can't break an area into common shapes, sketch the figure to scale on a grid and count the number of squares it takes up. Then multiply that number by the grid square's area.

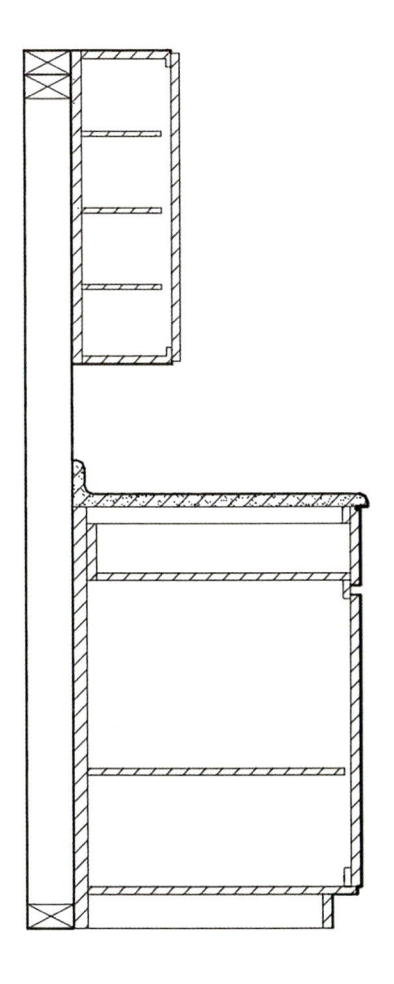

Figure 7.20
A section view through wall and base cabinets.

Figure 7.21
Cabinet elevation views and standard dimensions.

Front

Side

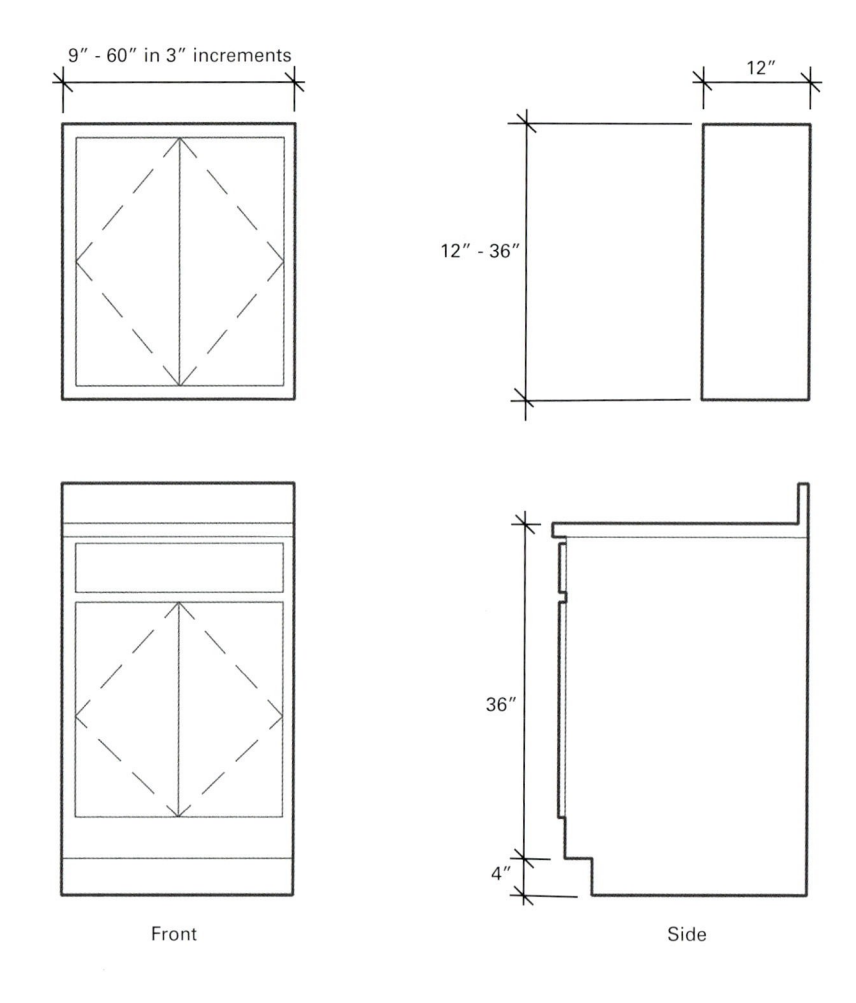

Figure 7.22
Formulae for calculating area. Break irregular shapes into basic ones.

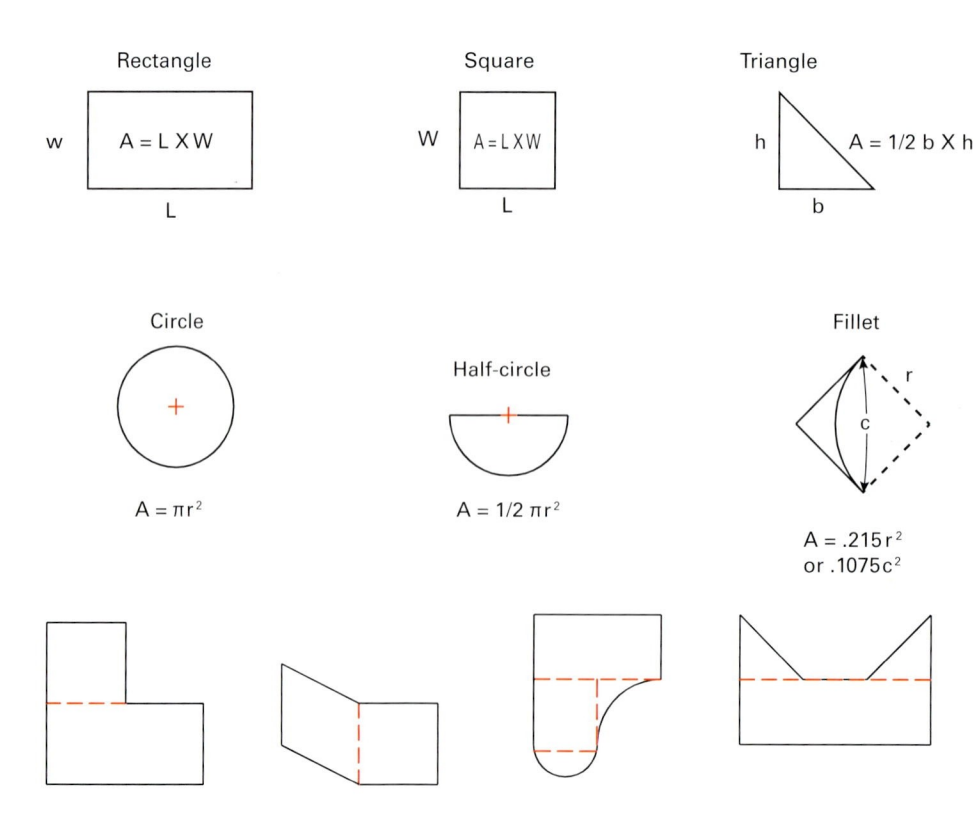

Rectangle
$A = L \times W$

Square
$A = L \times W$

Triangle
$A = 1/2\, b \times h$

Circle
$A = \pi r^2$

Half-circle
$A = 1/2\, \pi r^2$

Fillet
$A = .215 r^2$
or $.1075 c^2$

Misc. Shapes

176   CHAPTER 7   DIMENSIONING FLOOR PLANS AND ELEVATIONS

Summary

Dimensioning is the process of adding measurements to a drawing that describe the size and location of architectural spaces and features. The ANSI standard is used for architectural design drawings and the NKBA standard is used for more detailed kitchen and bath plans and millwork. Differently constructed buildings are also dimensioned differently. Designers must carefully add dimensions and notes to their drawings so that contractors can correctly and competently build them.

Classroom Activities

1. Obtain some floor plans from a magazine or other source and sketch extension and dimension lines on them.
2. Download appliance sheets from manufacturer websites and compare the actual sizes for appliances that are the same nominal size.
3. Measure, draw, and dimension your kitchen to NKBA standards.
4. Measure, draw, and dimension your classroom to AIA standards.

Questions

1. What two features of architectural spaces are dimensioned?
2. What construction type shows openings dimensioned to their centers?
3. True or False: Plumbing fixtures are dimensioned on a different stringer than cabinets.
4. What does the cabinet code "W1215-L" mean?
5. How are special features described on a drawing?

Keywords

- AFF (above finished floor)
- aligned
- American National Standards Institute (ANSI)
- annotation
- area
- base cabinet
- cast concrete
- CL (center line)
- clearance
- diameter
- dimension line
- dimension note
- dimensioning
- extension line
- location
- masonry
- milled lumber
- millwork
- radius
- shop drawing
- size
- stringer
- tick mark
- toe kick
- unidirectional
- wall cabinet
- wood frame

CHAPTER 8

Door and Window Symbols

Figure 8.0
Swing door flanked by sidelights, with transom and clerestory windows above. Courtesy Hy-Lite.

OBJECTIVES

Upon completion of this chapter you will be able to:

- Define door and window types
- Identify door and window symbols
- Analyze which type of door or window is appropriate for different locations
- Draft door and window symbols in plan and elevation views

Doors and windows are shown on floor plans, elevations, and section drawings. Each type has its own symbol.

What Do Door and Window Symbols Show?

These symbols represent hinge type—whether the components **swing** (push open from either side), **slide** (glide parallel to the wall), **fold** (creases in sections), or **spin** (revolve around a center point). The symbol also shows if they are **active**, **passive**, or **stationary**. The active panel is the hinged (operating) one. The passive panel may be opened only after the active panel is opened. A stationary panel is fixed; it doesn't move at all. All door and window frames have two **jambs** (sides), a **head** (top), and a **sill** (bottom).

Door and window construction and materials are described in detail drawings, specifications, and schedules. The arrangement of doors and windows in a building is called **fenestration**.

Door Types

Here are the different door types.

Swing (Figure 8.1) A swing door is hinged on one side and swings in or out of the room. Most open 90° to the wall; some open 180° to lie against the wall. Exterior doors have a **threshold** (the bottom of the door frame) symbol on the exterior side of the wall. Interior doors usually don't; exceptions are thresholds in commercial building restrooms. The dashed line in the elevation view is the hinge symbol; the apex points to the side the hinge is on and is drawn in the middle of the jamb and to the door's corners (not to the frame).

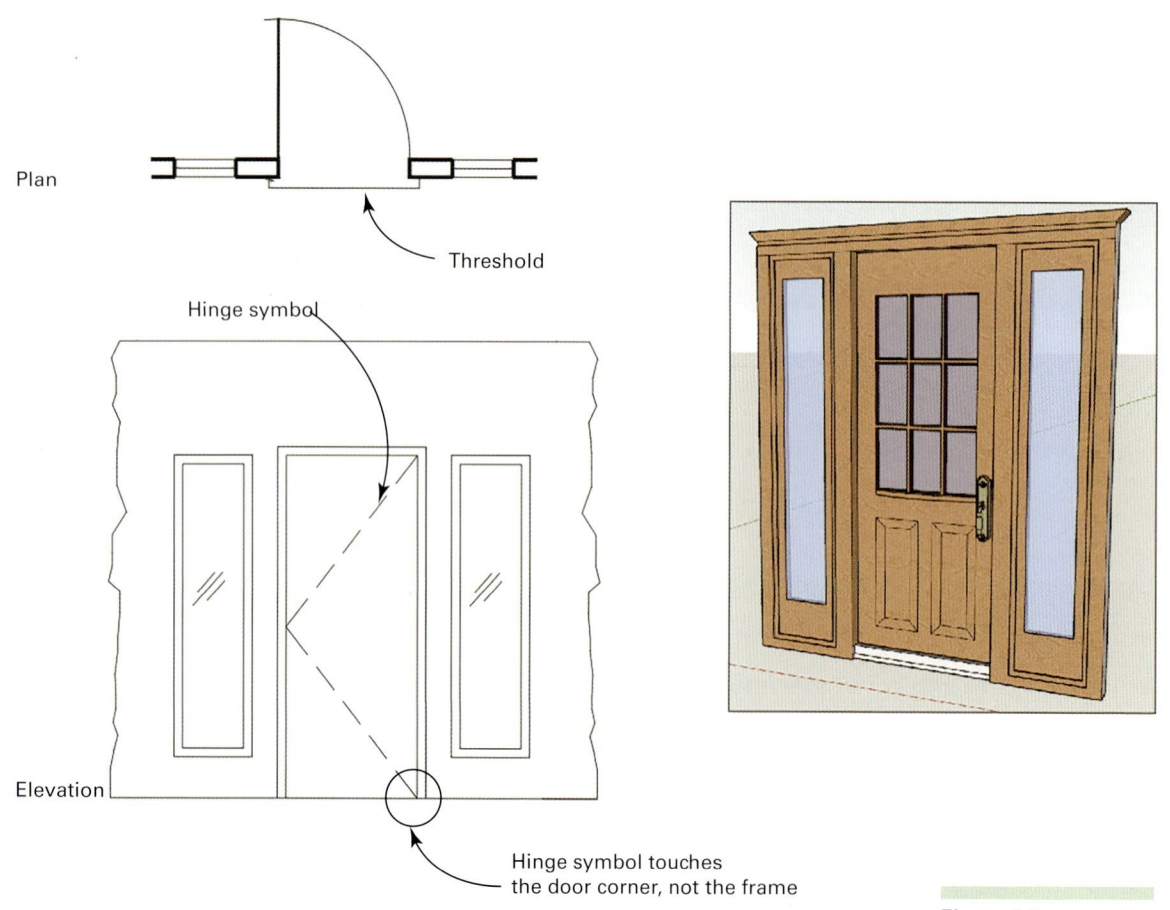

Figure 8.1
Residential swing door.

In residential construction, doors swing into the house and into the rooms (Figure 8.2). To maximize space, place doors in room corners and swing them against the wall. In commercial construction they swing out of the building and out of the rooms. This is to follow the direction of exit travel and to avoid people piling up against a door that opens inward in case of fire.

Figure 8.3 shows different ways to show a swing door symbol. Office standards vary, but whichever is chosen should be used consistently throughout a drawing set. Some designers draw 90° for new doors and 45° for existing doors.

Swing Door Hand   A swing door is classified by **door hand** (Figure 8.4), which describes which side it's hinged on and which way it swings. There are four hands, and design and building code criteria determines which is chosen. The hand must be known before buying a door because the hardware and predrilled holes for each is different.

To determine hand, stand in the hall or outside the building and face the door. A **left-hand** door has the hinge on your left and swings away from you. A **right-hand** door has the hinge on your right and swings away from you. A **left-hand reverse** has the hinge on your left and swings towards you. A **right-hand reverse** has the hinge on your right and swings away from you. If the door is between rooms, hence there is no hallway to stand in, put your back against the hinged side. A left-hand door swings to your left; a right-hand door swings to your right.

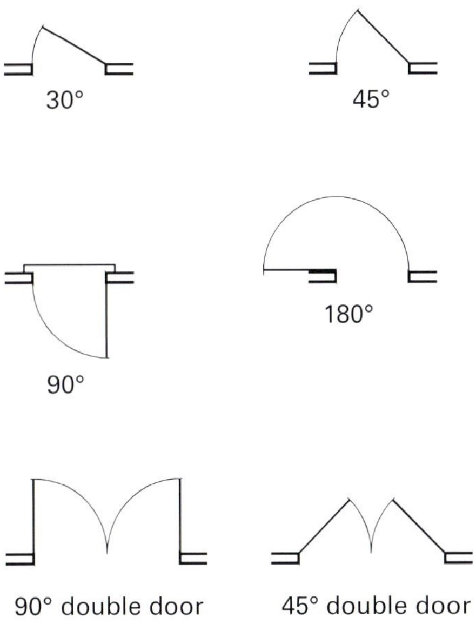

Exercise
Walk through your home and classroom building and identify door hands on different rooms.

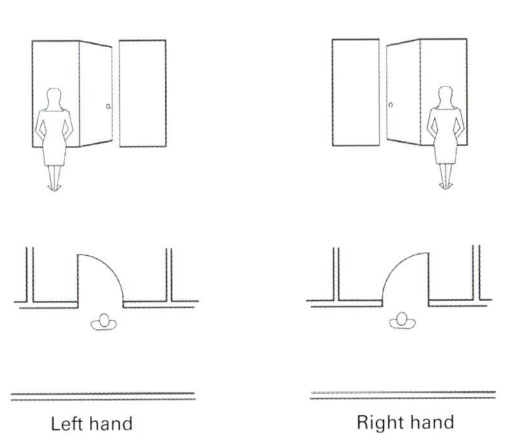

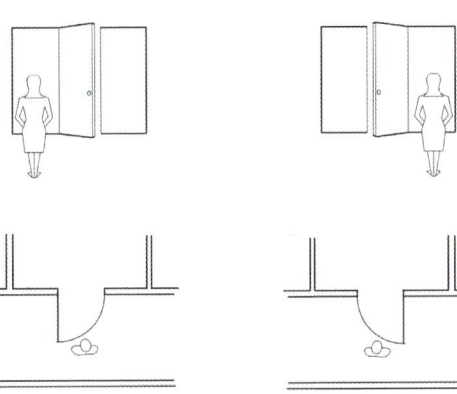

Figure 8.2 Residential doors swing into rooms.

30°   45°

90°   180°

90° double door   45° double door

Figure 8.3 Plan symbols for a swing door.

Left hand   Right hand   Left hand reverse   Right hand reverse

Figure 8.4 Door hands.

DOOR TYPES 181

Swing Door Variations These include:

Double two swing doors hung from opposite jambs in one opening, often used in entrances to large rooms (Figure 8.5). A **French** door is a double door with glass inserts.

Plan

Align this point with the wall line

Elevation

Figure 8.5
Double (French) door

Dutch a swing door with top and bottom **leaves** (panel halves; plural of **leaf**) that operate independently, often used in daycares (Figure 8.6).

Double-action a door that swings both ways (Figure 8.7), often used in restaurant kitchens.

Fire This swing door is part of a **fire door assembly**. That's a commercial construction door and frame that protects against the passage of fire. It is placed in areas where fire walls are required, such as areas of refuge and egress. It must have a latch and a self-closing device. Fire doors are rated how much time they can withstand an intense fire without failing or allowing the fire to penetrate.

Plan

Elevation

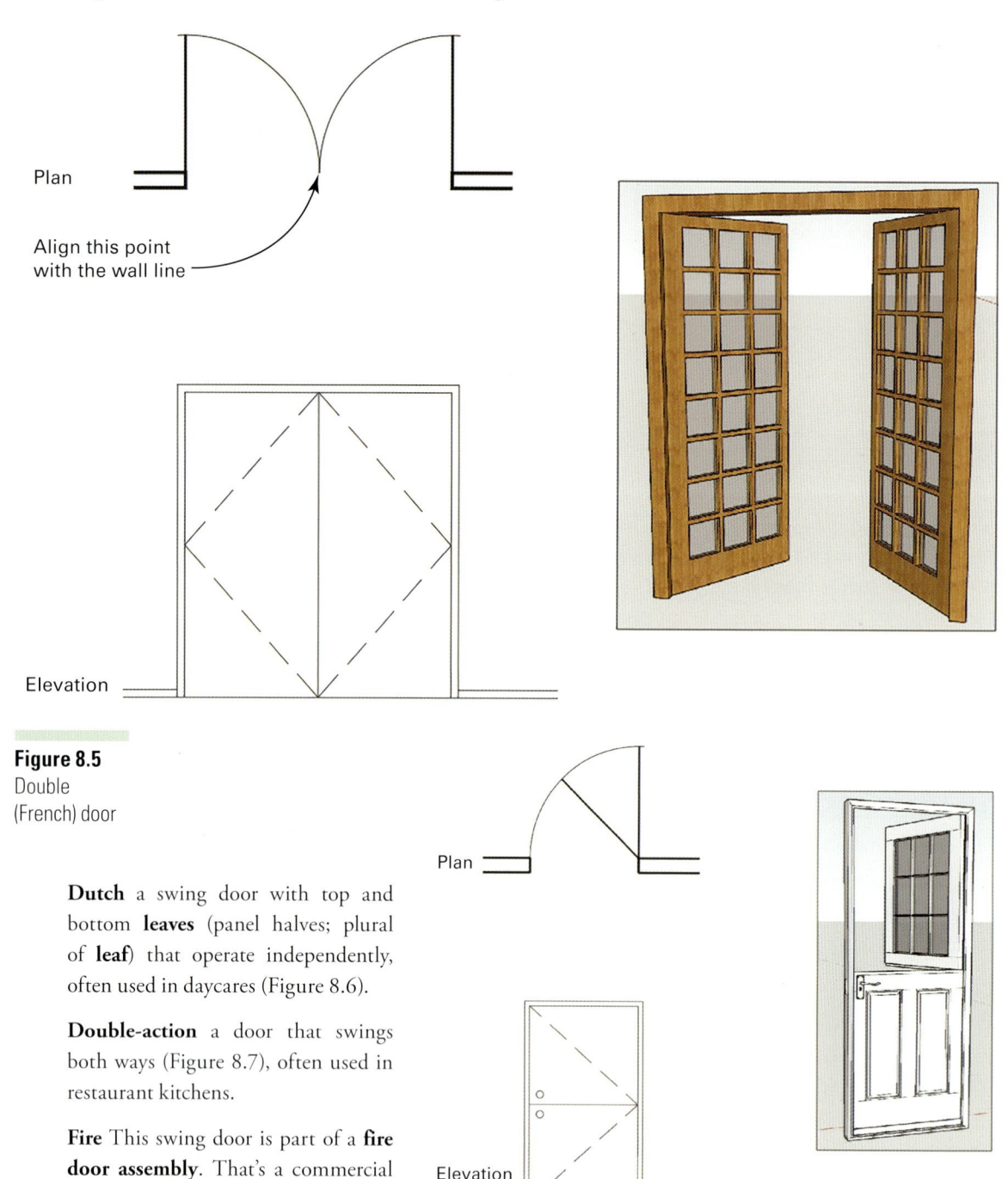

Figure 8.6
Dutch door.

How to Draw a Swing Door Symbol in Plan Figure 8.8 uses a circle template to draw a swing door symbol. Measure the door length from the corner it swings from; make it the same length as the opening; and use a circle whose radius is the same length as the door. To find that circle, line up its **quadrant markers**

182　CHAPTER 8　DOOR AND WINDOW SYMBOLS

Plan

Elevation

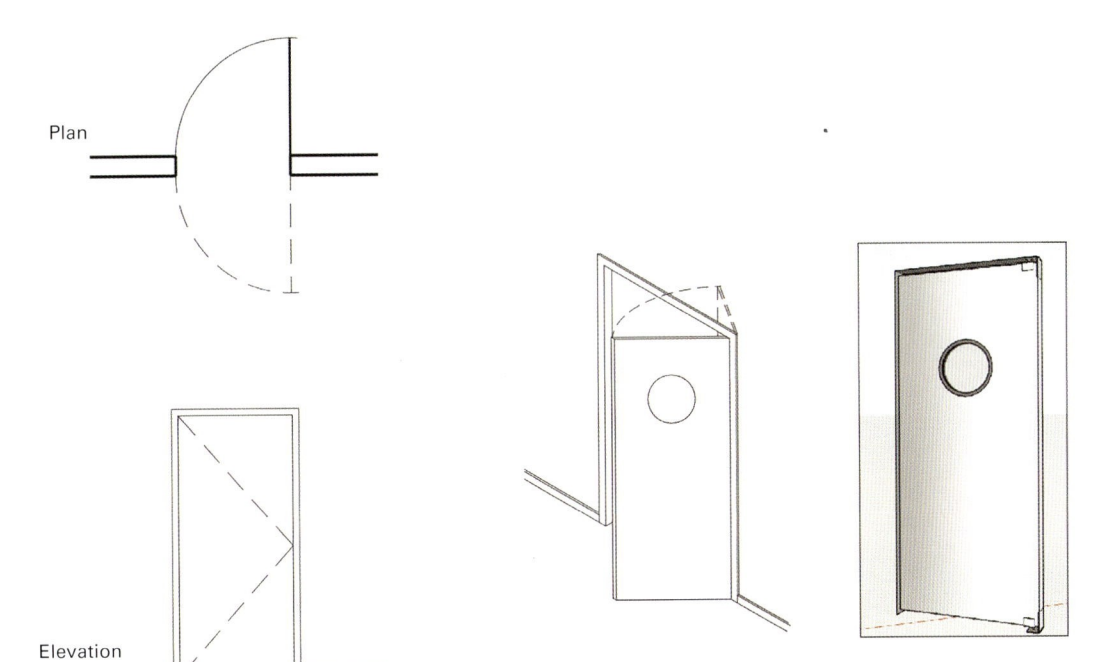

Figure 8.7
Double-action door.

(printed lines at the circle's four quarters) with the top of the door and with the jamb. If you don't have an appropriate size circle, use a compass. Note that the door, arc, and jamb (side of the door opening) all touch; there are no gaps between them. Include at least 3" of wall on both sides of any door.

If you use a door template (Figure 8.9), choose a size that matches the opening and align so its printed lines touch the end of the door and the jamb.

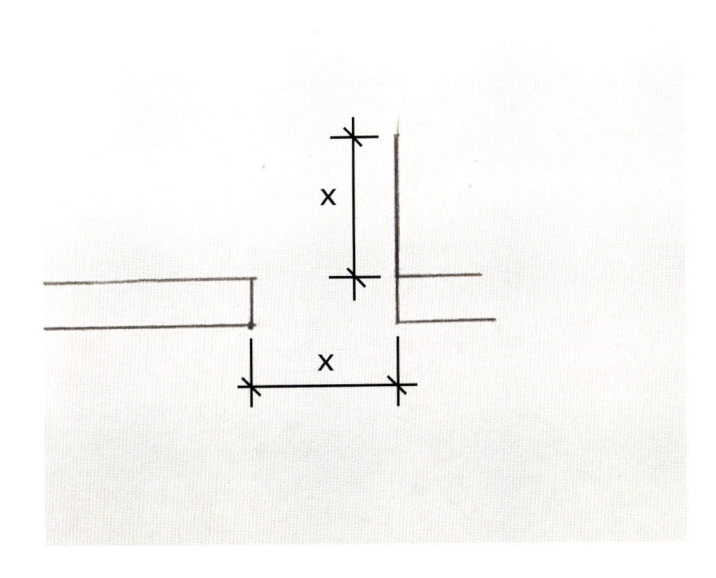

1.

2.

Figure 8.8
Align the circle's quadrant markers with the door and jamb to draw the arc.

DOOR TYPES 183

Figure 8.9
Draw a door swing symbol manually or with a template.

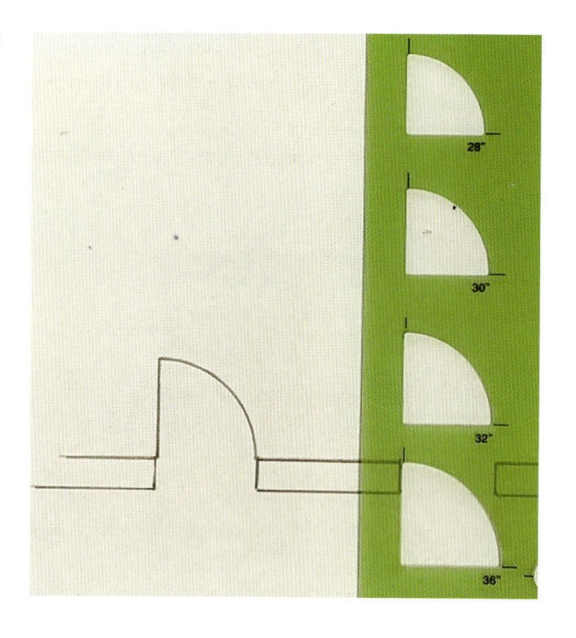

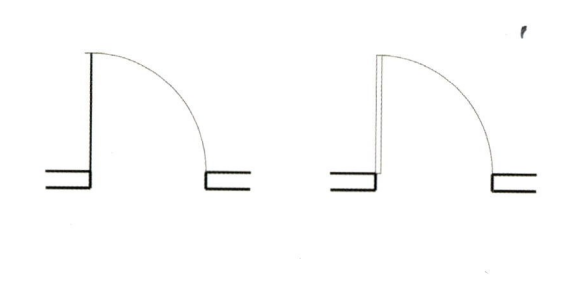

How to Draw a Swing Door Symbol in an Angled Wall Align a triangle with the wall. Brace the triangle against a second triangle. Draw along the first triangle's two edges to make lines parallel and perpendicular to the angled wall (Figure 8.10). The perpendicular line will be the door line.

Flush and Panel Doors
Flush doors are flat surfaced. **Panel doors** (Figure 8.11) have raised or recessed rectangular surfaces surrounded by the frame. The frame consists of **rails** (horizontal parts) and **stiles** (vertical parts). A separate frame called **trim** or **casing** is installed on the walls around the door to hide the joint where the door meets the walls. Show swing doors as flush or panel when drawing their elevation views.

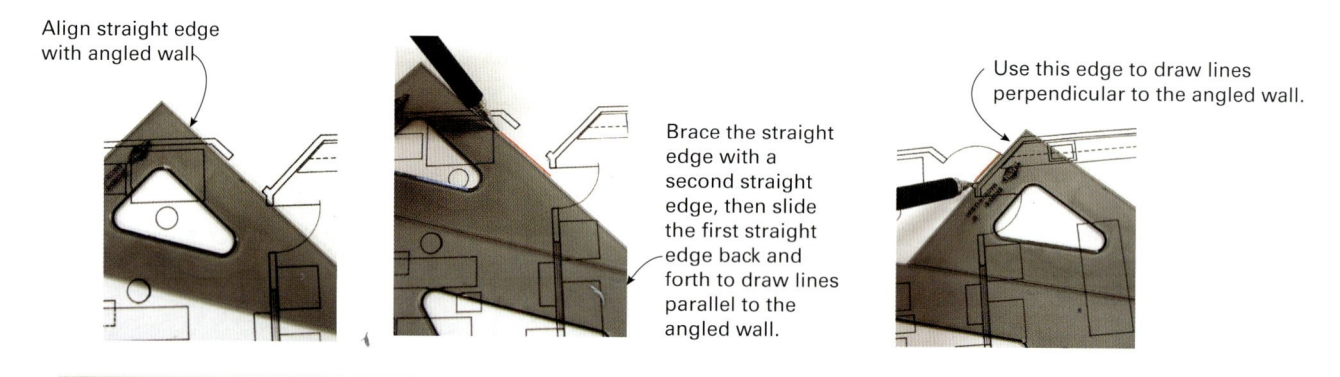

Figure 8.10
Drawing a door in an angled wall.

Narrow door stiles are 2" wide, medium door stiles are 4" wide, and wide door stiles are 5" wide. A narrow-stile door usually has a 4" bottom rail, and a medium- or wide-stile door has a 6" bottom rail. High bottom rails are 10". Both door types come in many decorative styles, and may have **lites**, or framed glass inserts.

Figure 8.11
Parts of a panel door.

184 CHAPTER 8 DOOR AND WINDOW SYMBOLS

Sliding This door (Figure 8.12) glides along a metal track. Double sliding doors usually have one movable door that slides in front of one **fixed** (non-moving) **panel**. Exterior doors glide along tracks in the frame head and sill; interior doors glide along a track in the frame head and have **guides** (hardware installed at the bottom) to keep them from swaying when closed. Note that the plan symbol for glass doors is drawn with three lines; wood doors are drawn with two lines.

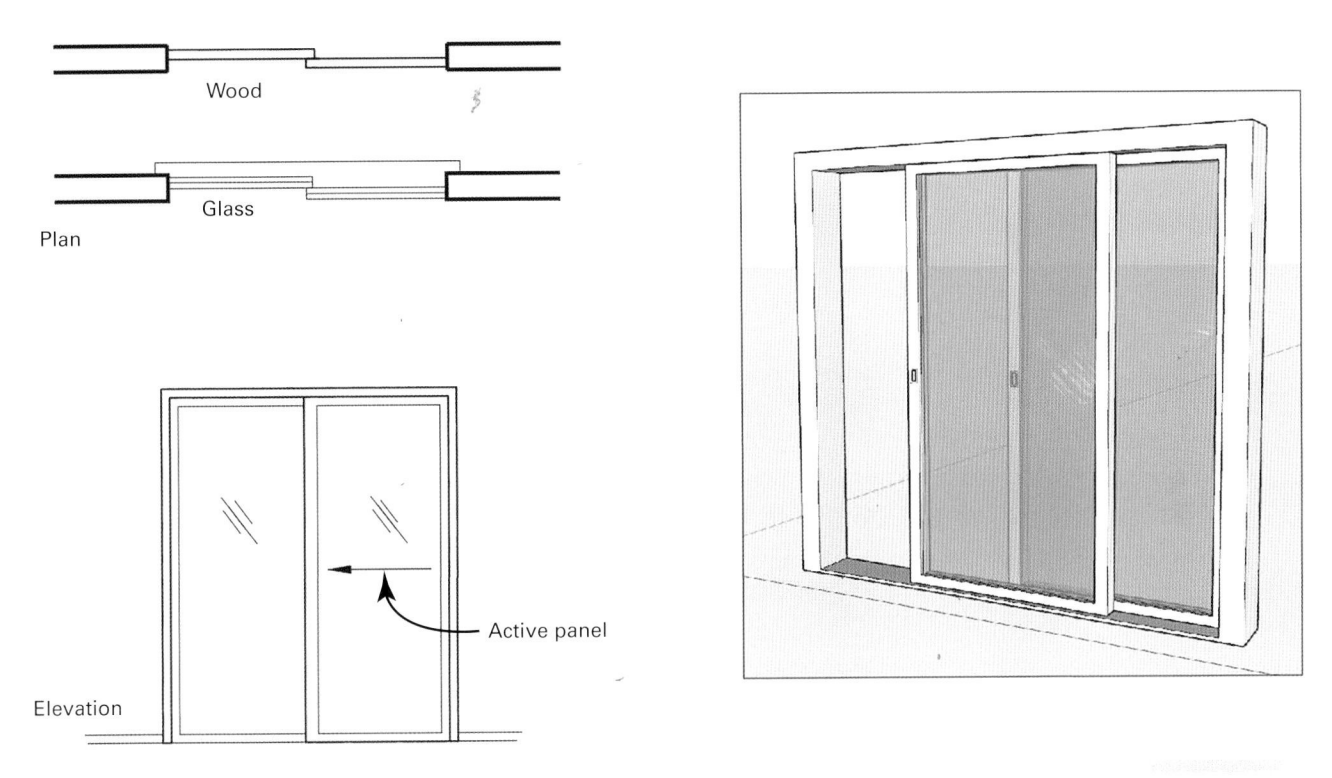

Figure 8.12
Sliding door.

Sliding Door Variations   Sliding doors include:

Pocket (Figure 8.13). This is a single sliding door that retracts into a wall cavity, useful for small spaces. Draw the opening, pocket, and door the same length (the door should appear to fill the opening if closed). Draft the wall at least 6" wide.

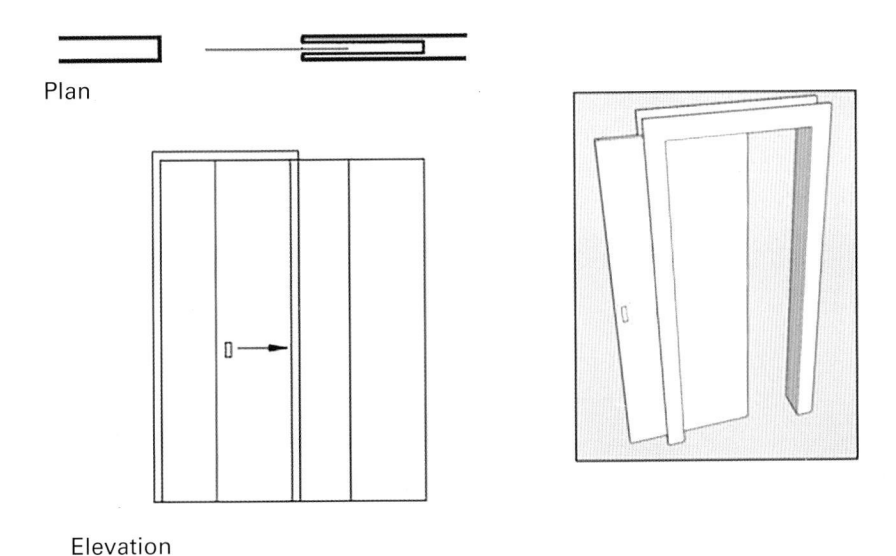

Figure 8.13
Pocket door.

DOOR TYPES **185**

Barn (Figure 8.14). This is a surface hung door; the track is mounted on the wall. It works for most any space plus large, open spaces. Barn doors are larger than the door opening.

Figure 8.14 Barn door.

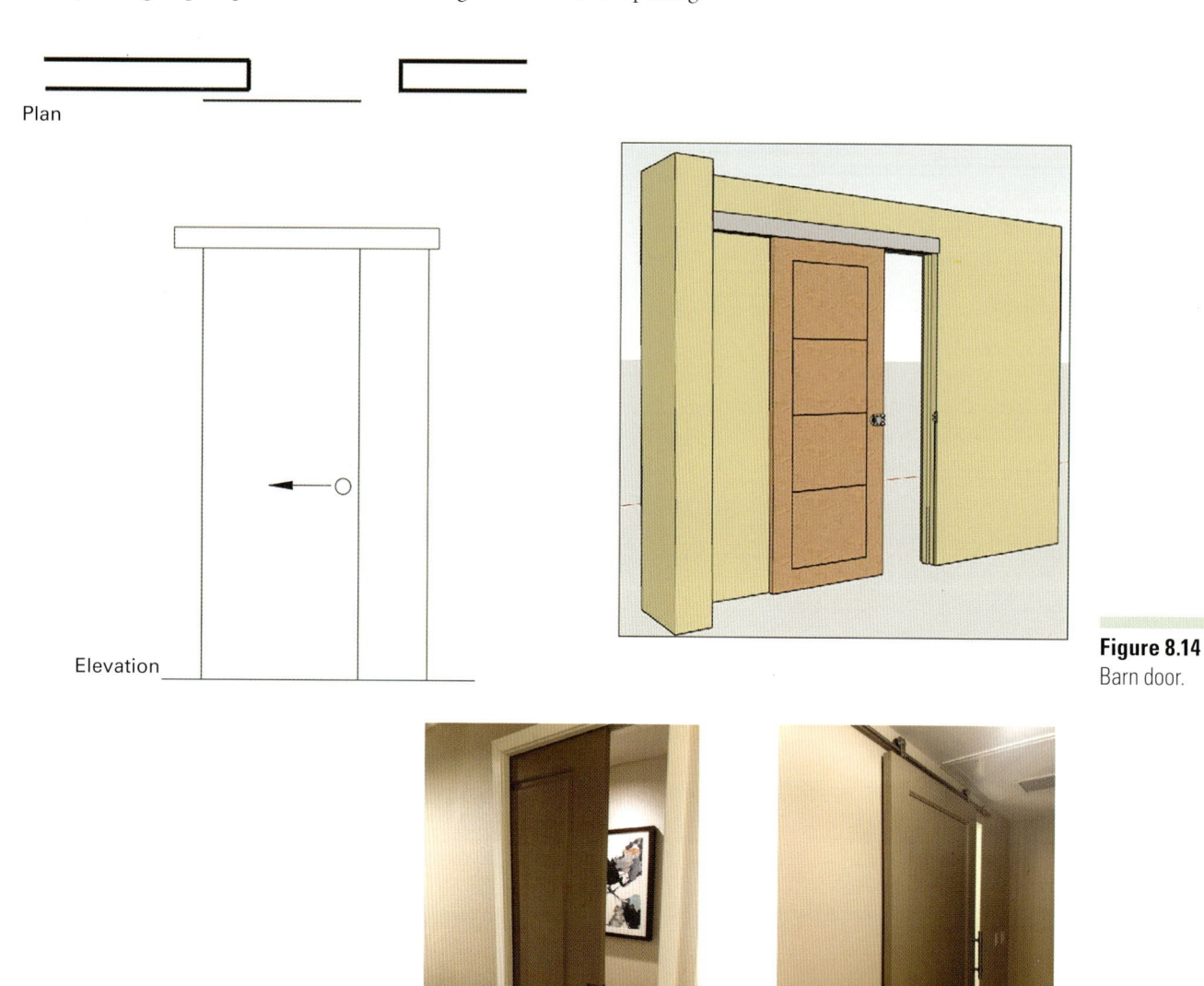

Figure 8.15 Pocket vs. barn door.

Figure 8.15 shows a pocket vs. a barn door.

Bifold This door is part hinged and part sliding. It's often used on closets, pantries and laundry areas. Two leaves are hinged together, and one is hinged to the jamb. Both leaves slide in a track. A single bifold has two leaves, and a double bifold has four (Figure 8.16). Figure 8.17 shows how to draft double bifold doors.

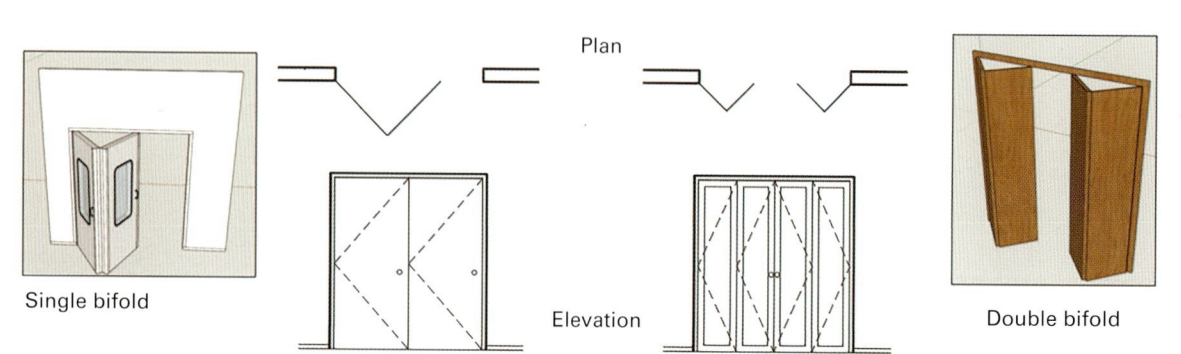

Figure 8.16 Bifold and double bifold doors.

186 CHAPTER 8 DOOR AND WINDOW SYMBOLS

1. Divide the opening into four equal parts

2. Position a triangle at the jamb and draw a leaf 1/4th the total length.

3. Flip the triangle and draw a second leaf. Note how it ends on the construction line.

4. Project the fold point over.

5. Repeat steps 2 and 3.

6. Bifold doors.

Figure 8.17 Drawing bifold doors.

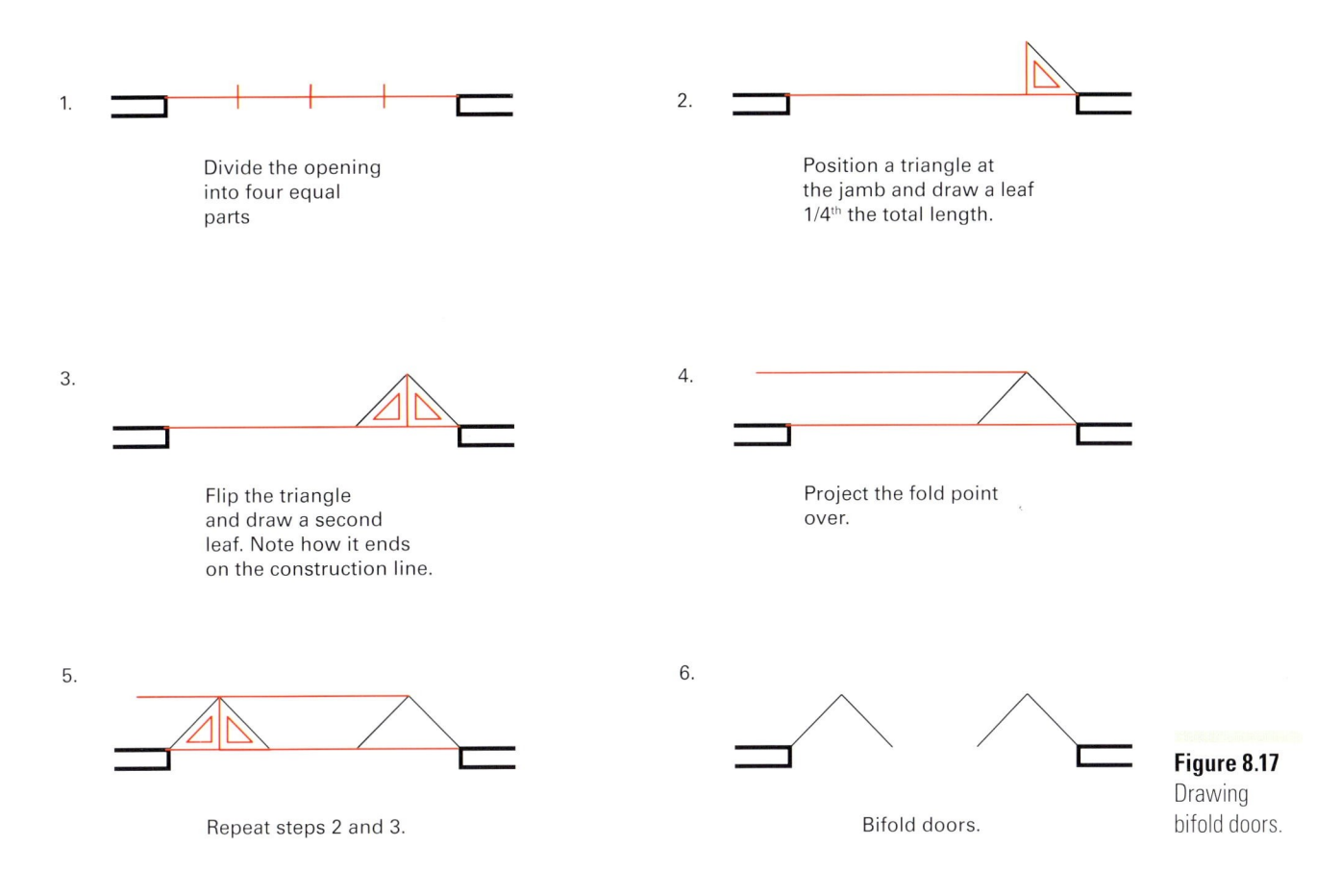

Accordion An **accordion door** operates like a folding door but has multiple leaves (Figure 8.18). When retracted, it can be, but is not always, hidden in a wall pocket. It's often used as a room divider in a large space or on a closet or pantry.

Plan

Elevation

Figure 8.18 Accordion door.

DOOR TYPES 187

Pivot A **pivot door** has one panel that rotates on a hidden spindle instead of on hinges (Figure 8.19). It is large (over 42" wide), heavy, and used in commercial buildings such as spas and exercise rooms, and in residences where an impressive entry is wanted.

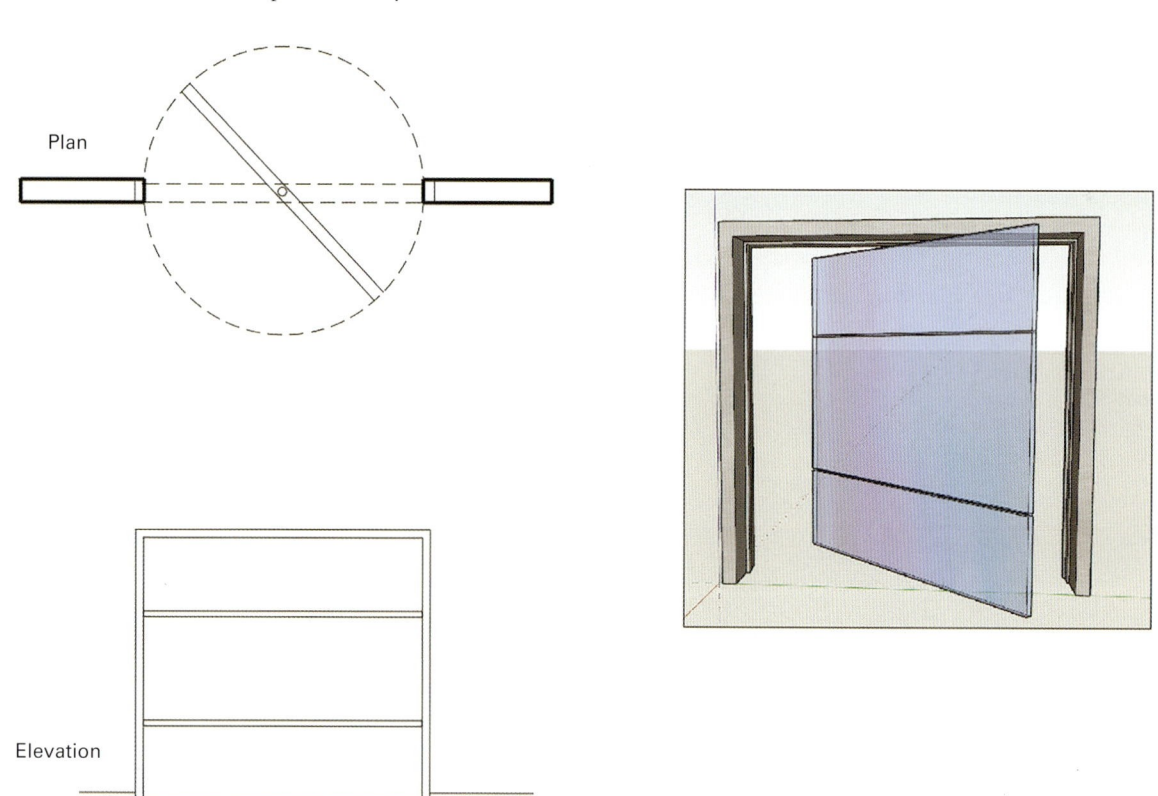

Figure 8.19
Pivot door.

Revolving A **revolving door** has multiple panels that rotate on pivot hardware. It's used in commercial and industrial buildings because it accommodates heavy traffic and minimizes drafts. Figure 8.20 shows a revolving door flanked by swing doors and a **transom window** (horizontal over the door).

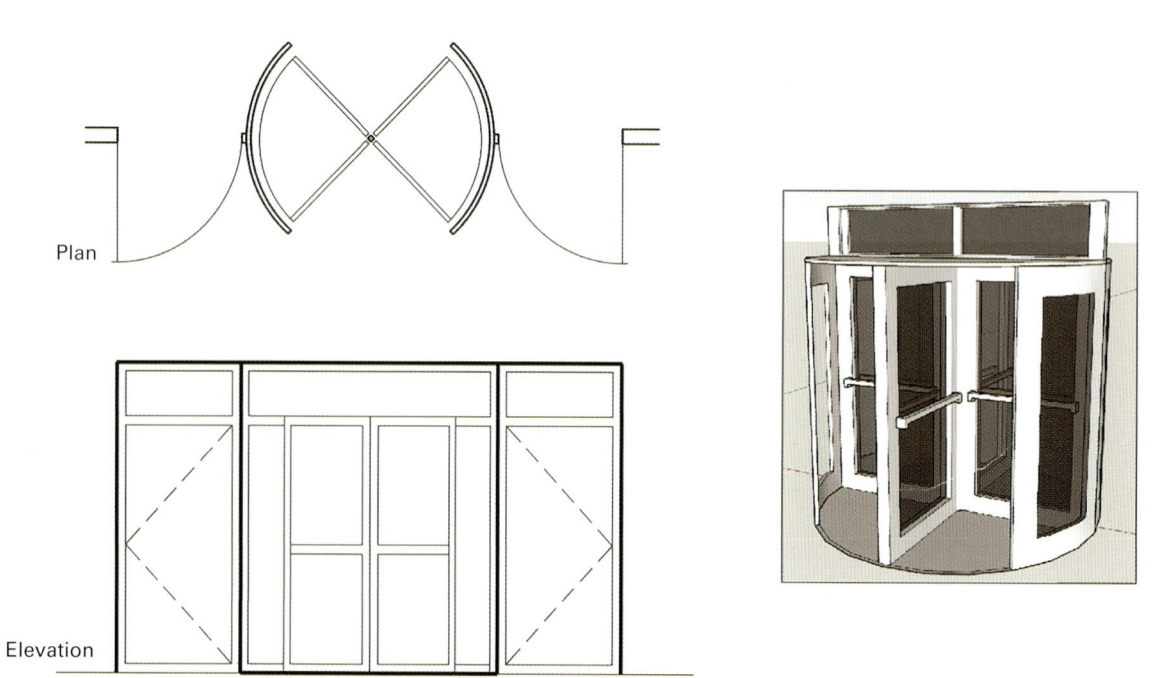

Figure 8.20
Revolving door.

188   CHAPTER 8   DOOR AND WINDOW SYMBOLS

Full Height Also called a **ceiling height** door, a **full height door** may have an overhead stationary panel or the whole hinged part may reach the ceiling (Figure 8.21). Full height doors can swing or slide. They're used to create an illusion of space and height.

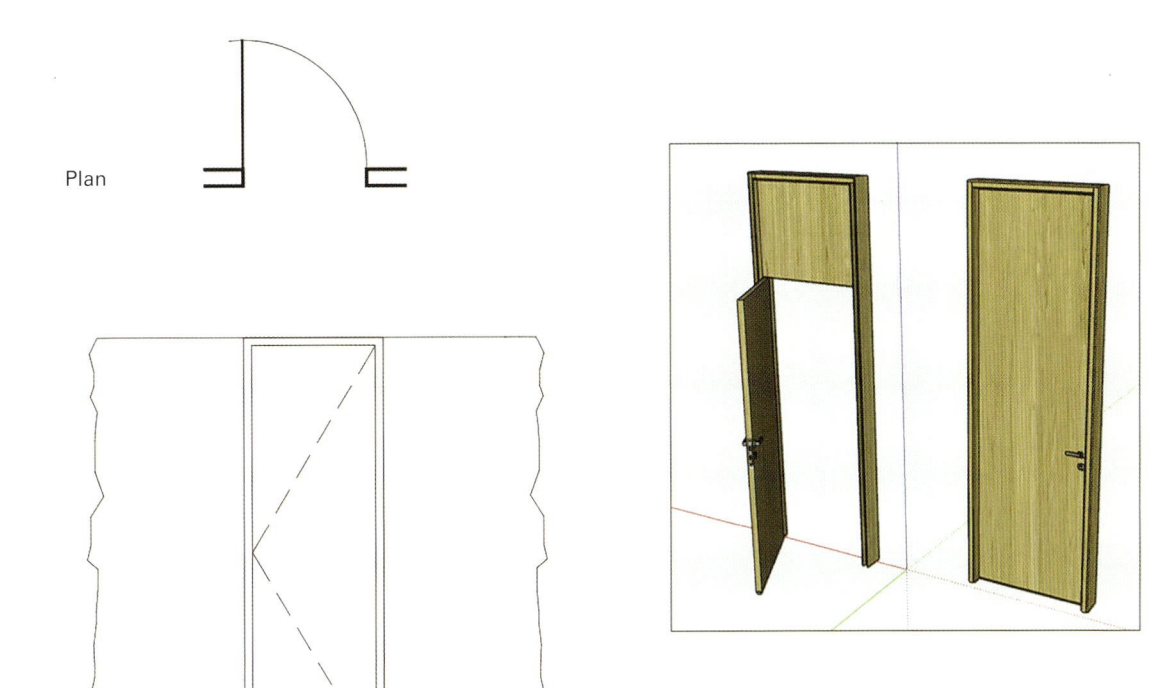

Figure 8.21 Full Height door.

Overhead Sectional This large door is assembled in sections and moves vertically along a track (Figure 8.22) via cables, pulleys, springs, and an opener. **Overhead sectional doors** are used on garages and warehouses. The typical residential sectional overhead door height is 6'-0" to 8'-0" in increments of 3", and 7'-0" (four sections) high, or 8'-0" (five sections) high. Standard widths range from 8'-0" to 18'-0". The rough opening is the same size as the door because the door fits against the opening from the inside.

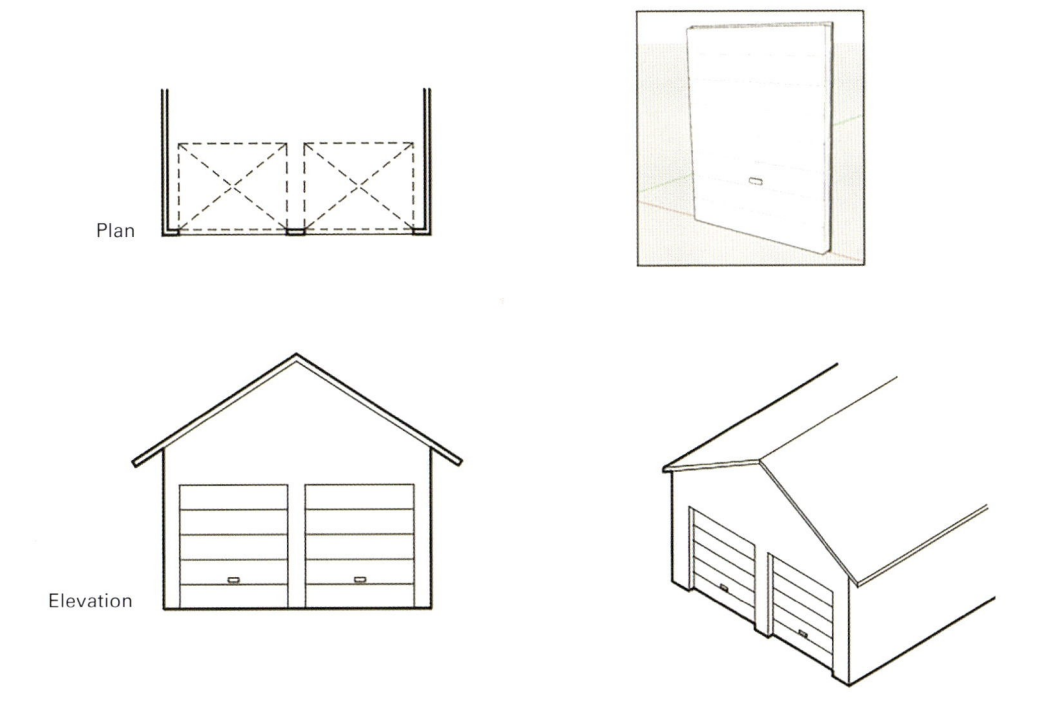

Figure 8.22 Overhead sectional door.

DOOR TYPES **189**

Cased Opening   This is an opening with no door (Figure 8.23). **Cased openings** may be door height or reach to the ceiling. The floor plan symbol looks the same whether the top is straight or arched because the arch is foreshortened in the plan view. Draw the hidden lines thinner than the solid lines and make sure they touch the solid lines at both ends.

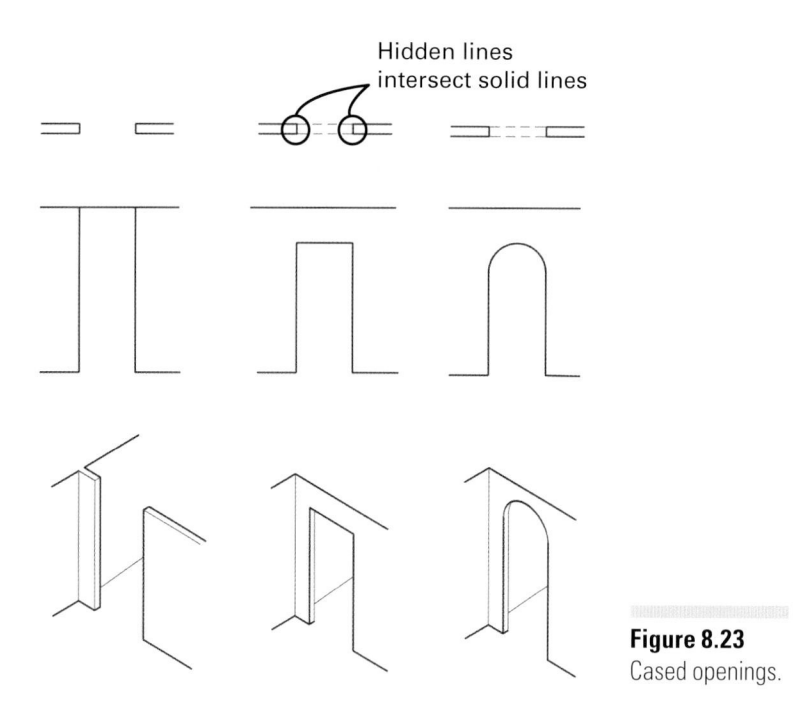

Figure 8.23
Cased openings.

Door and Rough Opening Sizes   The standard size for an exterior residential swing door is 6'-8" tall × 3'-0" wide. However, 8'-0" high and 30"–32" wide doors are also common in new construction. In wheelchair-accessible homes, 2'-8" minimum is needed; if the door is in a hall where wheelchair turning is needed, make the door 3'-0" wide.

The typical interior swing door is 6'-0" to 8'-0" tall. Standard widths range from 2'-0" to 3'-0" in increments of 2". The minimum width for a bedroom door should be 2'-6", but a 2'-8" or 2'-10" enables easy furniture moving. **Wheelchair-accessible doors** must be at least 2'-8" wide, but 3'-0" is preferable.

The **rough opening** (hole in the wall) for an interior door is usually framed 2 5/8" higher than the door height and 2 1/2" more than the door width to accommodate the frame. Rough opening applies to wood-framed buildings; **masonry opening** applies to concrete or brick buildings. Traditional sliding doors are 6' to 8' tall with a total unit width of 6'-0" or 8'-0".

Egress Requirements and Egress Doors

When planning spaces, designers must consider how occupants leave them. The **International Residential Code (IRC)** and **International Building Code (IBC)** have tables and calculations of number of exits needed, their sizes and locations, maximum hallway length to the exits, and more. Every dwelling unit needs at least two compliant forms of egress so that if one is blocked, there's another way out. All bedrooms, including basement apartments, must have two ways to reach the outside. Here are some code terms to know.

Egress: a clear, continuous, unobstructed travel path through a building to a public way (outside).

Means of egress: a continuous, unobstructed travel path from your location to the **public way** (outside public space such as a street). This path can include a window, door, staircase, or fire escape. Doors along this route cannot lock from the side people are leaving. A means of egress has three parts: **exit access**, **exit**, and **exit discharge**.

- Exit access: path from your location to an exit.
- Exit: the means of leaving the building. This is a door, or in a multi-story building, an enclosed exit stairs or fire escape (exterior stairs).
- Exit discharge: The path from the exit to the public way.

Egress Door An **egress door** facilitates escape from the building during an emergency or evacuation. Egress doors must be placed to provide that path. Codes and accessibility require them to be swing doors at least 6'-8" high and when open, have a minimum of 32" clear width (the distance between the jamb protrusions). A door leaf cannot be bigger than 4'-0" wide. Doors must swing in the direction of travel.

A 36" -wide door provides 34" of clear width. One door leaf may not be more than 4' wide. Therefore, if 60" of exit width are required, more than one door is needed. Two 36" doors provide 68" of clear width, which exceeds the requirement but is the closest increment. If 40" of exit width is needed, one 48" door works.

An egress door cannot reduce any required stair landing dimensions or corridor width by more than 7" when fully open, nor can it impinge on more than half of the required corridor width at any open position. Figures 8.24 and 8.25 show clearance requirements.

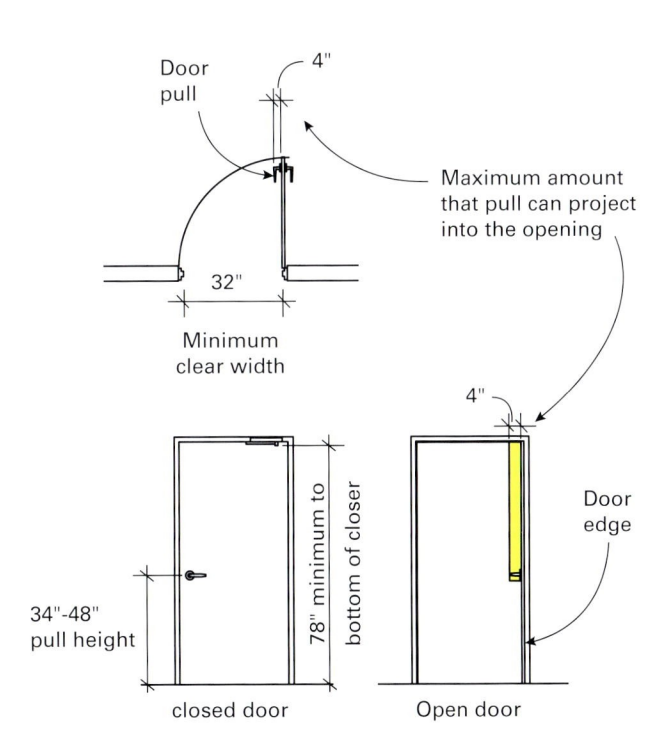

Figure 8.24 Hardware clearances for a means of egress door.

Half-Diagonal Rule If more than one exit is needed, at least two of the exits must be separated by a distance equal to or greater than one-half of the maximum diagonal dimension of the area served. Measure this distance in a straight line between the exits (Figure 8.25). If the building is sprinklered, one-third of the diagonal is permitted.

Tip
Windows may be used as emergency exits if they are operable from the inside without tools. The sill must be within 44" of the floors, with a minimum area of 5.7 square feet (ground level windows may have a minimum area of 5 square feet). They must have a net clear opening at least 20" wide and 24" high. Net clear opening is the actual free and clear space when the window is open, not the rough opening size or the glass panel size.

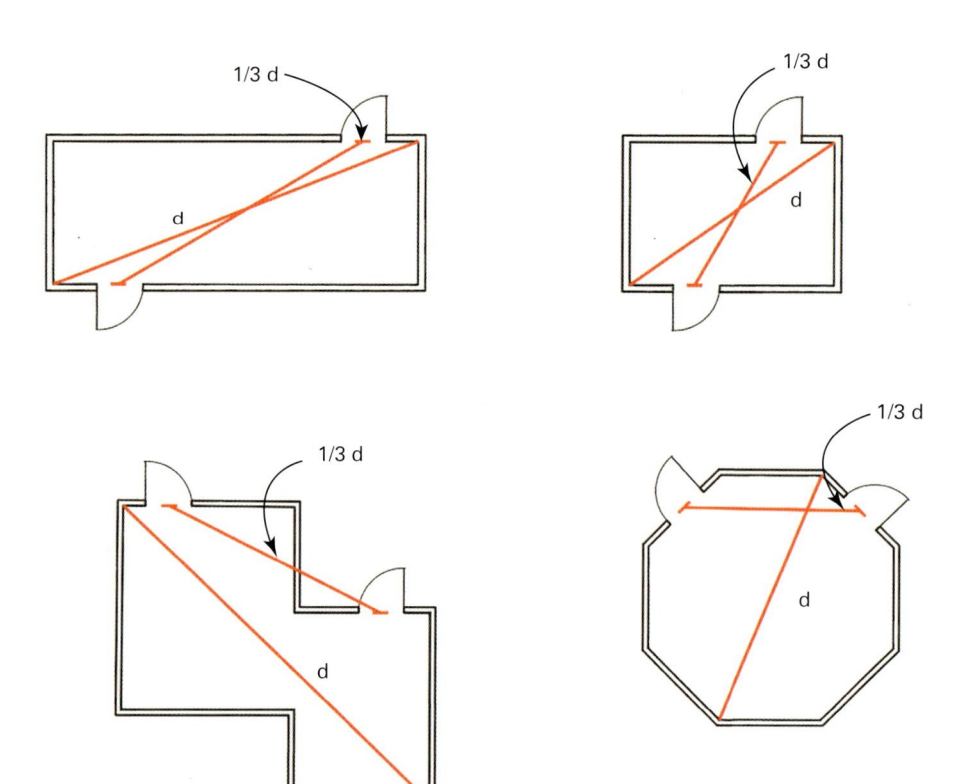

Figure 8.25 The **half-diagonal rule** for exits.

Door Hardware

Door hardware includes everything that operates and holds a door in place: screws, hinges, locks, handles, knobs, exit devices, bolts, closers, holders, stops, screens, kick plates, and pull/push plates. These are shown in construction detail drawings. Figure 8.27 shows some hardware and its sizes.

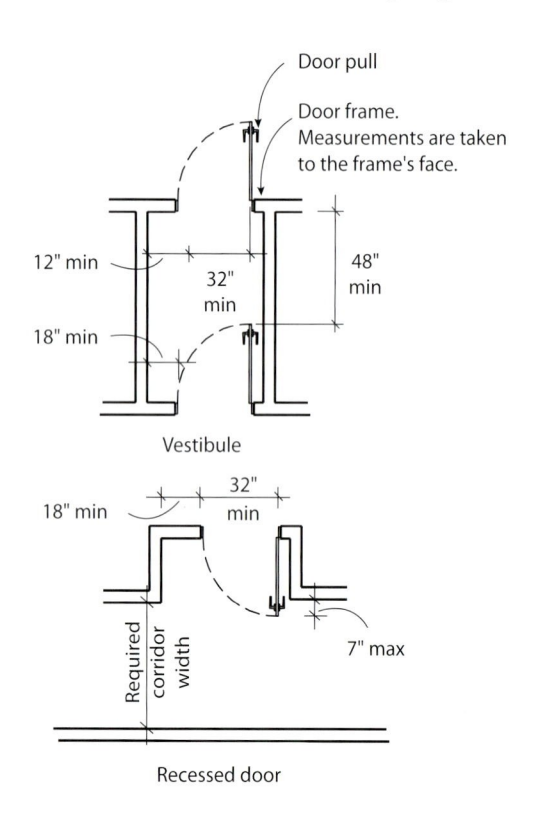

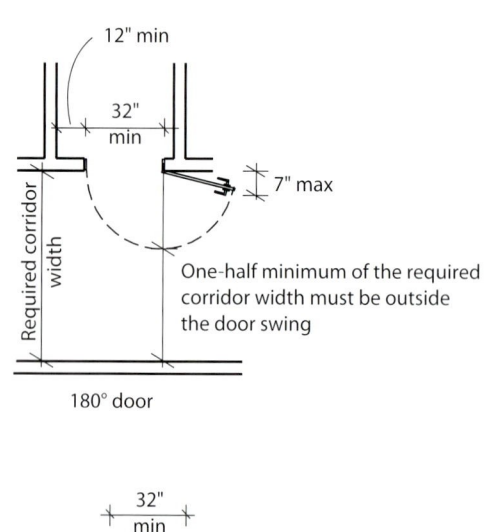

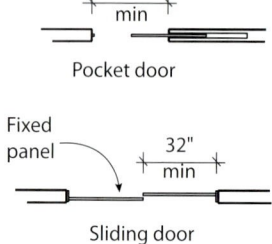

Figure 8.26 Opening and hallway clearances for a means of egress door.

192   CHAPTER 8   DOOR AND WINDOW SYMBOLS

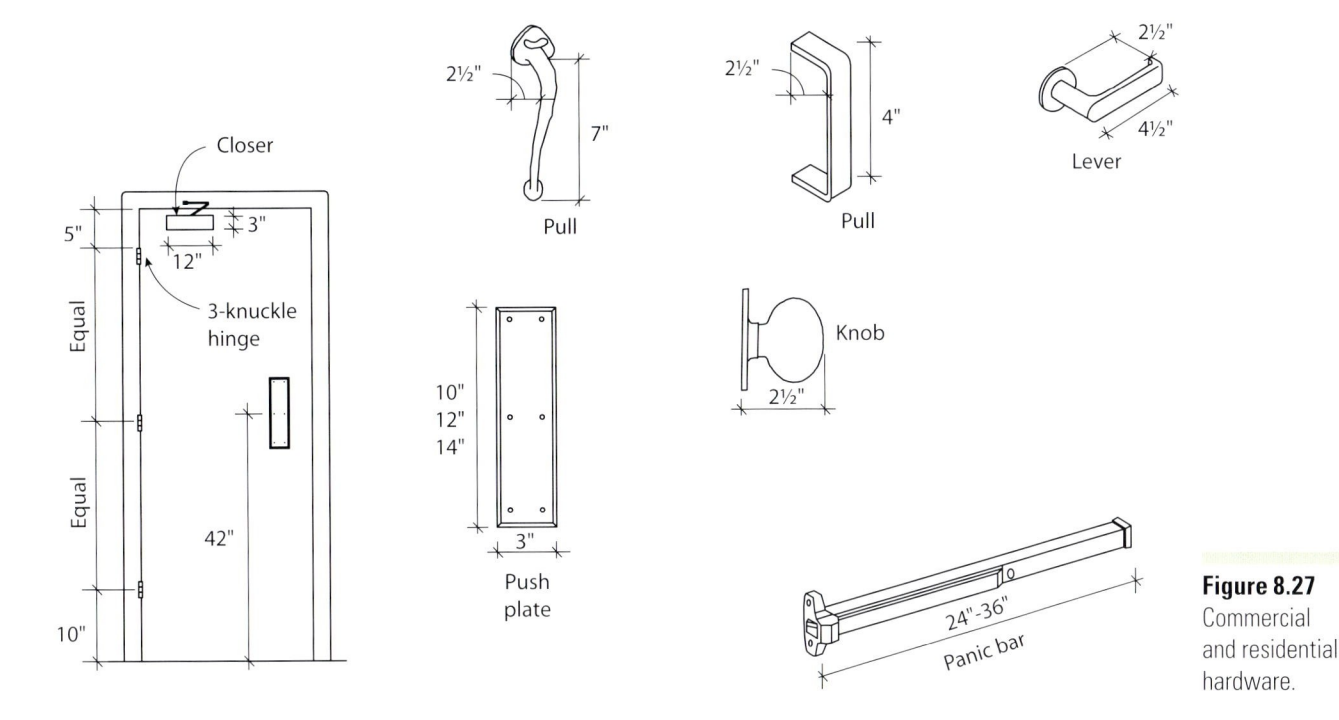

Figure 8.27 Commercial and residential hardware.

Codes regulate hardware, too. For instance, occupants in a commercial building must be able to exit a door in one operation with one simple maneuver, such as pushing a lever down. If a **deadbolt** (a lock that requires a key) is added, that extra step means the door will not meet code.

Hinges   **Hinges** are hardware that hold the door to the wall. The most common is the three knuckle, the knuckle being the hollow circular part that holds the pin.

Handle and Knob   Handles and knobs are hardware pieces that open a latch that opens the door. Lever handles are used in commercial buildings; knobs are only used in residential ones.

Lock   This is a security device with three components: a latch bolt, a lock strike, and a cylinder. There are five types of locks, named for how they are installed in the door: bored, preassembled, mortised, integral, and magnetic.

Bolt   A **bolt** provides supplemental security to a lock or holds the inactive half of a double door in place.

Exit Device   An **exit device** consists of a **panic (push) bar** and latch. The latch is released by pressing, and facilitates quick, easy egress. Codes vary in its placement; some require them to be placed between 30" and 44" above the finished floor; others require them to be placed at least 36" above the finished floor.

Closer   A **closer** shuts a door automatically after the door has been manually opened, It consists of a spring, a checker, and an arm.

Holder   A **holder** is hardware that is attached to the door and frame and keeps a door open.

Stop   A **stop** protects walls and equipment from the door as it swings open. There are many models, from the traditional metal springs with rubber bumpers to adhesive pads applied to the wall.

What Does a Window Symbol Show?

A window's plan symbol is made with a horizontal slice between 4' and 5' up from the floor, like the rest of the floor plan. The specific symbol represents its hinge type—swing, slide, or fold. In the symbol, the line in the center represents the **sash**, which is the glass and surrounding frame. One line represents one sash (such as a **fixed window**); two lines represent two sashes (such as a double-hung window).

A window located above the cutting plane is shown with hidden lines (Figure 8.28). If multiple windows are stacked vertically with some at and some above the cutting plane, draw the window symbol solid, because you're showing the cutting-plane–level window. Clarify the fenestration in an elevation drawing.

Figure 8.28
Window above the plan's cutting plane.

Window Types

The elevation symbols in the following illustrations are drawn as seen from the inside of the building, for use in interior elevation drawings. Some windows, such as the double-hung, look slightly different on the exterior side.

Double-Hung A **double-hung window** has two sashes that slide vertically in grooves inside a frame (Figure 8.29). A **single-hung window** has one sliding sash and one fixed sash, but its plan and elevations symbols look the same as the double-hung.

On the double-hung window's plan symbol, the two lines in the center are the two sashes. The bottom line is the wall edge and at the top is the sill. The sill goes on the outside of the building and it covers the outside wall edge, so a line for that edge isn't drawn. Sill and glass lines are drawn thinner than wall lines. On the elevation view, note the glass poché. It is drafted with a 45° triangle, not drawn freehand.

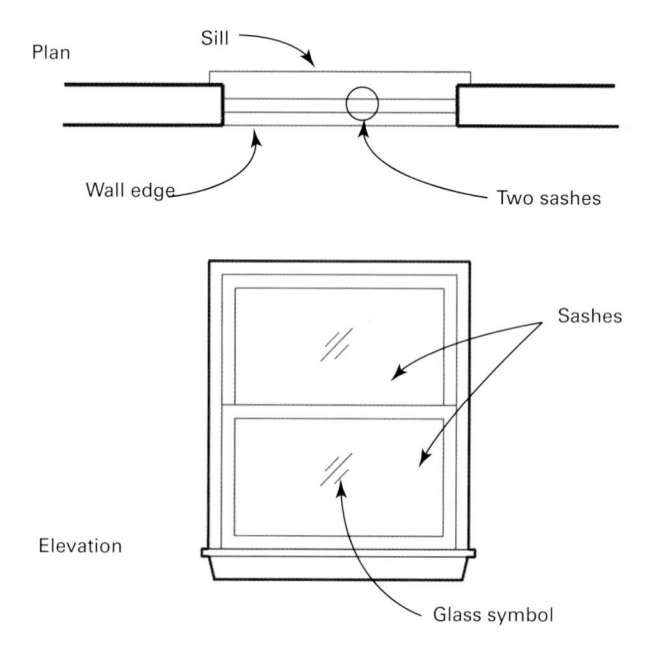

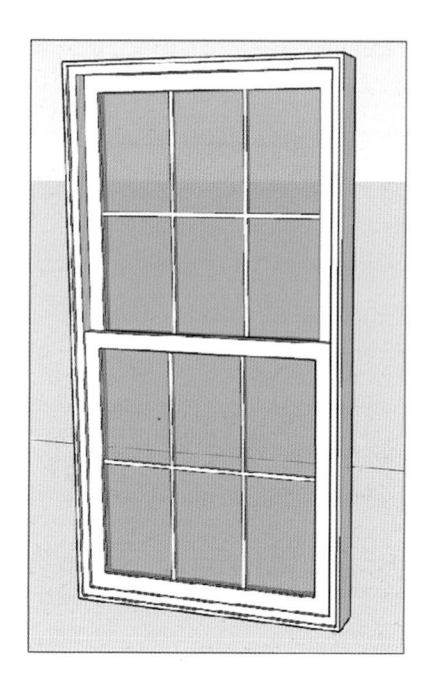

Figure 8.29
Double-hung window.

194 CHAPTER 8 DOOR AND WINDOW SYMBOLS

Fixed This window's symbol has one line to represent its one sash. The symbol looks the same whether the window is square, rectangular, trapezoidal, round, or half-circle. For wood frame construction draw a sill, but the sill in masonry construction is integral with the wall, so it isn't drawn (Figure 8.30). Since a fixed window does not open, screens, weather-stripping, and hardware are not needed. It can be bought as a preassembled unit or the glass can be custom cut on site and directly attached to the wall frame. A large fixed window is also called a **picture window** (Figure 8.31).

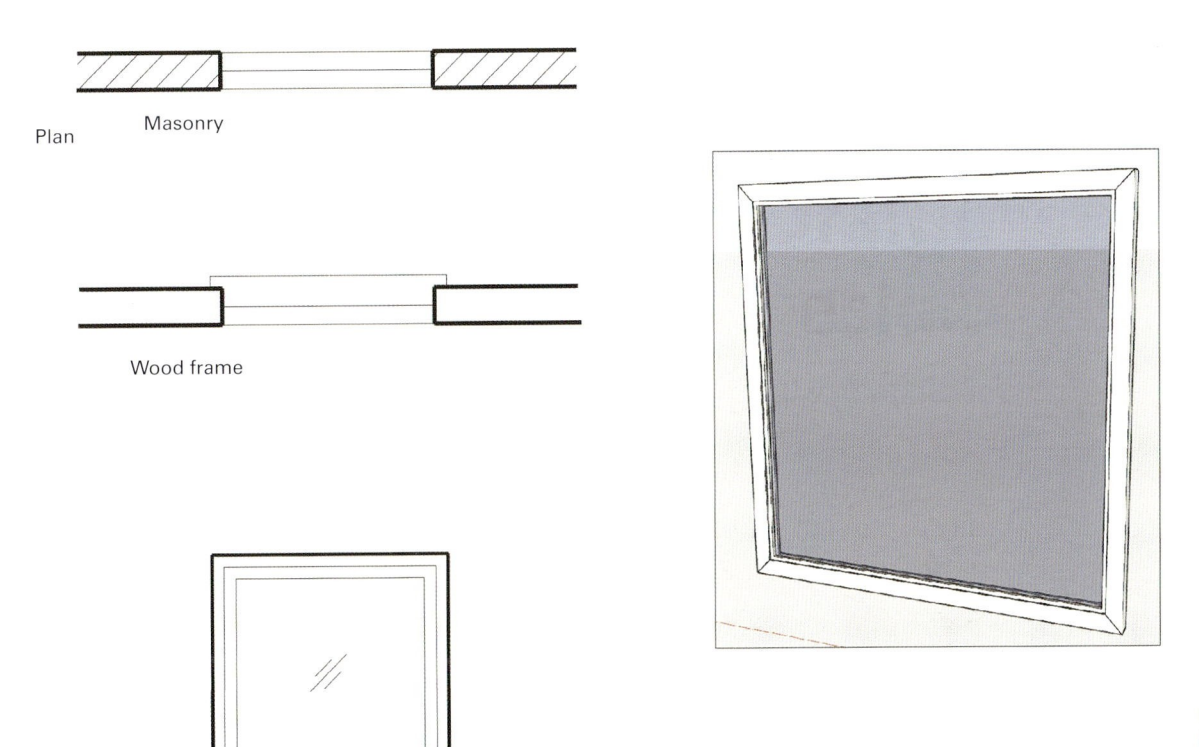

Plan — Masonry

Wood frame

Elevation

Figure 8.30
Fixed window.

Figure 8.31
Picture window. Courtesy of Hy-lite.

WINDOW TYPES   **195**

Sliding Also called a glider, a **sliding window** is a two-sash window that moves horizontally along a track (Figure 8.32). Both sashes may slide or one may be fixed. Draw an arrow on the active slider(s).

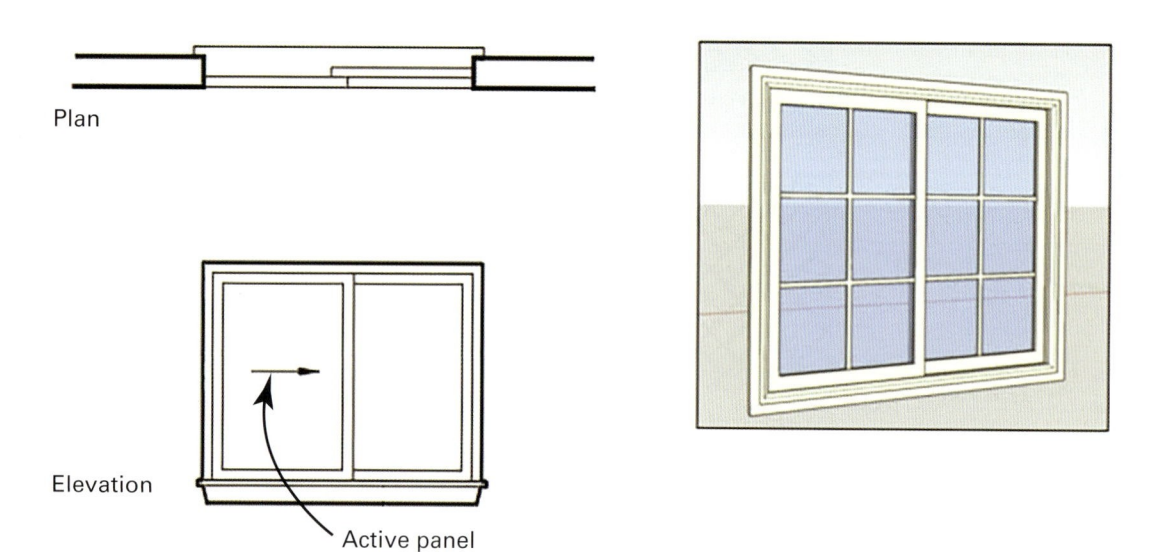

Figure 8.32
Sliding window.

Casement A **casement window** is hinged on the side, swings out, and is operated by a hand crank. It may have one sash or multiple sashes separated by a **mullion** (vertical bar between panes of glass). A pair of casement sashes may swing in the same direction or swing opposite each other (Figure 8.33). Draw the hinge line point to the corners of the sash, not to the frame.

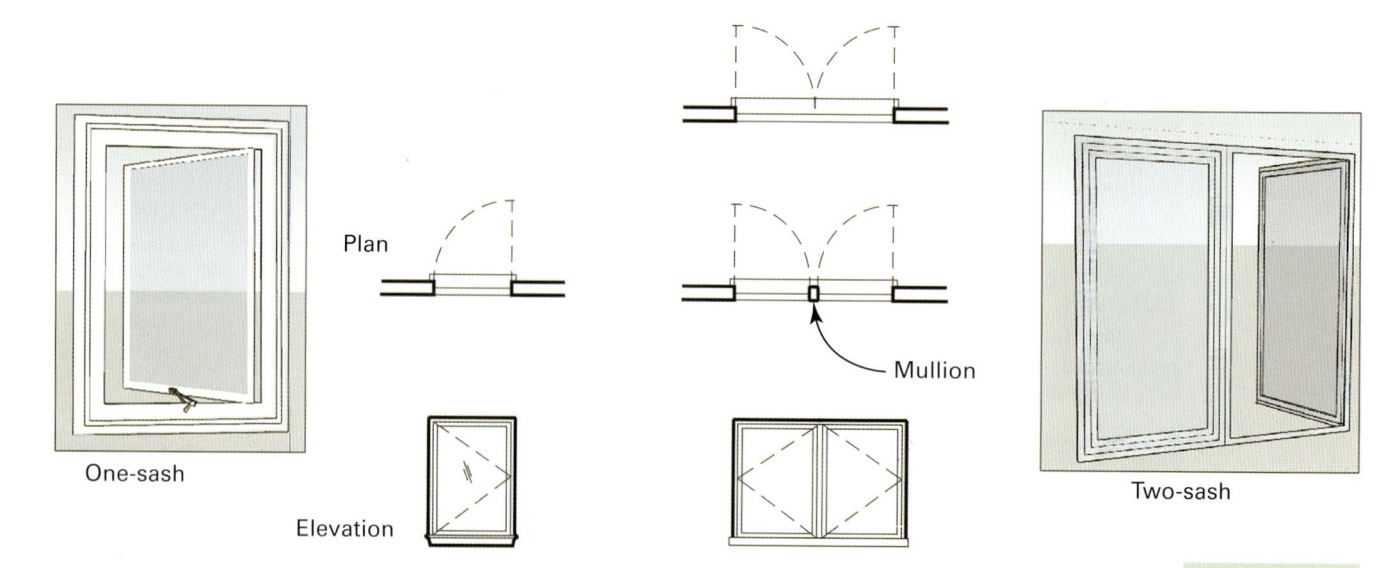

Figure 8.33
Casement window.

Awning An **awning window** is hinged at the top and swings out (Figure 8.34). It may have one sash or multiple sashes separated by a mullion (Figure 8.35). Multiple sashes may swing in the same direction or opposite directions.

196 CHAPTER 8 DOOR AND WINDOW SYMBOLS

Plan

Elevation

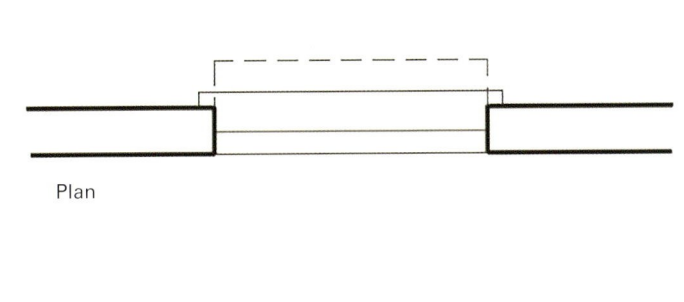

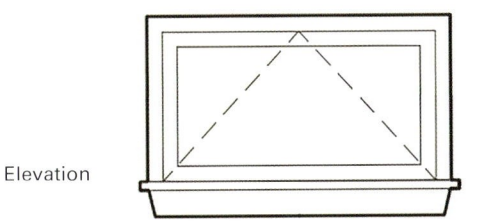

Figure 8.34
Awning window.

Figure 8.35
Two-sash awning window.
Courtesy of Hy-Lite.

Hopper This is hinged at the bottom and swings into the house (Figure 8.36, 8.37) via a handle at the sash top. Both awning and **hopper windows** open in the direction they do to keep rainwater out.

WINDOW TYPES **197**

Plan

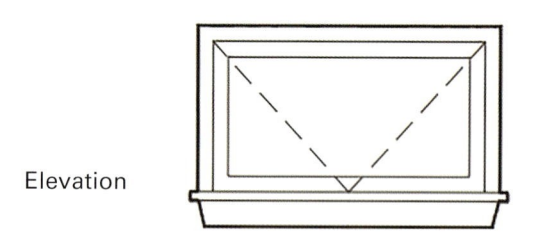

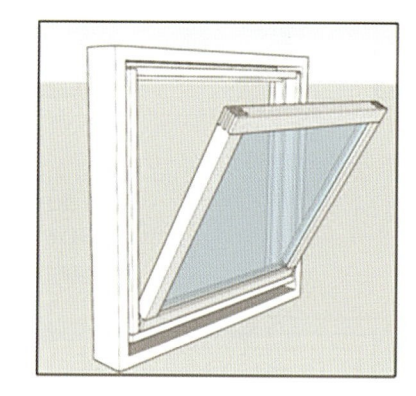

Elevation

Figure 8.36
Hopper window.

Figure 8.37
Hopper window.
Courtesy of Hy-Lite.

Jalousie Also called **louvered**, a **jalousie window** has rows of narrow, horizontal glass slats attached with clips to a frame (Figure 8.38). The slats open and close together like a mini-blind. Figure 8.39 shows how to draft five equally wide slats without measuring.

Plan

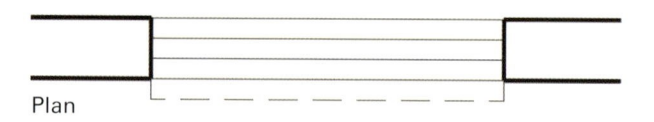

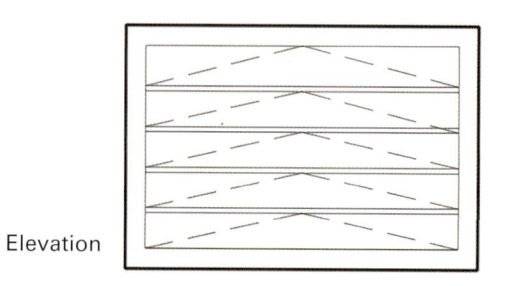

Elevation

Figure 8.38
Jalousie window.

198   CHAPTER 8   DOOR AND WINDOW SYMBOLS

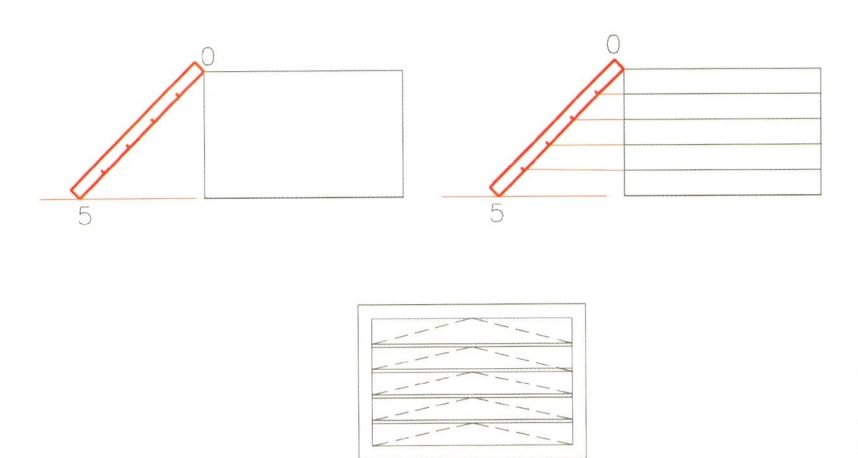

Figure 8.39
How to draft five slats of equal width without measuring.

Pivot Instead of hinges, this window has pivots (pins) on which it rotates (Figure 8.40). A **pivot window** may rotate horizontally or vertically.

Plan

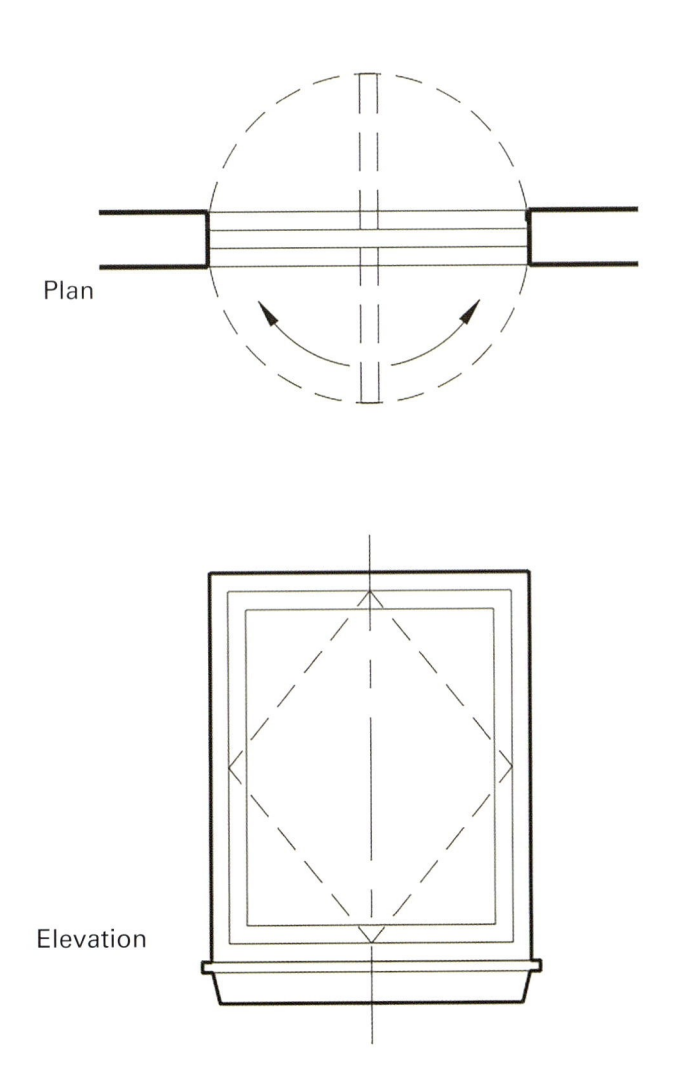

Elevation

Figure 8.40
Pivot window.

WINDOW TYPES **199**

Bay A **bay window** is a unit that can be installed on a flat wall (Figure 8.41). Its glass is typically fixed, double-hung or casement, or a combination. Its sides are angled 30° or 45° to the wall and the window projects out 18" to 24". Figure 8.42 shows how to draw one.

Plan

Elevation

Figure 8.41
Bay window.

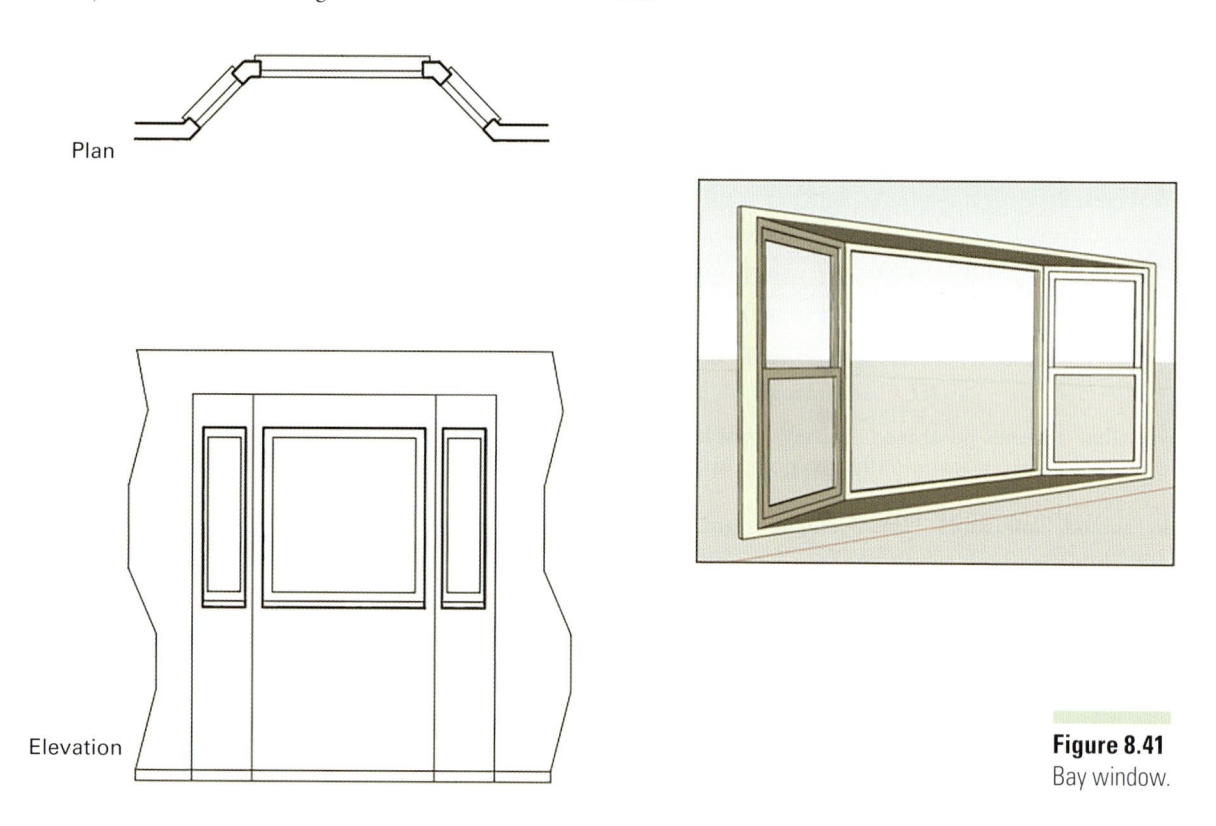

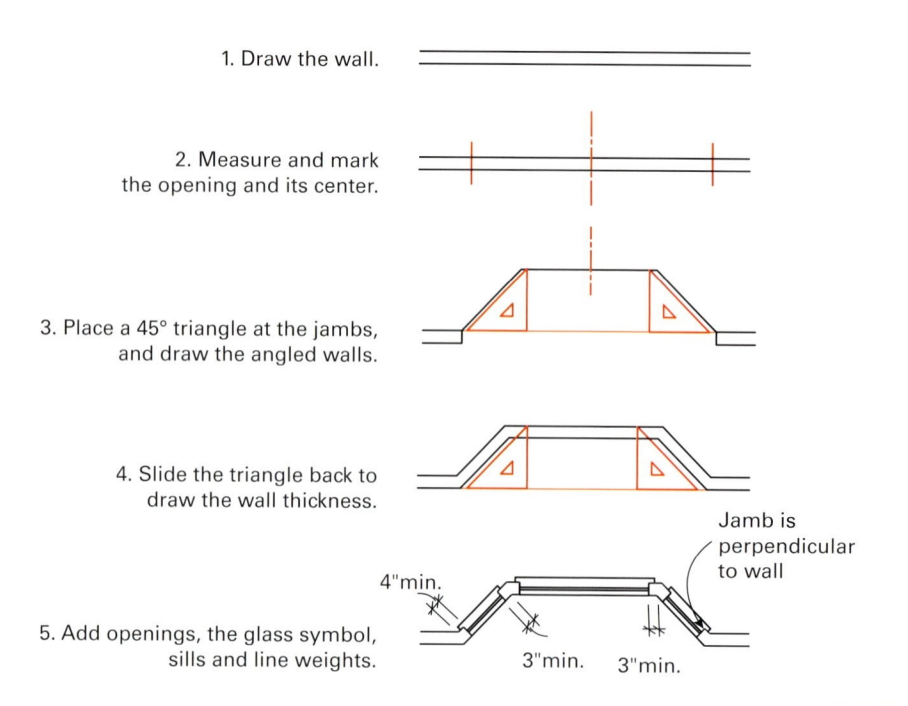

1. Draw the wall.

2. Measure and mark the opening and its center.

3. Place a 45° triangle at the jambs, and draw the angled walls.

4. Slide the triangle back to draw the wall thickness.

5. Add openings, the glass symbol, sills and line weights.

4"min.

3"min. 3"min.

Jamb is perpendicular to wall

Figure 8.42
How to draw a bay window.

200 CHAPTER 8 DOOR AND WINDOW SYMBOLS

Box Bay   Also called a **garden window**, a **box bay window** is a bay with sides that are 90° to the wall (Figure 8.43). It projects 12" to 18" from the wall.

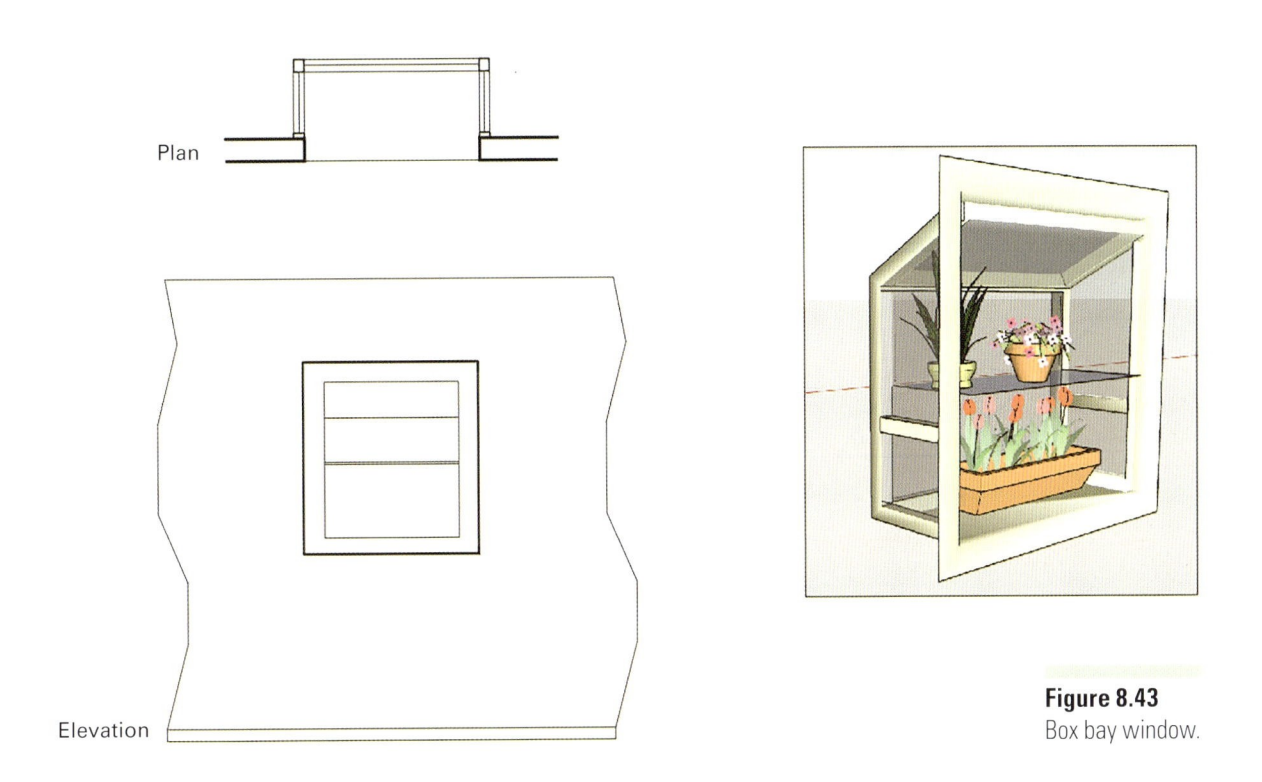

Figure 8.43
Box bay window.

Bow   A **bow window** projects from the wall and has narrower and more numerous glass panes than a bay window (Figure 8.44). The panes are installed to give the illusion of a curve. Bow windows can be fixed or operable. Figure 8.45 shows how to draw one.

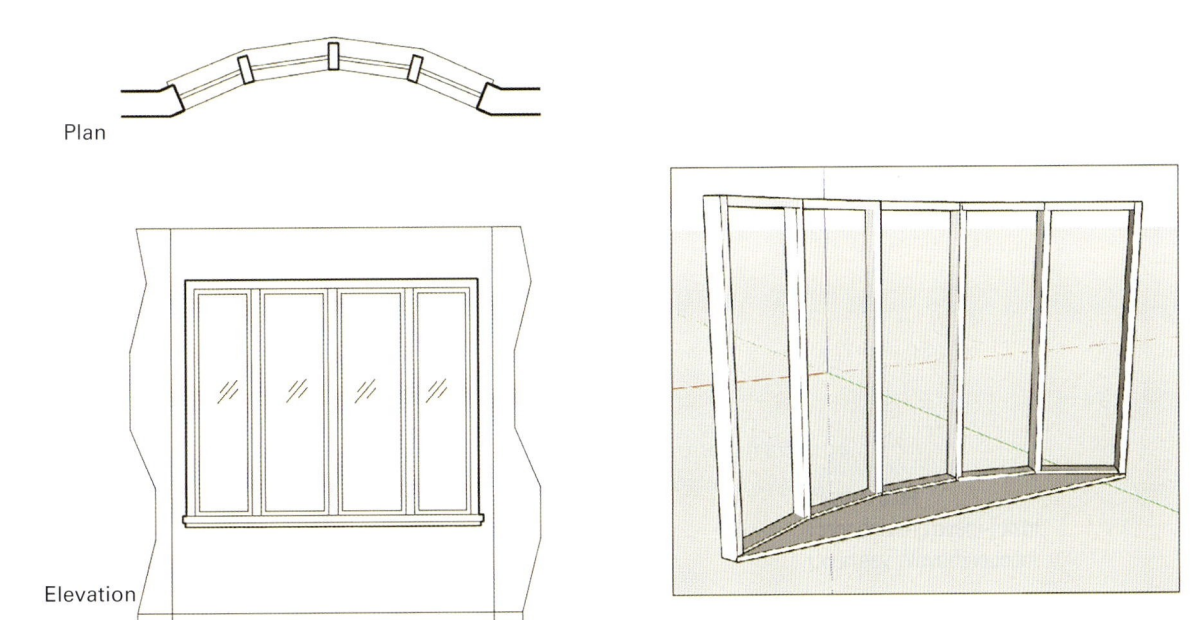

Figure 8.44
Bow window.

WINDOW TYPES   **201**

1. Measure and mark the wall opening. Position the compass by centering it on the opening and set its radius to the opening width. Swing an arc.

2. Move the compass's pivot point back the wall's thickness. Swing a second arc. Draw jambs.

3. Center a mullion on the arc. Measure and divide the window openings with a protractor.

4. Add glass symbol, sill, and line weights.

Figure 8.45 How to draw a bow window.

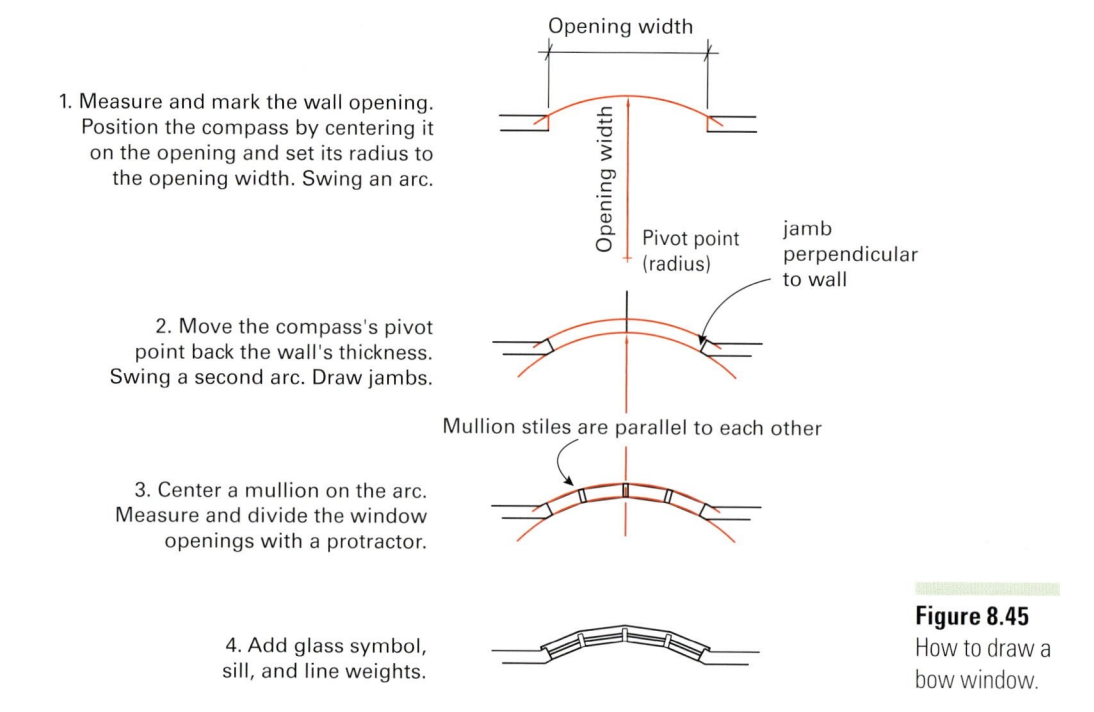

Specialty These windows are different shapes than any of the ones just discussed (Figure 8.46). They can be fixed or hinged, bought premade or custom cut.

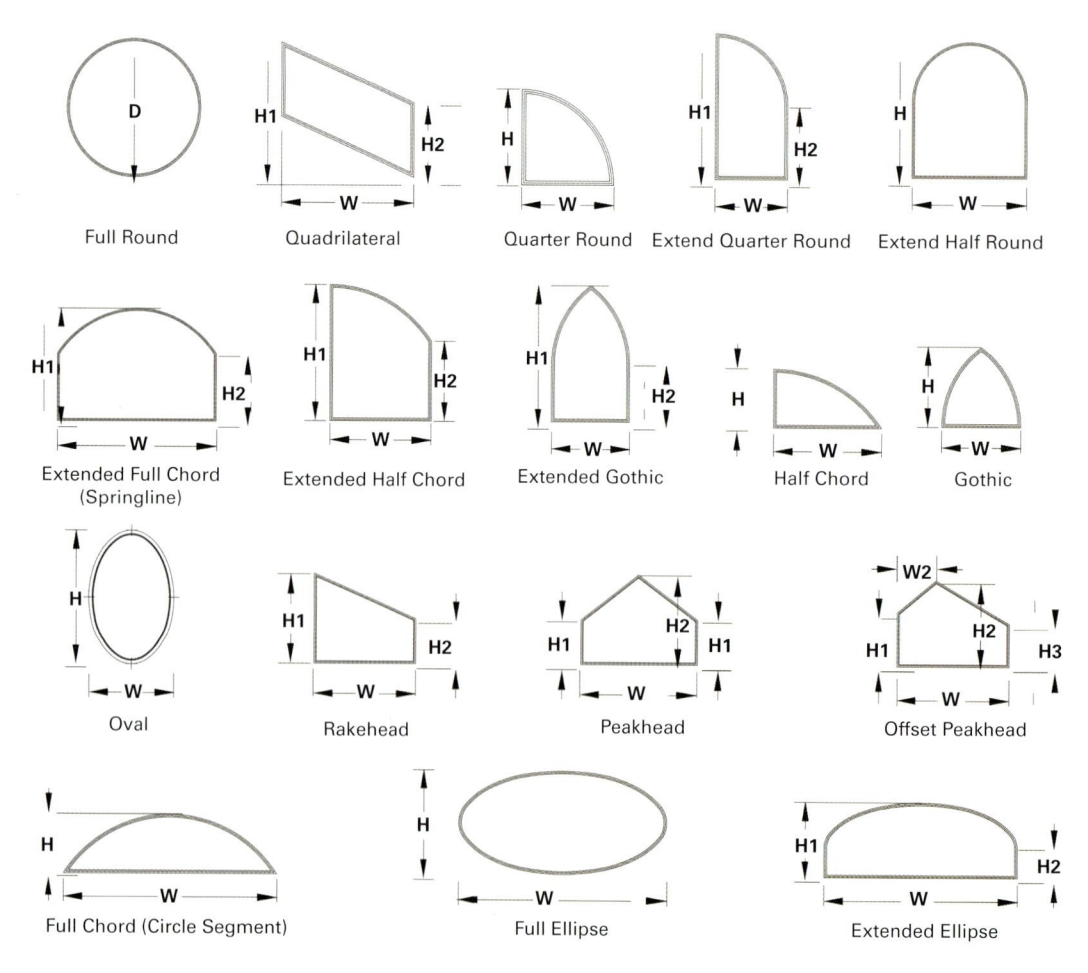

Full Round — Quadrilateral — Quarter Round — Extend Quarter Round — Extend Half Round

Extended Full Chord (Springline) — Extended Half Chord — Extended Gothic — Half Chord — Gothic

Oval — Rakehead — Peakhead — Offset Peakhead

Full Chord (Circle Segment) — Full Ellipse — Extended Ellipse

Figure 8.46 Specialty windows. Courtesy of Loewen Windows.

G2 Technical Guide Specialty Windows

202 CHAPTER 8 DOOR AND WINDOW SYMBOLS

Clerestory and Skylights **Clerestory** windows are windows set high in the wall (Figure 8.47). Their purpose is to admit light, not provide a view. **Skylights** are windows in the roof, and their purpose is also to provide light (Figure 8.48). They may also provide ventilation, a view and emergency egress. Outline a skylight on the floor plan with hidden lines and with solid lines on a **reflected ceiling plan** (plan of the ceiling and the components on it).

Figure 8.47
Clerestory windows in this mid-century modern house.

Figure 8.48
Skylights in a spa.

Window Placement in Elevation

Window placement in the wall varies with ceiling height and function. The sill (bottom) height is based on furniture, room arrangement, and view. The head (top) height may vary depending on ceiling height but is usually 6'-8" height above finished floor, same as the door heads. When deciding placement of a window in an elevation drawing, it is helpful to sketch a standing or sitting scale figure. A mistake beginning drafters often make is placing windows at unrealistic heights because they are not cognizant of eye level. Figure 8.49 shows eye levels, sill and head heights. Note that the door, window, and wall cabinet heads are all aligned.

Figure 8.49
Eye levels.

Window Definitions

Glazing A sheet of glass.

Trim (casing) A separate frame around the whole window that hides the joint where the window meets the walls.

Muntins Thin, non-structural bars that divide a large piece of glass into multiple panes (Figure 8.51). Draw them 2" wide.

Mullions Horizontal and vertical bars between individual windows (Figure 8.50).

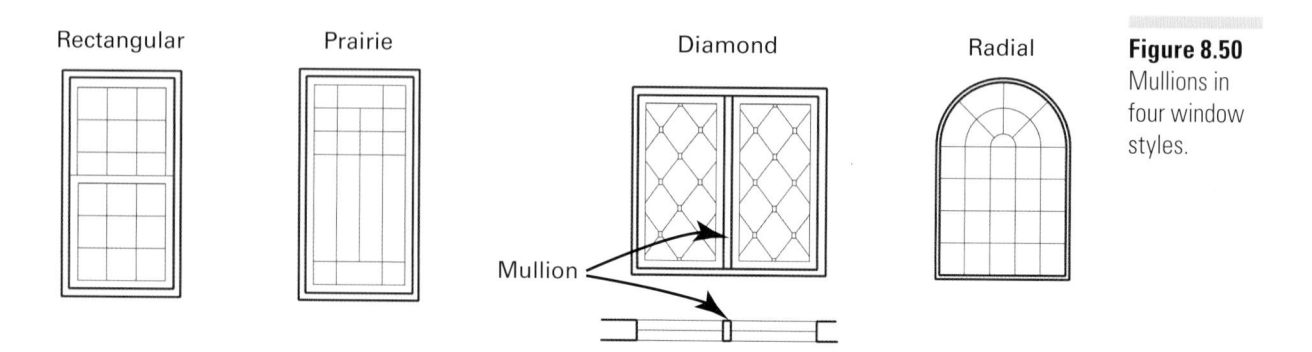

Figure 8.50
Mullions in four window styles.

Pane Framed sheet of glass within a window. If there are no muntins, the pane is the whole piece of glass, stile to stile, rail to rail.

Double-pane glass Two panes separated by an air space (Figure 8.51).

Unit size The overall dimensions of the window, including sashes, frame, weatherstripping, and hardware (Figure 8.52). Use **unit size** when drafting. Width is given first and height second.

Sash size The glass plus the frame that grips it. Sashes are measured as separate pieces; two sashes together are not measured as one.

Glass size The size of the glass alone.

Rough opening/masonry opening The hole in a framed or masonry wall.

Framed opening The hole with the frame installed. It's about 1/4" larger on all sides than the unit size.

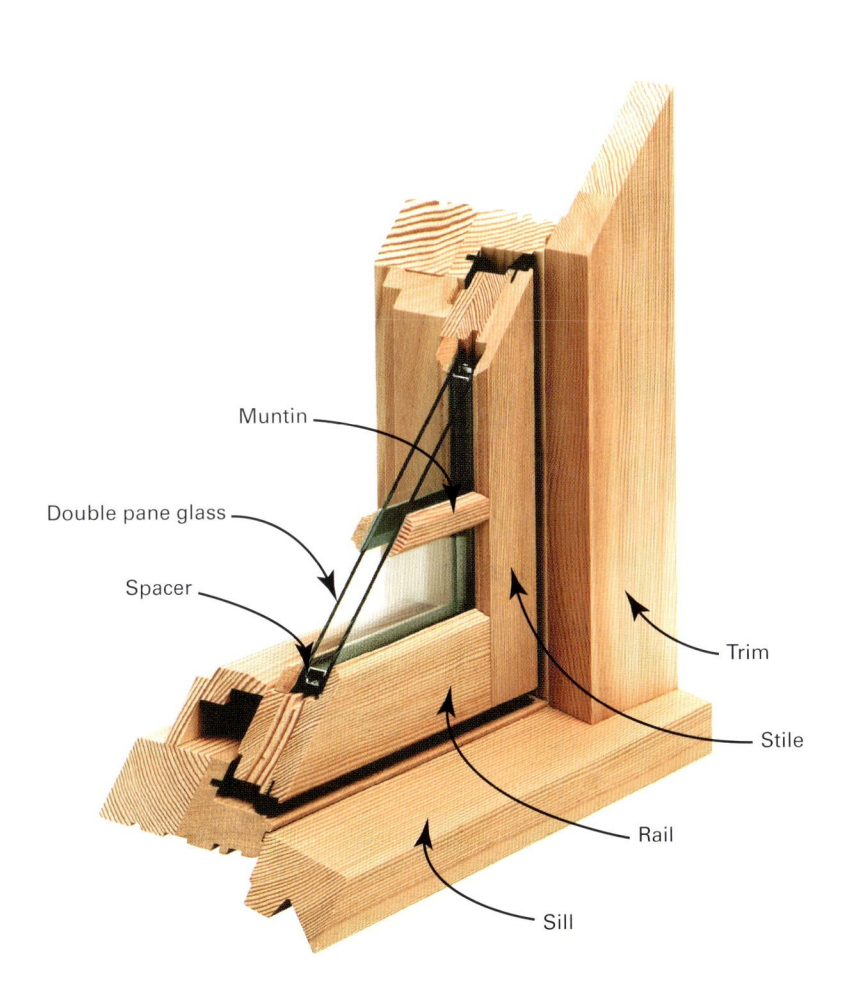

> **Exercise**
> Take photos of windows in elevation and trace them to familiarize yourself with their construction. Be detailed with the muntins, mullions, and trim.

Figure 8.51
Parts of a double-pane window. Courtesy Loewen Windows.

WINDOW DEFINITIONS   **205**

Window Sizes

Most windows range from 2'-0" to 12'-0" widths, in 6" increments. Consult product catalogs for specific brand sizes. Table 8.1 has generic window unit sizes for drafting reference.

Figure 8.52 Installing a double-hung unit inside a framed opening. Courtesy Pella Windows and Doors.

Table 8.1
COMMON WINDOW UNIT SIZES

Type	Width	Height	Notes
Double-Hung	1'-10", 2'-0", 2'-4", 2'-6", 2'-8", 2'-10", 3'-0", 3'-4", 3'-6", 3'-8", 3'-10", 4'-0", 4'-6"	2'-6", 3'-0", 3'-2", 3'-6", 4'-0", 4'-6", 4'-10", 5'-0", 5'-6", 5'-10", 6'-0"	
Horizontal Glider	3'-8", 4'-8", 5'-8", 6'-0", 6'-6"	2'-10", 3'-6", 4'-2", 4'-10", 5'-6", 6'-2"	For multiple units, overall scale is the sum of each unit plus 3/4" for each mullion used.
Casement	1'-6", 2'-0", 3'-0", 3'-4", 4'-0", 6'-0", 8'-0", 10'-0", 12'-0"	1'-0", 2'-0", 3'-0", 3'-4", 3'-6", 4'-0", 4'-6", 5'-0", 5'-4", 5'-6", 6'-0"	For multiple units, overall size is the sum of each unit plus 1/8" for each mullion used.
Fixed	1'-0", 1'-6", 2'-0", 2'-6", 3'-0", 4'-0", 5'-0", 5'-10", 6'-0"	4'-6", 4'-10", 5'-6", 6'-6", 7'-0", 7'-6", 8'-0"	
Hopper	2'-0", 2'-8", 3'-6", 4'-0"	1'-4", 1'-6", 1'-8", 2'-0", 4'-0", 6'-0"	
Louvered	1'-8", 2'-0", 2'-6", 3'-0"	3'-0", 4'-0", 5'-0", 6'-0"	
Bay/Bow	4'-0", 6'-0", 8'-0", 10'-0", 12'-0"	3'-0", 4'-0", 5'-0", 6'-0"	
Awning	2'-0", 2'-6", 2'-8", 3'-0", 3'-4", 4'-0", 5'-4", 6'-0", 6'-5", 8'-0", 10'-0", 12'-0"	1'-6", 2'-0", 3'-0", 3'-6", 4'-0", 5'-0", 6'-0"	For multiple units, overall size is the sum of each unit plus 1/8" for each stack used.
Fan	2'-0", 2'-4", 2'-6", 2'-10", 3'-0", 3'-6", 4'-0", 4'-8", 5'-0", 5'-4", 6'-0"	1'-6", 2'-0", 2'-8", 2'-10", 3'-0", 4'-0"	Height is chord height.

Figures 8.53 through 8.60 are charts of Loewen (Manitoba, Canada) window sizes. Find more of their windows and detailed information about them at https://www.loewen.com/professionals/literature/.

Casement
1 Wide Window Sizes

					Width						
Rough Opening					16 1/2 [419]	20 7/16 [519]	24 3/8 [619]	28 5/16 [719]	30 1/4 [769]	32 1/4 [819]	36 3/16 [919]
		Frame			15 3/4 [400]	19 11/16 [500]	23 5/8 [600]	27 9/16 [700]	29 1/2 [750]	31 1/2 [800]	35 7/16 [900]
Wood Exterior	**Metal Clad**		**Visible Glass**		9 3/4 [247]	13 11/16 [347]	17 5/8 [447]	21 9/16 [547]	23 1/2 [597]	25 1/2 [647]	29 7/16 [747]
25 1/16 [636]	24 3/8 [619]	23 5/8 [600]	17 3/8 [442]		CA1 0406	CA1 0506	CA1 0606	CA1 0706	CA1 7506		
29 [736]	28 5/16 [719]	27 9/16 [700]	21 5/16 [542]		CA1 0407	CA1 0507	CA1 0607	CA1 0707	CA1 7507	♦ CA1 0807	
30 15/16 [786]	30 1/4 [769]	29 1/2 [750]	23 5/2 [592]		CA1 0475	CA1 0575	CA1 0675	CA1 0775	♦ CA1 75757	CA1 0875	
32 15/16 [836]	32 1/4 [819]	31 1/2 [800]	25 1/4 [642]		CA1 0408	CA1 0508	CA1 0608	♦ CA1 0708	CA1 7508	♦ CA1 0808	♦ CA1 0908
36 7/8 [936]	36 3/16 [919]	35 7/16 [900]	29 3/16 [742]		CA1 0409	CA1 0509	CA1 0609	♦ CA1 0709	♦ CA1 7509	♦ CA1 0809	♦♦ CA1 0909
40 13/16 [1036]	40 1/8 [1019]	39 3/8 [1000]	33 1/8 [842]		CA1 0410	CA1 0510	♦ CA1 0610	♦ CA1 0710	♦ CA1 7510	♦♦ CA1 0810	♦♦ CA1 0910
42 11/16 [1084]	42 [1067]	41 1/4 [1048]	35 1/16 [890]		CA1 0411	CA1 0511	♦ CA1 0611	♦ CA1 0711	♦♦ CA1 7511	♦♦ CA1 0811	♦♦ CA1 0911
48 11/16 [1236]	48 [1219]	47 1/4 [1200]	41 [1042]		CA1 0412	CA1 0512	♦ CA1 0612	♦♦ CA1 0712	♦♦ CA1 7512	♦♦ CA1 0812	♦♦ CA1 0912
56 9/16 [1436]	55 7/8 [1419]	55 1/8 [1400]	48 7/8 [1242]		CA1 0414	CA1 0514	♦ CA1 0614	♦♦ CA1 0714	♦♦ CA1 7514	♦♦ CA1 0814	♦♦ CA1 0914
60 9/16 [1536]	59 13/16 [1519]	59 1/16 [1500]	52 13/16 [1342]		CA1 0415	CA1 0515	♦ CA1 0615	♦♦ CA1 0715	♦♦ CA1 7515	♦♦ CA1 0815	♦♦ CA1 0915
64 7/16 [1636]	63 3/4 [1619]	63 [1600]	56 3/4 [1442]		CA1 0416	CA1 0516	♦ CA1 0616	♦♦ CA1 0716	♦♦ CA1 7516	♦♦ CA1 0816	♦♦ CA1 0916
72 5/16 [1836]	71 5/8 [1819]	70 7/8 [1800]	64 5/8 [1642]		CA1 0418	CA1 0518	♦ CA1 0618	♦♦ CA1 0718	♦♦ CA1 7518	♦♦ CA1 0818	♦♦ CA1 0918
84 1/8 [2136]	83 7/16 [2119]	82 11/16 [2100]	76 7/16 [1942]		CA1 0421	CA1 0521	♦ CA1 0621	♦♦ CA1 0721	♦♦ CA1 7521	♦♦ CA1 0821	♦♦ CA1 0921

Height (left axis) | *Product Code* (right)

New oversized CA operator up to CA1 1224
- watch for sizes where temp, tri sash, sash stabilize etc. are read.
- rectangular
- roto only

Glass Size = Visible Glass + 15/16" (24mm)

Standard Sizes Shown. Additional sizes may be available. Custom sizes can be ordered.

Note:
- Available with Mission sash. For a unit with a Mission sash, subtract 13/16" (21mm) from the vertical visible glass measurement.
- SDL/Grille patterns are dependent on SDL/Grille type and window size. Please verify SDL/Grille patterns before confirming
- For Masonry opening information see page Section A
- Sizes are shown as double glazing, not all sizes may be available in triple glazing

- Egress calculations are based on a 20" width opening. Opening does not open a full 90°
  Meets US Egress: 20" in width 24" in height, with a total egress area of 5.7 Sq Ft.
  Meets Canadian Agress: 15" in width 15" in height, with a total egress area of 3.8 Sq Ft.
- Please check your local building codes for any variation in Egress Standards from those started above. For egress information, contact your Authorized Loewen Dealer.

B4 | Technical Guide   Casement Windows

Figure 8.53

Awning
1 Wide Window Sizes

Width

Rough Opening			20 7/16 [519]	24 3/8 [619]	28 5/16 [719]	30 1/4 [769]	32 1/4 [819]	36 3/16 [919]	40 1/8 [1019]	42 [1067]	42 [1067]	55 7/8 [1419]	59 13/16 [1519]	63 3/4 [1619]	71 5/8 [1819]
		Frame	19 11/16 [500]	23 5/8 [600]	27 9/16 [700]	29 1/2 [750]	31 1/2 [800]	35 7/16 [900]	39 3/8 [1000]	41 1/4 [1048]	47 1/4 [1200]	55 1/8 [1400]	59 1/16 [1500]	63 [1600]	70 7/8 [1800]
Wood Exterior	Metal Clad	Visible Glass	13 11/16 [347]	17 5/8 [447]	21 9/16 [547]	23 1/2 [597]	25 1/2 [647]	29 7/16 [747]	33 3/8 [847]	35 1/4 [895]	41 1/4 [1047]	49 1/8 [1247]	53 1/16 [1347]	56 15/16 [1447]	64 13/16 [1647]
17 3/16 [436]	16 1/2 [419]	15 3/4 [400]	9 1/2 [242]	AW1 0504	AW1 0604	AW1 0704	AW1 0504	AW1 0804	AW1 0904	AW1 1004	AW1 1104	AW1 1204	AW1 1404	AW1 1504	AW1 1604	AW1 1804
21 1/8 [536]	20 7/16 [519]	19 11/16 [500]	13 7/16 [342]	AW1 0505	AW1 0605	AW1 0705	AW1 0505	AW1 0805	AW1 0905	AW1 1005	AW1 1105	AW1 1205	AW1 1405	AW1 1505	AW1 1605	AW1 1805
25 1/16 [636]	24 3/8 [619]	23 5/8 [600]	17 3/8 [442]	AW1 0506	AW1 0606	AW1 0706	AW1 0506	AW1 0806	AW1 0906	AW1 1006	AW1 1106	AW1 1206	AW1 1406	AW1 1506	AW1 1606	AW1 1806
29 [736]	28 5/16 [719]	27 9/16 [700]	21 5/16 [542]	AW1 0507	AW1 0607	AW1 0707	AW1 0507	AW1 0807	AW1 0907	AW1 1007	AW1 1107	AW1 1207	AW1 1407	AW1 1507	AW1 1607	AW1* 1807
30 15/16 [786]	30 1/4 [769]	29 1/2 [750]	23 5/16 [750]	AW1 0575	AW1 0675	AW1 0775	AW1 0575	AW1 0875	AW1 0975	AW1 1075	AW1 1175	AW1 1275	AW1 1475	AW1 1575	AW1 1675	AW1* 1875
32 15/16 [1465]	34 1/4 [819]	31 1/2 [800]	25 1/4 [642]	AW1 0508	AW1 0608	AW1 0708	AW1 0508	AW1 0808	AW1 0908	AW1 1008	AW1 1108	AW1 1208	AW1 1408	AW1 1508	AW1 1608	AW1* 1808
36 7/8 [936]	36 3/16 [919]	35 7/16 [900]	29 3/16 [742]	AW1 0509	AW1 0609	AW1 0709	AW1 0509	AW1 0809	AW1 0909	AW1 1009	AW1 1109	AW1 1209	AW1 1409	AW1 1509	AW1 1609	AW1* 1809
40 13/16 [1036]	40 1/8 [1000]	39 3/8 [1000]	33 1/8 [842]	AW1 0510	AW1 0610	AW1 0710	AW1 0510	AW1 0810	AW1 0910	AW1 1010	AW1 1110	AW1 1210	AW1 1410	AW1 1510	AW1 1610	AW1 1810
42 11/16 [1084]	42 [1067]	41 1/4 [1048]	35 1/16 [890]	AW1 0511	AW1 0611	AW1 0711	AW1 0511	AW1 0811	AW1 0911	AW1 1011	AW1 1111	AW1 1211	AW1 1411	AW1 1511	AW1 1611	AW1 1811
48 11/16 [1236]	48 [1219]	47 1/4 [1200]	41 [1042]	AW1 0512	AW1 0612	AW1 0712	AW1 0512	AW1 0812	AW1 0912	AW1 1012	AW1 1112	AW1 1212	AW1 1412	AW1 1512	AW1 1612	AW1* 1812
56 9/16 [1973]	56 7/8 [1419]	55 1/8 [1400]	48 7/8 [1242]	AW1* 0514	AW1* 0614	AW1* 0714	AW1* 0514	AW1* 0814	AW1* 0914	AW1* 1014	AW1* 1114	AW1* 1214	AW1* 1414	AW1* 1514	AW1* 1614	AW1* 1814
60 1/2 [1536]	59 13/16 [1519]	59 1/16 [1500]	52 13/16 [1342]	AW1* 0515	AW1* 0615	AW1* 0715	AW1* 0515	AW1* 0815	AW1* 0915	AW1* 1015	AW1* 1115	AW1* 1215	AW1* 1415	AW1* 1515	AW1* 1615	
64 7/16 [1636]	64 3/4 [1600]	63 [1600]	56 3/4 [1442]	AW1* 0516	AW1* 0616	AW1* 0716	AW1* 0516	AW1* 0816	AW1* 0916	AW1* 1016	AW1* 1116	AW1* 1216	AW1* 1416	AW1* 1516	AW1* 1616	
75 5/16 [1836]	71 5/8 [1819]	70 7/8 [1800]	64 5/8 [1642]	AW1* 0518	AW1* 0618	AW1* 0718	AW1* 0518	AW1* 0818	AW1* 0918	AW1* 1018	AW1* 1118	AW1* 1218	AW1* 1418	AW1* 1518		

Height (left axis) — **Product Code**

Glass Size = Visible Glass + 15/16" (24mm)

Standard Sizes Shown. Additional sizes may be available. Custom sizes can be ordered.

Note:
- Available with Mission sash. For unit with a Mission shash, subtract 13/16" (21mm) from the vertical visible glass measurement.
- SDL/Grille patterns are dependent on SDL/Grille type and window size. Please verify SDL/Grille patterns before confirming your order.
- Available with Push Out hardware
- For Masonry opening information see page A24-A29.
- Sizes are shown as double glazing, not all sizes may be available in triple glazing

- Egress calculations are based on the unit being opened at 70° for washability hinge, 80° for egress hinge, and having minimum clear openings of:
  - ★ Meets US Egress: 20" in width, 24" in height, with a total egress area of 5.7 Sq Ft (Push Out Only)
  - ✦ Meets Canadian Egress: 15" in width, 15" in height, with a total egress area of 3.8 Sq Ft
- Please check your local building codes for any variation in Egress Standards from those stated above. For egress information, contact your Authorized Loewen Dealer.

C4 | Technical Guide   Awning Windows

Figure 8.54

Double/Single Hung Fixed Window Sizes

Rough Opening				Width						
Wood Exterior	Metal Clad	Frame	Visible Glass	16 1/2 [419] / 15 3/4 [400] / 11 5/48 [296]	20 7/2 [519] / 19 11/16 [500] / 15 9/16 [396]	24 3/8 [619] / 23 5/8 [600] / 19 1/2 [496]	28 5/16 [719] / 27 9/16 [700] / 23 7/16 [596]	30 1/4 [769] / 29 1/2 [750] / 25 7/16 [646]	32 1/4 [819] / 31 1/2 [800] / 27 3/8 [696]	36 3/16 [919] / 35 7/16 [900] / 31 5/16 [796]	40 1/8 [1019] / 39 3/8 [1000] / 35 1/4 [896]
13 1/4 [336]	12 9/16 [319]	11 13/16 [300]	7 11/16 [196]	PS1 0403	PS1 0503	PS1 0603	PS1 0703	PS1 7503	PS1 0803	PS1 0903	PS1 1003
17 3/16 [436]	16 1/2 [419]	15 3/4 [400]	11 5/8 [296]	PS1 0404	PS1 0504	PS1 0604	PS1 0704	PS1 7504	PS1 0804	PS1 0904	PS1 1004
21 1/8 [536]	20 7/16 [519]	19 11/16 [500]	15 9/16 [396]	PS1 0405	PS1 0505	PS1 0605	PS1 0705	PS1 7505	PS1 0805	PS1 0905	PS1 1005
25 1/16 [636]	24 3/8 [619]	23 5/8 [600]	19 1/2 [496]	PS1 0406	PS1 0506	PS1 0606	PS1 0706	PS1 7506	PS1 0806	PS1 0906	PS1 1006
29 [1364]	29 5/16 [719]	27 9/16 [700]	23 7/16 [596]	PS1 0407	PS1 0507	PS1 0607	PS1 0707	PS1 7507	PS1 0807	PS1 0907	PS1 1007
30 15/16 [786]	30 1/4 [769]	29 1/2 [700]	25 7/16 [646]	PS1 0475	PS1 0575	PS1 0675	PS1 0775	PS1 7575	PS1 0875	PS1 0975	PS1 1075
32 15/16 [836]	32 1/4 [819]	31 1/2 [800]	27 3/8 [696]	PS1 0408	PS1 0508	PS1 0608	PS1 0708	PS1 7508	PS1 0808	PS1 0908	PS1 1008
36 7/8 [936]	36 3/16 [919]	35 7/16 [900]	31 5/16 [796]	PS1 0409	PS1 0509	PS1 0609	PS1 0709	PS1 7509	PS1 0809	PS1 0909	PS1 1009
40 11/16 [1770]	40 1/8 [1770]	39 3/8 [1000]	35 1/4 [1896]	PS1 0410	PS1 0510	PS1 0610	PS1 0710	PS1 7510	PS1 0810	PS1 0910	PS1 1010
48 11/16 [1236]	48 [1219]	47 1/4 [1200]	43 1/8 [1096]	PS1 0412	PS1 0512	PS1 0612	PS1 0712	PS1 7512	PS1 0812	PS1 0912	PS1 1012
56 9/16 [1436]	56 7/8 [1419]	55 1/8 [1400]	51 [1296]	PS1 0414	PS1 0514	PS1 0614	PS1 0714	PS1 7514	PS1 0814	PS1 0914	PS1 1014
60 1/2 [1536]	59 13/16 [1519]	59 1/16 [1500]	54 15/16 [1396]	PS1 0415	PS1 0515	PS1 0615	PS1 0715	PS1 7515	PS1 0815	PS1 0915	PS1 1015
64 7/16 [1636]	63 3/4 [1619]	63 [1600]	587/8 [1496]	PS1 0416	PS1 0516	PS1 0616	PS1 0716	PS1 7516	PS1 0816	PS1 0916	PS1 1016
72 5/16 [1836]	71 5/8 [1819]	70 7/8 [1800]	66 [1696]	PS1 0418	PS1 0518	PS1 0618	PS1 0718	PS1 7518	PS1 0818	PS1 0918	PS1 1018
84 1/8 [2136]	83 7/16 [2119]	82 11/16 [2100]	78 9/16 [1996]	PS1 0421	PS1 0521	PS1 0621	PS1 0721	PS1 7521	PS1 0821	PS1 0921	PS1 1021
95 15/16 [2457]	95 1/4 [2419]	94 1/2 [2400]	90 3/8 [2296]		PS1 0524	PS1 0624	PS1 0724	PS1 7524	PS1 0824	PS1 0924	PS1 1024
119 9/16 [3036]	118 7/8 [3019]	118 1/8 [3000]	114 [2896]			PS1 0630	PS1 0730	PS1 7530	PS1 0830	PS1 0930	PS1 1030

Product Code

Glass Size = Visible Glass + 15/16" (24mm)

Standard Sizes Shown. Additional sizes may be available. Custom sizes can be ordered.
Note: • For Masonry opening information see Section A.

Figure 8.55

Double/Single Hung
1 Wide Window Sizes

				Width								
Rough Opening				22 9/16 [573]	26 9/16 [675]	30 9/16 [777]	32 9/16 [827]	34 9/16 [878]	36 9/16 [929]	38 9/16 [980]	42 9/16 [1081]	46 9/16 [1183]
		Frame		22 13/16 [554]	25 13/16 [656]	29 13/16 [758]	31 13/16 [808]	34 13/16 [859]	36 13/16 [910]	37 13/16 [961]	41 13/16 [1062]	45 13/16 [1164]
Wood Exterior	Metal Clad		Visible Glass	15 1/16 [382]	19 1/16 [484]	23 9/16 [586]	25 1/16 [636]	27 1/16 [687]	29 1/16 [738]	31 1/16 [789]	35 1/16 [890]	39 1/16 [996]
33 11/16 [856]	33 1/16 [839]	32 5/16 [820]	11 [280]	DH1 1612	DH1 2012	DH1 2412	DH1 2612	DH1 2812	DH1 3012			
37 11/16 [957]	37 [940]	36 1/4 [921]	13 [330]	DH1 1614	DH1 2014	DH1 2414	DH1 2614					
41 11/16 [1059]	41 [1042]	40 1/4 [1023]	15 [381]	DH1 1616	DH1 2016	DH1 2416	DH1 2616	DH1 2816	DH1 3016	DH1 3216	DH1 3616	DH1 4016
49 11/16 [1262]	49 [1245]	48 1/4 [1226]	19 [483]	DH1 1620	DH1 2020	DH1 2420	DH1 2620	DH1 2820	DH1 3020	DH1 3220	DH1 3620	DH1 4020
53 11/16 [1364]	53 1/16 [1347]	52 5/16 [1328]	21 [534]	DH1 1622	DH1 2022	DH1 2422	DH1 2622	DH1 2822	DH1 3022	DH1 3222	DH1 3622	DH1 4022
57 11/16 [1465]	57 [1448]	57 1/4 [1429]	23 [584]	DH1 1624	DH1 2024	DH1 2424	DH1 2624	DH1 2824	DH1 3024	DH1 3224	DH1 3624	DH1 4024
61 11/16 [1567]	61 [1550]	60 1/4 [1531]	25 [635]	DH1 1626	DH1 2026	DH1 2426	DH1 2626	DH1 2826	DH1 3026	DH1 3226	DH1 3626	DH1 4026
65 11/16 [1668]	65 [1651]	64 1/4 [1632]	27 [686]	DH1 1628	DH1 2028	DH1 2428	DH1 2628	DH1 2828	DH1 3028	DH1 3228	DH1 3628	DH1 4028
69 11/16 [1770]	69 [1753]	68 1/4 [1734]	29 [737]	DH1 1630	DH1 2030	DH1 2430	DH1 2630	DH1 2830	DH1 3030	DH1 3230	DH1 3630	DH1 4030
73 11/16 [1872]	73 1/16 [1855]	72 5/16 [1836]	31 [788]	DH1 1632	DH1 2032	DH1 2432	DH1 2632	DH1 2832	DH1 3032	DH1 3232	DH1 3632	DH1 4032
77 11/16 [1973]	77 [1956]	76 1/4 [1937]	33 [838]	DH1 1634	DH1 2034	DH1 2434	DH1 2634	DH1 2834	DH1 3034	DH1 3234	DH1 3634	DH1 4034
81 11/16 [2075]	81 [2058]	80 1/4 [2039]	35 [889]	DH1 1636	DH1 2036	DH1 2436	DH1 2636	DH1 2836	DH1 3036	DH1 3236	DH1 3636	DH1 4036
85 11/16 [2176]	85 [2159]	84 1/4 [2140]	37 [940]	DH1 1638	DH1 2038	DH1 2438	DH1 2638	DH1 2838	DH1 3038	DH1 3238	DH1 3638	DH1 4038

Glass Size = Visible Glass + 15/16" (24mm)

Standard Sizes Shown. Additional sizes may be available. Custom sizes can be ordered.

Note:
- Substitute SH1 for DH1 when ordering a Single Hung unit.
- SDL/Grille patterns are dependent on SDL/Grille type and window size. Please verify SDL/Grille patterns before confirming your order.
- For Masonry opening information see page A30-A31.
- Sizes are shown as double glazing, not all sizes may be available in triple glazing.
- Only available in Liberty (non-tilt).

- Egress calculations are based on the unit being opened 70° for washability hinge, 80° for egress hinge, and having minimum clear openings of:
  ★ Meets US Egress: 20" in width 24" in height, with a total egress area of 5.7 Sq Ft.
  ✦ Meets Canadian Egress: 15" in width 15" in height, with a total egress area of 3.8 Sq Ft.
- Please check your local building codes for any variation in Egress Standards from those started above. For egress information, contact your Authorized Loewen Dealer.

Figure 8.56

Double/Single Hung Fixed Window Sizes

Rough Opening				Width					
				42 9/16 [1081]	46 9/16 [1183]	50 9/16 [1285]	54 9/16 [1386]	62 9/16 [1589]	70 9/16 [1793]
		Frame		41 13/16 [1062]	45 13/16 [1164]	49 13/16 [1266]	52 13/16 [1367]	61 13/16 [1570]	69 13/16 [1774]
Wood Exterior	Metal Clad		Visible Glass	35 1/16 [890]	39 1/16 [992]	43 1/16 [1094]	47 1/16 [1195]	55 1/16 [1398]	63 1/16 [1602]
33 11/16 [856]	33 1/16 [839]	32 5/16 [820]	24 1/16 [611]	DF 3625	DF 4025	DF 4425	DF 4825	DF 5625	DF 6425
37 11/16 [957]	37 [940]	26 1/4 [921]	28 1/16 [712]	DF 3629	DF 4029	DF 4429	DF 4829	DF 5629	DF 6429
41 11/16 [1059]	41 [1042]	40 1/4 [1023]	32 1/16 [814]	DF 3633	DF 4033	DF 4433	DF 4833	DF 5633	DF 6433
49 11/16 [1262]	49 [1245]	48 1/4 [1226]	40 1/16 [1017]	DF 3641	DF 4041	DF 4441	DF 4841	DF 5641	DF 6441
53 11/16 [1364]	53 1/16 [1347]	52 5/16 [1328]	44 1/16 [1119]	DF 3645	DF 4045	DF 4445	DF 4845	DF 5645	DF 6445
57 11/16 [1465]	57 [1448]	56 1/4 [1429]	48 1/16 [1220]	DF 3649	DF 4049	DF 4449	DF 4849	DF 5649	DF 6449
61 11/16 [1567]	61 [1550]	60 1/4 [1531]	52 1/16 [1322]	DF 3653	DF 4053	DF 4453	DF 4853	DF 5653	DF 6453
65 11/16 [1668]	65 [1651]	64 1/4 [1632]	56 [1423]	DF 3657	DF 4057	DF 4457	DF 4857	DF 5657	DF 6457
69 11/16 [1770]	69 [1753]	68 1/4 [1734]	60 1/16 [1525]	DF 3661	DF 4061	DF 4461	DF 4861	DF 5661	DF 6461
73 11/16 [1872]	73 1/16 [1855]	72 5/16 [1836]	64 1/16 [1627]	DF 3665	DF 4065	DF 4465	DF 4865	DF 5665	DF* 6465
77 11/16 [1973]	77 [1956]	76 1/4 [1937]	68 1/16 [1728]	DF 3669	DF 4069	DF 4469	DF 4869	DF 5669	DF* 6469
81 11/16 [2057]	81 [2058]	81 1/4 [2039]	72 1/16 [1830]	DF 3673	DF 4073	DF 4473	DF 4873	DF* 5673	DF* 6473
85 11/16 [2176]	85 [2159]	84 1/4 [2140]	76 [1931]	DF 3677	DF 4077	DF 4477	DF 4877	DF* 5677	DF* 6477

Product Code

Glass Size = Visible Glass + 15/16" (24mm)

Standard Sizes Shown. Additional sizes may be available. Custom sizes can be ordered.

Note:
- SDL/Grille patterns are dependent on SDL/Grille type and window size. Please verify SDL/Grille patterns before confirming your order.
- For Masonry opening information see section A20-A31.
- Sizes are shown as double glazing, not all sizes may be available in triple glazing.
- Only available with a Grand Sash.

E6 | Technical Guide   Double/Single Hung Windows

Figure 8.57

Bow & Bay (Frame Wall) Window Sizes

Width

		34° Kitchen Bay			34° Bay			
Rough Opening		57 3/8 [1458]	65 1/4 [1658]	73 1/8 [1858]	94 1/8 [2390]	101 15/16 [2590]	109 13/16 [2790]	117 11/16 [2990]
	Frame	54 13/16 [1393]	62 11/16 [1593]	70 9/16 [1793]	92 1/2 [2325]	99 13/8 [2525]	107 1/4 [2727]	115 1/8 [2925]
		Projections: Framed Wall 2 x 4 = 10 9/16" (268mm) 2 x 6 = 10 9/16" (268mm)			Projections: Framed Wall 2 x 4 = 14 15/16" (380mm) 2 x 6 = 14 15/16" (380mm)			
Wood Exterior	**Metal Clad**	15 3/4 - 23 5/8 - 15 3/4 [400 - 600 - 400]	12 1/2 - 31 1/2 - 12 1/2 [400-800-400]	15 3/4 - 39 3/8 - 15 3/4 [400-1000-400]	23 5/8 - 47 1/4 - 23 5/8 [600-1200-600]	23 5/8 - 55 1/8 - 23 5/8 [600-1400-600]	23 5/8 - 63 - 23 5/8 [600-1600-600]	23 5/8 - 70 7/8 - 23 5/8 [600-1800-600]
36 7/8 [936]	36 3/16 [919]	35 7/16 [919]	BY3 1409	BY3 1609	BY3 1809			
40 13/16 [1036]	40 1/8 [1019]	39 3/8 [1000]	BY3 1410	BY3 1610	BY3 1810			
48 16 1/2 [1236]	48 [1219]	47 1/4 [1200]				BY3 2412	BY3 2612	BY3 2812	BY3 3012
56 9/8 [1436]	56 7/8 [1419]	55 1/8 [1400]				BY3 2414	BY3 2614	BY3 2814	BY3 3014
60 1/2 [1536]	59 13/16 [1519]	59 1/16 [1500]				BY3 2415	BY3 2615	BY3 2815	BY3 3015
64 7/16 [1636]	63 3/4 [1619]	63 [1600]				BY3 2416	BY3 2616	BY3 2816	BY3 3016
72 5/16 [1836]	71 5/8 [1819]	71 7/8 [1800]				BY3 2418	BY3 2618	BY3 2818	BY3 3018

Product Code

Glass Size = Visible Glass + 15/16" (24mm)

[1] Wood exterior:
Requires additional 3/4" (19mm) for optional head

[2] Metal Clad:
Requires additional 3/4" (19mm) for optional head
Requires additional 3/4" (19mm) for optional seat

Standard Sizes Shown. Additional sizes may be available. Custom sizes can be ordered.

Note:
- For glass sizes, see casement section.
- For Masonry opening information see section A.

Figure 8.58

Patio Door And Transom
2 Wide Door Sizes

			Width		
Rough Opening			59 13/16 [1519]	71 5/8 [1819]	95 1/4 [2419]
	Frame		59 1/16 [1500]	70 7/8 [1800]	94 1/2 [2400]
		Visible Glass	24 1/8 [612]	30 [762]	41 13/16 [1062]
			TRANSOM		
16 1/2 [419]	15 3/4 [400]	10 1/2 [266]	WPT2* 1504	WPT2* 1804	WPT2* 2404
24 3/8 [619]	23 5/8 [600]	18 3/8 [466]	WPT2* 1506	WPT2* 1806	WPT2* 2406
			DOOR		
80 1/8 [2035]	79 1/2 [2020]	70 1/2 [1790]	WP2 1520	WP2 1820	WP2 2420
81 7/8 [2080]	81 5/16 [2065]	72 1/4 [1835]	WP2 1568	WP2 1806	WP2 2468
83 1/4 [2115]	82 11/16 [2100]	73 5/8 [1870]	WP2 1521	WP2 1821	WP2 2421
86 9/16 [2199]	86 [2184]	76 15/16 [1954]	WP2 1570	WP2 1870	WP2 2470
95 1/16 [2415]	94 1/2 [2400]	85 7/16 [2170]	WP2 1524	WP2 1824	WP2 2424
98 [2489]	97 3/8 [2474]	88 3/8 [2244]	WP2 1580	WP2 1880	WP2 2480

Height (leftmost column) — Product Code

Glass Size = Visible Glass + 15/16" (24mm)

Standard Sizes Shown. Additional sizes may be available. Custom sizes can be ordered.

Note:
- SLD/Grille patterns are dependent on SDL/Grille type and window size. Please verify SDL/Grille patterns before confirming your order.
- O = Fixed, X = Operating
- Transom units.

J4 | Technical Guide   Sliding Patio & French Doors

Figure 8.59

Bifold Door Sizing Charts

Width

		34 11/16" [881]	46 1/2" [1181]	60 3/16" [1529]	66 3/4" [1695]	72" [1810]	79 7/8" [2029]
R.O.							
	2 Panel Frame	33 15/16" [862]	45 3/4" [1162]	59 7/8" [2259]	66" [2508]	71 1/4" [1810]	118 7/16" [3009]
R.O.		50 11/16" [1287]	68 3/8" [1737]	88 15/16" [2259]	98 3/4" [2508]	106 5/8" [2709]	118 7/16" [3009]
	3 Panel Frame	49 15/16" [1268]	67 5/8" [1718]	88 3/16" [2240]	98" [2489]	105 7/8" [2690]	117 11/16" [2990]
R.O.		66 5/8" [1693]	90 1/4" [2293]	117 11/16" [2989]	130 3/4" [3321]	141 15/16" [3589]	157 1/16" [3989]
	4 Panel Frame	65 7/8" [1674]	89 1/2" [2274]	116 15/16" [2970]	130" [3302]	140 9/16" [3570]	156 516" [3970]
R.O.		82 5/8" [2099]	112 3/16" [2849]	146 7/16" [3719]	162 3/4" [4134]	175 15/16" [4469]	195 5/8" [4969]
	5 Panel Frame	81 7/8" [2080]	111 7/16" [2830]	145 11/16" [3700]	162" [4115]	175 3/16" [4450]	194 7/8" [4950]
R.O.		98 5/8" [2506]	134 1/16" [3405]	175 3/16" [4430]	194 3/4" [4947]	210 9/16" [5349]	234 3/16" [5949]
	6 Panel Frame	97 7/8" [2486]	133 5/16" [3386]	174 7/16" [4430]	194" [4928]	209 13/16" [5330]	233 7/16" [5930]
R.O.		114 5/8" [2911]	155 15/16" [3961]	203 7/8" [5179]	226 3/4" [5760]	245 1/4" [6229]	272 13/16" [6929]
	7 Panel Frame	113 7/8" [2892]	155 3/16" [3942]	203 1/8" [5106]	226" [5741]	244 1/28" [6210]	272 1/19" [6910]
R.O.		130 9/16" [3317]	177 13/16" [4517]	232 5/8" [5909]	258 3/4" [6573]	279 7/8" [7109]	311 3/8" [7909]
	8 Panel Frame	129 13/16" [3298]	177 1/16" [4498]	231 7/8" [6639]	258" [6554]	279 1/8" [7090]	310 5/8" [7890]
R.O.		146 9/16" [3723]	199 3/4" [5073]	261 3/8" [6639]	290 13/16" [7386]	314 1/2" [7989]	349 15/16" [8889]
	9 Panel Frame	145 13/16" [3704]	199" [5054]	260 5/8" [6620]	290 1/16" [7367]	13 3/4" [7970]	349 3/16" [8870]
R.O.		162 9/16" [4129]	221 5/8" [5629]	290 1/8" [7369]	322 13/16" [8199]	349 3/16" [8869]	388 9/16" [9869]
	10 Panel Frame	161 13/16" [4110]	220 7/8" [5610]	289 3/8" [7350]	322 1/16" [8180]	348 7/16" [8850]	387 13/16" [9850]
R.O.		178 9/16" [4535]	243 1/2" [6185]	318 7/8" [8099]	354 13/16" [9012]	383 13/16" [9749]	427 1/8" [10849]
	11 Panel Frame	177 13/16" [4516]	242 3/4" [6166]	318 1/8" [8080]	354 1/16" [8993]	383 1/16" [9730]	426 3/8" [10830]
R.O.		194 1/2" [4941]	265 3/8" [6741]	347 5/8" [8829]	386 13/16" [9825]	418 7/16" [10629]	465 11/16" [11829]
	12 Panel Frame	193 3/4" [4922]	264 5/8" [6722]	346 7/8" [8810]	386 1/16" [9806]	417 11/16" [10610]	464 15/16" [11810]
R.O.		210 1/2" [5347]	287 5/16" [7297]	376 5/16" [9559]	418 13/16" [10638]	453 1/8" [11509]	504 4/16" [12809]
	13 Panel Frame	209 3/4" [5328]	286 9/16" [7278]	357 9/16" [9540]	418 1/16" [10619]	452 3/8" [11490]	503 9/16" [12790]
R.O.		226 1/2" [5753]	309 3/16" [7853]	405 1/16" [10289]	450 13/16" [11451]	487 3/4" [12389]	542 7/8" [13789]
	14 Panel Frame	225 3/4" [5734]	308 7/16" [7834]	404 5/16" [10270]	450 1/16" [11432]	487" [12370]	542 1/8" [13770]
R.O.		241 1/2" [6159]	331 1/16" [8409]	433 13/16" [11019]	482 13/16" [12264]	522 3/8" [13269]	581 7/16" [14768]
	15 Panel Frame	241 3/4" [6140]	330 5/16" [8390]	433 1/16" [11000]	482 1/16" [12245]	521 5/8" [13250]	580 11/16" [14750]
R.O.		258 7/16" [6565]	352 15/16" [8965]	462 9/16" [11749]	514 13/16" [13077]	557 1/16" [14149]	620" [15749]
	16 Panel Frame	257 11/16" [6546]	352 3/16" [8946]	461 13/16" [11730]	514 1/16" [13058]	556 6/16" [14130]	619 1/4" [15730]

				15 13/16" [402]	21 3/4" [552]	28 9/16" [726]	31 7/8" [809]	34 1/2" [876]	38 7/16" [976]
R.O.	Frame	Panel	Visible	7 3/16" [182]	13 1/16" [332]	19 15/16" [506]	23 3/16" [589]	25 13/16" [656]	29 3/4" [756]
81 5/6 [2066]	80 9/6 [2047]	76 1/8 [1933]	63 7/8 [1623]	BiFold	BiFold	BiFold	BiFold	BiFold	BiFold
83 1/8 [2111]	82 3/8 [2092]	88 7/8 [1978]	65 11/16 [1668]	BiFold	BiFold	BiFold	BiFold	BiFold	BiFold
87 13/16 [2230]	87 1/16 [2211]	82 9/16 [2097]	703/8 [1787]	BiFold	BiFold	BiFold	BiFold	BiFold	BiFold
96 5/16 [2446]	95 9/16 [2427]	91 1/16 [2313]	78 7/8 [2003]	BiFold	BiFold	BiFold	BiFold	BiFold	BiFold
99 5/16 [2520]	98 7/16 [2501]	94 [2387]	81 3/4 [2077]	BiFold	BiFold	BiFold	BiFold	BiFold	BiFold
100 7/8 [2562]	100 1/8 [2543]	95 5/8 [2429]	83 7/16 [2119]	BiFold	BiFold	BiFold	BiFold	BiFold	BiFold
108 1/8 [2746]	107 3/8 [2727]	102 7/8 [2613]	90 1116 [2303]	BiFold	BiFold	BiFold	BiFold	BiFold	BiFold

R.O = Frame Height + 3/4" [19mm] • Note: Bifold doors using standard TD panels will be 1 1/16" [27mm] higher than standard TD/FD units. If astragal is used for TwinPointlock add 1" frame width.

Note:
- The preparation of the rough opening for large openings such as those required by, but not limited to, LiftSlide, Bifold and MultiSlide doors have unique requirements. Structural that allow for deflection no greater than 1/8" along the unsupported length once the header is full loaded are required. Special care needs to be taken when including transoms above such large opening units.
- Loewen is not responsible for site measurements or determination of structural and architectural requirements for the installation of large opening products such as but limited to, LiftSlide, Bifold and MultiSlide doors. Site specifications are the responsibility of building professionals or engineers to determine.

K6 | Technical Guide   Bifold Doors

Figure 8.60

Summary

There are many types of doors and windows and accompanying hardware. All are represented with their own symbols in plan and elevation drawings. Designers must familiarize themselves with the different types to make appropriate selections in their floor plans. They must also know code requirements to competently place them in hallways and exits.

Classroom Activities

1. Measure the height, width and thickness of doors and frames in your classroom building and then draft them.
2. Visit door and window manufacturer websites and study their photos, drawings, hardware, and product descriptions.
3. Study the IRC and IBC sections on codes for doors.

Questions

1. Do residential entry doors swing into or out of the house?
2. Do commercial entry doors swing into or out of the building?
3. What does a door hand describe?
4. What kind of door retracts into a wall cavity?
5. What is an egress door?
6. Any window can be an egress window T F
7. What is the difference between muntins and mullions?
8. How tall and wide is a traditional residential interior door?
9. What is a window sash?
10. What is trim?

Further Resources

Author tutorial on drafting a bay window. https://www.youtube.com/watch?v=KofIC-qX57U

Author tutorial on drafting a double-hung window symbol. https://www.youtube.com/watch?v=q6oA0SjA788

Author tutorial on drafting swing and bifold doors. https://www.youtube.com/watch?v=Gc_JuB8U2KE

IRC website https://codes.iccsafe.org/content/IRC2018

Loewen windows www.loewen.com

Pella windows and doors www.pella.com

Keywords

- accordion door
- active panel
- awning window
- barn door
- bay window
- bifold door
- bolt
- bow window
- box bay window
- cased opening
- casement window
- casing

- ceiling height door
- clerestory
- closer
- deadbolt
- door hand
- double door
- double-action door
- double-hung window
- double-pane glass
- Dutch door
- egress
- egress door
- exit
- exit access
- exit device
- exit discharge
- fenestration
- fire door
- fire door assembly
- fixed panel
- fixed window
- flush door
- fold
- French door
- full height door
- garden window
- guides
- half-diagonal rule
- handle and knob
- head
- hinges
- holder
- hopper window
- International Building Code (IBC)
- International Residential Code (IRC)
- jalousie window
- jamb
- leaf
- leaves
- left-hand door
- left-hand reverse door
- lite
- lock
- louvered window
- masonry opening
- means of egress
- mullions
- muntins
- overhead sectional door
- pane
- panel door
- panic bar
- passive panel
- picture window
- pivot door
- pivot window
- pocket door
- public way
- quadrant markers
- rail
- reflected ceiling plan
- revolving door
- right-hand door
- right-hand reverse door
- rough opening
- sash
- sash size
- sill
- single-hung window
- skylight
- slide
- sliding window
- spin
- stationary panel
- stile
- stop
- swing door
- threshold
- transom window
- trim
- unit size
- wheelchair-accessible door

CHAPTER 9

Building Construction

Figure 9.0 Wood frame recreational vehicle garage under construction. Courtesy Rex Porter and Joan Organ Jenkins.

OBJECTIVES

Upon completion of this chapter you will be able to:

- Identify foundation, wood frame, masonry, concrete, and steel components used in residential and light commercial construction.
- Explain their purpose in the overall building design
- Recognize their plan, section, and detail drawings.

Interior designers benefit from knowing what is behind the walls as well as what's in front of them. Competent design requires knowledge of basic building construction and components.

Benefit of Building Construction Knowledge

A few examples of how basic building construction knowledge relates to interior design work are:

- Before hanging a heavy item on the wall or a heavy light from the ceiling, blocking must be installed to support it.
- Load-bearing and non-load-bearing walls must be identified during a renovation that involves moving walls.
- Locating fixtures on exterior walls or near columns may conflict with footings, structural cross-bracing, and stack pipe installation and make the fixtures' pipes vulnerable to freezing in cold climates.
- If a new floor material is installed and it's thicker or requires a subfloor (such as tile on concrete backer board), the operation of appliance doors may be impeded.
- Communicating with tradespersons is easier with a construction vocabulary.

The Foundation

A building has two main parts, the **substructure** (below-ground portion) and the **superstructure** (above-ground portion). The foundation is the substructure, and all walls and floors above the ground are the superstructure.

The **foundation** is the part of the building that provides a level surface to build on and transfers all loads to the earth. It also forms the basement walls, carries the building's **loads** (weights, such as furnishings, wind, snow), and keeps moisture-sensitive materials off the ground. It's usually made of **concrete**, which is a mixture of cement, water, and rock, and **rebar**, which are round steel bars that provide strength, or a steel mat called **woven wire mesh** (**WWM**). In some climates permanent wood foundations are built.

A **foundation plan** is a top-down view of the building's substructure, with the viewpoint at the top of the foundation wall, unlike the floor plan's 4' cut above the floor. It shows all structural components, such as walls, footings, grade beams, and **pilasters**.

> **Tip**
>
> *Concrete* and *cement* are not synonymous. Cement is a powdery grey material that is an ingredient in concrete. You have a concrete sidewalk or patio, not a cement one.

Structural vs. Partition Wall

A **load-bearing**, also called **structural**, component carries its own load plus the loads of other components such as the roof and upper floors. It also carries wind, snow and rain, and live loads (anything movable—people, partitions, furniture, equipment). Structural components can be columns, beams, and walls. Load-bearing walls can be exterior and interior. Figure 9.1 shows some load-bearing components.

A **partition wall** is **non-load-bearing**; it only carries its own weight. A partition wall's purpose is aesthetic or to divide larger spaces into smaller ones. It may be moveable. **Partial walls**, also called **knee walls**, are partitions that aren't full height. They're used to organize a large space, provide transitions from one use to another, and on staircases. Codes require they be at least 36" above the floor.

Structural System   This is the assembly of structural components that support and transmit loads to the ground. Structural materials are wood, masonry, concrete, and steel. Most buildings contain all but are primarily constructed of one.

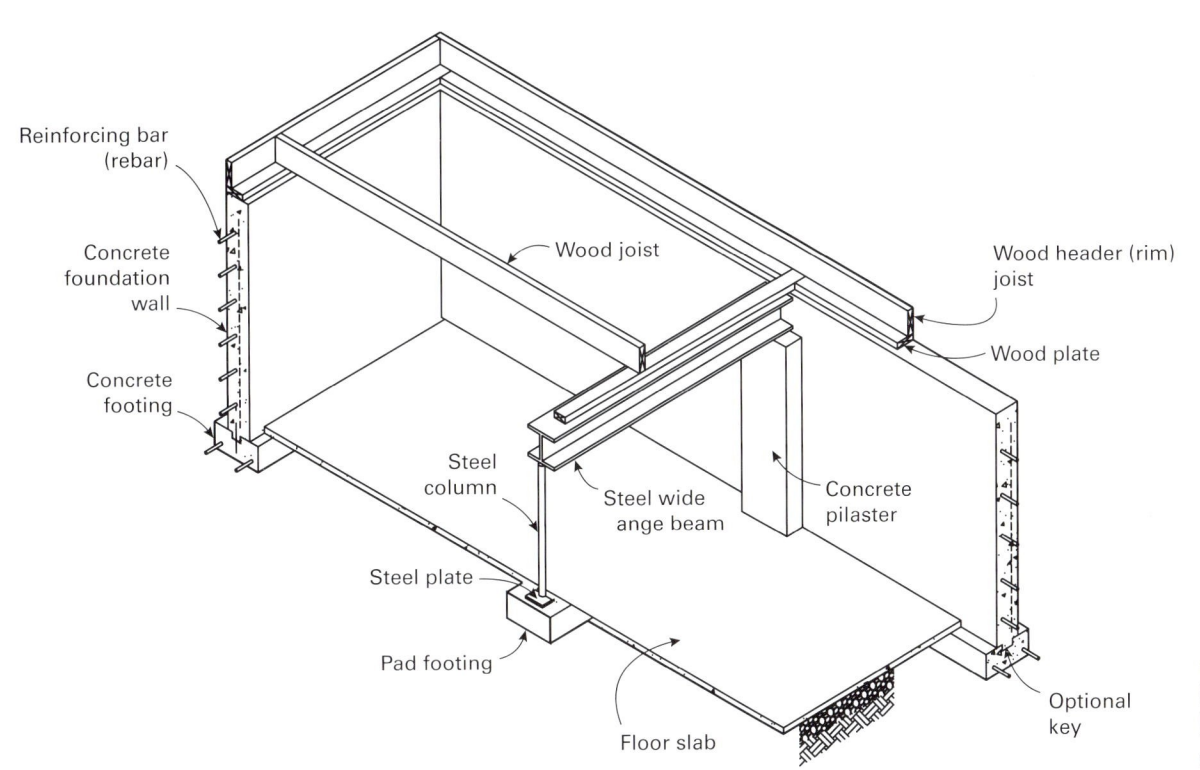

Figure 9.1 Load-bearing components.

Foundation Types

There are a lot of foundation types, but the ones most relevant to residential construction are the **slab-on-grade**, **spread footing**, and **post-on-pad**.

Slab-on-Grade This is a thick, monolithic piece of concrete on the ground (Figure 9.2). Concrete is poured into a shallow excavation in the ground or into an aboveground form. The slab is thickened around

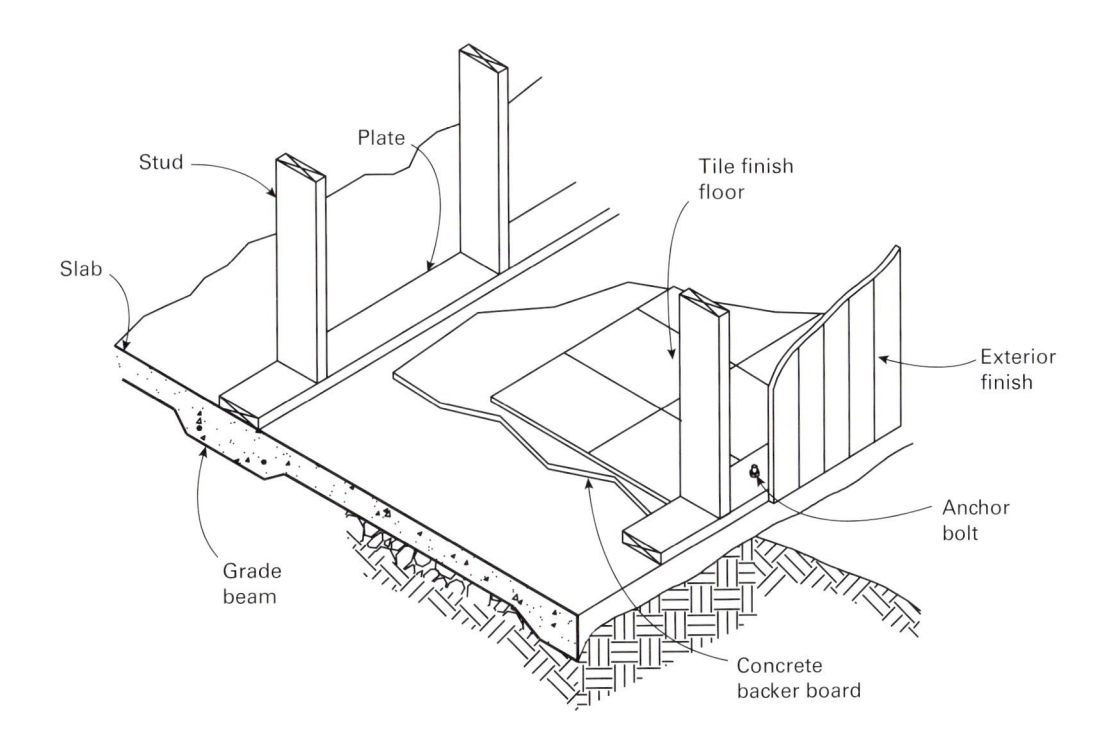

Figure 9.2 Slab-on-grade.

FOUNDATION TYPES 219

the perimeter to support the perimeter load-bearing walls (Figure 9.3). **Grade beams**, which are long, ground-level supports placed under load-bearing walls, may be integrated into the slab to support intermediate load-bearing walls. Figure 9.4 shows a slab-on-grade foundation plan with six grade beams and seven separate pours (slabs). Figure 9.5 shows multiple pours visible on a driveway, and the forms that the concrete is poured in.

Figure 9.4 also shows two section cuts; one at the grade beam and one at the ramp. Figure 9.6 shows **detail drawings**, which are slices through small parts of a building. Their purpose is to show how the parts go together.

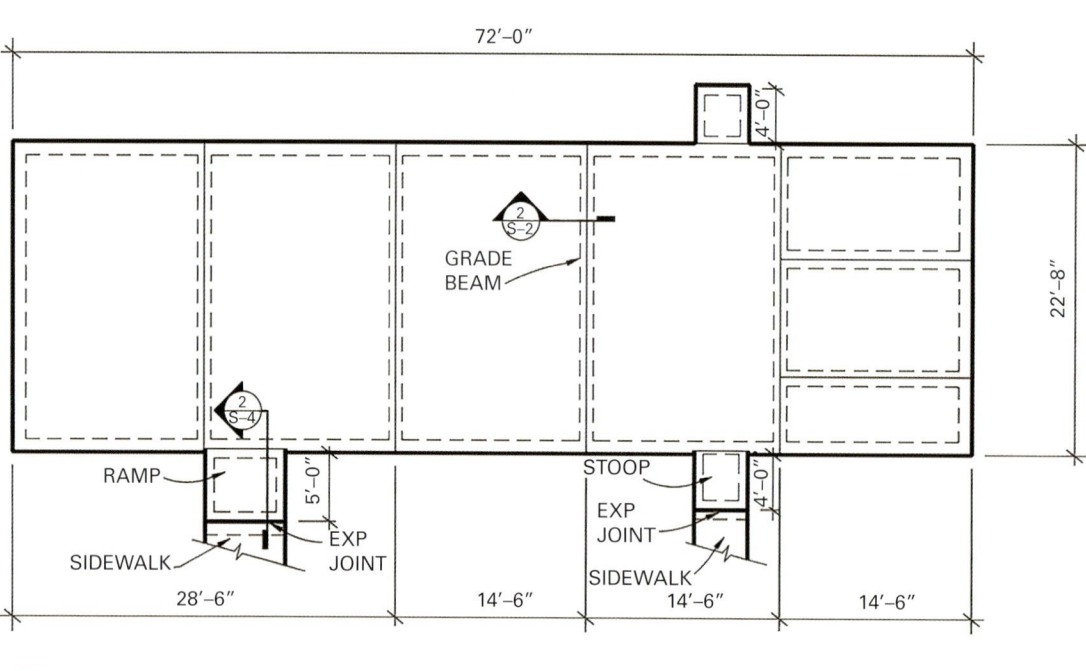

Figure 9.3 A slab's perimeter is thickened.

Figure 9.4 Foundation plan of a slab-on-grade shows grade beams and pours.

Spread Footing This is a wall built on top of a **footing**, which is a wide base under a wall. This system enables the construction of basements and crawlspaces. The footings are placed below the **frost line**, the level at which the ground no longer freezes. In a cold climate, this is about 36". Figure 9.7 shows a plan view of a spread footing and Figure 9.8 a **wall section**, which is a detail drawing of a wall made by slicing vertically through the foundation's perimeter. Figure 9.9 shows a spread footing under construction. Note the rebar around which the walls are poured.

Tip
Neither floor nor foundation plans annotate which walls are load-bearing. Although a structural engineer should be consulted before walls and beams are moved, you can overlay the floor and foundation plans to see which walls align with grade beams. Structural walls need the extra support grade beams provide.

CHAPTER 9 BUILDING CONSTRUCTION

Figure 9.5
The driveway shows multiple pours. Note the wood forms the concrete is poured into.

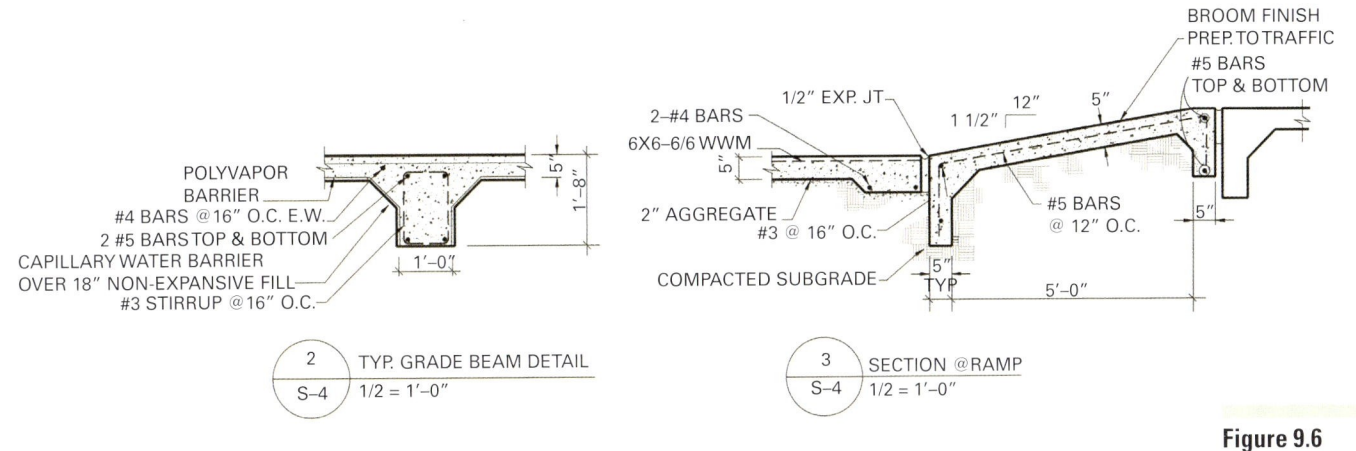

POLYVAPOR BARRIER
#4 BARS @16" O.C. E.W.
2 #5 BARS TOP & BOTTOM
CAPILLARY WATER BARRIER
OVER 18" NON-EXPANSIVE FILL
#3 STIRRUP @16" O.C.

2 / S-4 TYP. GRADE BEAM DETAIL 1/2 = 1'-0"

BROOM FINISH PREP. TO TRAFFIC
#5 BARS TOP & BOTTOM
1/2" EXP. JT.
2–#4 BARS
6X6–6/6 WWM
2" AGGREGATE
#3 @ 16" O.C.
COMPACTED SUBGRADE
#5 BARS @ 12" O.C.

3 / S-4 SECTION @ RAMP 1/2 = 1'-0"

Figure 9.6
Detail drawings of a grade beam and a ramp.

28'-0"
16"x10" CONCRETE FTG
11'-0"
2/S-2
FLOOR JOISTS 2 x 8 16" O.C.
4" DIA. STL. PIPE COLUMN
PILASTER 4" x 16"
3 – 2 x 12 GRIDER
12 x 12 PAD
18" x 18" x 10" FTG
9'-4" 9'-4" 9'-4"
22'-0"
UP
11'-0"
8"
1" x 3" CROSS BRIDGING

1 / S-2 FOUNDATION PLAN 1/4" = 1'-0"

Figure 9.7
Plan of a spread footing foundation.

FOUNDATION TYPES   221

Floor joist

Sill plate

Floor

Gravel

Rebar

Perforated drain pipe

Footing

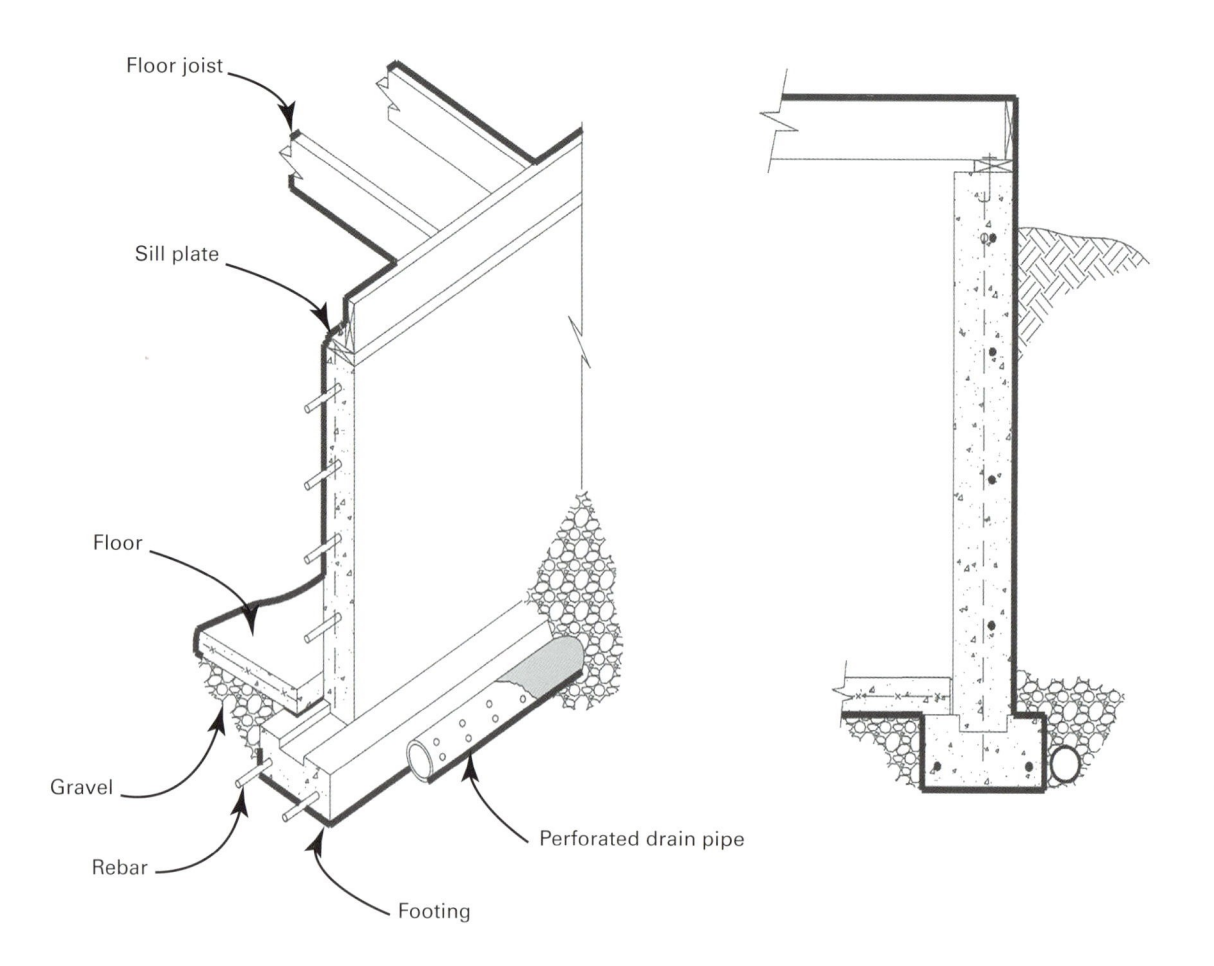

Figure 9.8 Wall detail of a spread footing foundation.

Figure 9.9 A spread footing foundation under construction.

Post-on-Pad This is a post (vertical load-bearing member that runs from level to level) on a thick concrete square. The plan in Figure 9.7 shows two posts on pads. Posts can be short, such as for under a deck or crawlspace (Figure 9.10), or be tall, such as in a basement supporting the floor above (Figure 9.11).

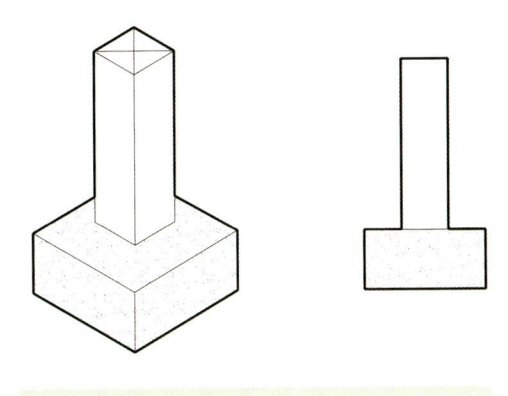

Figure 9.10
Short post-on-pad, as might be found under a deck.

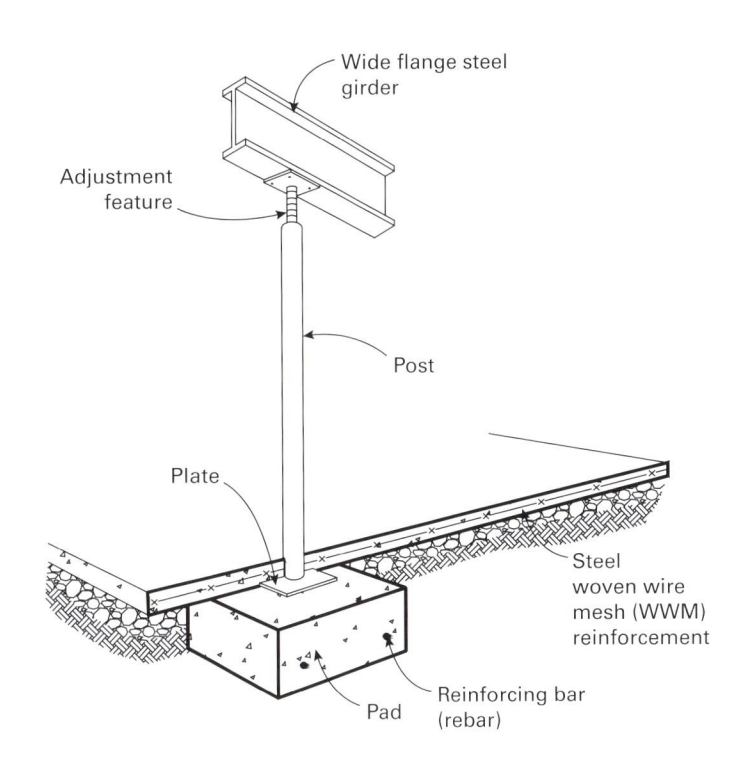

Figure 9.11
Tall post-on-pad, as might be found in a basement.

One house may utilize all three foundation types; for instance, a spread footing under the living space, a slab-on-grade under the garage, and a post-on-pad under a deck.

Other More foundation types include:

- **crawl space** foundation, in which concrete blocks or bricks sit on footings and support the walls above.
- **stepped footing**, in which the foundation stair-steps into a sloped site.
- **pile** (giant concrete shaft without footings) for high-rise buildings
- **strip footing**, which are long, horizontal beams that support load-bearing masonry walls
- **raft**, also called mat. There are no footings; the whole basement floor is the foundation. The building "floats" on this raft in the soil.
- **Insulated Concrete Forms (ICF)**. These are hollow foam blocks that serve as forms for below-grade walls and remain as insulation after the walls are poured (Figure 9.12). Rebar is integrated inside the blocks.

Figure 9.12
ICF foundation blocks. Courtesy foxblocks.com

FOUNDATION TYPES 223

Masonry

Masonry is units of concrete block, brick, glass block, stone, and structural clay tile. Stone is mainly used for veneers and accents. **Concrete blocks** are also called **concrete masonry units (CMU)**. They are precast products with one or more cavities, and are available in many shapes, sizes, and finishes (Figures 9.13, 9.14). All are based on multiples of 4". When designing with masonry, calculate wall lengths and heights based on that multiple.

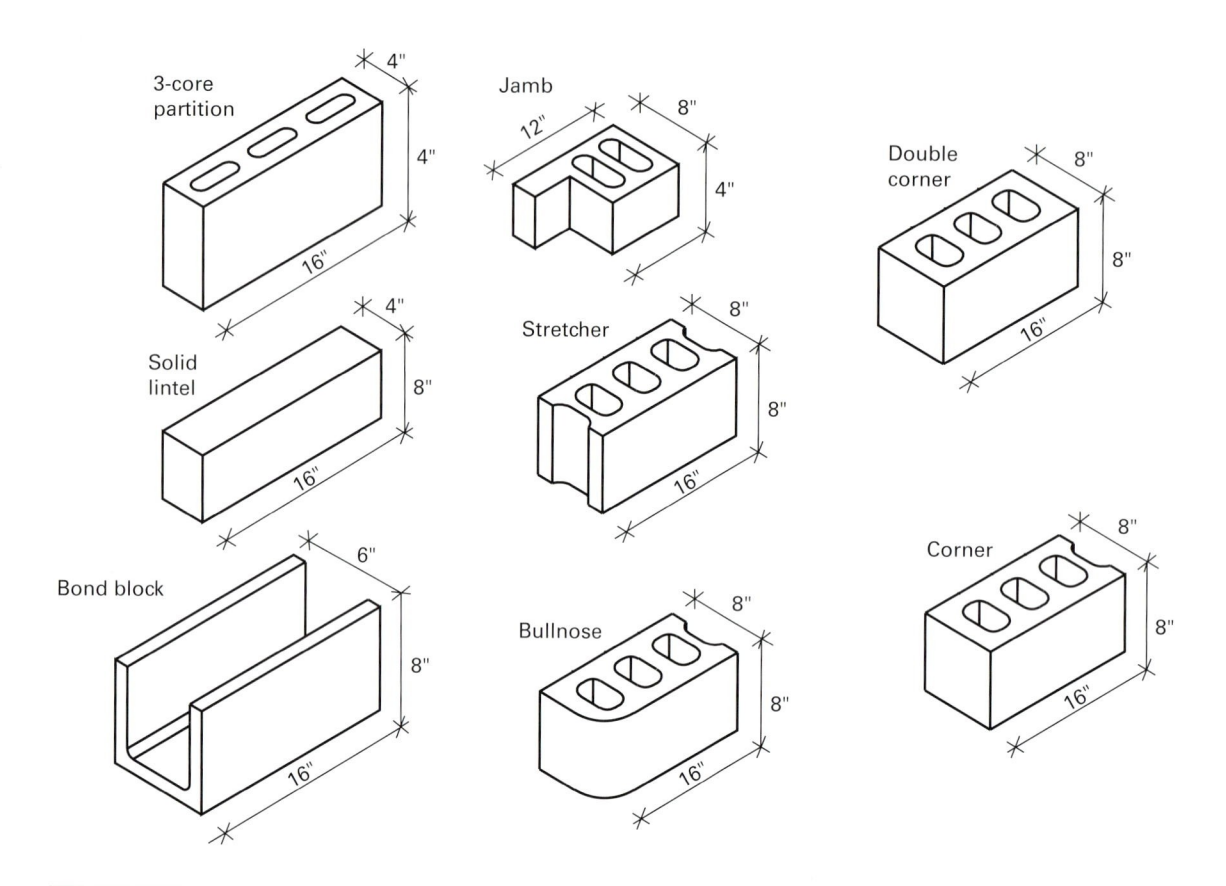

Figure 9.13

Concrete Block Types A **stretcher** is the most common and is laid with its length parallel to the face of the wall. A **double corner** is used any place where both block ends are visible. A **bullnose** is used as a corner block to achieve rounded edges. A **jamb** is combined with stretchers and corners around window openings and the casing is placed in the recess.

Masonry Sizes These units have **nominal** and **actual sizes**. Nominal size means *in name only* and is used for identification. Actual size is the manufactured dimensions. Masonry units are 3/8" smaller than their nominal sizes on each side to accommodate a **mortar joint**, which is the cement/sand filler in the spaces between blocks (Figure 9.15). Annotate an 8" × 8" × 16" block with its nominal size but draft its actual size of 7 5/8" × 7 5/8" × 15 5/8". Dimensions are written as width × height × length.

> **FYI**
> The use of masonry materials for construction is as old as civilization. In 3000 C.E. the Egyptians built the pyramids by mixing mud and straw to make bricks and mortar. The Chinese used cement-like materials to build the Great Wall and to hold bamboo boats together. The Romans used concrete to build the Appian Way, Coliseum, and Pantheon, and utilized sophisticated stone construction technologies in the last few centuries C.E.

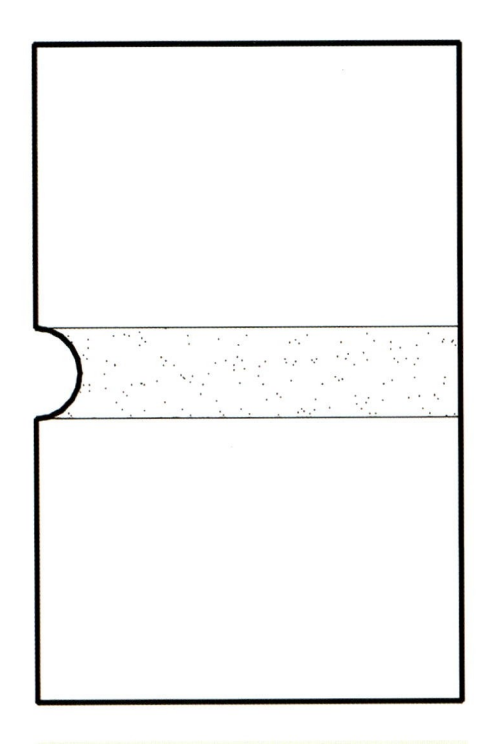

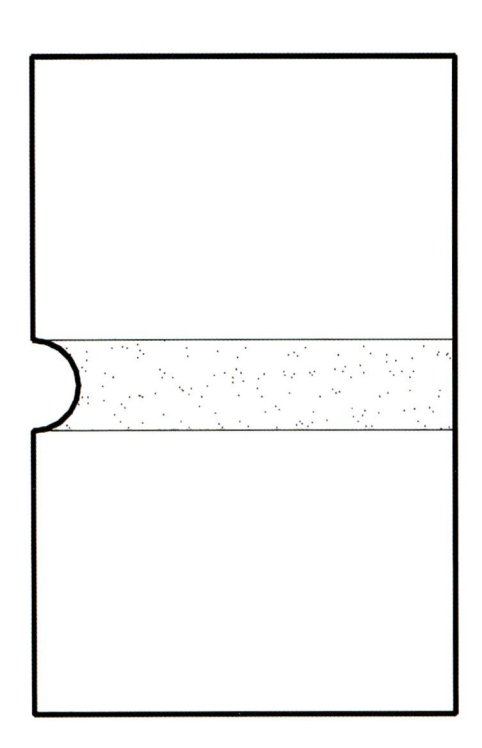

Figure 9.14
Bricks and blocks come in many sizes, shapes and finishes.

Figure 9.15
Concave mortar joint.

Masonry Wall Construction Masonry wall construction types include **solid** (consists only of concrete blocks built as one wythe, and is found mostly in pre-1950s homes), **veneer** (consists of concrete blocks with a façade of another material) and **cavity** (two wythes of concrete blocks with a space in between). Figure 9.16 shows a solid concrete block wall. Visible is a **bond** (U-shaped) block, a row of which helps hold the wall

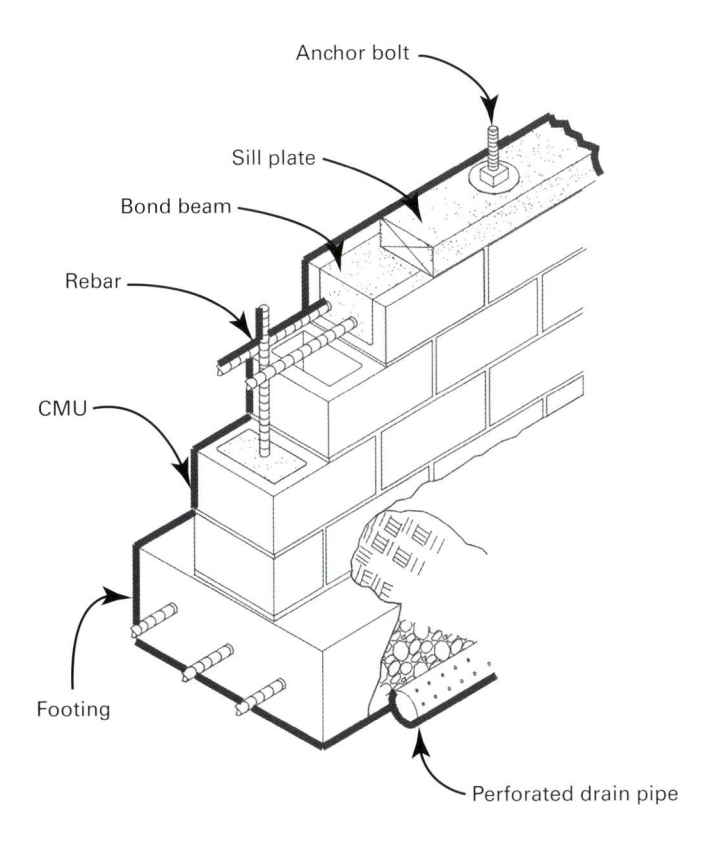

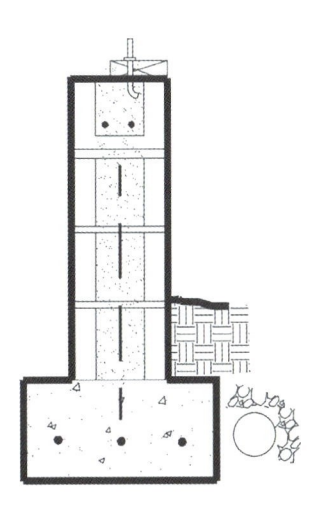

Figure 9.16
Concrete block wall.

MASONRY   225

together. Steel rebar runs horizontally through the **bond beam** and vertically through the **cells** (holes) in the blocks to make the wall stronger. Solid walls can also be walls of two materials tied together by the materials' positioning (Figure 9.17).

Figure 9.17
A one-wythe wall.

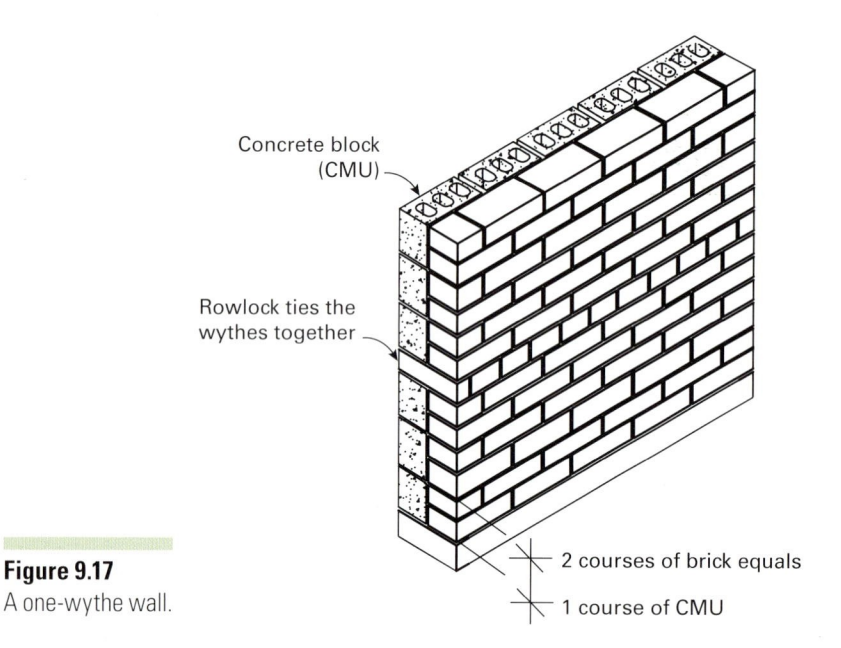

Figure 9.18 shows **cavity** and **veneer walls**. Cavity walls are two parallel walls separated up to 12" apart. They weigh less than solid masonry walls and provide better temperature and humidity insulation. Veneer walls are a structural wall and a separate, non-load-bearing aesthetic one, such as brick or stone on concrete. **Wythe** describes wall thickness; for instance, a solid wall is a one-wythe wall, and a cavity wall is a two-wythe (**double-wythe**) wall.

Brick veneer on concrete block (one wythe)

Brick veneer and concrete block (two-wythe)

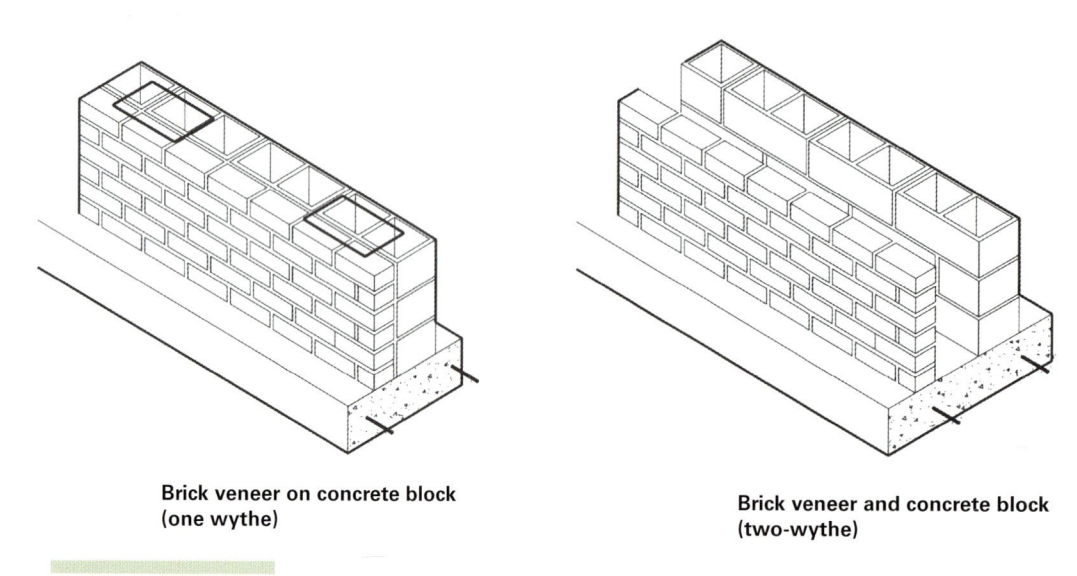

Figure 9.18
Veneer and cavity walls.

Masonry Bonds A **bond** is the arrangement of bricks and blocks in a wall. Bonds are named by the positions of those bricks and blocks (Figures 9.19, 9.20). Masonry units are laid in the walls in **courses**, which are horizontal rows. Bricks and blocks in a floor are called **pavers** (Figure 9.21).

Stone is laid as **ashlar** or **random rubble**. Ashlar is cut into rectangular pieces and laid in parallel courses. Random rubble is just stones surrounded with mortar; there is no bond pattern.

Running (5 X 12 bricks)

Common (5 X 12 bricks)

Horizontal stack (4 X 12 blocks)

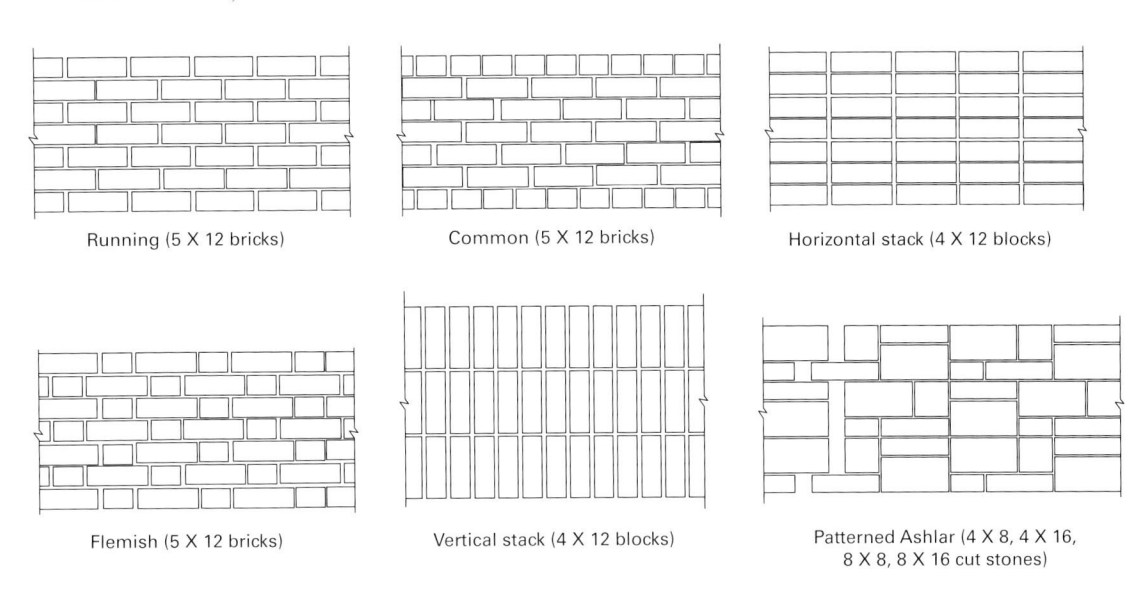

Flemish (5 X 12 bricks)

Vertical stack (4 X 12 blocks)

Patterned Ashlar (4 X 8, 4 X 16, 8 X 8, 8 X 16 cut stones)

Figure 9.19 Bond patterns.

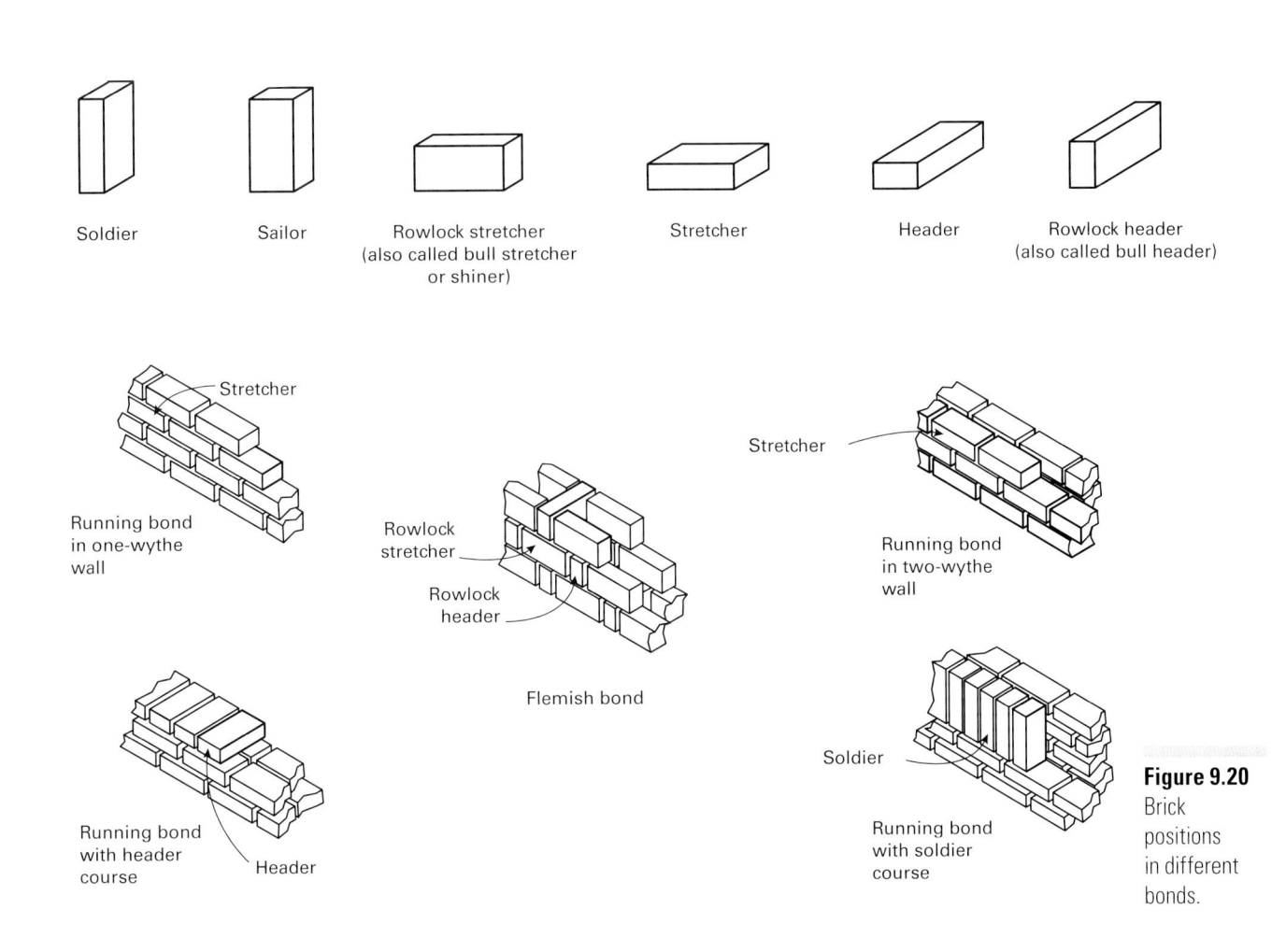

Soldier

Sailor

Rowlock stretcher (also called bull stretcher or shiner)

Stretcher

Header

Rowlock header (also called bull header)

Running bond in one-wythe wall — Stretcher

Flemish bond — Rowlock stretcher, Rowlock header

Running bond in two-wythe wall — Stretcher

Running bond with header course — Header

Running bond with soldier course — Soldier

Figure 9.20 Brick positions in different bonds.

MASONRY   227

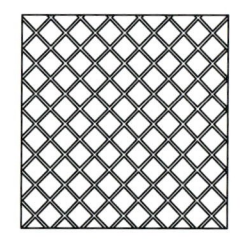

Diagonal stacking (4 X 4)

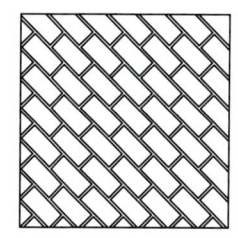

Diagonal (4 X 8)

Herringbone (5 X 12)

Diagonal basketweave (4 X 8)

Figure 9.21
Paver types.

Tilt-Up Construction

Tilt-up construction is both a building and a construction technique (Figure 9.22). Concrete panels are made on site for the exterior walls, tilted up with a crane, and temporarily braced until the rest of the structural components (floors, walls, roofs) are attached. Finish materials are then added to the panels' interior and exterior.

Figure 9.22
Tilt-up construction.

Wood Framing Components

Figure 9.23 shows the components of a wood frame structural system. Wood frame buildings are made of **dimensional lumber** and **engineered wood products (EVP)**.

Dimensional lumber is wood cut to standard sizes such as 1" × 4" or 2" × 4" (Figure 9.24). That dimension is the nominal size; the actual size is 1/2" smaller on all sides because it shrinks when kiln-dried. Annotate a wood component with its nominal size but draft its actual size. Wood components are dimensioned to their centers; 16" **o.c.** (on center) means the components are spaced 16" apart, center-to-center (Figure 9.24). Draw an X through their cross sections (Figure 9.25).

228   CHAPTER 9   BUILDING CONSTRUCTION

Figure 9.23
Components of a wood frame structural system.

WOOD FRAMING COMPONENTS **229**

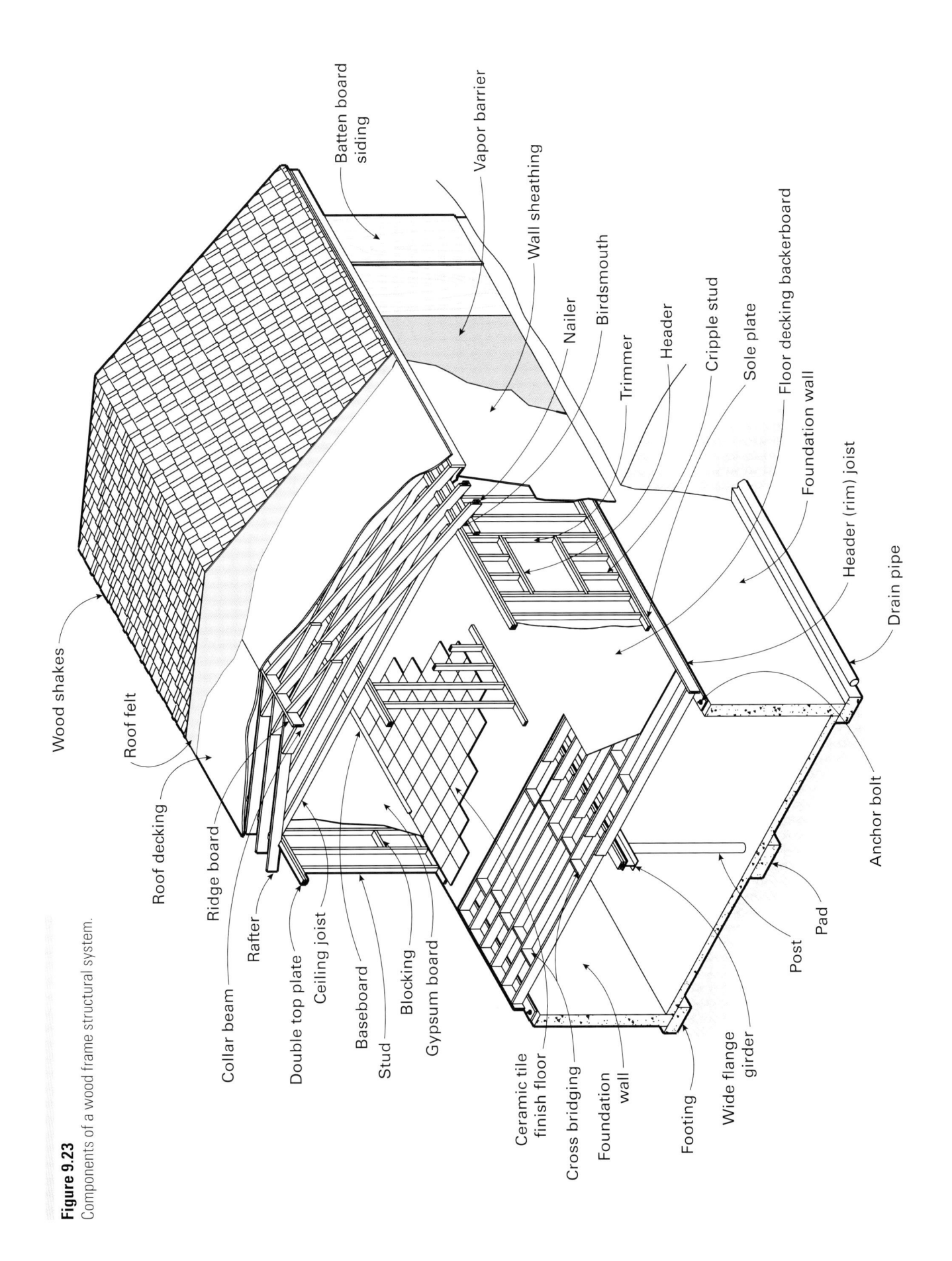

Engineered wood products are made of real wood, scrap wood, shredded wood fibers, sawdust, and manmade products glued together. They're used where their superior load bearing and spanning capabilities are needed. They may be more lightweight, stable, stronger, and provide greater **clear spans** (distance between two surfaces). Common EWP products are:

- **Glue laminated timber (Glulam)** This is multiple layers of wood held together with high-strength adhesive. It's used for straight and curved beams in both residential and commercial construction.
- **Laminated veneer lumber (LVL)** Also called **microlam**, this is made from peeled or sliced thin veneers laid parallel and bonded together under heat and pressure. It's used for beams, timber wall cladding, and columns.
- **Oriented strand board (OSB)** This is multiple layers of wood strands compressed and glued together in specific orientations. It's used for sub-surfaces and withstands water, moisture, and weather conditions well.
- **Medium density fiberboard (MDF)** This is wood fibers combined with wax and a resin binder. It's not water or weather resistant and is used to make panels that need a smooth finish.

Engineered wood products can be made into aesthetically interesting shapes. Figure 9.26 shows Glulam arches spanning a hockey rink. Figure 9.27 shows a **box beam**, which is a lightweight non-structural beam used to provide visual interest.

Figure 9.24
1" × 4" dimensional lumber boards.

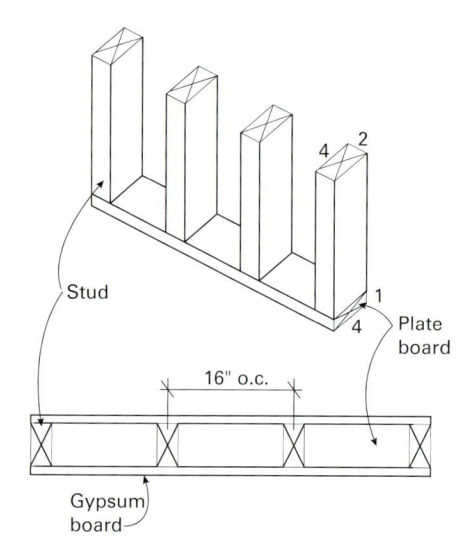

Figure 9.25
2" × 4" studs on a 1" × 4" plate.

Figure 9.26
Glulam arches spanning a hockey rink.
Courtesy of APA–The Engineered Wood Association.

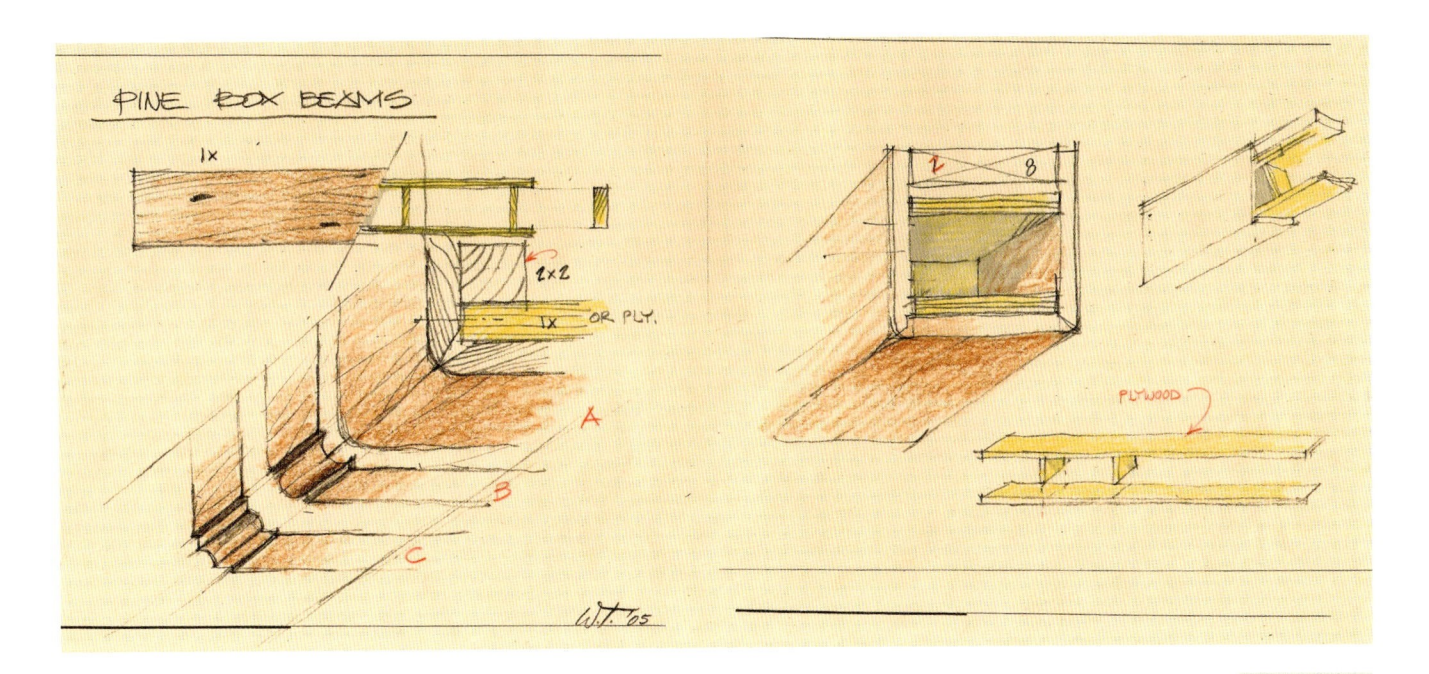

Figure 9.27
Box beam. Courtesy of wolfgangtrost.com

Wood Framing Definitions

There are many terms associated with wood framing, but here are the most common ones you'll see on a drawing set.

Stud Vertical load-bearing member inside a wall. **Cripple studs** are short and placed above or below a window.

Plate (Figure 9.28) Horizontal board. Bottom plates distribute loads placed on them; top plates tie studs together. Plates are named according to their location: **sole** plates are at ground level; **sill** plates are window bottoms; **double top** plates are at ceilings and above doors.

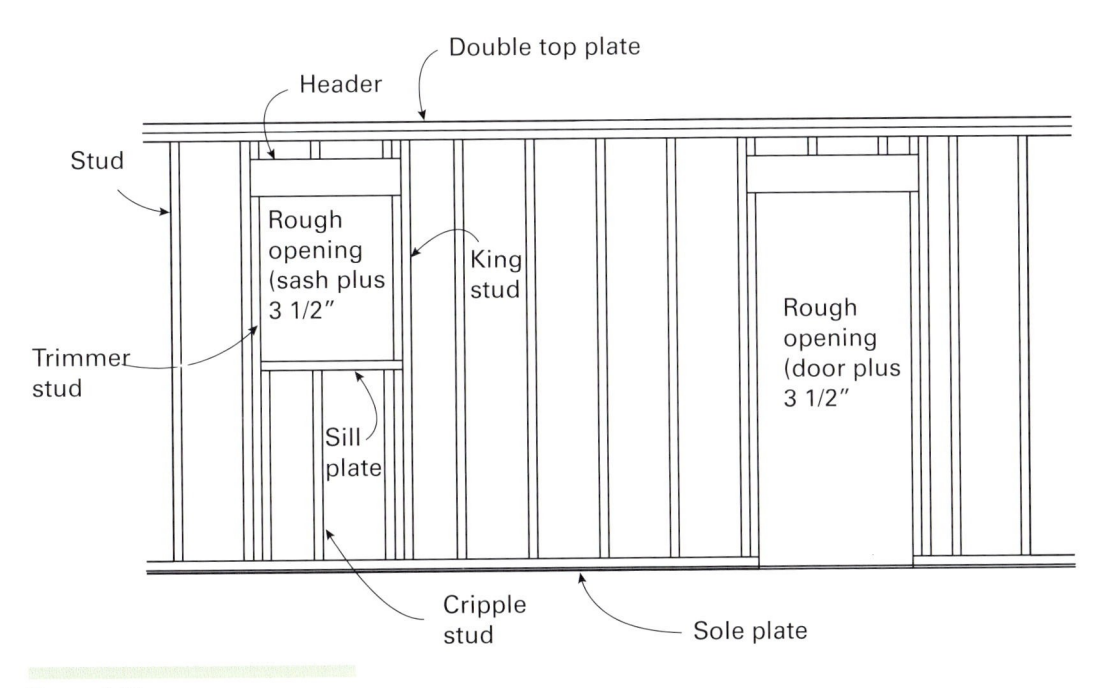

Figure 9.28
Headers, studs and plates in a wall.

WOOD FRAMING DEFINITIONS 231

Beam Horizontal load-bearing member. It has different names depending where it is located:

- **Joist**: horizontal beam in ceilings and floors
- **Girder**: large beam that supports smaller beams (Figure 9.29)
- **Lintel**: also called **header**, horizontal beam over a door, window, or fireplace (Figure 9.30).
- **Rafter**: inclined beam at the roof (Figure 9.31)
- **Trussed rafter (truss)**: a composite rafter (Figures 9.32, 9.33)

Sheathing Vertical boards on exterior walls and under the final finish.

Decking (underlayment) Horizontal boards on the floor and roof and under the final finish (Figures 9.34, 9.35).

Figure 9.29
A girder and the joists it supports.

Figure 9.30
Header and I-joist floor joists.
Courtesy Weyerhaeuser.

Figure 9.31
Rafters and sheathing.

232 CHAPTER 9 BUILDING CONSTRUCTION

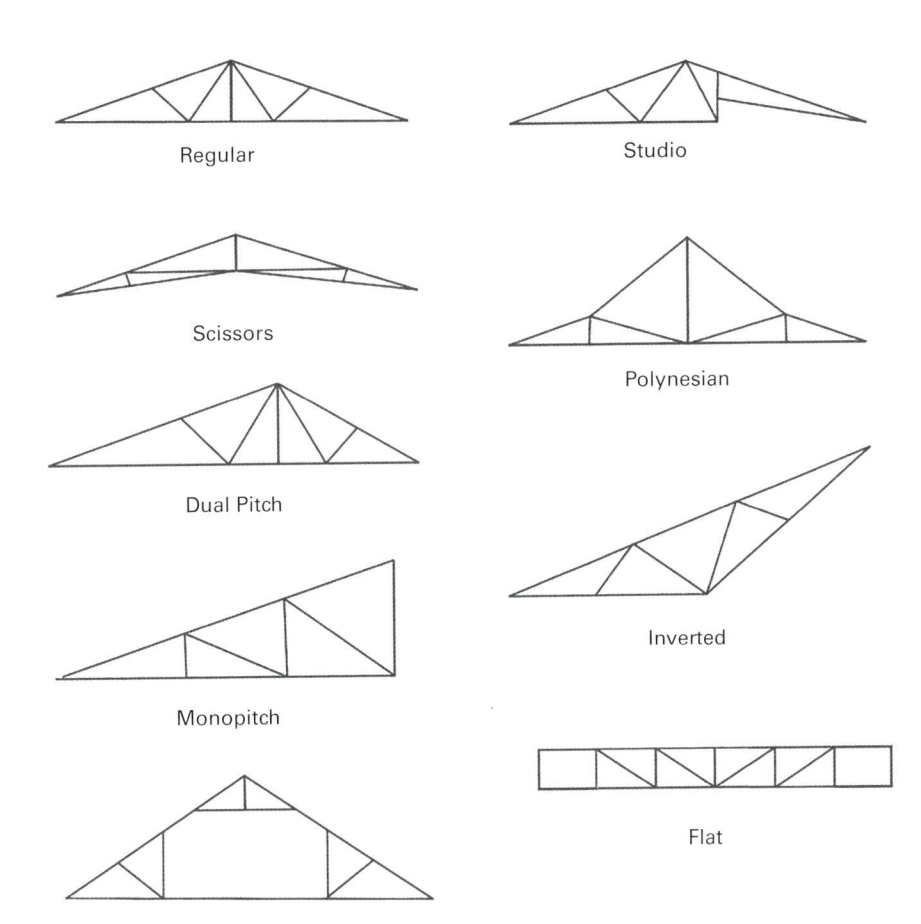

Regular

Studio

Scissors

Polynesian

Dual Pitch

Inverted

Monopitch

Flat

Attic

Figure 9.32
Common truss styles.

Figure 9.33
A truss being hoisted into place with a crane.
Courtesy Rex Porter and Joan Organ Jenkins.

Figure 9.34
Floor decking over i-joists.
Courtesy Weyerhaeuser.

WOOD FRAMING DEFINITIONS

Wood Frame Types

Two broad categories of wood framing are post and beam and skeleton.

Post-and-Beam, also called **timber framing**, has large, such as 6" × 6", wood structural components spaced far apart, such as 4'-0" o.c (Figure 9.36). Most of the building's weight is carried by the posts. This makes the walls **curtain walls**, which are non-load-bearing walls that simply shield and enclose. Post-and-beam construction enables larger windows and clear spans than other framing types.

Figure 9.35
Roof decking.

FYI
Post-and-beam framing was the method of wood construction used for 2,000 years. Trees were cut, hewn, and connected with mortise-and-tenon or dovetail notches. The art of joinery was a skilled craft. Over the centuries, different countries developed their own recognizable style of timber framing. Axe and adze marks, types of joints, and shapes of hewn logs identify a builder's home country and the year built of American log cabins.

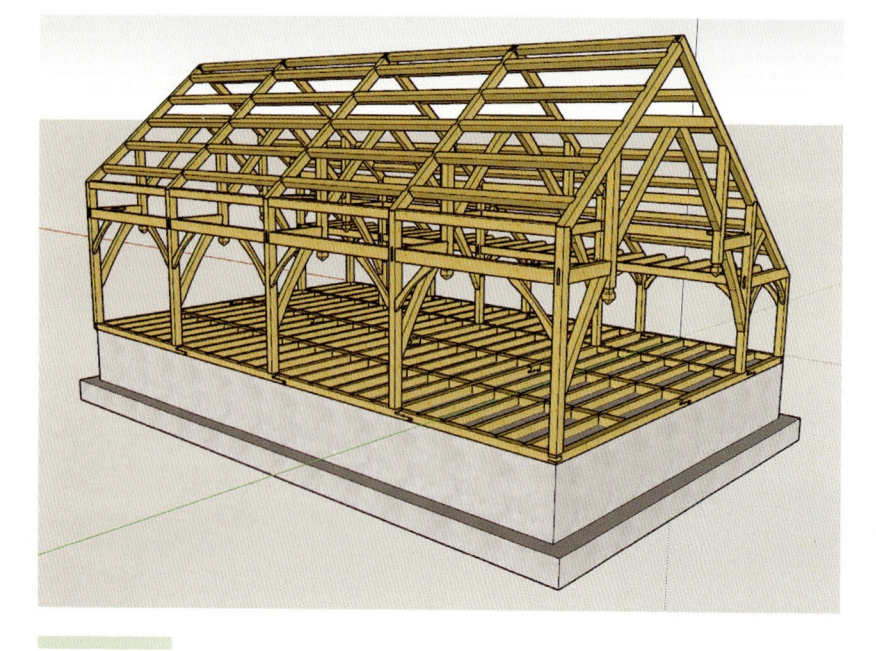

Figure 9.36
Post-and-beam framing.

Skeleton, also called **stick framing**, has small, such as 2" × 4", wood structural components spaced closely together, such as 16" **o.c.** Most of the building's weight is carried by load-bearing walls. There are two types of skeleton framing: balloon and platform.

- **Balloon framing** (Figure 9.37) is early skeleton framing, originating in the 1830s. It has long studs that run continuously from ground to roof. The second floor rests on a ribbon board attached to the studs. Although the balloon method is still used in occasional stucco or masonry veneer buildings, around 1930 platform framing began replacing it.
- **Platform,** also called **western framing** (Figure 9.38), is the current standard for most modern homes. Walls are built on top of the floors, not attached to them. The studs run from level to level, making the platform-framed wall independent for each floor.

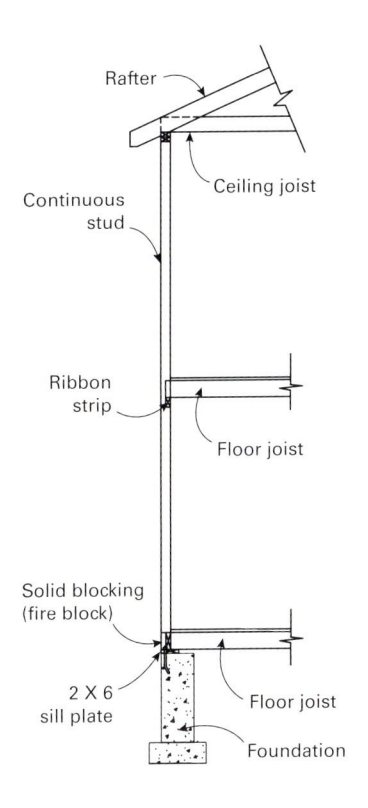

Figure 9.37 Balloon framing.

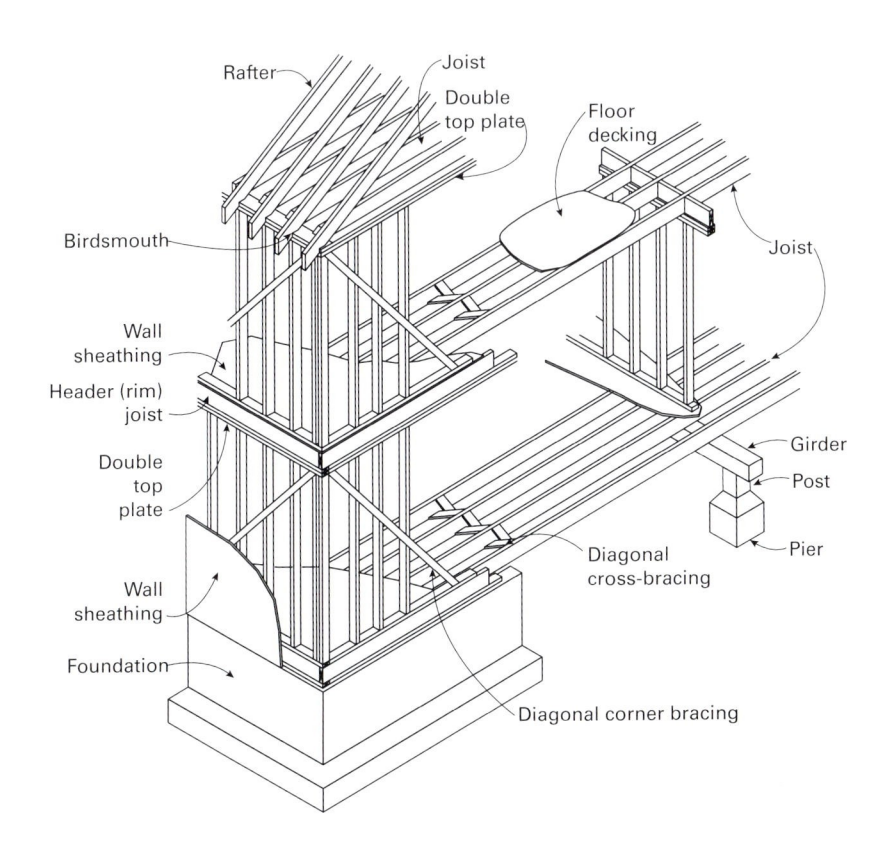

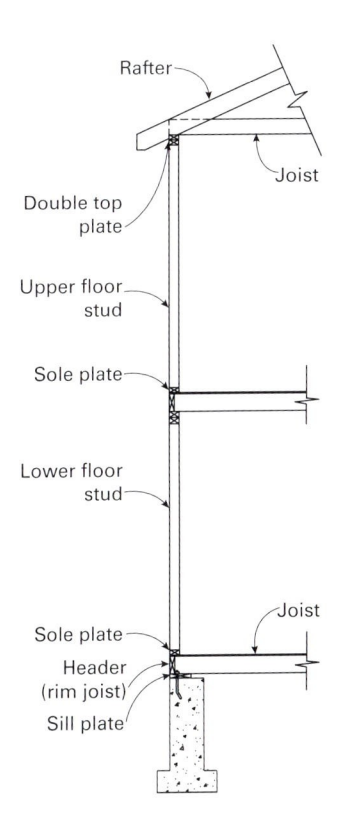

Figure 9.38 Platform framing.

WOOD FRAME TYPES 235

In both balloon and platform framing, wall studs, ceiling rafters, and floor joists are placed every 16" or 24" o.c., which enables the use of floor, ceiling, and wall materials manufactured in standard 4' widths.

Other Framing Items

Within the frame are **cross bracing**, which strengthens the walls; **cross bridging**, which strengthens the floors; and **blocking**, which is material that helps slow down the spread of fires. Material that strengthens walls and ceilings to support heavy objects is also called blocking. In the roof, there are **hip** and **valley** roof rafters (Figure 9.39). A hip rafter extends from the wall plate to the ridge and forms the angle of a hip roof. A valley rafter runs from a wall plate to the ridge along the valley, which is the "fold" where a roof changes direction. Hip rafters attach to a **ridge board**, which is the board to which the rafters are attached. Figure 9.40 shows a technique for making a round arch in a wood frame.

FYI

When machines and factories were built in the 19th-century Industrial Revolution, sawmills produced smaller, lightweight pieces of lumber in standardized sizes, and factories mass-produced nails. This suited the needs of the young, westward-growing United States better than post-and-beam, whose heavy timbers were harder to transport and required more skill to construct.

Figure 9.39
Hip and valley rafters.

Figure 9.40
Round arch in a wood frame.

Other Construction Plans

A **roof plan** is a bird's-eye view of the roof, and shows hips, valleys, skylights, and anything mounted on it (Figure 9.41). A **roof framing plan** is a top-down view that shows the rafter layout (Figure 9.42). A **demolition plan** is a top-down view that shows items to be removed or relocated. This is done when there are many such items (Figure 9.43).

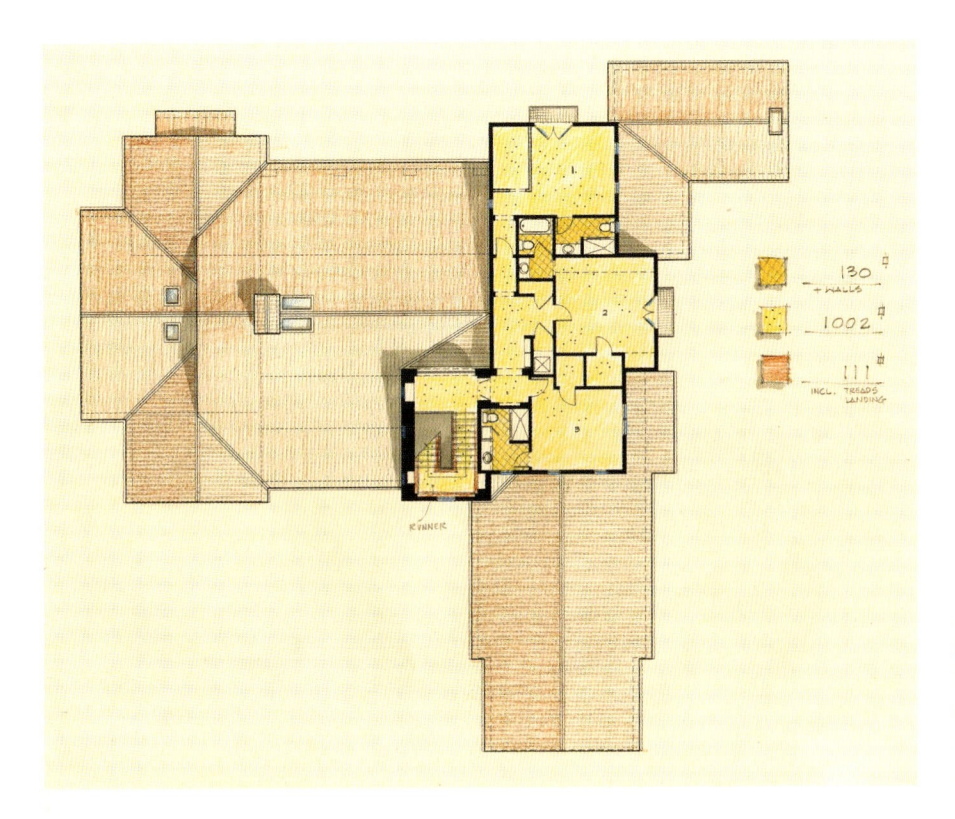

Figure 9.41
Roof plan and square footage calculations. Courtesy wolfgangtrost.com

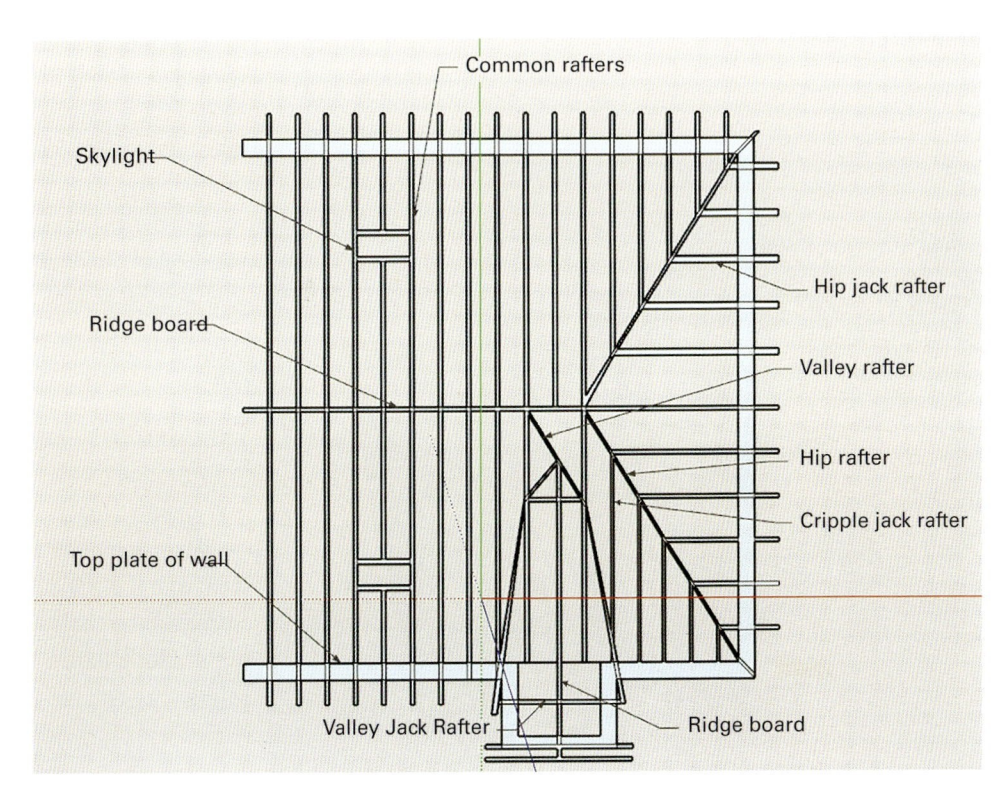

Figure 9.42
A partial roof framing plan.

OTHER CONSTRUCTION PLANS 237

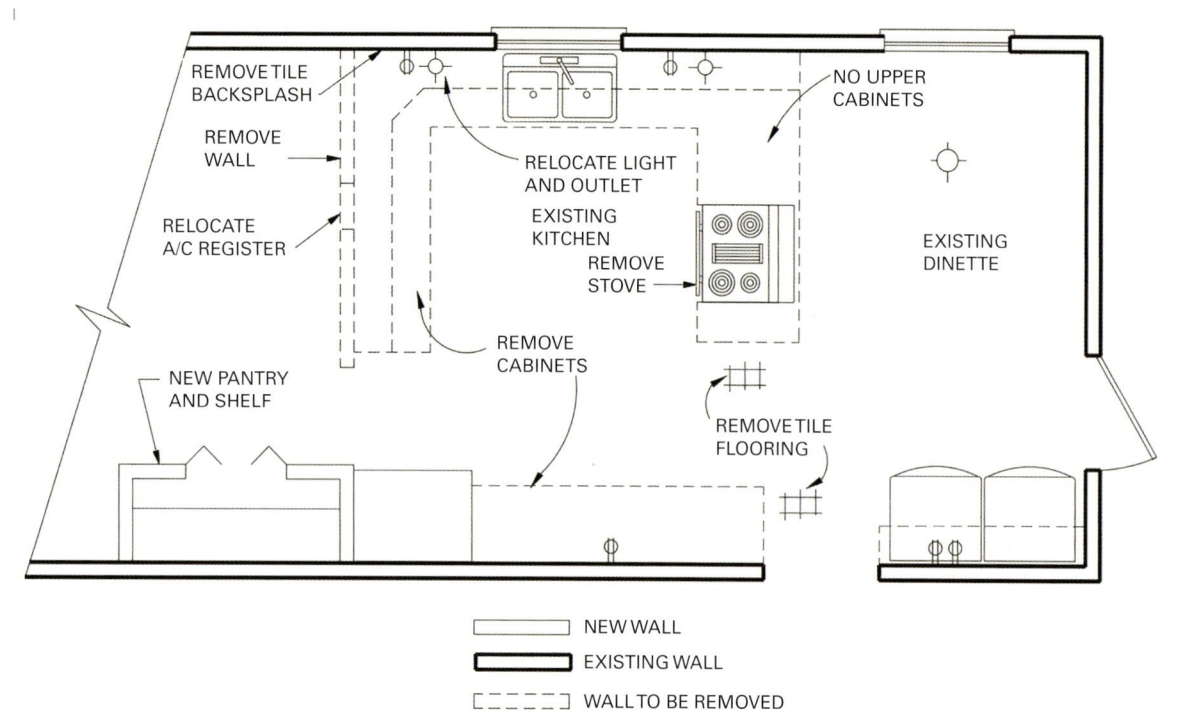

Figure 9.43
Demolition plan.

NEW WALL
EXISTING WALL
WALL TO BE REMOVED

Modular Construction

Also called **prefabricated**, **modular construction** (Figure 9.44) is a process in which standardized sized parts called modules (units) are mass-produced in a factory and shipped to the site for assembly. They can be mixed-and-matched into almost any configuration and stacked high. The modules are shipping container-sized: 40' × 20', with an 8' interior ceiling height. Everything inside can also be fabricated as drop-in modules, including the mechanical and electrical systems. Modular buildings use the same materials as site-built ones, the same codes and standards apply, and they can be constructed in about half the time.

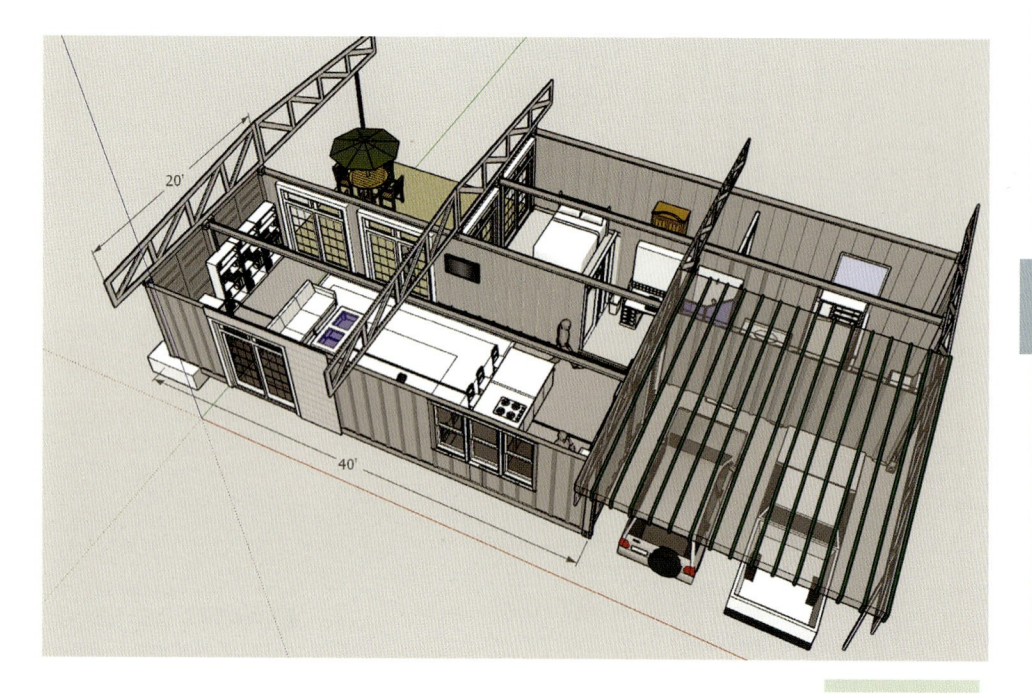

Tip
Here's how to draw a 4:12 roof pitch. Draw a 12" horizontal line. At its right endpoint, draw a 4" vertical line. Connect the horizontal line's left endpoint with the top of the vertical line. The resultant angled line will have a 4:12 pitch. Any scale can be used to do this.

Exercise
Walk around campus and identify steel building components, such as bar joists and trusses. Take photos of them for future reference.

Figure 9.44
A modular home.

3D Printing

This is a technology that turns a computer model (3D drawing) into a physical object. A printer reads the drawing and builds the item layer by layer in anything from plastic to metal to concrete (Figures 9.45, 9.46). 3D-printed homes can be constructed in as little as 24 hours, minus all the interior finish work.

Figure 9.45
A 3D printer and some concrete blocks it printed.

Figure 9.46
A 3D-printed planter.

Steel Frame Components

Steel-framed buildings can reach great heights and provide large (up to 150') clear spans. They're built by welding and bolting sections (Figure 9.47), which are load-bearing, premade parts. Sections are fabricated by stretching and heating thin strips and then shaping them in a hot-rolling process. Figure 9.48 shows a portable machine that makes steel components on the jobsite.

A wide flange beam that is 8" deep and weighs 20 pounds per lineal foot is notated *W8×20*. A channel shape beam that is 6" deep and weighs 15 pounds per lineal foot is notated *C6×15*.

Composite steel sections like **bar joists**, **arches**, and **rigid frames** provide greater load-bearing and spanning capabilities. They're used for large-span construction such as warehouse stores, sports arenas, convention

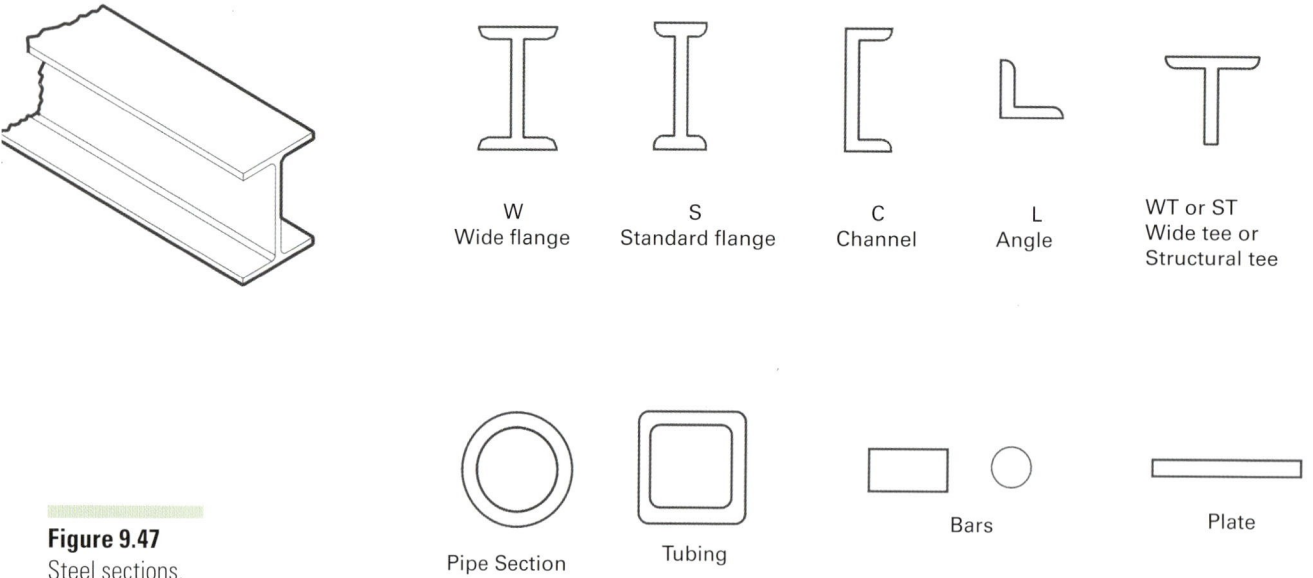

Figure 9.47
Steel sections.

W — Wide flange
S — Standard flange
C — Channel
L — Angle
WT or ST — Wide tee or Structural tee

Pipe Section
Tubing
Bars
Plate

Figure 9.48
A portable machine that makes steel parts onsite.

centers, and gyms. A bar joist is a flat truss (Figures 9.49, 9.50). It is annotated with a three-part code, such as *18K6*. This means that the bar joist is 18" deep and part of the *K* series (there are different series, each designed for specific span and load-bearing criteria) and is in *section 6* (this number refers the reader to a chart that describes how much bridging is needed between the joists).

CHAPTER 9 BUILDING CONSTRUCTION

Figure 9.49
Bar joist.

Elevation

Plan

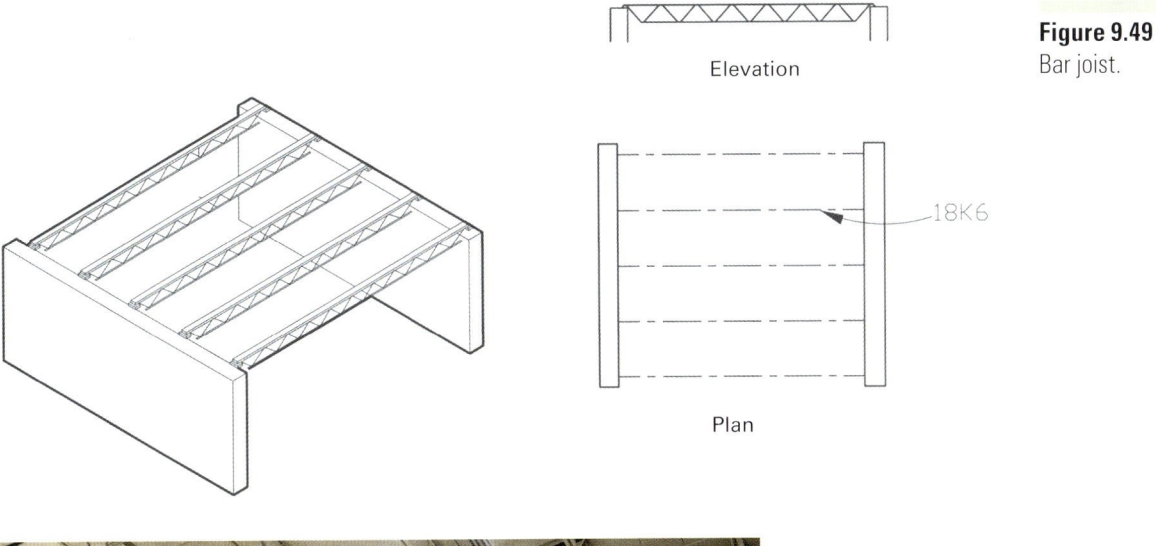

Figure 9.50
Bar joists span the ceiling in this warehouse store. The girders underneath them are made of standard flange sections bolted together.

Figure 9.51
Steel arches are the roof structure.

An arch is a curved symmetrical structure and a series of them form a building's roof (Figure 9.51) or its roof and sides. A rigid frame, also called a **bent**, is two **columns** (vertical load bearing member that runs from ground to roof) and a beam that spans between them (Figures 9.52, 9.53).

STEEL FRAME COMPONENTS   **241**

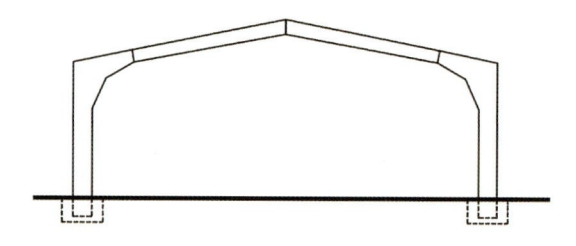

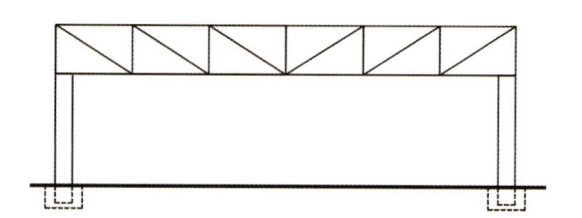

Figure 9.52
Rigid frames.

Figure 9.53
Warehouse made of rigid frames and curtain walls.

Steel Framing

Steel skeleton framing is done similarly to wood skeleton framing, just with steel sections. It is used in commercial construction and in residential where building codes require it for extreme weather, such as hurricanes. Decking is corrugated sheet steel (Figure 9.54). Figure 9.55 shows an apartment complex framed with channel sections as studs. The floors and roof are made of corrugated steel decking with concrete poured over it.

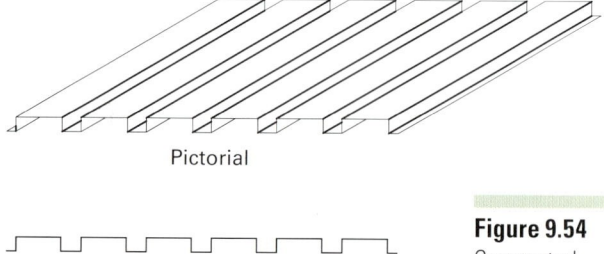

Figure 9.54
Corrugated steel decking.

Figure 9.55
Channel sections are used as studs in this apartment building.

242   CHAPTER 9   BUILDING CONSTRUCTION

Figure 9.56 shows a multi-bay steel structure and its framing plan. A **bay** is the space between four posts or between two rigid frames. It's the main division of a structure. All this structure's components are wide flange sections of different sizes. Figure 9.57 is a partial section drawing. It shows a steel arch, bar joists, and corrugated decking.

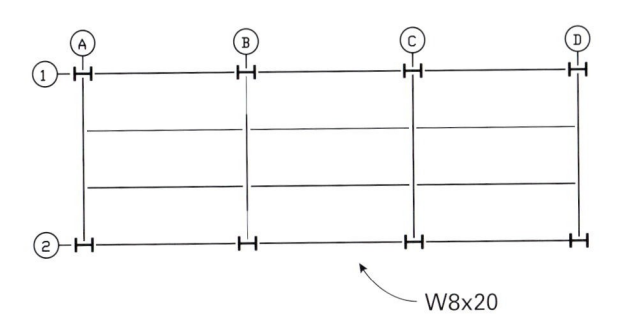

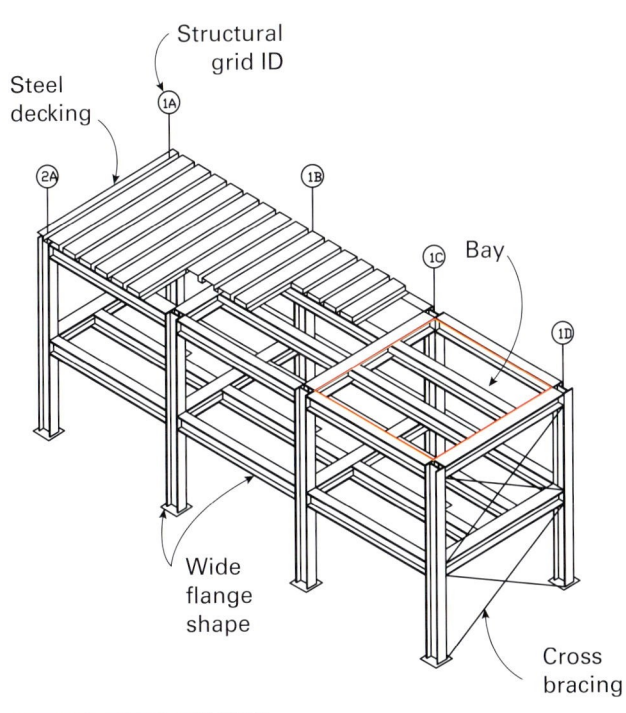

Figure 9.56
A multi-bay, steel structure.

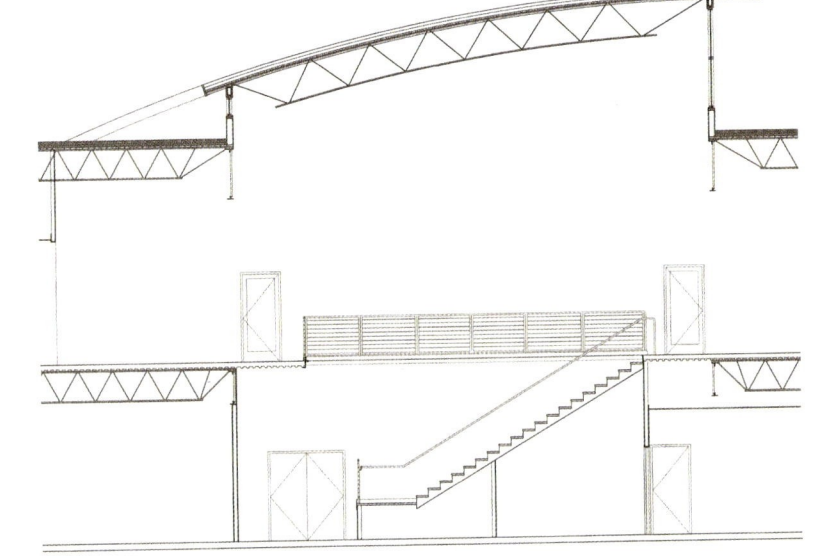

Figure 9.57
Partial section with steel components.
Courtesy mkerrdesign.com.

FYI

Iron and steel have been used since 250 C.E., but commercial-grade steel became possible in 1855 with the invention of the Bessemer blast furnace. Steel's strength, ductility, and relatively cheap price changed the look of commercial architecture. Masonry buildings' heights are limited, as each level up requires a thicker foundation. Steel's ability to support heavier loads without a proportionately thicker foundation enabled buildings to ascend. In 1885 the first skyscraper, the Home Insurance Building in Chicago, went up. Since then, height records have been broken every decade. The Burj Khalifa is currently the highest building in the world at slightly over half a mile tall.

STEEL FRAMING

Finish Materials

The structural system is covered with layers of other materials to make the interior and exterior walls, floors, and ceilings (Figure 9.58). The most common interior wall finish is painted **gypsum board** (Sheetrock® and Drywall® are brand names for this material). Gypsum board is gypsum powder compressed between two pieces of construction paper. It's attached to masonry walls with **furring strips**, which are thin vertical boards nailed to the masonry wall. Furring strips must be an appropriate thickness to clear any piping or ductwork planned between the walls.

Figure 9.58
Tile is laid over a cement substrate.
Courtesy Johnny's Midwest Painting/Remodeling Inc.

There are many material choices for interior surfaces, and new ones are added each year. It's a designer's responsibility to keep current on them all.

Detail Drawings

The structural system, finishes, and all other parts of a building are described via **detail drawings**, which are vertical and horizontal slices through those parts. **Wall sections** are details made by slicing vertically from foundation to roof (Figure 9.59). Figure 9.60 shows a foundation wall detail. Look at each material in the pictorial and find it in the orthographic drawing. Study them to understand how 2D and 3D relate. Figure 9.61 shows section pochés for such details and Figure 9.62 shows wall pochés and thicknesses.

Figure 9.59
Wall section.

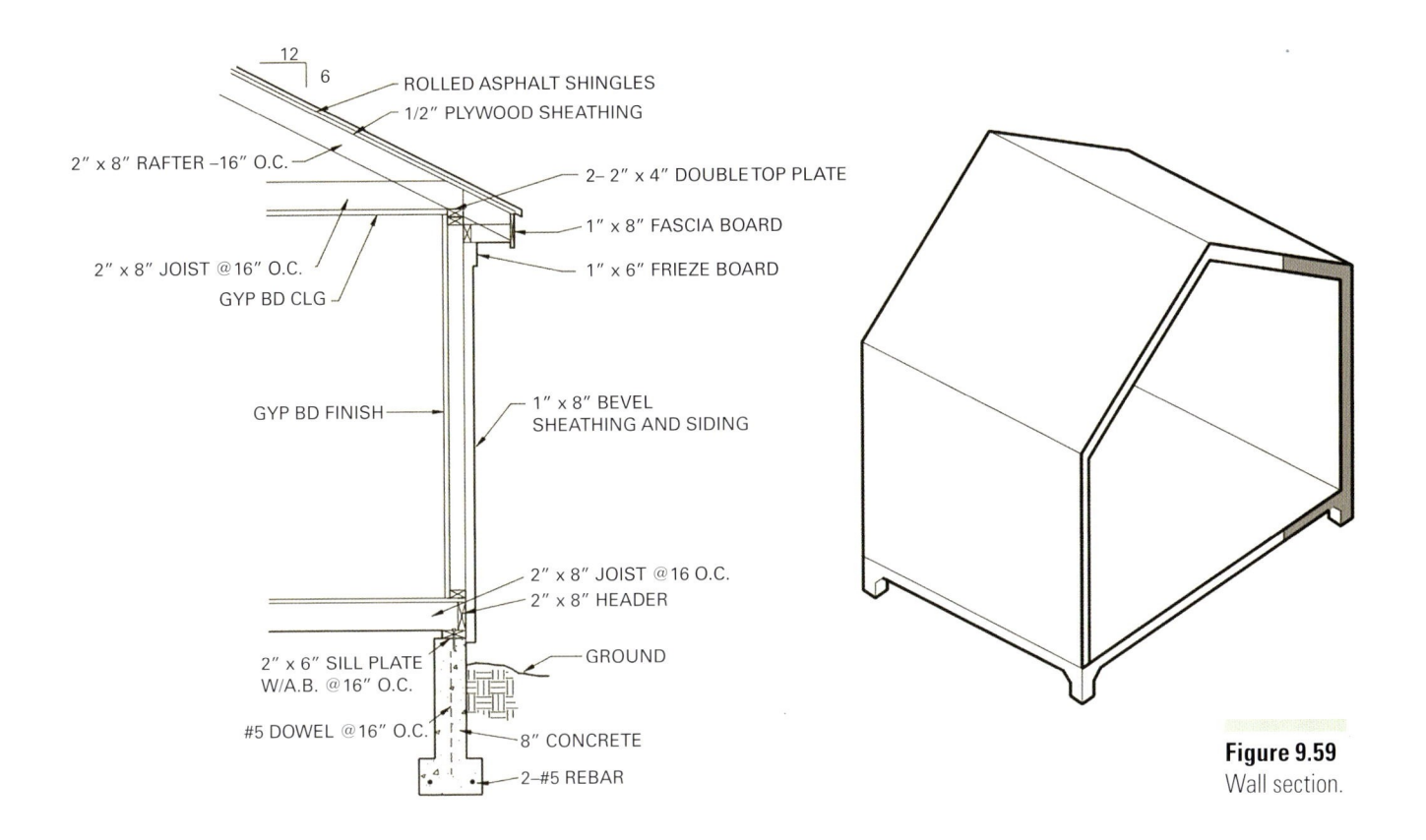

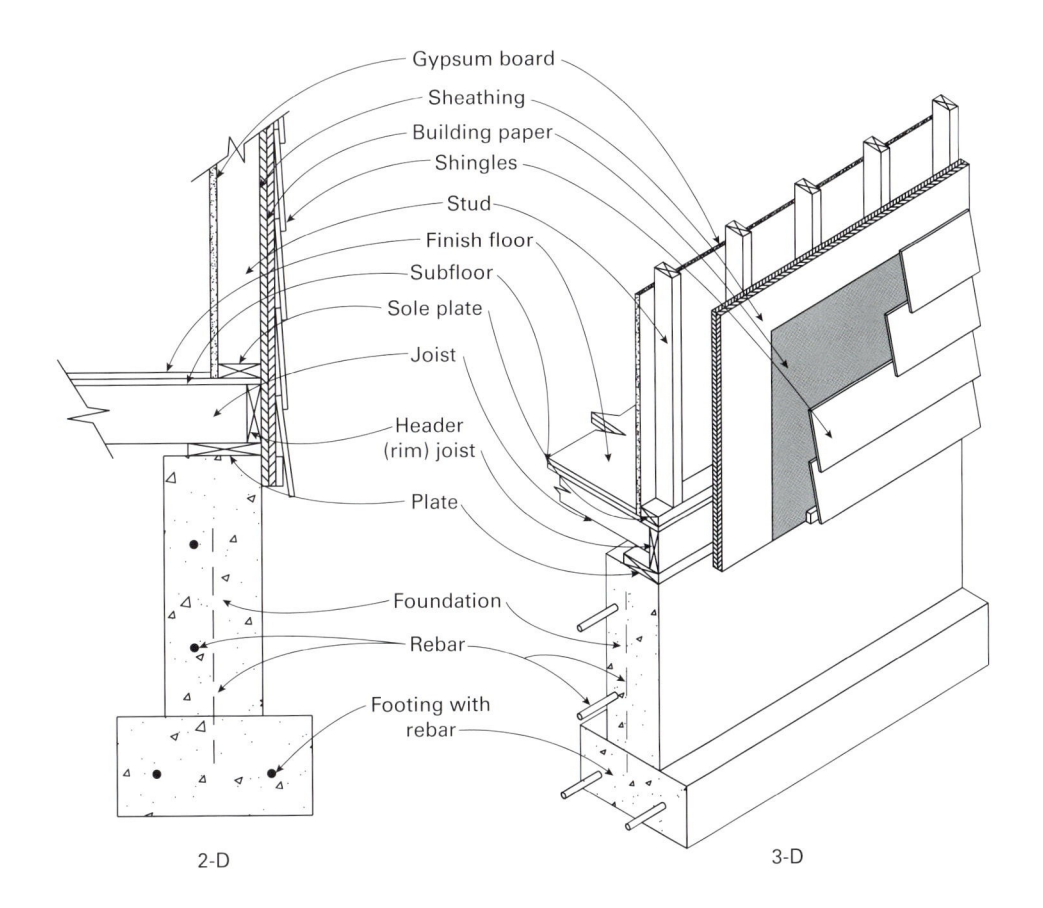

Figure 9.60
Foundation wall detail.

DETAIL DRAWINGS **245**

BATT/LOOSE FILL INSULATION

RIGID INSULATION

METAL LATH AND PLASTER

PLASTER

SAND/MORTAR

GRAVEL

EARTH/COMPACT FILL

TERRAZZO

CARPET ON PAD

ACOUSTICAL TILE

CONCRETE BLOCK

CAST CONCRETE

DIMENSION LUMBER

FINISH BOARD

ASHLAR STONE

RUBBLE STONE

FACE BRICK

FIRE BRICK

METAL

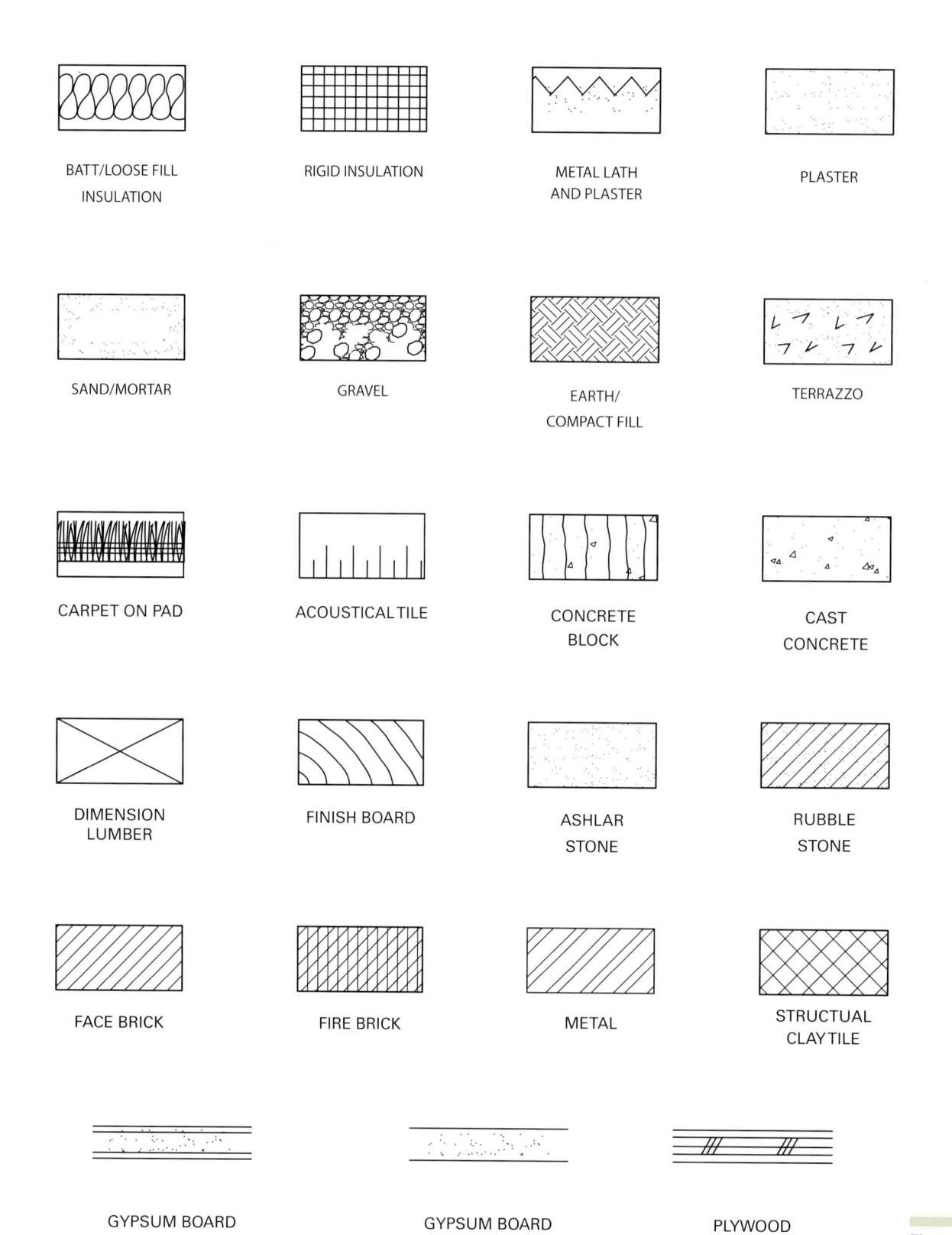

STRUCTURAL CLAY TILE

GYPSUM BOARD (LARGE SCALE)

GYPSUM BOARD (SMALL SCALE)

PLYWOOD

Figure 9.61
Section pochés.

Note: Composite walls have:
4" brick veneer
6" stone veneer
8" concrete block
1/2" gypsum board
2" air space

6" chase wall with wall and floor mounted toilets

Existing wall to remain

or

Existing wall to remove

Existing opening to enclose

Wood frame with siding (exterior) — 6"

Plaster on wood frame (exterior) — 6"

Stone on frame (exterior) — 10"

Brick on frame (exterior) — 10"

Spindles on knee wall (interior) — 6"

Wood studs w/gyp bd on both sides (interior) — 5"

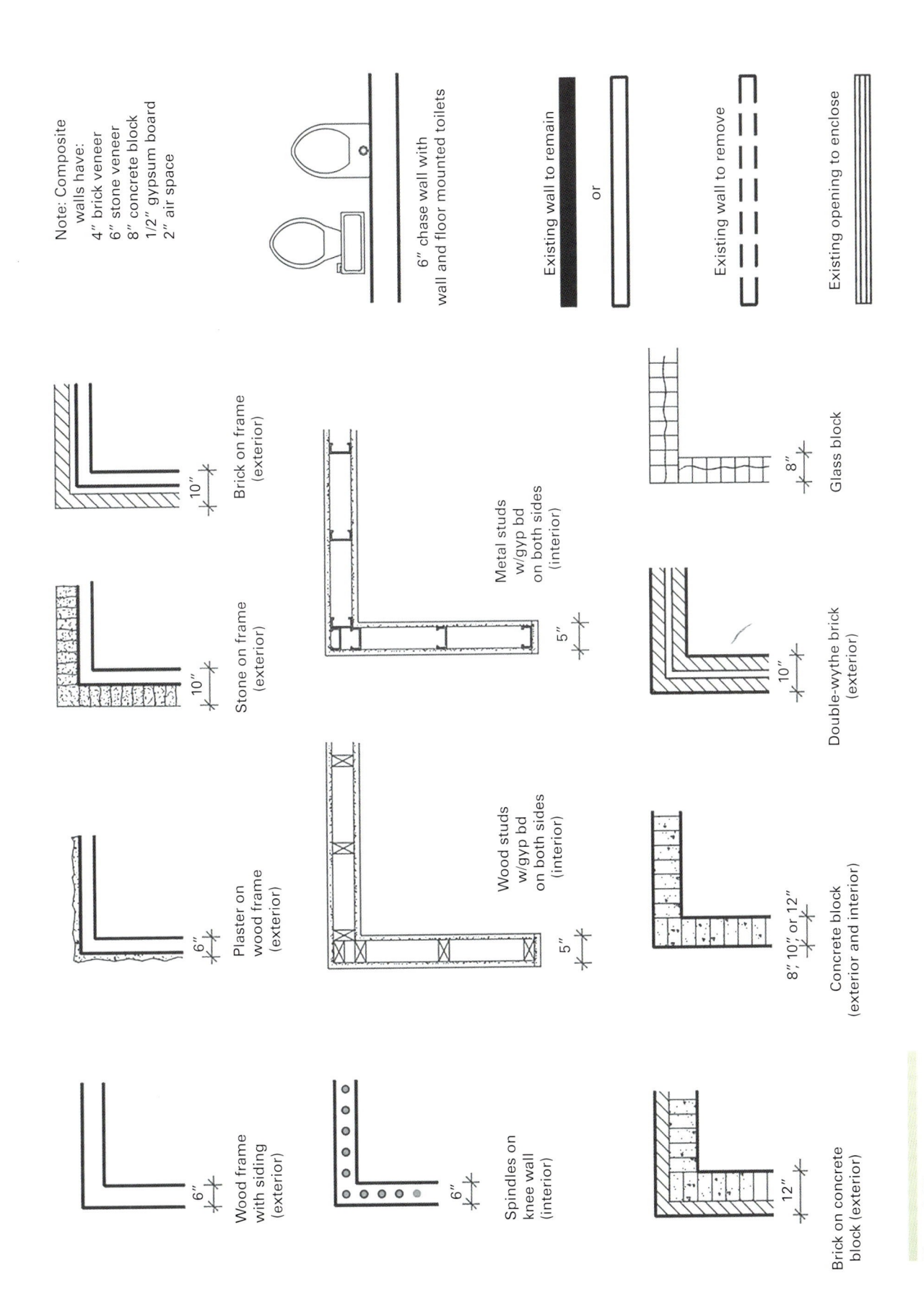

Metal studs w/gyp bd on both sides (interior) — 5"

Glass block — 8"

Concrete block (exterior and interior) — 8", 10", or 12"

Double-wythe brick (exterior) — 10"

Brick on concrete block (exterior) — 12"

Figure 9.62
Wall pochés and thicknesses.

DETAIL DRAWINGS **247**

Door and Window Details

Brand-specific drawings are typically supplied by the manufacturer, but some drafter customization may be needed to show their installation.

Figures 9.63 and 9.64 are pictorials of door and window framing. With doors, the rough opening is framed in the wall, the frame is attached to the rough opening, and the door is attached with a hinge to the frame. With windows, the window is placed in the rough opening and then nailed to the frame. **Casing** (**trim**) is applied afterward to conceal the joint between the window and the wall.

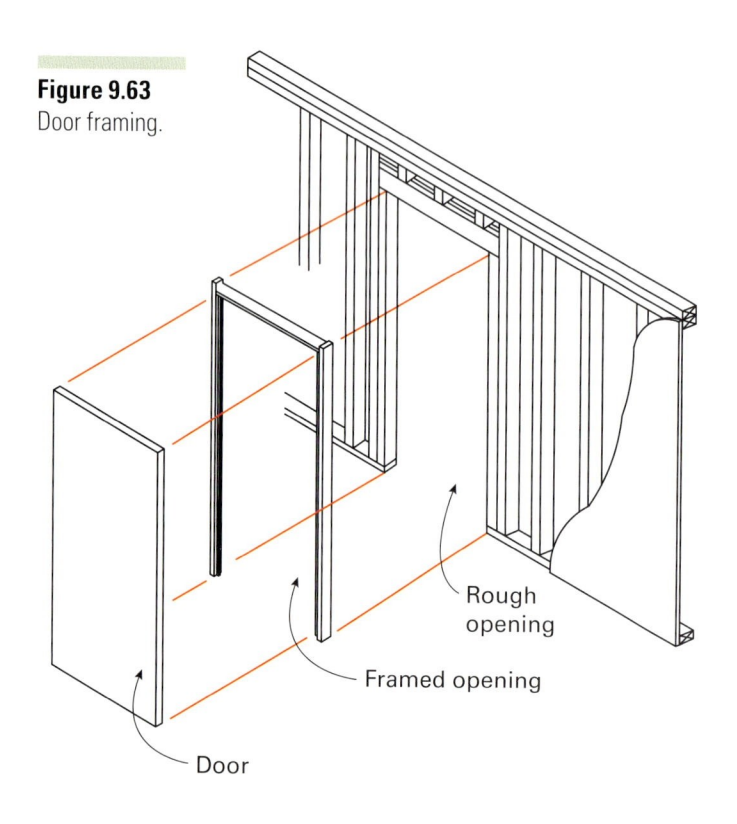

Figure 9.63
Door framing.

Figure 9.64
Window framing.

248   CHAPTER 9   BUILDING CONSTRUCTION

Door and window details consist of the **head**, **jamb**, and **sill**, and are typically drawn at the 1/2" = 1'-0" or 3/4" = 1'-0" scale. The head is the top, the jambs are the sides, and the sill is the bottom. Draw a head detail by slicing vertically through the top of the door or window. Draw a jamb detail by slicing horizontally through the side. Draw a sill detail by slicing vertically through the bottom. Figure 9.65 shows head, jamb, and sill details of an interior wood stud wall with a gypsum board finish on both sides. It is standard practice to rotate the jamb drawing 90° and draw the head, jamb, and sill lined up vertically. An explanation of what we're looking at follows.

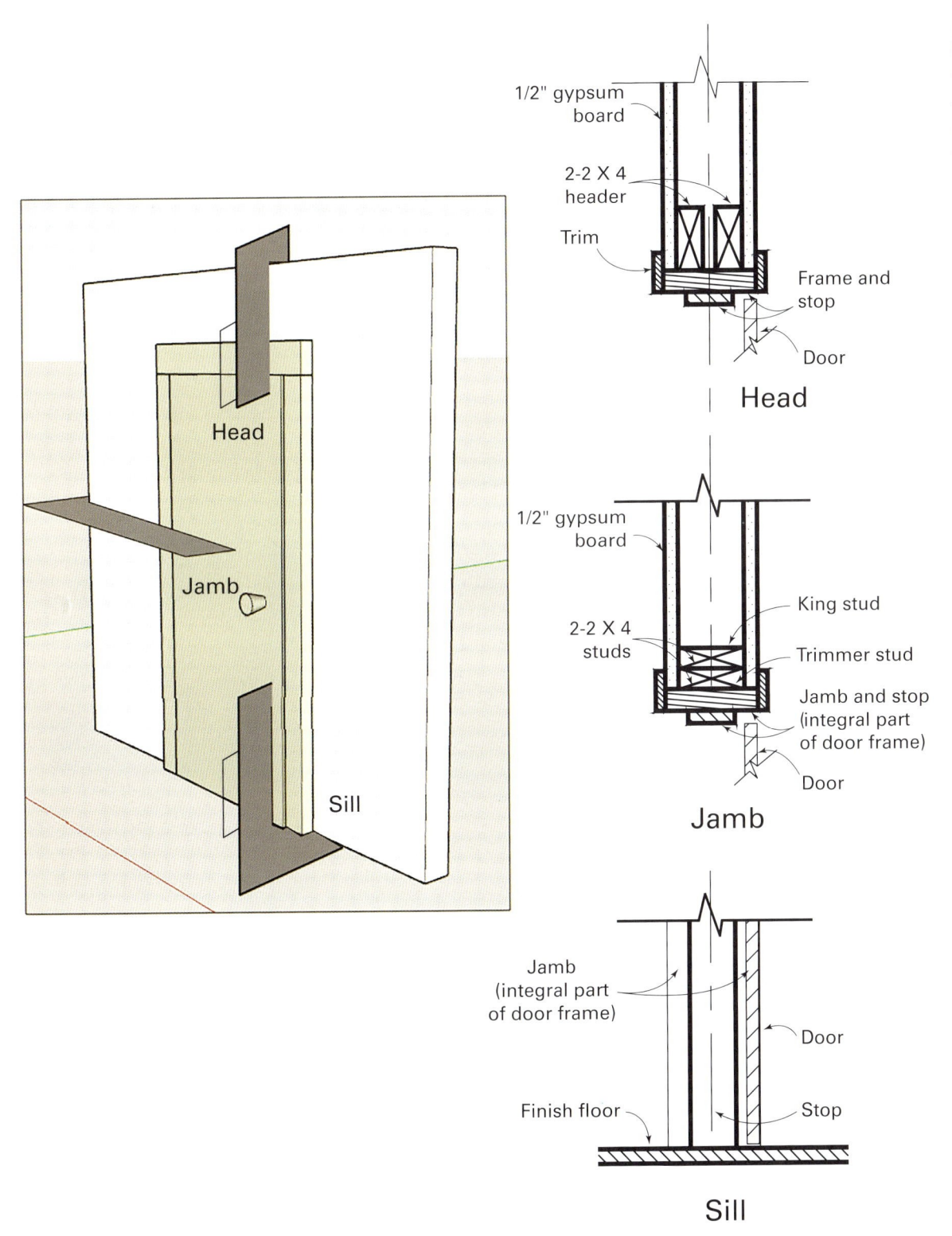

Figure 9.65
Cutting planes and the resultant door details.

DETAIL DRAWINGS 249

Head We're cutting vertically through the top of the door and wall over it. The two *X*s represent the header, two 2 × 4s. The gypsum boards on both sides of the header are the finished wall surfaces. The door frame is attached to the header. The top part of the door is also visible.

Jamb We're cutting horizontally through the side and looking down. Here the two *X*s represent the studs on each side of the opening. Again, we see the gypsum board, frame, and part of the door.

Sill We're making a vertical cut in the middle of the doorway. Even though this is a section view, most of what is seen here is in elevation.

Figure 9.66 shows a cutaway pictorial of a casement window in a vinyl-sided exterior wall, and its head, jamb, and sill details. This level of detail comes from a manufacturer-provided drawing.

Figure 9.66
Casement window and details.

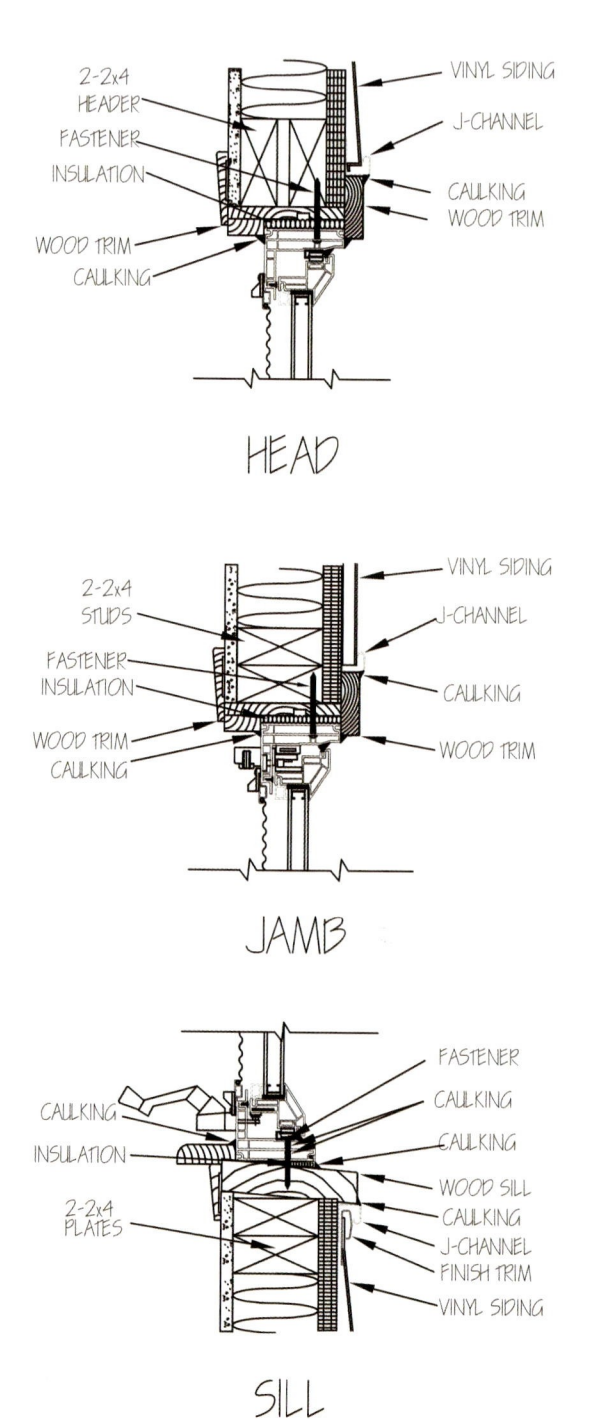

250   CHAPTER 9   BUILDING CONSTRUCTION

Fireplace

A **fireplace** is a framed opening in a chimney that holds an open fire and expels gas and smoke through a chimney or **vent** (opening that allows air or gas to pass out). Few fireplaces are site-built; most are prefabricated metal boxes installed into a framed opening (Figure 9.67)

The **firebox** is the combustion chamber where the fire is contained. The **surround** is the immediate border around the firebox. The **mantel** is the shelf over the firebox. The **hearth** is the floor inside and immediately outside the fireplace. The **chimney** is the vertical, freestanding structure that carries smoke and gas out of the room. The **flue** is a vertical duct inside the chimney. Each fireplace needs its own flue, but multiple flues may share a chimney (Figure 9.68). Some fireplaces are **direct vent**, meaning air for combustion is drawn from the outdoors and waste gasses are exhausted to the outdoors (Figure 9.69).

> **Exercise**
> Stand in a doorway and sketch possible head, jamb, and sill details.

Figure 9.67
Prefab fireplace in a framed opening.

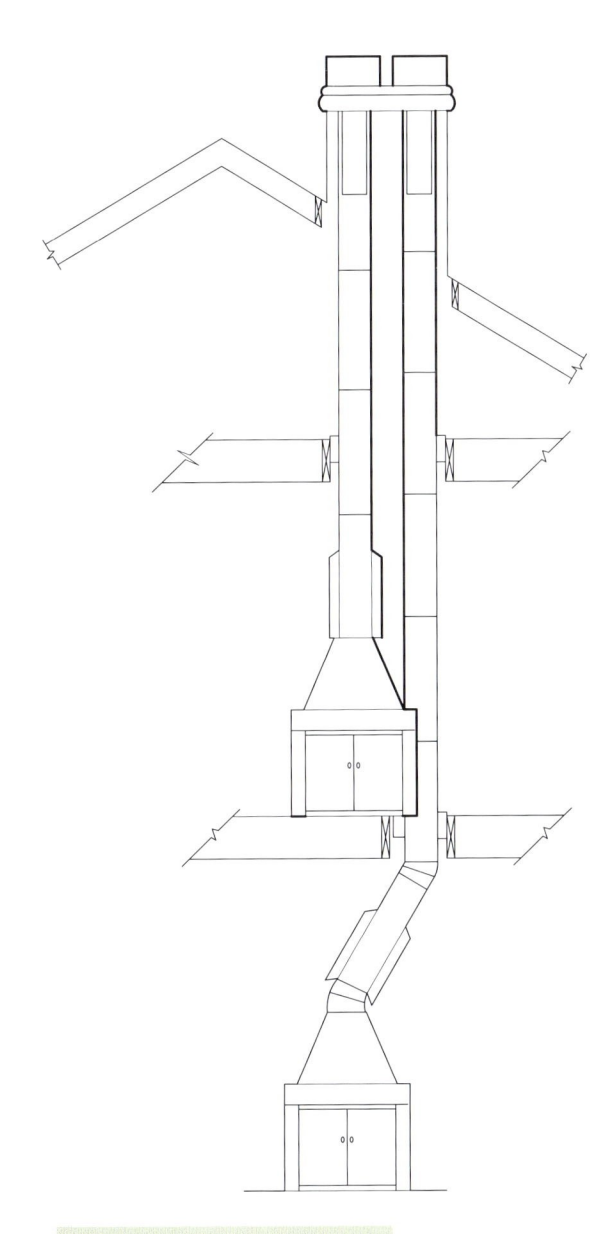

Figure 9.68
Each fireplace needs its own flue.

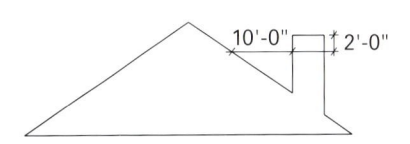

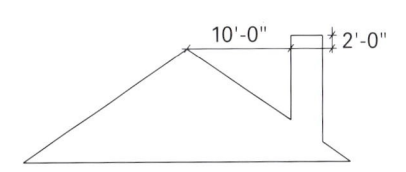

Figure 9.69
Direct vent fireplace.

FIREPLACE 251

Traditional Fireplace This has one, two, or three open faces (Figure 9.70). It can be flush with or protrude from a wall and have a flat or raised hearth. Figure 9.71 shows orthographic views of a one-face fireplace. Note the proportions. The depth, width, and wall angles of the firebox are designed to channel smoke back and up.

How to Draw a Traditional Fireplace Templates for the plan view of one-face fireplaces are available, and Figure 9.72 shows steps for drafting one.

A firebox with an opening of six square feet needs a hearth to collect sparks. Codes require that it be of non-combustible material and extend at least 16" in front of the firebox and 8" on either side. Larger openings require a hearth that extends 20" into the room and 12" on each side. Modern gas-only fireplaces with fixed glass fronts don't require a hearth extension; the manufacturers require a 36" clear space in front. Wood burning fireplaces with gas inserts require the hearth extension.

Draw the back and side walls of the firebox at least 12" thick, which is the minimum needed to fit a flue. Draw any wood trim at least 8" from the firebox opening. A raised hearth is typically 11" to 16" above the floor.

Firebox Size The firebox opening size should be 1/30th of the area of small rooms and 1/65th of the area of large rooms. A common size is 36" wide by 26" high. Table 9.1 is a chart of suggested openings for room sizes:

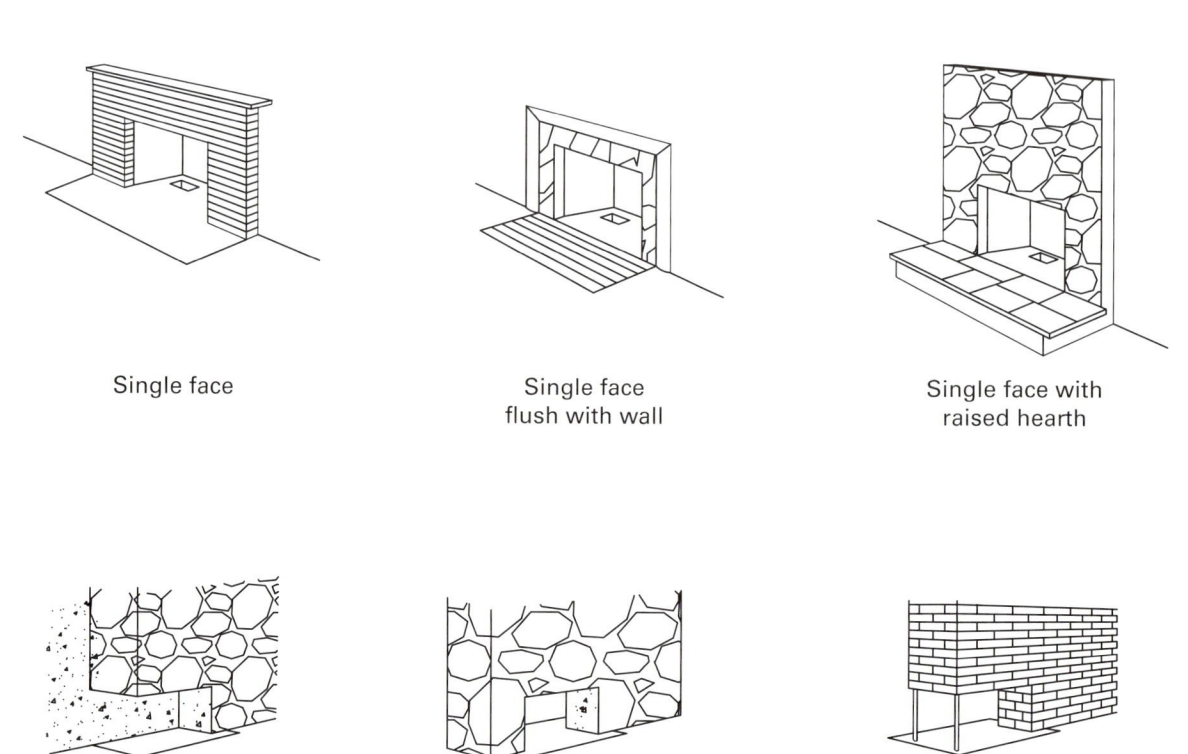

Single face

Single face flush with wall

Single face with raised hearth

Two-face adjacent (projecting corner)

Two-face, opposite

Three-face

Figure 9.70 Traditional fireplace styles.

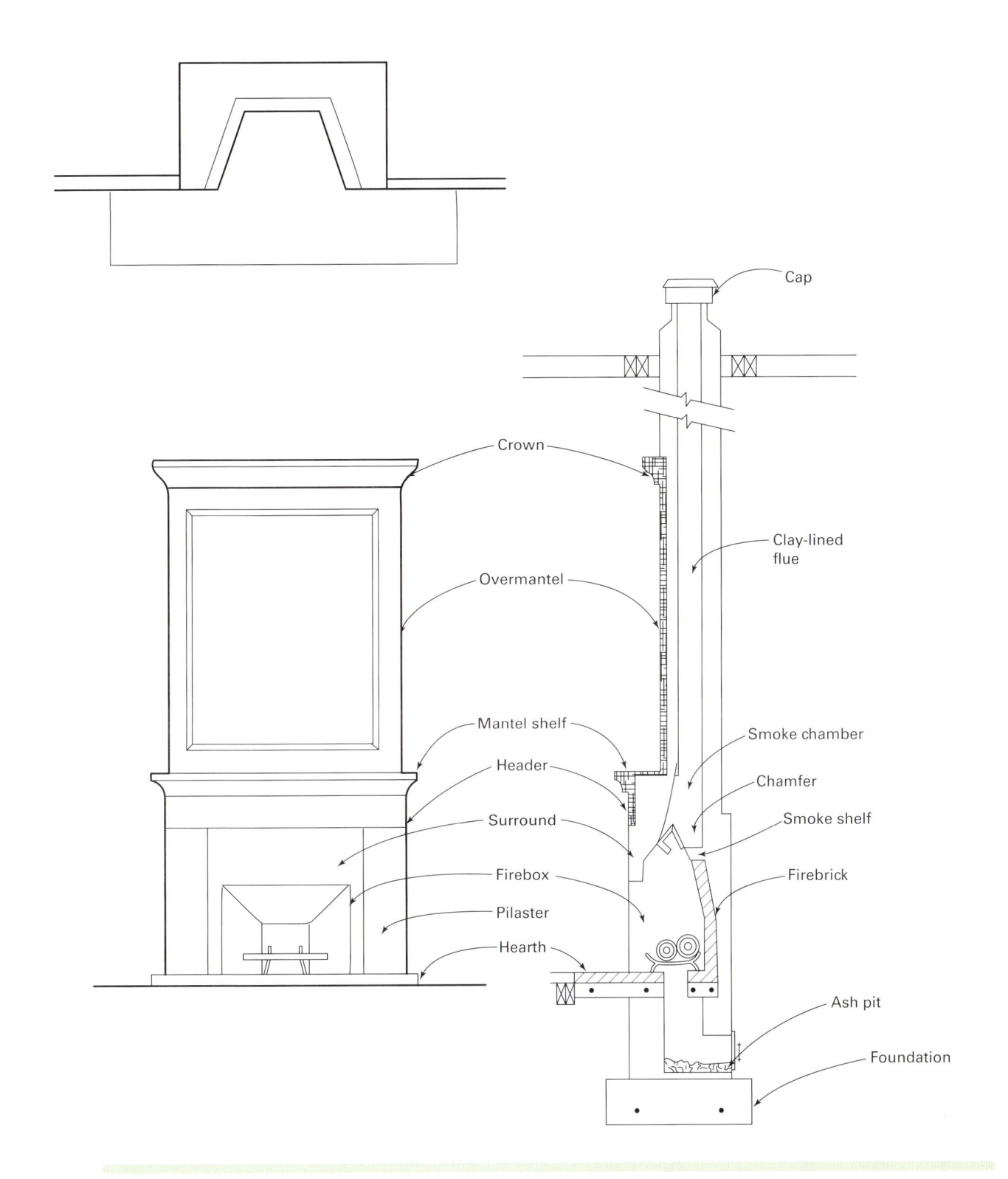

Figure 9.71
Orthographic views of a one-face fireplace

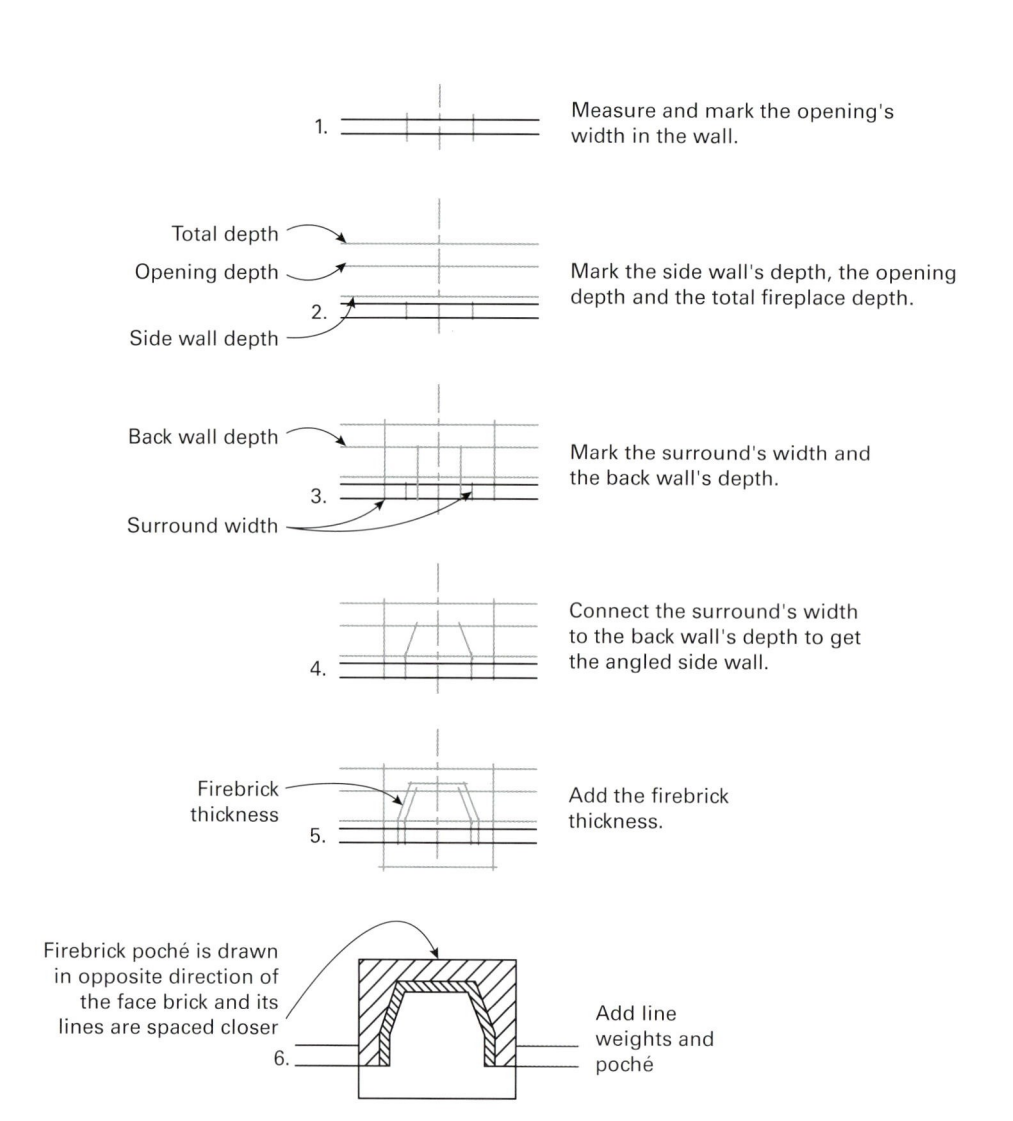

1. Measure and mark the opening's width in the wall.

Total depth
Opening depth
Side wall depth

2. Mark the side wall's depth, the opening depth and the total fireplace depth.

Back wall depth
Surround width

3. Mark the surround's width and the back wall's depth.

4. Connect the surround's width to the back wall's depth to get the angled side wall.

Firebrick thickness

5. Add the firebrick thickness.

Firebrick poché is drawn in opposite direction of the face brick and its lines are spaced closer

6. Add line weights and poché

Figure 9.72
Steps for drafting a fireplace's plan view.

Table 9.1
SUGGESTED WIDTH OF FIREPLACE OPENING IN RELATION TO ROOM SIZE

Room Size	Placed in Short Wall	Placed in Long Wall
10′ × 14′	24″	24″–32″
12′ × 16′	28″–36″	32″–36″
12′ × 20′	32″–36″	36″–40″
12′ × 24′	32″–36″	36″–48″
14′ × 28′	32″–40″	40″–48″
16′ × 30′	36″–40″	48″–60″
20′ × 36′	40″–48″	48″–72″

Non-Traditional Fireplace There are many innovative design fireplaces on the market, and they have different profiles and requirements than traditional ones (Figure 9.73). Obtain specific models' construction drawings from their manufacturers. Figure 9.74 shows drawings for a linear fireplace.

Figure 9.73 Electric fireplace with simulated wood fire. No chimney or venting is needed.

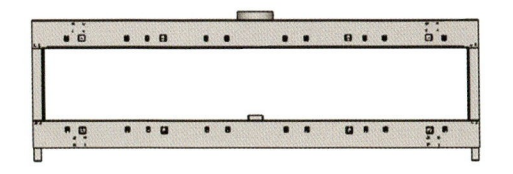

Figure 9.74 Linear fireplace.

Millwork

Also called woodwork, **millwork** is a general term for architectural products manufactured in a lumber mill. Mill items may be made of wood, engineered wood products, or polyurethane (plastic). Hundreds of designs and cross sections exist. Typical mill items are trim (finish woodwork), **molding** (decorative length of wood that hide the joints at floor/wall and floor/ceiling intersections; **base moldings** are at the floor, **crown moldings** are at the ceiling), **chair rails** (running trim 36" above the floor to protect walls from chair backs), **wainscots** (a protective surface on the lower 4' of an interior wall; Figure 9.75), and **case goods** (non-upholstered pieces of furniture that provide storage, such as cabinets, dressers, and desks).

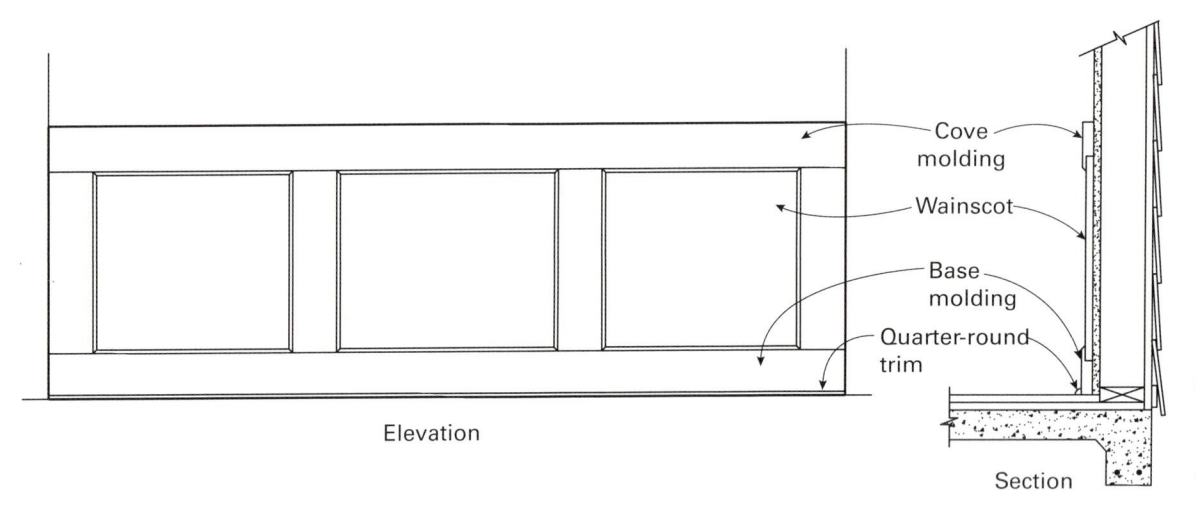

Figure 9.75 Paneled wainscot.

Cabinets These are cupboards with drawers or shelves for storage and display. Base and wall cabinets are made to standardized sizes discussed in Chapter 7. **Corner cabinets** are configured in a non-standard way to maximize usable space (Figure 9.76). **Blind cabinets** have one door and contain largely inaccessible space (Figure 9.77).

Framed vs. Frameless Cabinets may be constructed **framed** or **frameless** (Figure 9.78). Framed cabinets have **rails** (horizontal framing pieces) and **stiles** (vertical framing pieces) on the inside of their doors; unframed ones don't.

Cabinet Door Types (Figure 9.79) Doors are either **full** or **standard overlay**, which describes the amount of front frame covered by the door and drawer. A full overlay has a maximum 1/8" **reveal** (exposed portion of the frame) around each door and drawer and between the doors. A standard overlay cabinet has a minimum of 1" reveal around the door and drawer perimeters. They may also be **tambour**, **inset**, or **lipped**. A tambour rolls up. An inset door sits within the cabinet frame, flush with the cabinet box's front. A lipped door has a **rabbet** (groove) on the back that allows part of the door to go back into the cabinet and the remaining part rests on the cabinet's face frame. When viewed from the front, a lipped door looks like an overlay door.

Cabinet Door Styles This is the door's appearance. There are two major styles: **slab** and **panel** (Figure 9.80). A slab door looks like one solid piece. A panel style has either a raised or a recessed profile. A recessed panel has a wood frame around a middle panel like a picture frame. A raised panel protrudes forward. Both have hundreds of variations.

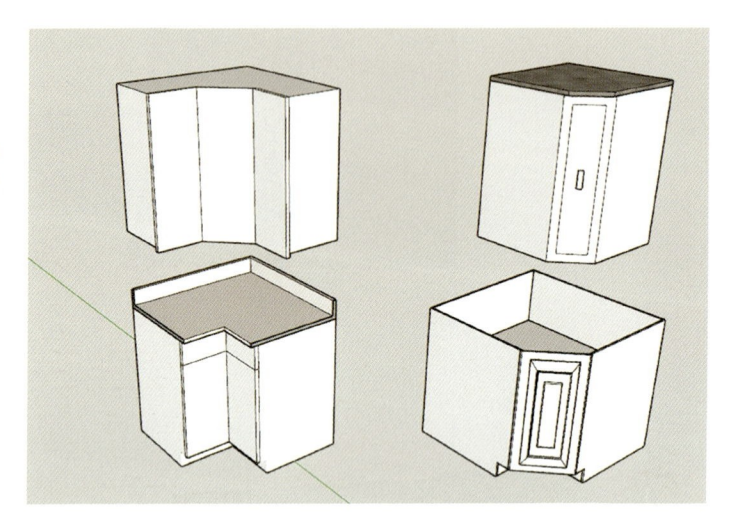

Figure 9.76 Corner cabinets are configured to maximize corner space.

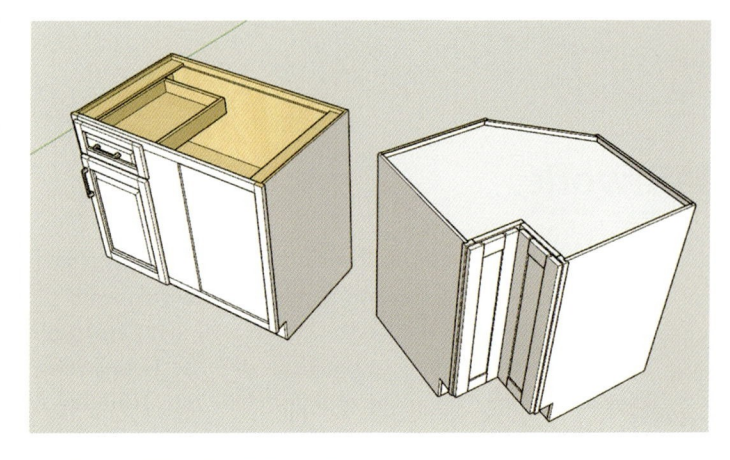

Figure 9.77 Blind cabinets have inaccessible space.

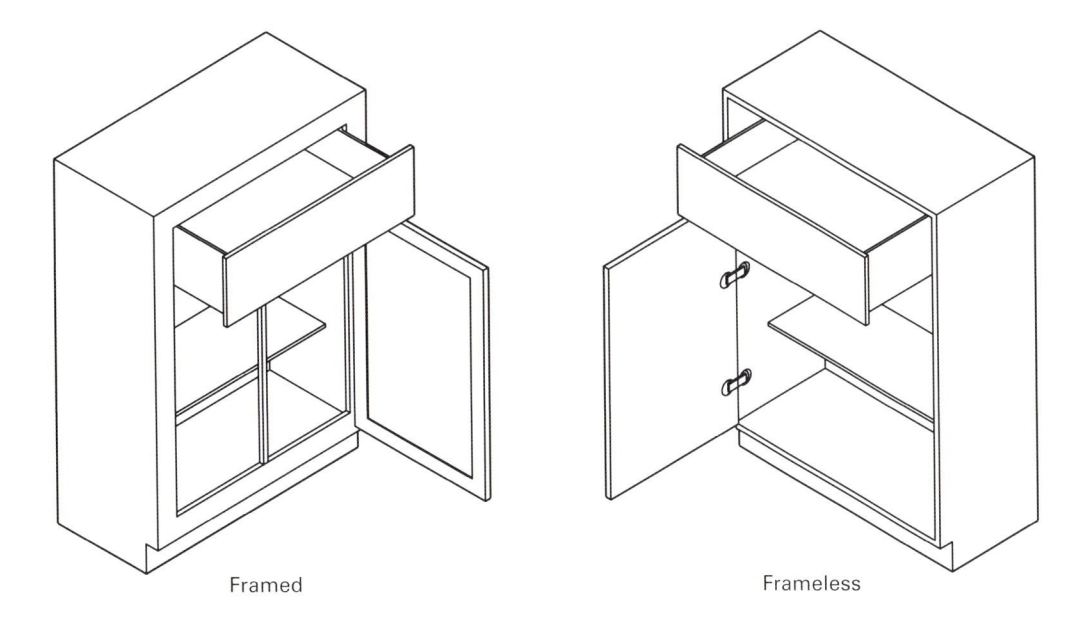

Figure 9.78 Framed and frameless cabinets.

Framed

Frameless

Tambour

Face frame
Cabinet door

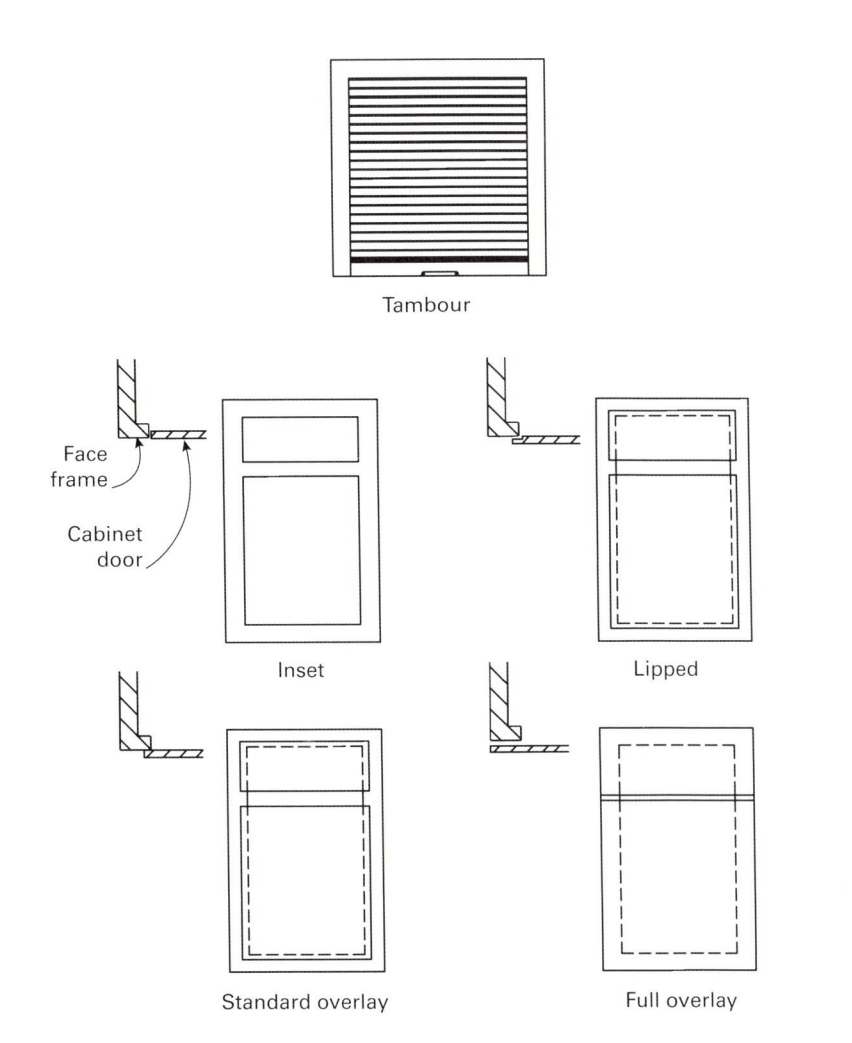

Inset

Lipped

Standard overlay

Full overlay

Figure 9.79 Cabinet door types. The hidden lines show the cabinet frame and are not included in an elevation drawing.

MILLWORK **257**

Figure 9.80
Slab and panel cabinet doors.

Range Hoods **Range hoods** are appliances that have fans to remove airborne cooking by-products such as grease, fumes, smoke, and steam. Some have **ducts** (passageways) that remove those by-products to the outdoors, and others are ductless, which just filter the by-products. Hoods are built into wall cabinets (Figure 9.81); **downdraft vents** (Figure 9.82) are built into base cabinets next to cooktops or integrated in the cooktop. Both hoods and downdrafts have many variations.

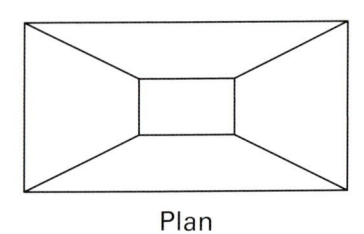

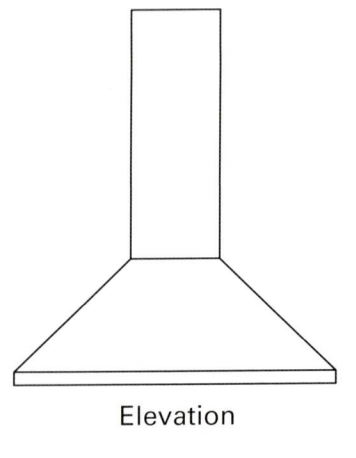

Figure 9.81
Range hood. Courtesy Bosche Home Appliances.

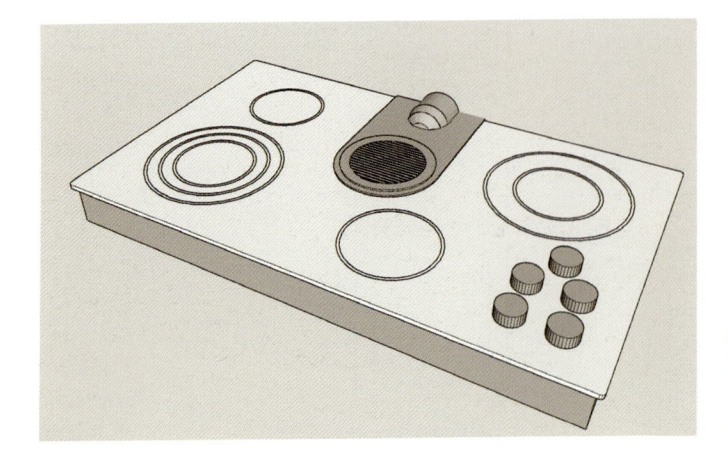

Figure 9.82
Cooktop with downdraft vent.

258   CHAPTER 9   BUILDING CONSTRUCTION

MasterFormat

MasterFormat is a categorization system written by the Construction Specifications Institute, a trade group of design and construction professionals. It's an industry standard used by government and private organizations for project manuals, bid documents, cost estimates, and specifications. Hence, it's useful to understand its numbers. Subjects are arranged into 49 divisions, with each division containing three levels. Each title has a six-digit number. For example:

12 56 33.23 Classroom Computer Furniture

The first two digits (12) are the division number; the next two (56 33) are for specific items in that division. Sometimes there's a dot and another pair of numbers if the level of detail requires yet another classification. Figure 9.83 shows the divisions. Figure 9.84 shows a page from Division 9.

MasterFormat GROUPS, SUBGROUPS, AND DIVISIONS

PROCUREMENT AND CONTRACTING REQUIREMENTS GROUP
Division 00 – Procurement and Contracting Requirements
   Introductory Information
   Procurement Requirements
   Contracting Requirements

SPECIFICATIONS GROUP

GENERAL REQUIREMENTS SUBGROUP
Division 01 – General Requirements

FACILITY CONSTRUCTION SUBGROUP
Division 02 – Existing Conditions
Division 03 – Concrete
Division 04 – Masonry
Division 05 – Metals
Division 06 – Wood, Plastics, and Composites
Division 07 – Thermal and Moisture Protection
Division 08 – Openings
Division 09 – Finishes
Division 10 – Specialties
Division 11 – Equipment
Division 12 – Furnishings
Division 13 – Special Construction
Division 14 – Conveying Equipment
Division 15 – Reserved for Future Expansion
Division 16 – Reserved for Future Expansion
Division 17 – Reserved for Future Expansion
Division 18 – Reserved for Future Expansion
Division 19 – Reserved for Future Expansion

FACILITY SERVICES SUBGROUP
Division 20 – Reserved for Future Expansion
Division 21 – Fire Suppression
Division 22 – Plumbing
Division 23 – Heating, Ventilating, and Air Conditioning (HVAC)
Division 24 – Reserved for Future Expansion
Division 25 – Integrated Automation
Division 26 – Electrical
Division 27 – Communications
Division 28 – Electronic Safety and Security
Division 29 – Reserved for Future Expansion

SITE AND INFRASTRUCTURE SUBGROUP
Division 30 – Reserved for Future Expansion
Division 31 – Earthwork
Division 32 – Exterior Improvements
Division 33 – Utilities
Division 34 – Transportation
Division 35 – Waterway and Marine Construction
Division 36 – Reserved for Future Expansion
Division 37 – Reserved for Future Expansion
Division 38 – Reserved for Future Expansion
Division 39 – Reserved for Future Expansion

PROCESS EQUIPMENT SUBGROUP
Division 40 – Process Interconnections
Division 41 – Material Processing and Handling Equipment
Division 42 – Process Heating, Cooling, and Drying Equipment
Division 43 – Process Gas and Liquid Handling, Purification, and Storage Equipment
Division 44 – Pollution and Waste Control Equipment
Division 45 – Industry-Specific Manufacturing Equipment
Division 46 – Water and Wastewater Equipment
Division 47 – Reserved for Future Expansion
Division 48 – Electrical Power Generation
Division 49 – Reserved for Future Expansion

Figure 9.83
Page from the 2020 CSI MasterFormat. © 2018 The Construction Specifications Institute, Inc. (CSI). MasterFormat® excerpt and trademarks used under permission from CSI.

| Number | Title | Explanation |

DIVISION 09

DIVISION 09 – FINISHES

09 00 00   **Finishes**   May be used as Division level Section title.

    09 01 00   **Maintenance of Finishes**

                Includes: maintenance, repair, rehabilitation, replacement, restoration, preservation, etc. of finishes.

                Level 4 Numbering Recommendation: Level 4 titles for finishes are provided as examples for user's creation of more detailed titles for other subjects as required. Following Level 4 numbering is recommended:
.51-.59 for maintenance.
.61-.69 for repair.
.71-.79 for rehabilitation.
.81-.89 for replacement.
.91-.99 for restoration.

                See: 02 83 33 for lead-based paint removal and disposal.

      09 01 20   Maintenance of Plaster and Gypsum Board
          09 01 20.91   Plaster Restoration
      09 01 30   Maintenance of Tiling
          09 01 30.91   Tile Restoration
      09 01 50   Maintenance of Ceilings
          09 01 50.91   Ceiling Restoration
      09 01 60   Maintenance of Flooring
          09 01 60.91   Flooring Restoration
      09 01 70   Maintenance of Wall Finishes
          09 01 70.91   Wall Finish Restoration
      09 01 80   Maintenance of Acoustic Treatment
      09 01 90   Maintenance of Painting and Coating
          09 01 90.51   Paint Cleaning
          09 01 90.52   Maintenance Repainting
          09 01 90.53   Maintenance Coatings
          09 01 90.61   Repainting
          09 01 90.91   Paint Restoration
          09 01 90.92   Coating Restoration
          09 01 90.93   Paint Preservation

    09 03 00   **Conservation Treatment of Period Finishes**
      09 03 21   Conservation Treatment for Period Plaster Assemblies
                May Include: requirements for gypsum plastering, Portland cement plastering, veneer plastering, and plaster fabrications and details.

      09 03 25   Conservation Treatment for Period Plastering
          09 03 25.23   Conservation Treatment for Period Lime-Based Plaster
      09 03 30   Conservation Treatment for Period Tile

197   Master List of Numbers, Titles, and Explanations

Figure 9.84 Page from Division 9 © 2018 The Construction Specifications Institute, Inc. (CSI). MasterFormat® excerpt and trademarks used under permission from CSI.

Site Plan

This is an aerial view of the property and is prepared by a civil engineer or landscape architect. It shows the site's physical characteristics. Site plans vary greatly based on project complexity, but all will include the following:

North Arrow   Graphic that indicates the direction of north.

Engineer Scale   Tool for measuring large distances. A site plan is often too large for the architect's scale. *1" = 10'* and *1" = 20'* are common engineer's scales used.

Footprint   The outline of the building's shape, size, and orientation. It may depict an outline, the roof, or a floor plan to show the relationships between the interior and exterior. Outbuildings, such as sheds or unattached garages, are also included.

Property Lines   These enclose the property's physical boundaries.

Hard Surfaces   These include walks, entries, driveways, access roads, patios, retaining walls, terraces, sport courts, the septic system, and any other features within the property lines. Their surface materials are pochéd.

Utility Lines   These include gas, electricity, water, and sewer lines that serve the building, and the connection of the house sewer to the public system.

Vegetation   Trees and shrubs. Trees are drawn at their trunk base locations, at a diameter that represents the trunk size.

Setbacks and Easements   A setback is the building's distance from the property line. Local zoning laws regulate the exact distance, which may differ between the front, side, and back property lines. Setback lines are drawn within, and parallel to, the property lines to show the total allowable building area. An easement is a waterway, walkway, or street that lies within the property lines but is designated for public use.

Contour Lines   These show ground elevations. They connect points on the land surface that are the same elevation above a benchmark or reference point. Closely spaced contour lines describe a steeply sloping site; lines spaced far apart describe a gradually sloping one. Dashed lines indicate "before" status; the solid ones indicate "after." Some site plans have *spot elevations* instead of contour lines. These are elevations at key points and are used when the property is relatively flat.

Dimensions and Bearing Angle   Overall lot dimensions and the length and compass direction of the property lines.

Figure 9.85 shows a site plan done for client presentation. Figure 9.86 shows some site plan symbols.

Figure 9.85
Residential site plan. Courtesy wolfgangtrost.com

SITE PLAN   **261**

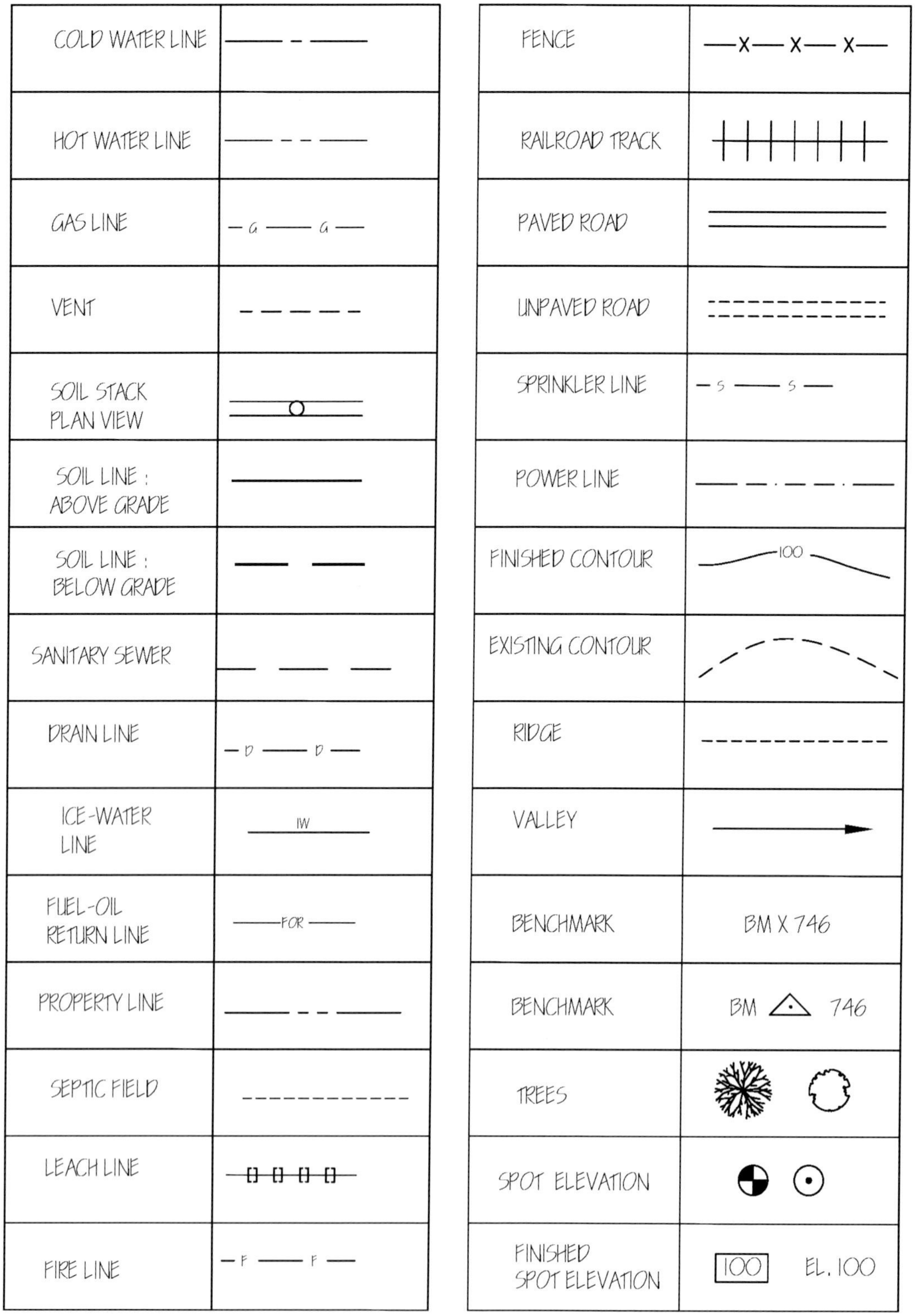

COLD WATER LINE	— - — - —	FENCE	—X—X—X—
HOT WATER LINE	— - - — - - —	RAILROAD TRACK	┼┼┼┼┼┼
GAS LINE	— G — G —	PAVED ROAD	═══════
VENT	— — — — —	UNPAVED ROAD	═ ═ ═ ═ ═
SOIL STACK PLAN VIEW	──○──	SPRINKLER LINE	— S — S —
SOIL LINE: ABOVE GRADE	──────	POWER LINE	— - · — - · —
SOIL LINE: BELOW GRADE	— — —	FINISHED CONTOUR	⌒100
SANITARY SEWER	— — —	EXISTING CONTOUR	⌒ (dashed)
DRAIN LINE	— D — D —	RIDGE	- - - - - - -
ICE-WATER LINE	— IW —	VALLEY	──→
FUEL-OIL RETURN LINE	— FOR —	BENCHMARK	BM X 746
PROPERTY LINE	— — - - —	BENCHMARK	BM △ 746
SEPTIC FIELD	- - - - - - - -	TREES	🌳 ◯
LEACH LINE	—▯▯▯▯—	SPOT ELEVATION	⊕ ⊙
FIRE LINE	— F — F —	FINISHED SPOT ELEVATION	[100] EL. 100

Figure 9.86
Site plan symbols.

Summary

Buildings range from simple to complex, but all rest on a foundation designed for their size, type, and location, and all contain load-bearing and non-load-bearing components. They're made of many different materials, with wood, concrete, steel, and masonry being the primary structural ones. Their complexity of construction is explained in section and detail drawings. Competent design, drafting, and interpretation of details is facilitated by understanding basic building construction.

Classroom Activities

1. Sketch a foundation plan for a floor plan you've drawn or seen in a magazine.
2. Look for brick paving on courtyards and sidewalks and sketch the patterns.
3. Look at brick walls and identify the bonds and exposed portions of the brick.
4. Visit a big-box building supply store and examine different components used in construction. Note standardized sizes.
5. Obtain manufacturer catalogs for different fireplace types and examine their design data, photographs, and drawings.
6. Sketch a site plan of a residence or commercial building. Study how the ground is sloped and include some contour lines.

Questions

1. What is a box beam?
2. What is a load-bearing wall?
3. How are balloon and platform framing different?
4. Why is the actual size of a brick smaller than its nominal size?
5. How wide should you draw a wood stud wall with gypsum board on the inside and wood siding on the outside?
6. What do closely spaced contour lines on a site plan indicate?
7. How far should the hearth extend in front of, and to the side of, a 6' square wood firebox opening?
8. What is a steel section?
9. Name three common foundation types.
10. What is a demolition plan?

Further Resources

ArchDaily article describing different types of engineered wood products. https://tinyurl.com/y4qtsvlp

CertainTeed website. Information on their products, design, and interior space planning and relevant codes. www.certainteed.com.

Ching, Francis DK. 2020. Building Construction Illustrated, 6th ed. Hoboken, NJ: Wiley.

Fireplace manufacturers. Information on their products and downloadable CAD files. hearthnhome.com, flarefireplaces.com, heatnglo.com

The Engineered Wood Association. Information about engineered wood products. www.apawood.org.

Weyerhaeuser. Information on their products, such as the i-joist system. weyerhaeuser.com

Keywords

- actual size
- arch
- balloon framing
- bar joist
- base molding
- bay
- beam
- bent
- blind cabinet
- blocking
- bond
- bond beam
- case goods
- casing
- chair rail
- chimney
- clear span
- column
- concrete block (cmu)
- corner cabinet
- course
- crawl space
- cripple stud
- cross bracing
- cross bridging
- crown molding
- curtain wall
- decking
- demolition plan
- detail drawing
- dimensional lumber
- direct vent
- double top plate
- downdraft vent
- duct
- engineered wood product
- firebox
- fireplace
- flue
- footing
- framed cabinet
- frameless cabinet
- full overlay
- furring strip
- girder
- Glulam
- grade beam
- gypsum board
- head
- header
- hearth
- hip rafter
- inset door
- Insulated Concrete Forms (ICF)
- jamb
- joist
- knee wall
- lintel
- lipped door
- load-bearing
- mantel
- MasterFormat
- microlam
- millwork
- modular construction
- molding
- mortar joint
- nominal size
- non-load bearing
- panel door
- partition wall
- paver
- pilaster
- pile
- plate
- platform (western) framing
- post and beam framing
- post-on-pad
- prefabricated
- rabbet
- raft
- rafter
- rail
- range hood
- rebar
- reveal
- ridge board
- rigid frame
- roof framing plan
- roof plan
- sheathing
- sill
- skeleton framing
- slab door
- slab-on-grade
- sole plate
- spread footing
- standard overlay
- steel
- stepped footing
- stick framing
- stile
- strip footing
- stud
- substructure
- superstructure
- surround
- tambour door
- tilt-up construction
- timber framing
- trim
- trussed rafter (truss)
- valley
- valley rafter
- vent
- wainscot
- wall section
- wythe

CHAPTER 10

Utility Systems

Figure 10.0
The utility lines are incorporated into the architectural design.

OBJECTIVES

Upon completion of this chapter you will be able to:

- Identify electrical, reflected ceiling, plumbing, and HVAC (Heating Ventilation and Air Conditioning) drawings
- Describe their relevance to interior design
- Establish basic code requirements for outlet, switch, and vent placement.
- List vent sizes and ways to integrate them into a design.

The ability to interpret utility systems drawings enables an interior designer to approach a project **holistically**, that is, to coordinate and integrate all its parts. This is especially useful when she or he is responsible for the whole project.

Benefit of Utilities Systems Knowledge

Examples of how understanding plumbing, HVAC, and electrical drawings benefits interior design work are:

- You'll be able to compare electrical switch and outlet locations with the space plan to ensure they aren't inaccessible behind furniture.
- You can verify plumbing pipe locations to ensure selection of a bathtub with a drain on the end that connects to those pipes.
- You'll be aware that electrical conduit may need to be cut and relocated when creating holes in walls for new doors or windows.
- You can overlay electrical drawings onto floor plans to place outlets and **thermostats** (devices that regulate furnace and air conditioner temperature) appropriately.
- You can coordinate the door swings of front-loading **appliances** (devices attached to electric and/or gas lines) and their piping connections so that the doors will open opposite each other.
- You'll be aware that kitchen remodels might require new or relocated ductwork, and that ductwork has to be designed into the remodel.

Electrical Drawings

These include **electrical**, **lighting**, **reflected ceiling**, **automated systems**, and **power**, also called **data**, **plans**. Electrical plans show everything powered by a plug and wire–appliances, lights, switches, outlets, etc.—and the connecting wires between them. Lighting plans show all lighting fixtures, switches, and their connecting wires. Commercial drawing sets have separate electrical and lighting plans; residential often combine both onto one. Reflected ceiling plans show everything on the ceiling—lights, air registers, smoke detectors, sprinklers, cornice mouldings, etc. Automated systems plans show electronic devices and the smart home components they operate—thermostats, lights, landscape sprinklers, entertainment, cooking, and security systems. These systems are controlled with a computer or phone app. Power plans show video cables, data, phone, Voice over Internet Protocol (VoIP), and fiber optics lines, and data ports.

Electrical Components

Residences and commercial buildings contain **service panels**, **circuits**, **outlets**, **switches**, **permanently attached light fixtures**, **equipment** (furnaces, water heaters, garbage disposals), and **appliances** (stoves, cooktops, dishwashers, refrigerators).

Service Panel   Also called a **breaker box**, this is a steel box with **circuit breakers** through which electricity enters the house. Circuit breakers are switches that shut down overloaded circuits—wires through which electricity is distributed throughout the house. 110-volt circuits power most outlets, switches, and permanently installed light fixtures (Figure 10.1). 220-volt circuits power high-heat generating appliances like air-conditioners and dryers. Dedicated circuits supply appliances that need their own circuits, such as sump pumps, washers, dryers, furnaces, and water heaters.

Outlets   Outlets are connectors for plug-in devices (Figure 10.2). Most are **duplex** (two plugs). **Single plug** outlets are for dedicated circuits. **GFCI outlets** shut down when they detect water. **Split wire** outlets look like duplexes, but one plug is always hot and the other activates with a switch (you might plug a lamp in one and a TV in the other). Most circuits are 110V; high-heat generating appliances such as dryers and air conditioners need 220V. Some outlets have a T-shaped slot that allows higher **amperage** (electric current strength) devices to be plugged in. **Grounded outlets** (three holes) have an additional wire that provides extra safety and are code-required. **Non-grounded outlets** (two holes) are found in older homes.

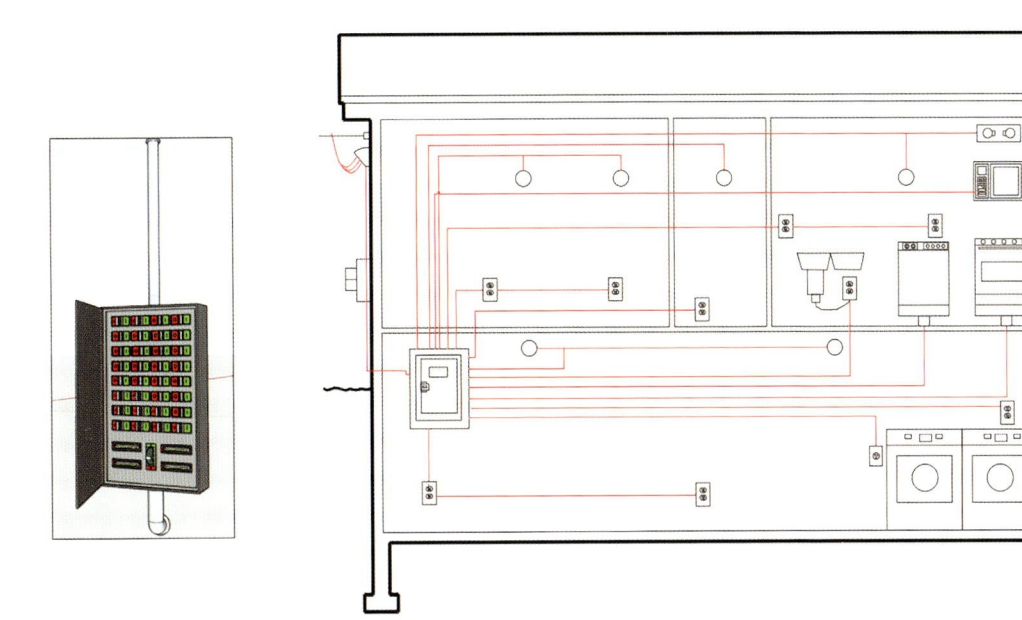

Figure 10.1
(Left) Service panel. (Right) Schematic drawing showing circuits emanating from the service panel to power the house.

Figure 10.2
Phone (left) and electrical (right) lines.

Switches These are devices that open and close a circuit (Figure 10.3). They can be:

- **Toggle**: an angled, two-position lever used for lights. A *single-pole* means one switch operates one light fixture (or group of fixtures); a *three-way* means two switches operate one light fixture; and a *four-way* means three switches operate one fixture. A **gang** is multiple switches on one plate.

Figure 10.3
Outlets and switches.

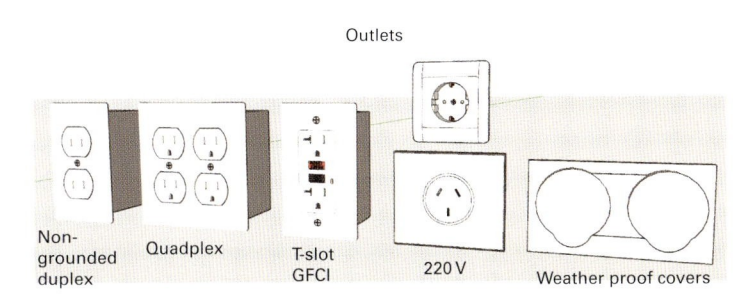

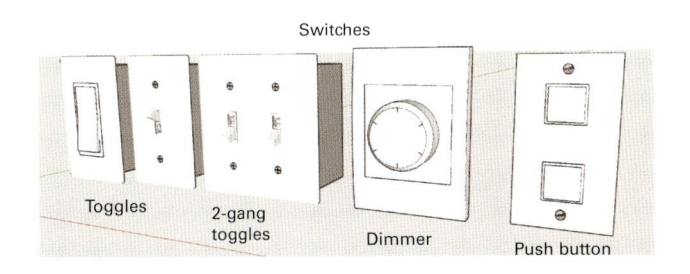

Outlets: Non-grounded duplex, Quadplex, T-slot GFCI, 220 V, Weather proof covers

Switches: Toggles, 2-gang toggles, Dimmer, Push button

FYI
How does a GFCI outlet work? A GFCI outlet monitors the amount of current going to the appliance and compares it to the current coming back. If the two are equal, it allows electricity to flow. However, if the amount of returning current is less than it should be, it shuts the current off, assuming that if the current is not traveling via the wire it must be traveling somewhere else, such as through a person (which is how electrocution takes place).

ELECTRICAL COMPONENTS 267

- **Push-button**: a press- and-release button used for doorbells and garage door openers.
- **Dimmer**: a rotating knob that adjusts light brightness.

Permanently Attached Light Fixtures   A light fixture is an electric unit with a power source and a **lamp**—the replaceable component known as a light bulb (Figure 10.4). Lamps can be:

- **incandescent** a light globe with a wire filament inside that heats until it glows.
- **fluorescent** a lamp consisting of a gas-filled glass tube through which an electrical current passes to produce uniform, glareless light.
- **compact fluorescent (CFL)** an energy-saving light and compact fluorescent tube designed to replace an incandescent light bulb
- **halogen** an enhanced incandescent lamp that is brighter and lasts longer.
- **light emitting diode (LED)** a semiconductor that emits light when a current passes through it. Lasts longer, and is cooler and more efficient than incandescent lamps.

The fixtures can be **track** (multiple bulbs on a rail), **recessed** (installed above the ceiling with an opening flush with the ceiling), **under-cabinet** (task lights mounted under wall cabinets), **chandelier** (suspended from the ceiling with light directed upward), **wall sconces** (surface mounted with light directed up or down), **surface** (mounted directly to the ceiling), **pendant** (suspended from the ceiling with light directed down), **cove** (placed high on the wall in a recess, with light directed to the ceiling), or **wall wash**, also called **soffit light** (attached to the wall with light radiating downward). Figure 10.5 shows some of these.

Figure 10.4
Lamps: Incandescent, fluorescent, CFL, halogen, LED.

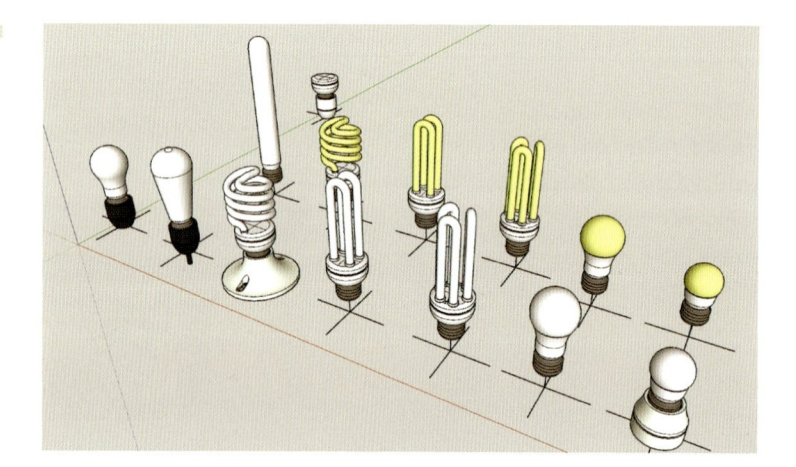

Electrical Plans

These are typically done by electrical engineers or lighting designers and governed by the National Electric Code (NFPA 70). Figures 10.6, 10.7, and 10.8 show electrical plans for residential rooms. Each electrical component has its own symbol. The curved, dashed lines are circuits and show which switches operate which components. Outlets dedicated for a specific appliance should be marked, for example *DW* for dishwasher and *GD* for garbage disposal. When exact locations of electrical components are needed, dimension to the center of each light fixture, outlet, and switch and to their distance from the edge of sinks and appliances.

Figure 10.9 shows an electrical plan of a whole house. The circled drawings in red are pictorial explanations, not part of a construction document. Figures 10.10 and 10.11 show common electrical symbols.

Pendant

Cove

Chandelier

Track

Sconce

Under Cabinet

Recessed

Wall Wash

Surface

Figure 10.5
Light fixtures.

Figure 10.6
Living room.

Figure 10.7
Hall and stairs.

ELECTRICAL PLANS **269**

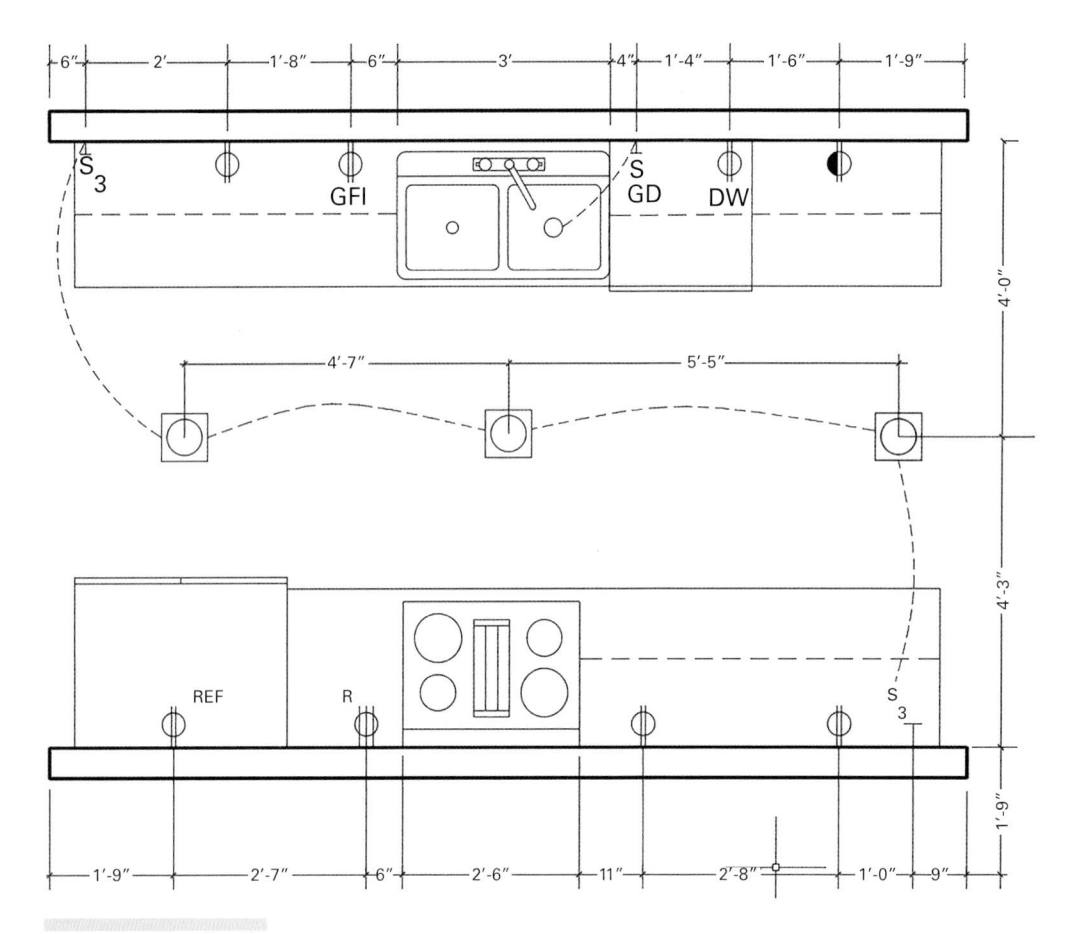

Figure 10.8
Dimensioned lighting plan of a galley kitchen.

Electrical Plan Design   Here are some code and design recommendations:

- Each room should have three outlets minimum, and one should be a split-wire.
- Every fixture, device, and appliance in the plan needs an outlet.
- Place duplex outlets:
    - 12' apart maximum on walls. 6' apart is ideal.
    - on every wall over 2' long.
    - 12" to 18" above the floor
    - 6' maximum from an opening.
    - 6' maximum from a room corner unless a door or built-in item occupies that space
    - on any wall between doors
    - near desks, tables, fireplaces and work areas.
    - in hallways that are over 10' long
    - on any counter more than 1' long
    - every 2' on kitchen countertops, with the maximum being 4' apart
    - 6" above the countertop (48" above the floor line).
    - near patios.

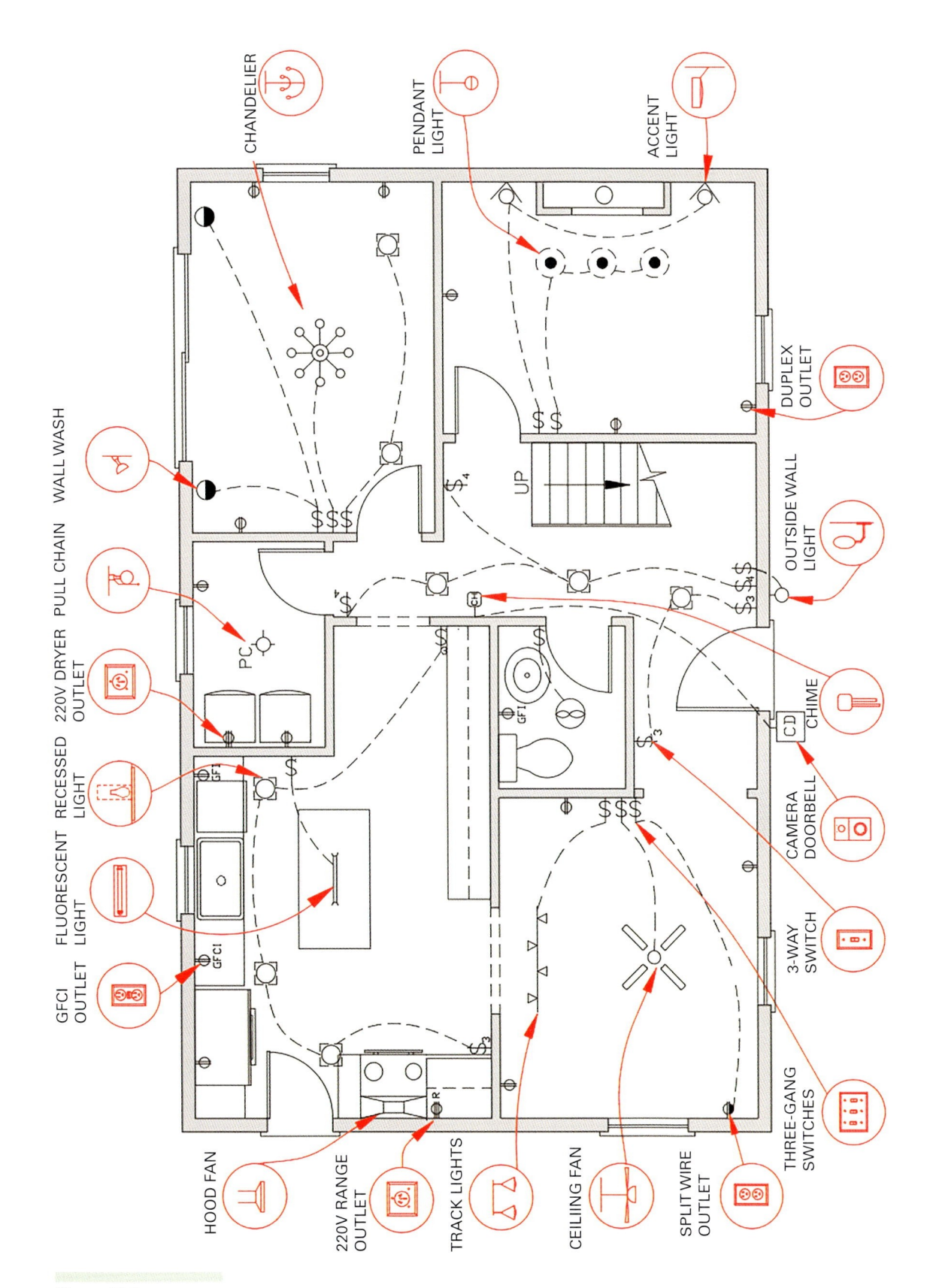

Figure 10.9
Electrical plan, symbols and pictorials.

ELECTRICAL PLANS 271

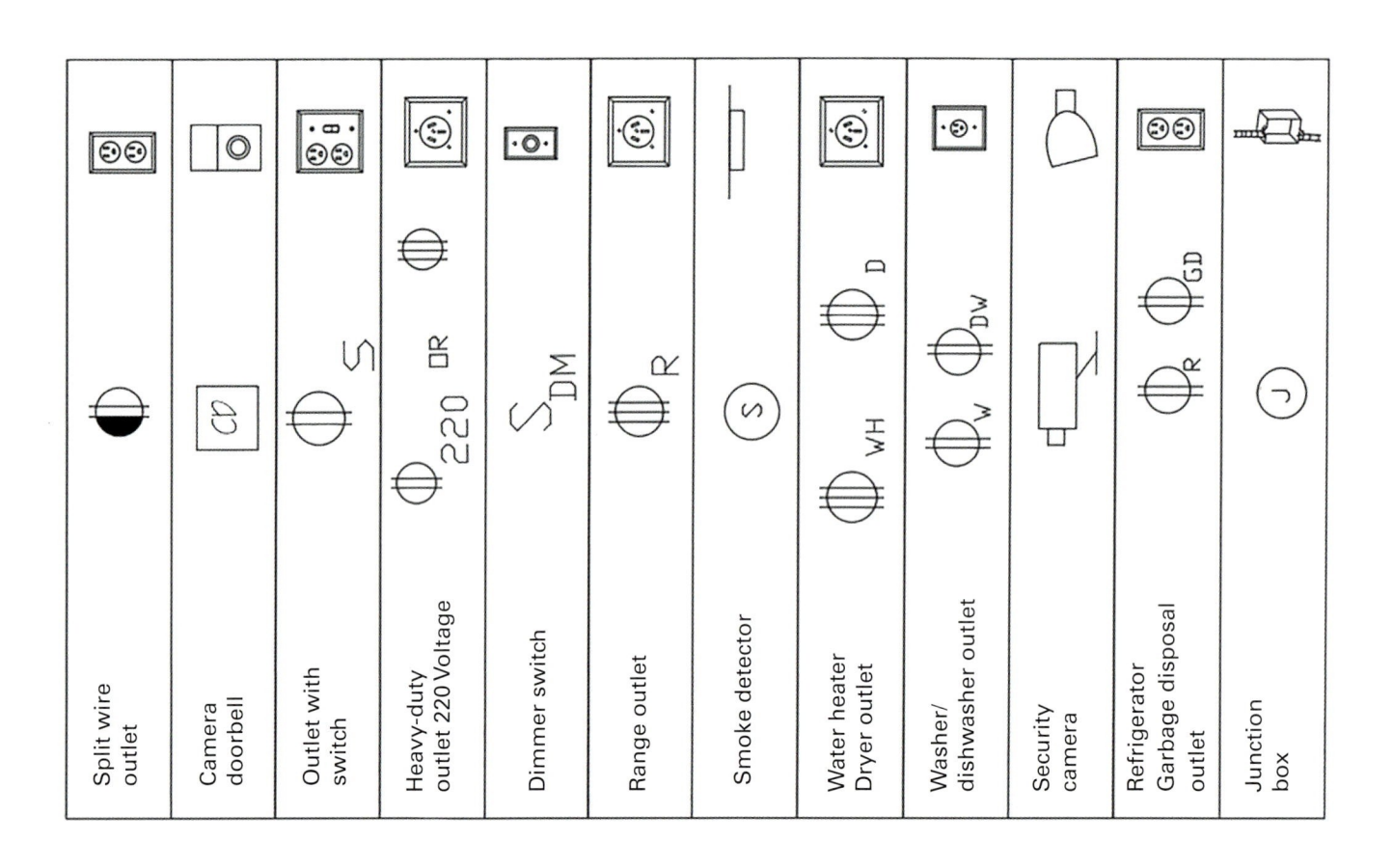

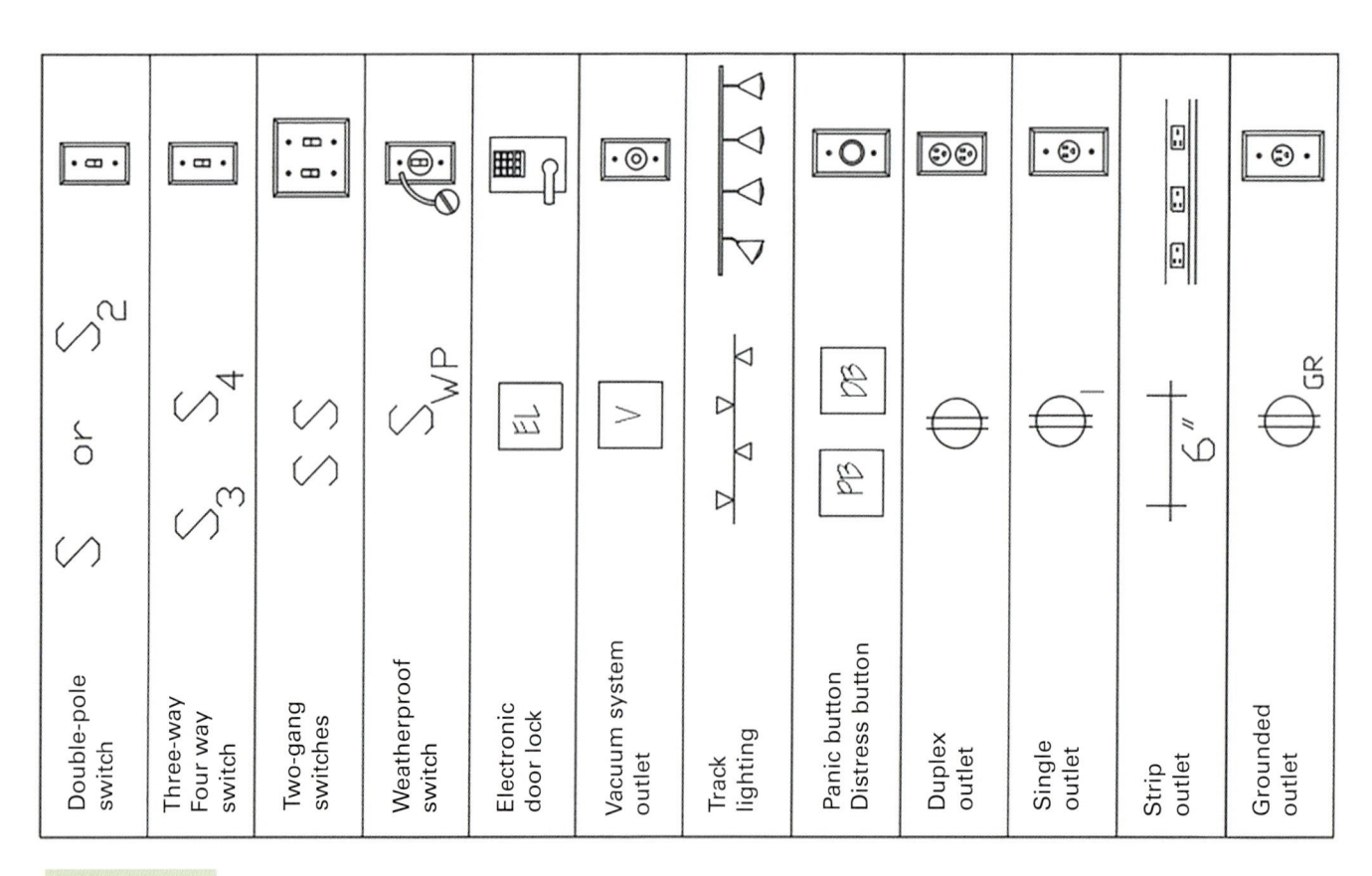

Figure 10.10
Electrical symbols.

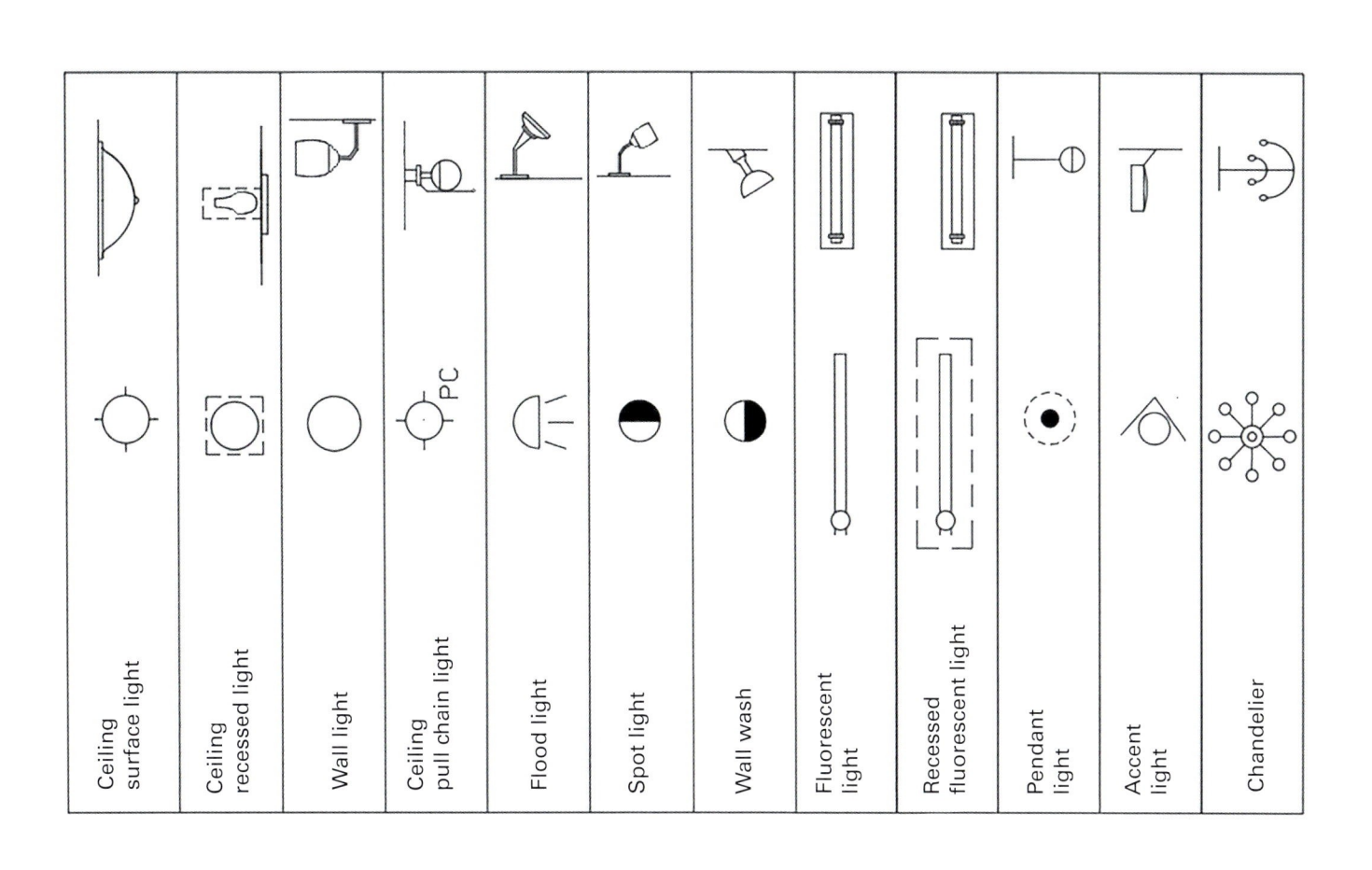

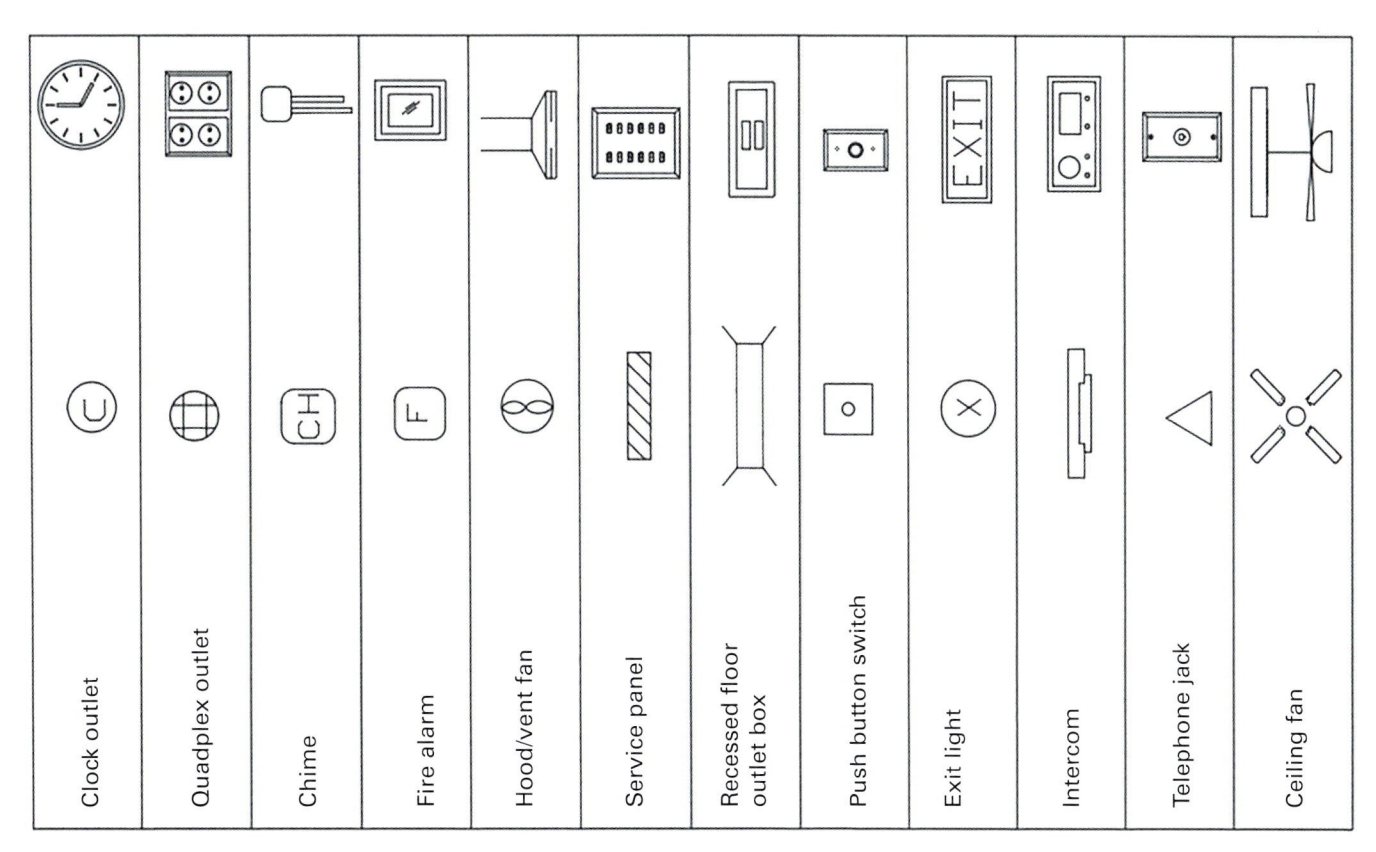

Figure 10.11
Electrical symbols.

ELECTRICAL PLANS   **273**

- Place at least 2 GFCI outlets in bathrooms; one above each countertop or vanity.
- Place a GFCI outlet
  - within 18" of water
  - near sinks
  - garages
  - outdoors
  - crawlspaces
  - unfinished basements
- Place weatherproof GFCI outlets
  - outdoors
  - at entries
  - near lighting
- Place switches
  - on the door latch side
  - 48" above the floor; 30" to 40" for wheelchair users
  - 2 1/2" minimum from the trim.
  - away from bathtubs or showers.
- In a 4-switch gang, the closest switch to the door is the most general; the furthest away is the most specific.
- Place four-way switches in rooms that have more than two exits.
- Place three-way switches
  - in all large rooms that have two exits
  - at hallway ends
  - at staircase ends
  - in garages.

Drafting an Electrical Plan Overlay tracing paper on a floor plan that shows walls, major appliances, and built-in features. Show everything attached to a wire—switches, outlets, phone jacks, circuits, permanently attached light fixtures, cameras, the service panel, fire and smoke detectors, vacuum system outlets, doorbells, chimes, and exhaust fans. Identify 220V, split wire, weatherproof, and other special use outlets. Symbols for automated systems may be included or drawn on a separate electronics plan.

Align alphanumeric symbols (such as *S3*) as per Figure 10.12. As a guide, draw lights twice the size of switches and outlets; 1/2" in diameter for lights and 1/4" in diameter for outlets. Draw the *S* for switch 1/4" tall.

After drawing the symbols, draw circuit lines with a French curve, not freehand. Don't draw them horizontal or vertical. On residential drawings the circuits are dashed; draw those dashes evenly sized and spaced. On commercial plans they're solid. Connect switches with the devices

> **Exercise**
> Sketch a lighting plan. First, measure and sketch a room to scale on grid paper. Then mark workspaces, countertops, shelves, furniture, and wall art. Make the room's focal point the start of your plan and radiate outward from it. If there is no focal point, start your lighting plan at the room's center.

> **Tip**
> Calculate general room lighting placement by dividing the height of the ceiling by two. The result is the amount of space to leave between each light. For example, recessed lighting spacing for a 10'-high ceiling is 5' feet between each light. Ideal lighting differs in each room and for different tasks. Make the room's focal point the start of your layout plan and build outward from there. If there is no focal point, start the lighting plan from the room's center.

they operate; for example, connect push-button switches with their door chimes or door openers, and toggle switches with their lights or split-wire outlets. Figure 10.13 shows correct and incorrect ways of drawing circuits. These lines aren't the wires' actual positions; they just show the control link. The electrician chooses where to run the physical wires.

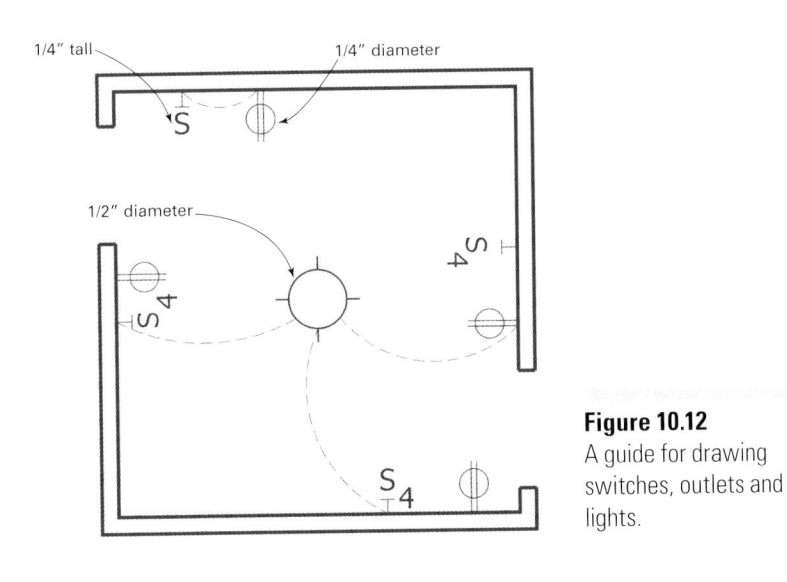

Figure 10.12
A guide for drawing switches, outlets and lights.

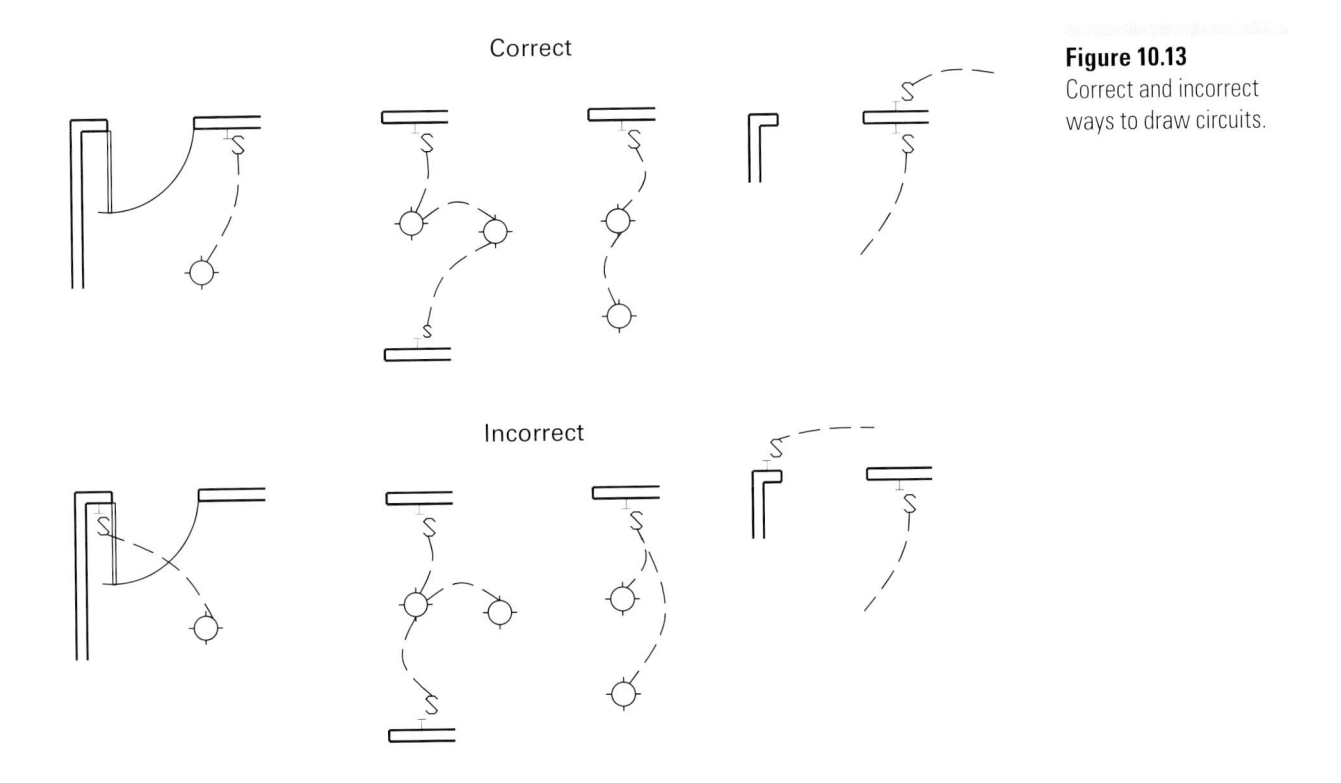

Figure 10.13
Correct and incorrect ways to draw circuits.

Power Plans

Some of the information on a power plan may overlap with the electrical plan or be combined with it. But if a project is large and complex, a separate plan for this information is usually prepared (Figures 10.14, 10.15).

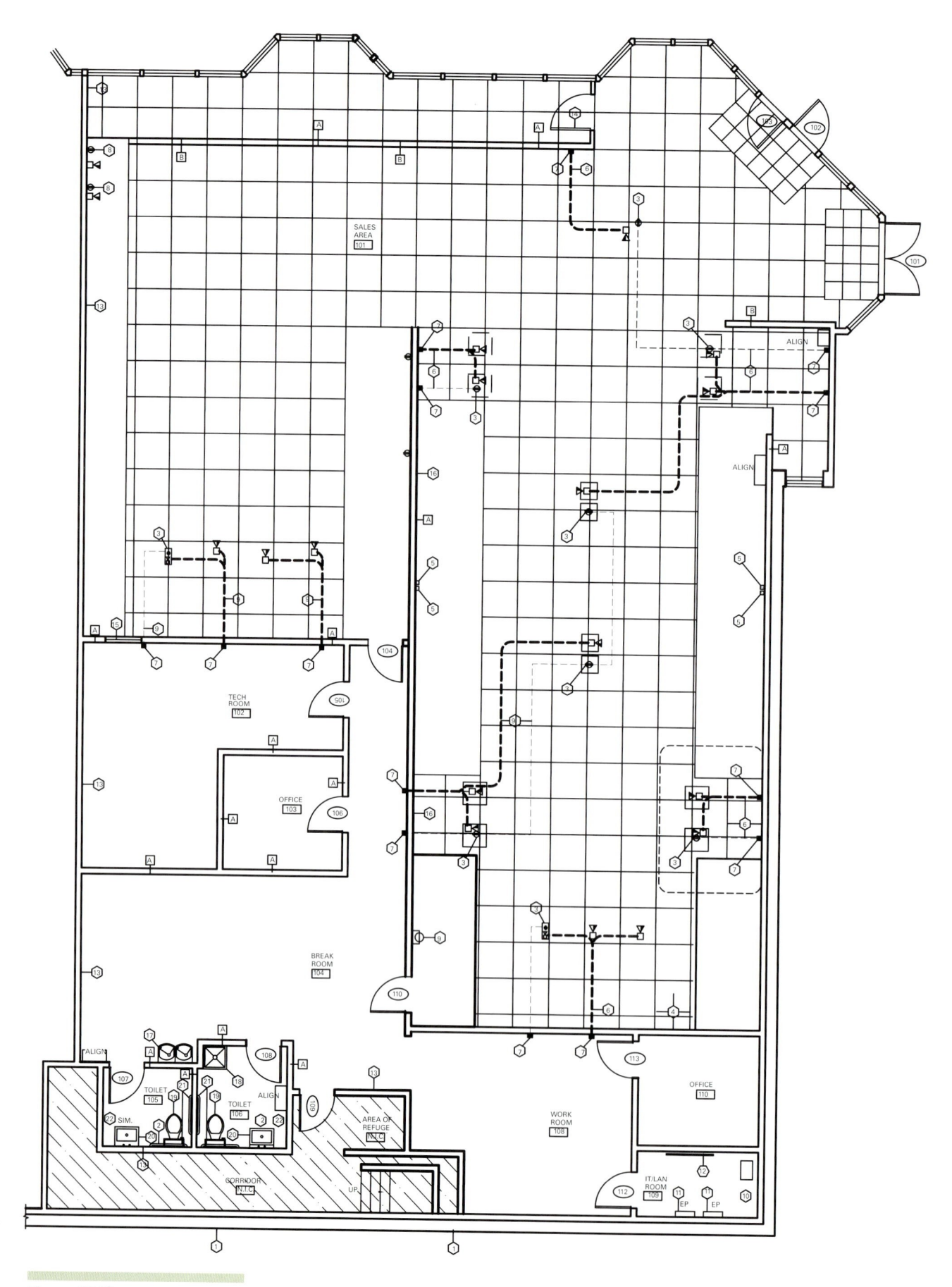

Figure 10.14
Power plan for an office building.

276 CHAPTER 10 UTILITY SYSTEMS

① HATCH REPRESENTS THE AREA OF LITTLE OR NO WORK.
② PROVIDE FIBER REINFORCED PANELS 4'-0" WIDE X 4'-0" HIGH A.F.F. INSTALLED ON WALL. (TYP. ALL WET AREAS) RE: SPECIFICATIONS
③ DIMENSIONAL LOCATION OF ELECTRICAL OR DATA/PHONE FLOOR FITTING. RE: ELECTRICAL FOR ADDITIONAL INFORMATION.
④ LINE OF CARPET TILE SEAM STARTING POINT, TYP. GC SHALL LAYOUT CARPET TILE PATTERN ON FLOOR PRIOR TO ROUGH IN OF WALL BOXES AND FLAT WIRE INSTALLATION TO COORDINATE FLAT WIRE AND CARPET SEAMS. RE: FINISH PLAN.
⑤ DIMENSIONAL LOCATION FOR JUNCTION BOX AT 18' O.C. AFF. RE: ELECTRICAL FOR INFORMATION.
⑥ UNDER CARPET FLATWIRE LOCATION FOR DIMENSIONAL REF. ONLY RE: A3/A-1 AND ELECTRICAL DRAWINGS FOR ADD. INFORMATION.
⑦ DIMENSIONAL LOCATION FOR ELECTRICAL WALL BOX AT 10" AFF. FOR DIM REF. ONLY. RE: ELECTRICAL DRAWINGS FOR ADD. INFO.
⑧ DIMENSIONAL LOCATION FOR ELECTRICAL AT 12" AFF. RE: ELECTRICAL FOR ADD. INFO.
⑨ SURFACE MOUNTED FIRE EXTINGUISHER LOCATION. (5 LBS. TYPE ABC. MANUFACTURER - KIDD, GRANCEIR OR EQUAL) MUST HAVE CURRENT INSPECTION TAGS ATTACHED.
⑩ NEW IT/LAN RACK.
⑪ NEW ELECTRICAL PANEL LOCATION. RE: ELECTRICAL DRAWINGS FOR ADD. INFORMATION.
⑫ NEW 48" X 48" PLYWOOD PHONE BOARD. RE: ELECTRICAL DRAWINGS FOR ADD. INFO. PAINT TO MATCH WALL.

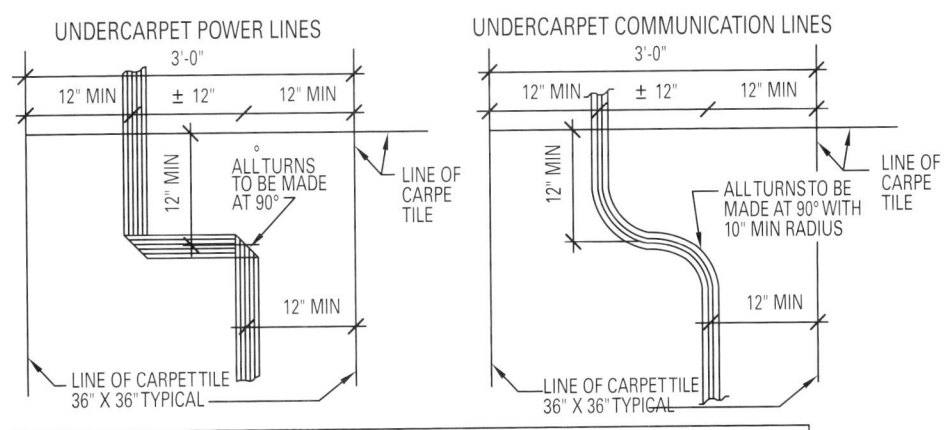

UNDERCARPET POWER LINES — UNDERCARPET COMMUNICATION LINES

MANUFACTURER: AMP CONNECT - UL #E73212; FCC FITTING #E73213
NOTE: REFER TO MANUFACTURER SPECIFICATIONS FOR ADDITIONAL INFORMATION

NOTE:
SEE UNDERCARPET DIMENSION PLAN ON SHEET A-1 AND ADDITIONAL INFORMATION ON SHEET E-3. GC IS RESPONSIBLE FOR ALL COORDINTATION OF UNDERCARPET WIRING AND CARPET TILE. NO RAISED CARPET TILE SEAMS WILL BE ACCEPTED.

- DATA/ELEC
- DUPLEX OUTLET
- QUADPLEX OUTLET
- FLATWIRE TRANSITION BOX
- ELECTRICAL FLOOR FITTING
- WALL MOUNTED FIRE EXTINGUISHER
- EP ELECTRICAL PANEL
- E   EXISTING TO REMAIN

- NEW WALL
- EXISTING WALL TO REMAIN
- ELECTRICAL FLATWIRE
- DATA/PHONE FLATWIRE
- LEASE LINE

Figure 10.15 Legend for the power plan.

Reflected Ceiling Plan

This is done by an architect, interior designer, or lighting designer. A reflected ceiling plan (RCP) is a plan viewed as if it were reflected onto a floor mirror. Alternatively, think of it as a plan drawn as if looking down on it from above the ceiling (Figure 10.16). You project the ceiling down to overlay the floor plan. All features of the RCP overlap the features of the floor plan directly below it. For example, features in the RCP's upper-right corner overlay the features in the plan's upper right-hand corner. One RCP is needed for each floor level.

Figure 10.16
Draw the RCP as if looking down on it from above the ceiling.

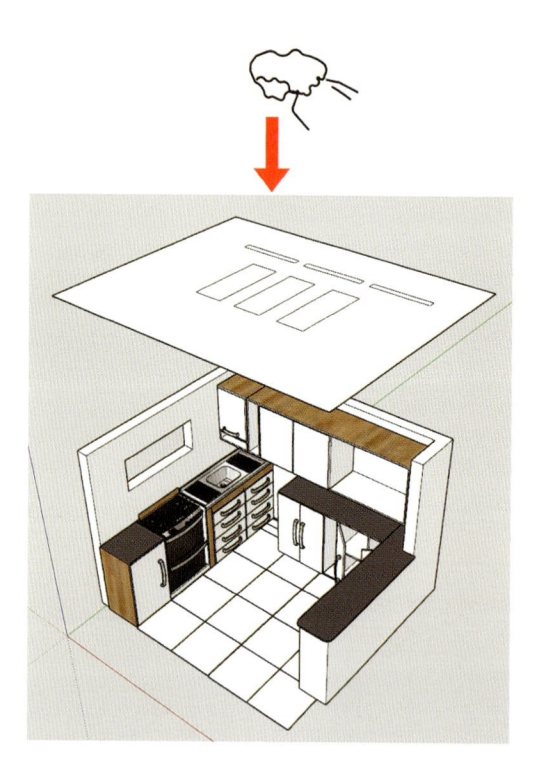

RCPs show ceiling materials, exposed beams and ductwork, ornamentation (Figure 10.17), cornice moulding, skylights, changes in ceiling heights, light fixtures, exit lights, sprinkler heads, air supplies and returns, access panels, speakers, projectors, and anything else that is on or touches the ceiling. Walls and partitions that extend to and above the ceiling are shown, and a ceiling height note. Commercial plans include the structural grid.

Figure 10.17
All details of this richly ornamented ceiling are drafted in the RCP.

Exercise
Sketch a reflected ceiling plan of the picture in 10.17

CHAPTER 10 UTILITY SYSTEMS

Figure 10.18 is of a residential RCP. It shows a gypsum board ceiling and recessed, surface and pendant lights, and **air supply vents**, the openings through which treated air enters a room. Although air supplies and exposed ducts are shown on the HVAC drawings, they are also on the reflected ceiling plan to make their coordination easier and ensure that the designer fully understands what the ceiling will look like. RCPs may also have section symbols directing the reader to ceiling details.

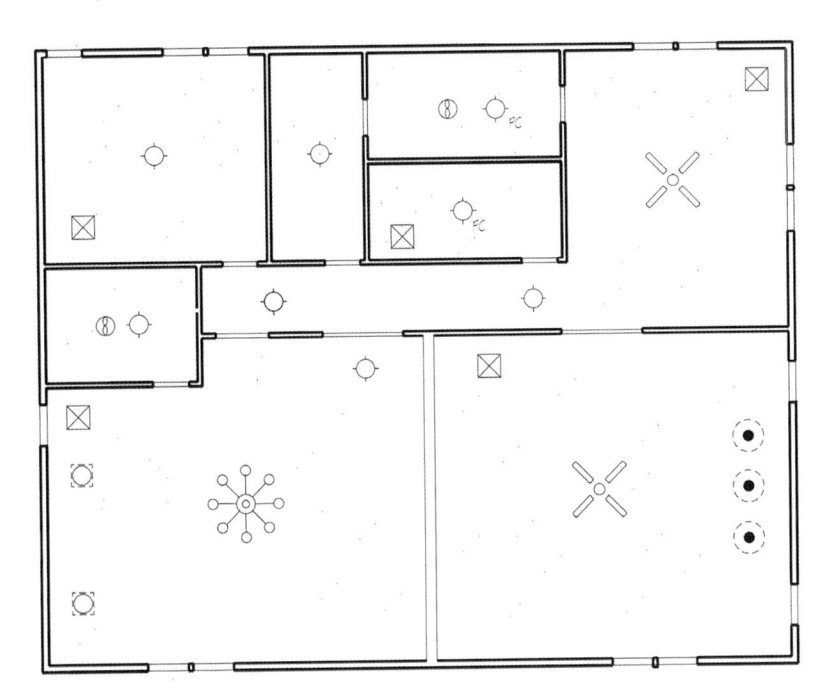

Figure 10.18
A residential RCP.

Figure 10.19 shows a commercial RCP and legend. It's a **drop-in**, also called **suspended**, ceiling. A grid is suspended from the underside of the roof with wires and acoustic tiles and lights are dropped into it. The space between the ceiling and roof is the **plenum** (Figures 10.20, 10.21).

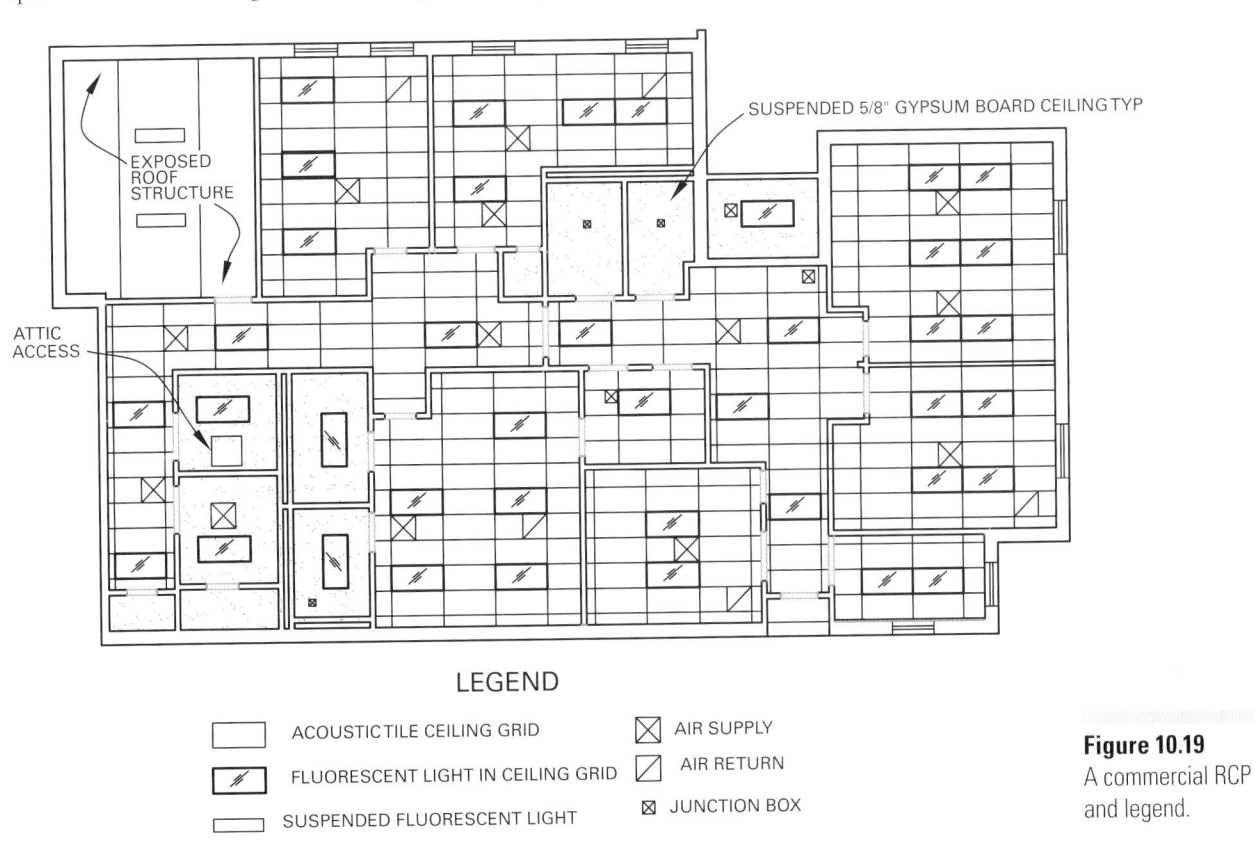

LEGEND

- ACOUSTIC TILE CEILING GRID
- FLUORESCENT LIGHT IN CEILING GRID
- SUSPENDED FLUORESCENT LIGHT
- AIR SUPPLY
- AIR RETURN
- JUNCTION BOX

Figure 10.19
A commercial RCP and legend.

REFLECTED CEILING PLAN   **279**

Figure 10.20
A drop-in ceiling. The space between the ceiling and roof is the plenum. Courtesy certainteed.com

Figure 10.21
Plenum space between the ceiling and roof. Note the gypsum board wall on the right and the wires tying the acoustic tile grid to the steel wide flange joist above.

Dimension any component whose placement isn't clear. For instance, you can locate lights in a drop-in ceiling by counting the tiles, but not in a gypsum board ceiling. As with electrical plans, dimension light fixtures to their centers.

Drafting a Drop-In Ceiling

1. Choose an acoustic tile size and type of light. Round canned lights and fluorescents that are the grid size will both work (Figure 10.22). Common ceiling tile sizes are 12" × 12", 12" × 24", 24" × 24", and 24" × 48". Determine if interior walls in the floor level you're drawing the RCP for are partitions that stop under the ceiling or if they go to the roof. If they're ceiling-height partitions, draw one continuous RCP. If the walls go to the roof, treat each space as separate when drawing the RCP. Here we'll assume the interior wall goes to the ceiling.

2. Draw vertical and horizontal center lines in each space. The ceiling grid will be centered on these lines.

3. Mark vertical lines the tile size apart, starting at the center.

4. Mark horizontal lines the tile size apart, starting at the center.

5. Add line weights.

6. Add details. Darken in the perimeters of the lights and add the glass symbol to them. Add any other details on the ceiling, such as air vents.

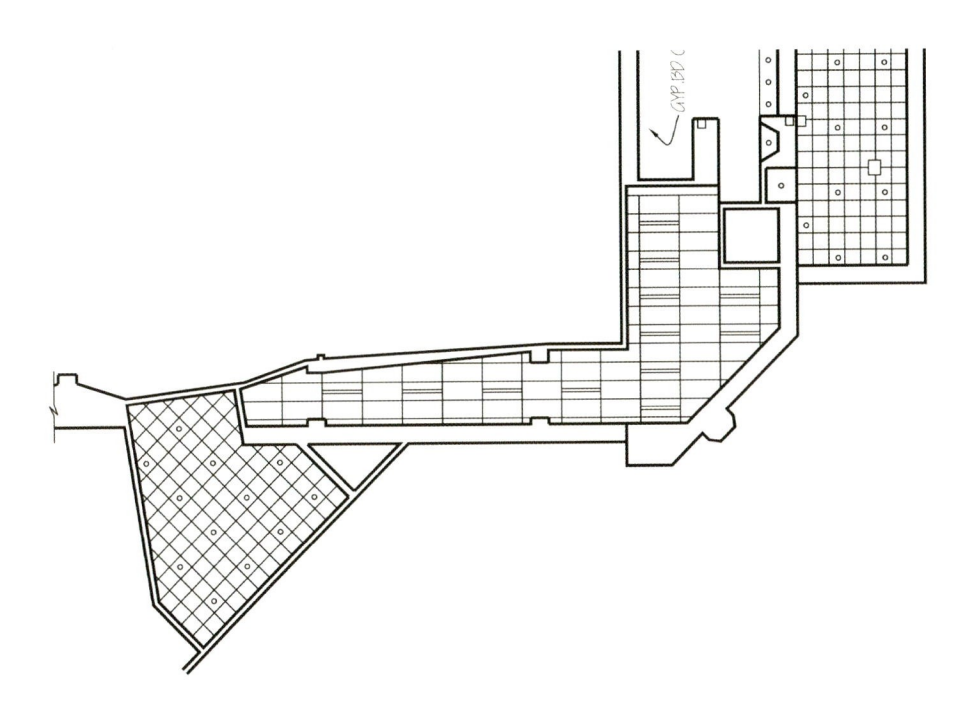

Figure 10.22
This commercial reflected ceiling plan has both canned and fluorescent lighting.

Figure 10.23
How to draw an acoustic tile reflected ceiling plan.

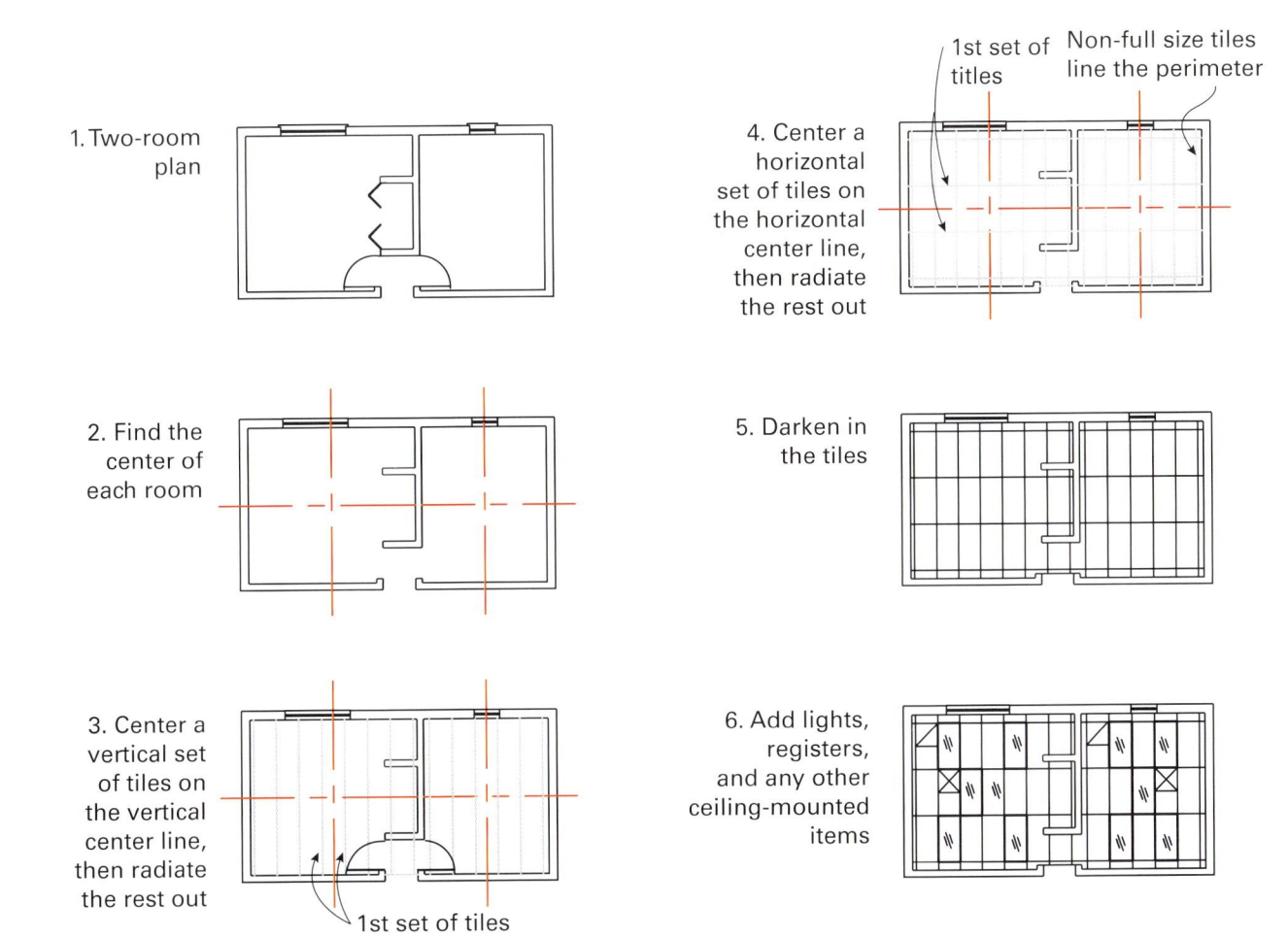

1. Two-room plan

2. Find the center of each room

3. Center a vertical set of tiles on the vertical center line, then radiate the rest out

   1st set of tiles

4. Center a horizontal set of tiles on the horizontal center line, then radiate the rest out

   1st set of titles | Non-full size tiles line the perimeter

5. Darken in the tiles

6. Add lights, registers, and any other ceiling-mounted items

DRAFTING A DROP-IN CEILING 281

Heating, Ventilation, and Air-Conditioning (HVAC) Systems

A building's **HVAC** system, also called **climate control system**, circulates pleasant, clean air. It regulates heat, coolness, freshness, humidity, and temperature. The most common HVAC systems are **forced air**, **radiant heat**, and **solar**.

Forced Air Also called **central air**, this system uses air to transfer heat and cold. A **furnace** (heat-producing appliance) draws air from outside, heats it up, and blows the heated air through **ducts** and out air supply vents in each room (Figures 10.24, 10.25).

Figure 10.24
Furnace and air conditioner.

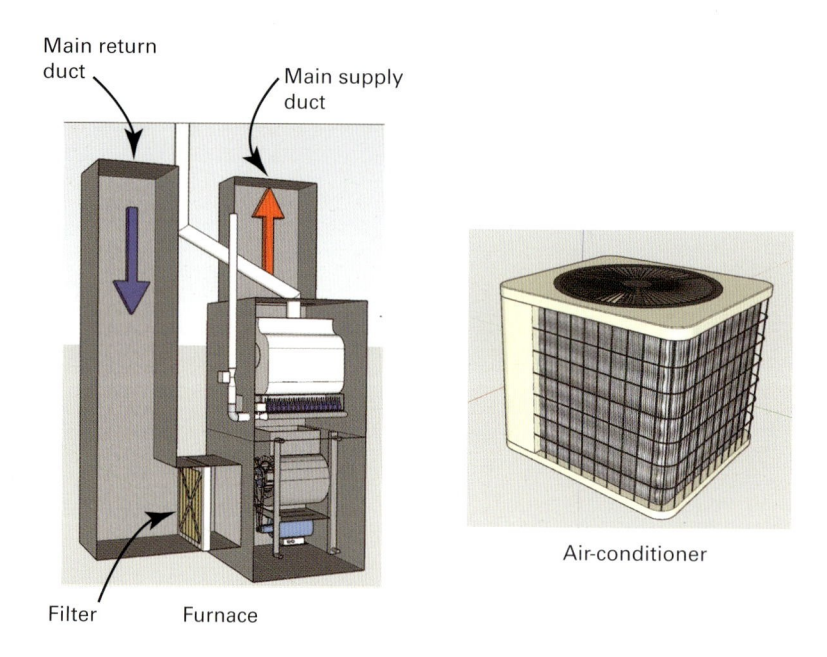

Figure 10.25
Pictorial of a forced-air heating system.

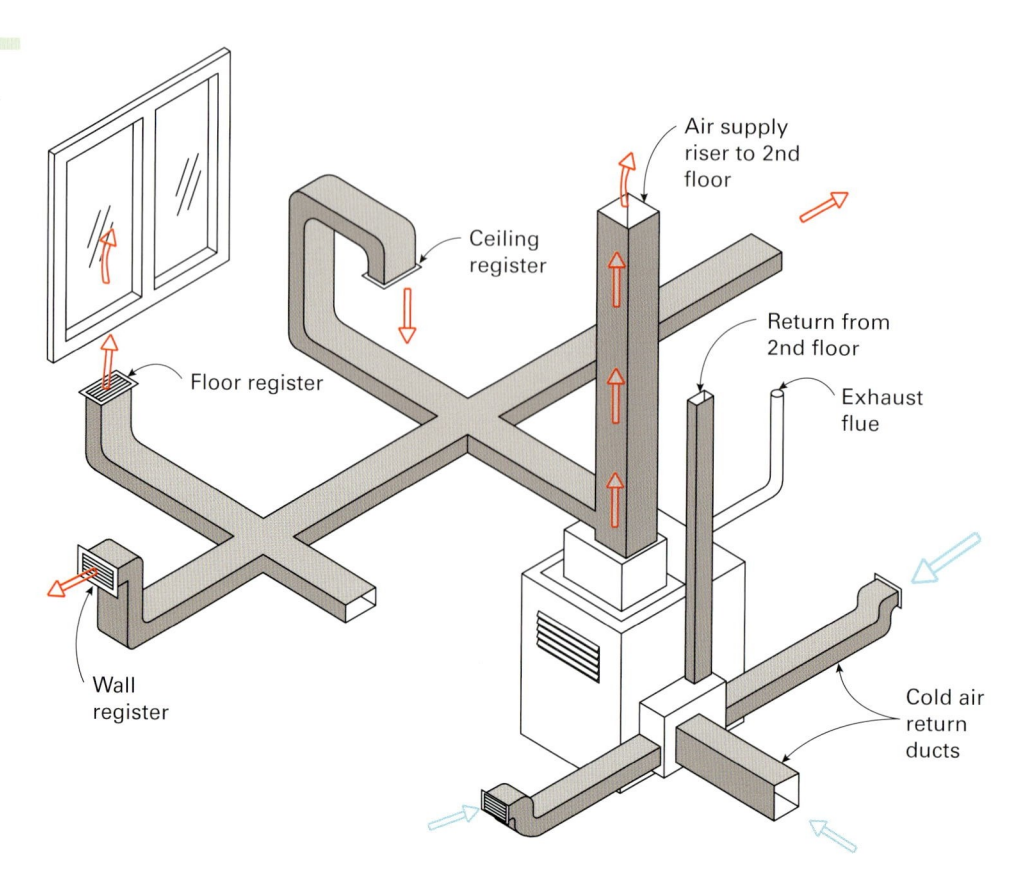

Ducts are sheet metal passageways and can be round or square. A **main supply duct** is attached to the furnace, and **branch ducts** run from the main duct to individual rooms (Figure 10.26). A filter and **humidifier** (optional appliance that adds moisture to the air) is installed on the main supply duct. A vertical duct that fits between the wall studs is a **wall stack**.

LEGEND

▬▬▬▬ SUPPLY DUCT   ⊠ SUPPLY REGISTER

▭ ▬ ▭ RETURN DUCT   ⊠ RETURN REGISTER

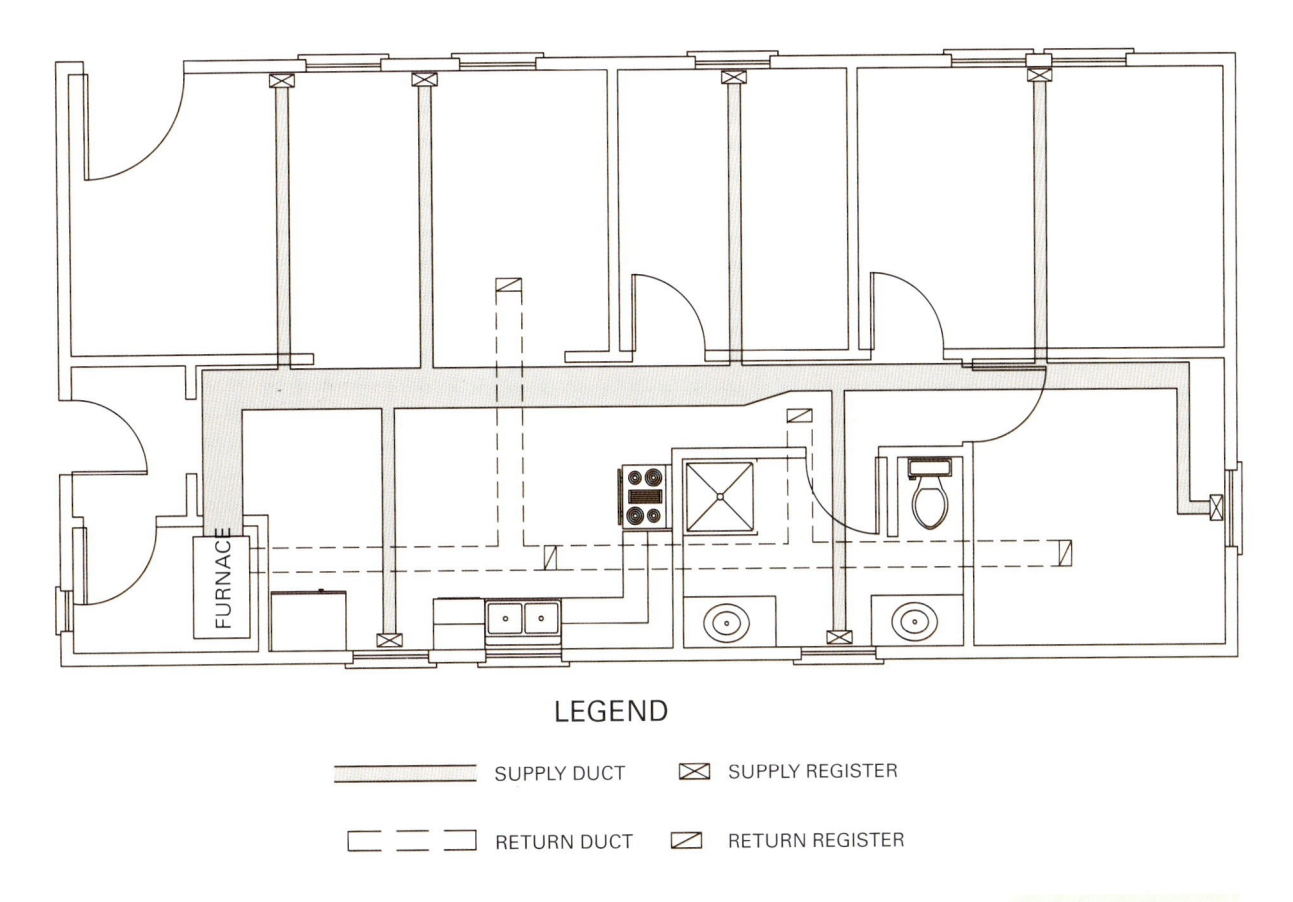

Figure 10.26
Plan showing furnace, main supply, branch and return ducts, and return and supply vents.

Filters and humidifiers can be put on the house's main duct, where the air passes before being distributed to branch ducts.

There are separate supply ducts and return ducts. Treated air is blown through supply ducts and supply vents into the space. Cooler room temperature air is drawn back into separate return vents and ducts back to the furnace, where it is reheated and sent out through the supply ducts again. **Air conditioners** (air chilling appliance) draw in outside air, chill it, and send it through the house using the same ducts and the furnace's filter and blower motor.

Figure 10.27 shows a creative installation of a furnace installed horizontally on steel channels suspended from the roof. This saved floor space, minimized duct work, requires shorter air intake and exhaust pipes, and is closer to the outside gas supply tank. Note the ductwork, air intake, and exhaust pipes and large air supply vent.

Figure 10.27
Horizontally installed furnace.
Courtesy Rex Porter and Joan Organ Jenkins.

Supply vents are in the floors and ceilings; return vents are on the walls. Supply vents have louvers behind the grille for directing airflow and are smaller than return vents. Decorative **grilles** (cover plates) are available in many styles and materials to integrate with the interior design (Figure 10.28).

Vent sizes are based on room size. Supply vents are typically 4" × 10" or 4" × 12"; return vents are 16" × 20" or larger. Place one supply vent in every room that is 100 square feet or smaller; place two or three in larger rooms. Place one return vent in each room. Both should be unobstructed by furniture, as that inhibits air flow and pressure. Supply vents are placed in front of patio doors and windows to spot-treat that area.

Figure 10.28
This wood grille was custom made to fit under a furniture piece.

Mini-Split A **mini-split** is a ductless heat pump that both heats and cools a room (Figure 10.29). It's installed high on a wall and connected to an outdoor condenser. It works by heating or cooling the interior air and then blowing room-temperature air outside. Mini-splits are used in remodels and spaces where there is no ductwork and it's impractical or cost-prohibitive to install it.

Radiant Floor Heating **Radiant floor heating** uses electric coils or water-filled tubes under the floors to heat the floors (Figure 10.30). The heat then radiates out into the space. It can be used in whole houses, parts of houses, and embedded into driveways in cold climates. There are no fans or ducts, which increases energy efficiency because there's no chilled moving air or energy loss during transportation. However, radiant heat cannot be combined with air-conditioning, humidifiers, or air filters.

Figure 10.30
Radiant floor heating coils.

Figure 10.29
A mini-split unit will heat and cool the office in this shipping container home.

Solar Heating There are two kinds of solar systems: **passive** and **active**. Passive uses a house's design to collect the sun's energy. South-facing windows and a specially designed roof and walls soak up the sun all day (Figure 10.31). When the temperature cools at night, the stored heat is released directly into the space. No mechanical components are needed. Passive systems only heat air.

Active systems collect and store energy in solar panels. Pumps or fans transport the heated air through pipes to a fluid-filled **heat exchanger**, which is a device that passes heat from its fluid directly to the air inside the house or to the household water tank (Figure 10.32). Solar heating is a back-up to a forced-air system because it doesn't work on cold or overcast days.

Evaporative Cooler Also called a **swamp cooler**, an **evaporative cooler** cools air through water evaporation. It's mounted on the roof and uses wet pads and a fan to blow cool air into a home. It can use a home's existing ductwork or blow air directly in through a roof vent. Evaporative coolers only work in climates with low humidity.

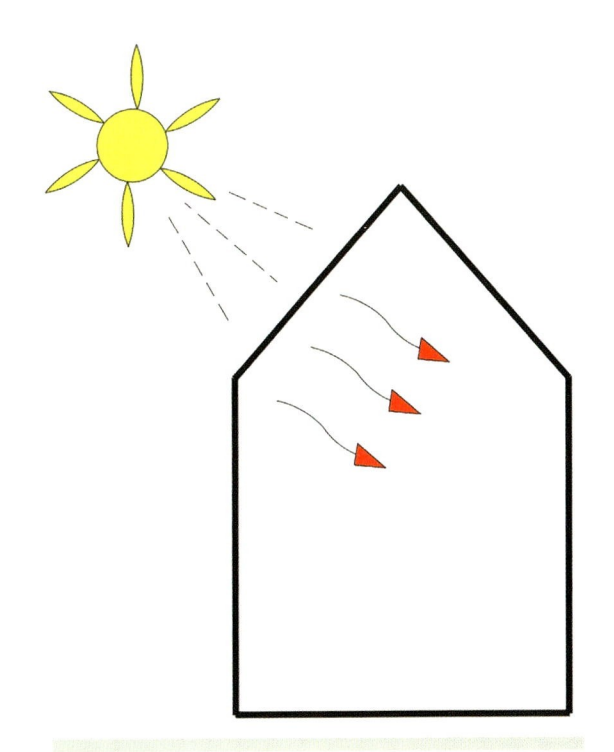

Figure 10.31
Passive solar system.

HEATING, VENTILATION, AND AIR-CONDITIONING (HVAC) SYSTEMS

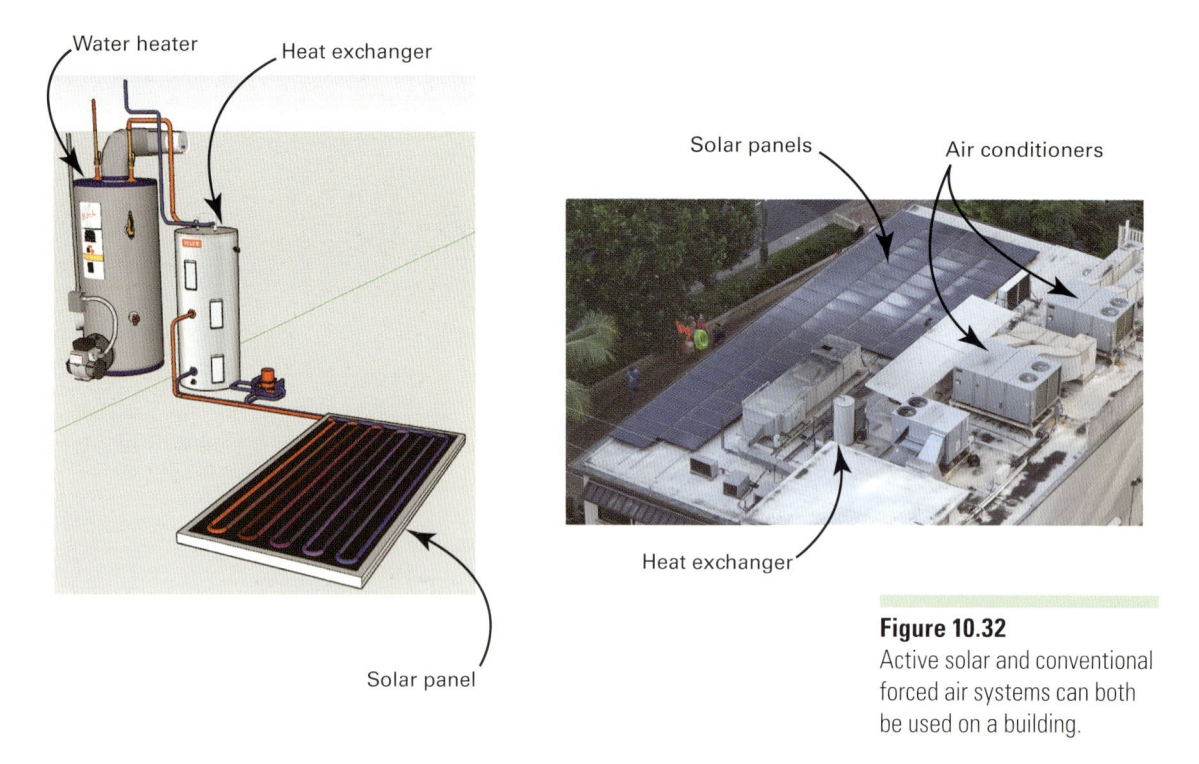

Figure 10.32
Active solar and conventional forced air systems can both be used on a building.

HVAC Drawings

These are called HVAC, mechanical, or **equipment plans**. They're done by a mechanical engineer and governed by the **International Residential Code (IRC)** or **Uniform Mechanical Code**. An **HVAC/mechanical plan** shows the furnace, air conditioner, water heater, ducts, filters, humidifiers, pipes, control devices, and vents. An equipment plan just shows the equipment used for heating and cooling and their complexity is based on the building's complexity. For instance, in a factory or other large commercial building there are pumps and valves that control systems other than heating. A restaurant has pumps and motors for motorized lifts and sewage pumps. Multi-story buildings have elevator motors with hydraulic pumps. Figure 10.33 shows a residential HVAC plan.

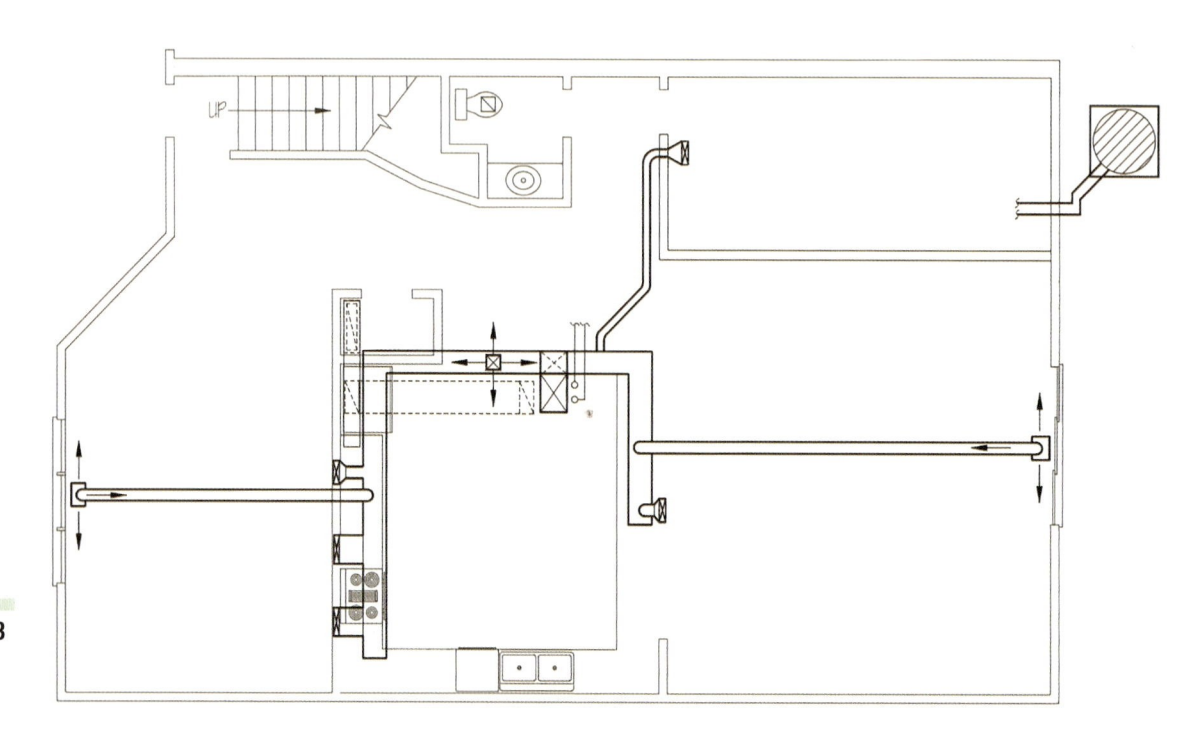

Figure 10.33
A residential HVAC plan.

286 CHAPTER 10 UTILITY SYSTEMS

Reading an HVAC Plan Walls and built-in architectural features are lightly outlined, and the rest of the drawing shows mechanical items. The type, size, and location of all heating, cooling, ventilating, humidification, and air-cleaning equipment, pipes, ducts, registers, inlets, and thermostats are shown.

The thermostat's location is especially important for space planning purposes. For example, furniture shouldn't go in front of it. Heat-generating items like lamps shouldn't be near it, as they'll send false signals to its sensor, making the HVAC system inefficient. It should be placed between 36" and 48" above the floor. Thermostats are available in programmable electronic models (Figure 10.34) and in Braille.

HVAC drawings are **schematic**, meaning they utilize simple line types and symbols that don't resemble the physical appearance of the items they represent like architectural symbols do. Horizontal ducts are shown with their width outlined; vertical ducts are shown with diagonal lines. Air return ducts are drawn with a hidden line. Pipes are drawn with single lines. Heating units are drawn connected to their supply ducts or hot water supply pipes. Figure 10.35 shows common HVAC plan symbols.

Figure 10.34
A programmable thermostat. This one has an Eco setting for energy efficiency.

Exercise
Identify these parts of the HVAC system in your house or other building: furnace, air conditioner, registers, humidifier, and thermostat.

Ceiling supply outlet		Radiator	RAD
Floor register		Convector	CONV
Duct lowering		Room airconditioner	RAC
Duct rising		Furnace	FURN
Duct return		Fuel-oil tank	OIL
Duct supply		Humidistat	H
Heat transfer surface		Heat pump	HP
Warm air supply		Pump	
Cold air return		Humidification line	H
Second floor supply		Forced convection	
Second floor return		Hydronic-radiant panel coil	
Gas outlet		Hot water heating return	
Heat outlet		Hot water heating supply	
Heat register		Thermostat	T

Figure 10.35
HVAC symbols

HVAC DRAWINGS **287**

Plumbing System

A **plumbing system** is a network of pipes and controllers that delivers a building's fresh water and removes dirty water and waste. It's designed by a mechanical engineer and governed by the Uniform Plumbing Code.

Water from a river or lake is routed to a public utility. The utility filters and treats the water and pumps it through an underground supply pipe to its consumers. This delivery is chiefly gravity-driven; some municipalities store water in high towers to provide the pressure needed to push the water through the pipes. A water tower typically holds enough water to serve a community's needs for one day.

After going through a usage meter, some water is routed to a **hose bib** (outdoor faucet) or sprinkler system. The rest enters the house through the house's main supply pipe, then passes through filters and a **softener**, an optional appliance that removes hard minerals from the water. Then it is routed to a **water heater**, an appliance that heats and stores water. Figure 10.36 shows a conventional water heater. A gas flame heats up the water and the heated water rises to the top, where it is drawn off. Figure 10.37 shows a tankless heater, which heats instantaneously on demand, cutting energy costs.

Separate cold and hot lines emanate from the water heater. These lines run parallel to each other, 6" apart, and the hot water line is located to the left of the cold-water line. The water is then delivered to all the appliances, sinks, **water closets** (toilets), **lavatories** (sinks), tubs, and showers. **Wastewater** (dirty water) starts at each fixture through a drain hole (Figure 10.38) and is pushed via gravity through a **branch** (horizontal) **pipe** that connects to a **stack** (vertical) **pipe** and then down to the house's sewer line (Figure 10.39, 10.40). The house sewer line connects to the municipality's sewer line and the wastewater goes to a treatment plant, where it is cleaned and released as **reclaimed** (recycled) **wastewater** into streams or used for agricultural and municipal irrigation.

> **FYI**
> Here's how a tank water heater works. A dip tube carries the cold water to the tank's bottom so cold water doesn't get poured on top of the hot water. Hot and cold-water supply pipes are attached to the top of the heater. The outbound hot water pipe takes the water from the tank's top. As the hot water is drawn, cold water enters the tank to replace it. After all the hot water is used, cold water is delivered.

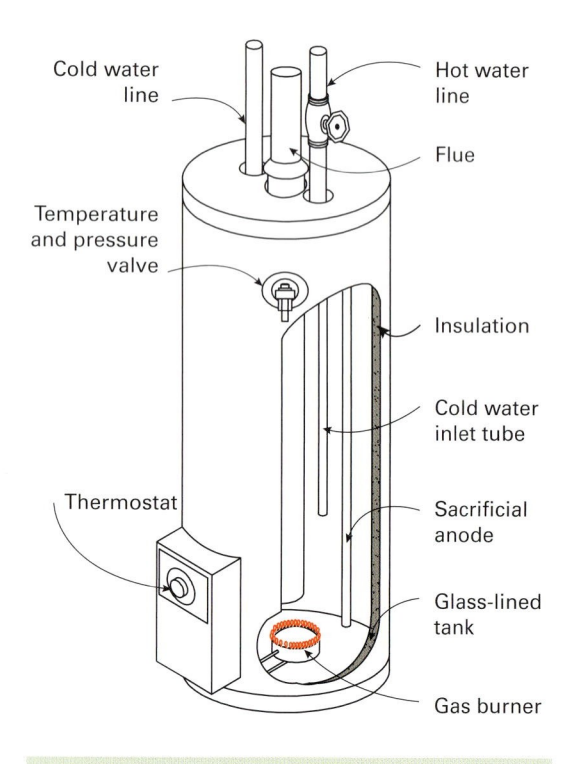

Figure 10.36
A conventional hot water heater.

Figure 10.37
A tankless hot water heater.

Figure 10.38
Each fixture has a drain hole.

Figure 10.39
Stack and branch pipes in a laundry area.

Rural areas often aren't served by city utilities due to cost. Those consumers get their water from a **well**, which is a hole drilled in the ground to access groundwater. Dirty water is routed into a **septic tank**, which is a buried container that leaches liquids into a drain field around it. A septic system has its own drawings on a plumbing set and is shown on the site plan.

Figure 10.41 shows schematic drawings of freshwater supply and wastewater removal. A **gate valve** (shut-off) is on the main freshwater line to stop the water supply to the whole house for repairs. Each fixture has its own gate valve so it can be worked on without shutting off water to the whole house. In the wastewater removal schematic notice the **P-trap** under each sink, a curved piece of pipe that prevents sewer gases from bubbling back into the bowl. Every plumbing **fixture** (device attached to a water pipe) has a different kind of trap. Toilets are self-trapped (Figure 10.42). **Cleanout valves** are plugged holes in the house's main sewer line and at the base of all waste stack pipes for accessing clogs. The **vent stack** is a wastewater pipe that rises above the roof, allowing sewer gases to escape. Figure 10.43 is a pictorial of bathroom branch and stack waste pipes.

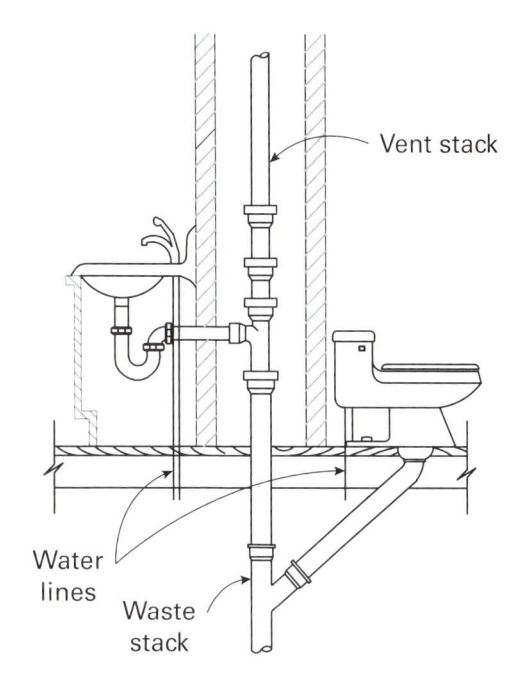

Figure 10.40
A stack pipe inside a chase wall.

PLUMBING SYSTEM   289

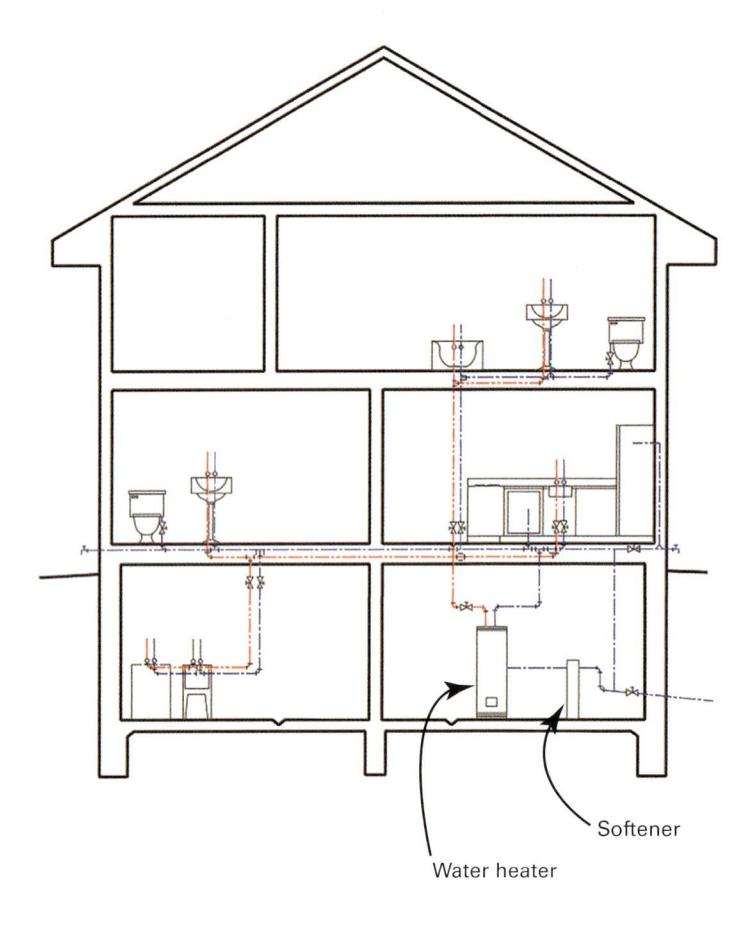

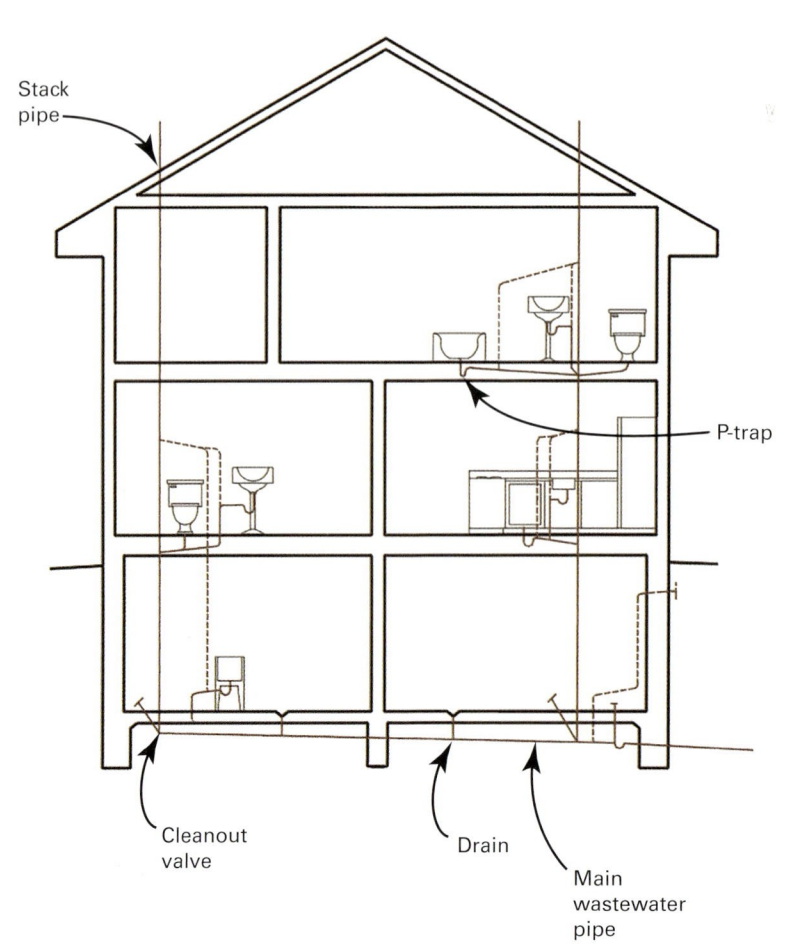

Figure 10.41
Schematic drawings of freshwater supply and wastewater removal.

290 CHAPTER 10 UTILITY SYSTEMS

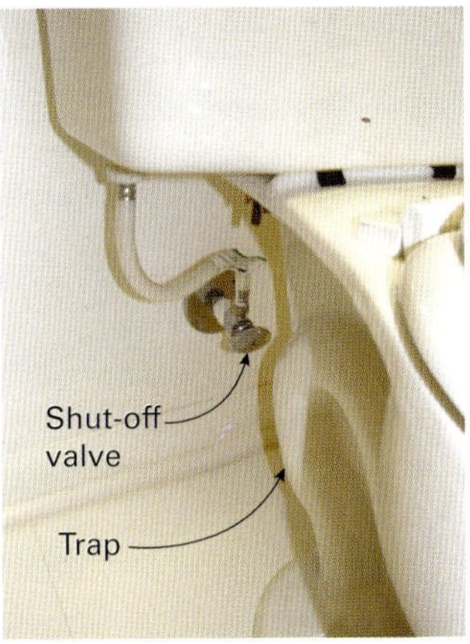

Figure 10.42
Toilets are self-trapped.

Figure 10.43
Bathroom branch and stack waste pipes.

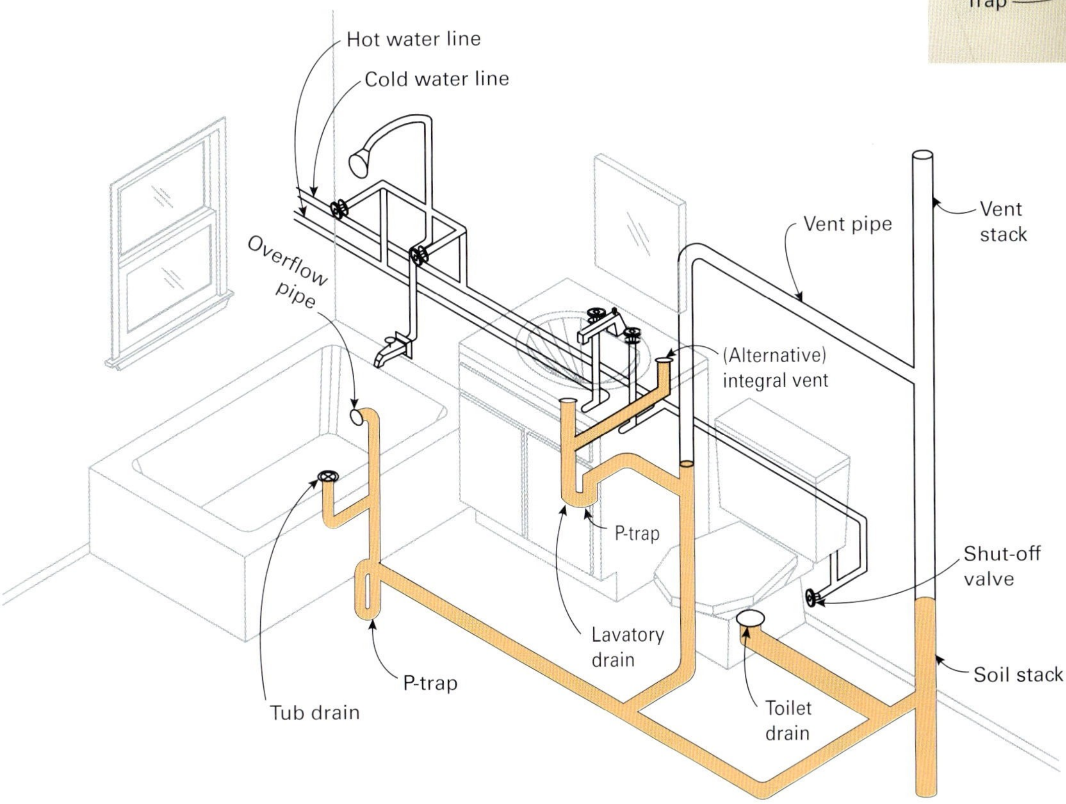

Plumbing Plan

The **plumbing plan** shows gas appliances, supply and waste removal pipes, fixtures, hose bibs, floor drains, roof vents, valves, pipe intersections, and any built-in vacuum systems, radiant heating systems, and control devices. They're all connected with plumbing line symbols to the water supply, waste discharge, gas line and stack pipe that serves them. Hot water lines connect to the house main, to the water heaters, and to each fixture that requires them. Shutoff valve symbols are drawn on those lines. Wastewater lines are larger than supply lines, so they are drawn thicker. A trap symbol is shown at each fixture. Stack and other vertical pipe locations are shown with a circle inside walls and partitions. The house main and sewer lines are connected to the stack symbols, clean-outs, and house trap. Typically, the only dimensions given are for walls in which stack pipes will be placed before the foundation is poured and underground clean-out valves. Pipe types and sizes are noted.

Figure 10.44 shows a simple plumbing plan with hot and cold water, wastewater and gas pipe lines. Figure 10.45 shows common plumbing symbols.

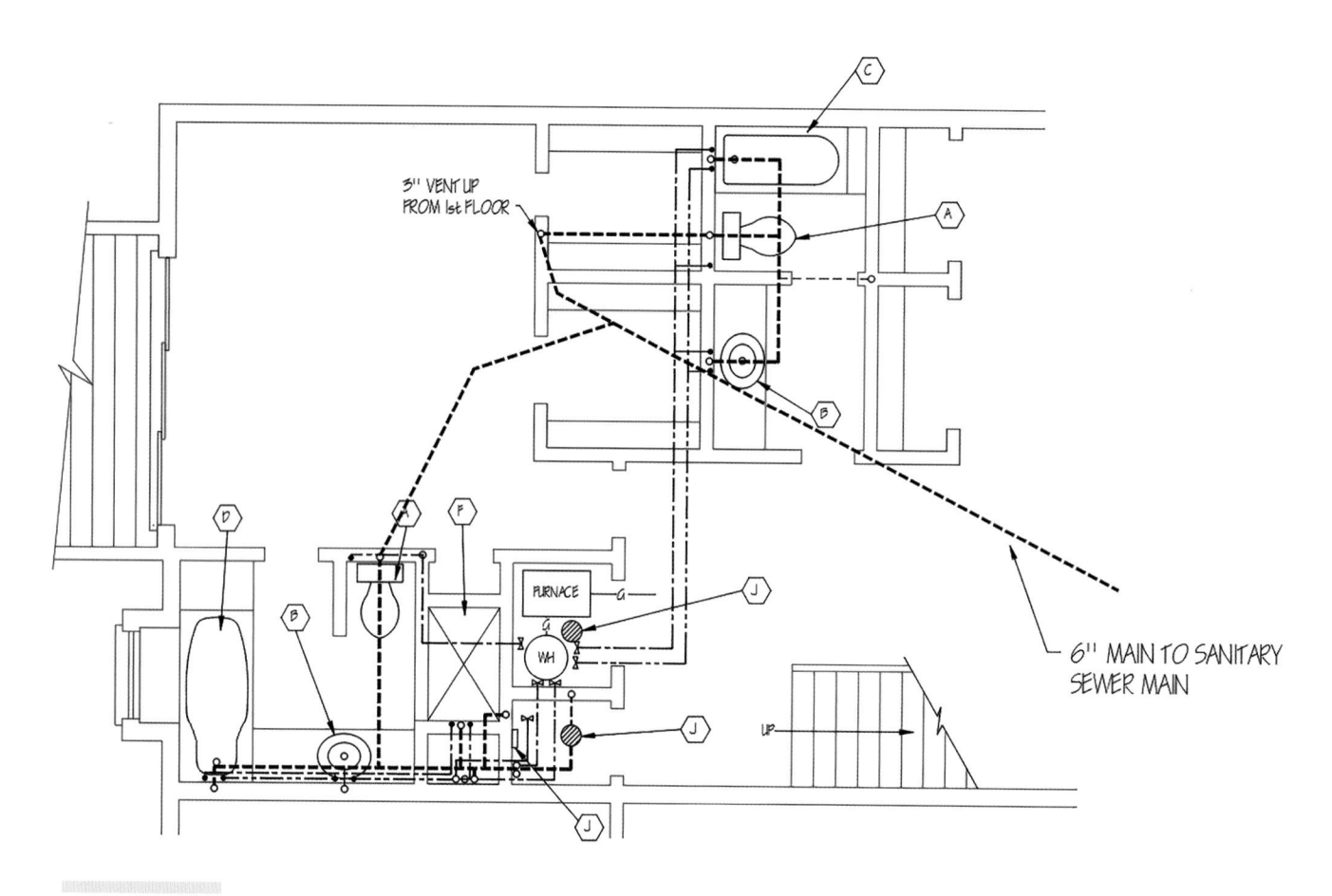

Figure 10.44
Plumbing plan.

Double branch elbow	⊤	Solid waste pipe	———	Gas line	— G —	Sprinkler line	—○——○—
Chilled drinking water	— DWS —	Vent pipe	— — —	Down elbow pipe outlet	—⊙—	Flanged fitting	—‖—
Gate valve	—⋈—	Clean out lateral or Y		Water meter	—(M)—	Welded fitting	—✕—
Hand valve	⊗	Cold water supply	— — —	Fire service	— F —	Sprinkler	—⊕—
Combination waste & vent	— CWV —	Hot water supply	— — —	90 degree elbow	⌐	Expansion joint	—⊟—
Water heater	WH	Floor drain	⊘	P-trap	⌐	Vent	○

Figure 10.45
Plumbing symbols.

The plans only show horizontal positions and dimensions of pipes, valves, and fixtures. **Riser diagrams**, which are elevation drawings of the pipes, show the pipes' vertical heights, angles above the floor, and the flow of freshwater and waste between levels.

> **Exercise**
> Identify the pipes in an unfinished basement or other space. Look for hot and cold-water lines, soil stacks, shutoff valves, and fixture traps.

Summary

Electrical, reflected ceiling, mechanical, and plumbing plans show how utilities are delivered to, from, and throughout a building. Although the interior designer may not draw all of these plans, the ability to read them is important in order to recognize features that must be considered when designing spaces.

Classroom Activities

1. Draw a reflected ceiling plan of the ceiling in your classroom.
2. Sketch an electrical plan of the lights, switches, and outlets in your classroom.
3. Collect, compare, and contrast manufacturer literature on different plumbing and HVAC products.
4. Visit the plumbing department of a big box home improvement store and examine different components used in residential plumbing.

Questions

1. What is an electrical plan?
2. What is a reflected ceiling plan?
3. Name three components that may be found on a reflected ceiling plan.
4. What is an HVAC plan?
5. Name three types of HVAC systems.
6. What is a plumbing plan?
7. How many outlets minimum should a room have?
8. How far above the countertop should outlets and switches be?
9. What is a 3-way switch?
10. What should be considered when locating a thermostat?

Further Resources

National Fire Protection Agency. Obtain the NFPA 70 electrical code here. https://www.nfpa.org/

Stiebel-Eltron website. Contains information on their tankless water heaters. https://www.stiebel-eltron-usa.com/water-heating

Keywords

- active solar
- air conditioner
- air supply vent
- amperage
- appliance
- automated systems plan
- branch duct
- branch pipe
- breaker box
- central air
- chandelier
- circuit
- circuit breakers
- cleanout valve
- climate control system

- compact fluorescent
- cove light
- data plan
- dimmer switch
- drop-in ceiling
- duct
- duplex outlet
- electrical plan
- equipment
- equipment plan
- evaporative cooler
- fluorescent
- forced air system
- furnace
- gang
- gate valve
- GFCI outlet
- grille
- grounded outlet
- halogen
- heat exchanger
- holistically
- hose bib
- humidifier
- HVAC
- HVAC plan
- incandescent
- International Residential Code (IRC)
- lamp
- lavatory
- light emitting diode (LED)
- lighting plan
- main supply duct
- mechanical plan
- mini-split
- non-grounded outlet
- outlet
- pendant
- permanently attached light fixtures
- plenum
- plumbing fixture
- plumbing plan
- plumbing system
- power plan
- P-trap
- push-button switch
- radiant heat
- recessed lighting
- reclaimed wastewater
- reflected ceiling plan
- riser diagram
- schematic
- septic tank
- service panel
- single plug outlet
- soffit light
- softener
- solar heating
- split wire outlet
- stack pipe
- suspended ceiling
- swamp cooler
- switch
- thermostat
- toggle switch
- track lighting
- under-cabinet lighting
- Uniform Mechanical Code
- vent stack
- wall sconce
- wall stack
- wall wash
- wastewater
- water closet
- water heater
- well

CHAPTER 11

Stairs

Figure 11.0
This spiral stair at a Caribbean resort makes a dramatic architectural statement.

OBJECTIVES

Upon completion of this chapter you will be able to:

- Identify different staircase types
- Identify staircase parts
- Name IRC-required dimensions of interior staircases and steps.
- Calculate the number and size of steps needed between floors.
- Prepare staircase drawings.

A **staircase**, also called a **flight**, is a set of steps that leads from one floor of a building to another. It may be **enclosed** (surrounded with walls), partially enclosed, **open** (no surrounding walls), and constructed of wood, concrete, or metal. The International Residential Code (IRC), International Building Code (IBC), Occupational Safety and Health (OSHA), Americans with Disabilities Act (ADA), Architectural Barriers Act (ABA), and local codes govern commercial stairs. We'll use IRC requirements here.

> **Exercise**
> Download the free file *Visual Interpretation of the IRC* at www.stairways.org (Stairways Manufacturers Association). Write down five requirements for stair construction from it.

Staircase Parts and Sizes

There are many components to a staircase (Figures 11.1, 11.2). The largest is the **stringer**, which is the diagonal support structure for the steps. A staircase has at least two stringers, and sometimes more. The **well** is the hole in the upper floor framing. A **stairwell** is the vertical shaft in which the staircase is built.

Treads and Risers   Stair steps are made of **treads** (horizontal boards) and **risers** (vertical boards). While the exact riser height and tread width depends on the staircase type and distance between floors, here are some guidelines. Residential treads must be at least 36" long and 10" deep (11" is ideal). In a commercial building with over 50 occupants, treads must be 44" long and between 11" and 14" deep. If a tread is less than 11" deep, a **nosing**, which is a riser overhang, must be included. Nosings protrude 1 1/4" beyond the riser face. Risers for both residential and commercial must be between 4" and 7 3/4" tall (7" is ideal). **Service stairs**, which access roof and equipment rooms, may have risers 8" high, which is too steep for general public use.

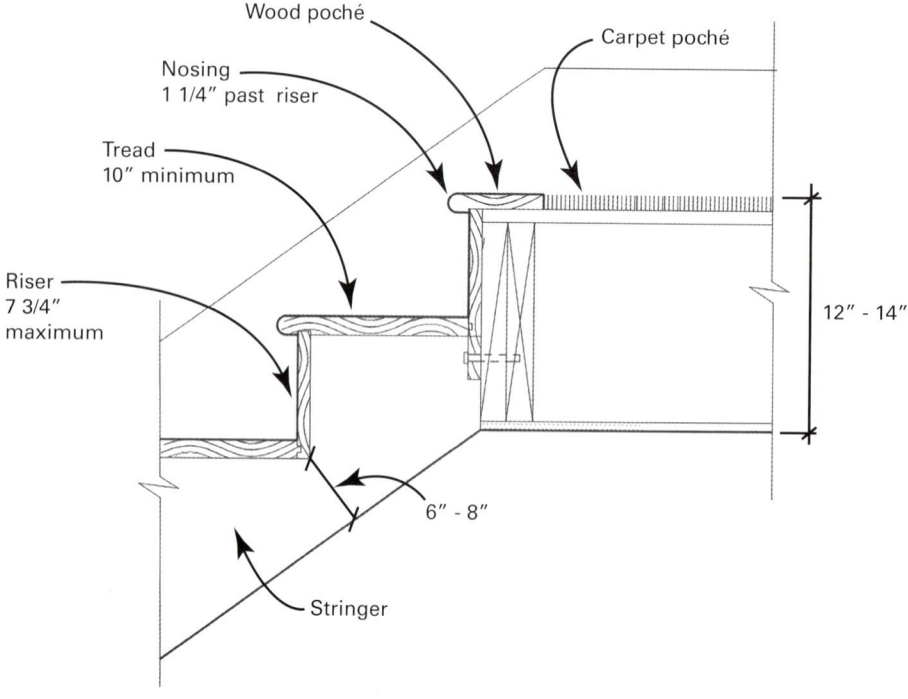

Figure 11.1
Parts of a step.

Figure 11.2
Parts of a staircase.

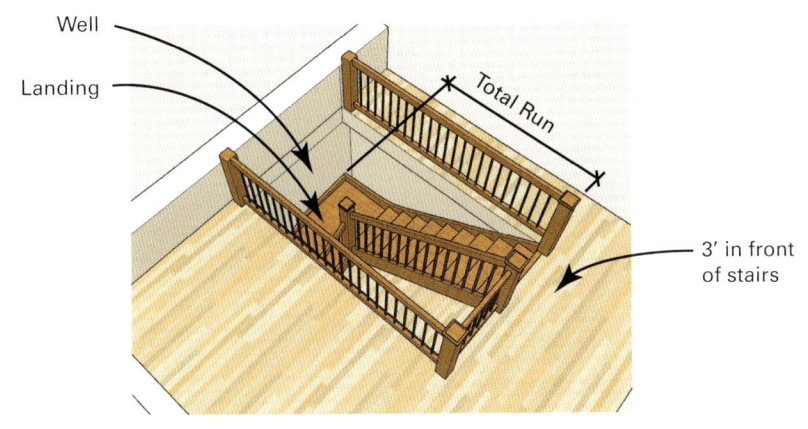

Tip
Avoid single steps to sunken rooms, as they are tripping hazards. Use at least two steps.

Baluster, Guard, and Handrail A **baluster** is a post under a **handrail**, the horizontal bar on top of it. A series of balusters is a **balustrade**. The first, and sometimes last, baluster may be a **newel post** (Figure 11.3), a larger, more ornamental baluster. Newels are often topped with **finial caps**, or decorative knobs. Together, the handrail, balustrade, and newel post are called the **guard**.

A guard must be at least 36" (residential) or 42" (commercial) from the floor to the top of the rail. It's needed when the stair elevation is 30" (four risers) or higher and there's no adjacent wall. It is placed between 30" and 38" above the tread. A handrail must be continuous for the whole length of the staircase and can't be more than 30" away from any place on a staircase. Therefore, place an intermediate rail every 60" of stair width. A handrail's cross-section must be between 1 1/4" and 2 3/4" wide for easy grasping. A **volute** is a handrail's curved end (Figure 11.4).

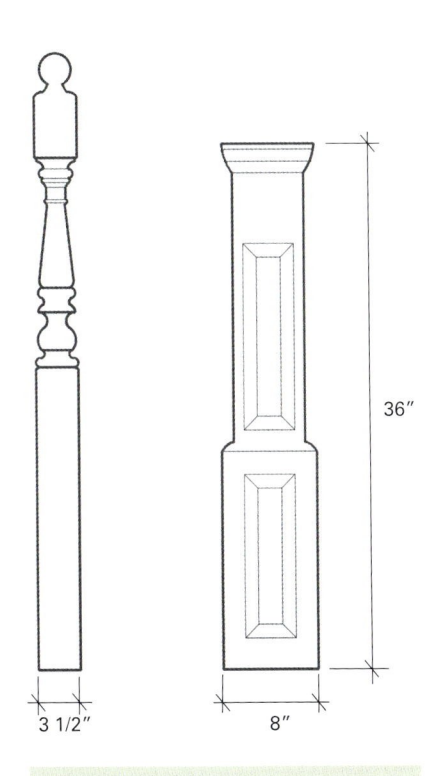

Figure 11.3
A baluster and newel post.

STAIRCASE PARTS AND SIZES

The space between balusters cannot allow passage of a 4" diameter sphere, and the space between the balustrade and treads cannot allow passage of a 6" diameter sphere (Figures 11.5, 11.6). A sphere is referenced instead of a circle because a sphere is the same size from any direction, as opposed to a flat, linear measurement.

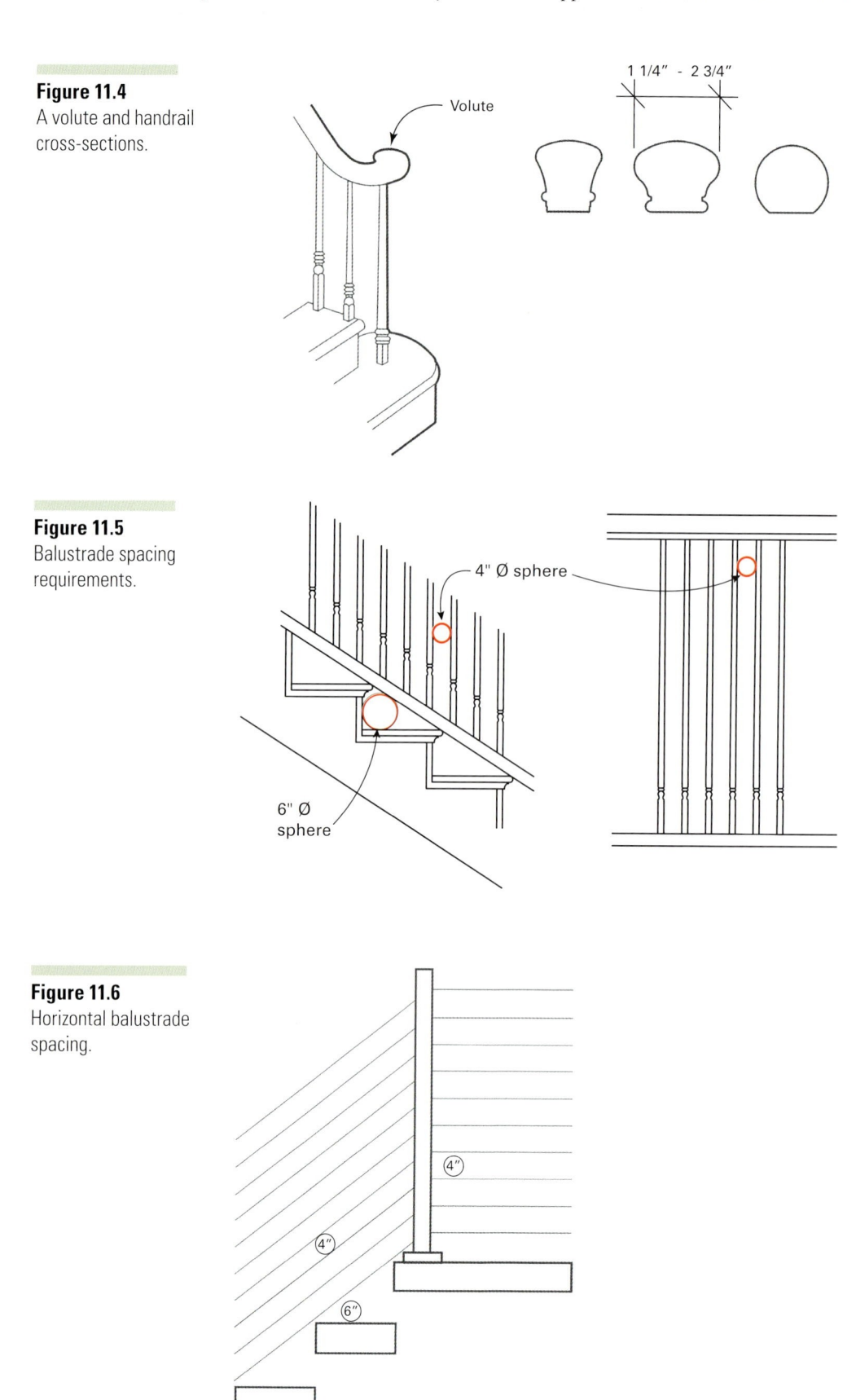

Figure 11.4
A volute and handrail cross-sections.

Figure 11.5
Balustrade spacing requirements.

Figure 11.6
Horizontal balustrade spacing.

298   CHAPTER 11   STAIRS

Wall Rail   A **wall rail** is a handrail mounted to the wall and is required on at least one side (Figure 11.7). If the steps are wider than 44", both walls need one.

The clear width between one wall rail and the opposite wall must be 31 1/2" minimum. The clear width between two wall rails, or a wall and handrail, must be 27" minimum (Figure 11.7). In a residence, wall rails must be 34–38" inches from the top of each tread to the top of the rail; in a commercial building, it is 42". Wall rails should be 1 1/2" from the wall and cannot project more than 4 1/2" on either side. On residential stairs the endpoints turn into the wall. On commercial stairs, ADA compliance requires that both wall rails and handrails extend 12" past the riser's nosing at each end (Figure 11.8).

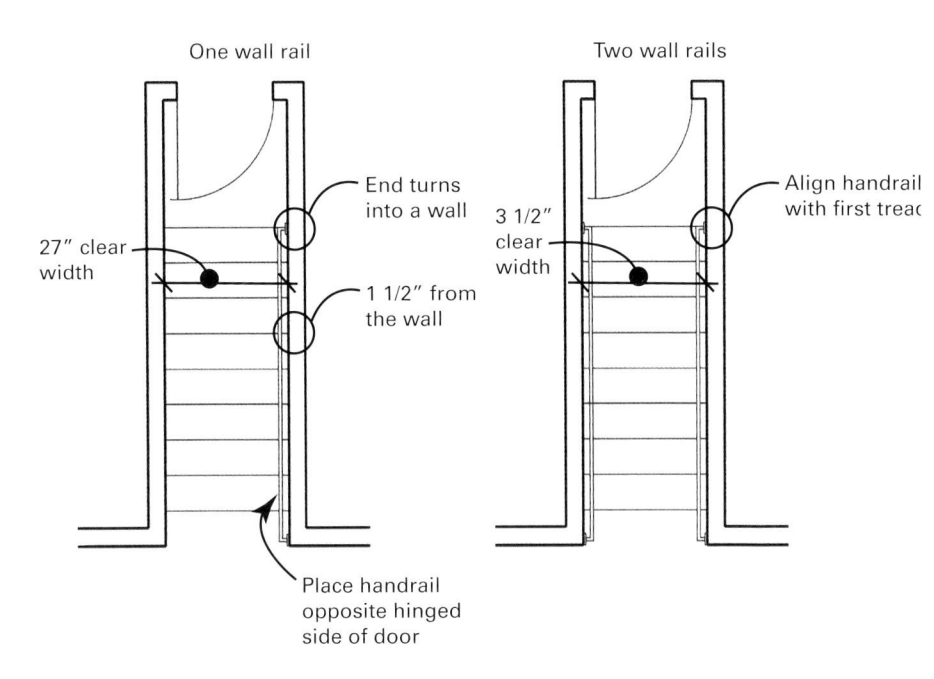

Figure 11.7
Wall rail requirements.

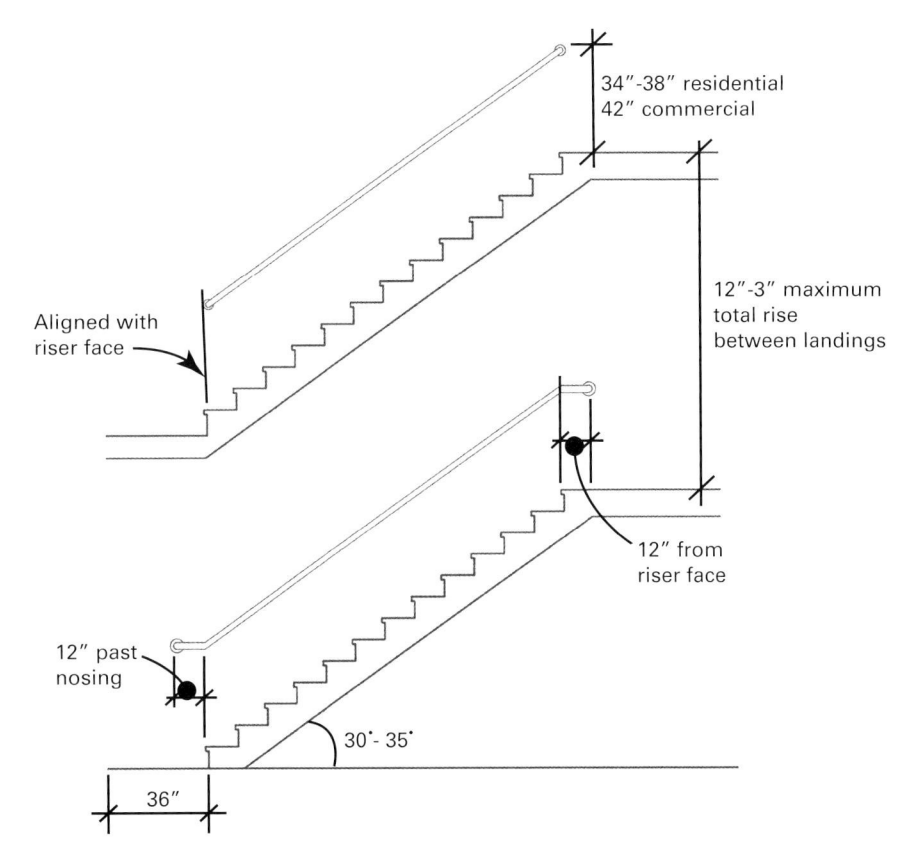

Figure 11.8
Staircase dimensions.

STAIRCASE PARTS AND SIZES   299

Headroom   Headroom is the clear vertical distance between a stair tread and the ceiling. The minimum headroom allowed is 6'-8", and the bottom of any ceiling-mounted item, such as an exit light or ductwork, must be at that 6'-8" height.

Landing   A **landing** is the short, level rest area on a staircase and is needed if the staircase's **total rise** (distance from finished floor to finished floor) is greater than 12'-3". Landings must be the same width as the stairs and at least 36" deep. 36" of floor space are needed at the top and bottom of a staircase. If a door is at the head of the stairs, it should not swing over them; at least 36" of landing is needed (Figure 11.9). When a door is fully open it should not project into the clear exit width more than 7". It cannot impede on more than half the required exit width at any open position.

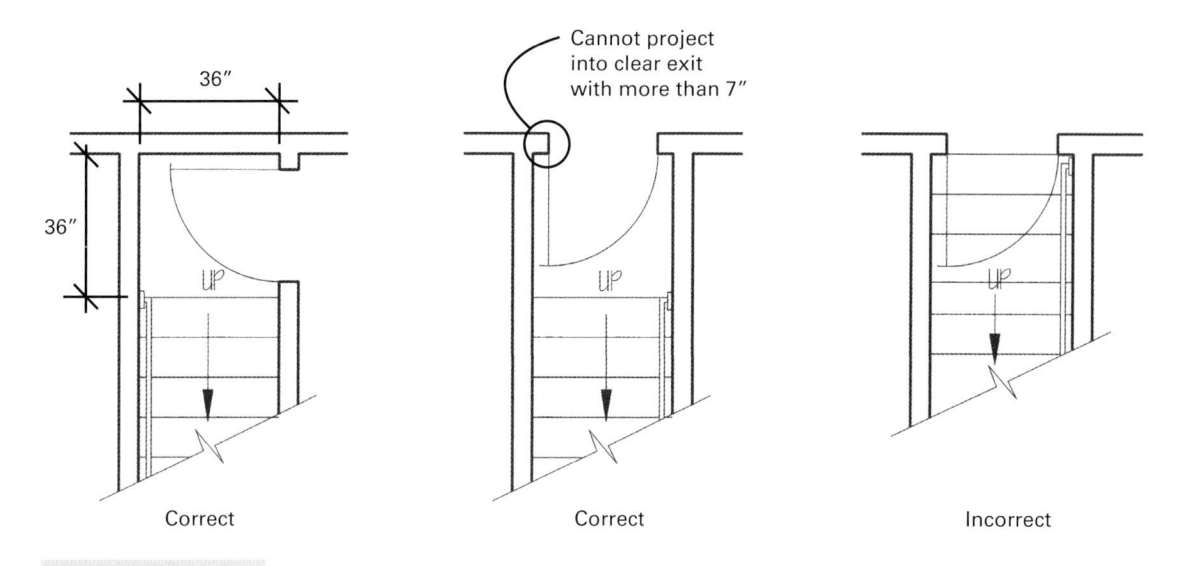

Figure 11.9
Correct/incorrect door and stairs placement.

How to Draft Stairs   Draw half of the complete stair on each floor plan more than half is needed to show a stair's defining characteristics (Figure 11.10). The second floor is a mirror and/or rotated image of the first floor. An alternative on the second floor is to draw the full flight, since you're looking down at it.

Position the full flight in the same location on both levels and leave enough room for the full flight on both levels even if you don't draw it all out. If you overlap the two floor plans and see one complete stairs plan, you've drawn the flights correctly. Draw an *UP* arrow on all flights; reverse the arrow on the second-floor flight and change the note to *DN* (down).

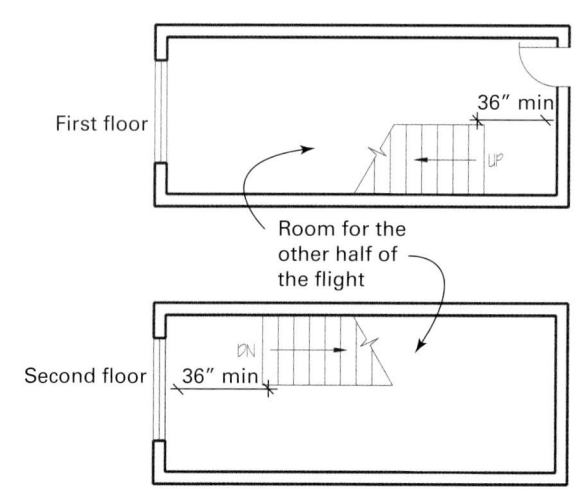

Figure 11.10
Draw half the flight on each level and leave enough room for the full flight.

To draw the elevation, draw the full flight in plan first and then project all the treads down (Figure 11.11). You'll need to know the riser height; calculating it is done later in this chapter.

Stair Types

The major stair types are straight run, L, turning, U, spiral, and circular. All are used in residential and commercial construction. The specific one you choose depends on space available, aesthetics desired, and code requirements.

Straight Run A **straight run** has no turns and requires a long **horizontal run** (total floor space of the stairs including landings). Figure 11.12 shows one with a landing. Figure 11.13 shows their floor plan symbols. Note the stacked stairs. This is flights arranged on top of each other in a multi-story building.

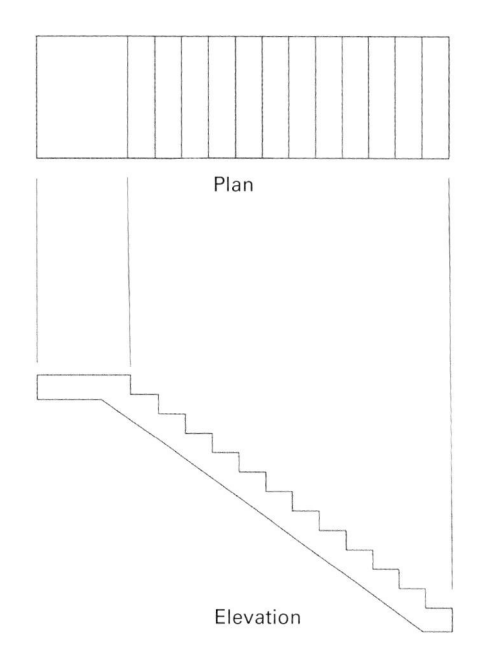

Figure 11.11
Project points from the plan down.

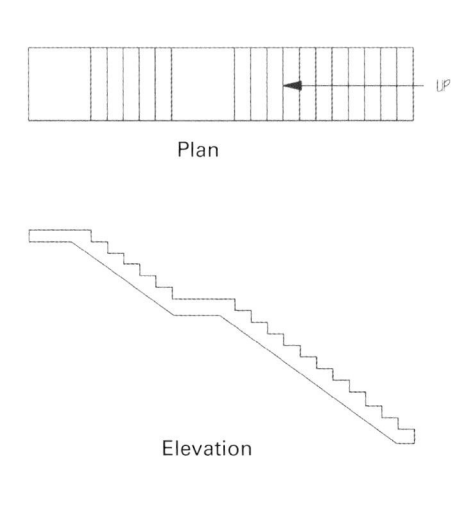

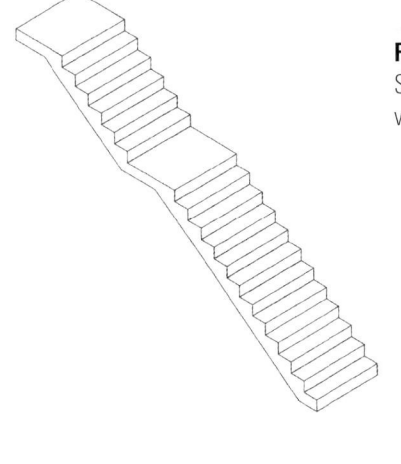

Figure 11.12
Straight-run staircase with a landing.

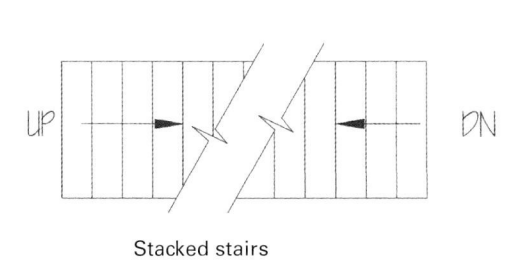

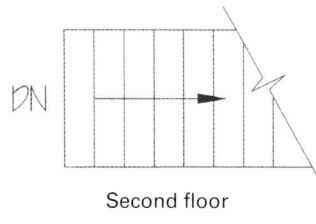

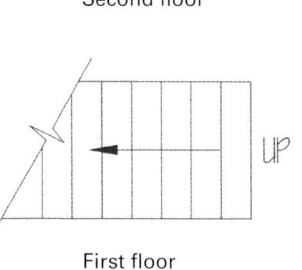

Figure 11.13
Straight-run stair symbols.

L An **L** (Figures 11.14, 11.15) has a landing and a turn. It's used when there isn't enough space for a straight run. An L may be equal leg, meaning the turn is at the midpoint, or one leg may be longer than the other one. L-shaped stairs are good for residences and low-traffic areas but cannot be fire exit stairs.

Figure 11.14
L-stair

Plan

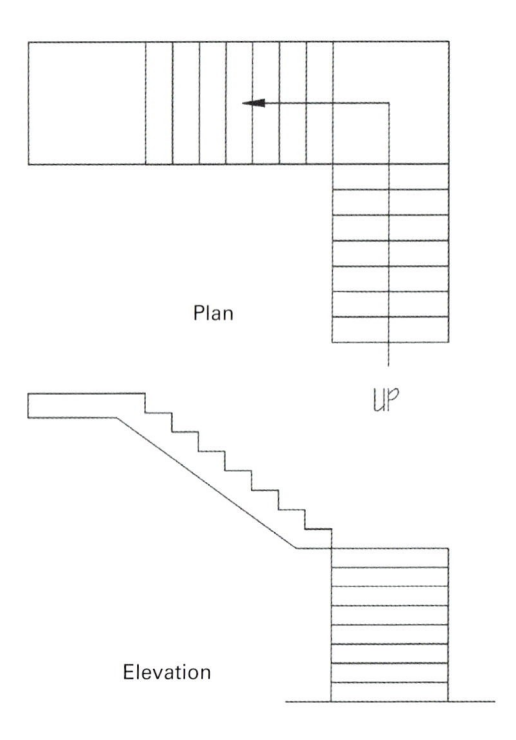

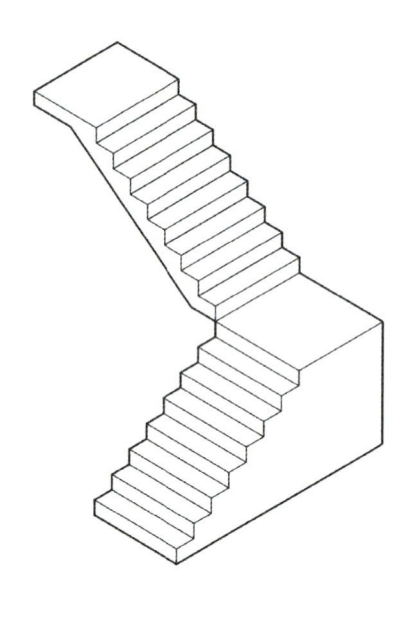

UP

Elevation

Figure 11.15
L-stair symbols.

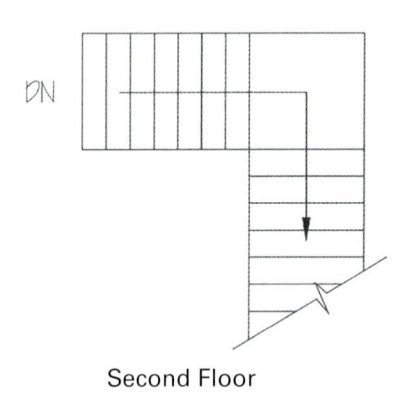

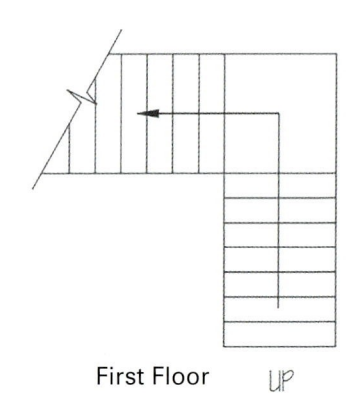

DN

Second Floor

First Floor   UP

302   CHAPTER 11   STAIRS

Turning   A **turning** (Figure 11.16, 11.17) is an L-stair that uses **winder treads** at the turn. Winder treads are **trapezoidal**, meaning one set of edges is non-parallel and used when there isn't enough space for a level landing. The midpoint of the winder treads must be equal to the tread width of the normal steps, and the narrowest portion must be at least 6" wide.

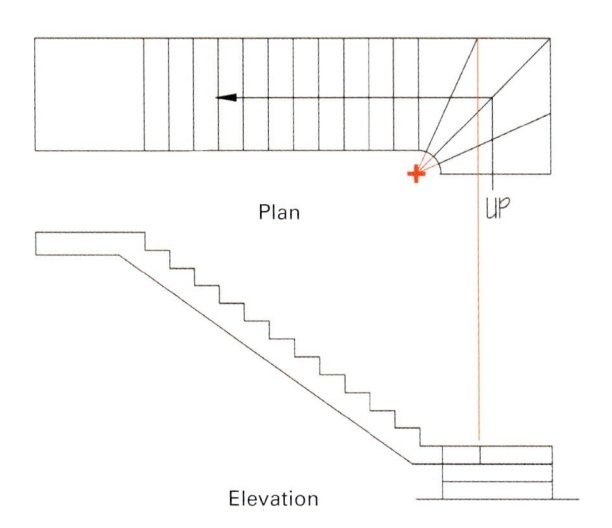

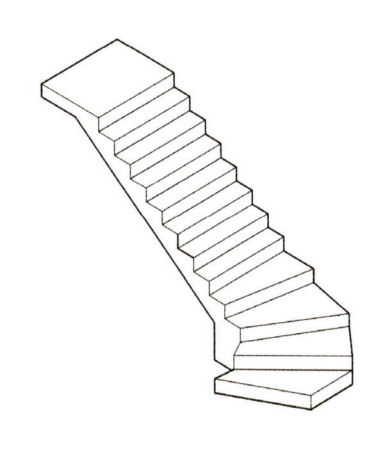

Figure 11.16
Turning stair

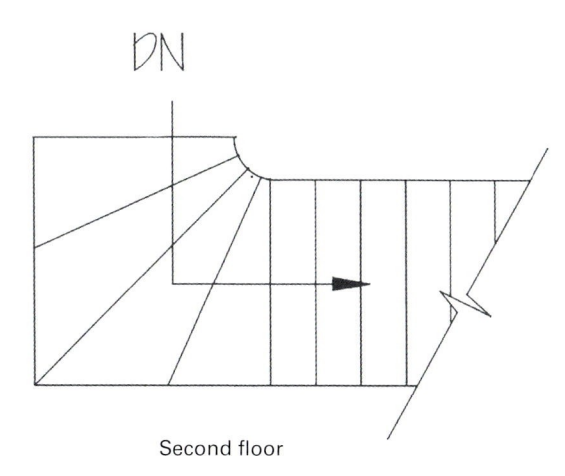

Figure 11.17
Turning stair symbols.

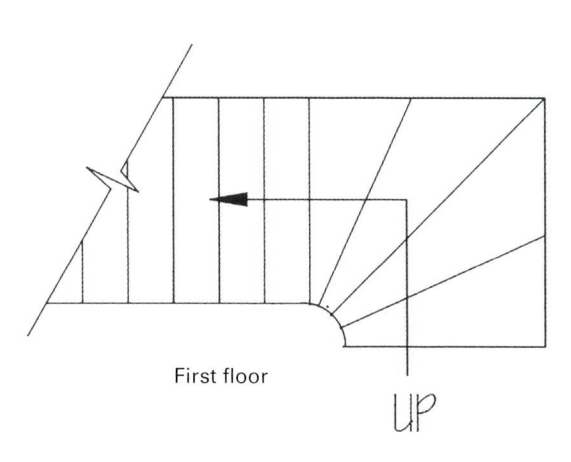

STAIR TYPES   **303**

U A **U** has two parallel flights (Figures 11.18, 11.19, 11. 20) and may be used when space for a straight-run isn't available. A **narrow U** has a small space between flights; a **wide U** has a **well hole** or large space. A **scissors** is two interlocking U-stairs with two separate means of egress within one enclosed stairwell (Figure 11.21). They're often used in small apartment buildings that require two vertical means of egress.

Figure 11.18
U stair.

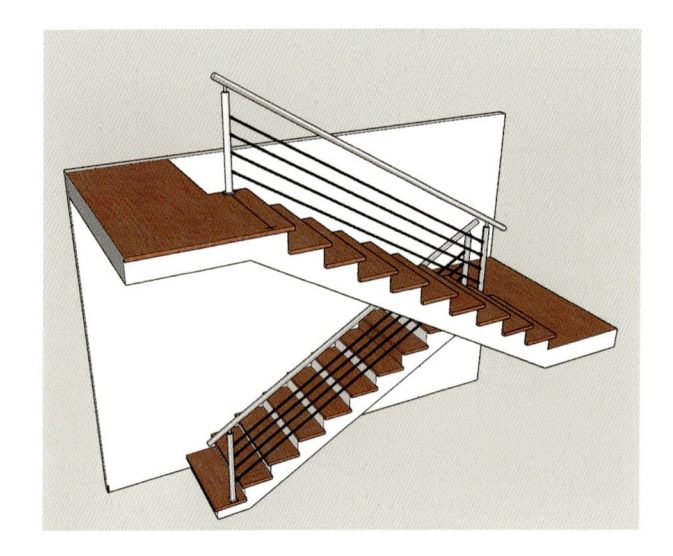

Figure 11.19
U stair.

Plan

Elevation

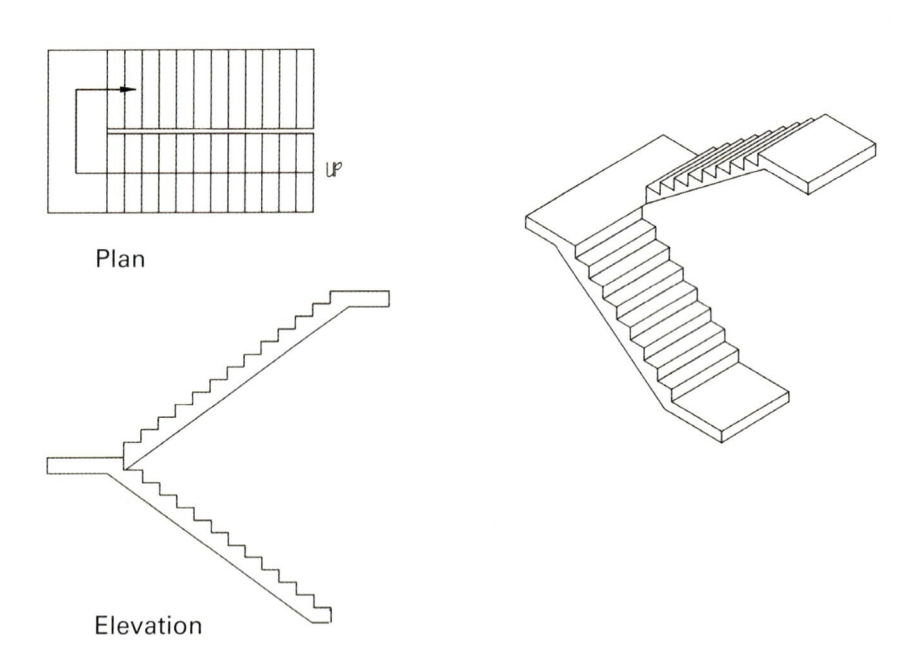

Figure 11.20
U-stair symbols.

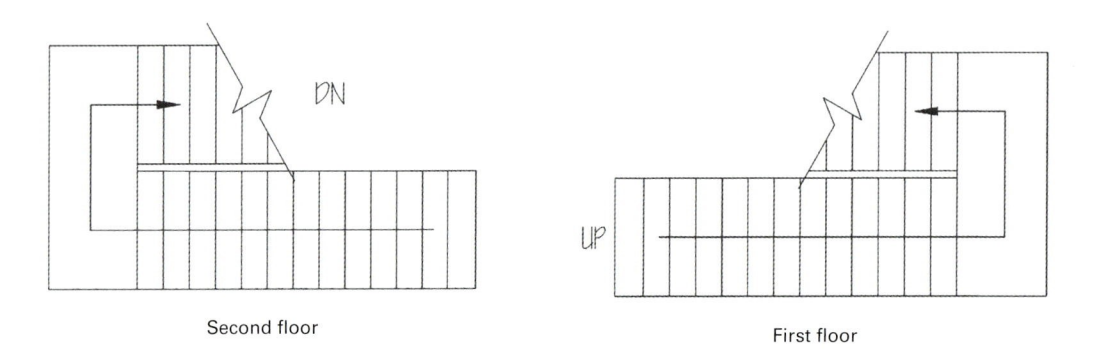

Second floor

First floor

304 CHAPTER 11 STAIRS

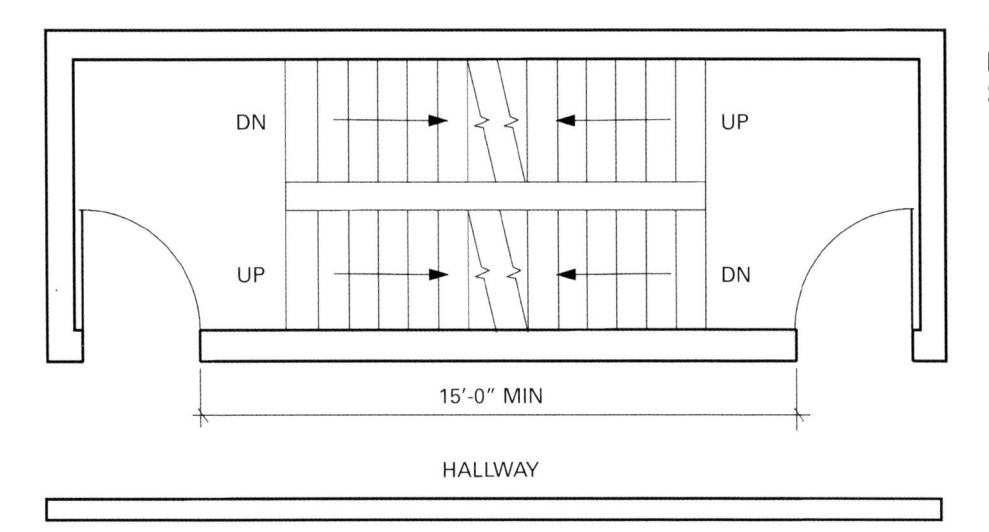

Figure 11.21
Scissors stair.

Spiral A **spiral** (Figure 11.22, 11.23, 11.24) rises in a complete circle. Its treads are trapezoidal and connected to a center pole. The treads must have a clear width of 26" measured from the outer edge of the center pole to the inner edge of the handrail, and the tread's narrowest part must be at least 7 1/2". The maximum riser height allowed is 9 1/2". A headroom of 6'-6" is allowed. Spiral stairs are used where there is little horizontal space.

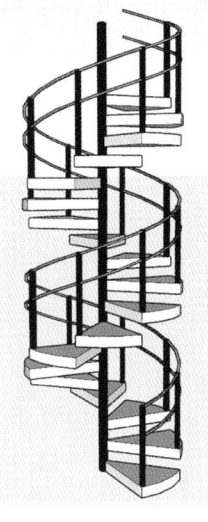

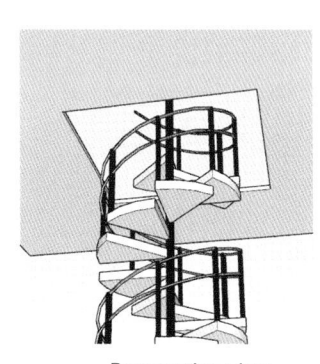

Figure 11.22
Spiral stair.

STAIR TYPES **305**

Plan

Elevation

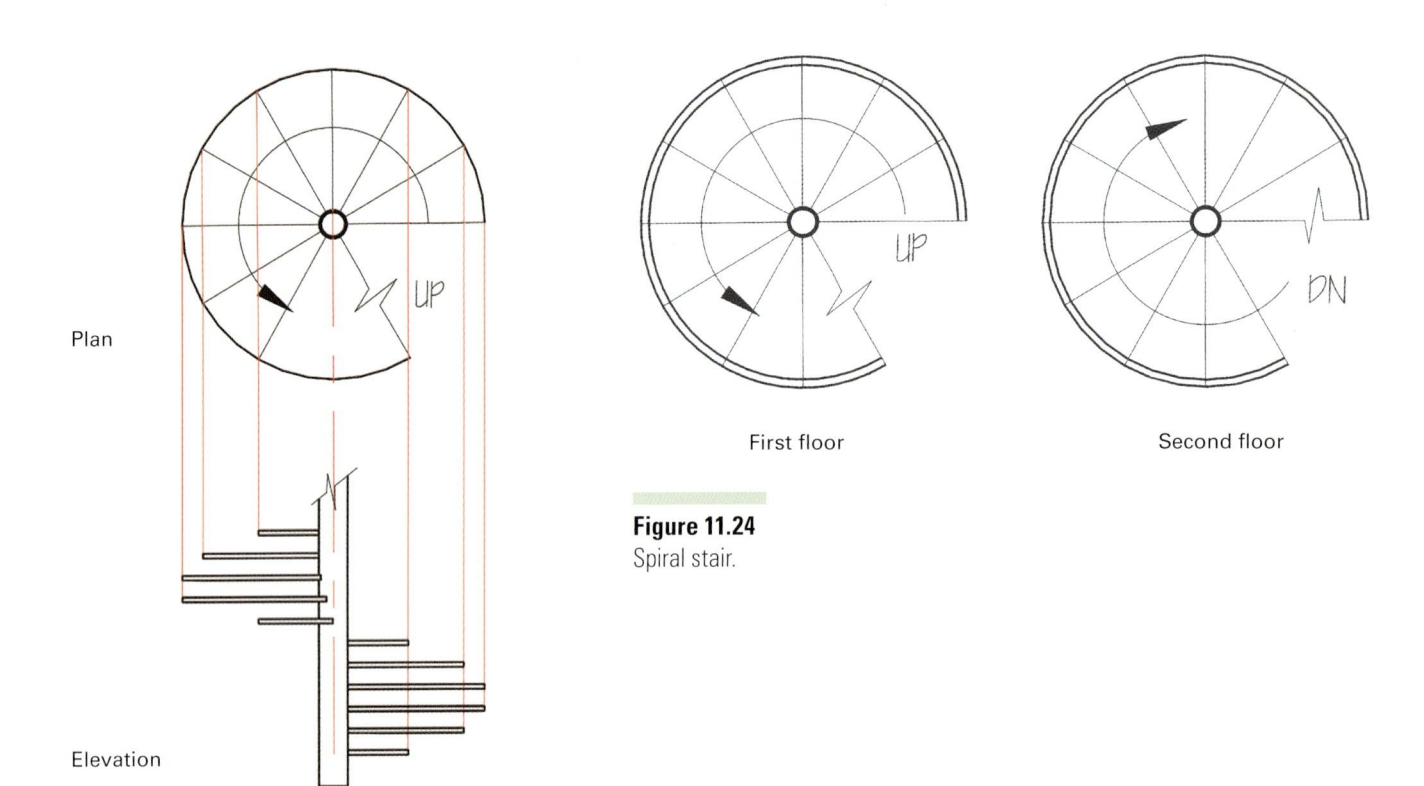

Figure 11.23
Spiral stair.

First floor

Second floor

Figure 11.24
Spiral stair.

Circular Also called **helical**, a **circular** staircase rises in a continuous "C," oval, or elliptical shape (Figure 11.25, 11.26). It has trapezoidal steps, a larger radius than a spiral and no central pole. It requires a lot of horizontal space.

Figure 11.25
Perspective view of a circular stair.

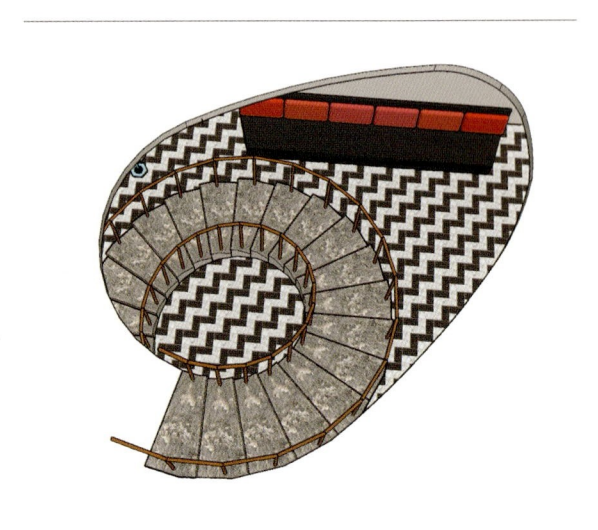

Top

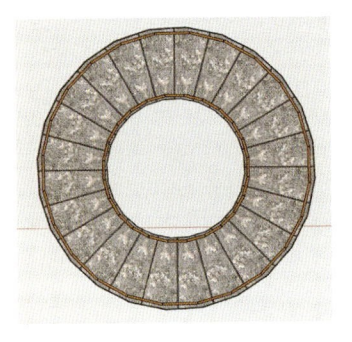

Front   Side

Figure 11.26
Orthographic views of the circular stair in Figure 11.25.

Exercise
Find buildings with interesting staircases and sketch plan and elevation views of them.

Other Configurations Combinations of basic stair types are often designed to suit specific building and traffic needs (Figure 11.27). For example, a **T**, also called a **bifurcated stair**, is used in large entrance halls in public buildings and residences.

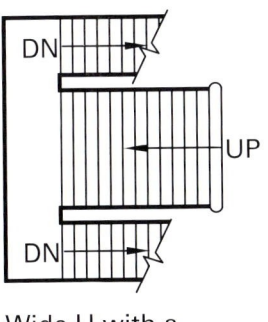

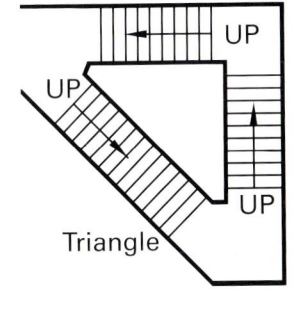

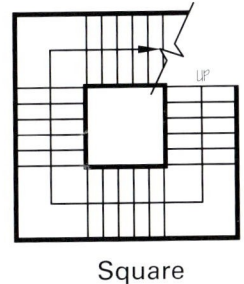

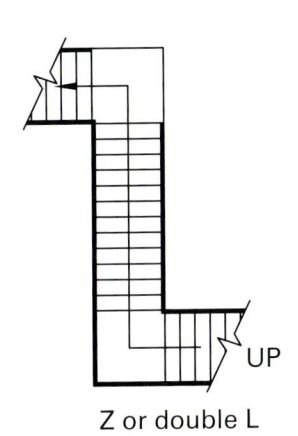

Figure 11.27
Combination staircase configurations.

Wide U with a middle return flight

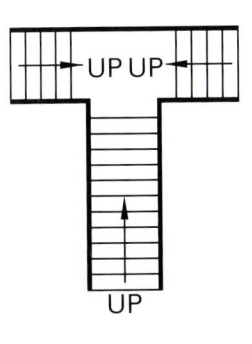

Triangle

T

Square

Z or double L

STAIR TYPES   **307**

Exit Stairs

Codes require commercial buildings to contain **exit stairs**, which consists of the stairs, a protected enclosure of fire-rated walls, and any doors. Most stair types may be part of the exit access. The minimum width of an exit access must be at least 36". Stairs must be wide enough so that two people can descend side by side. Interior exit stairs must be separated by at least 30'-0" or at least one-fourth of the length of the building's maximum overall diagonal dimension/areas to be served, whichever is less. Exit stairs doors must swing in the direction of the exit discharge. Therefore, all doors swing into the stairway except at the ground level, where the door swings toward the exit discharge or public way.

Exit stairs may also require an **area of refuge**, which is a safe waiting place for emergency assistance (Figure 11.28). Areas of refuge are usually provided next to, or inside, exit stairwells or at elevator lobbies. An area of refuge located next to an exit stair must be wheelchair accessible, and the number of wheelchair spaces that must be provided is determined by the floor's occupant load. The most common requirement is one space for every 200 occupants.

Figure 11.28
Staircases combined with areas of refuge.

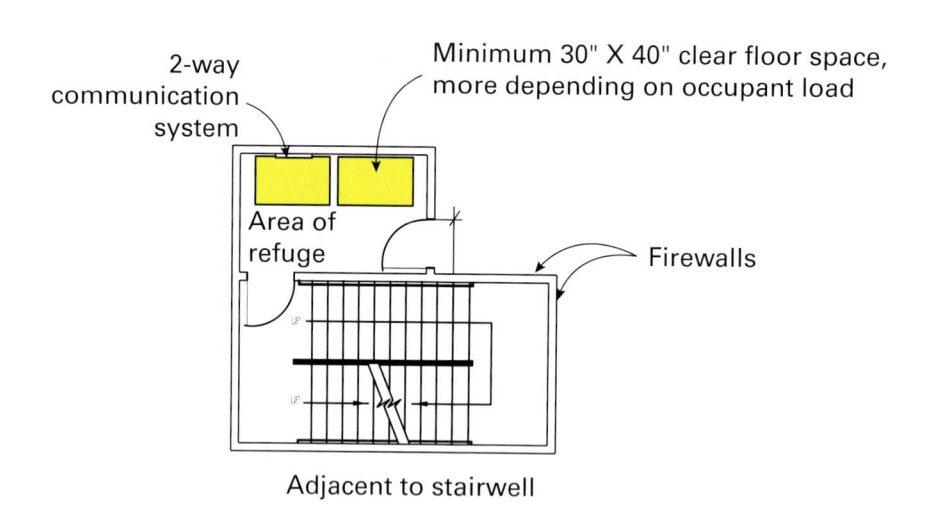

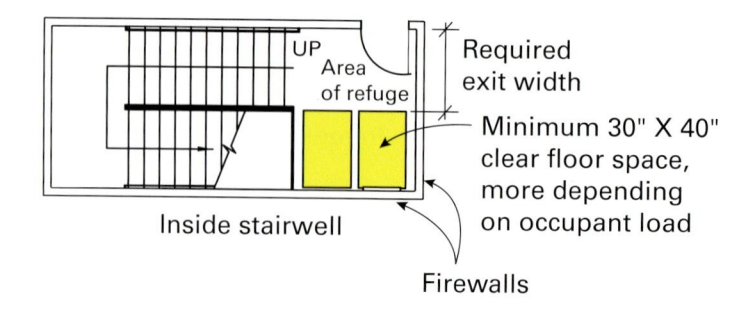

Exterior Stairs

Exterior stairs (Figure 11.29) are usually designed with smaller riser heights and wider treads than interior stairs, such as a 6" riser paired with an 11" tread. There should be a landing every 16 risers on continuous stairs. There is no riser requirement for wood stairs, so the staircase may be **open riser**, that is, one without riser boards, as long as a 4" sphere can't fit between the riser openings. If the staircase is shorter than 30", the 4" sphere rule doesn't apply.

Draw a 10° angle on the riser of a concrete stair. You don't need to dimension it because the exact angle will depend on the forms the builder uses. However, this angle cannot be greater than 30°.

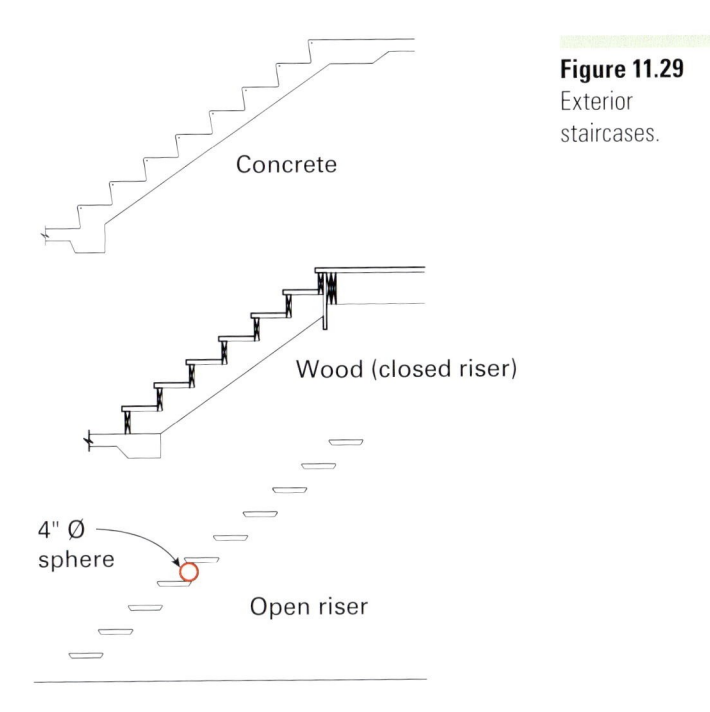

Figure 11.29
Exterior staircases.

How to Calculate Riser Height, Tread Width, and Total Run

The average staircase has 14 or 15 risers between two floors. Here's how to calculate how many risers and treads are needed, and their sizes. You need to know the **total rise** (distance from finished floor to finished floor). If working on an existing building, find that information on dimensioned drawings or measure it. In this example we'll use a total rise of 11'-6".

1. Convert the total rise into inches. Multiply 11' × 12" (because there are 12 inches in a foot) and add the remaining 6". The result is a total rise of 138".

2. Calculate the number of risers needed and their height. Assume an ideal riser height of 7". Then divide the total rise of 138" by 7. This gives us 19.71 risers. We need an even number—either 19 or 20. Divide them both into 138".

    138"/19 = 7.26"

    138"/20 = 6.90"

    Both results are very close to 7" so we can choose. Let's use 19 risers, each 7.26" tall.

3. Calculate the number of treads needed and their depth. There is one less tread than riser because the last tread is the second floor. So, 18 treads. Now use this formula: R + R + T = 25. This will yield a staircase angle between 30° and 35°, which is code compliant. Plug in the riser height (R) and solve for tread depth (T).

    7.26" + 7.26" + T = 25"

    T = 25" − 7.26" − 7.26"

    T = 10.48"

HOW TO CALCULATE RISER HEIGHT, TREAD WIDTH, AND TOTAL RUN   **309**

So, we need 19 risers that are 7.26" tall each, and 18 treads that are 10.48" deep each. If you need a wider tread, choose 20 risers at 6.90" tall each. Now calculate how much **total run** (horizontal distance) the staircase needs. In our example, multiply 18 treads × 10.48" for a result of 188.64". Add another 36" clearance at the top and bottom, and now we need 260.64". This is almost 22'-0". If the total rise was more than 12'-3", an intermediate landing is needed, too, which adds to the total run. If that much space is unavailable, choose a staircase with a smaller footprint, such as a U.

> **Exercise**
>
> Measure the height of an existing space for total rise, and then calculate the number and size of steps needed to reach the second floor.

Use a Grid to Draft a Staircase's Elevation View

When you know the total rise, total run, and number of risers and treads, you can draw the staircase's elevation view. Earlier in the chapter we showed how to draw the full flight in plan first and then project all the treads down. Alternatively, draw a grid with rectangles the height of each riser and the depth of each tread. Here are steps for drawing a grid with 19 risers and 18 treads.

1. *Draw a rectangle* with the length of the total run and the height the total rise (Figure 11.30).
2. *Divide one vertical line* into the number of risers needed (in this case, 19). Do this by selecting a scale—any scale—where you can place the 0 at one corner as shown and the 19 at the other corner (Figure 11.31).
3. *Mark off all 19 increments* and draw a horizontal line through each (Figure 11.32).
4. *Divide a horizontal line* into the number of treads needed (in this case, 18). Do this by placing the 0 where shown and the 18 anywhere on the vertical line (Figure 11.33).
5. *Mark off all 18 increments* and draw a vertical line through each (Figure 11.34).
6. Darken the risers and treads (Figure 11.35).

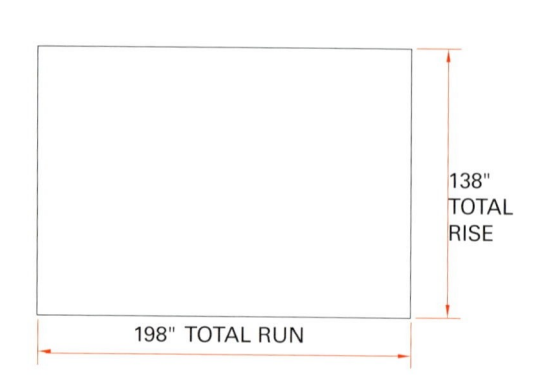

Figure 11.30
Step 1

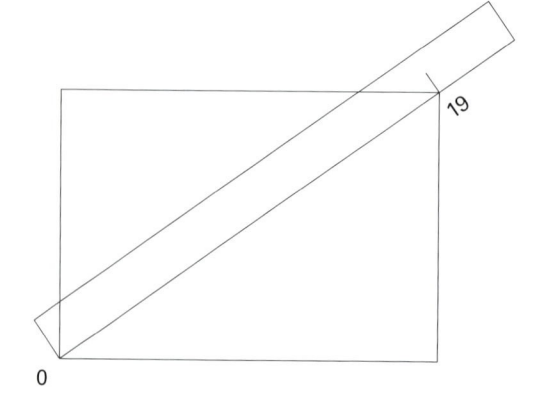

Figure 11.31
Step 2

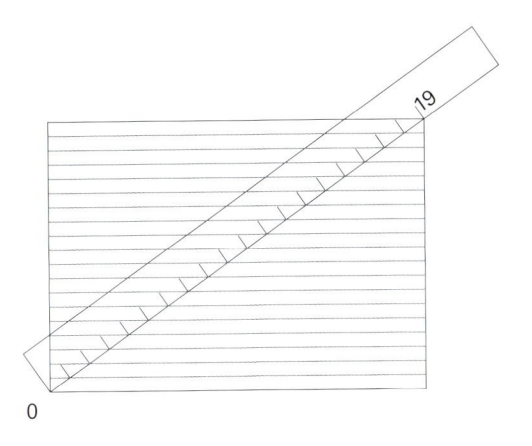

Figure 11.32
Step 3

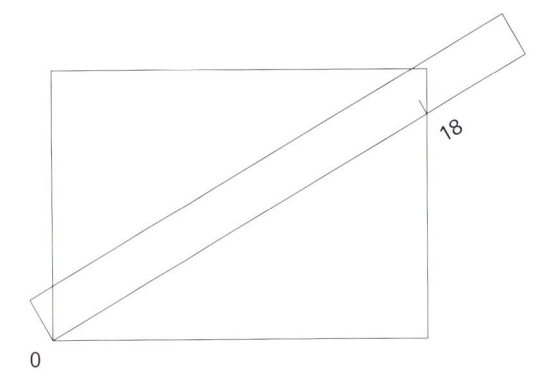

Figure 11.33
Step 4

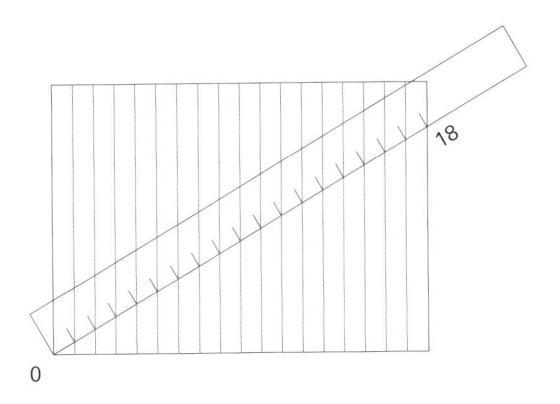

Figure 11.34
Step 5

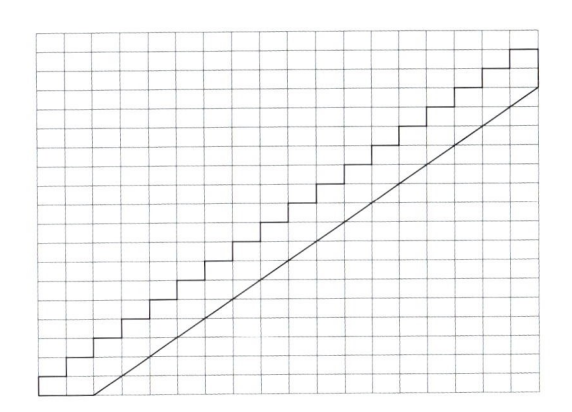

Figure 11.35
Step 6

Ramp

A **ramp** is a sloped surface that makes a building accessible to wheelchair users (Figure 11.36). It should have a nonslip surface and protection from rain, snow, and ice where possible. The IRC–recommended slope for both residential and commercial ramps is 1:8 (1 vertical unit for every 8 horizontal units), with 1:12 being the maximum. 30'-0" is the maximum length for a ramp. Landings are required at the top and bottom, and their length must consider any adjacent doors. Extend an entry platform 18" beyond the door's handle side to facilitate wheelchair use. Ramps longer than 30'-0" must be separated by a landing the same width as the ramp and at least 5'-0" long.

A ramp's minimum allowed width is 31 1/2", as measured to the inside of the handrail. 36" is needed for wheelchair accessibility, and 4'-0" is recommended. Any ramp that exceeds a 1:12 ratio must have at least one handrail, with a height between 34" and 38". Guards 36" high are needed when there is no adjacent wall and the ramp's overall rise is higher than 30".

RAMP **311**

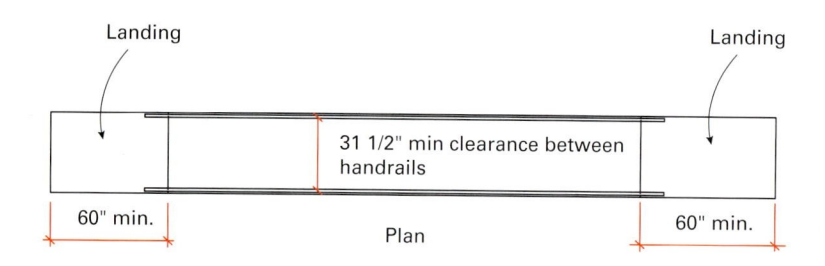

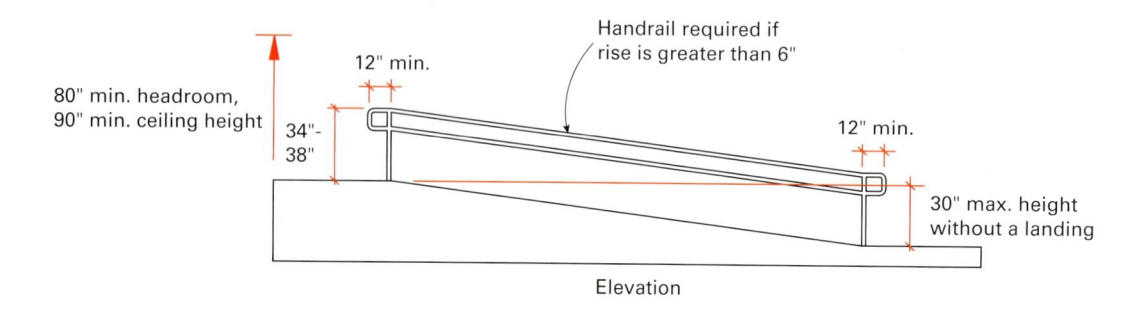

Figure 11.36
Ramp dimensions.

Elevator

An **elevator** is a vertical transport vehicle and is designed by a consultant with a mechanical or electrical engineering backgrounds. Elevator design and construction is governed by the IBC, ADA, and local building codes. There are **passenger** elevators (for general public use), **service elevators** (for access to equipment and storage rooms), and **occupant evacuation elevators** (for self-evacuation in fire emergencies).

Most elevators are 6'-6" to 7'-0" wide and 6'-0" deep. Minimum ADA compliance requires a 36" opening width, 51" cab depth, 68" cab width, and 80" cab width with center-opening doors. Figure 11.37 shows dimensions for occupant evacuation elevators.

Figure 11.37
Occupant evacuation elevators.

312 CHAPTER 11 STAIRS

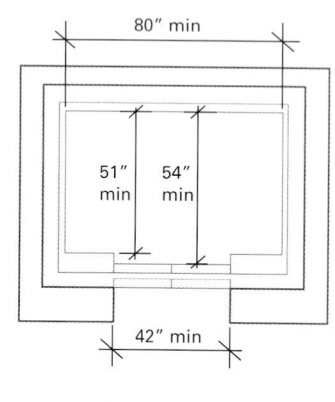

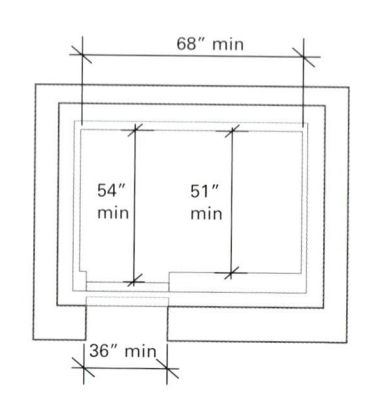

Copying Images into Plans

You can copy any of the staircase plan views here into your own drawings, assuming they have the correct number of treads and are scaled properly. Following is how to scale the staircase plans. This technique works for any image you want to scale and copy such as ones for design presentation boards. We'll refer to the staircase plan view as a "footprint," and assume the floor plan you're drawing is at the 1/4" = 1'-0" scale.

1. Figure out how big the footprint must be in your floor plan at the scale you're using. For example, your spiral staircase has 3'-0" wide treads on both sides of a 6" pole. So, 6'-6" in diameter.
2. Draw a 6'-6" diameter circle at the 1/4" = 1'-0" scale.
3. Use the true scale (1" = 1") to measure the footprint you just drew. This is the "size you want" dimension. Figure 11.38 shows it as 1 1/2".
4. Use the true scale (1" = 1") to measure the size of the footprint that you're copying. This is the "size you have" dimension. Figure 11.38 shows it as 2".

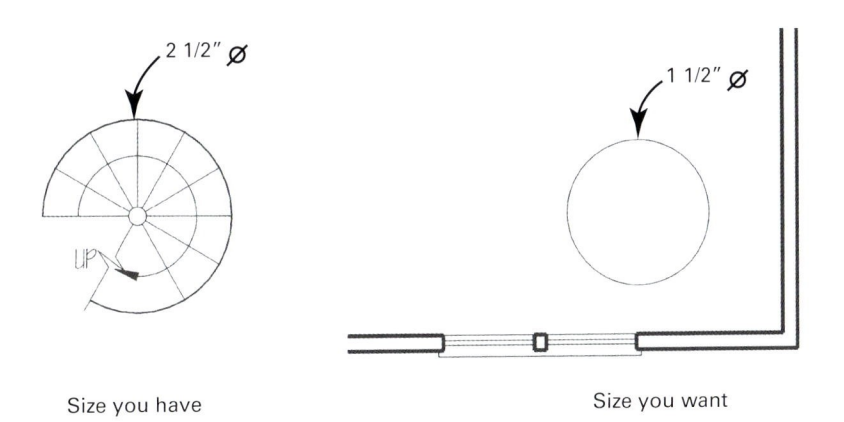

Figure 11.38 Measure the two footprints with a true scale to get numbers to plug into a proportional scale.

Size you have

Size you want

5. Plug the two numbers into an online **proportional scale**, such as the one at https://tinyurl.com/372ssbm (Figure 11.39). This is a tool that calculates percentage amounts for resizing images. Then set the zoom function on a copier to that.

Figure 11.39 An online proportional scale.

Summary

Staircases are vertical passageways from one level to another. They can be configured into different designs to accommodate the space available and aesthetic goals. As with all other aspects of building construction, they are regulated by various codes: IRC, IBC, ADA, and local. Hence, the drafter must know accessibility requirements and maximum and minimum dimensions for them and all their many components.

Classroom Activities

1. Search for prefabricated stair manufacturers online, download their product literature and study the photos, detail drawings, vocabulary, and other information about their products.
2. Arrange a field trip to a building under construction, perhaps on your campus, to study a staircase being built.

Questions

1. What is the ideal tread width?
2. What is the ideal riser height?
3. What must the clear width between two wall rails be?
4. What is the minimum height for a residential guard?
5. What does total rise mean?
6. Discuss the spacing between balusters and between the balusters and treads.
7. What is an area of refuge?
8. How many risers does the average staircase have?
9. What is the required minimum depth of a landing?
10. How high is a guard?

Further Resources

Information on ADA compliance https://tinyurl.com/y97lprl9

Stairways and Manufacturers Association https://stairways.org/

Video tutorial for how to use the proportional scale https://tinyurl.com/y2clnsth

Keywords

- area of refuge
- baluster
- balustrade
- bifurcated stair
- circular stair
- elevator
- enclosed staircase
- exit stairs
- finial cap
- flight
- guard
- handrail
- headroom
- helical
- horizontal run
- L stair
- landing
- narrow U stair
- newel post
- nosing
- occupant evacuation elevator
- open riser
- open staircase
- passenger elevator
- proportional scale
- ramp
- riser
- scissors stair
- service elevator
- service stairs
- spiral stair
- staircase
- stairwell
- straight run stair
- stringer
- T stair
- total rise
- total run
- trapezoidal
- tread
- turning stair
- U stair
- volute
- wall rail
- well
- well hole
- wide U stair
- winder stair
- winder tread

CHAPTER 12

Legends and Schedules

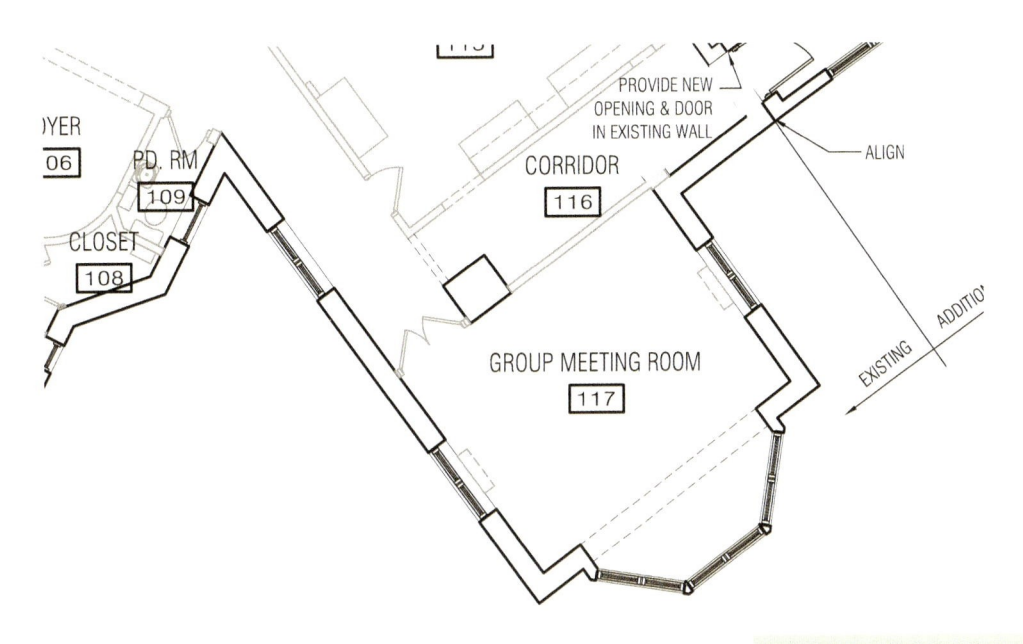

Figure 12.0
Callouts in the plan link each room to a finish schedule. Courtesy mkerrdesing.com.

OBJECTIVES

Upon completion of this chapter you will be able to:

- Define what a legend, schedule, and key is and describe how they're used
- Identify the information that legends, schedules, and keys convey
- Create and design legends, schedules, and keys

Legends, also called **keys**, and schedules are lists and charts of data about repeating components used in a design. Their purpose is to clarify symbols and describe features. Putting this data on separate drawings keeps the drawings from becoming overly cluttered.

What Are Legends and Schedules Used For?

These lists and charts give details about repeating building components. For example, architects make them for doors, windows, and columns; engineers make them for pipes, plumbing, and mechanical equipment; interior designers make them for room finishes, furniture, equipment, hardware, and lighting fixtures.

Callouts

Also called **key numbers**, **key letters**, or **marks**, these are symbols that link legends, schedules, and keys to their components on the plans (Figure 12.1, 12.2) and elevations. The symbols are geometric shapes—circles, squares, hexagons, rectangles, ellipses, triangles, and diamonds, with a number or letter inside. The same numbers are then used to identify those components on the legend or schedule.

Each component type needs a different symbol. Within the component types, duplicates get the same letter or number, but any difference requires a different number or letter. For instance, you might draw all window callouts with a hexagon, but double-hung windows would be labelled "1," fixed windows "2," etc. Attach callouts to the component's center with a line, or if that's not possible, near the component (Figure 12.2). Don't obscure other lines on the drawing.

> **Tip**
> Find a template of geometric shapes to aid drawing callout marks.

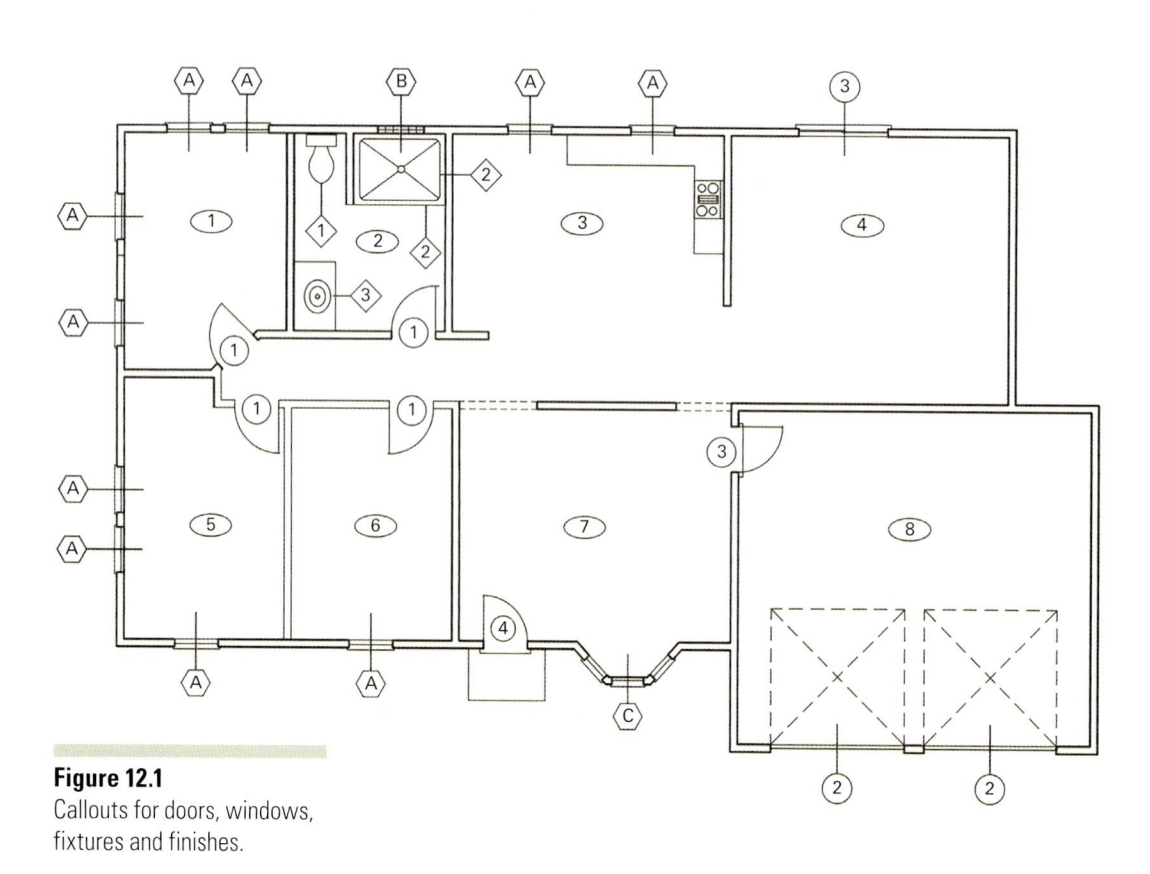

Draw all symbols 1/4" tall and letter the labels inside 1/8" tall. Letter and numbers always read upward, no matter the direction of the callout.

Figure 12.1
Callouts for doors, windows, fixtures and finishes.

316   CHAPTER 12   LEGENDS AND SCHEDULES

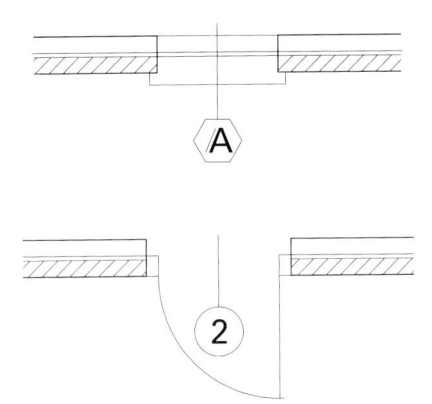

Figure 12.2
Window and door callouts.

Legends

These are charts that clarify floor plan symbols. A **legend** tells the reader what the symbol represents. Figure 12.3 shows callouts and a legend for wall types and material pochés; Figure 12.4 shows callouts and a legend for electrical components. Legends are usually drawn as a list, but may be arranged in a tabular grid, too. They are placed near the plan and typically don't have their own **ID label**.

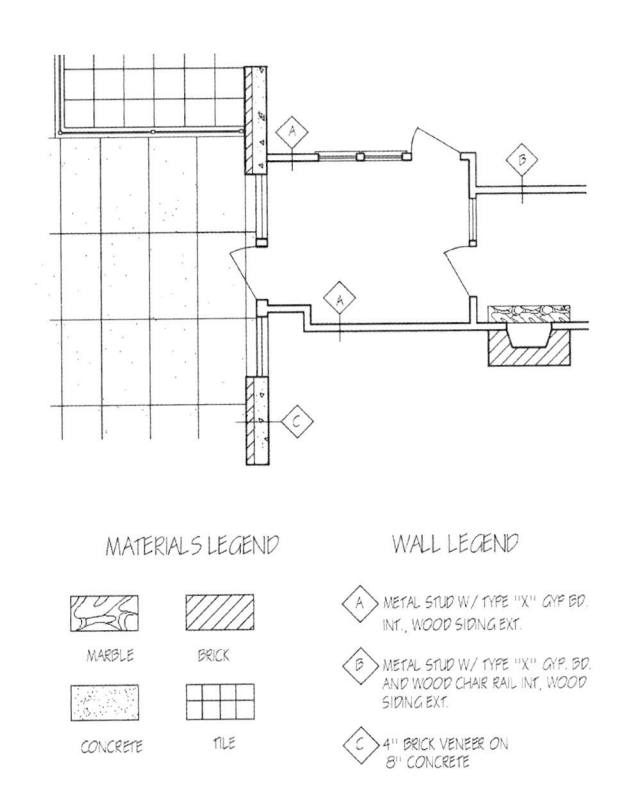

Figure 12.3
Callouts and legend for material pochés and wall types.

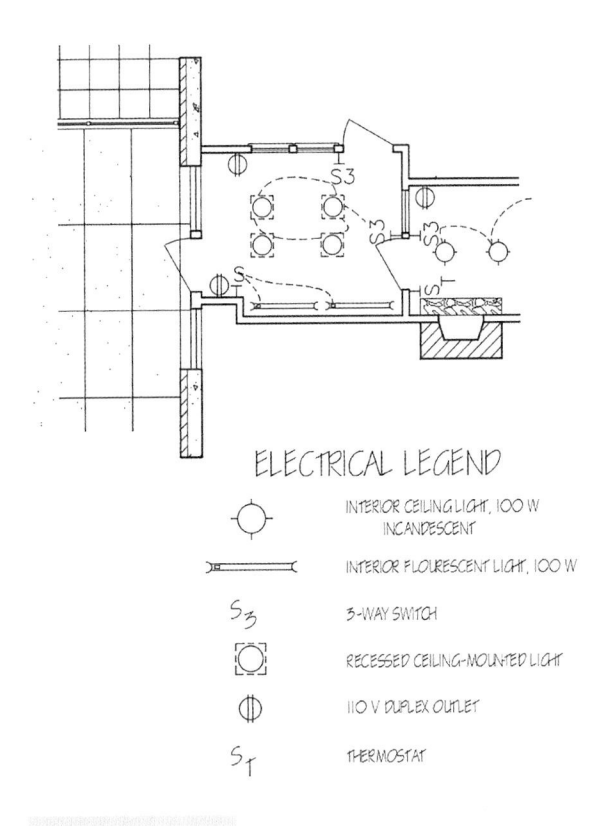

Figure 12.4
Callouts and legend for electrical components.

LEGENDS 317

Legends for furniture are usually called **keys**. A **furniture key** lists every piece of furniture and sometimes even non-permanently attached lamps, that is shown on the floor plan. Start with number 1 and label them in consecutive order. Include annotations to explain special features. Figure 12.5 shows a plan and key for a residential project.

furniture plan key

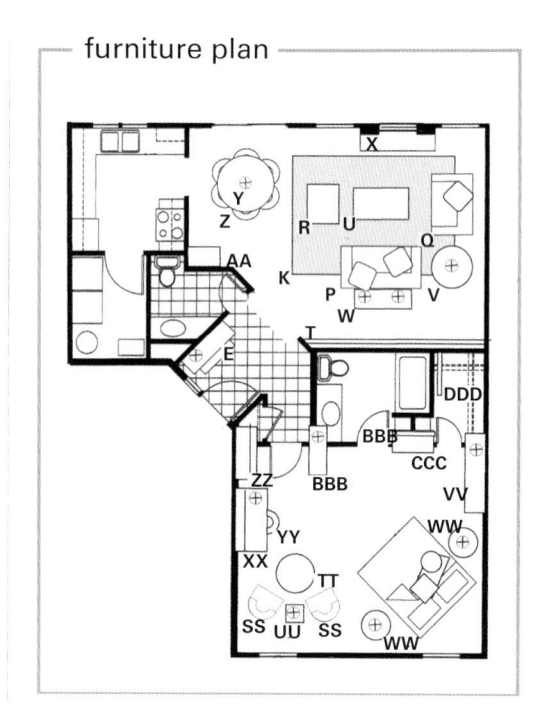

E.	OTTOMAN UNDER CONSOLE TABLE
K.	AREA RUG
P.	SOFA
Q.	LOVESEAT
R.	OTTOMAN
T.	SHELVING FOR CHINA OVER CABINETRY/STERO
U.	COFFEE TABLE
V.	ENDTABLE
W.	SOFA TABLE
X.	WALL-MOUNTED PLASMA TV
Y.	DINING TABLE
Z.	6 ARMLESS DINING CHAIRS
AA.	WINE BAR.SERVER
OO.	KING BED W/HEADBOARD
SS.	PAIR OF SIDE CHARIS
TT.	OTTOMAN
UU.	SIDE TABLE
VV.	DRESSER
WW.	PAIR OF NIGHTSTANDS
XX.	DESK
YY.	DESK CHAIR
ZZ.	BOOKCASE OVER LATERAL FILE CABINET
AAA.	SHELF/DIVIDER
BBB.	WALL-MOUNTED LCB TV
CCC.	CHEST
DDD.	CLOSET WALL MIRROR

Figure 12.5
Furniture key for a residential project. Courtesy Kacy Childs-Winston.

Schedules

Schedules are charts with detailed information about components. They're not as detailed as the project specifications but should be detailed enough for a builder to know what to provide. Specific details vary based on what the designer thinks is most important to communicate. There is no one way to lay out a schedule, as it depends on the complexity of the items and the amount of information given. A correctly done schedule is organized, easy to read, and has all necessary information for the person who will be reading it.

A schedule is usually in tabular form but may be pictorial if the components are custom-made or otherwise warrant it. Schedules may be included on the drawing sheets or bound separately (Figure 12.6).

Types of Schedules Interior designers create schedules for furniture, fixtures, equipment, doors, lighting fixtures, and hardware. A simple furniture schedule describes all the furnishings, such as sofas, chairs, and table or floor lamps. A **furniture fixtures and equipment (FFE)** schedule describes all moveable equipment; bookcases, filing cabinets, computers, electronic equipment, partition walls, etc. A **door schedule** describes all of a building's doors. A **hardware schedule** describes the hinges, handles, locks, knobs, etc., on the doors. A **lighting schedule** describes all the permanently attached lighting fixtures.

Furniture Fixture and Equipment schedules are done for commercial projects. The callouts are usually codes instead of simple numbers or letters. For instance, *C* means chair and *T* means table. The code *C-1* means the first type of chair; *T-3* is the third type of table; *CR-1* the first type of credenza. *T-1/100* is the first type of table in Room 100 (Figure 12.7).

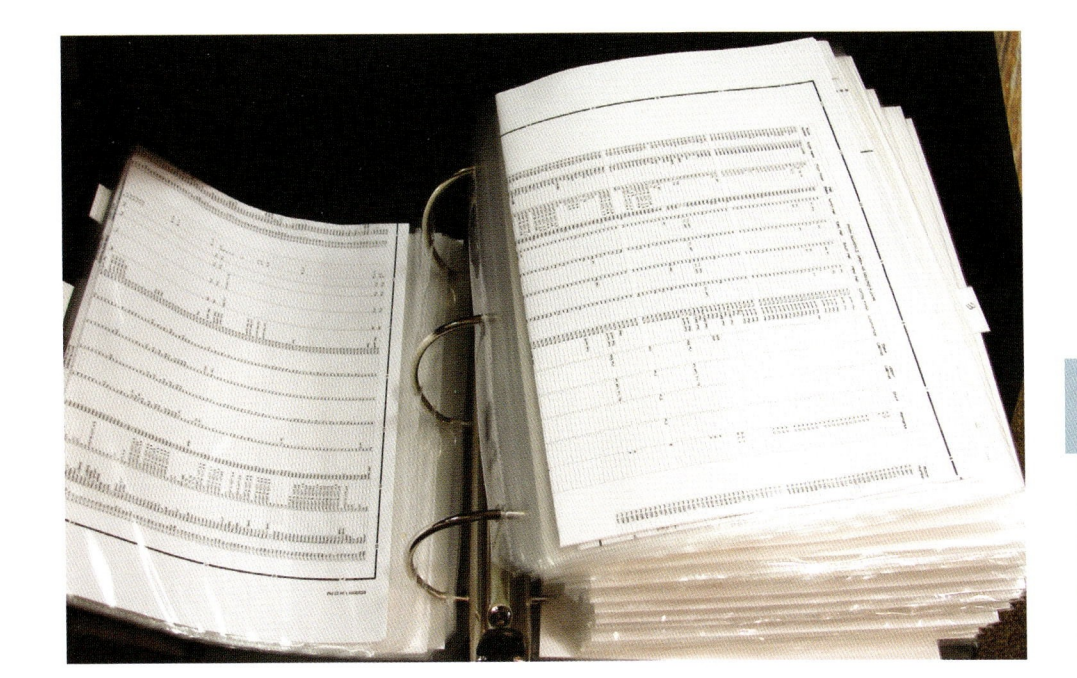

Figure 12.6
Schedules for a care clinic project.

> **Exercise**
> Look around your classroom and write a schedule for the furniture, fixtures and equipment in it.

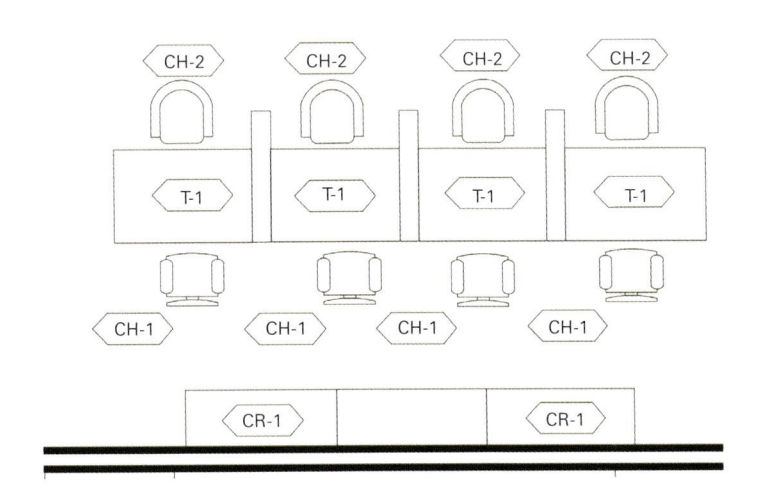

FFE SCHEDULE							
MARK	QTY	MFGR CATALOG #	DESCRIPTION	FABRIC	FINISH	REMARKS
C-1	42	SAFCO-MAYLINE	OFFICE CHAIR 25"W x 21"D x 16"H	C-1	DARK BLUE	CLASS A FLAME SPREAD FABRIC
C-2	38-1	SAFCO-MAYLINE	OFFICE CHAIR 25"W x 21"D x 16"H	C-1	DARK BLUE	CLASS A FLAME SPREAD FABRIC
T-2	42	SAFCO-MAYLINE	OFFICE CHAIR 60"W x 36"D x 32"H	SOLID CHERRY TOP	NATURAL STAIN	BRUSHED CHROME BASE
CR-1	10	HERMAN MILLER	CREDENZA 72"W x 18"D x 32"H	SOLID CHERRY TOP	NATURAL STAIN	BRUSHED CHROME BASE

Figure 12.7
FFE callouts and schedule.

SCHEDULES **319**

Hardware schedules should include:

- Type, style, function, size, label, hand, and finish of each door hardware item.
- Manufacturer of each item.
- Fastenings and other pertinent information.
- Location of door hardware set, cross-referenced to Drawings, both on floor plans and in door and frame schedule.
- Explanation of abbreviations, symbols, and codes contained in schedule.
- Mounting locations for door hardware.
- Door and frame sizes and materials.

On commercial projects it's common to write a hardware set for each door and then reference that set in a column on the door schedule. For example, Door 101 might have the set shown in Figure 12.8. The schedule would reference it next to the door it applies to.

Figure 12.8
Hardware set.

HARDWARE SET #1			
QTY	ITEM	MODEL/SIZE	MFGR
6	Heavyweight Hinge	HTA786 x NRP 4 1/2" x 4 1/2"	McKinney
2	Magnetic Lock	M680BDX	Securitron
2	Push Plate	70C	Rockwood
2	Pull Plate	110 x 70C	Rockwood
2	Automatic Operator	5910/5930 (as required)	Norton
2	Armor Plate	K1050 F 36" x 2" LDW 4BE	Rockwood
2	Silencer	608	Rockwood
1	Sensor	DPS	Dorma Door Controls
1	Push Button	EEB3N	Securitron
1	Power Supply	BPS-24-2	Securitron
1	Card Reader	Provided by Owner	

Creating a Schedule   You can use a 3D modelling program, a 2D drawing or spreadsheet program, or draft one manually. The most efficient way is with modelling software. All 3D programs, whether they're BIM or not, have a Tables function that hot-links components to a chart. The information you enter about a component in the drawing layers automatically populates the chart. When you change those components, the table updates.

Hand-entering information into a Microsoft Excel spreadsheet or into AutoCAD's Tables tool is another way to make a schedule. Make the grid row depths whatever size is needed to fit the information. They don't all have to be the same depth. If you're hand drafting a schedule, letter the information within the grid lines, not on them, and use 1/8" letters. Place ID labels under schedules as with other drawings.

Figures 12.8–12.16 show different schedule types and layouts. Note the different information in the header boxes, such as mounting height above finish floor; compass direction of walls; and manufacturer/model numbers. You can combine the information from any of them into a schedule that suits your own needs.

> **Tip**
> Autodesk Revit has different types of schedules: Basic (List and Quantify all Elements), Sheet+View Lists, Material Takeoff (calculate materials), and Note Block (organize plan notes). Familiarize yourself with them for ideas on what to include on your own schedules.

FURNITURE SCHEDULE

Mark	Item	Mfgr	Finish	Dimensions	Qty	Product #

EQUIPMENT SCHEDULE

Mark	Item	Mfgr	Finish	Dimensions	Qty	Product #

Figure 12.9 Furniture and equipment schedule layouts.

FURNITURE SCHEDULE

Key	Item	Mfgr	Finish	Model #	Qty	Fabric
A	Low Console	Century	Old World Mahogany	779-706	3	
B	Rug	Hellenic	Linear, Gold/Camel	TC04	3	
C	Sofa	Baker	Java	211-84-9	3	Vanguard 150347
D	Lounge Chair	Baker	Java	657	12	Kravet 19992
E	Club Chair	Baker	Fruitwood	445	6	Kravet 25649-19
F	Bar Stools	Kincaid	Fruitwood	96-081	12	Sentinel Fabic 12431
G	Side table	Accents Beyond	Fruitwood	F-1090	4	

Figure 12.10 Furniture schedule.

SCHEDULES

ROOM FINISH SCHEDULE							
NO	ROOM NAME	FLOOR	BASE	WALLS	CEILING	HEIGHT	NOTES

Floor columns: Sealed Concrete, Carpet, Sheet Vinyl, Vinyl Composition Tile
Base column: Vinyl Base
Walls columns: Painted Block, Epoxy Painted Block, Painted Drywall, Exposed Brick
Ceiling columns: Exposed Structure-Pnt, 2x4 Lay-In, 5/8" Drywall Suspended, 5/8" Drywall on Framing

NO	ROOM NAME	Floor	Base	Walls	Ceiling	HEIGHT	NOTES
300	ENTRY	Sheet Vinyl	Vinyl Base	Painted Drywall	2x4 Lay-In	8'-0"	
301	LOBBY	Sheet Vinyl	Vinyl Base	Painted Block, Epoxy Painted Block	2x4 Lay-In	8'-6"	PAINTED GYP BD
302	HALLWAY	Sheet Vinyl	Vinyl Base	Painted Block	2x4 Lay-In	8'-6"	
303	WOMEN'S BATHROOM	Sheet Vinyl	Vinyl Base	Painted Drywall	5/8" Drywall Suspended	8'-0"	
304	MEN'S BATHROOM	Vinyl Composition Tile	Vinyl Base	Painted Drywall	5/8" Drywall Suspended	8'-0"	
305	OFFICE	Vinyl Composition Tile	Vinyl Base	Painted Block	2x4 Lay-In	8'-6"	
306	STORAGE	Vinyl Composition Tile	Vinyl Base	Painted Block	5/8" Drywall Suspended	8'-0"	
307	LIBRARY	Sealed Concrete	Vinyl Base	Painted Block	2x4 Lay-In	8'-6"	
308	MECHANICAL	Vinyl Composition Tile		Painted Drywall	Exposed Structure-Pnt	—	
309	HALLWAY	Vinyl Composition Tile	Vinyl Base	Painted Block	2x4 Lay-In	8'-6"	
310	HALLWAY	Vinyl Composition Tile	Vinyl Base	Painted Block	5/8" Drywall Suspended	8'-6"	
311	MAINTENANCE	Sealed Concrete		Painted Block	5/8" Drywall Suspended	9'-0"	
312	GARAGE	Sealed Concrete		Painted Block	Exposed Structure-Pnt	9'-0"	PAINTED GYP BD
313	STORAGE	Sealed Concrete		Painted Block	5/8" Drywall on Framing	9'-0"	
314	STORAGE	Sealed Concrete		Painted Block	5/8" Drywall on Framing	9'-0"	
315	OFFICE			Painted Block	2x4 Lay-In	9'-0"	
316	CONFERENCE ROOM	Vinyl Composition Tile	Vinyl Base	Painted Block	2x4 Lay-In	9'-0"	
317	HALLWAY	Vinyl Composition Tile	Vinyl Base	Painted Block	2x4 Lay-In	8'-6"	
318	OFFICE	Carpet	Vinyl Base	Painted Block	2x4 Lay-In	9'-0"	
319	OFFICE	Carpet	Vinyl Base	Painted Block	2x4 Lay-In	9'-0"	
320	KITCHEN	Vinyl Composition Tile	Vinyl Base	Painted Drywall	5/8" Drywall Suspended	8'-6"	
321	BATHROOM	Sheet Vinyl	Vinyl Base	Painted Drywall	5/8" Drywall Suspended	8'-0"	
322	UTILITY CLOSET	Vinyl Composition Tile	Vinyl Base	Painted Drywall	5/8" Drywall Suspended	8'-0"	

Figure 12.11
Room finish schedule.

ROOM FINISH SCHEDULE								
Room #	Room Name	Floor	Walls				Notes	
			N	S	E	W	Matl	Ht

Figure 12.12
Room finish schedule.

DOOR SCHEDULE

#	SIZE	THK	DOOR MAT'L	FRAME MAT'L	GLASS	LOCKSET	FIRE RATING	FINISH	HW SET	COMMENTS
200	2'8" x 6'8"	1 3/8	WD	WD		CLASSROOM		STAIN	1	
201	3'0" x 6'8"	1 3/8	H.M.	WD	1" INSULATED	ENTRY			2	1'-8" SIDELIGHT
202	3'0" x 6'8"	1 3/8	H.M.	WD		ENTRY			2	
203	PR 2'4" x 6'8"	1 3/8	H.M.	WD		NONE			3	BI-PASS HARDWARE
204	2'6" x 6'8"	1 3/8	HM	HM		PRIVACY	20 min		4	
205	PR 2'6" x 6'8"	1 3/8	WD	WD		NONE		STAIN	3	
206	2'0" x 6'8"	1 3/8	HM	HM		PRIVACY	20 min		5	
207	(4) 1'0" x 6'8"	1 3/8	WD	WD		NONE		STAIN	5	BI-FOLD HARDWARE
208	2'0" x 6'8"	1 3/8	WD	WD		PRIVACY		STAIN	6	
209	2'0" x 6'8"	1 3/8	WD	WD		PRIVACY		STAIN	6	
210	2'6" x 6'8"	1 3/8	WD	WD		PRIVACY		STAIN	6	
211	2'0" x 6'8"	1 3/8	HM	HM		PASSAGE	20 min		7	
212	2'6" x 6'8"	1 3/8	WD	WD		PRIVACY		STAIN	6	
213	2'6" x 6'8"	1 3/8	WD	WD		PASSAGE		STAIN	6	

Figure 12.13 Door Schedule.

BATHROOM ACCESSORIES SCHEDULE

CALL-OUT	A	B	C	D	E	F	G	H	I	J	K	L	M	N
	TOWEL DISPENSER/ WASTE RECEPTACLE	TOWEL DISPENSER	MIRROR 24" x 36"	FOLDING SHELF	2 COMPARTMENT SANITARY DISPOSAL	1 COMPARTMENT SANITARY DISPENSER	SANITARY RECEPTACLE	TOILET PAPER DISPENSER	GRAB BAR	DOUBLE ROBE HOOK	TOWEL PIN	RECESSED SOAP DISPENSER	RECESSED SOAP DISH	CURTAIN ROD
MOUNTING HEIGHT A.F.F.	5'-6"	5'-0"	6'-0"	4'-0"	2'-6"	2'-6"	6'-0"	2'-6"	3'-0"	5'-0"	5'-0"	3'-6"	4'-0"	6'-9"

ROOM #	ROOM NAME	A	B	C	D	E	F	G	H	I	J	K	L	M	N
106	WOMEN'S NO. 1	1			2	1			2	2					
107	MEN'S NO. 1	1							1	2					
108	MEN'S DRESSING RM. 1			1	1				1						
109	WOMEN'S DRESSING. RM. 1			1	1				1						
121	TOILET NO. 1			1	1				1						
129	STAR DRESSING RM.			1	1				1		1	1		1	1
130	WOMEN'S NO. 2								1		1	1			
132	TOILET NO. 2		1	1					1						

Figure 12.14 Bathroom accessories schedule.

⬡	WIDTH	HEIGHT	MATERIAL	GLASS	TYPE	NOTES
			WINDOW SCHEDULE			
A	3'-0"	3'-0"	METAL-CLAD	3/8" INSULATED	FIXED	STAINED GLASS INSERT
B	3'-0"	3'-0"	D.O.	D.O.	FIXED	STAINED GLASS INSERT
C	D.O.	D.O.	D.O.	D.O.	FIXED	
D	D.O.	D.O.	D.O.	D.O.	DOUBLE-HUNG	
E	D.O.	D.O.	D.O.	D.O.	DOUBLE-HUNG	
F	5'-0"	5'0"	D.O.	1" INSULATED	DOUBLE-HUNG	
G	6'-0"	5'-0"	D.O.	1" INSULATED	DOUBLE-HUNG	
H	6'-0"	6'-8"	D.O.	1" TEMPERED INSULATED	SLIDING GLASS	

Figure 12.15
Window schedule

Leilani Lynn Residence Maui, Hawaii

Type	Description	Manufacturer/ Model Number	Lamp and Voltage
Pendant CF/PD/S/Y 	Residential Ceiling Pendant Light Fixtures with Medium (E26) Base Yellow Sona	Sunlite 88730	35 watt, 120 volt
Under Cabinet LI24T4/CW 	Fixture with Plug White Finish, (Bulb Included) 34.5 inch	Sunlite 53056	24 watt fluorescent 120 volt
Vanity Lamp B318/CH 	Globe Style Fixture, Chrome Finish	Sunlite 45025	40 watt, voltage as determines by electrical engineer
Wall Sconce Fixture 	Rectangle Incandescent, Black Powder Finish	Sunlite 46092	35 watt, 120 volt

Figure 12.16
Lighting fixtures schedule

Summary

Legends and schedules are charts of information that provide information about symbols and components in the drawings. They're linked to the plan with callout symbols. This gives readers information on those subjects at a glance while keeping the drawings from becoming overly cluttered. Schedule layout and detail needed varies with project complexity.

Classroom Activities

1. Find a floor plan in a magazine and create separate schedules for its finishes, furniture, and lighting fixtures.
2. Create a room finish schedule for a floor plan you already have. Draw callouts in the floor plan to link it to the schedule.

Questions

1. What is a legend?
2. What is a schedule?
3. What is a callout?
4. Name three categories on a room finish schedule.
5. Name three categories on a lighting schedule.

Keywords

- callout
- door schedule
- furniture fixture and equipment (FFE) schedule
- furniture key
- furniture schedule
- hardware schedule
- ID label
- key
- key letter
- key number
- legend
- lighting fixture schedule
- mark
- schedule

CHAPTER 13

Hand Drafting Isometric and Perspective Pictorials

Figure 13.0
3-point interior perspective view of a living room.

OBJECTIVES

Upon completion of this chapter you will be able to:

- Manually construct an isometric pictorial drawing from a plan and elevations
- Manually construct one and two-point interior perspective drawings from a plan and elevations
- Choose which type of drawing is most appropriate for a space

Isometric and perspective pictorials are 3D drawings used for both presentation and construction purposes.

Why Construct 3D Drawings by Hand?

The discipline of drafting a 3D drawing by hand strengthens the impromptu sketching skills that are so important during the conceptual design phase. This thought process helps a designer better understand the space. It also facilitates making and using a perspective grid, because those require prior knowledge of perspective drawing rudiments, such as what portion of the space will appear in the pictorial and how to construct heights. Finally, being able to manually create a pictorial enables communicating precise intent, as opposed to working with software limitations.

What Is an Isometric Pictorial?

This is a **paraline drawing**, meaning all line sets are parallel, or equidistant, to each other. An isometric drawing has one vertical axis and two axes that are skewed 30° to the horizontal. All axes have the same scale, and the resultant drawing is measurable. In fact, the word "isometric" means "equal measure." It doesn't show objects and space exactly as we see them but is a good approximation. Isometric is the most common drawing type for technical illustration, as many drafters prefer its trade-off of realism vs. time saved.

What Is a Perspective Pictorial?

This is a drawing whose sets of parallel lines converge to points on the horizon, which causes objects to appear smaller and closer together the farther away they are. It is a realistic drawing, as it shows space how we really see it. But because it doesn't show space as it exists, it is not measurable and is more time-consuming and difficult to construct than an isometric.

Draft an Isometric Pictorial of a Gable House

Figure 13.1 shows views of a gable roof house. Following are steps for drafting it. Use a divider for the measurements, since numeric lengths aren't needed. All vertical lines are parallel to the z-axis and to each other. All horizontal lines are parallel to the x- and y-axes and to each other. Draw non-isometric lines by finding and connecting their endpoints.

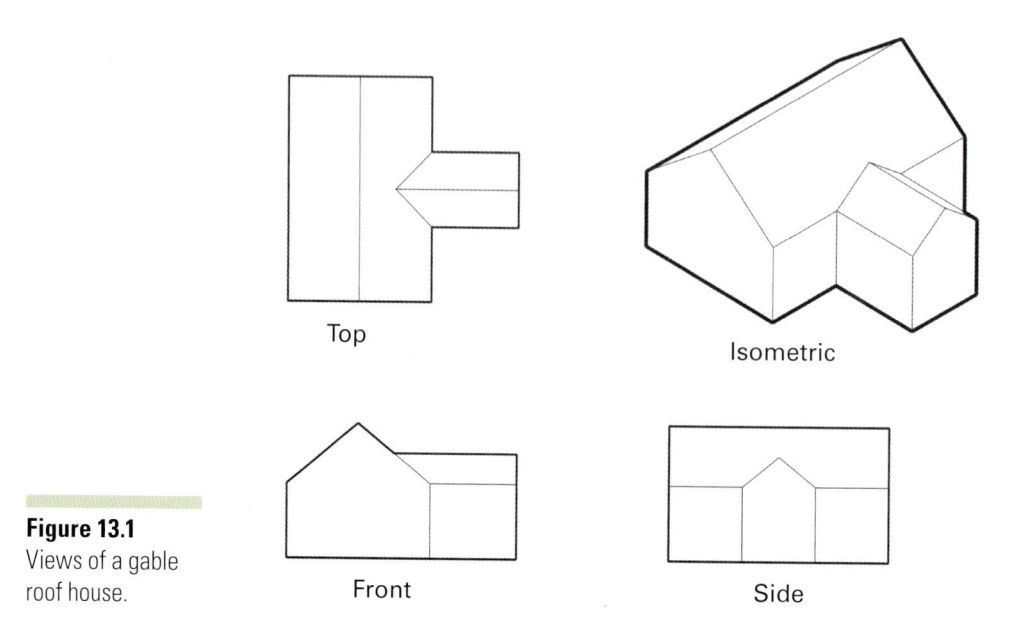

Figure 13.1
Views of a gable roof house.

1. Draw the x-, y-, and z-axes (Figure 13.2). The z-axis is vertical; make all height measurements on it. The x- and y-axes are drawn 30° to the horizontal. Draw all the house's horizontal lines on them. Then draw the perimeter of the main part of the house on them, using lengths obtained from the plan view.
2. *Draw the wall height* (Figure 13.3). Project vertical lines up from each corner and mark the wall height, which is obtained from the elevation view.
3. Find the roof ridge height. (Figure 13.4) First, find the center of the short wall by connecting its corners with diagonal lines. Where those lines intersect is the wall's center. Draw a vertical line through that center up to the ridge height, which is obtained from the elevation view. Then draw roof lines connecting to it as shown.
4. Draw the roof (Figure 13.5a, b, c). Align the triangle edge with the top of the wall, slide it through the roof ridge point and then through the opposite wall point.

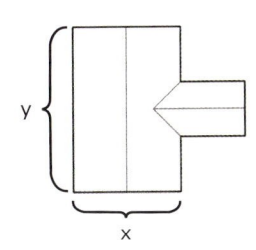

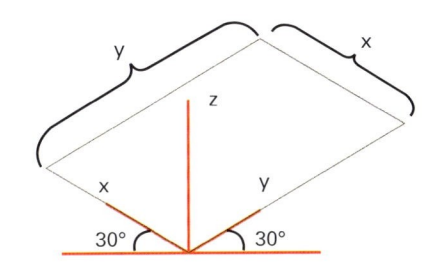

Figure 13.2
Step 1 Draw the axes.

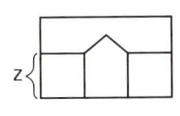

Figure 13.3
Step 2 Draw the wall height.

Figure 13.4
Step 3 Find the roof ridge height.

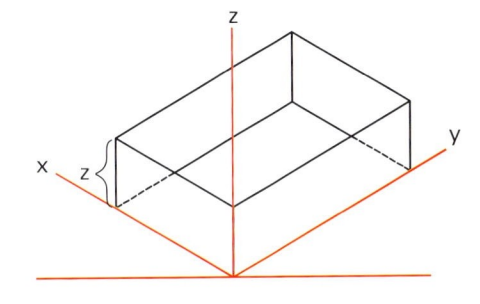

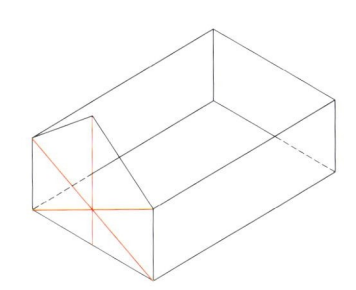

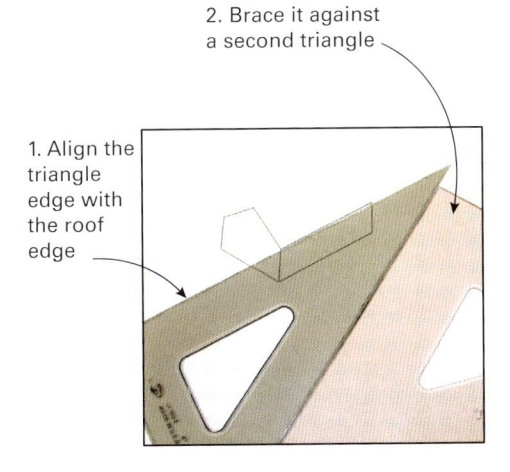

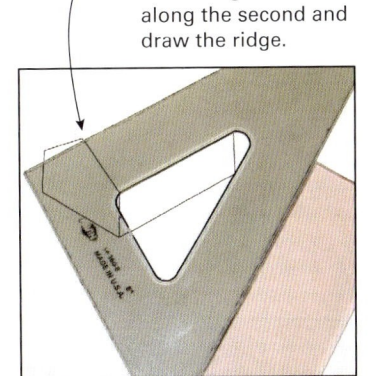

1. Align the triangle edge with the roof edge

2. Brace it against a second triangle

3. Slide the first triangle along the second and draw the ridge.

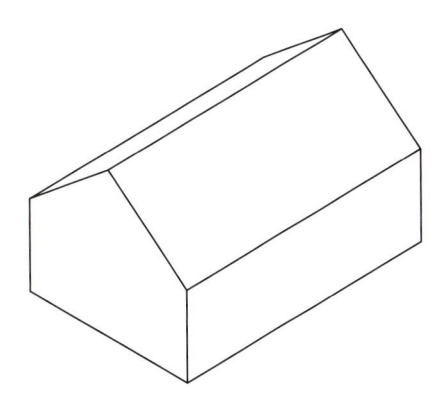

Figure 13.5 a,b,c
Step 4 Draw the roof.

DRAFT AN ISOMETRIC PICTORIAL OF A GABLE HOUSE

5. Draw the wing (Figure 13.6). Measure its lengths on the floor plan and mark them along the isometric axes.

6. Draw the wing's wall height and ridge height (Figure 13.7). Project the wall height up from each corner. Draw diagonal lines through the corners to find the wing's center as shown. Then draw a vertical line through that center up to the ridge point, obtaining its height from the elevation view.

7. Construct the wing's roof (Figure 13.8). Draw the wing's roof ridge, obtaining its length from the plan view. Connect the walls to it to form the roof.

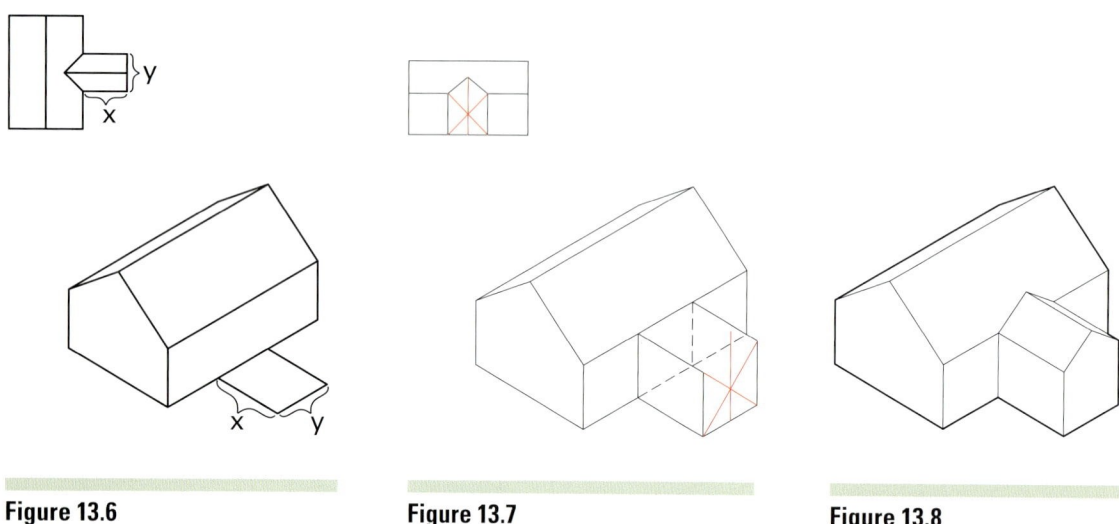

Figure 13.6
Step 5 Draw the wing.

Figure 13.7
Step 6 Draw the wing's wall and ridge height.

Figure 13.8
Step 7 Construct the wing's roof.

Isometric Circle   A circle appears as an **ellipse** in isometric (Figure 13.9), which is a foreshortened circle. You can see why by holding a circle template up and slowly rotating it. The circles on the template narrow (foreshorten) and become ellipses. When the template is rotated 90° to your eye, the circles become straight lines. Being able to manually construct one with a compass is useful if a template isn't available.

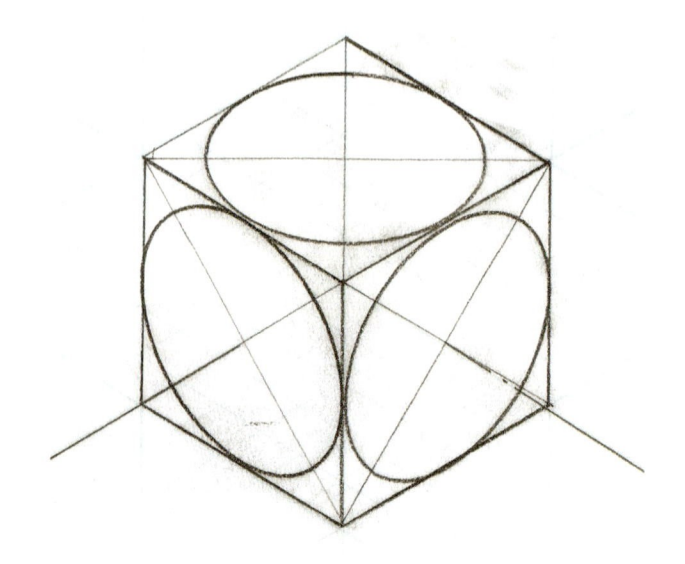

Figure 13.9
Circles in isometric appear as ellipses.

330   CHAPTER 13   HAND DRAFTING ISOMETRIC AND PERSPECTIVE PICTORIALS

Draft an Isometric Circle

Figure 13.10 shows views of a circular table. Following are steps for drafting it.

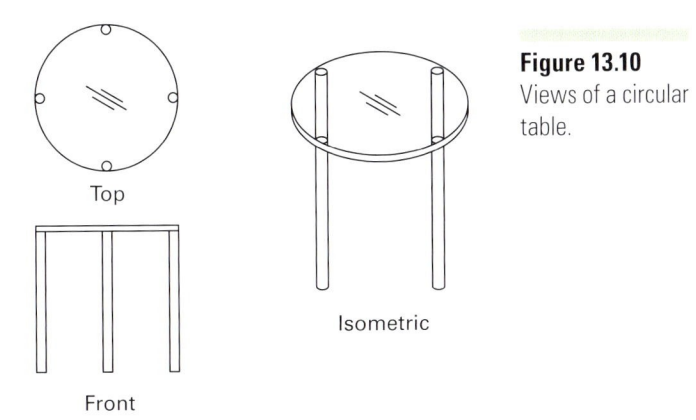

Figure 13.10
Views of a circular table.

1. *Draw an isometric square* with its length and width equal to the tabletop's diameter (Figure 13.11). Then draw lines inside it as shown.

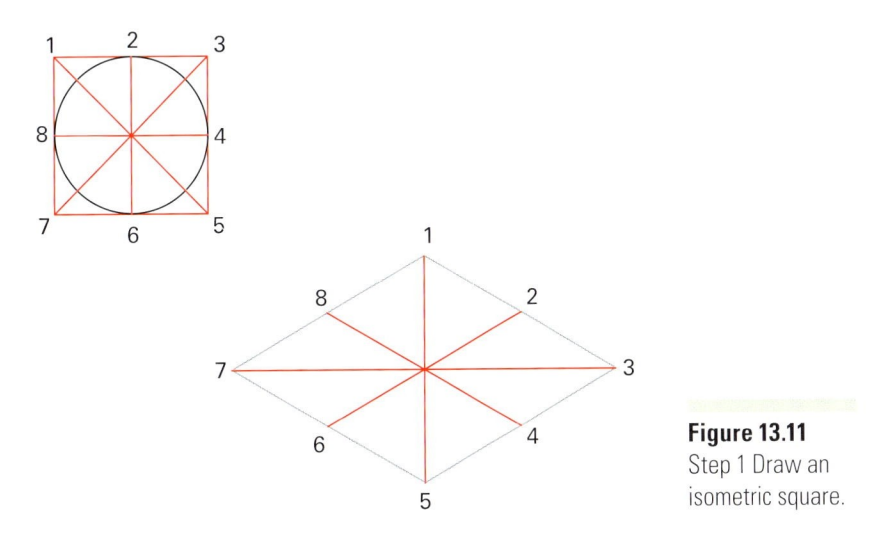

Figure 13.11
Step 1 Draw an isometric square.

2. *Connect the points as shown* (Figure 13.12). The circled intersections are the first two compass swing points

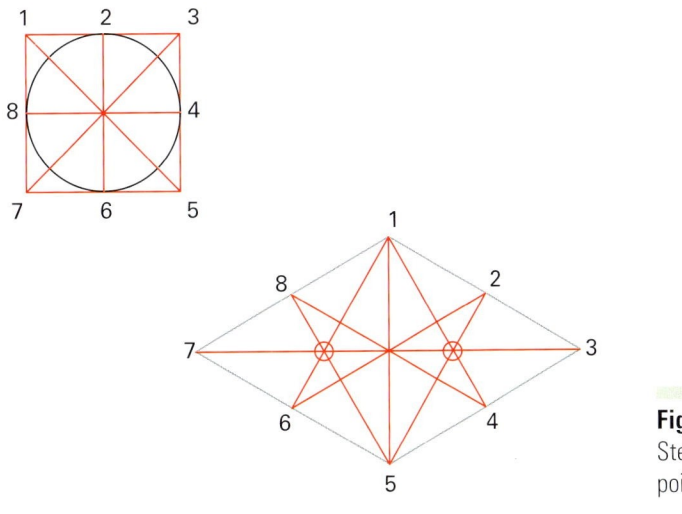

Figure 13.12
Step 2 Connect the points.

DRAFT AN ISOMETRIC PICTORIAL OF A GABLE HOUSE   **331**

3. *Swing two arcs* at the circled swing points as shown (Figure 13.13).

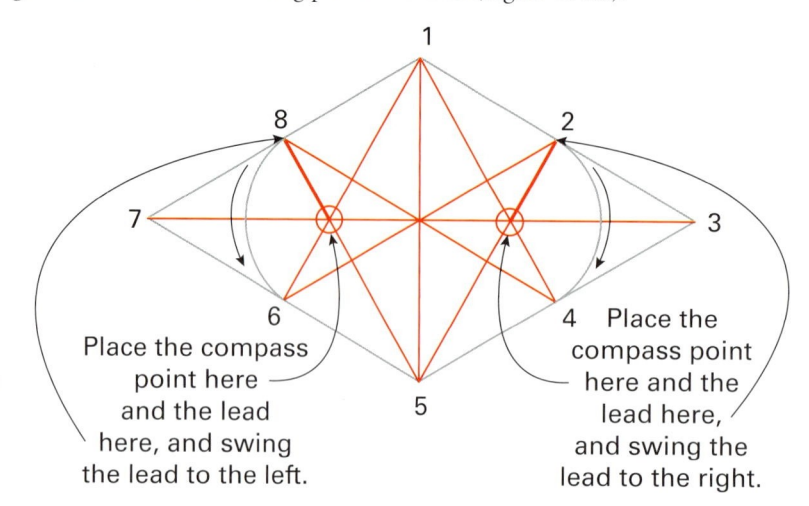

Figure 13.13 Step 3 Swing two arcs.

Place the compass point here and the lead here, and swing the lead to the left.

Place the compass point here and the lead here, and swing the lead to the right.

4. *Swing a second pair of arcs* at the circled swing points as shown (Figure 13.14).

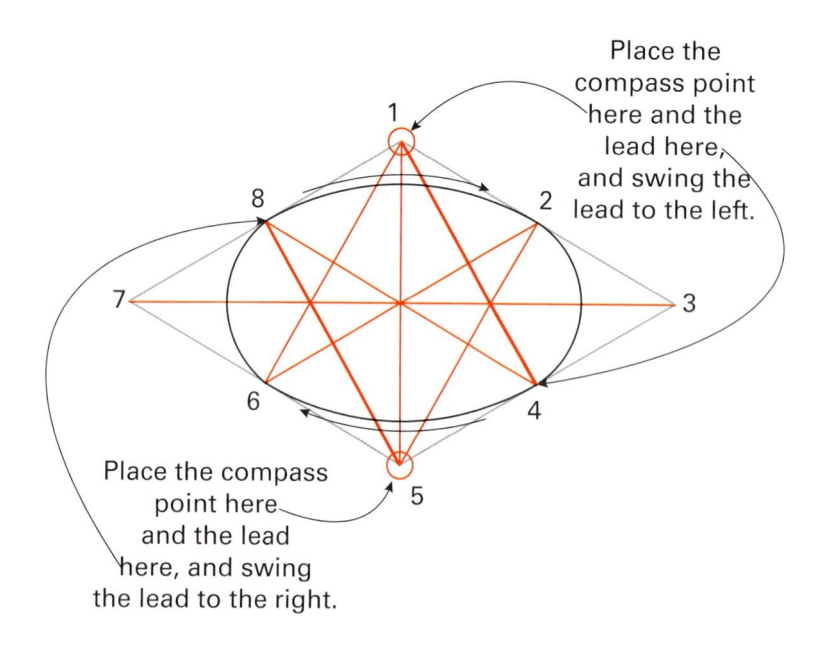

Place the compass point here and the lead here, and swing the lead to the left.

Place the compass point here and the lead here, and swing the lead to the right.

Figure 13.14 Step 4 Swing a second pair of arcs.

5. *Draw the legs and tabletop thickness* (Figure 13.15). Draw the leg tops parallel to the ellipse axis. Leave some space between them and the ellipse edge to represent the glass thickness. Draw the legs the length shown in the elevation view. Finally, add thickness to the tabletop with an irregular curve; move it around to copy the ellipse's curve (Figure 13.15).

Leave space to represent tabletop thickness

Tabletop thickness

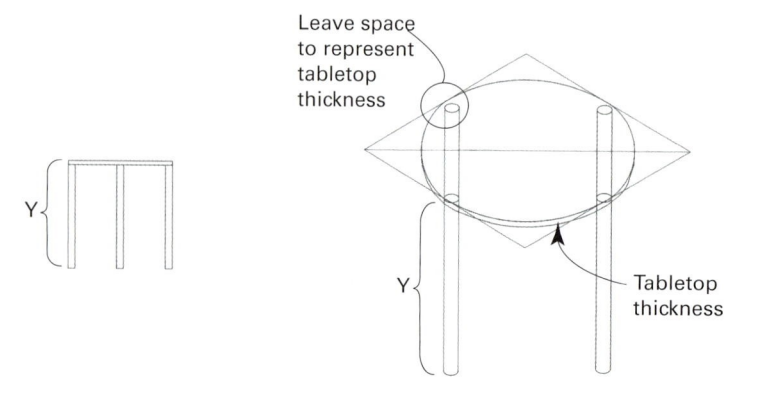

Figure 13.15 Step 5 Draw the legs and tabletop thickness.

332 CHAPTER 13 HAND DRAFTING ISOMETRIC AND PERSPECTIVE PICTORIALS

Cutaway Isometric

This drawing omits or partially removes walls and roof to show what is behind them. It's a useful drafting technique when you want to show a 3D space but don't want to construct a perspective.

Draft a Cutaway Isometric Room Let's draw a room using the plan and exterior elevations in Figure 13.16.

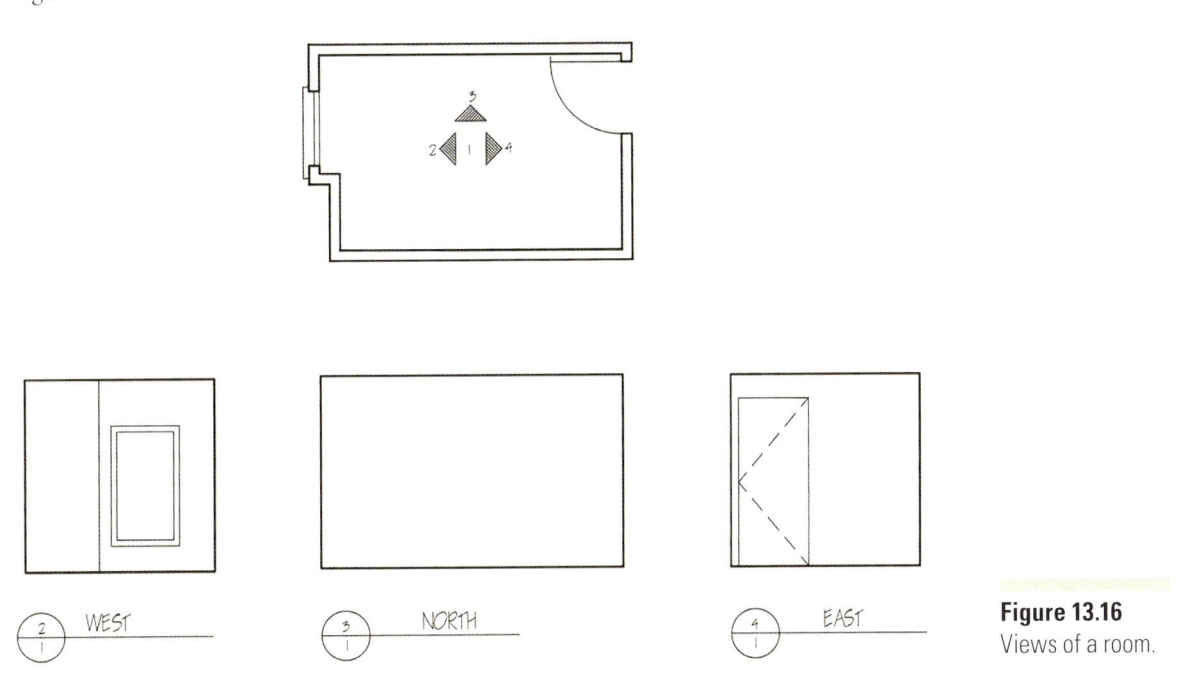

Figure 13.16
Views of a room.

1. *Draw the plan's perimeter along the isometric axis* (Figure 13.17). Draw the whole rectangle first to ensure the overall dimensions are correct, and then mark off the niche.

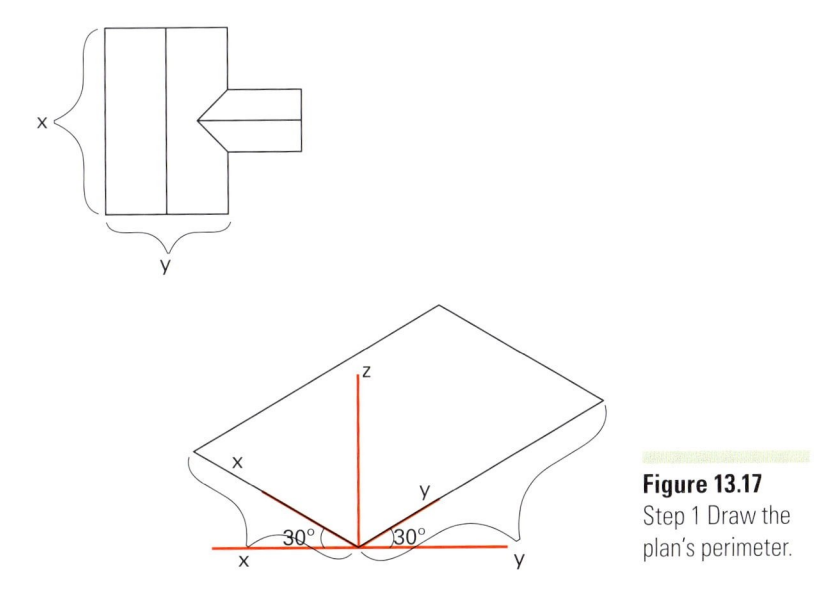

Figure 13.17
Step 1 Draw the plan's perimeter.

2. *Project the wall, window, and door heights up from the plan.* (Figure 13.18).

3. *Cut away the walls* as needed to show interior detail (Figure 13.19). Angle the cuts so they're not confused with the wall lines. Add line weight.

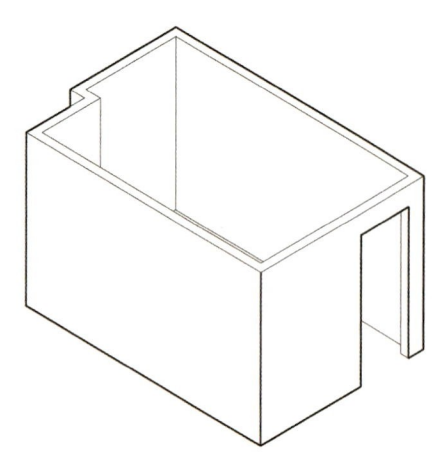

Figure 13.18
Step 2 Project heights up.

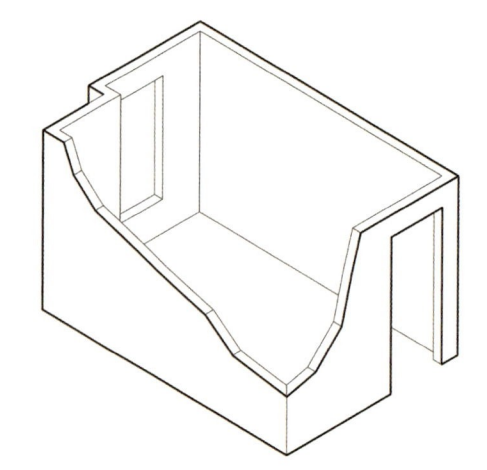

Figure 13.19
Step 3 Cut away the walls.

Draft a Cutaway Isometric of a Floor Plan Following are steps for drafting a cutaway isometric of the residential floor plan in Figure 13.20.

Exercise
Sketch and then draft a floor plan and cutaway isometric of a room in your home. Include all furniture.

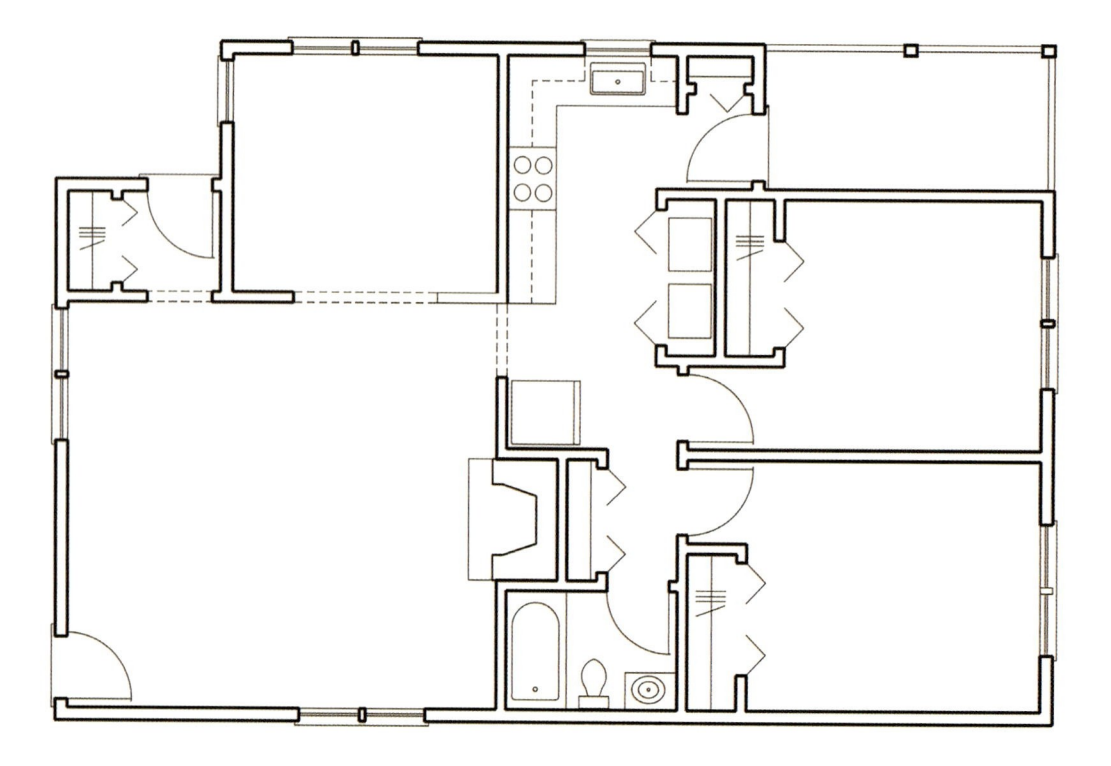

Figure 13.20
Residential floor plan.

1. *Draw the plan along the isometric axis* (Figure 13.21). Start at corner A and then measure and mark the perimeter. Then measure and mark the interior walls.

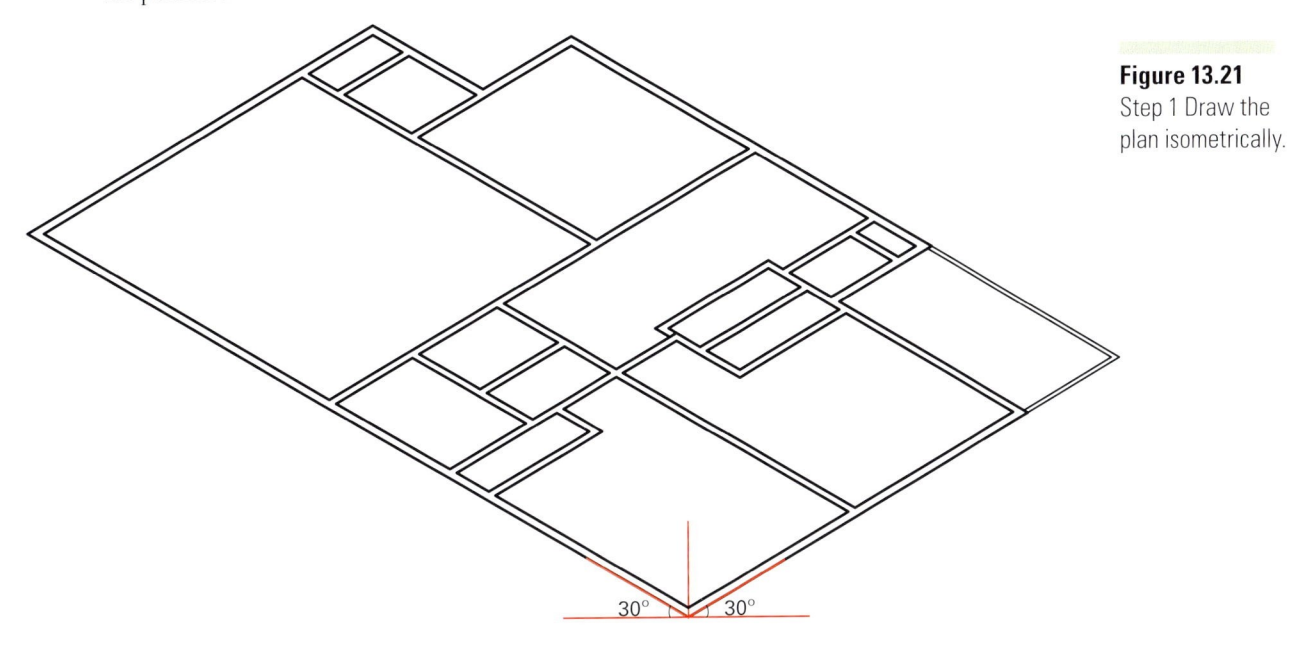

Figure 13.21
Step 1 Draw the plan isometrically.

2. *Measure and mark the window and door openings* (Figure 13.22).

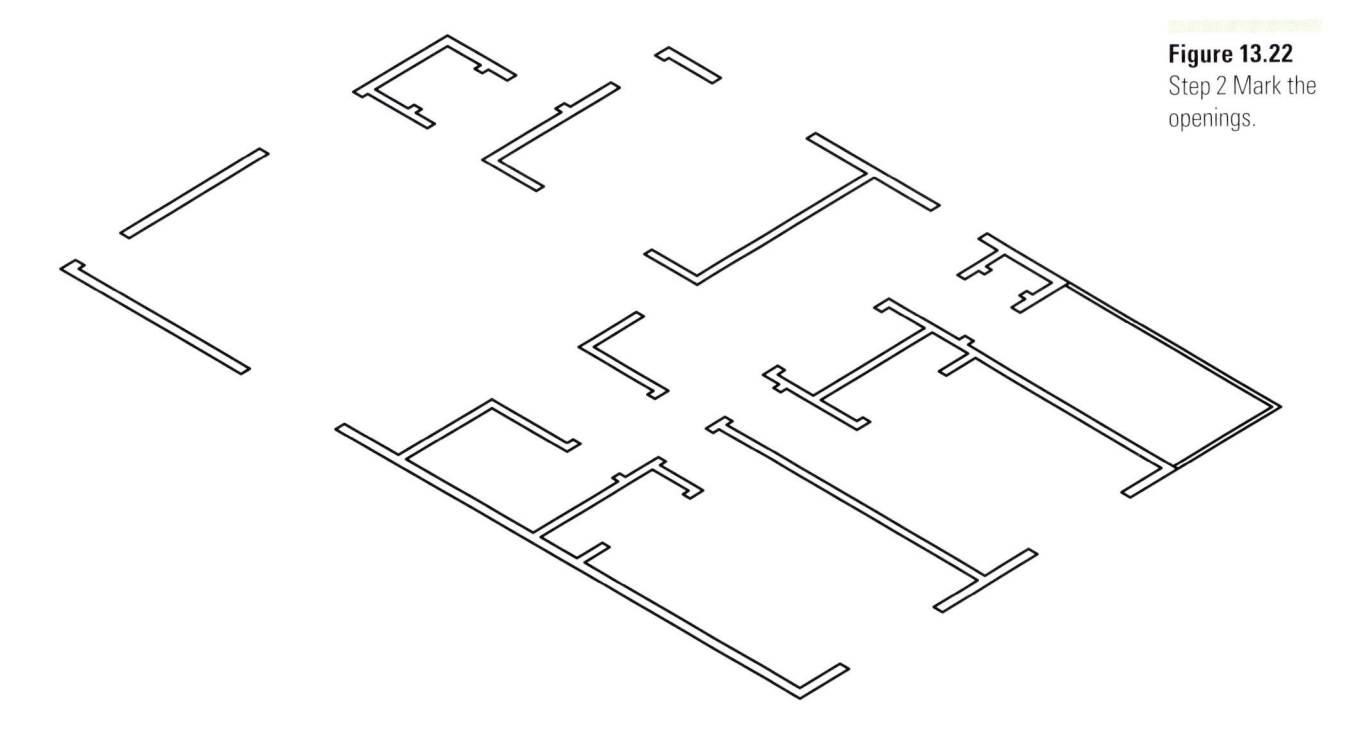

Figure 13.22
Step 2 Mark the openings.

3. *Project the walls up.* Measure them up from the floor line and check different walls to ensure they are a consistent height. Mark the window and door opening locations (Figure 13.23).

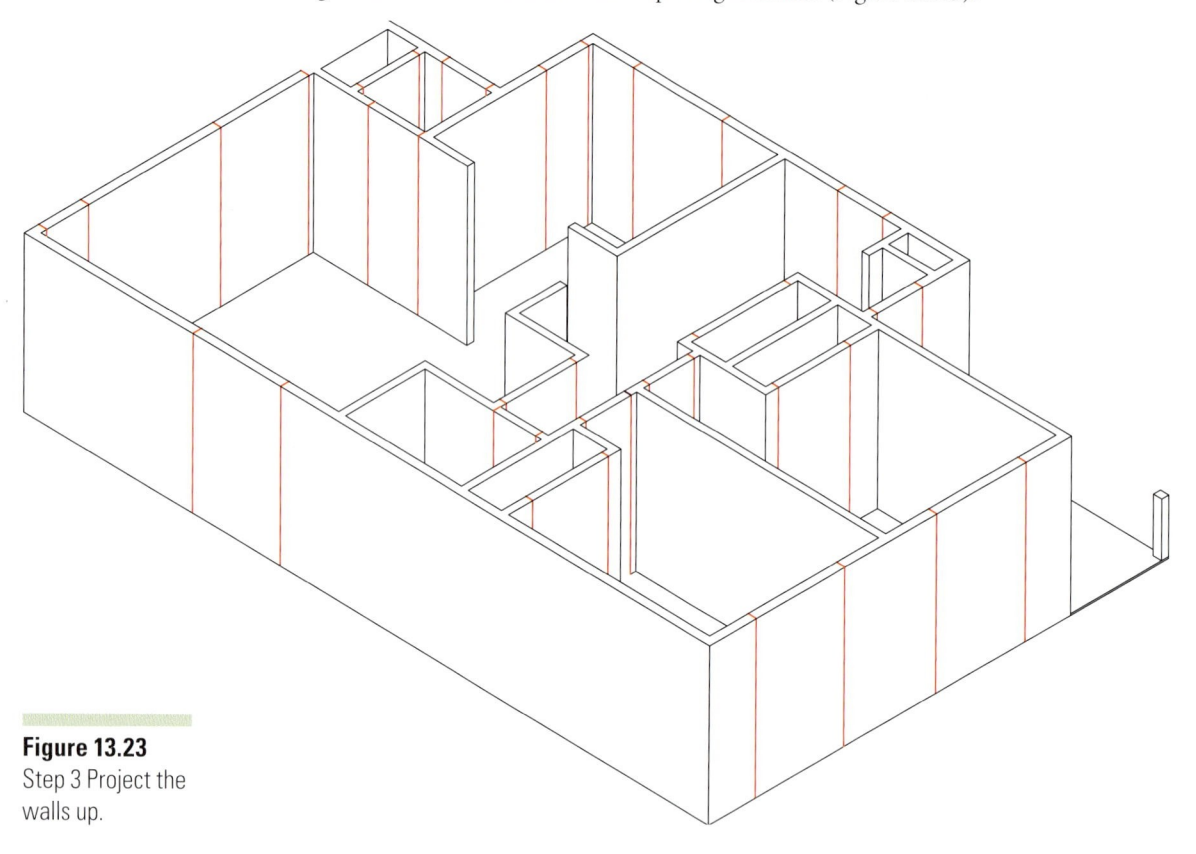

Figure 13.23
Step 3 Project the walls up.

4. *Measure and mark the window and door heights* (Figure 13.24).

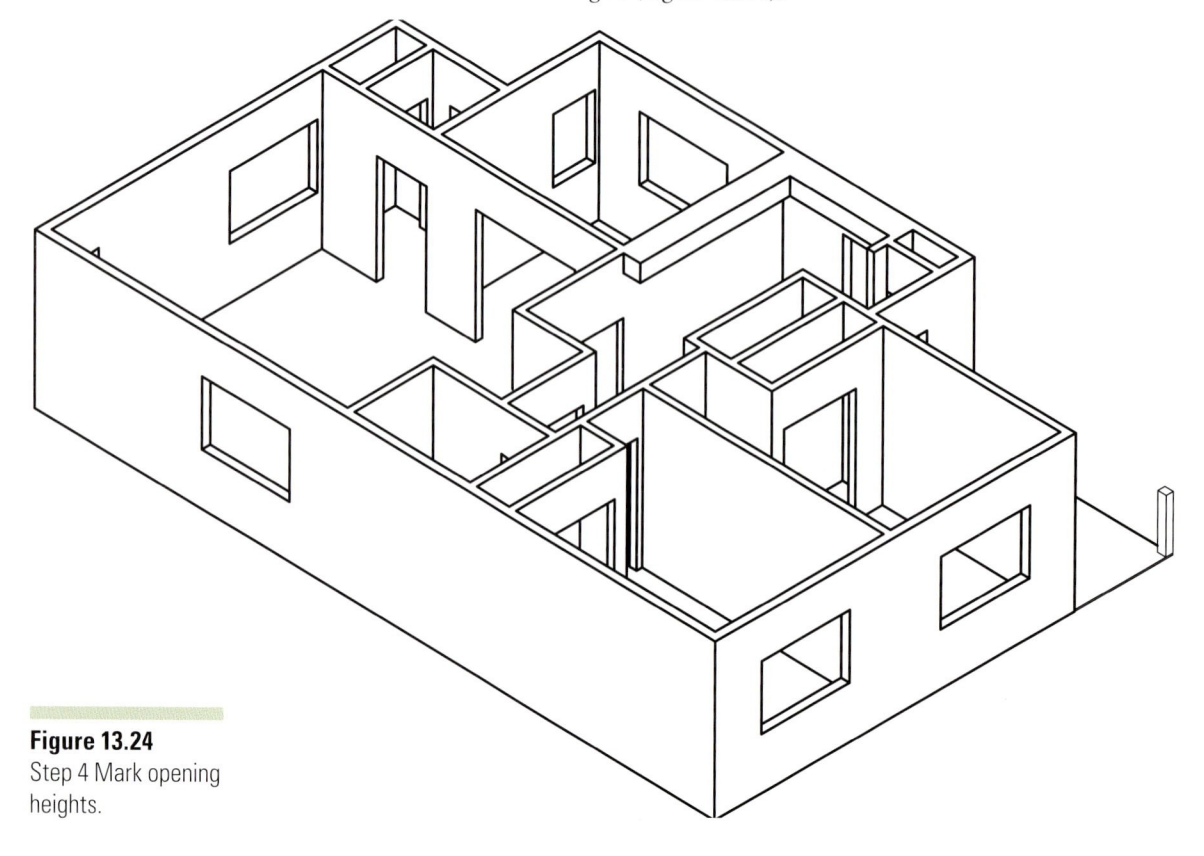

Figure 13.24
Step 4 Mark opening heights.

5. *Add furniture, cabinetry* and other relevant interior details (Figure 13.25).

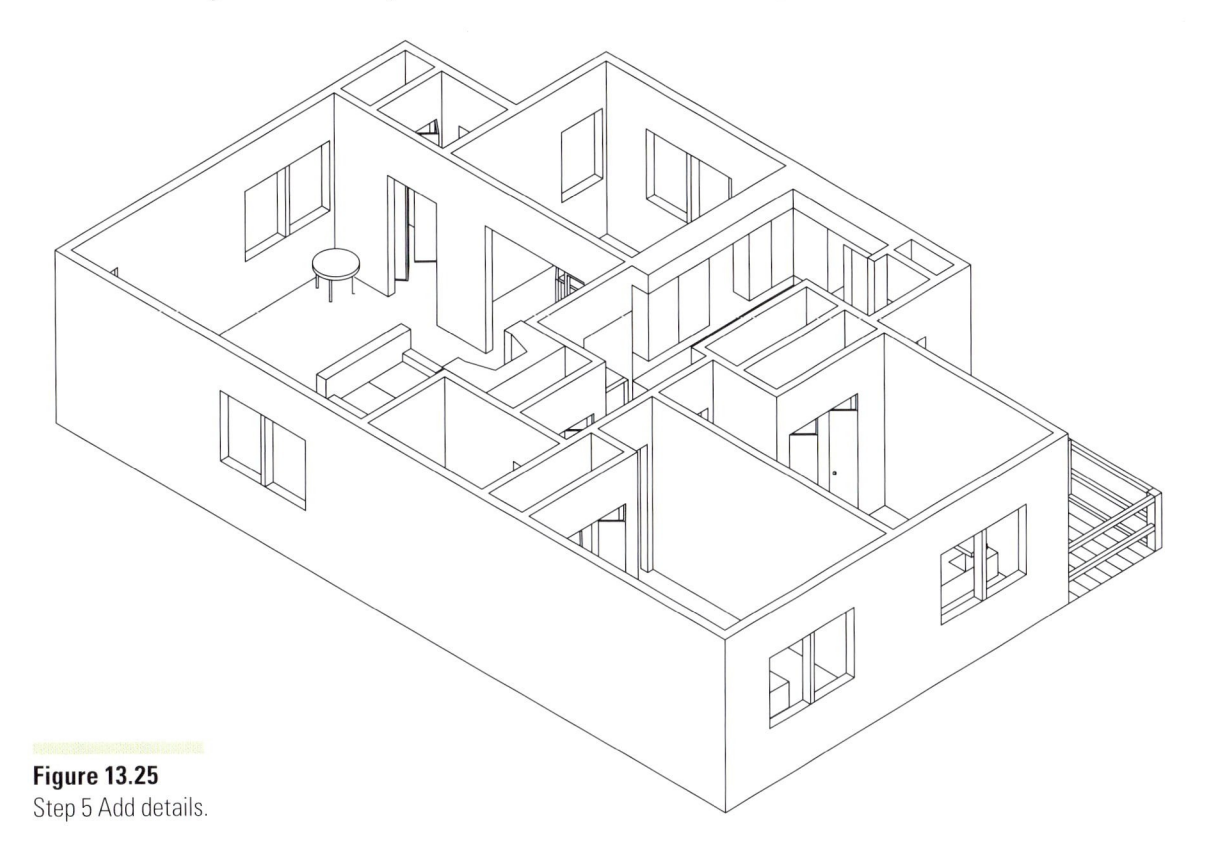

Figure 13.25
Step 5 Add details.

6. *Cut away walls* in front of important features to display them (Figure 13.26). Angle the cuts so they are not confused with the wall lines.

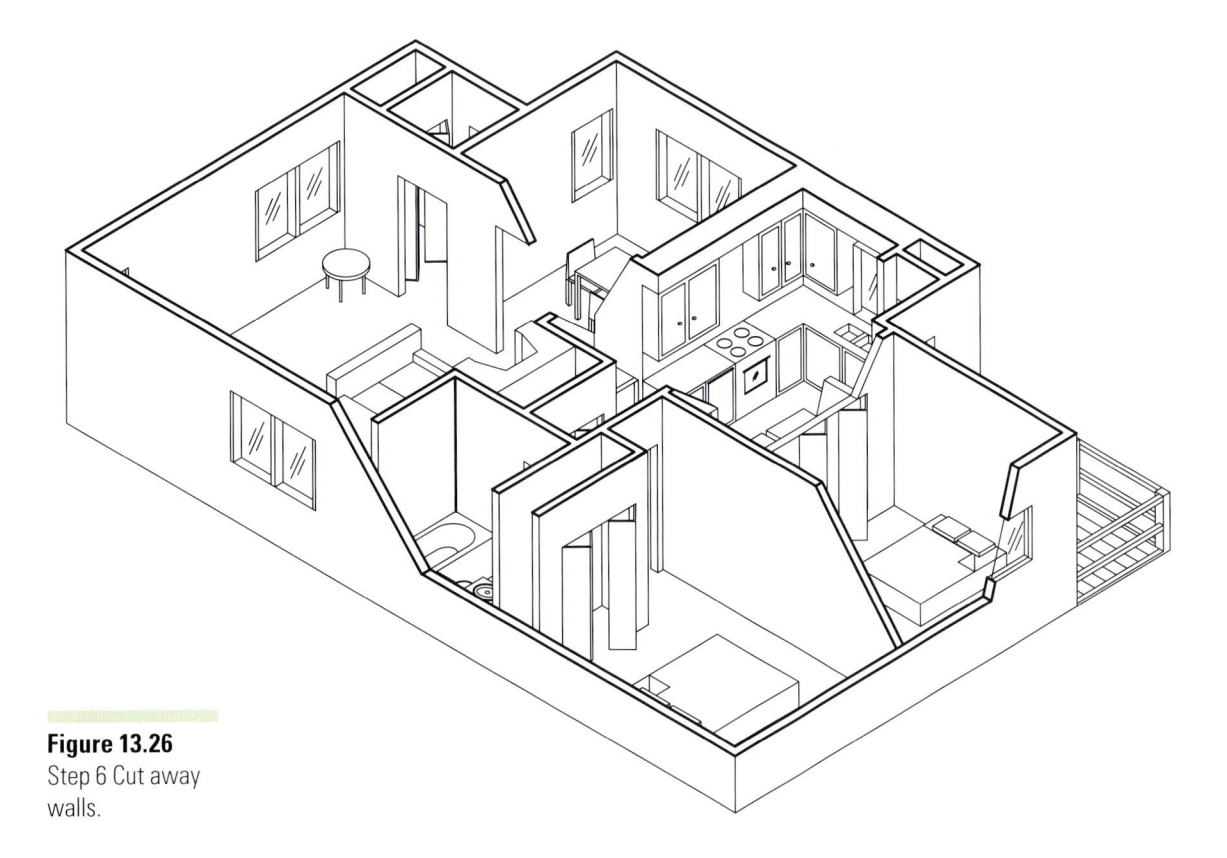

Figure 13.26
Step 6 Cut away walls.

CUTAWAY ISOMETRIC **337**

Perspective Drawing Manipulation

Exploit perspective drawings to emphasize specific features and angles (Figure 13.27). For instance, the **station point**, which is viewer location, can be inside or outside the room. Standing outside shows more of the space and standing inside takes the viewer deep into it. Standing in the center emphasizes the whole room, while moving to the side emphasizes one wall. The latter technique enables one side of the drawing to display product, and the former enables it to display depth.

FYI
Modern perspective construction theory was developed during the Renaissance when Italian artists applied Euclidean geometry to their art. They perfected perspective drawing as a means of showing item size and shape as a direct function of distance and viewpoint.

Figure 13.27
The top view emphasizes a wall; the bottom emphasizes the whole room.

Understanding how different station points create different views is critical to creating good perspective drawings. For instance, a station point too far away gives a viewed-from-afar look, and one too deep in the room shows just a small portion. Develop station point selection skills by photographing existing spaces. Some tips:

- Stand in different places and analyze the resultant photos. Which cause nearby furniture to loom large, blocking out other things?
- Photograph a room at different eye levels by standing on stairs or crouching.
- Hold the camera level with the floor to avoid the vertical line tilting that results when the camera is pointed up or down.
- We see within a 60° **cone of vision** (COV). This is an angle that defines the limits of our view. To understand the cone of vision, hold a 30°/60° triangle parallel to the floor, with the 60° corner touching your nose. The triangle's edges define the cone of vision. You see what lies between the edges. Items immediately outside the edges appear in your peripheral vision.
- Stand in one place, choose a view, and sketch what you see (Figure 13.28). Lock your gaze in that direction; do not swivel your head because a perspective drawing shows only one view. Unlike real space, you cannot look around to build up an image.

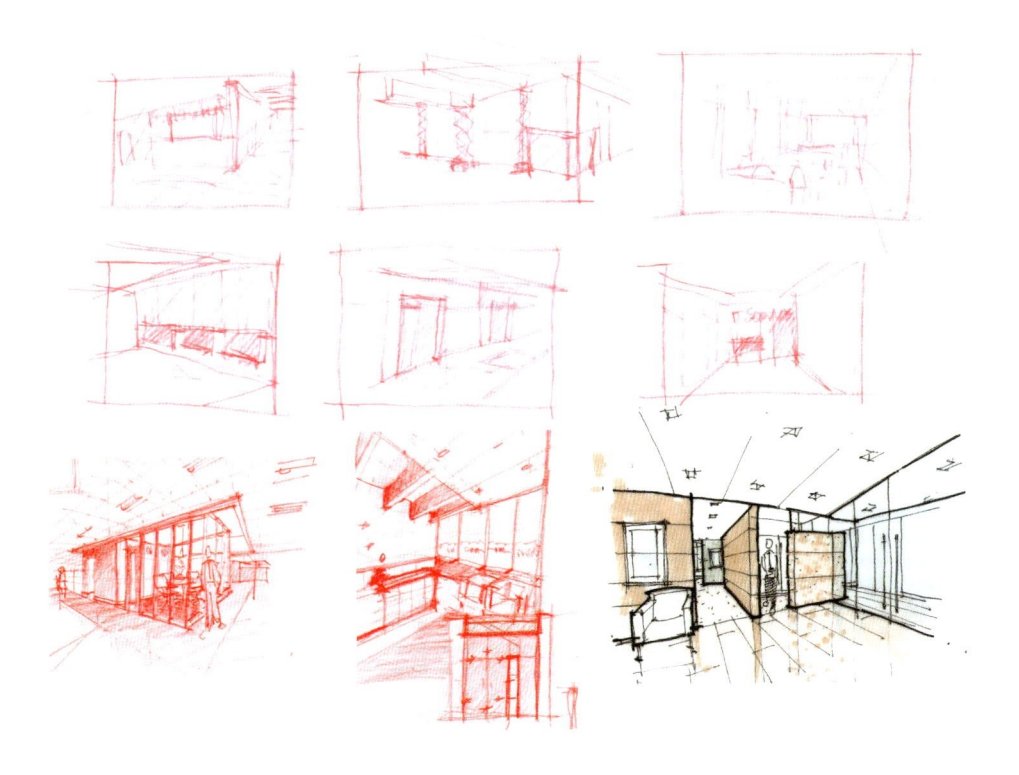

Figure 13.28
Sketch different views before committing to the final one.

Perspective drawings are complex, but all apply four basic concepts of construction. Once those concepts are understood, their complexity is easily managed. Look at Figure 13.29. Notice that:

1. Items appear smaller the farther they are from the viewer, and their shapes change.
2. Parallel horizontal lines converge to a **vanishing point**, the location in which they meet. All sets of parallel lines have their own vanishing points.
3. All horizontal lines below eye level slope up; all horizontal lines above eye level slope down.
4. All vertical lines remain vertical.

PERSPECTIVE DRAWING MANIPULATION

Figure 13.29
One-point perspective.
Courtesy mkerrdesign.com

Perspective pictorials are either **one-**, **two-**, or **three-point** (Figure 13.30). Architectural perspective drawings may be **exterior**, showing the outside of the building, or **interior**, showing the inside. A scaled, hard-line plan and elevations are needed to draw one. The smaller the plan, the smaller the final drawing, so don't use one smaller than 1/4" = 1'-0".

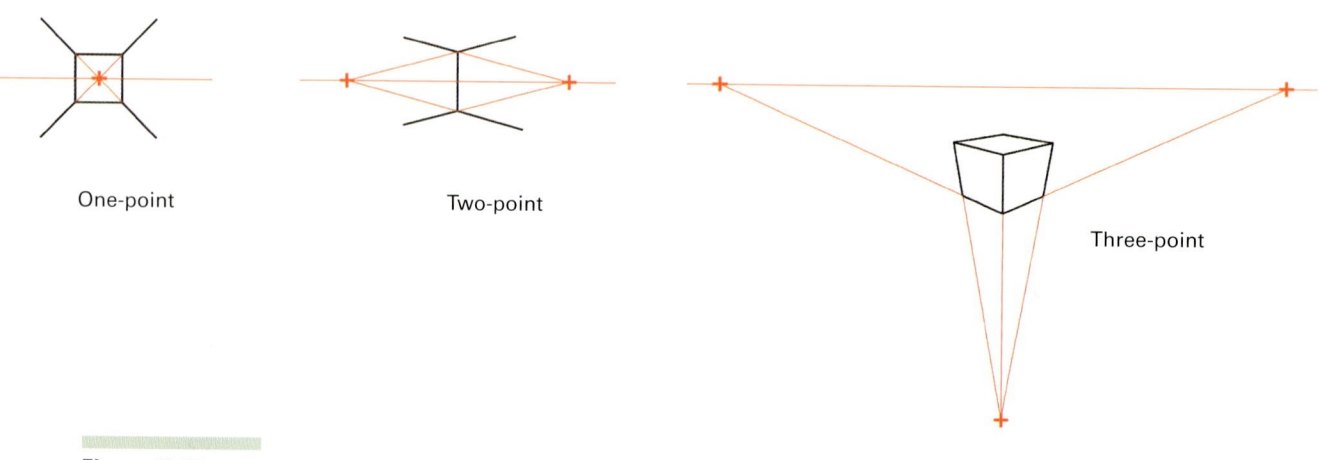

One-point   Two-point   Three-point

Figure 13.30
Perspective types.

340   CHAPTER 13   HAND DRAFTING ISOMETRIC AND PERSPECTIVE PICTORIALS

Following are steps for constructing a two-point interior perspective using the **rotated plan** method. It's called that because the plan is rotated on the drawing board during the setup process. We'll use the living room floor plan and elevations in Figure 13.31.

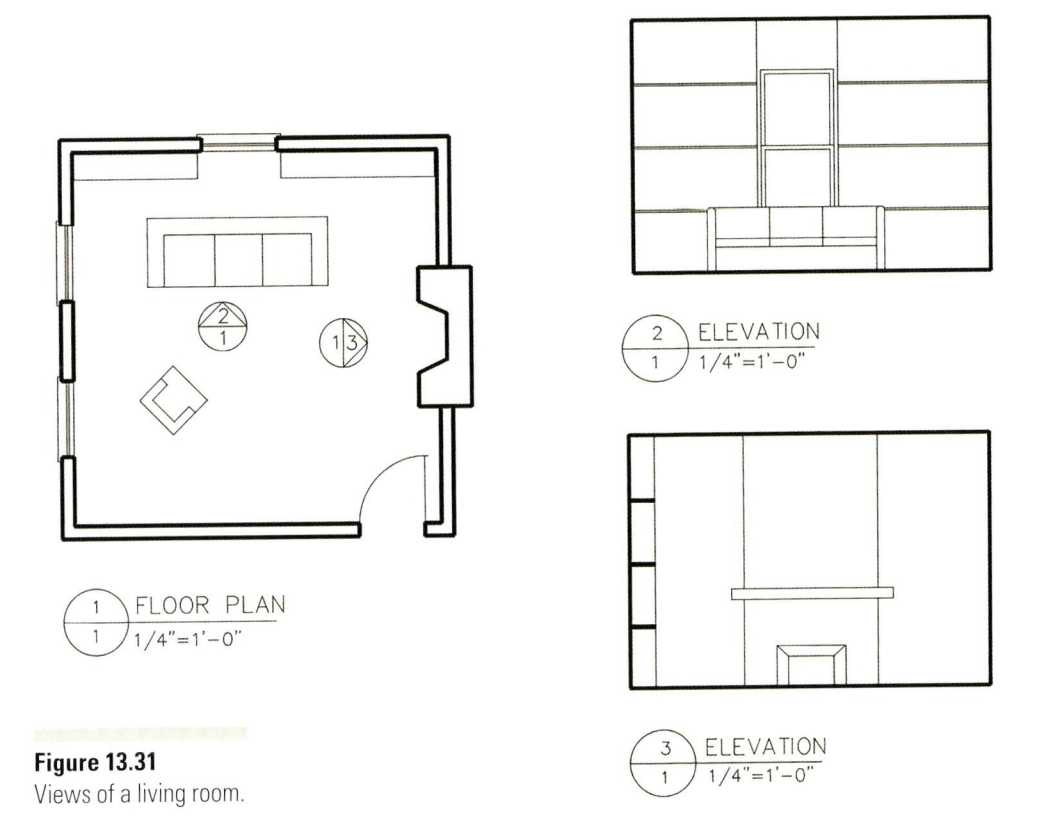

Figure 13.31
Views of a living room.

Draft a Two-Point Interior Perspective Using the Rotated Plan Method

1. *Select a station point and direction of view* (Figure 13.32). Choose a station point (SP) on the floor plan. Then place the 60° corner of a 30°/60° triangle on it and swivel the opposite end. Everything between the triangle's edges is inside the COV and will appear in the drawing. Alternatively, you can use a 90° COV, which will include more of the room but look somewhat distorted.

Figure 13.32
Step 1 Select a station point and direction of view.

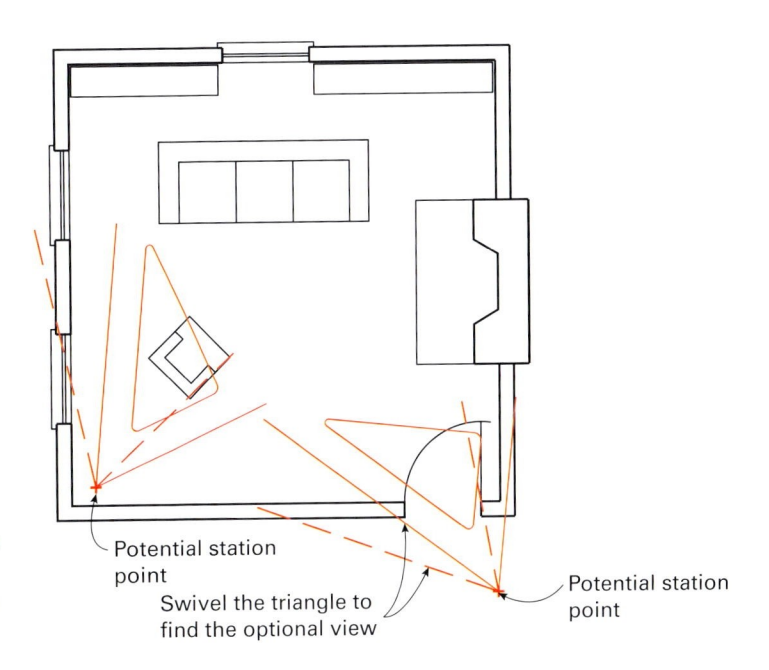

PERSPECTIVE DRAWING MANIPULATION

2. *Choose the best view* and then draw an arrow in the center of the cone of vision. Mark the SP with a cross, and then draw the COV lines and an arrow in the cone's center (Figure 13.33).

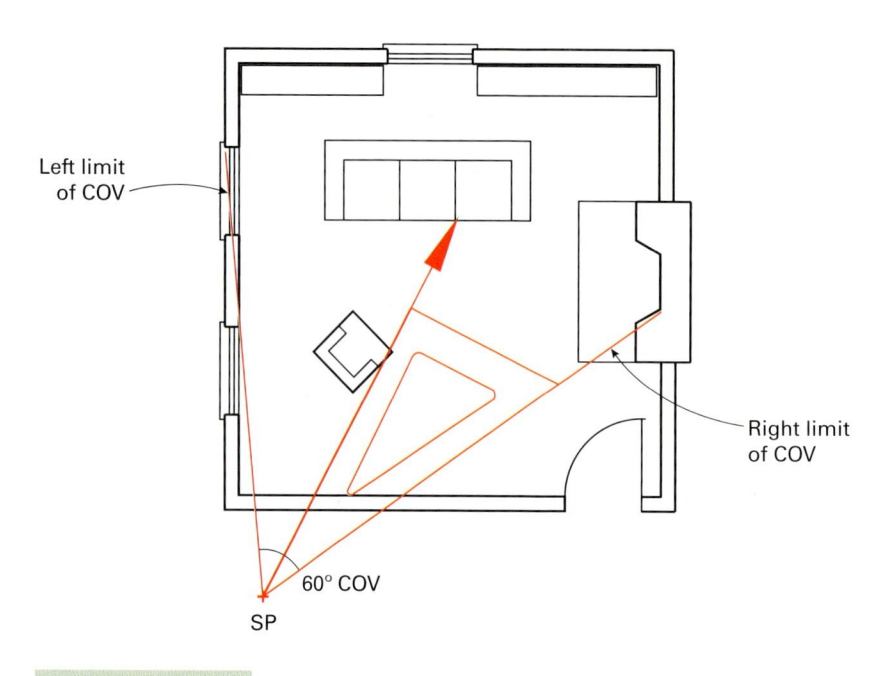

Figure 13.33
Step 2 Choose the best view.

3. *Physically pick up and revolve the paper* so the arrow points straight up (Figure 13.34). Then tape the paper to the board, placing it high enough so there's enough room to construct the perspective drawing below it.

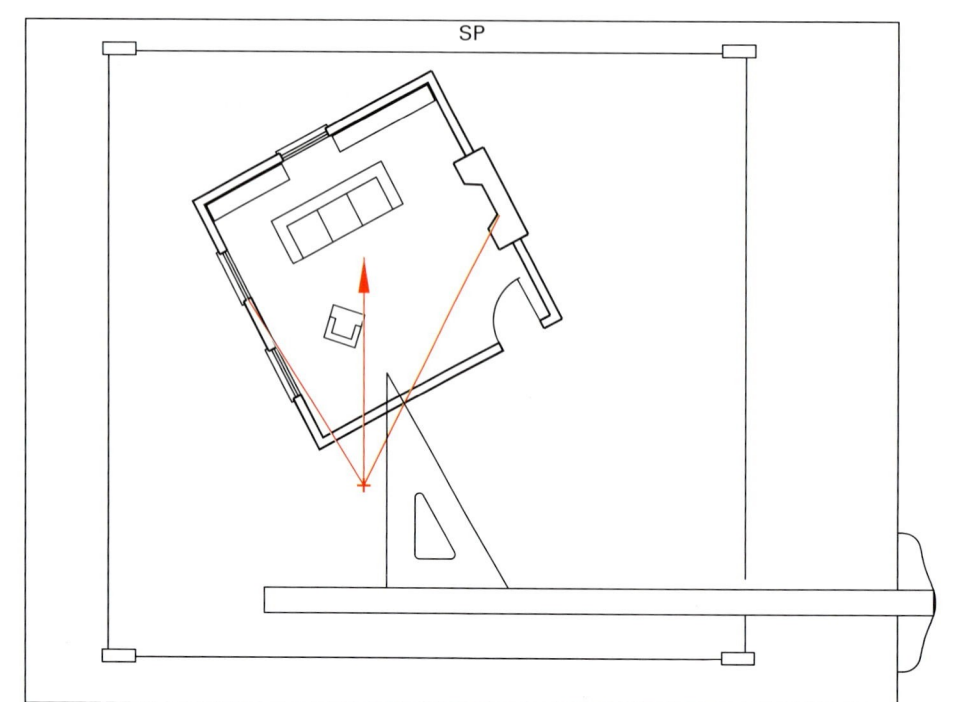

Figure 13.34
Step 3 Physically pick up and revolve the paper.

4. *Draw the picture plane (PP) and horizon line (HL), vanishing points and cone of vision limits* (Figure 13.35). Placing the PP in the plan's farthest-back corner usually gives the best view. Make sure you place it through the *interior* corner of the plan, not the exterior corner. Next, draw a horizontal line anywhere under the floor plan. This is the horizon line and will be the center of the pictorial drawing.

   Each set of parallel walls has its own vanishing point (VP). This rectangular room has two sets of walls, hence two vanishing points. Draw a line through the SP and parallel to the walls until the line reaches the PP. Where the lines intersect the PP, project down to the horizon line, mark with crosses, and label them "LVP" (left vanishing point) and "RVP" (right vanishing point).

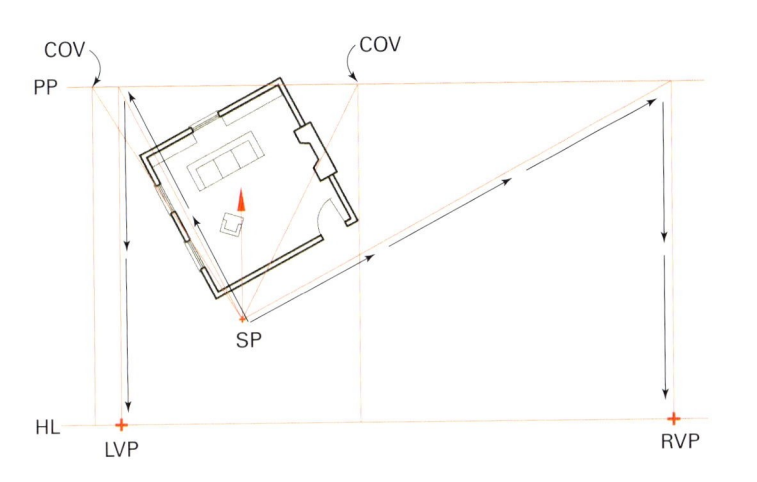

Figure 13.35
Step 4 Draw the picture plane (PP) and horizon line (HL), vanishing points and cone of vision limits.

   Extend both sides of the cone of vision up to the picture plane. Where the COV lines intersect the PP, draw vertical lines down. Those vertical lines define the perspective drawing's left and right sides. These COV lines are construction lines and not part of the final drawing.

5. *Draw the* **height**, *also called the true line* (Figure 13.36). This is the line on which you mark all measurements. Where the PP and room corner intersect, draw a vertical line down to the horizon line. Choose a 5' eye level for the view and measure and mark that distance down from the horizon line. From that mark, measure up the ceiling height (we'll use 9') and make another mark. Darken in the vertical line between those two marks. This is the height line. It is also the room's first corner.

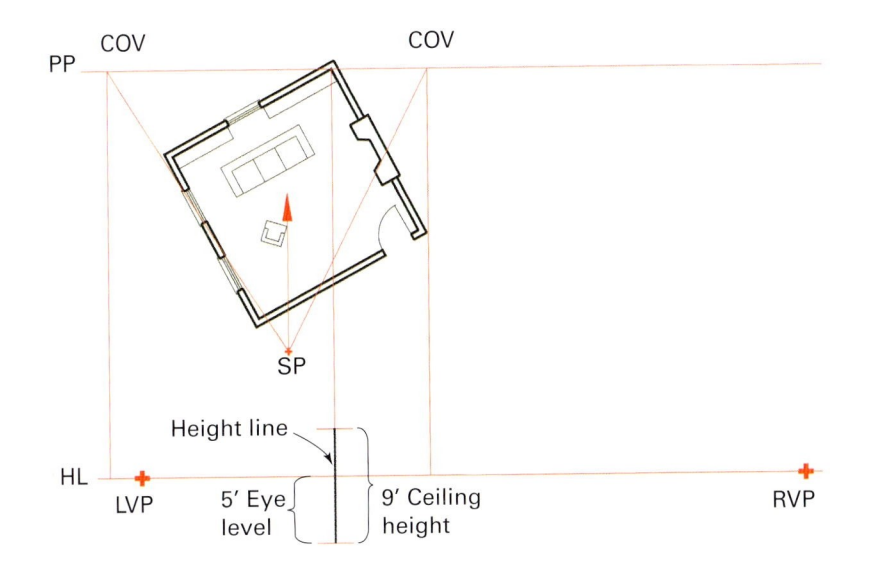

Figure 13.36
Step 5 Draw the height line.

PERSPECTIVE DRAWING MANIPULATION   343

6. *Draw the ceiling and floor lines* (Figure 13.37). Align a straightedge so that it touches the RVP and the top of the height line. Draw a line forward from the height line until it reaches the left COV's line. That is now the room's ceiling line. Next, align a straightedge with the RVP and the bottom of the height line. Draw a line forward from the height line until it reaches the COV's left line again. That's the floor line. Repeat this procedure from the LVP.

Figure 13.37
Step 6 Draw the ceiling and floor lines.

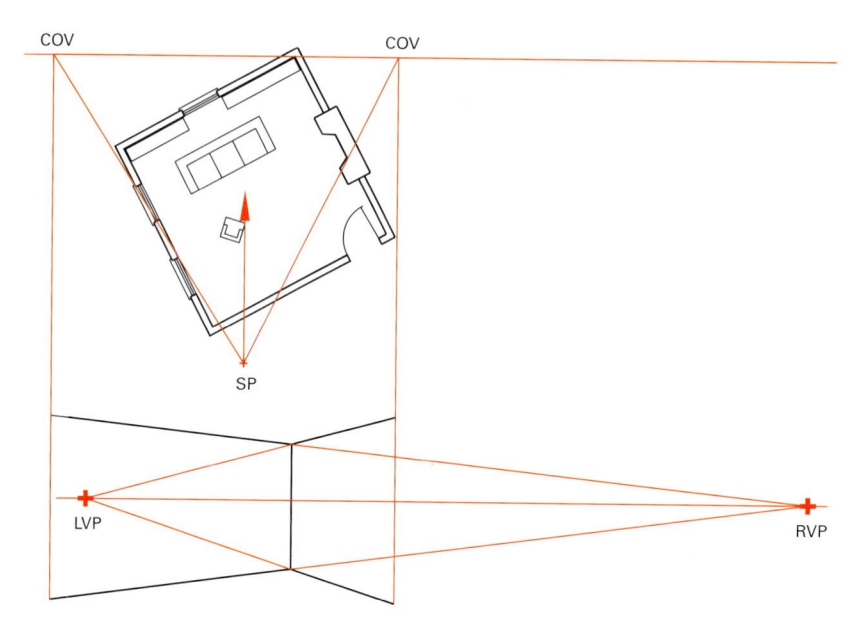

7. *Draw the room's second corner* (Figure 13.38). In the plan view, draw a **sight ray** (construction line) from the SP through the room's second corner. Extend the sight ray to the picture plane and then drop it down to intersect the ceiling and floor lines. The vertical line between the floor and ceiling lines is the second corner. If this corner didn't fall within the COV you wouldn't draw it; the ceiling and floor lines would simply continue to the left COV line and then stop.

Figure 13.38
Step 7 Draw the room's second corner.

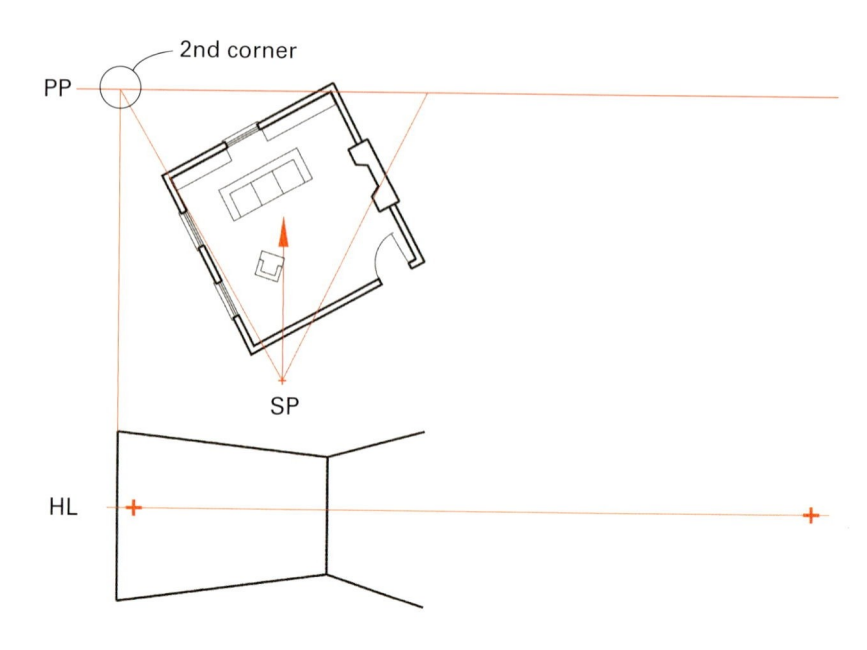

8. *Draw the third wall* (Figure 13.39). Do this by drawing lines from the LVP through the top and bottom of the second corner. Now there are three walls and two corners, matching what is inside the COV location selected in Step 1.

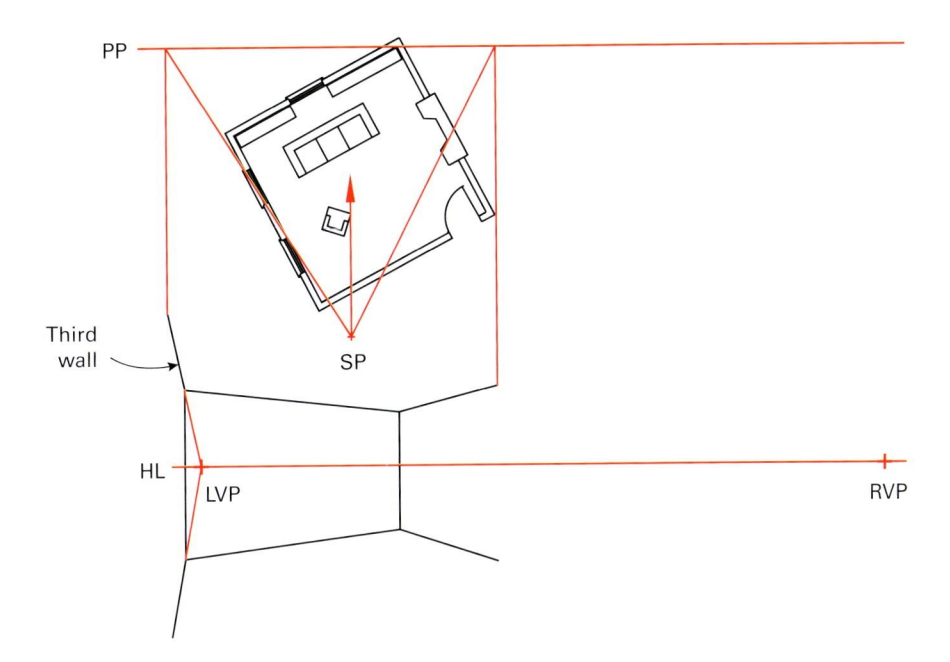

Figure 13.39
Step 8 Draw the third wall.

9. *Draw the window* (Figure 13.40). Draw sight rays from the SP through the window's left and right *interior* corners and project them up to the PP. Where the rays intersect the PP, draw vertical lines down to the wall. Those are the window's sides. Measure and mark the window bottom and top on the height line, and then draw lines from the RVP through those marks until they intersect the window sides. The resultant rectangle is the window. To divide the window in half (Figure 13.41), connect its corners and draw a line through the lines' intersection to the RVP.

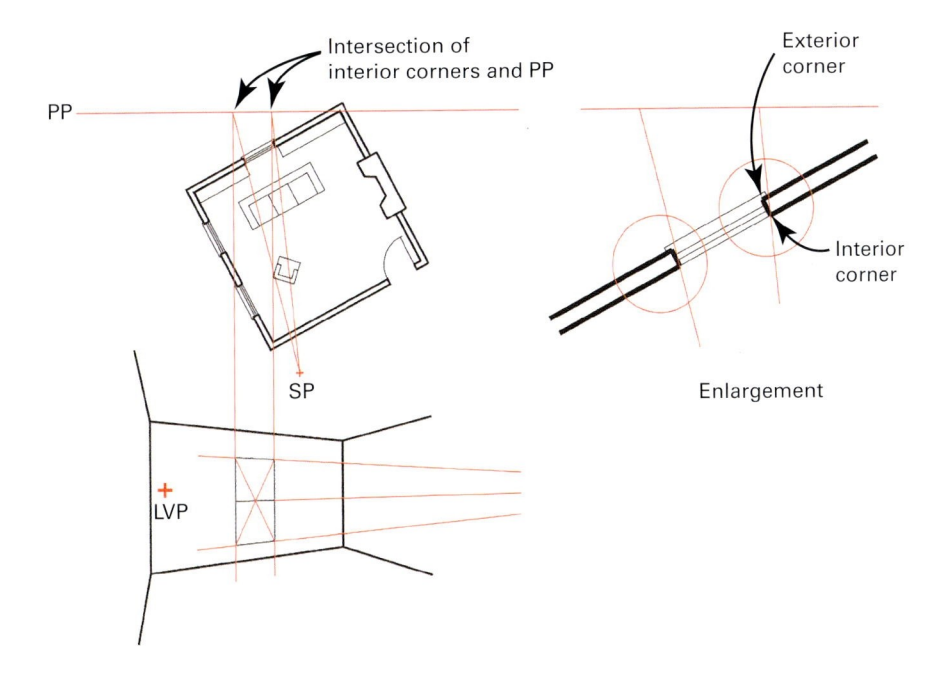

Figure 13.40
Step 9 Draw the window.

PERSPECTIVE DRAWING MANIPULATION

10. *Add thickness to the window* by running a sight ray from the SP through the window's *exterior* corner and projecting it down. Draw a line from the window's lower-right corner to the LVP and draw another line from that intersection to the RVP for the sill.

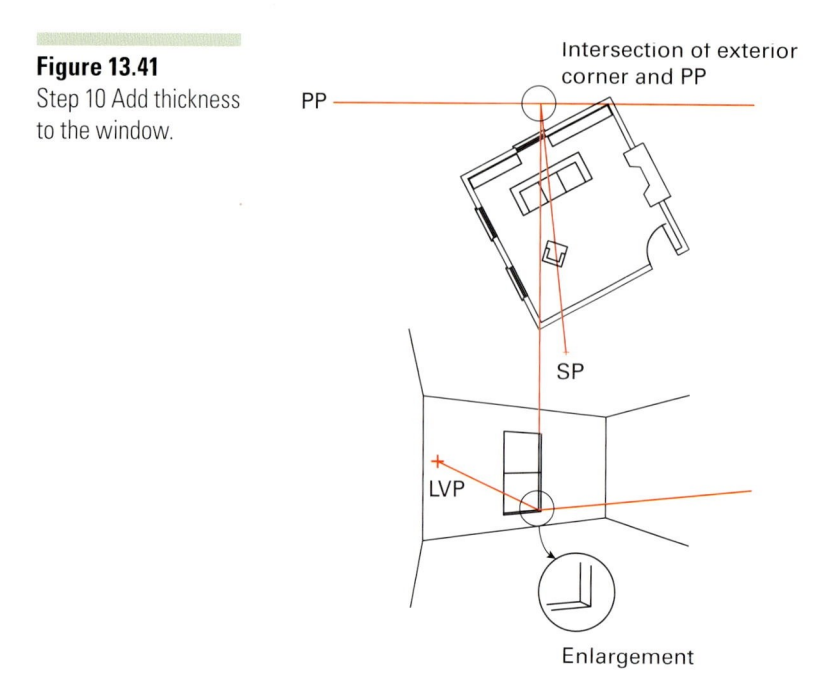

Figure 13.41
Step 10 Add thickness to the window.

11. *Draw the fireplace front* (Figure 13.42). Draw a sight ray through the intersection of fireplace and wall and then up to the PP. Drop a vertical line at the PP intersection down to intersect the floor in the pictorial. Draw another sight ray to the fireplace's front corner, take it to the PP, and then drop that intersection down to the pictorial. Note that only half the fireplace is inside the COV.

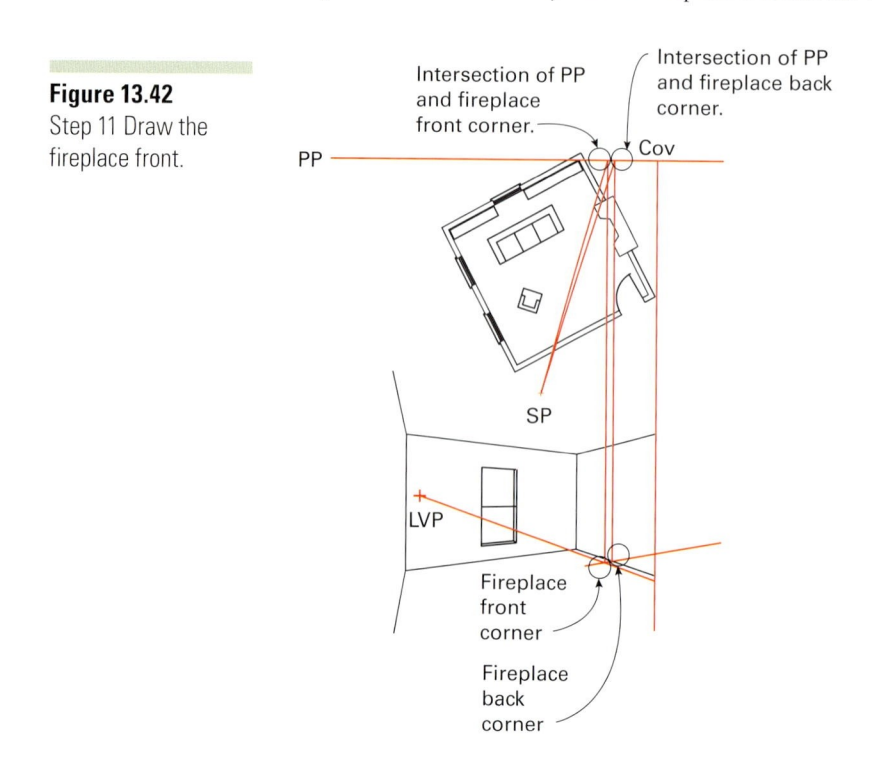

Figure 13.42
Step 11 Draw the fireplace front.

346   CHAPTER 13   HAND DRAFTING ISOMETRIC AND PERSPECTIVE PICTORIALS

12. *Draw the fireplace back* (Figure 13.43). In the pictorial, draw a line from the RVP to the fireplace's back corner/floor intersection and continue it forward until it intersects the vertical line that's the fireplace front corner. Connect that front corner to the LVP and draw a line that stops at the right COV. Draw the fireplace the full wall height (as it is shown in the elevation view) and then repeat this construction process at the ceiling line.

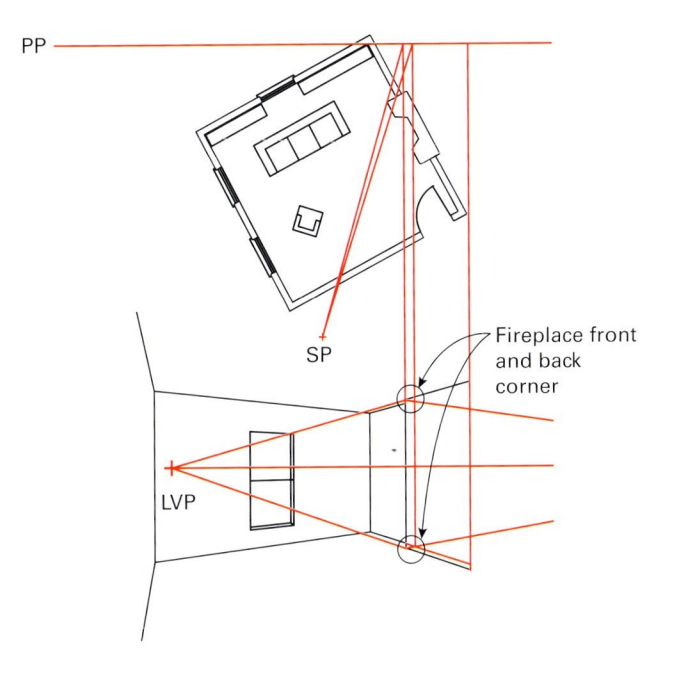

Figure 13.43
Step 12 Draw the fireplace back.

13. *Draw the firebox's angled walls* by finding and connecting their endpoints as shown (Figure 13.44).

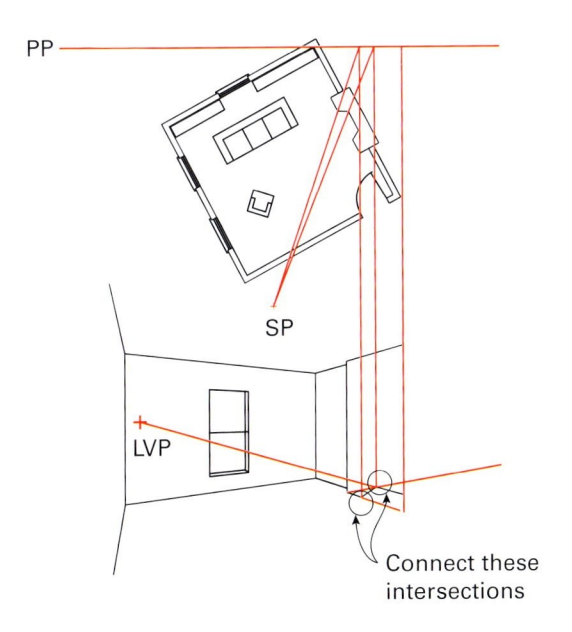

Figure 13.44
Step 13 Draw the firebox's angled walls.

PERSPECTIVE DRAWING MANIPULATION **347**

14. *Draw the firebox height* (Figure 13.45) Mark the firebox height on the height line and draw a line to it from the LVP. Since the fireplace is not flush with the wall, "zigzag" the line from the back wall to the front of the fireplace as shown.

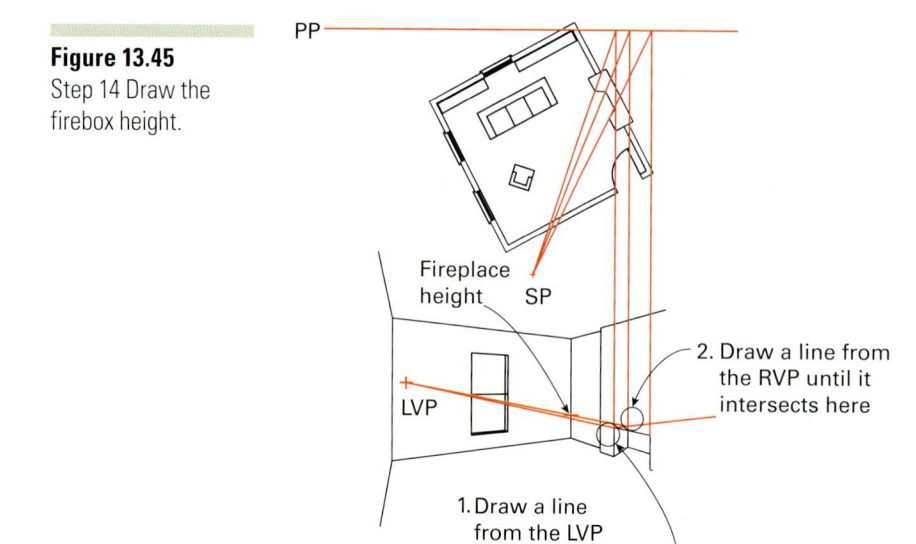

Figure 13.45
Step 14 Draw the firebox height.

15. *Draw furniture adjacent to the wall: the bookcases* (Figure 13.46). Draw two sight rays to the bookcases' back corners and up to the PP. Then draw a third sight ray through one front corner, up to the PP, and down to the pictorial. The bookcases go to the ceiling, as shown in the elevation.

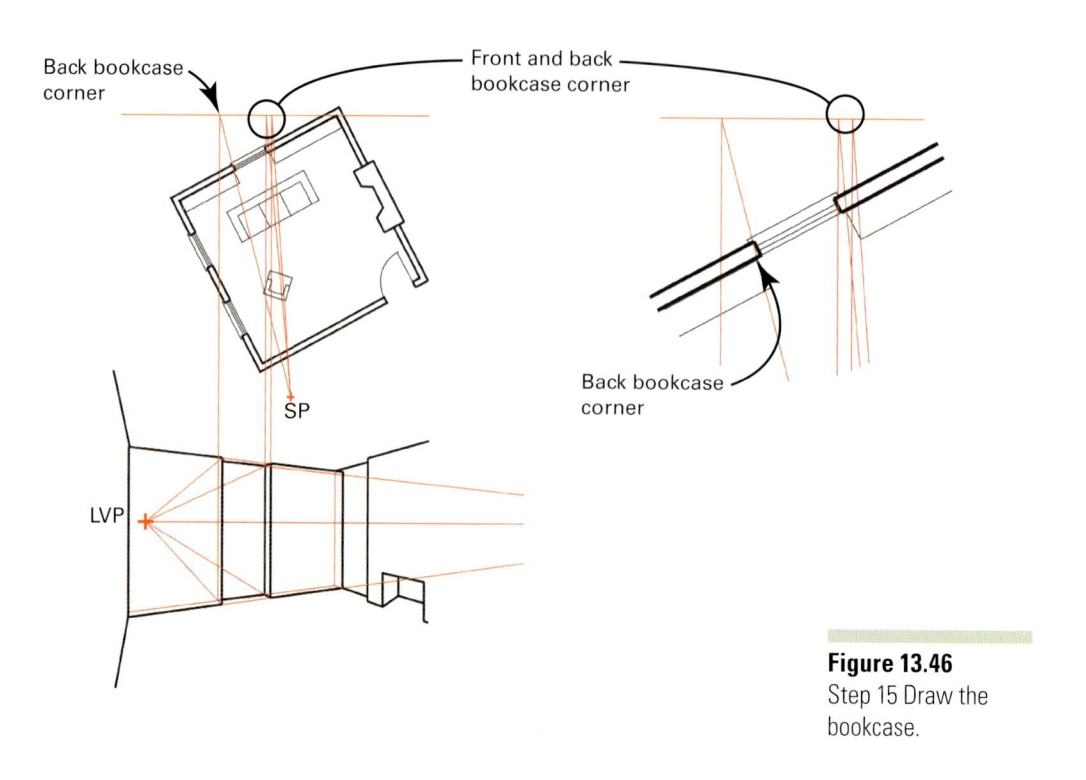

Figure 13.46
Step 15 Draw the bookcase.

348 CHAPTER 13 HAND DRAFTING ISOMETRIC AND PERSPECTIVE PICTORIALS

16. *Draw furniture not adjacent to a wall: the couch* (Figure 13.47). In the plan view, project the couch's edges to the back wall as shown. Draw sight rays through those edge/wall intersections up to the PP and drop them down to the floor line in the pictorial. In the pictorial, draw lines from the LVP through the intersections you just drew. We'll use the vertical line on the right later to construct the couch's height.

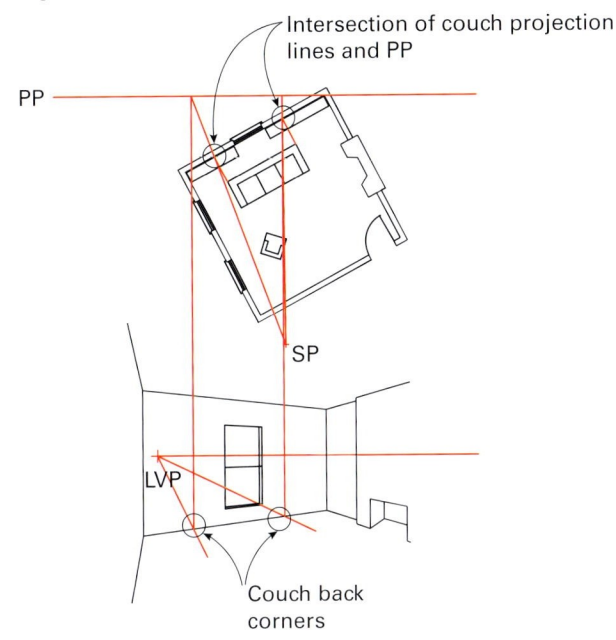

Figure 13.47
Step 16 Draw the couch.

17. *Construct the couch's* **footprint**, *or outline* (Figure 13.48). Project the actual locations of three couch corners up to the picture plane, then down to the pictorial.

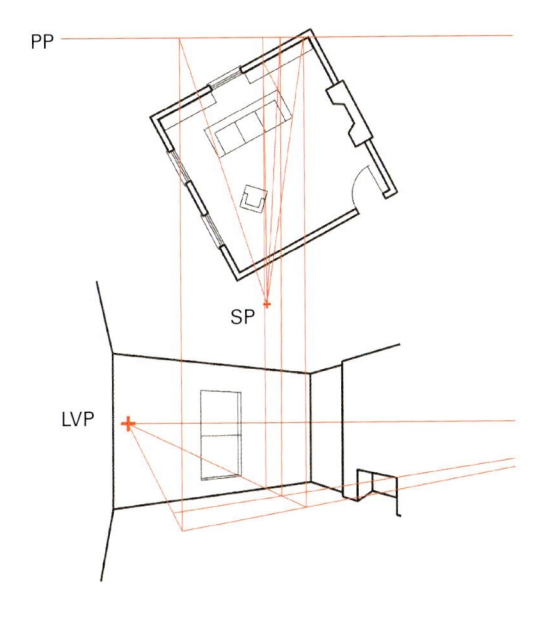

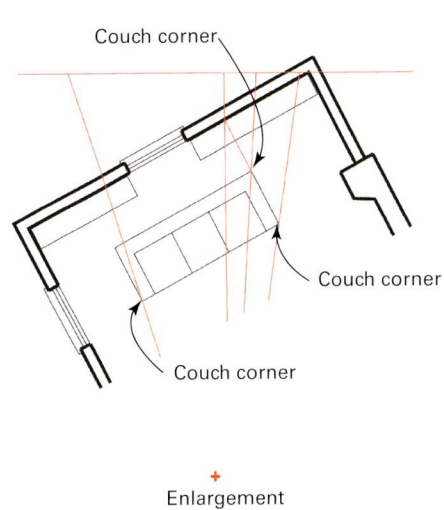

Figure 13.48
Step 17 Construct the couch's footprint.

PERSPECTIVE DRAWING MANIPULATION

18. *Construct the couch's form* (Figure 13.49). Mark the couch's height on the height line and draw a line through it from the RVP until it intersects the vertical line shown. At that intersection, use the LVP to bring another line forward that intersects with a vertical line emanating from the footprint's right front corner. Construct the rest of the couch's rectangular form with vertical lines emanating from the other three corners. Whittle the resulting rectangular mass into a couch, looking at catalog photos for guidance. Make sure to draw all lines on it to the left or right vanishing points.

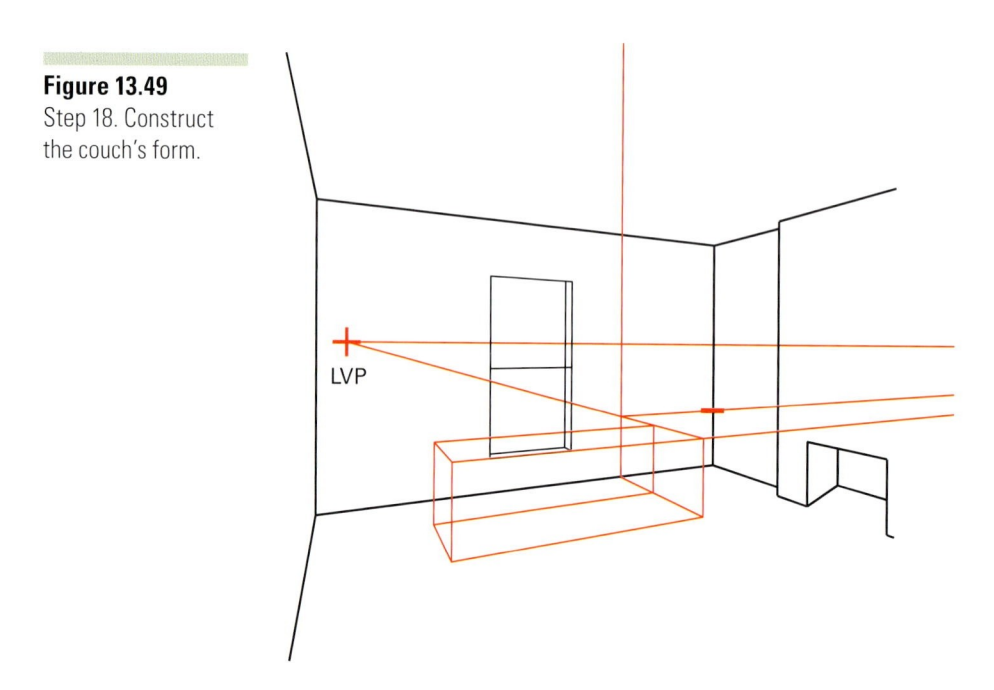

Figure 13.49
Step 18. Construct the couch's form.

19. *Draw caddy-corner furniture: the chair* (Figure 13.50). Furniture not parallel to the walls requires its own vanishing points. Find the chair's vanishing points by drawing lines parallel to its edges through the SP. Label its vanishing points with the name of the chair (CLVP and CRVP).

Figure 13.50
Step 19 Draw the chair.

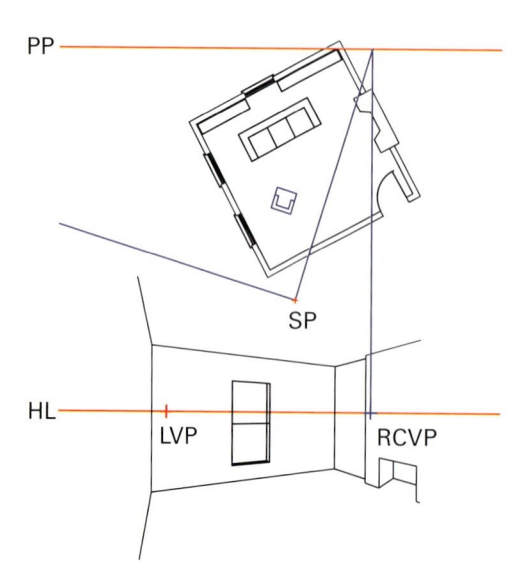

Tip
Use color pencils for furniture to distinguish their lines from the other construction lines.

350   CHAPTER 13   HAND DRAFTING ISOMETRIC AND PERSPECTIVE PICTORIALS

20. *Locate one chair corner in the pictorial* (Figure 13.51). Draw a sight ray from SP through the chair corner that is closest to the SP (1). Take it up to the picture plane, then straight down to the pictorial until it intersects the floor line (2). Draw a line from the room's LVP through that intersection until you reach the vertical line shown (3). This is the chair corner. Connect this corner to the CRVP (4).

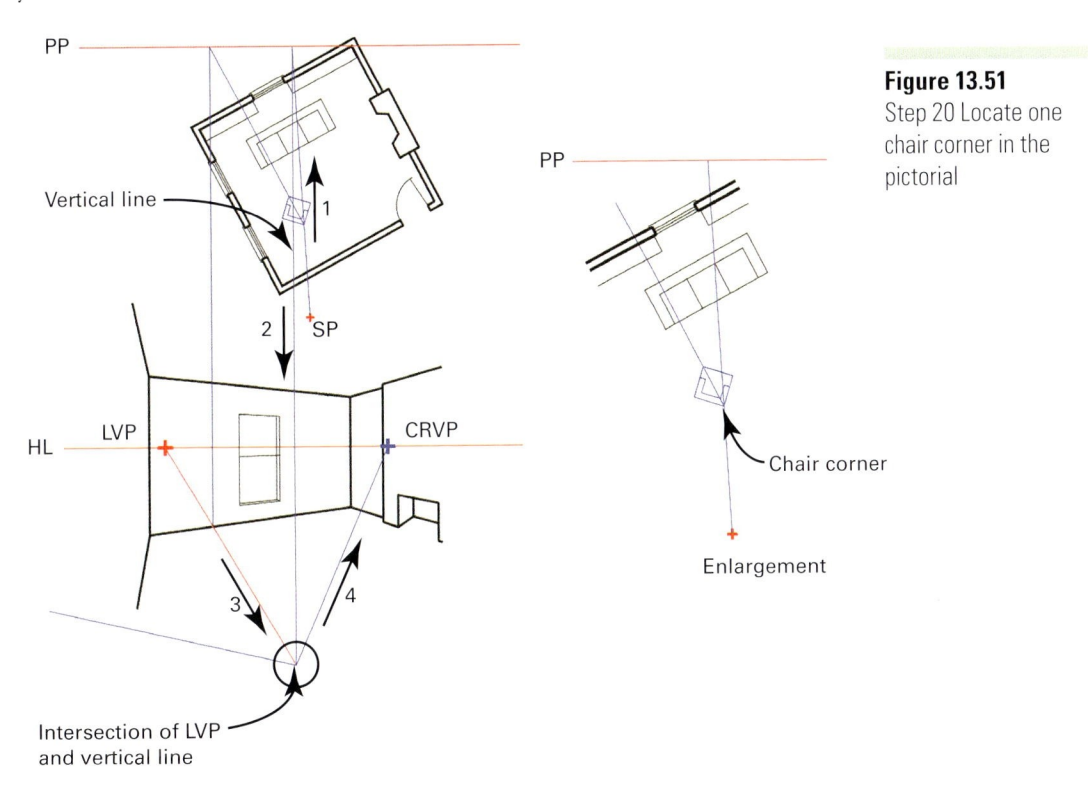

Figure 13.51
Step 20 Locate one chair corner in the pictorial

21. *Construct the chair's footprint* (Figure 13.52). Draw sight rays through two more chair corners up to the picture plane and then down to the pictorial to create the chair's footprint (Figure 13.54). Use the chair's vanishing points.

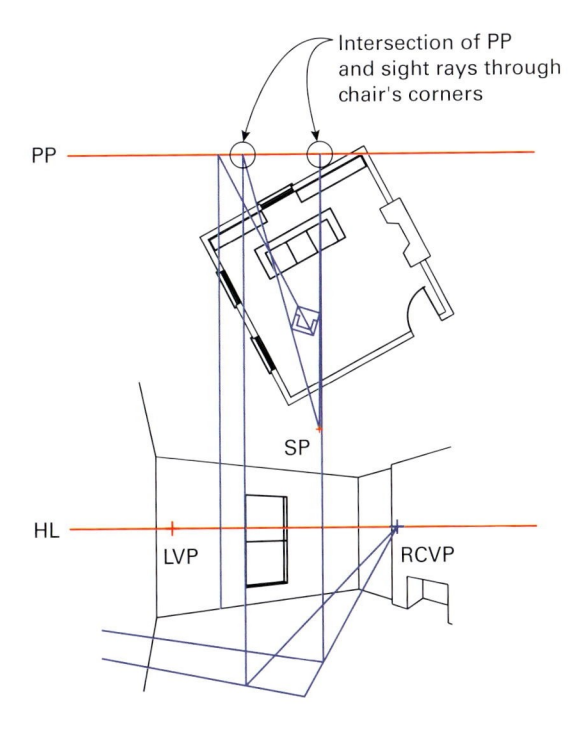

Figure 13.52
Step 21 Construct the chair's footprint.

PERSPECTIVE DRAWING MANIPULATION 351

22. *Construct the chair's height* (Figure 13.53). Mark the chair height on the room's height line. Draw a line from the room's RVP through that mark until it intersects the vertical line used to find the first chair corner. Then draw a line from the room's LVP through that intersection until it hits the vertical line emanating from that first corner in the footprint. This intersection is the chair height. Draw lines from the top of that height to the chair's vanishing points. Project the remaining corners up to finish the chair's form. Figure 13.54 shows the completed living room.

Figure 13.53
Step 22 Construct the chair's height.

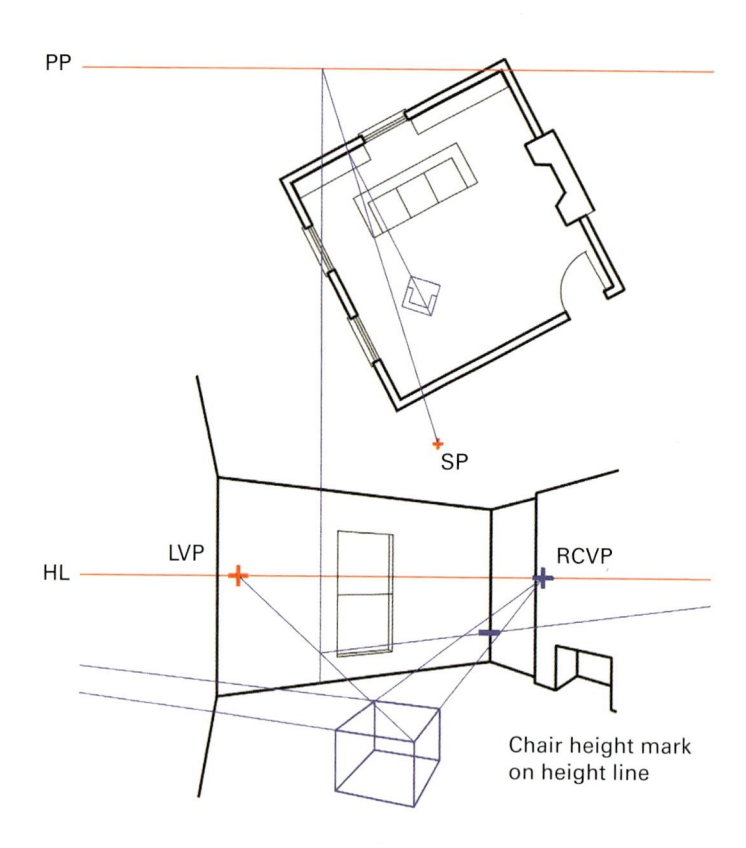

Figure 13.54
Pictorial view of the living room.

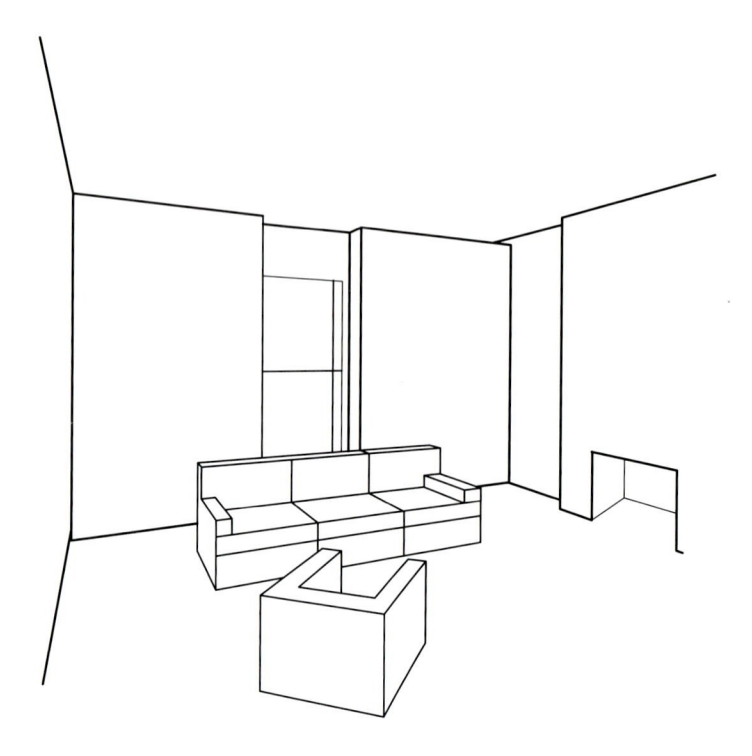

352 CHAPTER 13 HAND DRAFTING ISOMETRIC AND PERSPECTIVE PICTORIALS

Construct a Scale Figure See Figure 13.55. Mark a person's height on the height line and draw a line through it from the RVP. Any vertical line on the wall between that line and the floor represents that height. To bring the figure into the room, draw lines from the LVP through the top and bottom of a vertical line drawn anywhere on the wall. The closer the figure is to the viewer, the larger it will appear.

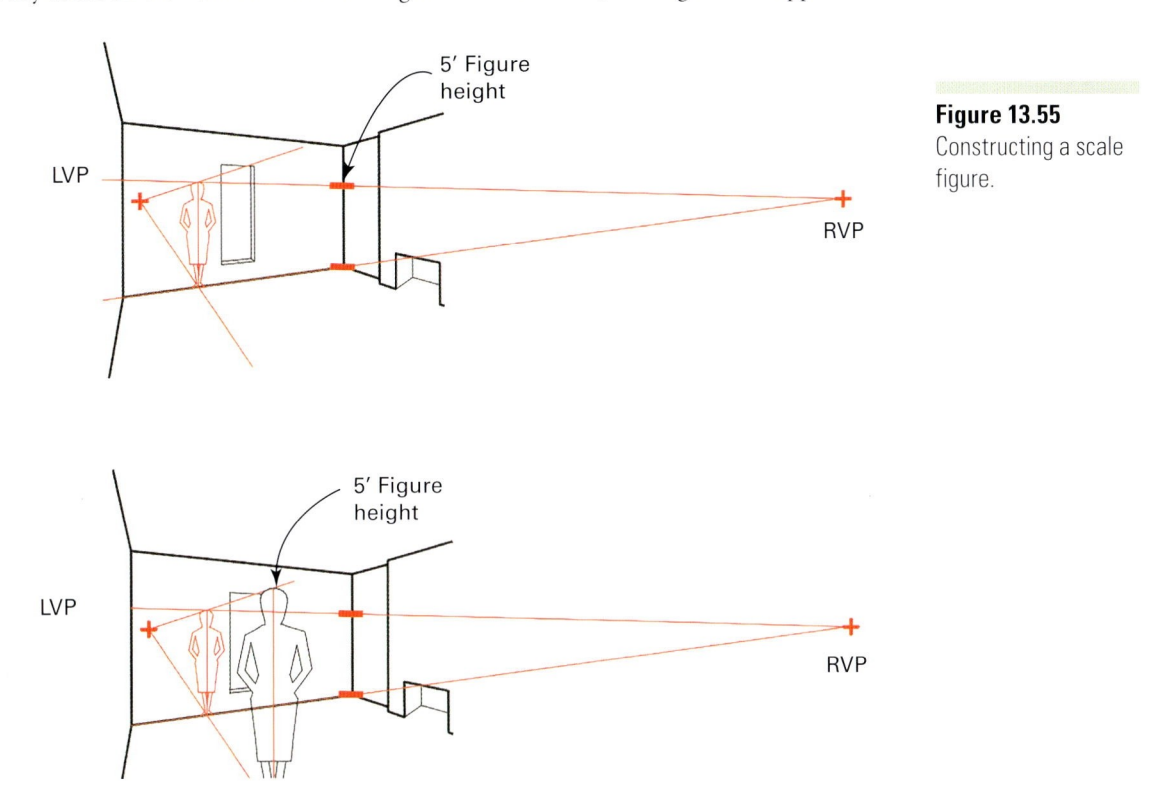

Figure 13.55
Constructing a scale figure.

Draft a Two-Point Perspective of a Circle Circles in perspective appear as ellipses. Construct one by enclosing the circle in a square and then drawing the square in perspective. Mark points on the circle, project each point to the PP and down to the top and bottom lines of the square in the pictorial. The more points you plot, the more accurate the curve will be. Connect the dots with a French curve (Figure 13.56).

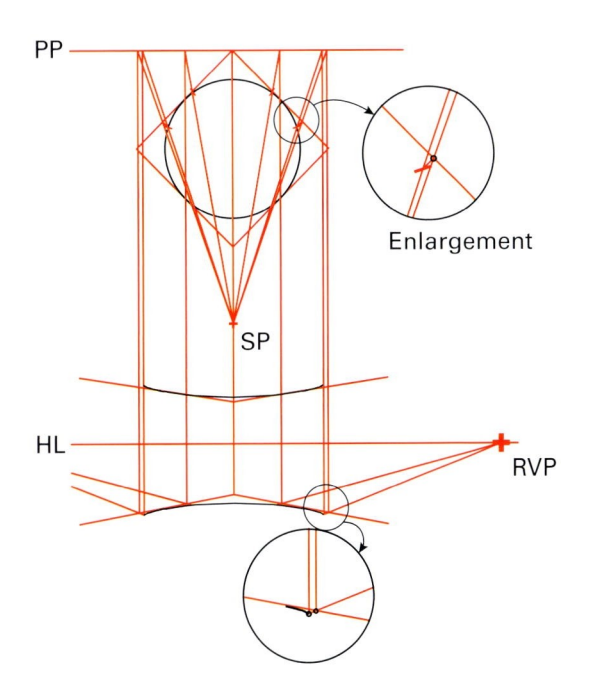

Figure 13.56
Constructing a circle.

PERSPECTIVE DRAWING MANIPULATION **353**

Draft a One-Point Perspective In a **one-point perspective** the floor plan is oriented so that the whole back wall is on the picture plane, making it an elevation, and there is only one vanishing point. You can make measurements anywhere on that wall. Let's draw a one-point perspective of the living room in Figure 13.57.

Figure 13.57
Plan and elevation of a living room.

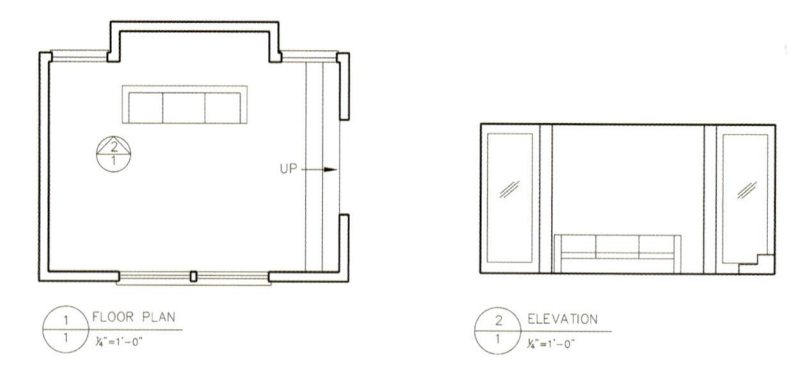

1. Set up the drawing (Figure 13.58). Align the back wall with the parallel bar and draw the picture plane along that back wall. Draw a horizon line below the plan, choose a station point, and draw the cone of vision.

Figure 13.58
Step 1 Set up the drawing.

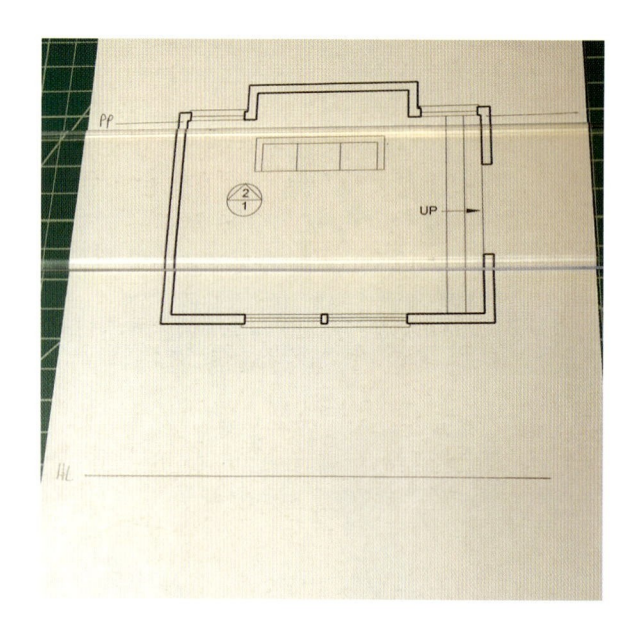

354 CHAPTER 13 HAND DRAFTING ISOMETRIC AND PERSPECTIVE PICTORIALS

2. Draw the back wall (Figure 13.59) by projecting the wall's edges down. From the horizon line measure down the viewer's eye level to mark the floor, and from there measure up the total room height up to mark the ceiling. Then project the windows and offset down to the pictorial. Find the vanishing point by projecting the SP straight down to intersect the HL. There's no second vanishing point because a line parallel to the second set of walls will never intersect the PP.

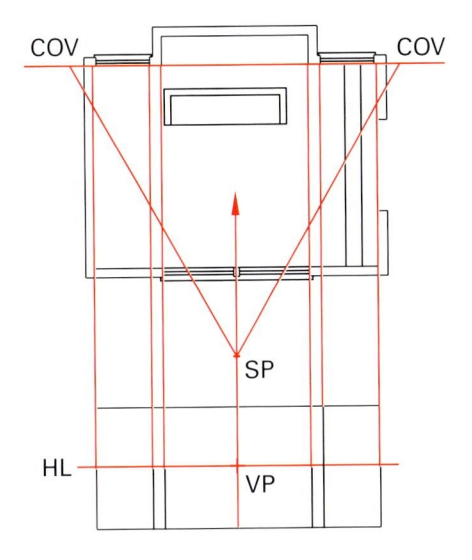

Figure 13.59
Step 2 Draw the back wall.

3. Draw the adjacent walls (Figure 13.60). Radiate lines from the vanishing point through the back wall's upper and lower corners. Stop drawing those lines when they reach the COV.

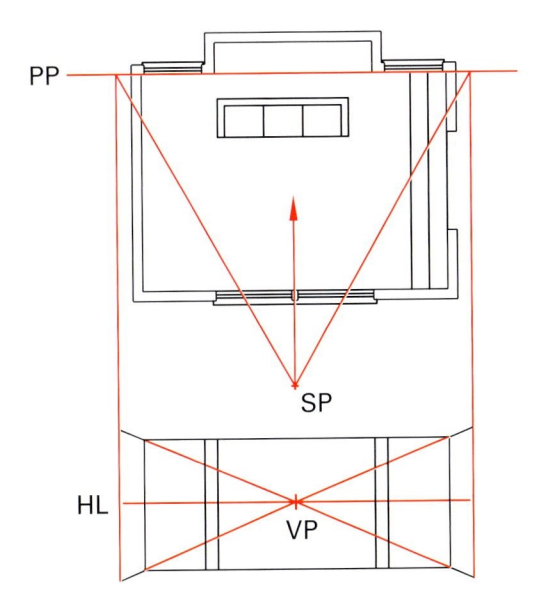

Figure 13.60
Step 3 Draw the adjacent walls.

4. Draw the offset (Figure 13.61). The offset is not aligned on the picture plane, so it will appear in perspective, not in elevation. Draw sight rays from the SP to the offset's back corners. At their intersections with the PP, draw vertical lines down to the pictorial. Radiate lines in the pictorial from the vanishing point to the offset's four corners. At the intersection of the radial and vertical lines, draw horizontal lines to represent the offset's floor and ceiling. Leave a construction line where the original wall was, as you'll need it for the couch and stairs construction.

Figure 13.61
Step 4 Draw the offset.

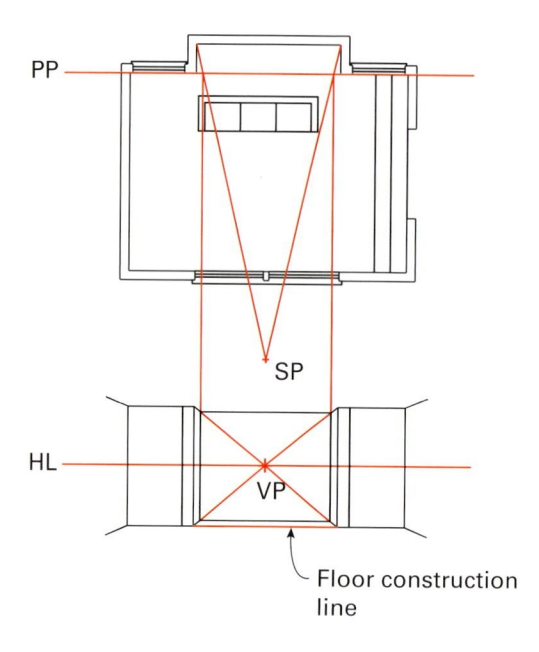

5. *Draw the couch's back edge* (Figure 13.62). Project the couch's two back corners from the plan down to the pictorial until they intersect the floor line. At those intersections draw two lines radiating from the vanishing point.

Figure 13.62
Step 5 Draw the couch's back edge.

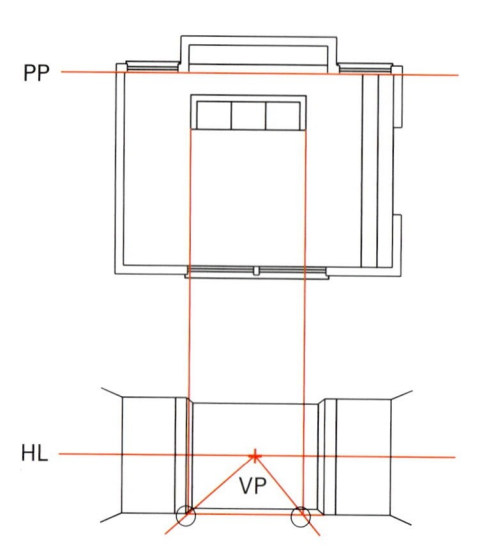

6. *Draw the couch's footprint* (Figure 13.63). In plan, draw sight rays through three couch corners, up to the PP, and then down to intersect the radiating lines. Connect the couch's front two corners to finish the footprint.

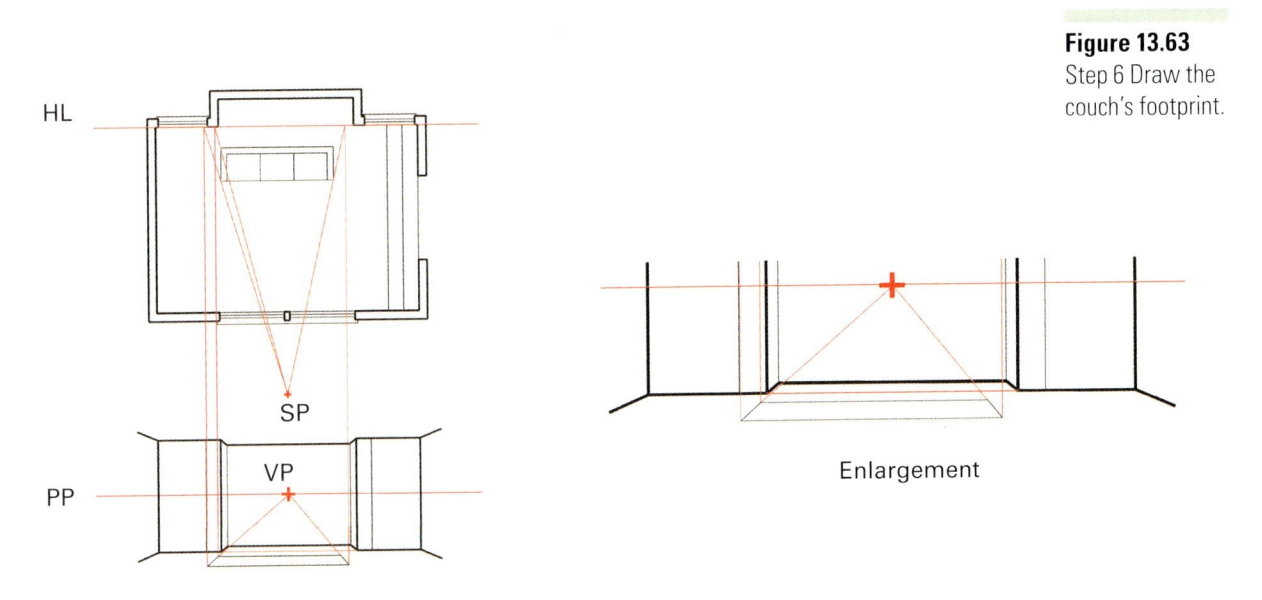

Figure 13.63
Step 6 Draw the couch's footprint.

Enlargement

7. *Draw the couch height* (Figure 13.64). Mark the couch's height on the room's back wall and extend a horizontal line from that mark until it intersects the vertical line as shown. Radiate a line from the vanishing point through that intersection and forward until it intersects a vertical line emanating from the couch's front corner. Draw the other three corners similarly to create the couch's rectangular form.

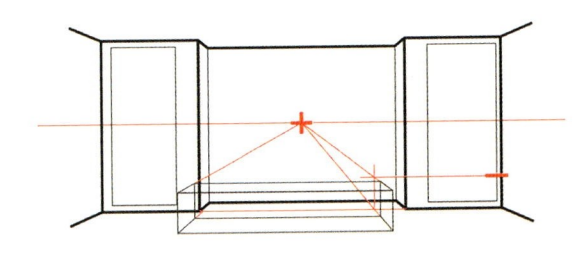

Figure 13.64
Step 7 Draw the couch height.

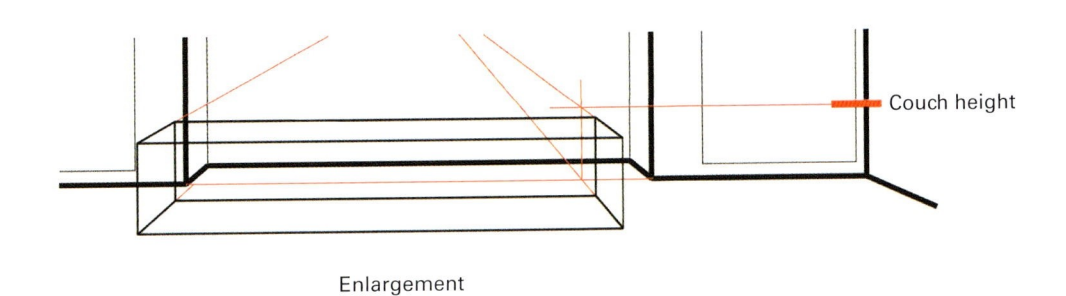

Enlargement

PERSPECTIVE DRAWING MANIPULATION

8. Draw the stairs' *footprint* (Figure 13.65). Project the tread widths down from the plan to intersect the pictorial's floor line. Draw lines from the vanishing point through those intersections and forward into the room.

Figure 13.65
Step 8 Draw the stairs' footprint.

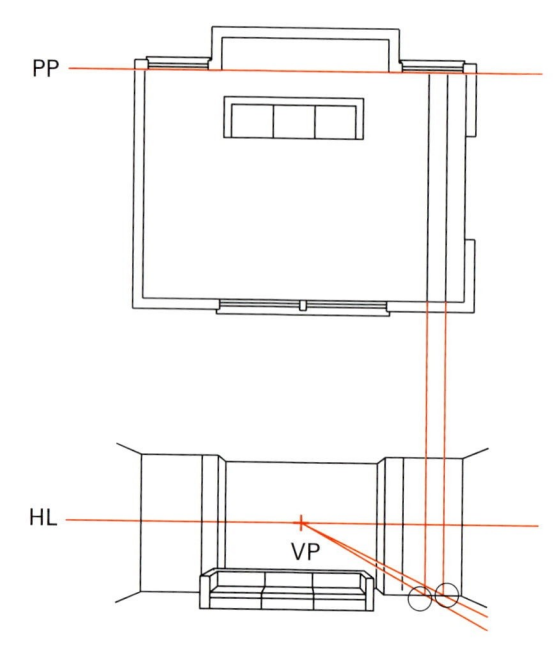

9. *Construct the steps* (Figure 13.66) Mark the riser heights and project them horizontally to intersect with vertical lines emanating up from each tread. Then draw lines from the LVP through those intersections and forward into the room. Complete the stairs by drawing horizontal tread lines.

Figure 13.66
Step 9 Construct the steps.

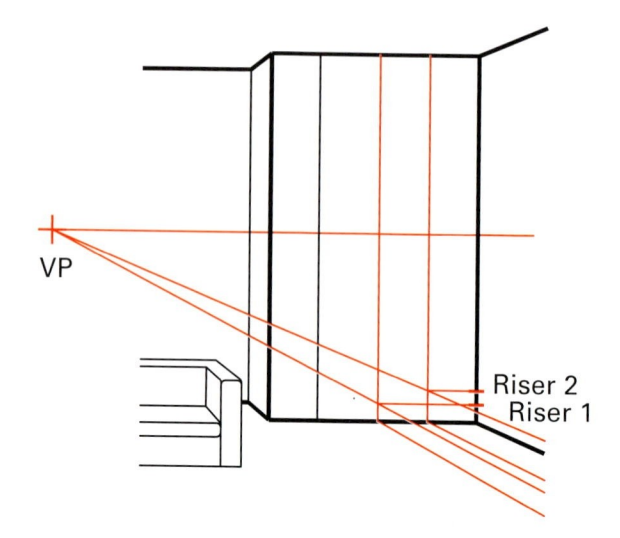

Exercise
Download a floor plan from your favorite design site, scale it inside a software program, print, and sketch one-point perspectives of its rooms.

Construct a Sloped Ceiling

Let's add a **gable ceiling**, which is a ceiling sloped on two sides (Figure 13.67). The hidden line in the plan is the ridge (intersection of two slopes).

1. *Draw sight rays* through the ridge's endpoints, project them up to the picture plane and down to the pictorial.
2. *Draw the gable line* at its ridge height above the wall.
3. *Radiate construction lines* from the vanishing point through the gable line endpoints until they intersect the ridge endpoints projected down in plan and connect those intersections with the corners of the shorter wall. The resultant angle of those lines represents the gable ceiling.

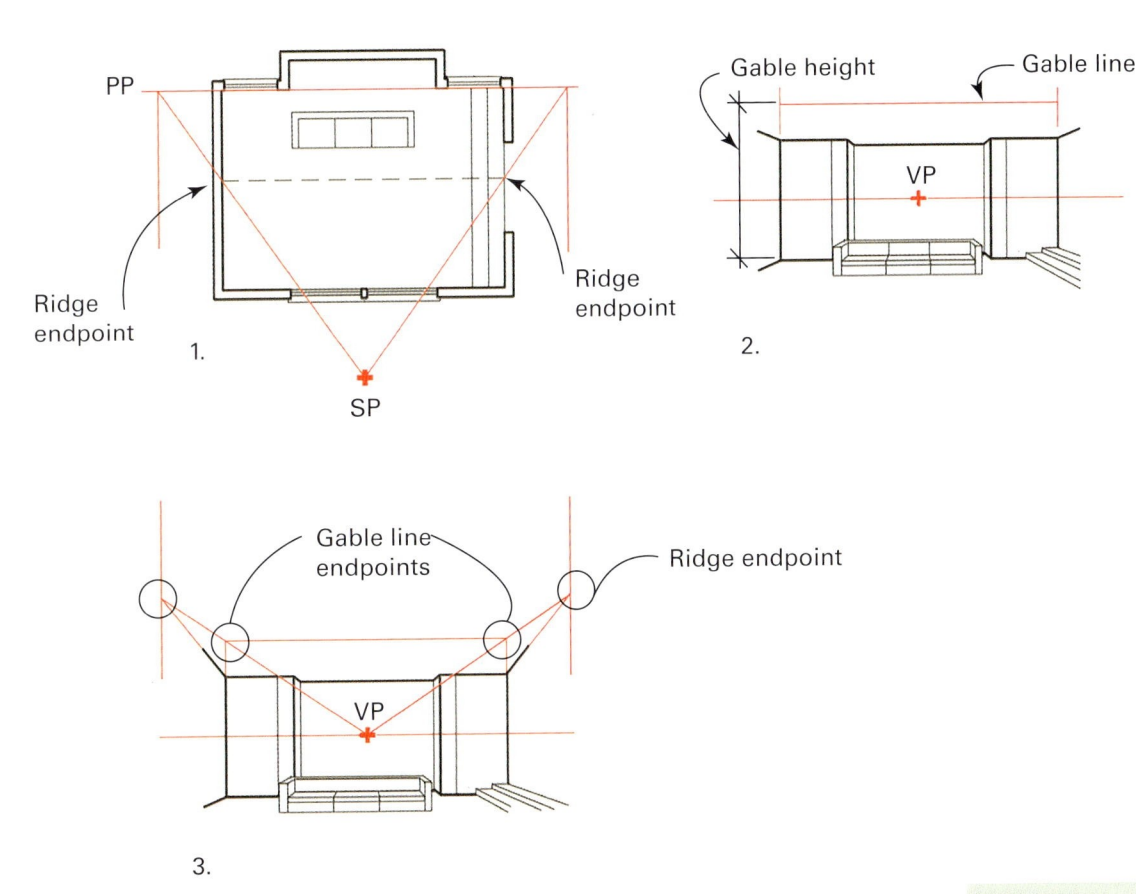

Figure 13.67
Construct a gable ceiling.

Troubleshooting

Perspective drawings have many steps and it is easy to make mistakes. If a drawing looks wrong, retrace all steps from the beginning, as that's where mistakes often start. Common ones are:

- Inadvertently using the vanishing point locations on the picture plane instead of the vanishing point locations on the horizon line
- Using a picture plane/cone of vision intersection as a vanishing point
- Not constructing vanishing points through the station point

- Not constructing vanishing points parallel to the walls
- Using an incorrect height line, such as a corner or edge of the plan that doesn't touch the picture plane
- Not projecting freestanding furniture perpendicular to the wall first, or projecting it at a non-90° angle
- Trying to draw object heights in the pictorial before constructing their footprints
- Not drawing lines to the two vanishing points; drawing them to random points instead.
- Darkening construction lines instead of object lines

Summary

Isometric and perspective drawings are 3D representations of space and objects. Both are used in architectural drafting to communicate ideas. Perspective drawings show spaces and objects as we see them but are complicated to create and are not measurable. Isometric drawings appear somewhat distorted but are easier to create and are measurable. Knowing how to mechanically draft them facilitates good sketching, understanding of space, and perspective grid use.

Classroom Activities

1. Draft a floor plan and elevation of your classroom and use them to draft a perspective drawing.
2. Copy a magazine floor plan and do several perspectives sketches of it, experimenting with cone-of-vision angles and eye-level heights.
3. Study pieces in a furniture catalog to understand how they appear in perspective.
4. Draft three views of the same space using perspective and paraline drawing techniques.

Questions

1. Can objects be measured on an isometric drawing?
2. Can objects be measured on a perspective drawing?
3. What does an isometrically drawn circle look like?
4. What is the benefit of a cutaway isometric drawing?
5. What is the cone of vision?
6. What is the station point?
7. What is a vanishing point?
8. What is the height line?
9. What is drawn first, an object's footprint or height?
10. What one extra step must be taken when drawing freestanding furniture that doesn't need to be done for furniture adjacent to a wall?

Further Resources

Brehm, Matthew. 2016 Drawing Perspective: How to See It and How to Apply It. Hauppauge, NY: BES

Cline, Lydia Sloan. 2011 Drafting and Presentation for Interior Designers, Upper Saddle River, NJ: Pearson Prentice Hall

Montague, J. 2013. Basic Perspective Drawing: A Visual Approach. 6th ed. Hoboken, NJ: Wiley.

Keywords

- cone of vision
- cutaway isometric
- ellipse
- exterior perspective
- footprint
- gable ceiling
- height line
- interior perspective
- isometric pictorial
- one-point perspective
- paraline drawing
- perspective pictorial
- rotated plan
- sight ray
- station point
- station point
- three-point perspective
- true line
- two-point perspective
- vanishing point

CHAPTER 14

Digital Drafting

Figure 14.0
This laser scanner was used to create a floor plan during a kitchen remodel.

OBJECTIVES

Upon completion of this chapter you will be able to:

- Identify industry-dominant software programs
- Explain the capabilities of modeling software
- Describe emerging technologies for visualization and prototyping
- Understand what generative design is.

Today's drafting and design technologies are sophisticated and continually evolving. Hand work, CAD work, generative design software, machine learning, virtual reality headsets, 3D cameras, and 3D printing machines can all be combined in one project **workflow** (series of steps for completing a task) for maximum efficiency and productivity.

2D Drawing Programs

These mimic the board drafting process. You click icons to create lines, circles, polygons, text, etc., and the cursor is your electronic pencil (Figure 14.1). The thought process is like that of manual drafting. You consider the process for drafting a picture on a board and then mimic that process on the computer. Perspective and isometric drawings done in the 2D environment are simply a collection of geometric pieces arranged to create an illusion of a 3D object. Popular 2D drafting and drawing programs include **AutoCAD** and **Illustrator.**

Figure 14.1
2D drafting on AutoCAD.

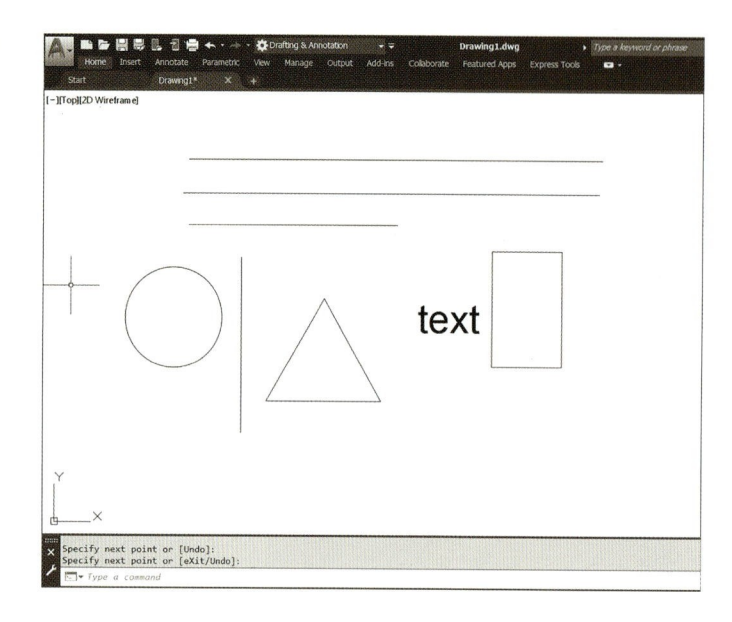

3D Drawing Programs

These are **modelling** programs. They create **digital models**, which are true 3D drawings. You can view a model from any location and angle by **orbiting** (moving around in a circle) the mouse (Figure 14.2). 2D drawings are generated from the model, not separately created (Figure 14.3). Modeling programs also include some 2D drawing capabilities.

Figure 14.2
View a model from any location by orbiting around it.

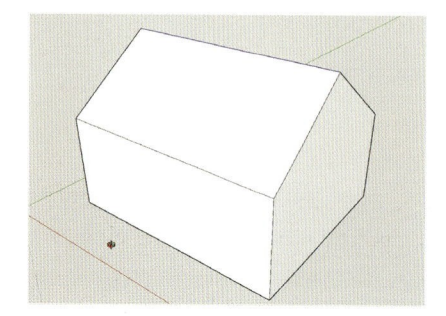

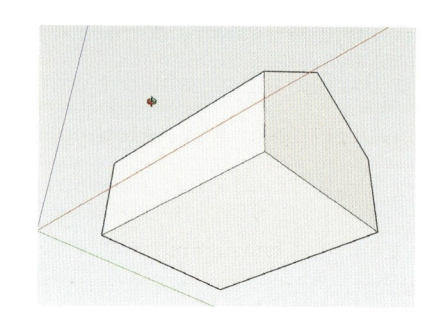

Figure 14.3
2D views are generated from a model.

3D modeling programs are **direct**, **parametric**, or **BIM** (Building Information Management). With direct modeling you create drawings from lines and curves. It is the simplest, most straightforward type of drafting, but if you update a portion of the model, you must manually update everything around it. Parametric drafting is more complex, requiring the use of features and constraints. However, when you update a portion of the model, everything around it updates, which is more efficient when working on a complicated model. BIM software is collaborative; the model is hosted on a server and everyone works directly on it. Materials quantities can be stored and retrieved, making cost estimates and other calculations easy. This differs from the old-school workflow of everyone creating separate drawings that the project manager compiles. BIM software also tracks a project through its whole lifecycle. **AutodeskRevit, SketchUp, Rhinoceros 3D, Chief Architect**, and **20/20** are the dominant programs in the architecture and interior design fields.

Which Program Is Best?

2D and 3D programs have different prices, subscription plans, and capabilities. For instance, Autodesk offers free licenses for its programs to students and faculty; Trimble offers low-cost ones; others just offer 30-day free trials. 20/20 is for kitchen and bath designers and has a robust library of brand products and prices. Revit is capable of incredibly detailed commercial building models and documentation drawings. SketchUp's strength is in conceptual design problem-solving and presentation drawings. It can also produce robust residential documentation drawings.

Other software selection considerations include learning curve; subscription cost; hardware needed; what the client or contractor use; specialized apps and premade models for download available; and ability to import into other programs in the project workflow. It is the designer's responsibility to research which program is most suitable for a project.

Other Programs

Popular non-drafting programs include **Adobe Photoshop, Irfanview, GIMP, Acrobat DC**, and **Excel**. Photoshop is a **digital imaging** program, software that manipulates pictures. Use it to clean up drawings, adjust brightness, change resolution, convert file type, embellish, apply art media filters, make entourage libraries and presentation boards (Figure 14.4). Irfanview and Gimp are free digital imaging programs with some of those capabilities. Acrobat DC is a PDF reader and editor. Excel is a spreadsheet program for creating reports and schedules.

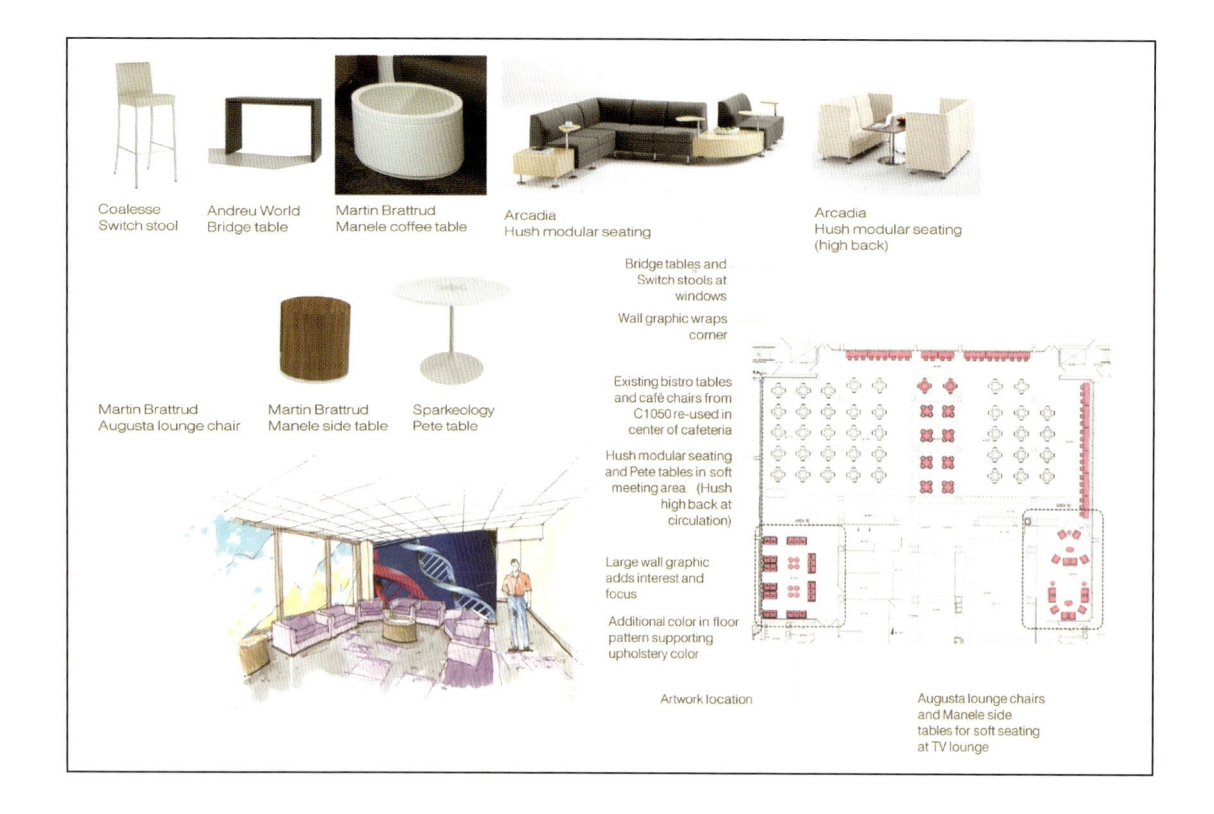

Figure 14.4
Digital presentation board made by combining images in Photoshop. Courtesy mkerrdesign.com

> **FYI**
>
> Digital files are **vector** or **raster**. Vector files are mathematical lines and curves that can be scaled up on the screen or printed without **pixelation** (jagged appearance). Their file sizes are relatively small. Common vector files are **DWG**, **RVT**, and **PDF**. Raster, also called **bitmap**, files are made of pixels. A **high-resolution** file has at least 300 dpi (dots, or pixels, per inch). A **low-resolution file** has fewer dots spread throughout the image, so it appears jagged. Common raster files are **JPG**, **GIF**, and **PNG**. A PDF can also be raster, depending on the software that makes it. Click anywhere on it; if it turns blue, it's a scanned drawing (raster).

Hardware

All programs mentioned except Revit run on both the PC and Mac operating systems, but the PC is dominant in architecture and interior design. A tower or laptop computer is needed; tablets provide limited drawing and editing capabilities and are usually just good for viewing. A Google search returns the exact hardware and operating system requirements but invest in the highest speed processor and powerful graphics card you can afford, because all drafting and modeling programs are very graphics intensive.

Mouse

A two-button mouse with a scroll wheel is needed to fully exploit a program's capabilities. The typical one-button Mac mouse will not work as well. A **3D mouse** (Figure 14.5) is even better, as it deftly navigates the 3D environment with dedicated keys for pan, zoom, and other functions, and has programmable buttons for your most-used commands. There are many different models; 3Dconnexion and Logitech make popular ones. The 3D mouse is used in tandem with the conventional one.

Figure 14.5
A 3D mouse is built to navigate a 3D modeling environment.

Exercise
Do an internet search for 3D mice and compare and contrast different models' features.

Plotter

Plotter is a large, freestanding inkjet printer that can print large drawings to scale (Figure 14.6). However, a not-to-scale copy of a large drawing can be printed on any desktop printer. Its paper tray can also be loaded with thin art paper, vellum, or plastic film for drawings intended for client presentation.

Figure 14.6
A plotter can print large drawings to scale.

PLOTTER 367

Workflow Techniques

Hand and computer work can be combined, and so can programs in a workflow. Examples are:

- Export a SketchUp model as a JPG file, print it on bond paper, add marker color for a hand-sketched look and some glows and shadows in Photoshop (Figure 14.7).
- Import SketchUp models into Revit. This enables you to do the conceptual design work in SketchUp and use Revit's robust tools to develop it.
- Import AutoCAD floor plans into SketchUp and model them.
- Think through a 3D space problem with a pencil and tracing paper and clean it up in Photoshop (Figure 14.8).
- Photograph subjects relevant to your project and turn them into entourage files with Photoshop. Add shadow filters for a different aesthetic. (Figure 14.9).
- Import a not-to-scale floor plan into SketchUp or AutoCAD, scale and trace it. A tutorial video is listed at the end of this chapter.
- Model a project to help visualize the space when hand drafting 2D drawings (Figure 14.10)

Figure 14.7
Color studies done with SketchUp, Photoshop, and manual work. Courtesy mkerrdesign.com

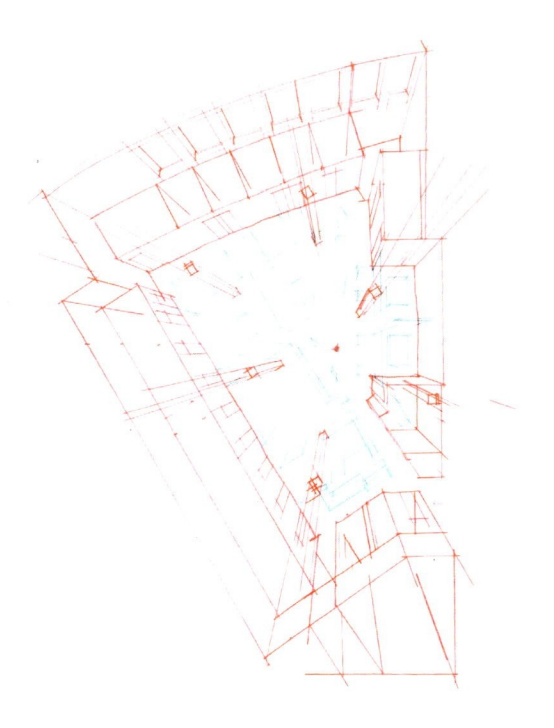

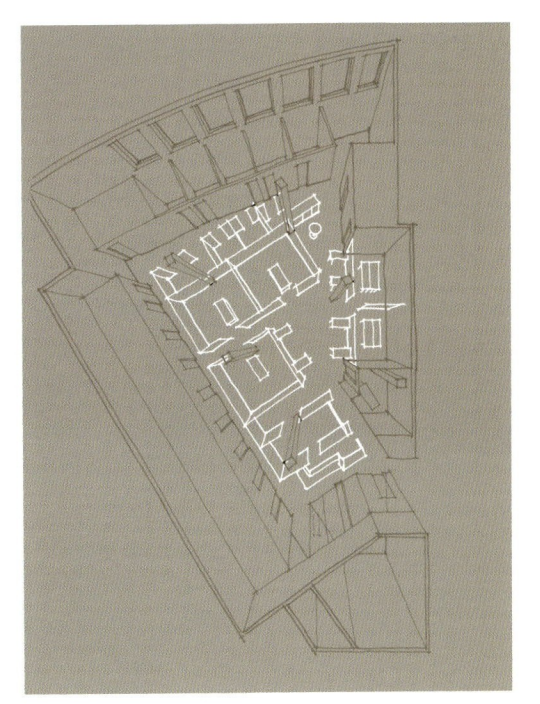

Figure 14.8
This one-point aerial perspective was drawn on tracing paper with colored pencils to study a complicated space. Then it was traced with pen, imported into Photoshop and cleaned up for a client presentation.

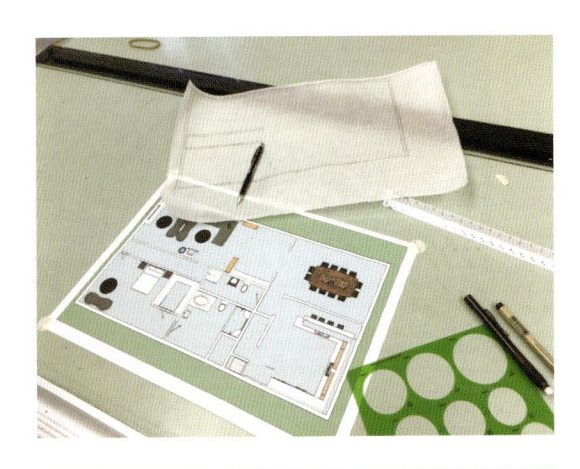

Figure 14.10
A manual drafting student digitally modeled a class project and generated 2D views to help visualize how to draft them.

Figure 14.9
Turn pictures into entourage figures with Photoshop.

WORKFLOW TECHNIQUES   **369**

Virtual Reality Headset

A **virtual reality headset** is a head-mounted device that creates an immersive reality experience for the wearer in a completely different environment (Figures 14.11, 14.12). These sets are used with video games, test simulators, and modeling software. An accompanying hand-held controller lets you "walk" through the model, enabling you to judge space, scale, and proportions as close to first-hand as possible. Popular headset brands include Oculus Rift and Google Cardboard.

Figure 14.11
A VR headset provides an immersive experience into a different environment.

Figure 14.12
Google Cardboard and a smartphone make an affordable VR headset.

Augmented Reality

Augmented Reality is an app that inserts models of a proposed design onto an existing background displayed on a mobile device. So, you combine virtual architectural designs with the reality of the job site. Popular apps include Reality Composer and Unity. The iPhone 12 Pro has **LiDAR** capabilities, which is a method for measuring distances with a laser light and sensor. LiDAR improves AR implementation.

3D Printing

3D printing is a manufacturing process that converts a digital model into a physical one. Most consumer 3D printers use a process called **Fused Deposition Modeling (FDM)** or **Fused Filament Fabrication (FFF)**. In this operation, an extruder heats up plastic filament that is wound on a spool and deposits it onto a build plate in the form of the digital model (Figure 14.13). 3D printing is popular for prototyping all kinds of consumer products from furniture (Figure 14.14) to lighting fixtures to prosthetics. During the COVID-19 pandemic, volunteers printed personal protective equipment (PPE) to donate to first responders and health workers. Figure 14.15 shows some face shield files available for download from thingiverse.com, an online repository for 3D-printable models.

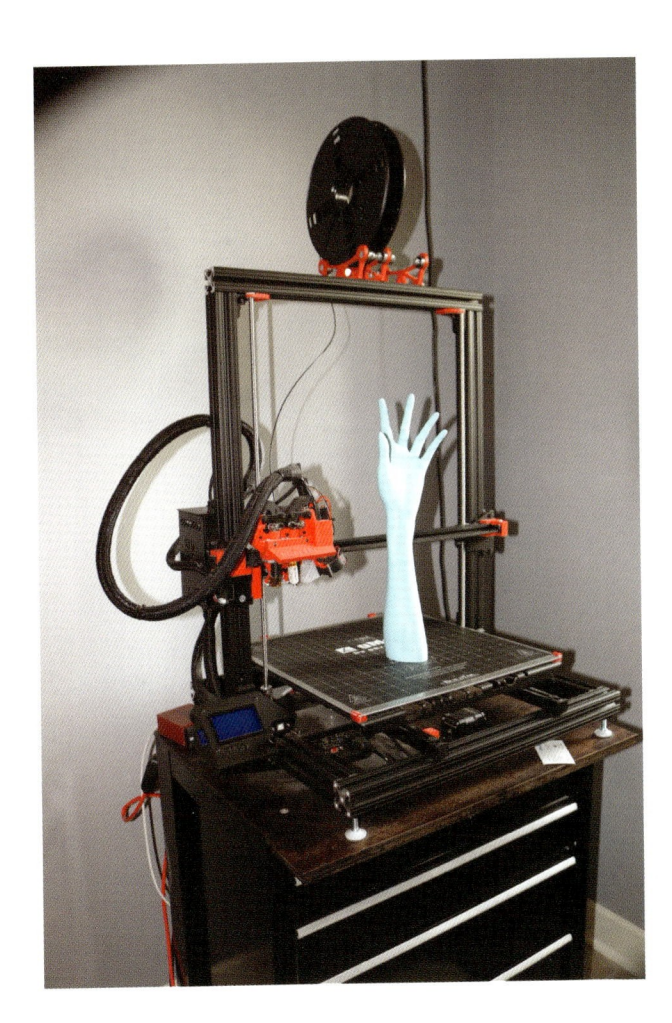

Figure 14.13
A 3D printer and jewelry holder printed on it.

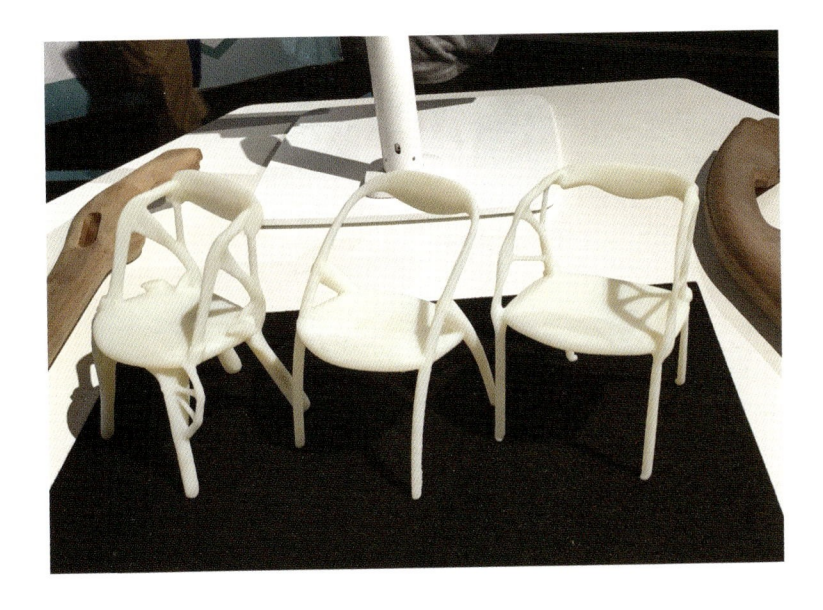

Figure 14.14
Small prototypes of chairs were 3D printed and one was chosen for a full-scale fabrication.

3D PRINTING

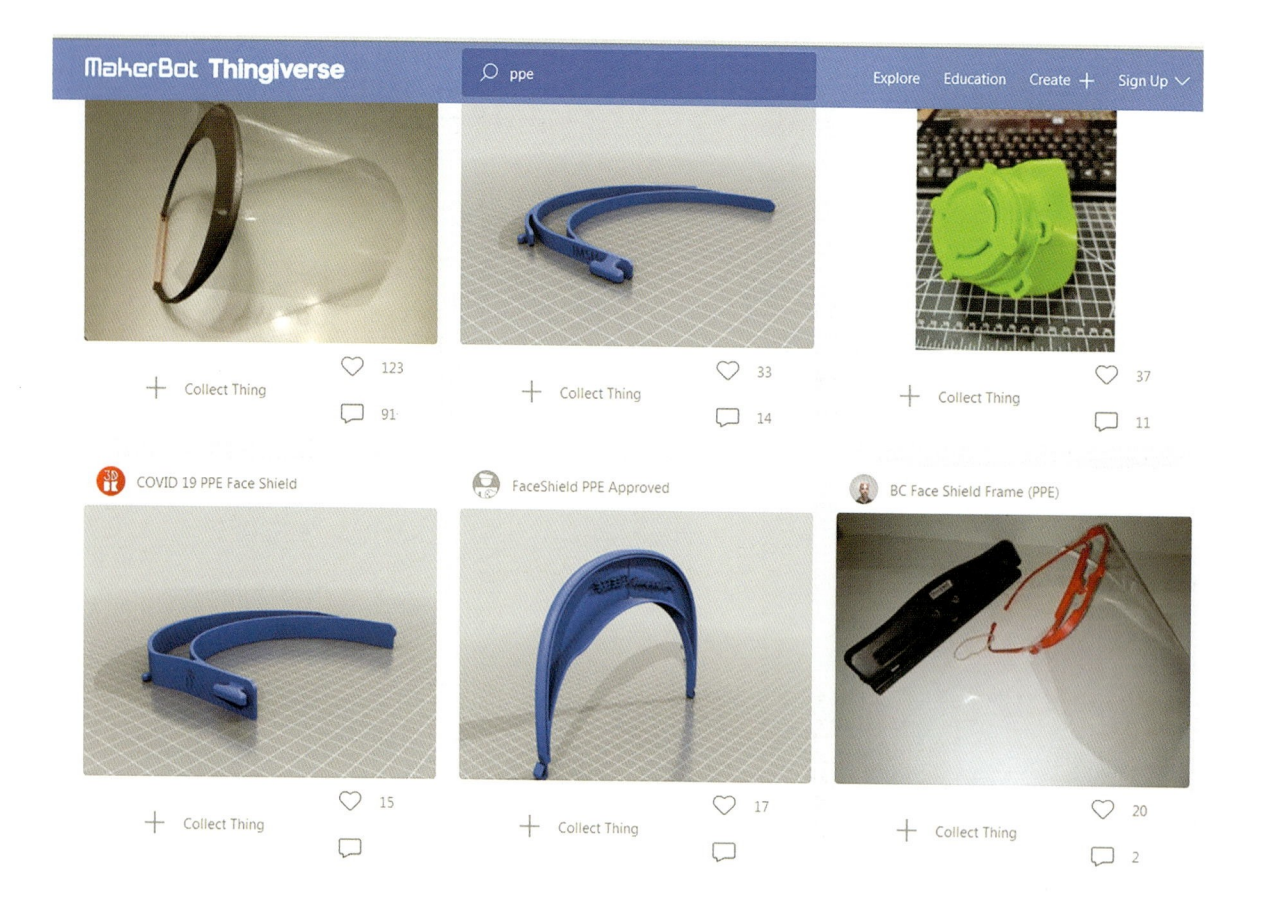

Figure 14.15
Face shield model files available for download at thingiverse.com.

Generative Design

This is a design exploration process done with cloud computing and artificial intelligence software. Instead of you designing a space or object, you input your design goals and parameters—size, weight, strength, manufacturing methods, cost constraints, etc. The software then designs and offers choices. If you don't like the choices, tell the software why and input new parameters. The software then offers different choices after an analysis process called **machine learning**. Figure 14.16 shows wheel rims for a Volkswagen Microbus that were generatively designed and displayed at **Autodesk University**, a yearly software conference.

Figure 14.16
Generatively designed wheel rims.

Generative design is not a specific program but rather a feature inside a program. Autodesk's Fusion 360, PTC's Creo, and Siemen's NX modeling programs have it. Designers have produced buildings, products, and new construction materials with generative design. Software that is not only generative, but intuitive and empathic is also being developed.

Photogrammetry

Photogrammetry is a process that creates a digital model from photographs. You fly a drone-mounted camera around the space or move a tripod-mounted scanner through it, taking photos at regular intervals. At many offices, this has replaced the old-school method of manually measuring each wall and drawing a floor plan. Figure 14.17 shows commercial use scanners. ReCap Pro and Recap Photo are professional photogrammetry programs.

Countertop installers use photogrammetry to measure and draft (see the chapter opener photo), as do product designers who want quick models of existing objects to design edits on. Figure 14.18 shows a handheld 3D camera suitable for such models and Figure 14.19 shows a project made with it.

Figure 14.17
Scanners on tripods used for photogrammetry.

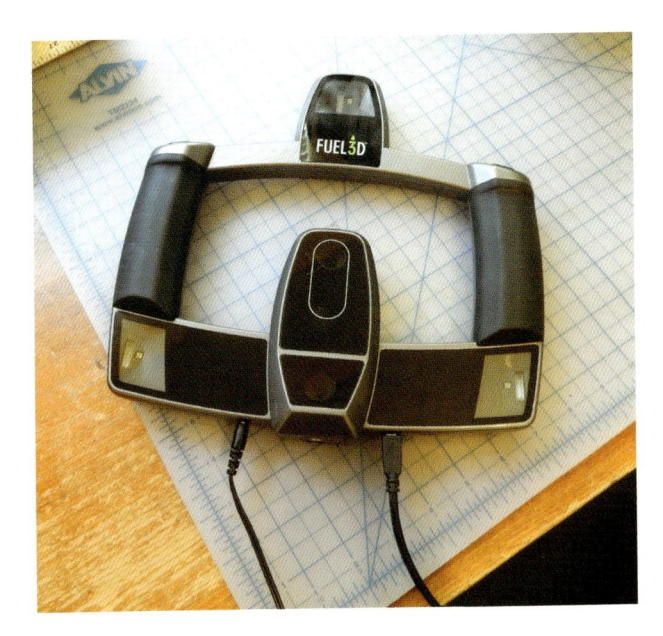

Figure 14.18
A 3D camera for making models of small objects.

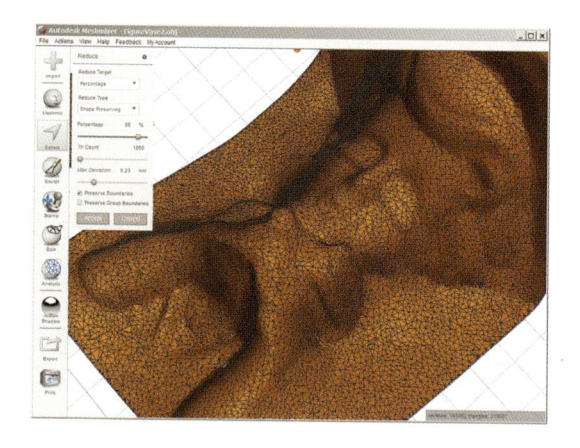

Figure 14.19
The porcelain vase was photographed with a 3D camera. The file was imported into Autodesk Meshmixer, where it was edited into a jewelry holder and then 3D printed.

374 CHAPTER 14 DIGITAL DRAFTING

Photogrammetry can also be done with an ordinary smartphone. Take photos in a circle around the object and upload them to a program like ReCap Photo for processing. Figure 14.20 shows a model of a glass head made this way.

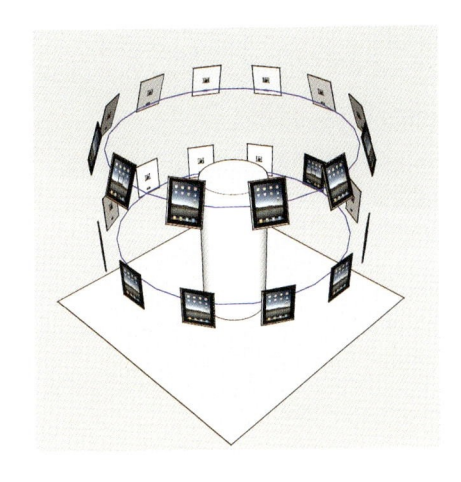

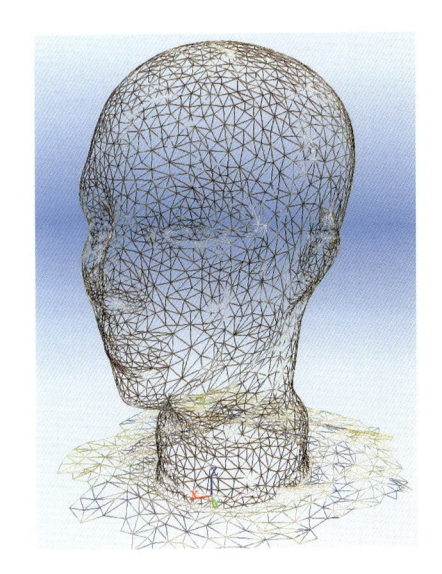

Figure 14.20
A model of a glass head made with a smartphone and ReCap Photo software.

Summary

Designers use 2D and 3D software for construction and presentation drawings. There is no one best program, as all have their own strengths. Drawings can be created inside one program or with multiple programs to take advantage of each's strengths. Emerging technologies in virtual and augmented reality and generative design help designers visualize their work as well as work smarter, more efficiently, and productively.

Classroom Activities

1. Build an entourage library from magazine or personal photos. Find people singly and in groups, shopping, bicycle riding, talking on their phones, etc., for specific projects.
2. Go to thingiverse.com and shapeways.com and browse files available for 3D printing.
3. Obtain and experiment with a Google Cardboard set.

Questions

1. What is a digital model?
2. What is generative design?
3. Why is a 3D mouse better for modeling than a traditional mouse?
4. What DPI makes a high-resolution image?
5. What is parametric drafting?

Further Resources

30 Day Free trial of 20/20 software https://www.2020spaces.com/2020products/2020design/

3D Sourced article on best photogrammetry apps. https://3dsourced.com/3d-software/best-photogrammetry-software/

Autodesk Project Dreamcatcher about generative design. https://autodeskresearch.com/projects/dreamcatcher

Autodesk University. Website for its yearly conference. Watch hundreds of recorded classes about their software and practical uses. https://www.autodesk.com/autodesk-university/

Cline, Lydia Sloan 3D Printing and CNC Fabrication with SketchUp McGraw-Hill NYC 2015

Cline, Lydia Sloan: 3D Printer Projects for Makerspaces McGraw-Hill NYC, 2017

Download student license Autodesk software https://www.autodesk.com/education/free-software/featured

Download student priced Adobe Creative Suite products. https://www.adobe.com/creativecloud/buy/students.html

Free digital imaging software www.gimp.org, www.irfanview.com

Google Cardboard. https://arvr.google.com/cardboard/get-cardboard/

Lydia's YouTube channel. Has manual drafting and SketchUp tutorials. https://www.youtube.com/user/ProfDrafting

Share and download BIM content https://www.bimobject.com/en-us/product

Share and download Revit models www.revitcity.com

Share and download SketchUp content https://3dwarehouse.sketchup.com/?hl=en

Tutorial video: Import and scale a floor plan in SketchUp for tracing. https://www.youtube.com/watch?v=WX9byO0j_Dw

Keywords

- 20/20
- 3D mouse
- 3D printing
- Acrobat DC
- Adobe Illustrator
- Augmented Reality app
- AutoCAD
- Autodesk Revit
- Autodesk University
- BIM modeling
- bitmap
- Chief Architect
- digital imaging
- digital model
- direct modeling
- DWG file
- Excel
- Fused Deposition Modeling (FDM)
- generative design
- GIF file
- GIMP
- high-resolution
- Irfanview
- JPG file
- LiDAR
- low resolution
- machine learning
- modeling
- orbiting
- parametric modeling
- PDF file
- photogrammetry
- Photoshop
- pixelation
- plotter
- PNG file
- raster
- Rhinoceros 3D
- RVT file
- SketchUp
- vector
- virtual reality headset
- workflow

APPENDIX I

Floor Plans for General Use

Suggestions:
- Download from **STUDiO** and import into CAD programs to trace and model.
- Space plan with template furniture.
- Draft a one-point or two-point perspective.
- Draft an isometric pictorial.
- Trace with ink.
- Add dimension lines.

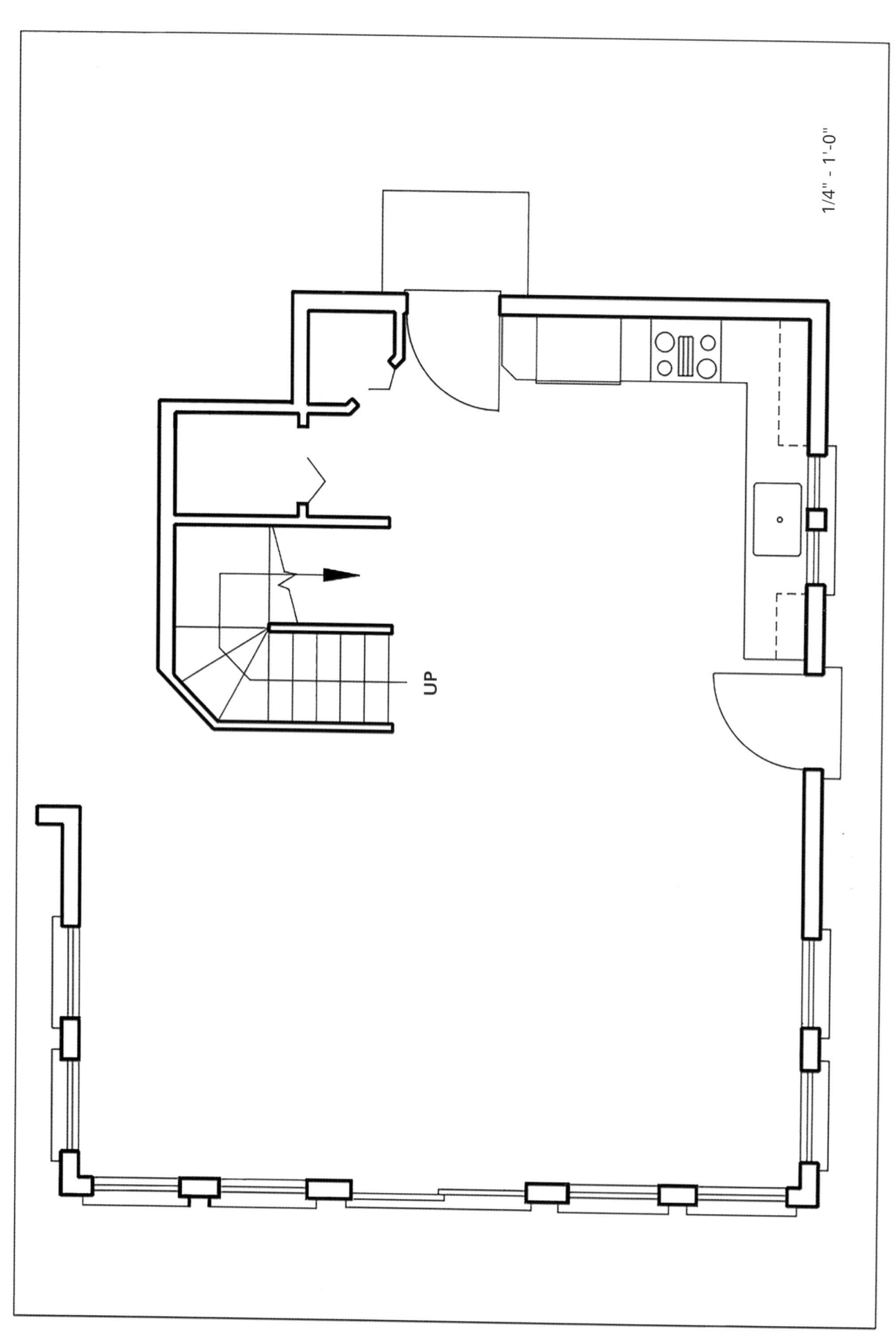

1/4" - 1'-0"

378 APPENDIX I

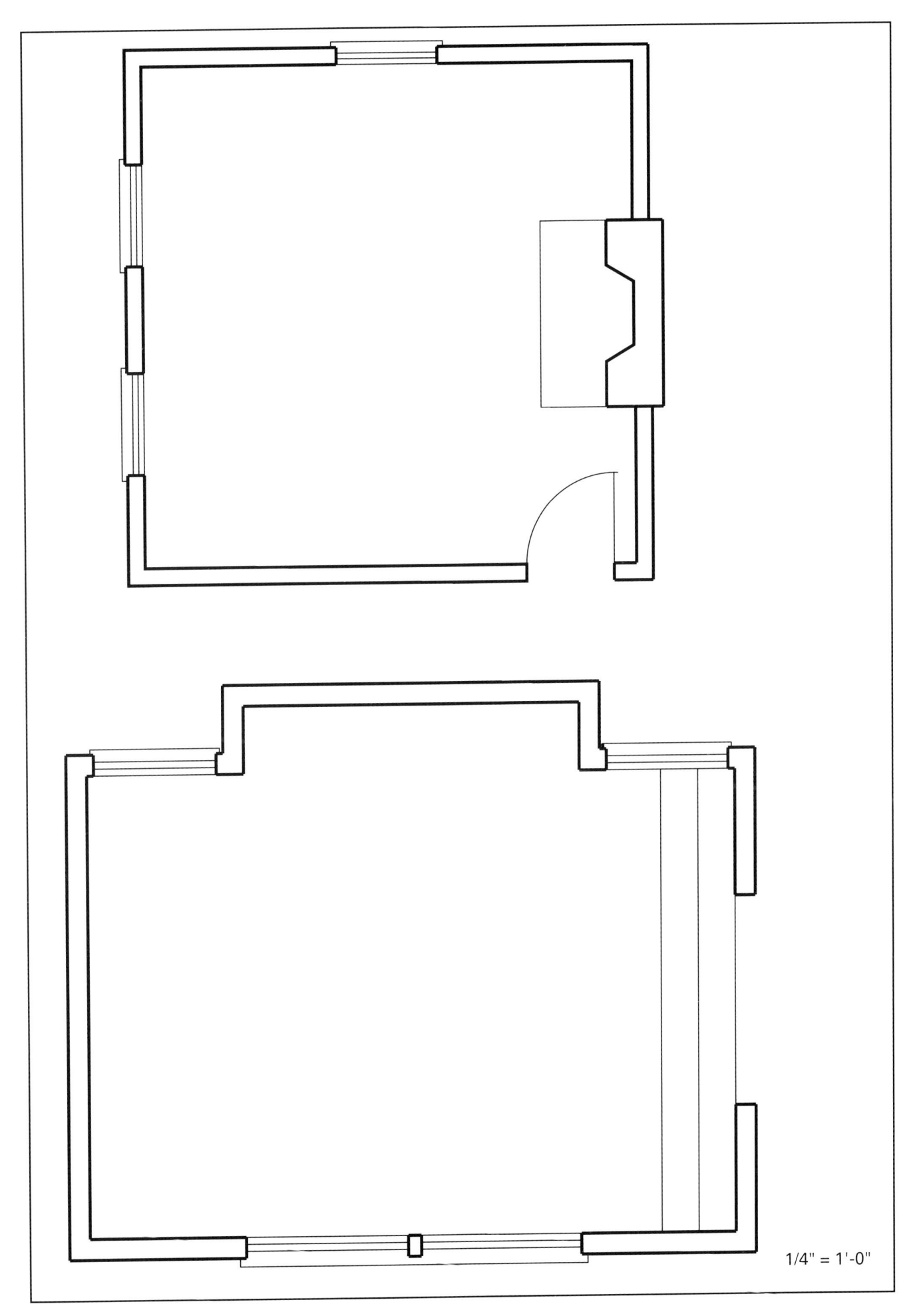

1/4" = 1'-0"

APPENDIX I

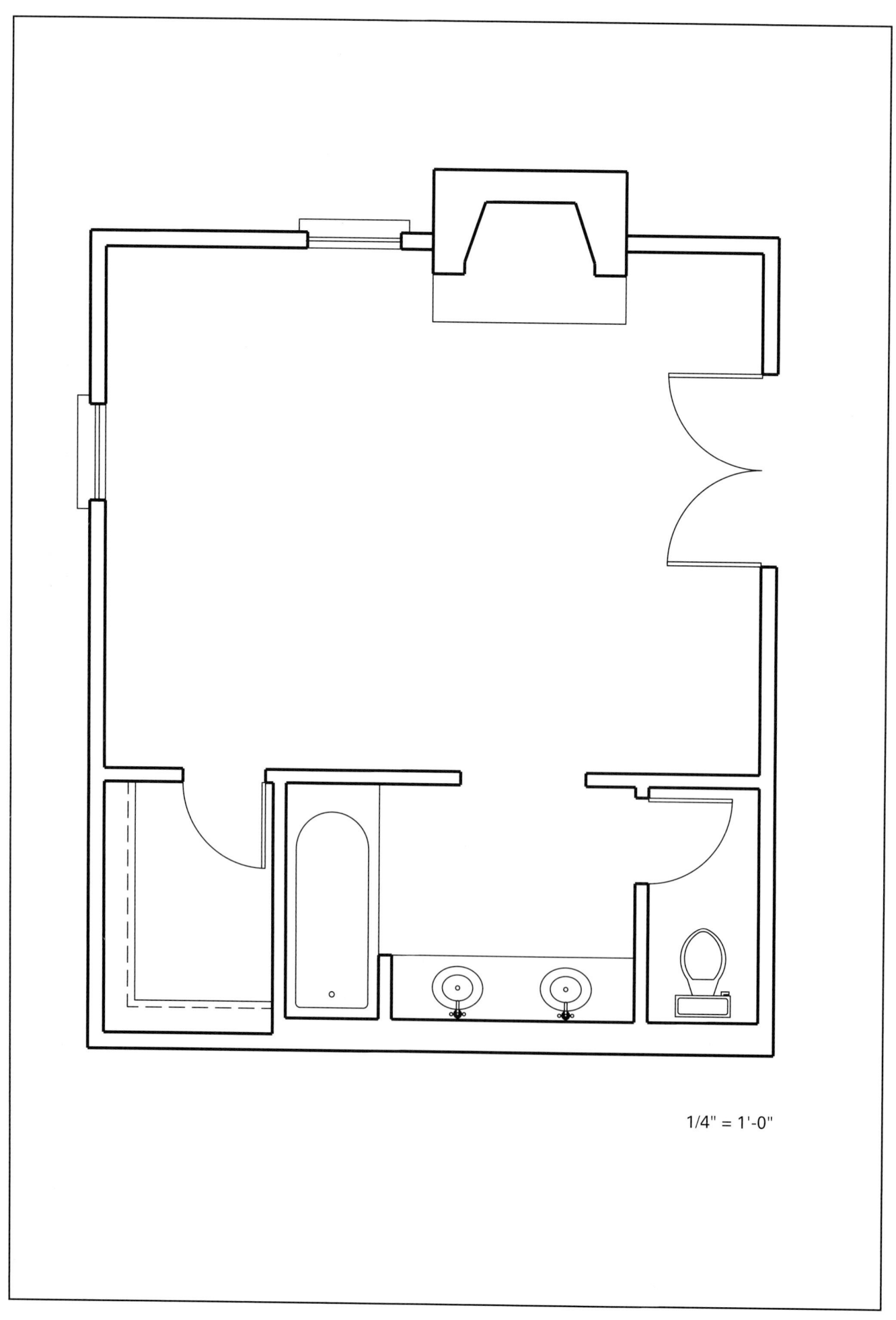

1/4" = 1'-0"

UP

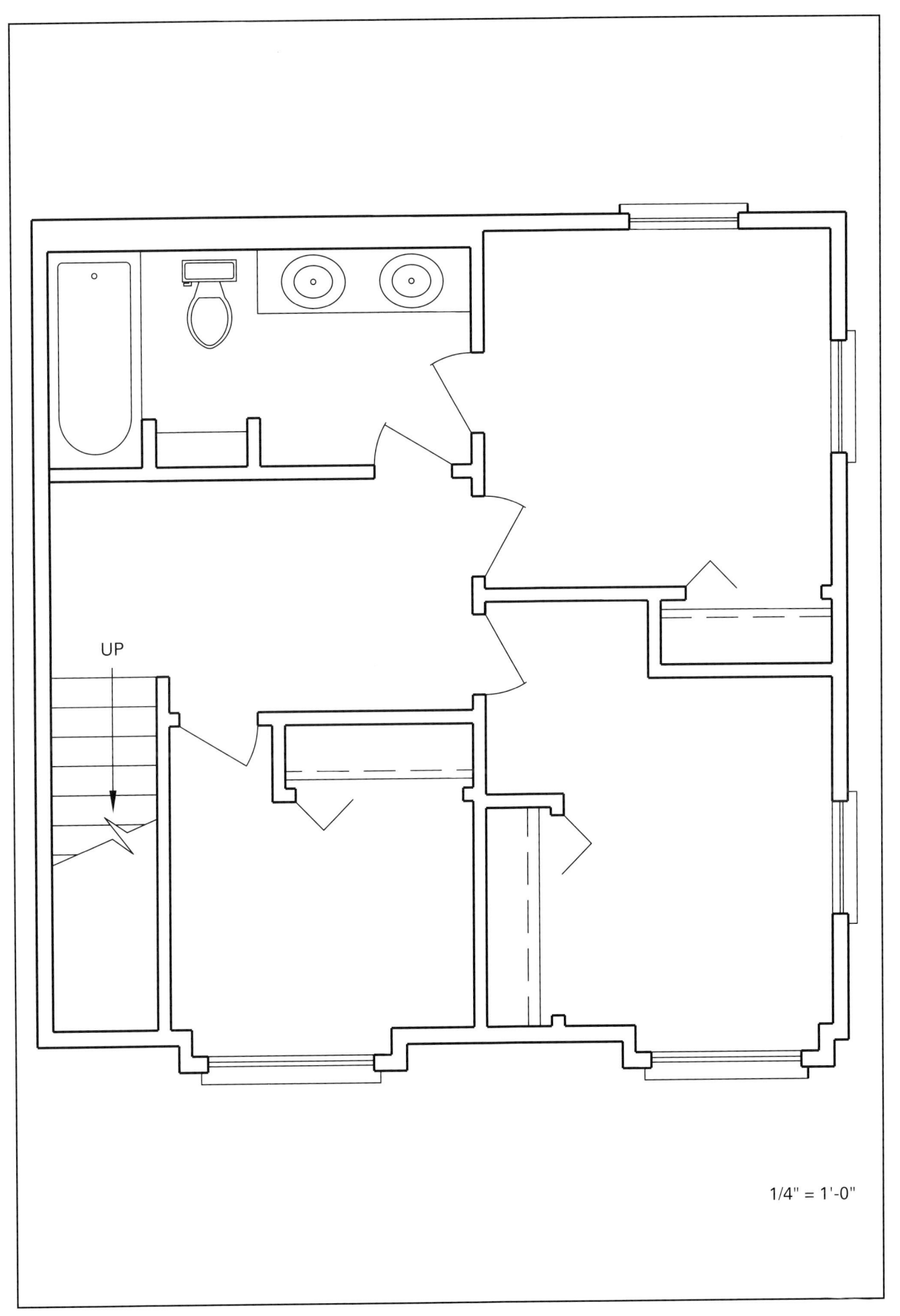

1/4" = 1'-0"

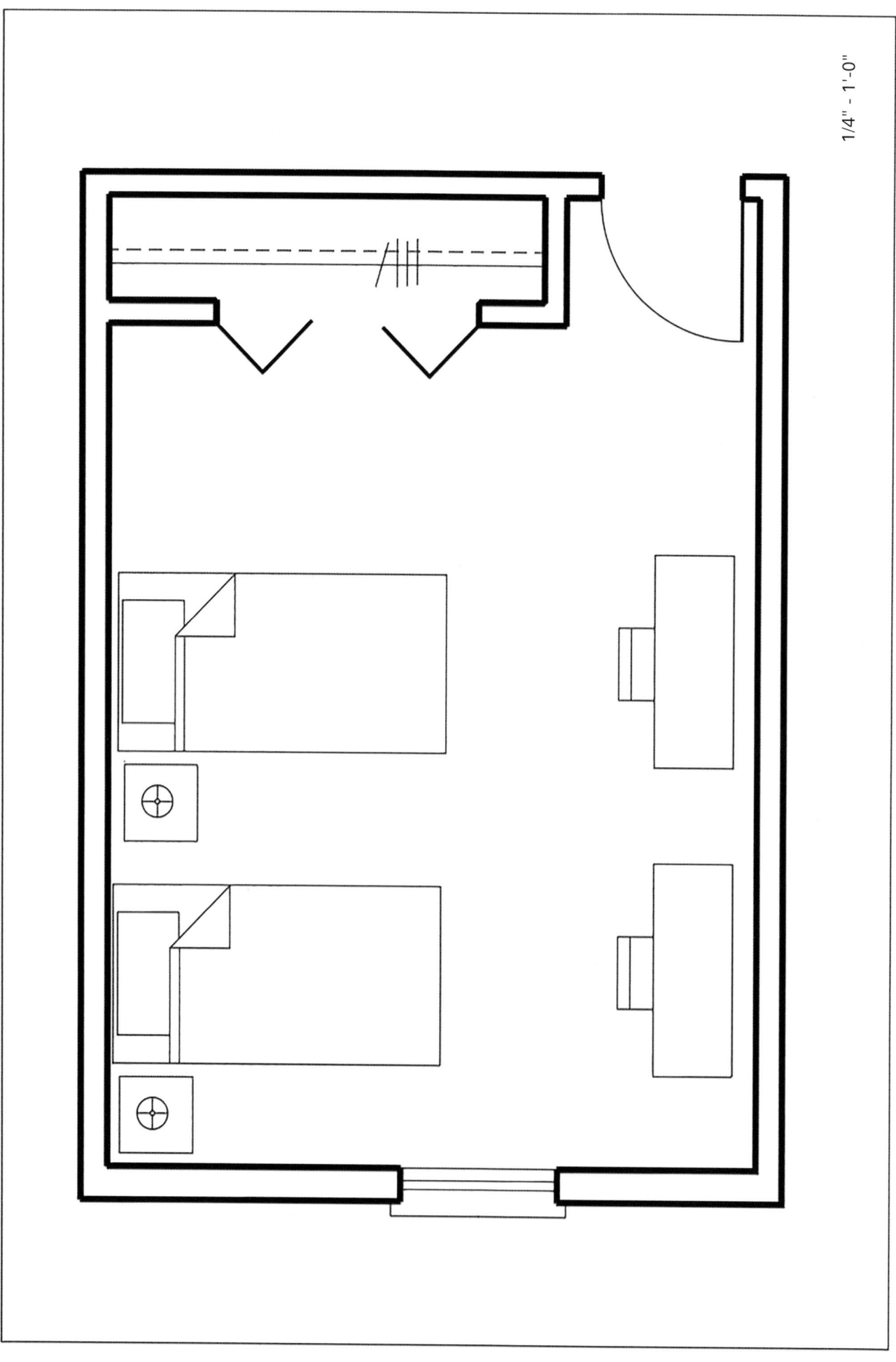

1/4" - 1'-0"

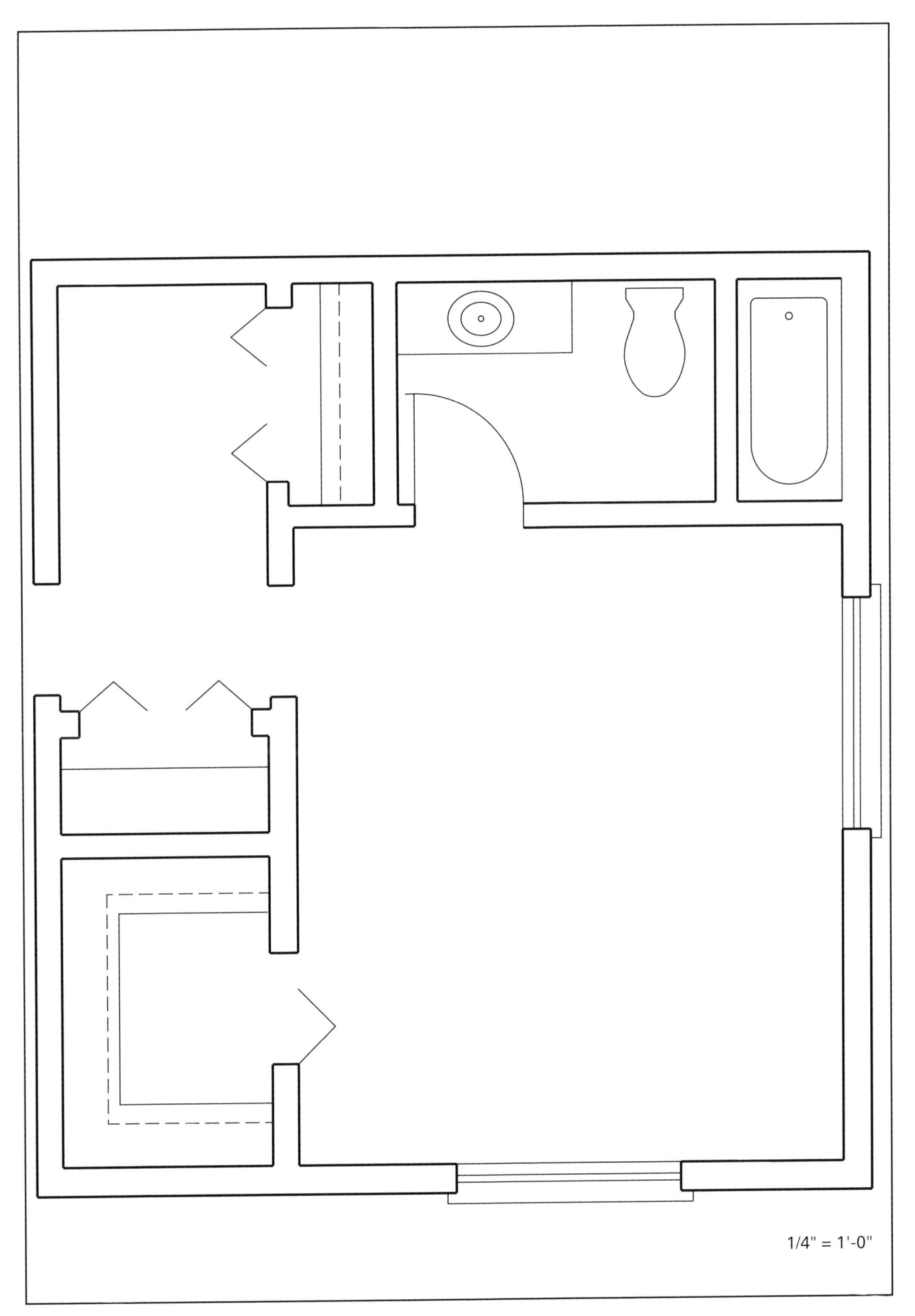

1/4" = 1'-0"

APPENDIX I   **383**

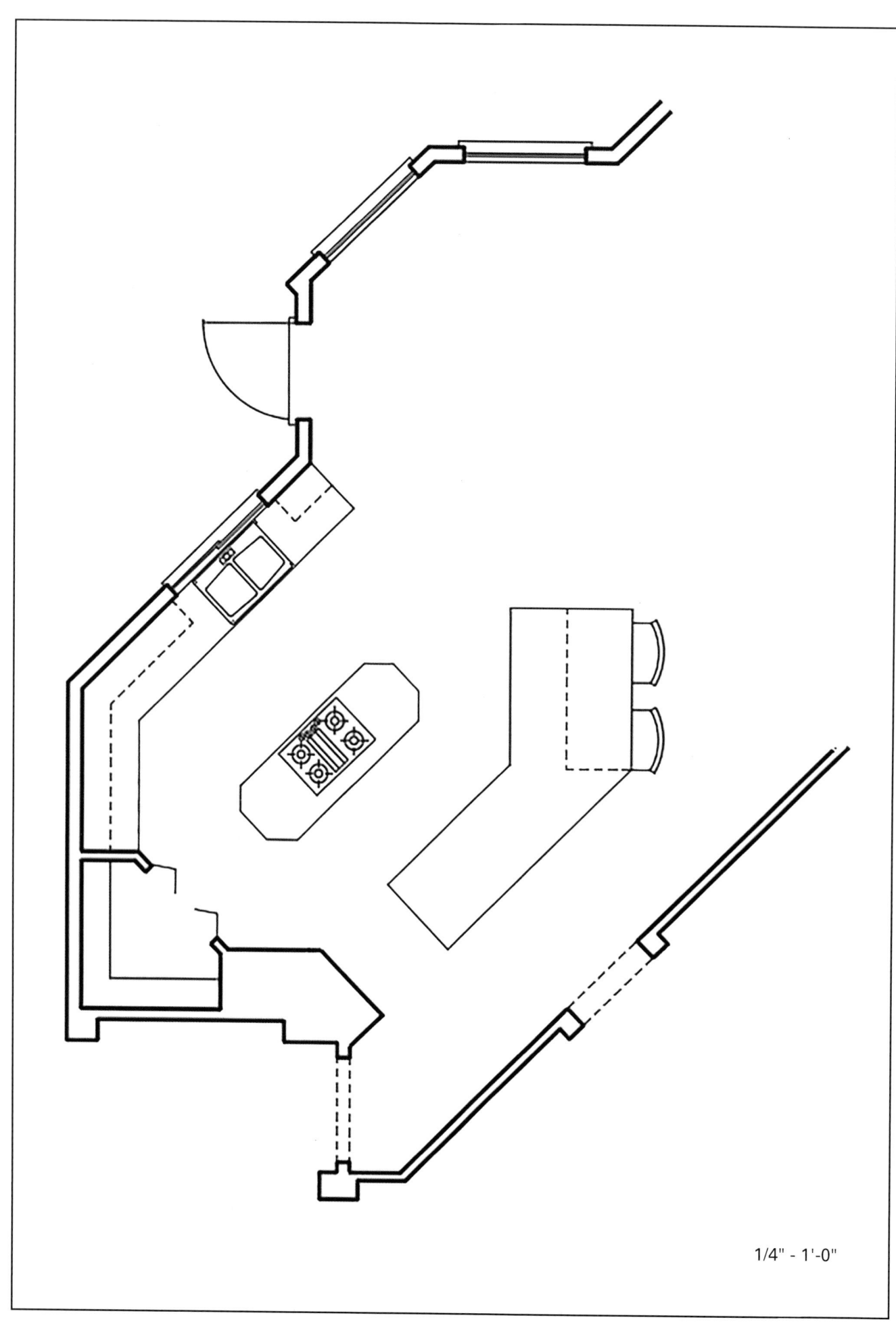

1/4" - 1'-0"

UP

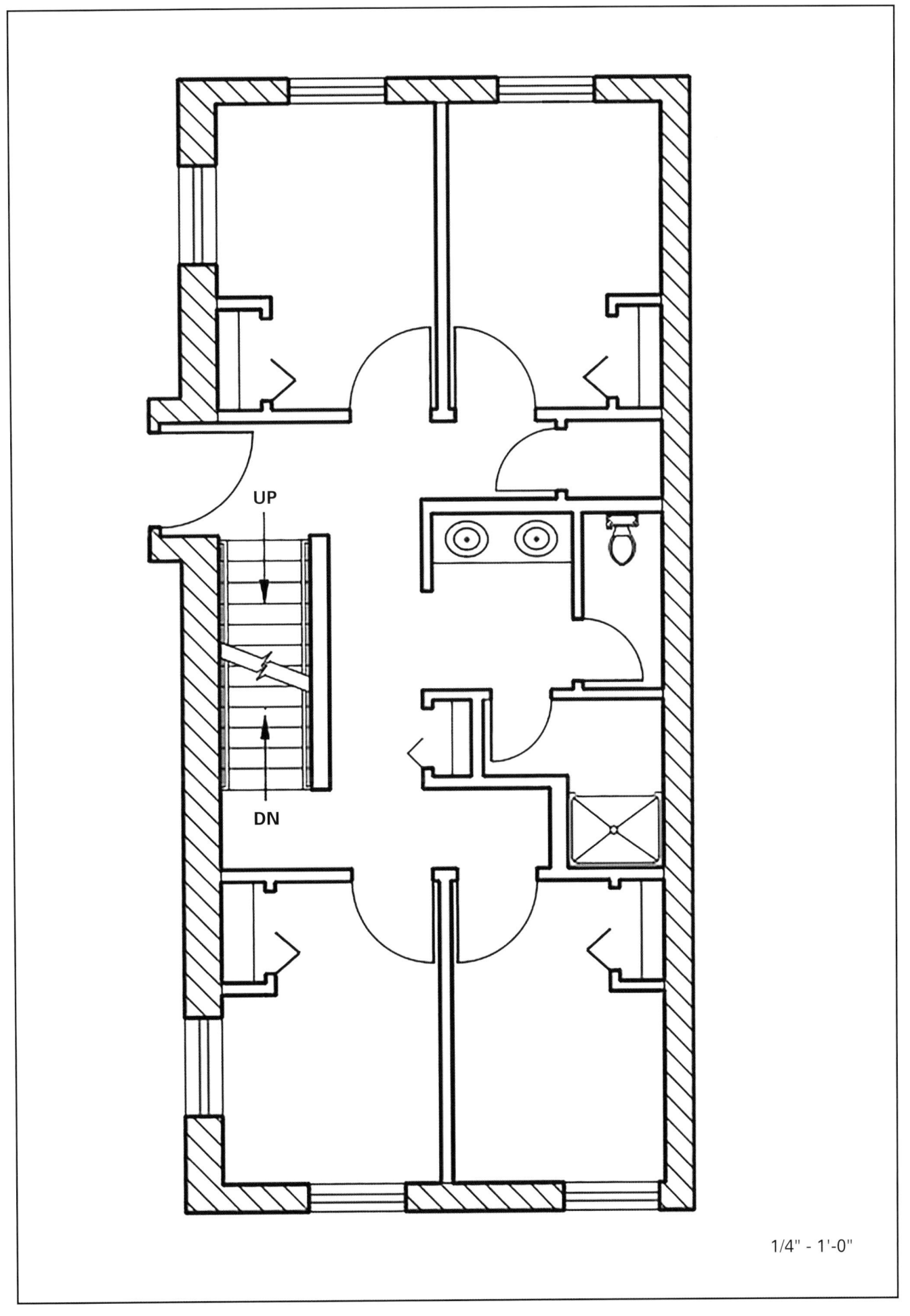

DN

1/4" - 1'-0"

APPENDIX I

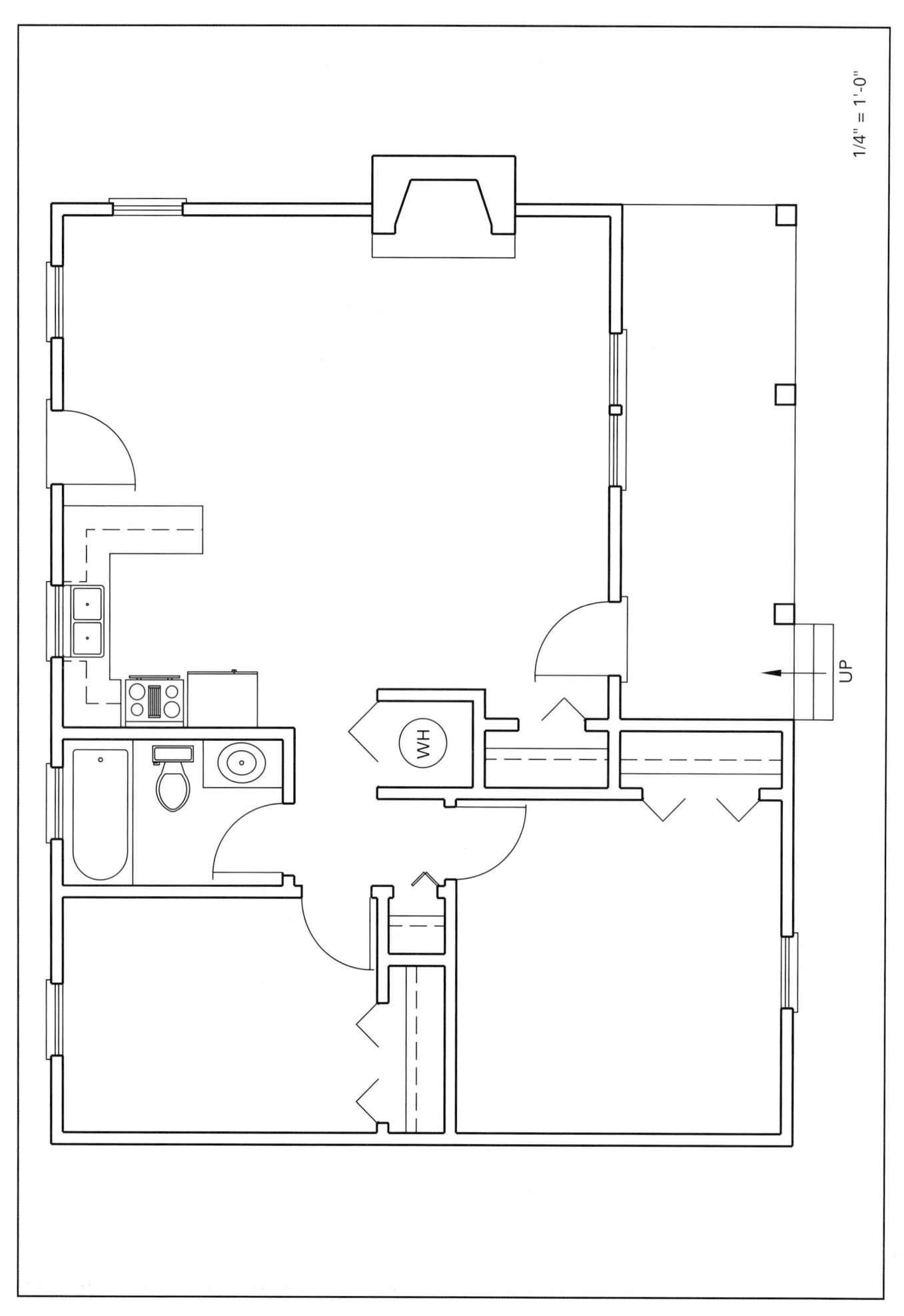

1/4" = 1'-0"

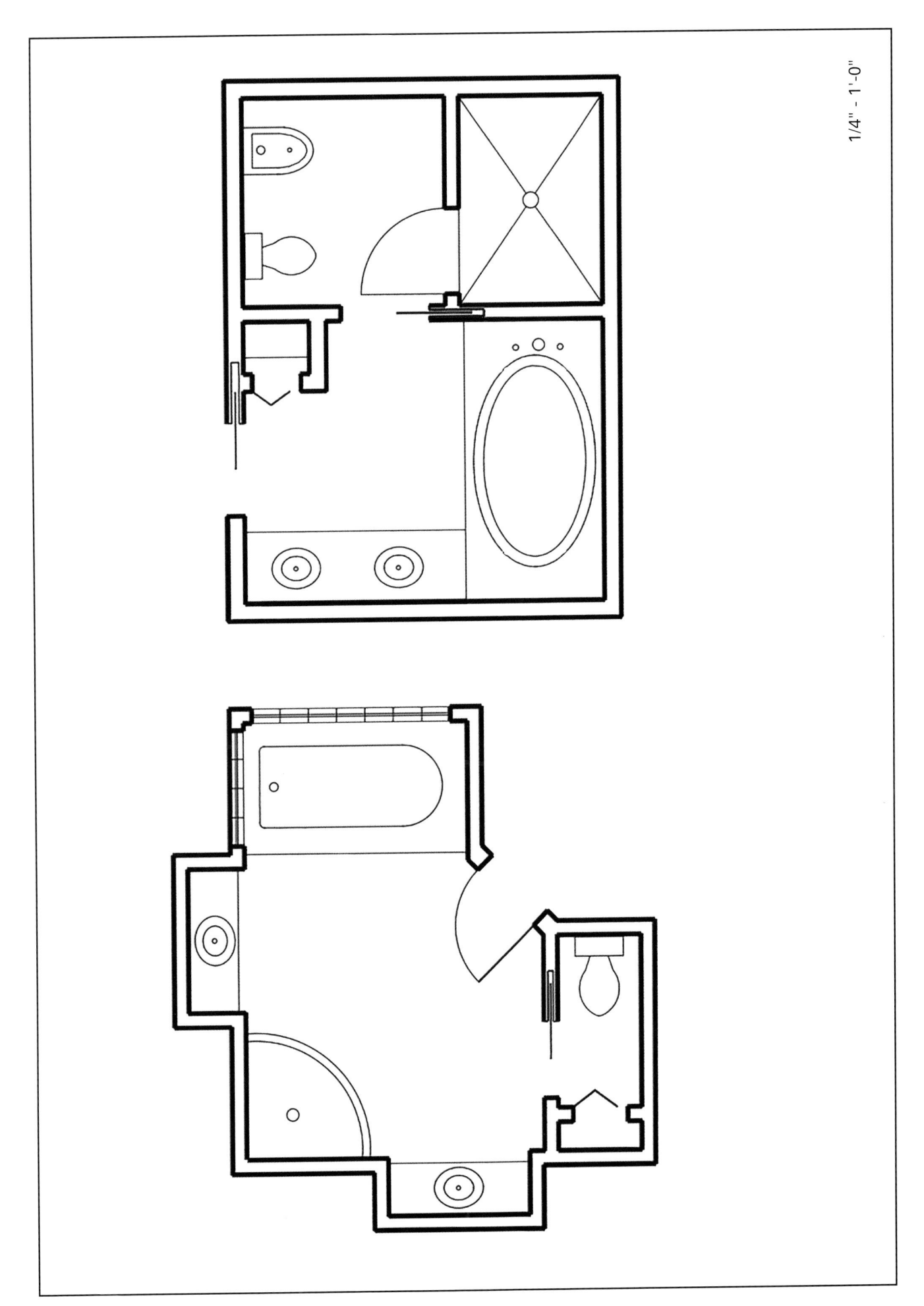

1/4" = 1'-0"

APPENDIX I

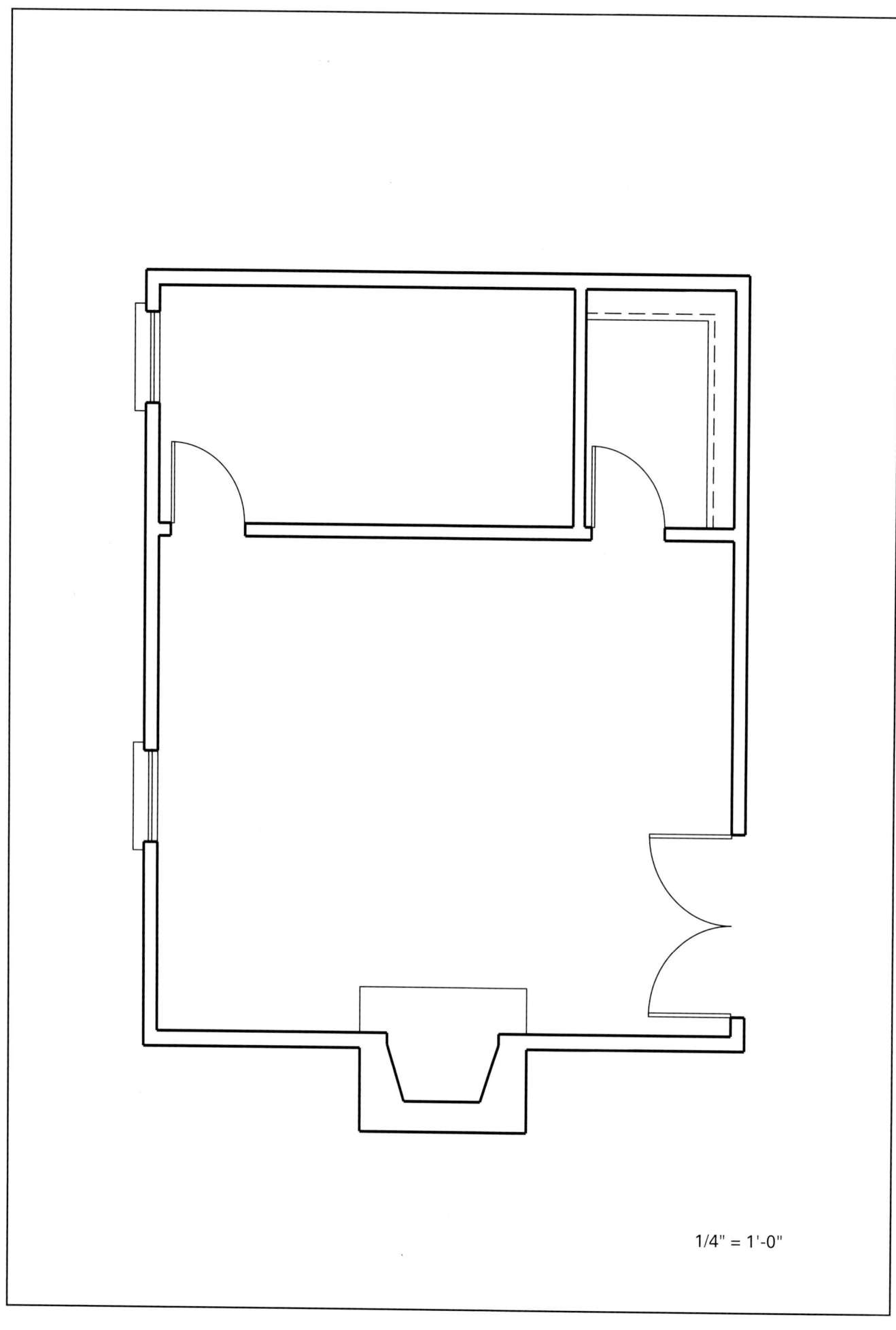

1/4" = 1'-0"

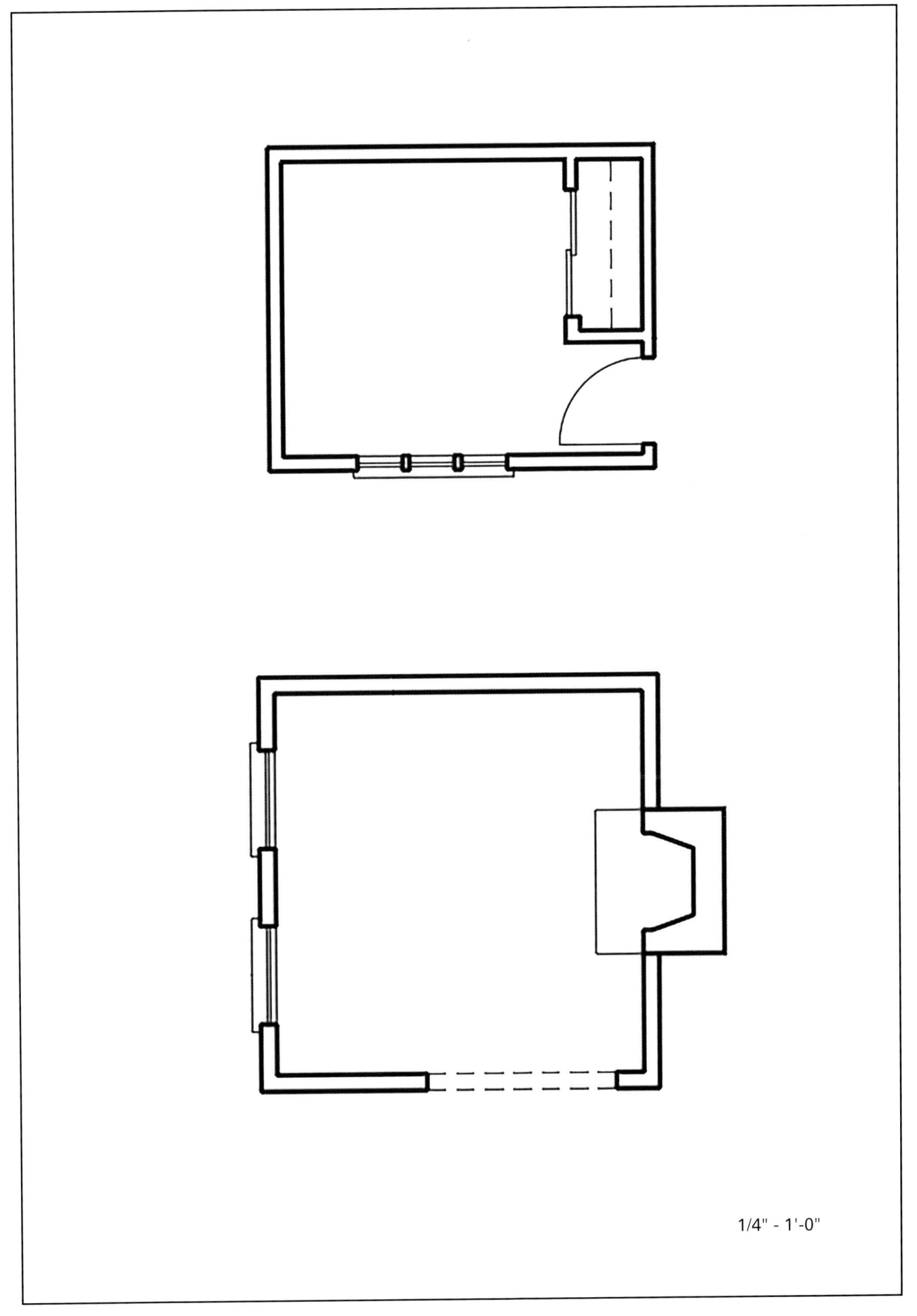

1/4" - 1'-0"

APPENDIX I

1/8" - 1'-0"

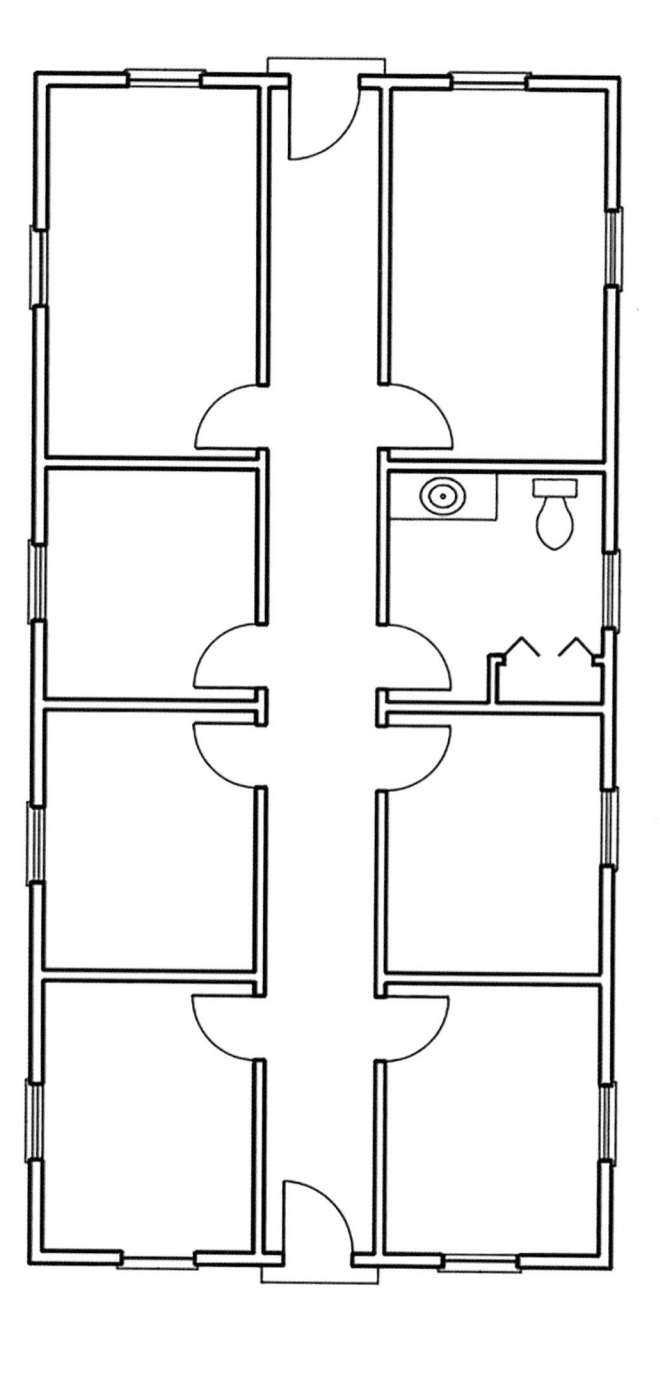

1/8" = 1'-0"

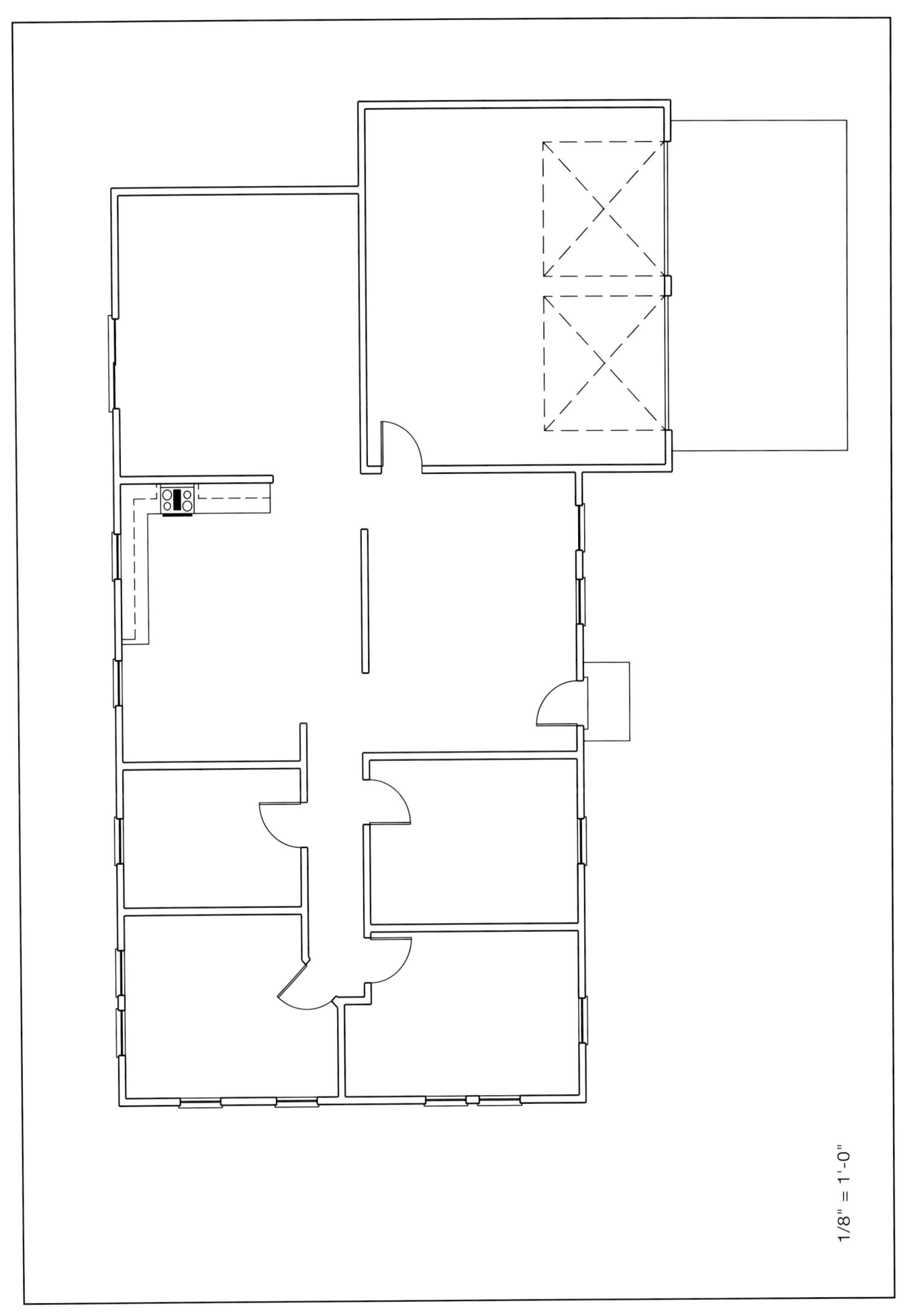

APPENDIX I

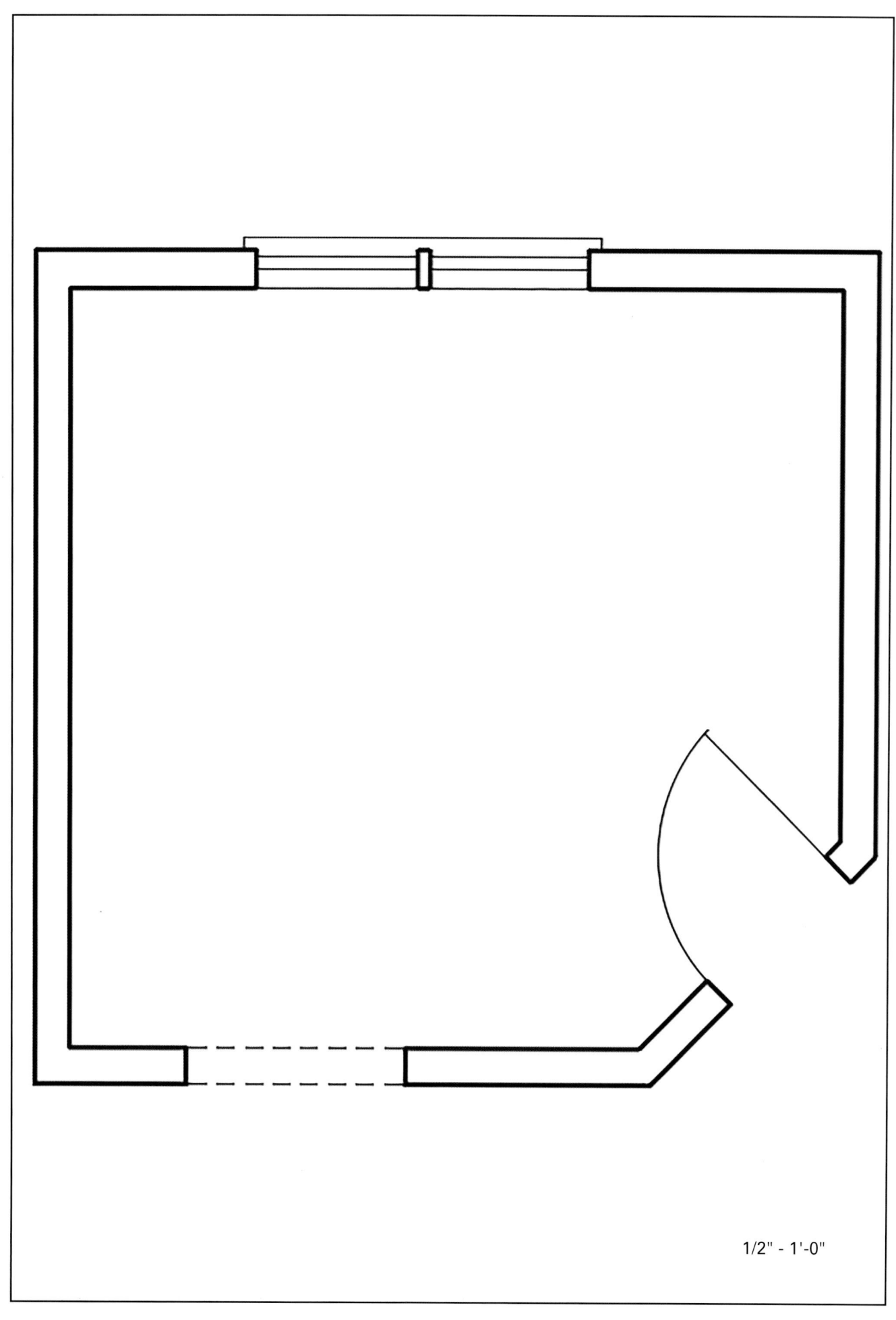

1/2" - 1'-0"

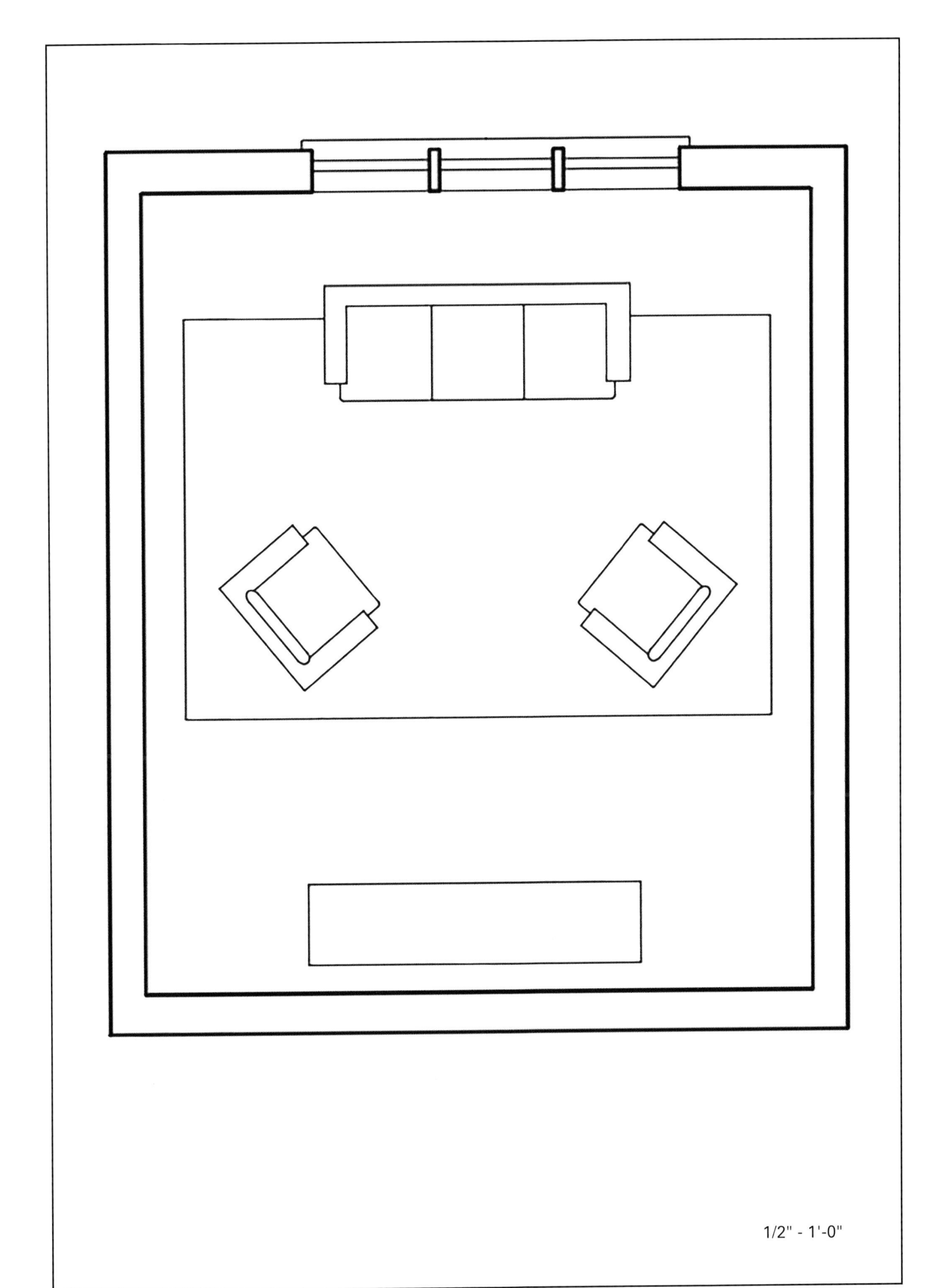

1/2" - 1'-0"

APPENDIX I   **393**

APPENDIX II

Worksheets

These are scaled worksheets. To print the PDFs from this book's **STUDIO** correctly, follow these steps:

1. Right-click and save to the computer.
2. Open with Adobe Reader.
3. Click File/Print. From the Page/Sizing options, choose "Actual Size." Don't fit or shrink the worksheet.
4. If an "Autocenter" option exists, click it to center the graphics on the paper.

The hard copy should look exactly like the textbook page, with same-size graphics. Overlay your print on the textbook worksheet to verify that your hard copy is printed correctly.

Measure the lines with the architect's scales given. Letter the answer above the line.

1. |—————————————————|
   1/4" = 1'-0"

2. |——————————————————————————|
   1/8" = 1'-0"

3. |————————————————|
   1" = 1" (true scale)

4. |———————|
   1/2" = 1'-0"

5. |—————————————————————|
   1 1/2" = 1'-0"

6. |———————————————|
   3" = 1'-0"

7. |————————————————|
   1" = 1" (true scale)

8. |——————————————————|
   1" -1'-0"

9. |——————————————————————————————|
   3/4" = 1'-0"

10. |——|
    3/32" = 1'-0"

NAME	PROJECT
CLASS	SECTION

APPENDIX II

Measure the lines in meters with the metric scales given. Letter the answer above the line.

1.
1:100

2.
1:200

3.
1:300

4.
1:30

5.
1:20

6.
1:10

7.
1:1

8.
1:40

9.
1:500

10.
1:1

NAME	PROJECT	
CLASS	SECTION	

APPENDIX II

Measure the lines in feet with the engineer's scales given.

1. |—————————————————|
   1" = 10'

2. |———————————————————————|
   1" = 20'

3. |————————————————|
   1" = 30'

4. |————————|
   1" = 40'

5. |————————————————————|
   1" = 50'

NAME	PROJECT
CLASS	SECTION

APPENDIX II

Write the length above each dimension line.

1.

2.

3.

4.

5.

6.

NAME	PROJECT	
CLASS	SECTION	

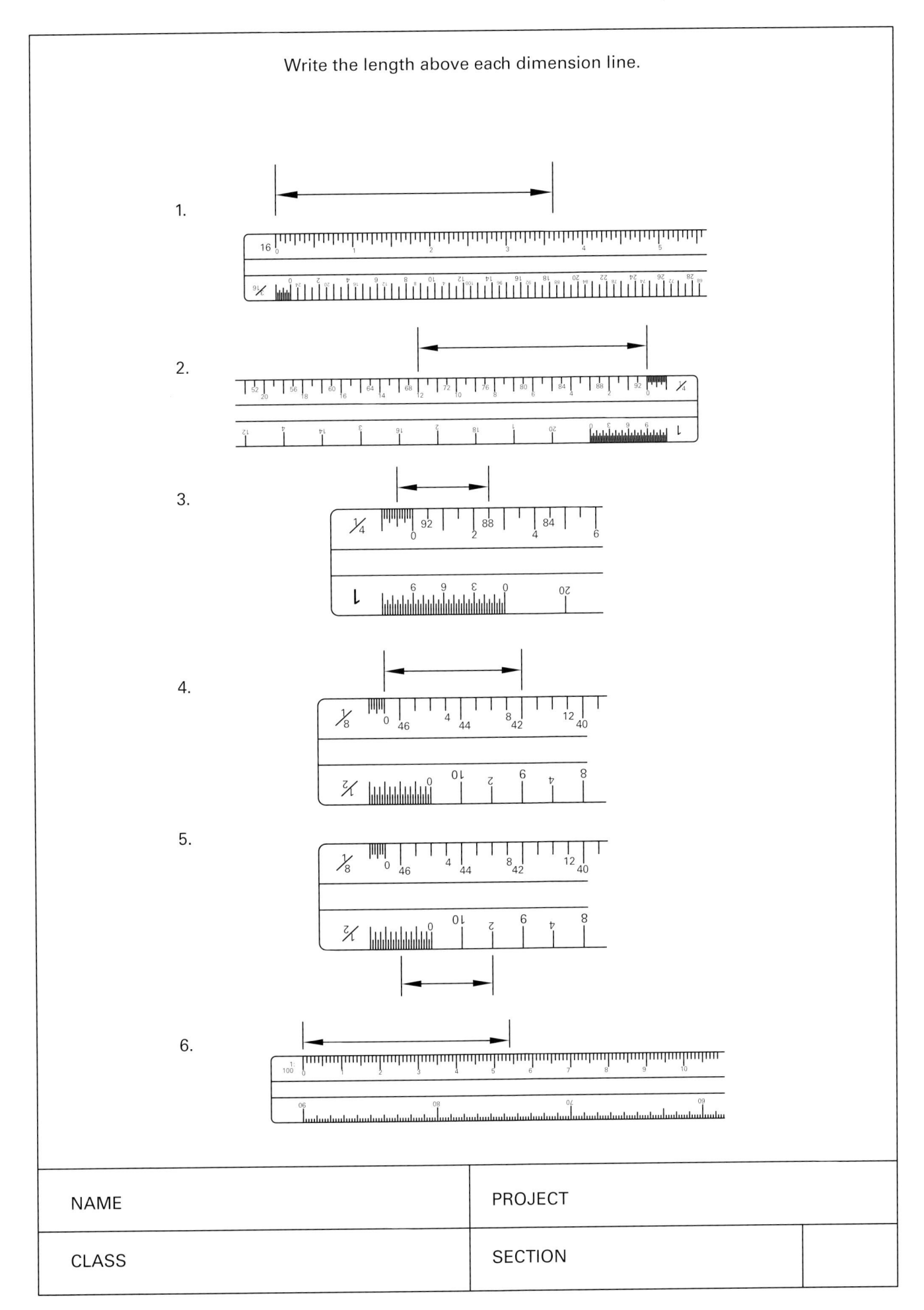

Measure walls A, B, and C at the scales given.

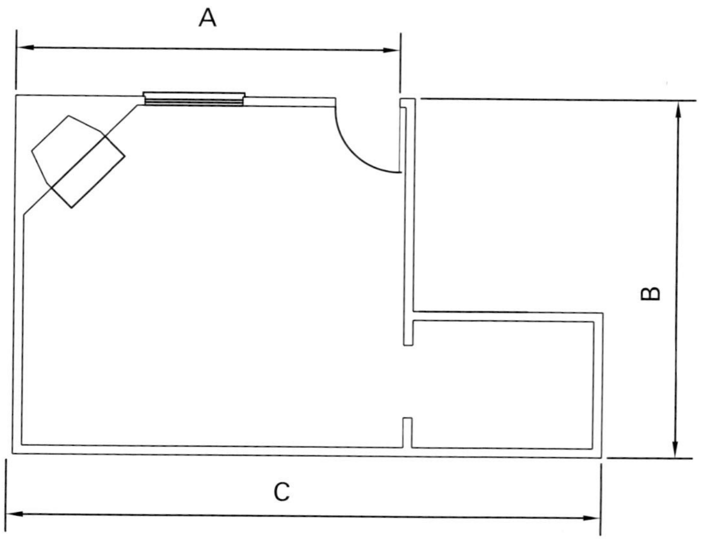

SCALE	A	B	C
1/4" = 1'-0"			
3/16" = 1'-0"			
1/8" = 1'-0"			
3/4" = 1'-0"			
1 1/2" = 1'-0"			

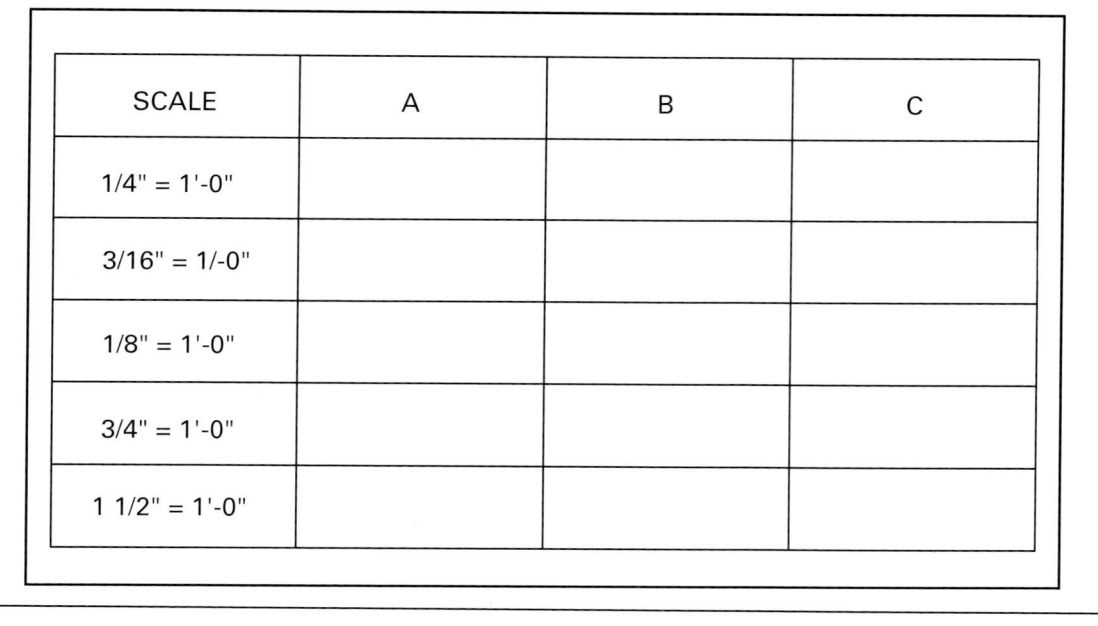

NAME	PROJECT
CLASS	SECTION

400 APPENDIX II

Use a compass or a circle template to draw circles the sizes shown.

1. 1/2" Ø

2. 1 13/16" Ø

3. 1 1/2" Ø

4. 3/4" Ø

NAME	PROJECT	
CLASS	SECTION	

APPENDIX II

Redraw this bed twice its size by using the dividers tool and straight edges.

1/4" = 1'-0"

NAME	PROJECT
CLASS	SECTION

Geometric Construction

Hints

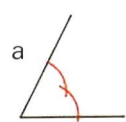

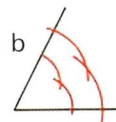

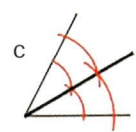

a) Swing two arcs about 2/3rds the way up one leg. b) Swing two more arcs at the arc/leg intersections. c) Draw a line through the intersections.

1. 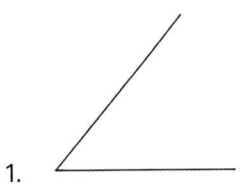 Bisect this angle.

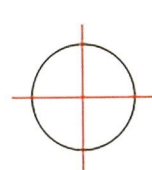

Use the 30-60 triangle.

2. Inscribe a hexagon.

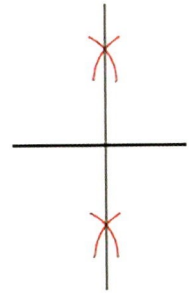

Swing arcs from the lines' endpoints.

3.  Draw lines perpendicular to these lines.

4. 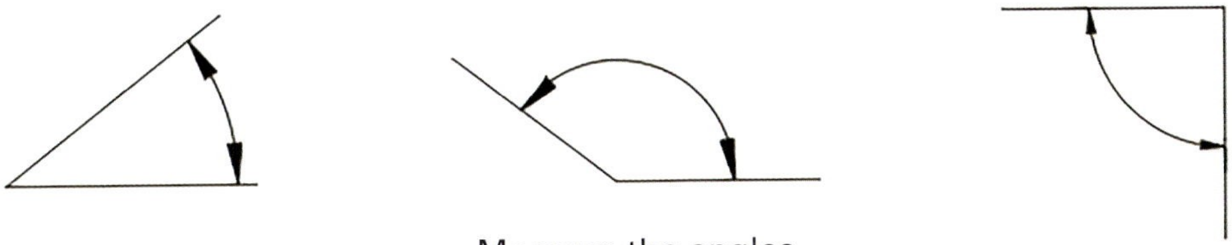 Measure the angles.

NAME	PROJECT
CLASS	SECTION

Use a circle template or compass to draw circles tangent to the lines.

Tangent: Making contact at a single point or along a line; touching but not intersecting.

2" Ø

1" Ø

1 1/4" Ø

1/2" Ø

NAME	PROJECT
CLASS	SECTION

16'-10"

18'-4"

8'-3"

8'-1"

1'-6"

Redraw this at 1/2"=1'-0" scale.

9'-0"

1/4"=1'0"

10'-0"

6'-10"

NAME	PROJECT	
CLASS	SECTION	

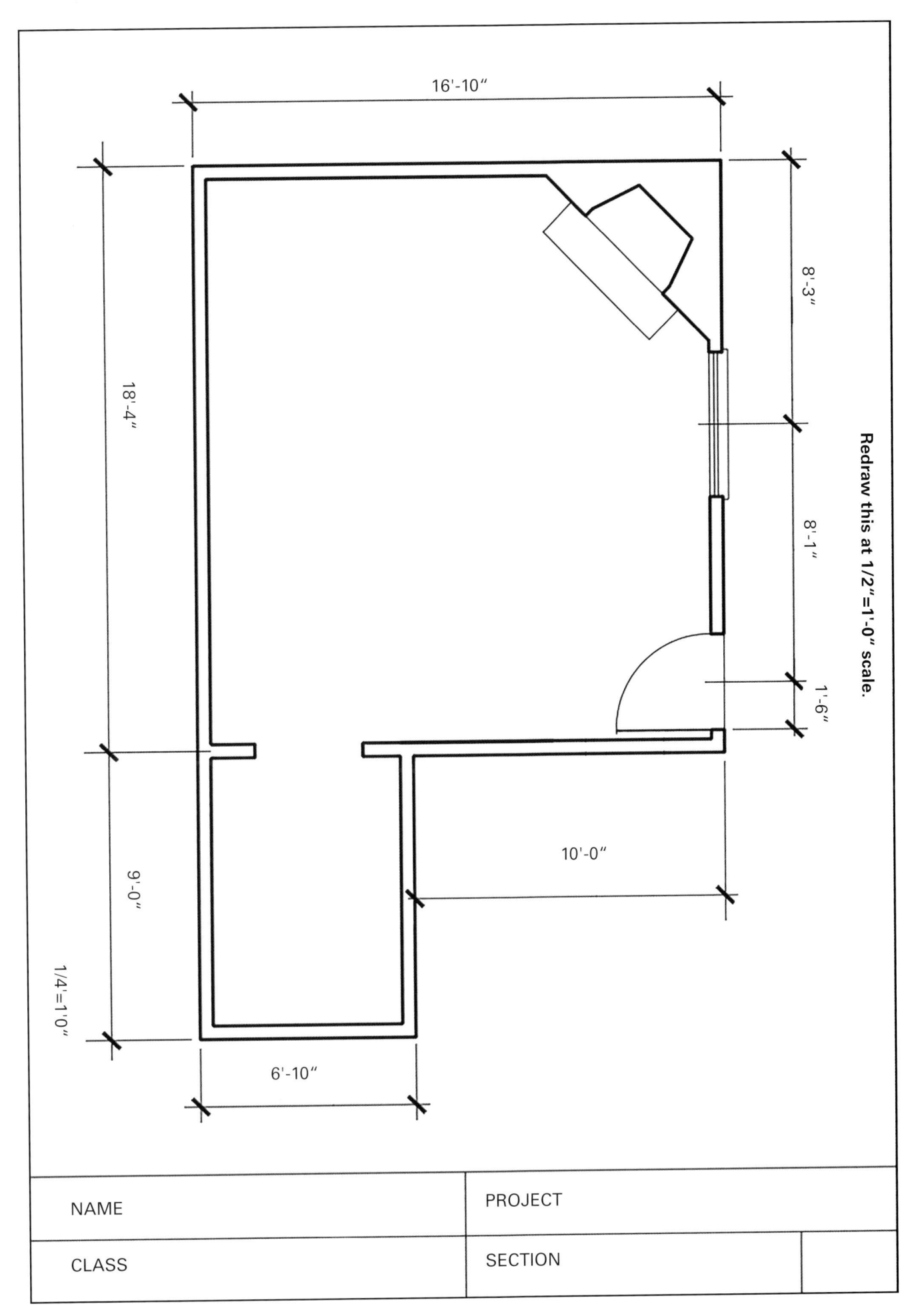

APPENDIX II

Sketch orthographic views of an object of your choice.

NAME	PROJECT
CLASS	SECTION

Sketch orthographic views of an object of your choice.

NAME	PROJECT	
CLASS	SECTION	

Sketch an isometric pictorial of an object of your choice.

NAME	PROJECT
CLASS	SECTION

Sketch a space of your choice.

NAME	PROJECT
CLASS	SECTION

APPENDIX II

Sketch a space of your choice.

NAME	PROJECT
CLASS	SECTION

410 APPENDIX II

Sketch a space of your choice.

NAME	PROJECT	
CLASS	SECTION	

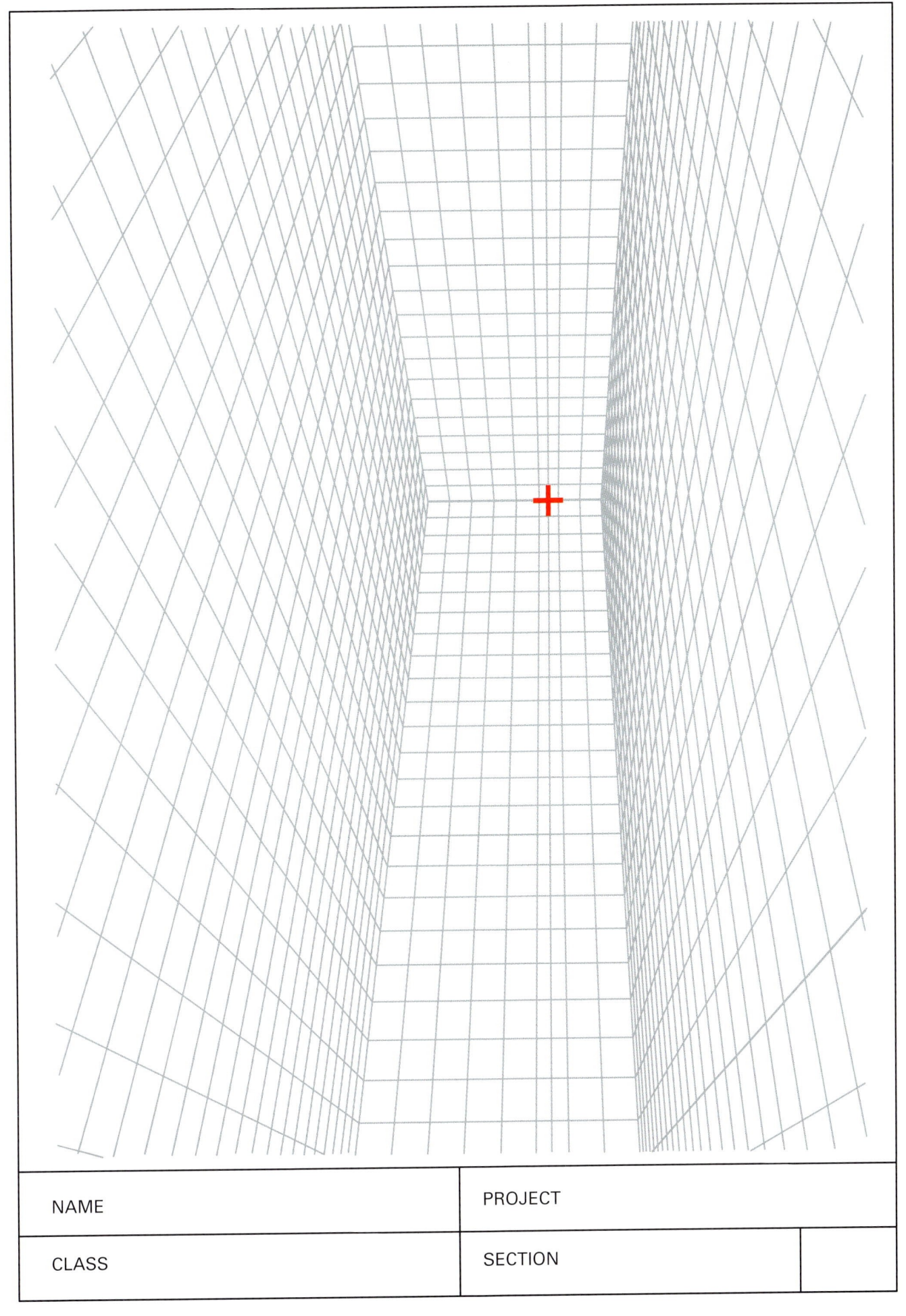

APPENDIX II

Sketch front and side orthographic views @ 1"=1'-0":

NOT TO SCALE

22" deep x 65" long x 21" tall
Legs are 6" tall.

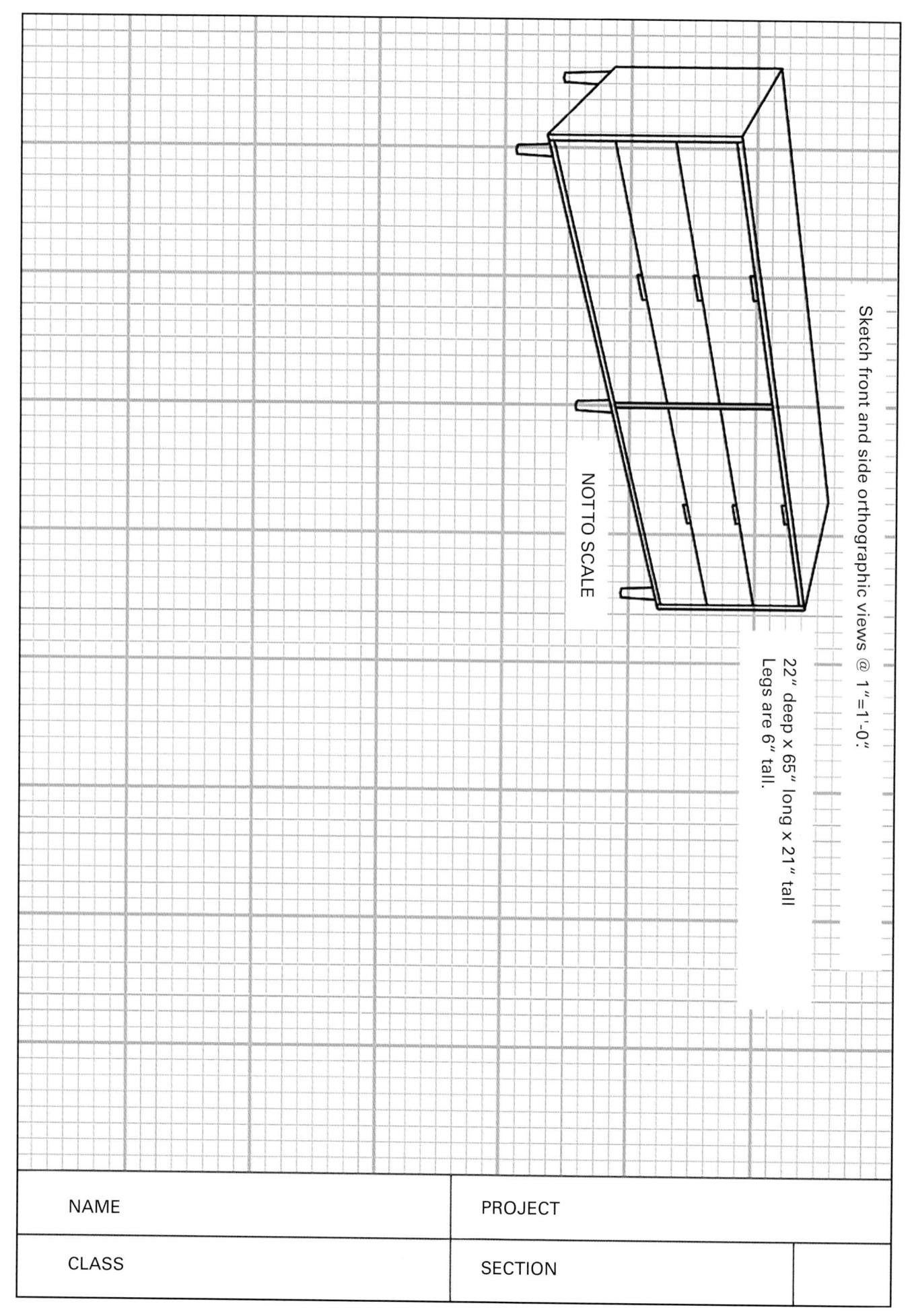

NAME	PROJECT	
CLASS	SECTION	

Sketch front and side orthographic views @ 1"=1'-0".

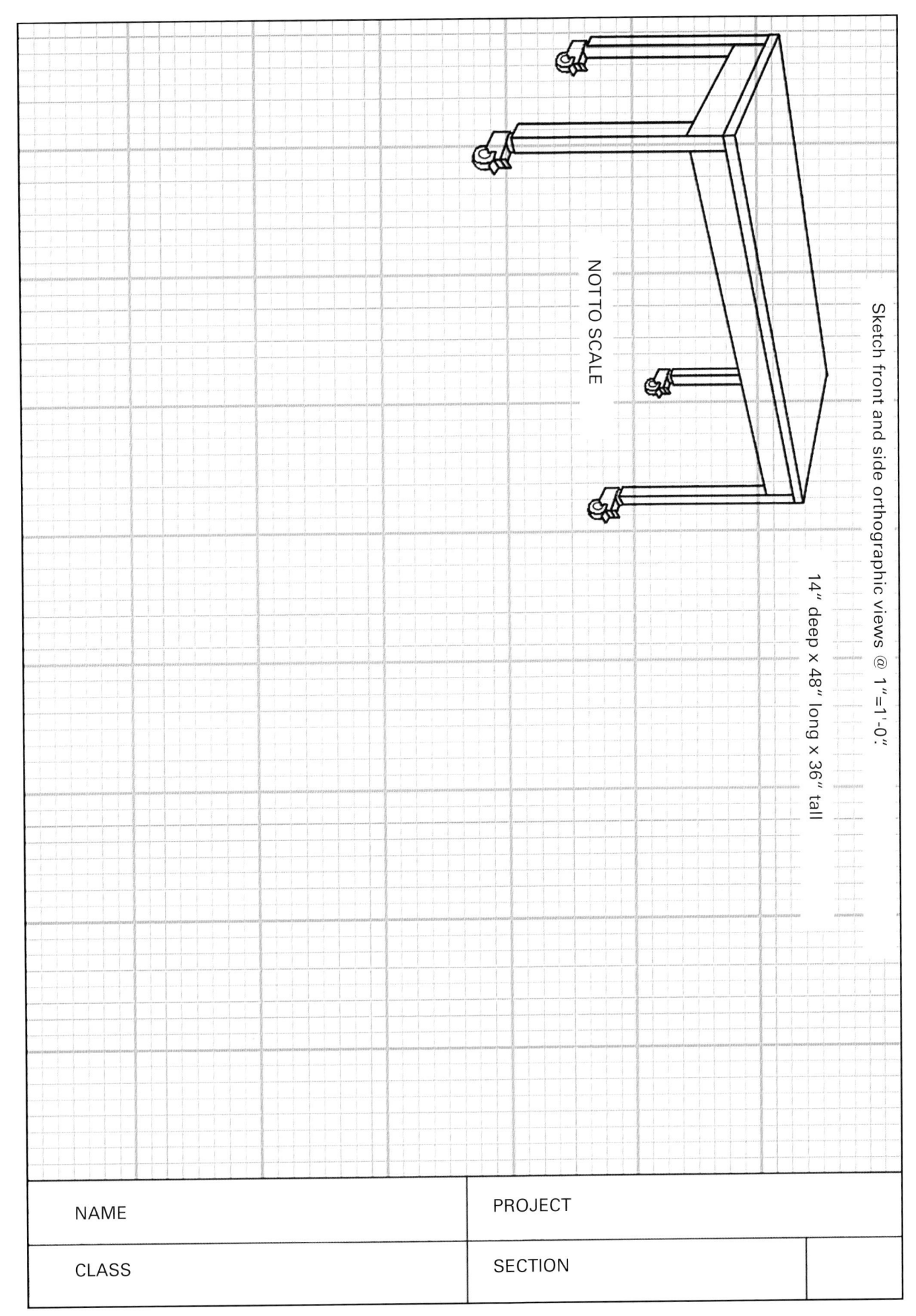

14" deep x 48" long x 36" tall

NOT TO SCALE

NAME

CLASS

PROJECT

SECTION

Sketch front and side orthographic views. 1"=1'-0".

Not to scale

17"
13"
25"
21"

NAME	PROJECT	
CLASS	SECTION	

414 APPENDIX II

1.

2.

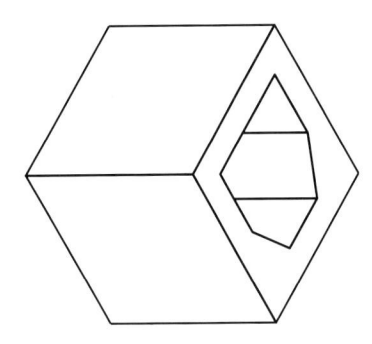

Draw top, front, and side views.

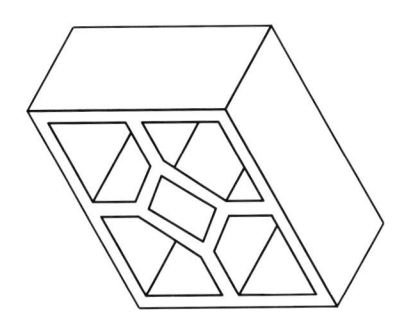

NAME	PROJECT	
CLASS	SECTION	

Draft top, front, and side views.

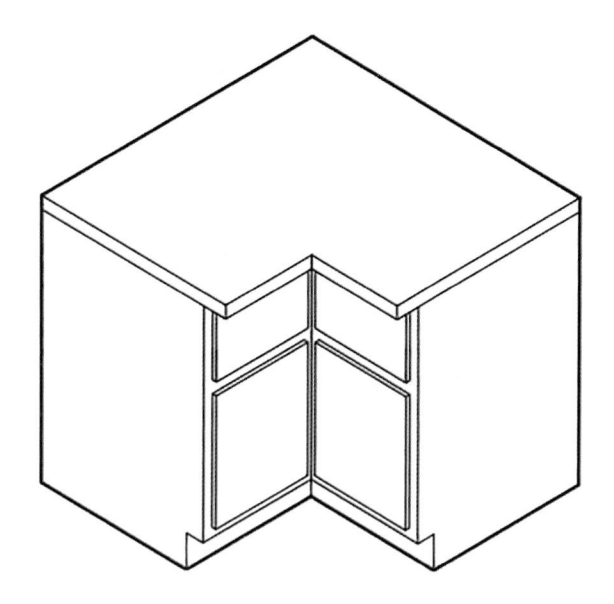

1/2"-1-0"

NAME	PROJECT
CLASS	SECTION

APPENDIX II

1/2" = 1'-0"

Start here.

Draw a cabinet oblique pictorial.

NAME	PROJECT	
CLASS	SECTION	

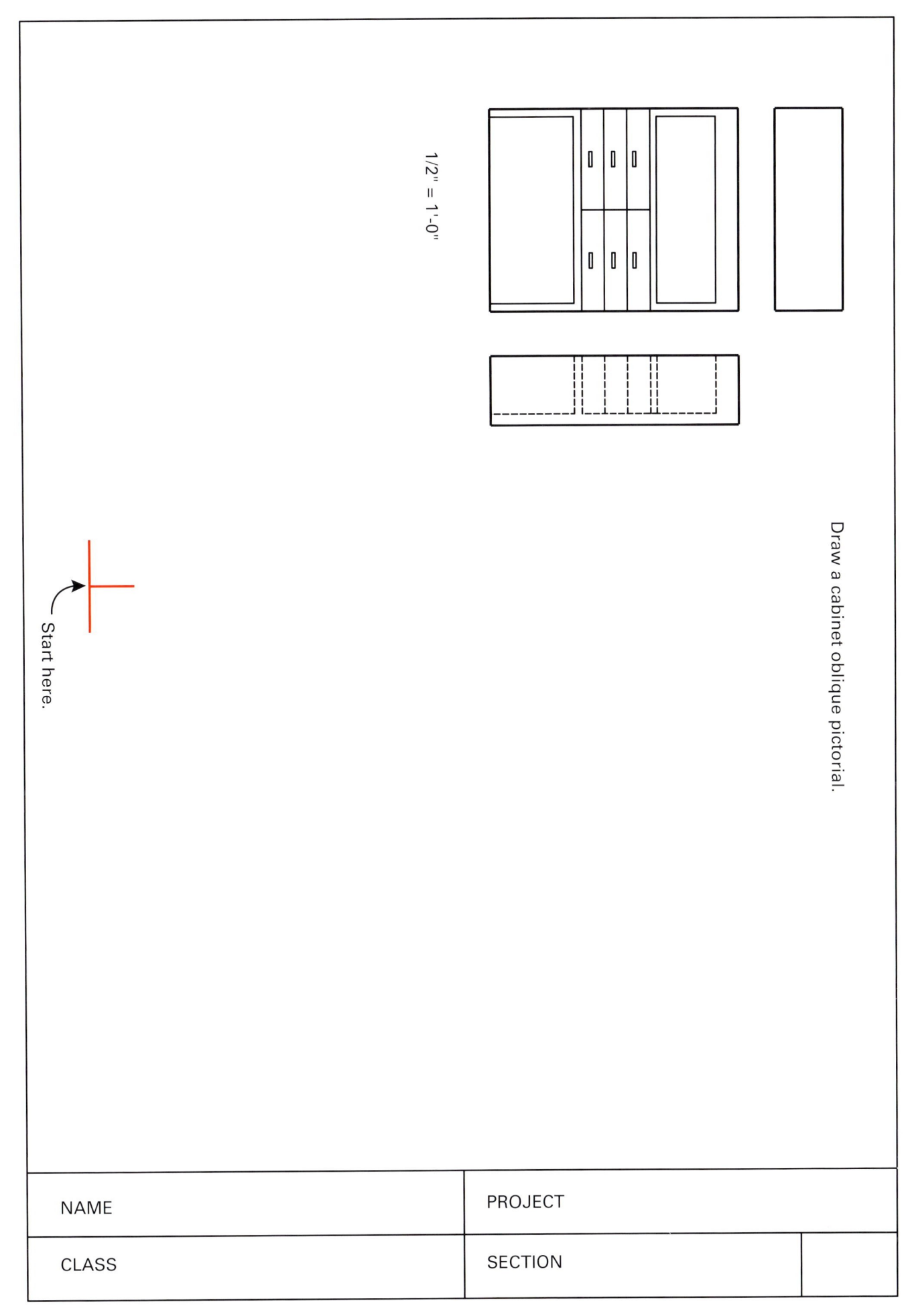

APPENDIX II

1/2" = 1'-0"

Draw top, front, and side views.

NAME	PROJECT
CLASS	SECTION

1/2" = 1'-0"

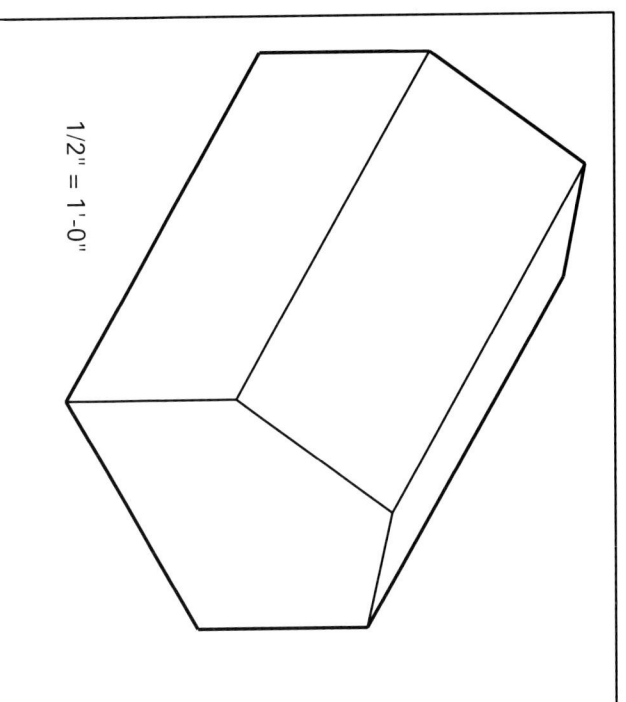

Draw top, front, and side views.

NAME	PROJECT	
CLASS	SECTION	

1/2" = 1'-0"

Draw top, front, and side views.

NAME	PROJECT
CLASS	SECTION

1/2" = 1'-0"

Draw an isometric pictorial.

NAME	PROJECT
CLASS	SECTION

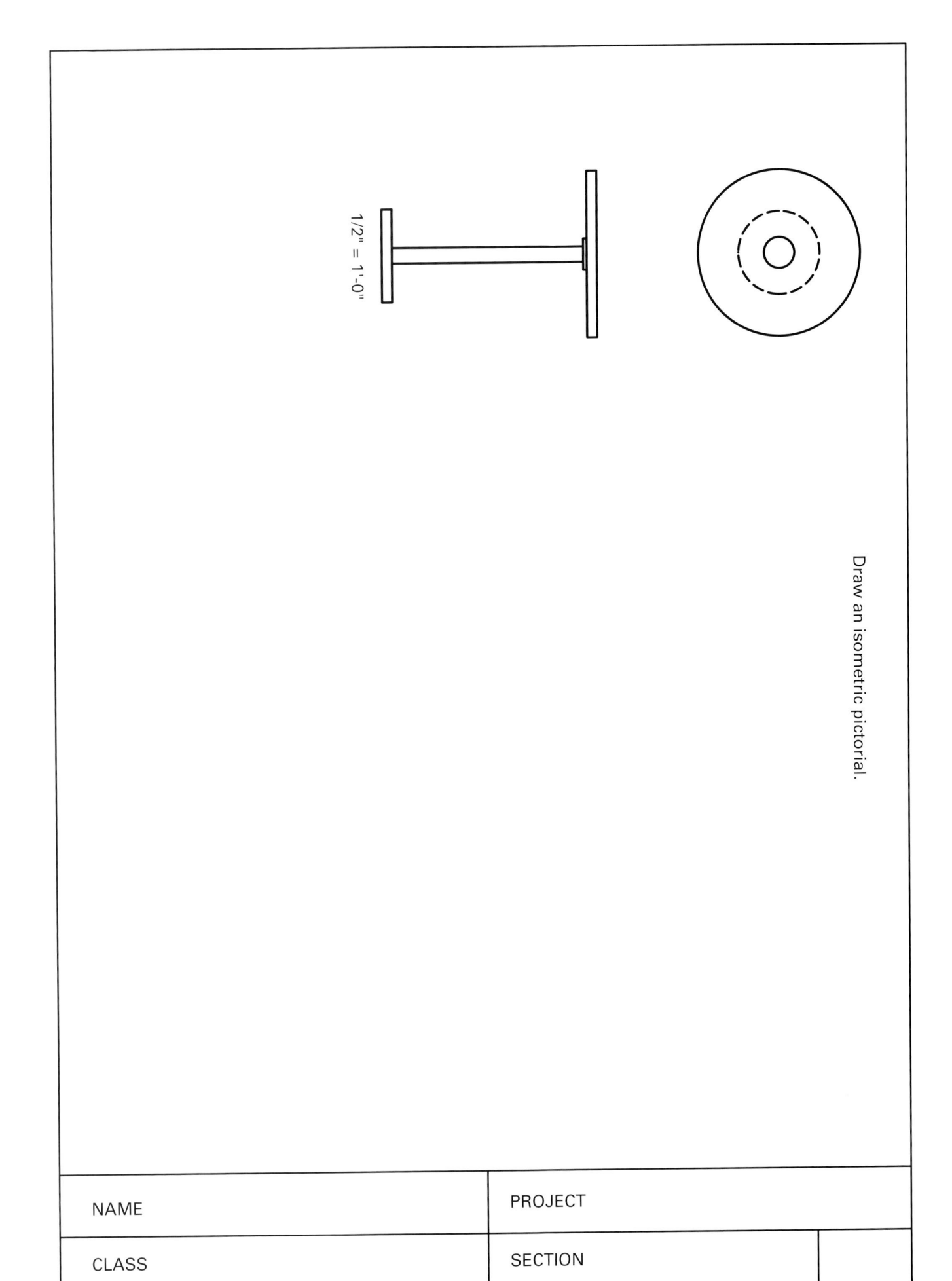

Below is the top view of a 6'-0" tall bookcase.
Draw a front view. 1/2" = 1'- 0".

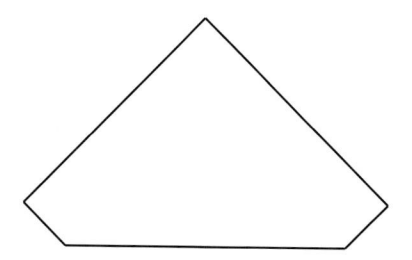

NAME	PROJECT	
CLASS	SECTION	

1 1/2" overhang

Not to scale

26"

45"

3 X 3 toe kick

36"

Draw top, front, and side views at 1 1/2" = 1'-0"

Tip:

NAME	PROJECT
CLASS	SECTION

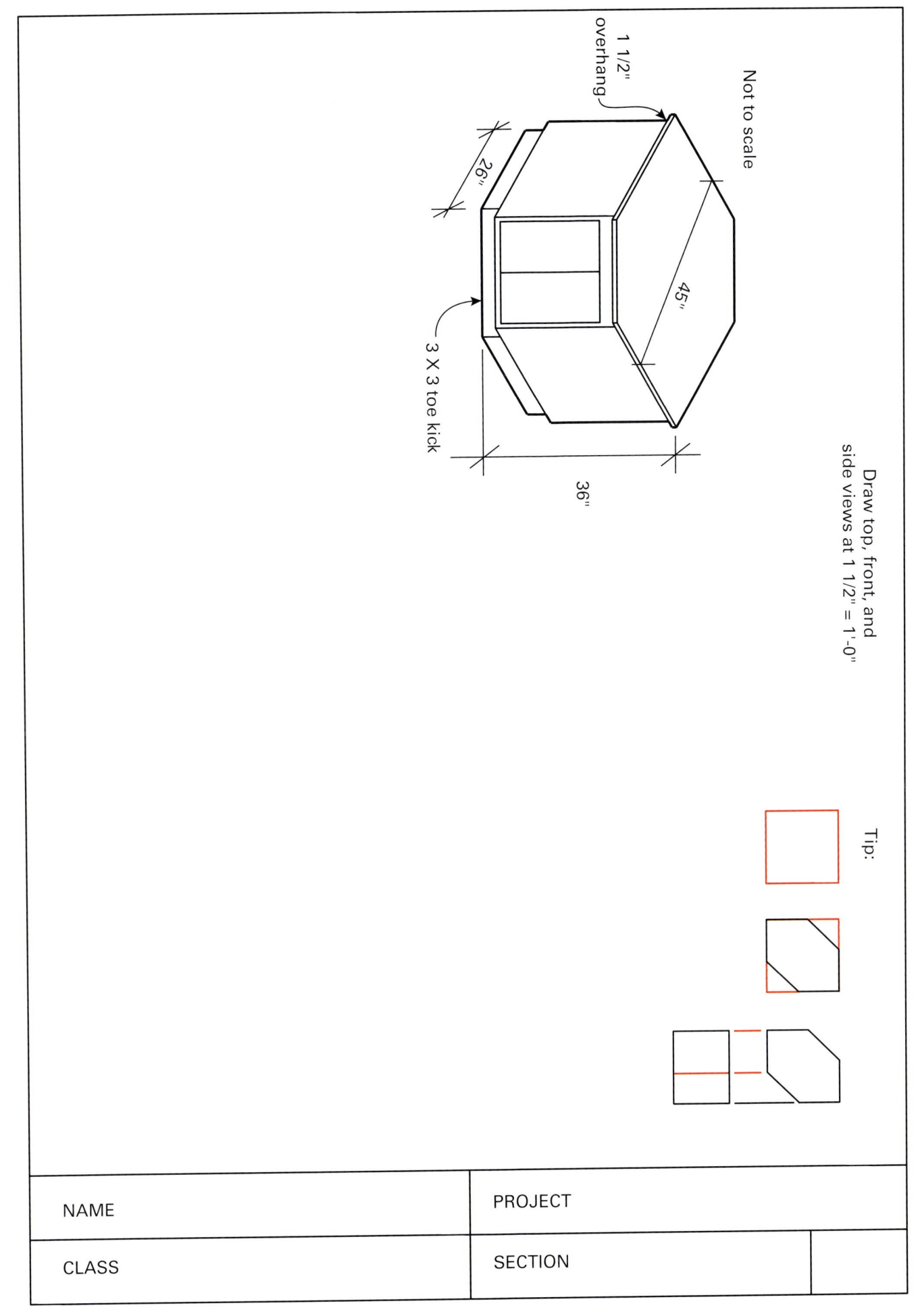

Complete the top and front views.
Then draw a side view. 3/4" = 1'-0"

| NAME | PROJECT |
| CLASS | SECTION |

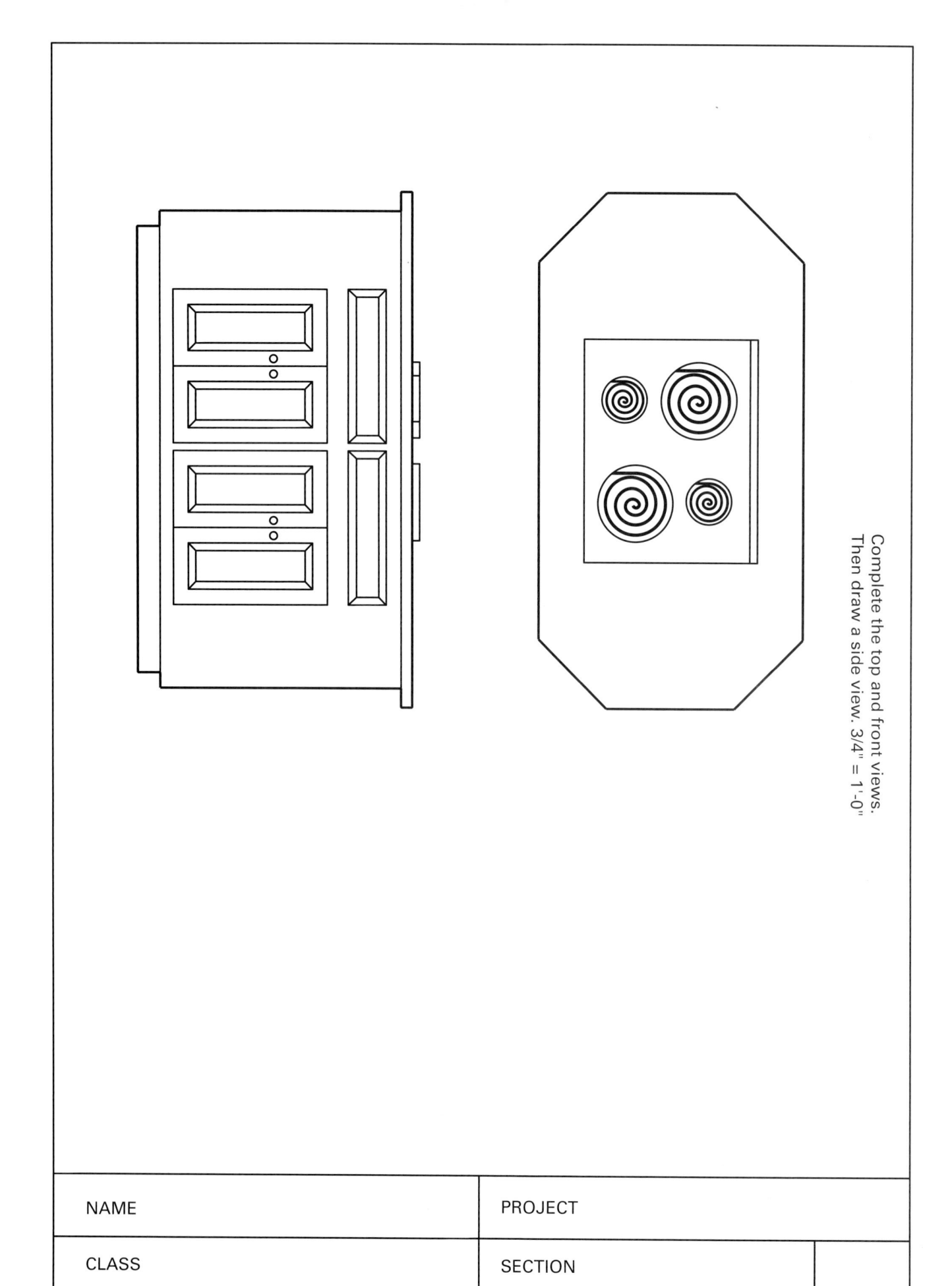

424 APPENDIX II

Draw a border and this title block true size on an 8 1/2" x 11" vellum.

- 8 1/2" x 11" vellum
- 1/2" border
- title block

Not to Scale

Sheet Number:	Date:	Drawn by:	Project Name Project Address City, State

1 1/2"

1" 1" 1" 4 1/2"

NAME	PROJECT	
CLASS	SECTION	

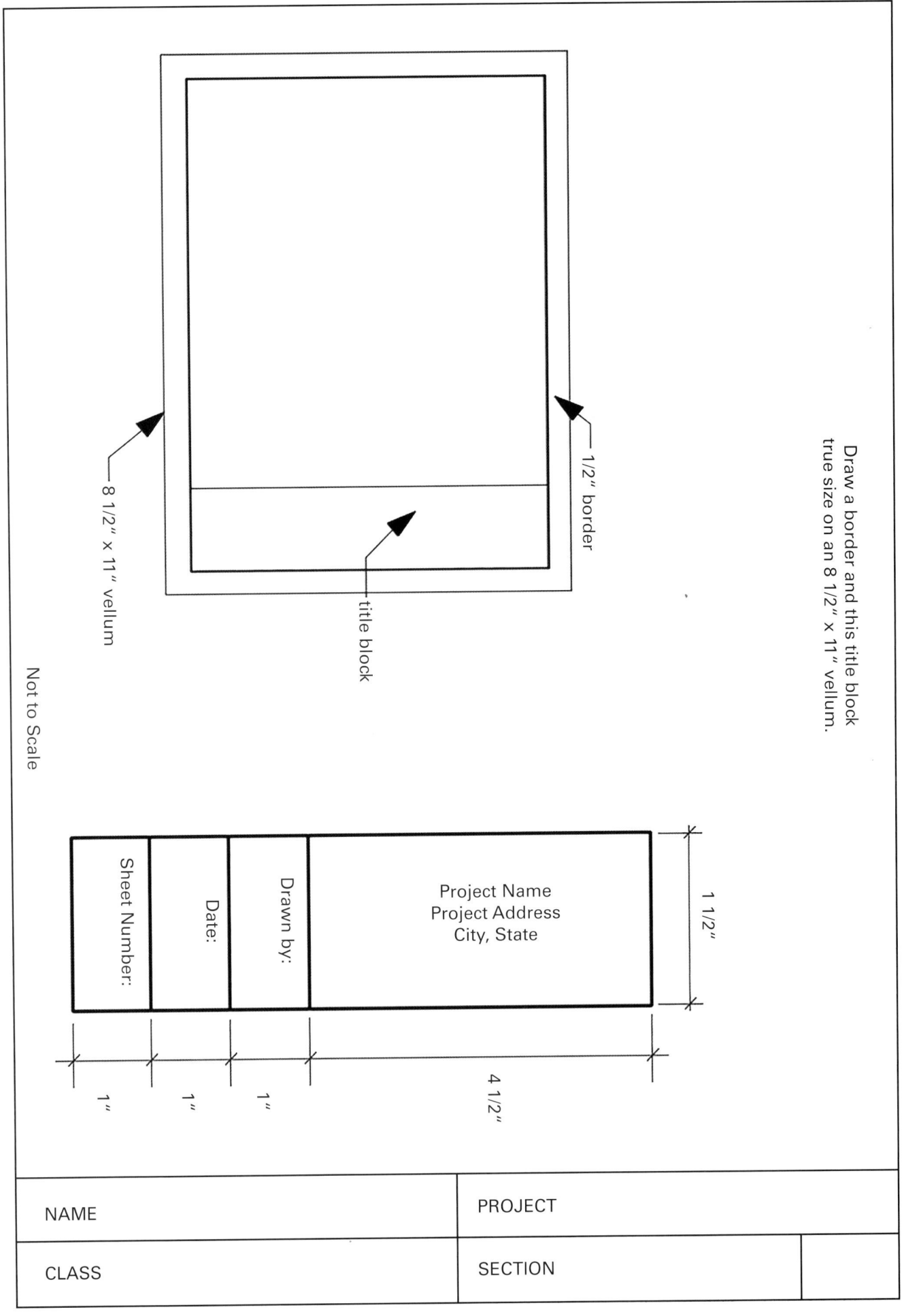

Draw the title blocks true size.

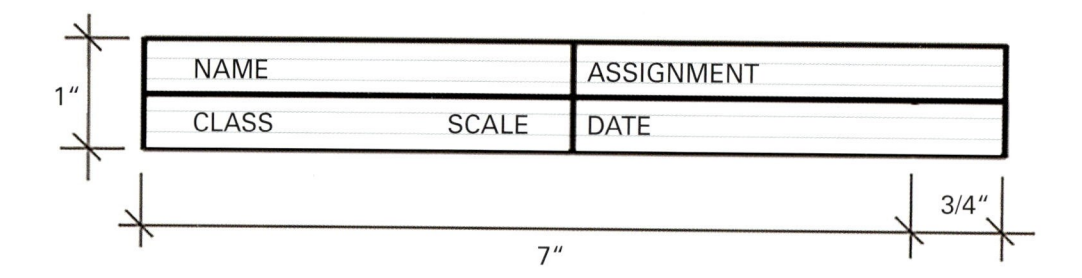

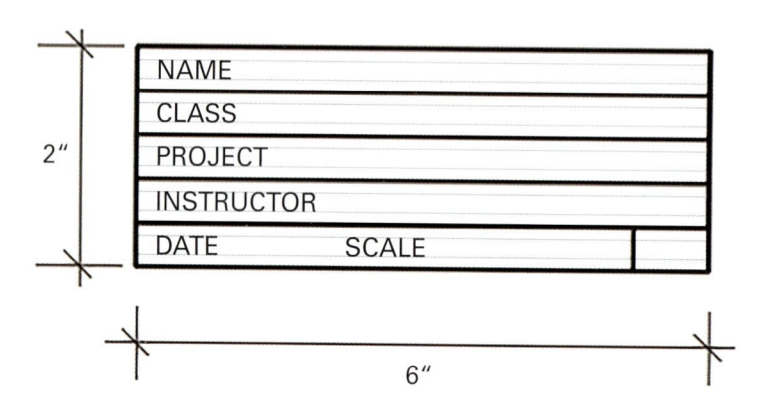

NAME		ASSIGNMENT
CLASS	SCALE	DATE

1" 7" 3/4"

NAME
CLASS
PROJECT
INSTRUCTOR
DATE   SCALE

2" 6"

Not to Scale

NAME	PROJECT	
CLASS	SECTION	

426 APPENDIX II

Draw a section symbol, two elevation callouts and an ID label. Place them appropriately.

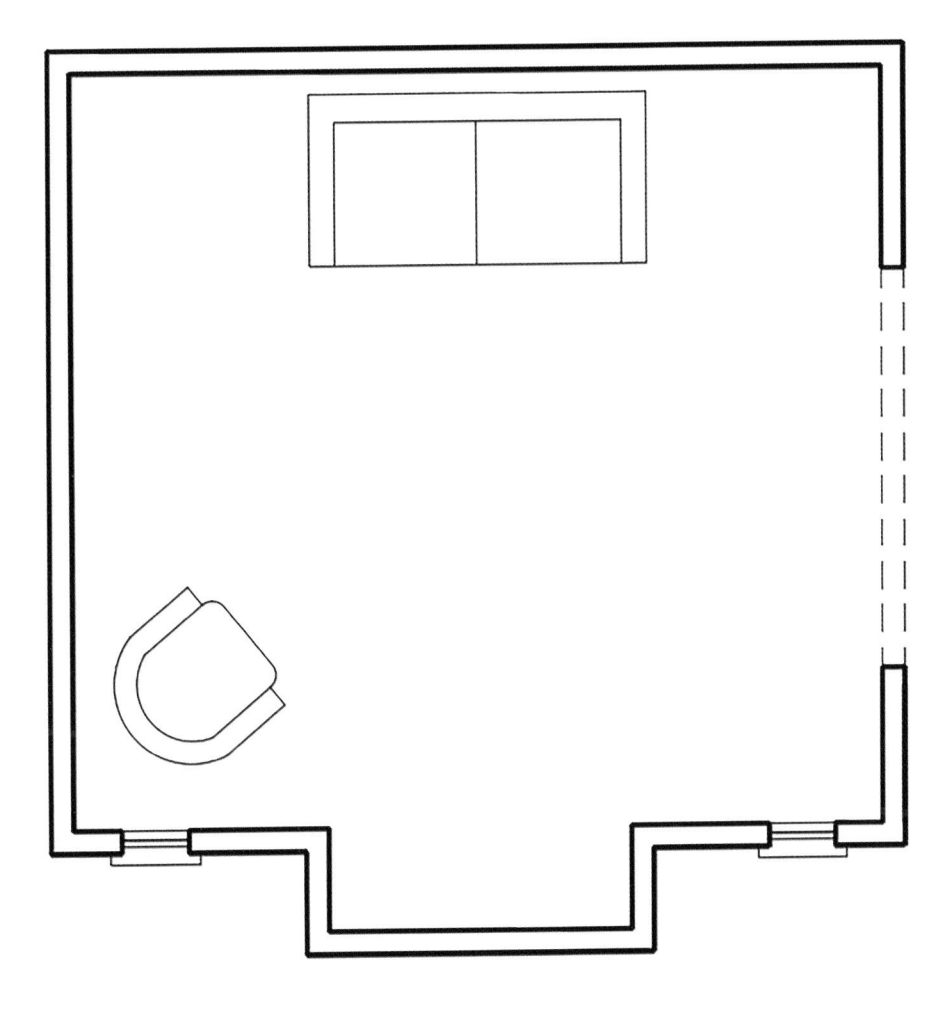

1/4"= 1'0"

APPENDIX II

Line types: Draw hidden lines where indicated for the wall cabinets and dishwasher, Draw horizontal center lines through the columns. Draw a break line and arrow through the stairs.

DW

COLUMNS

1/4" = 1'-0"

NAME	PROJECT
CLASS	SECTION

428 APPENDIX II

Use the Ames lettering guide and a hard lead to draw five sets of guidelines the heights indicated. Then fill the guidelines with letters and numbers.

1/8"

5/32"

1/4"

NAME	PROJECT
CLASS	SECTION

Center and ink this plan on vellum, using appropriate line weights.
Use templates for the fixtures. Include an ID label and title block.

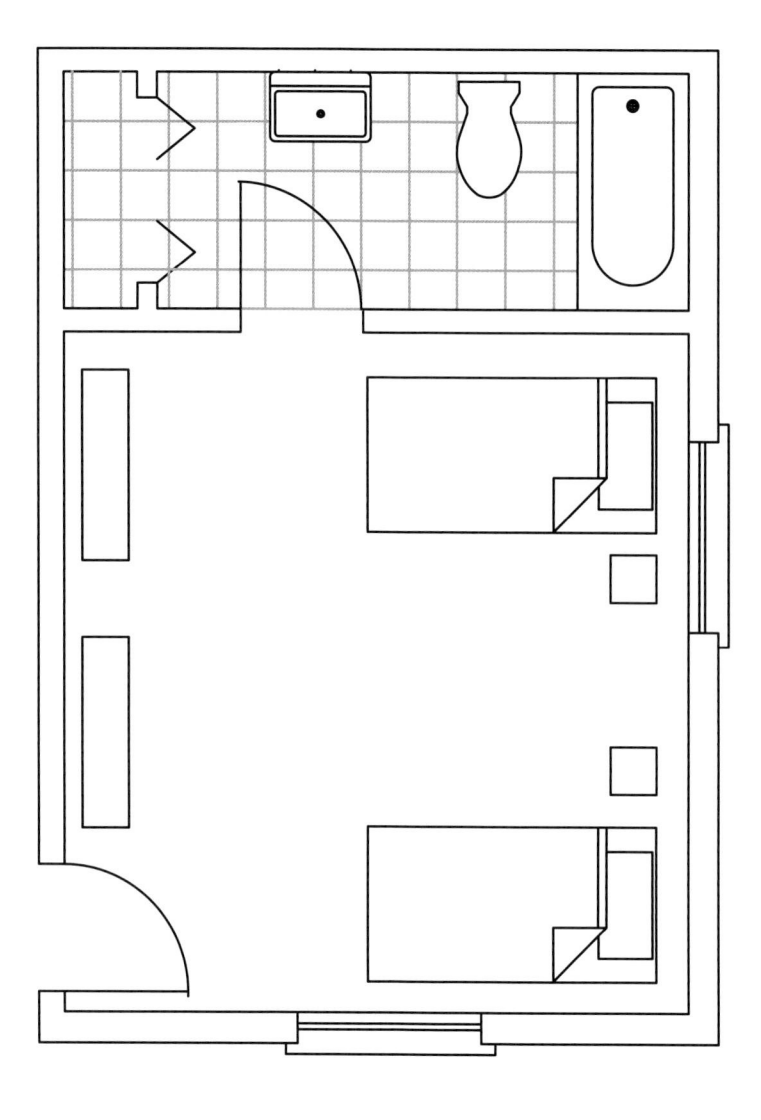

NAME	PROJECT
CLASS	SECTION

430 APPENDIX II

Line types: Offset a section line 1/4" inside the perimeter. Offset a center line 1/2" from the perimeter. Offset a hidden line 3/4" from the perimeter.

NAME	PROJECT	
CLASS	SECTION	

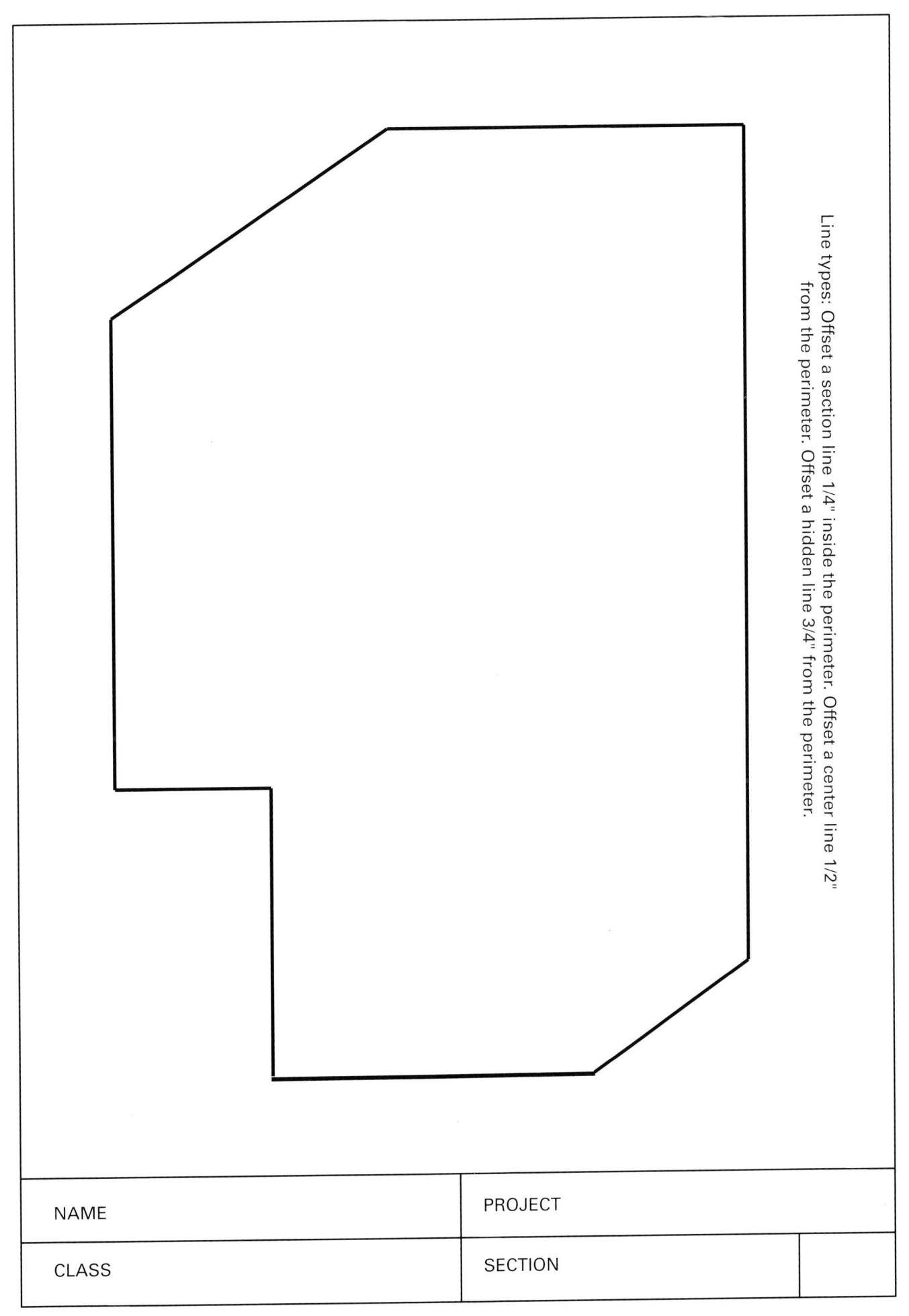

Lay out a bathroom and living room, using sizes and clearances from the text.

NAME	PROJECT
CLASS	SECTION

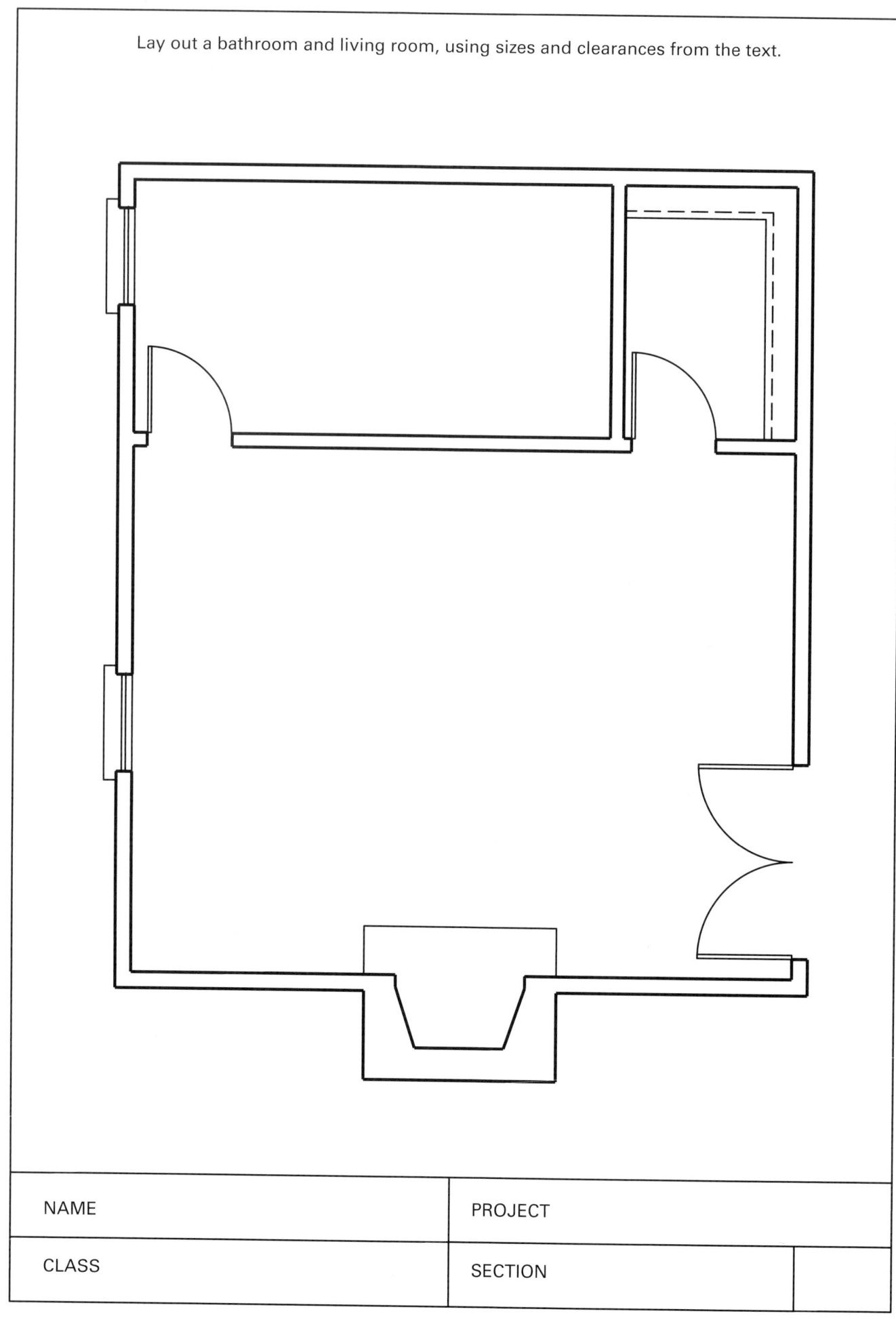

Draw the wall type, poché, and thickness indicated. 1/2" = 1'-0"

- 8" concrete block
- 2 X 6 wood frame
- Plaster on 2 X 4 wood frame
- 4" brick on 8" concrete block
- 2 X 6 wood frame with gypsum board on both sides.
- 4" brick, 2" air space, 2 X 4 wood frame
- 8" glass block
- 8" stone on 2 X 4 wood frame

NAME	PROJECT	
CLASS	SECTION	

APPENDIX II 433

DN

Ink this plan on film or vellum, and use templates to add furniture. 1/4" = 1'-0"

NAME	PROJECT
CLASS	SECTION

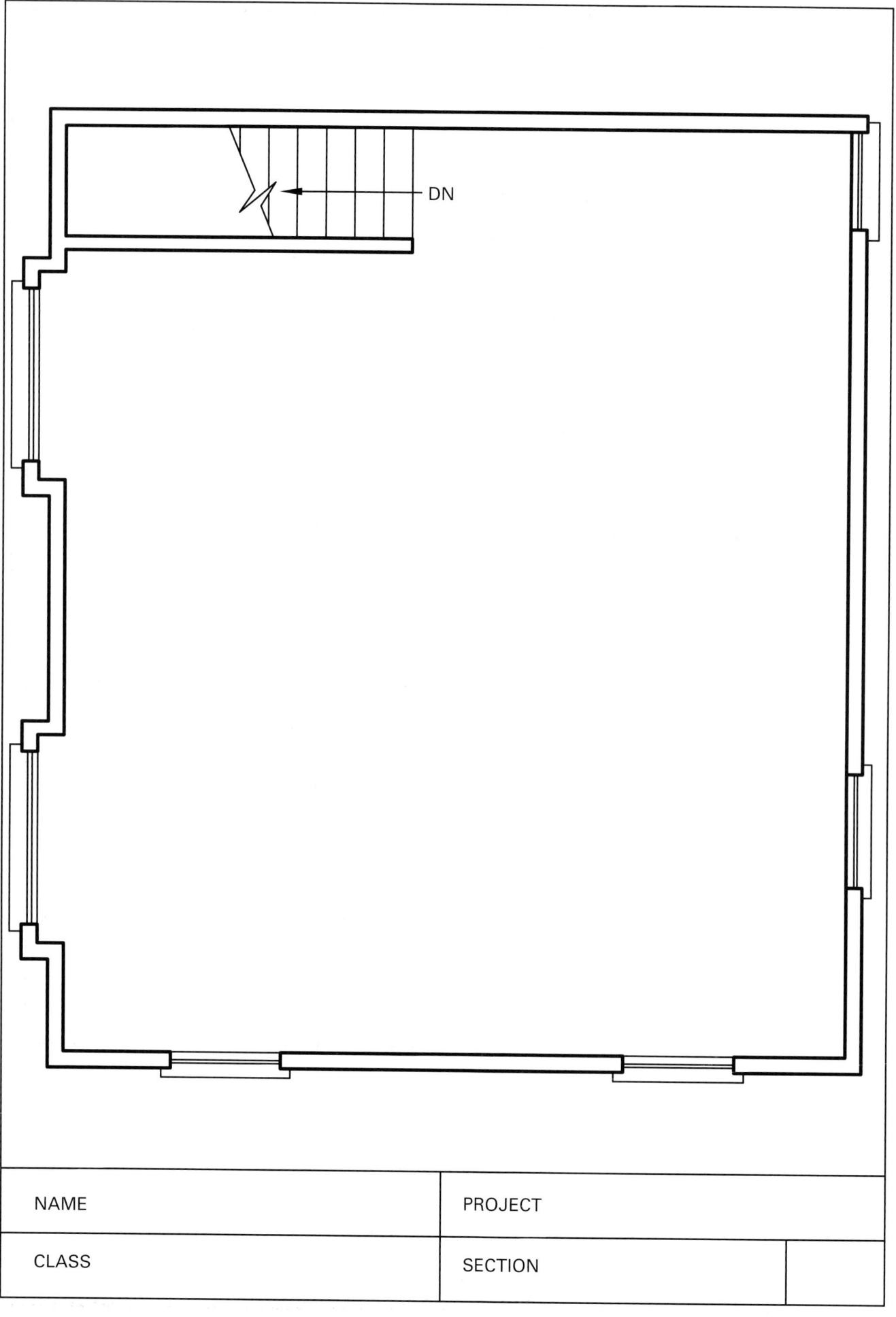

Draw the floor pochés

HONEYCOMB TILE	WOOD PLANKS	CONCRETE
GRANITE	CARPET	WOOD STRIPS
STONE	RUG	BRICK BASKETWEAVE
DIAMOND	12" TILE	6" TILE W/ GROUT

APPENDIX II **435**

DN

Ink this plan on film or vellum, and use templates to add furniture. 1/4" = 1'-0"

NAME	PROJECT
CLASS	SECTION

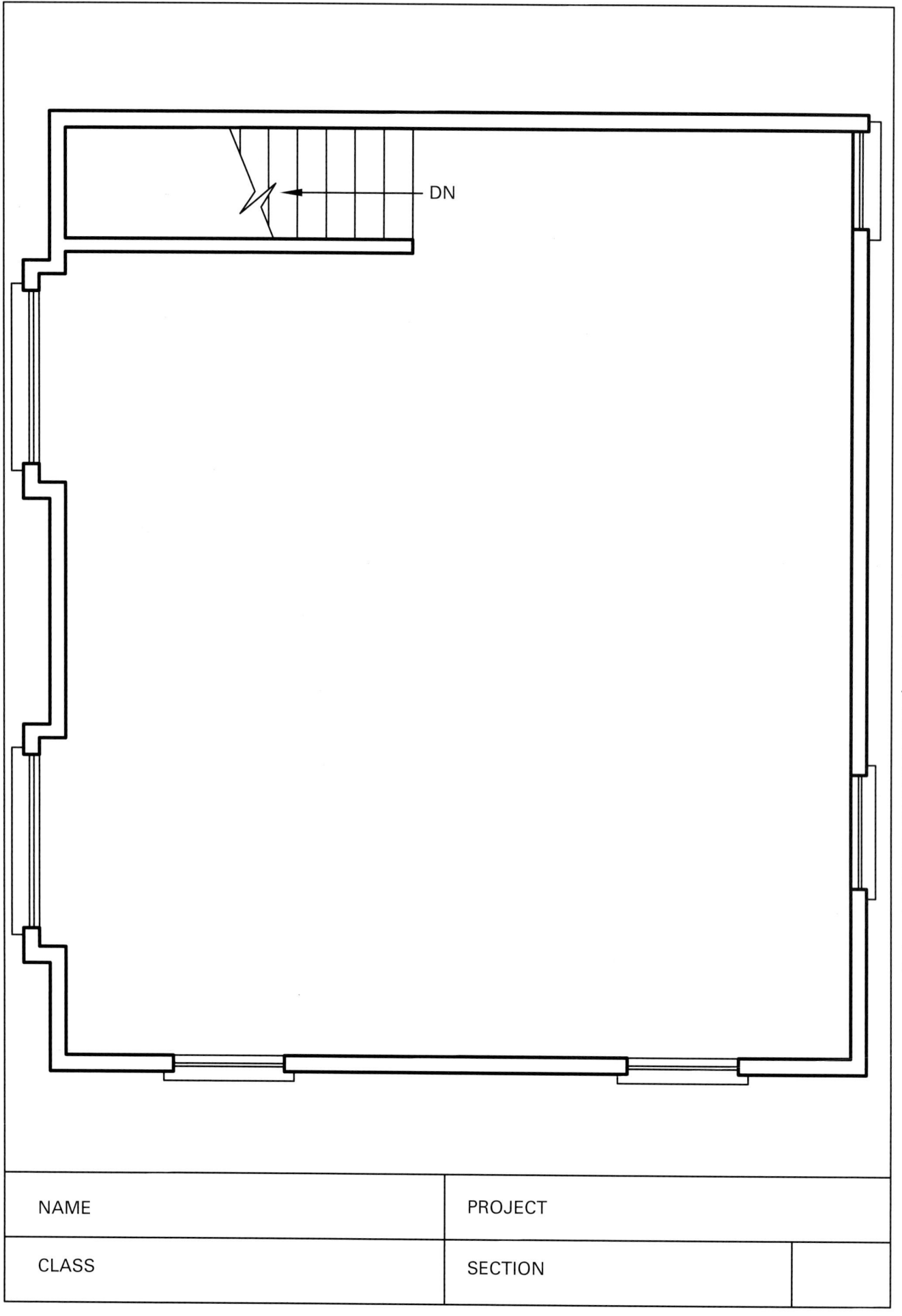

APPENDIX II

Sketch a floor plan of this house.

NAME	PROJECT
CLASS	SECTION

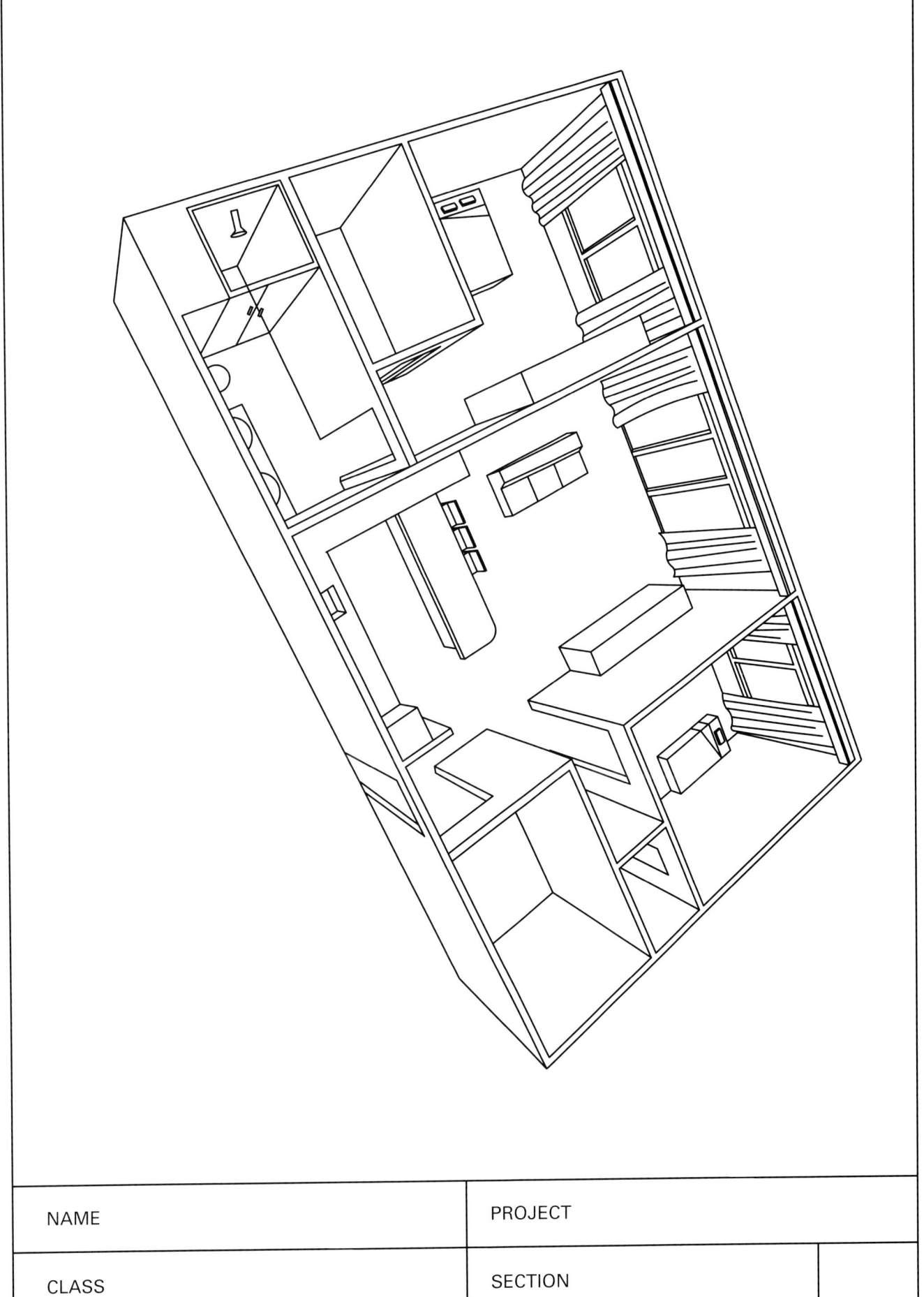

NAME	PROJECT
CLASS	SECTION

Draw a section at the cutting plane location. Assume a 9'-0" ceiling and flat roof.

1/8" = 1'0"

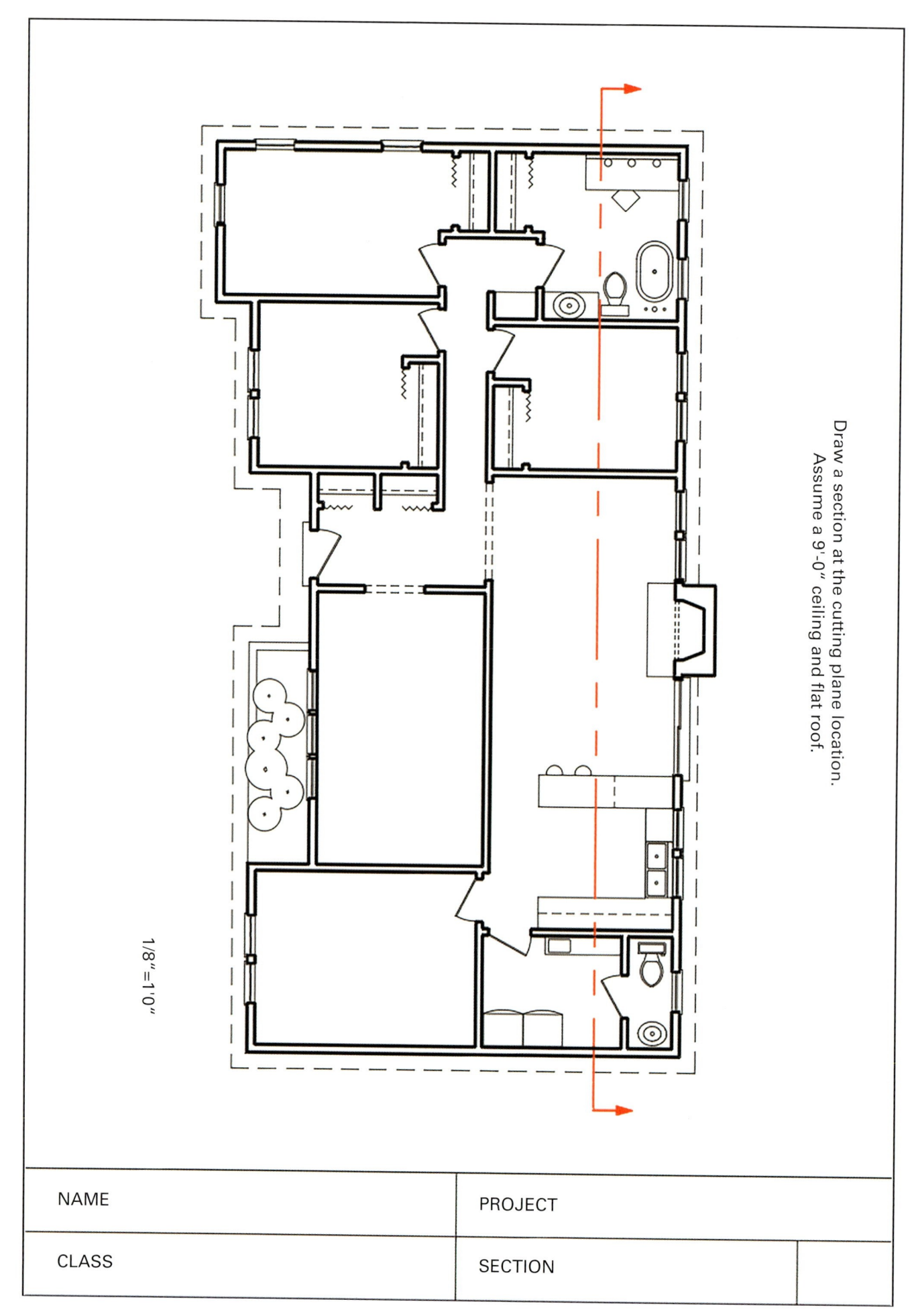

438   APPENDIX II

Draw interior elevations where indicated.
Ceiling is 9'-0" high.

1/4" = 1'-0"

NAME	PROJECT	
CLASS	SECTION	

APPENDIX II **439**

Draw an interior elevation where indicated.
Ceiling is 9'-0" high.

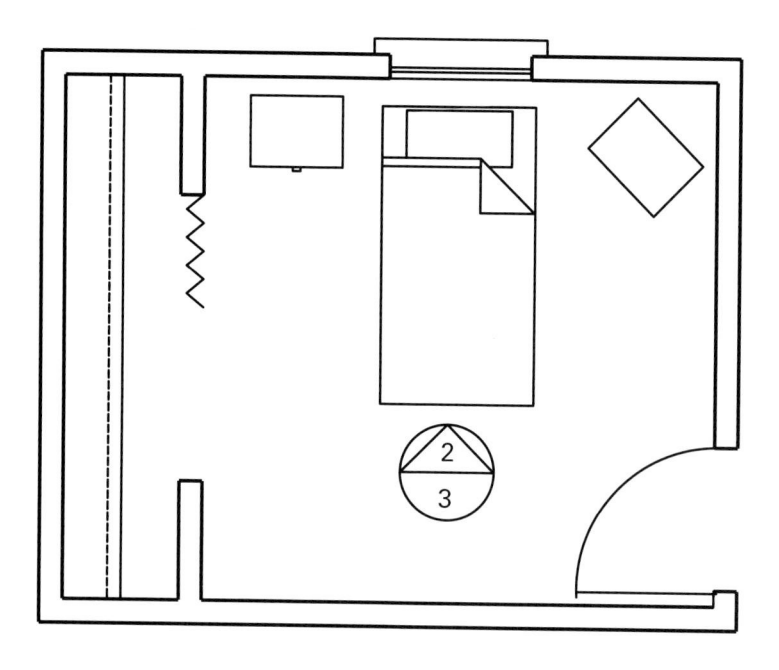

1/4" = 1'-0"

NAME	PROJECT
CLASS	SECTION

440 APPENDIX II

Draw an interior elevation where indicated.
Ceiling is 9'-0" high.

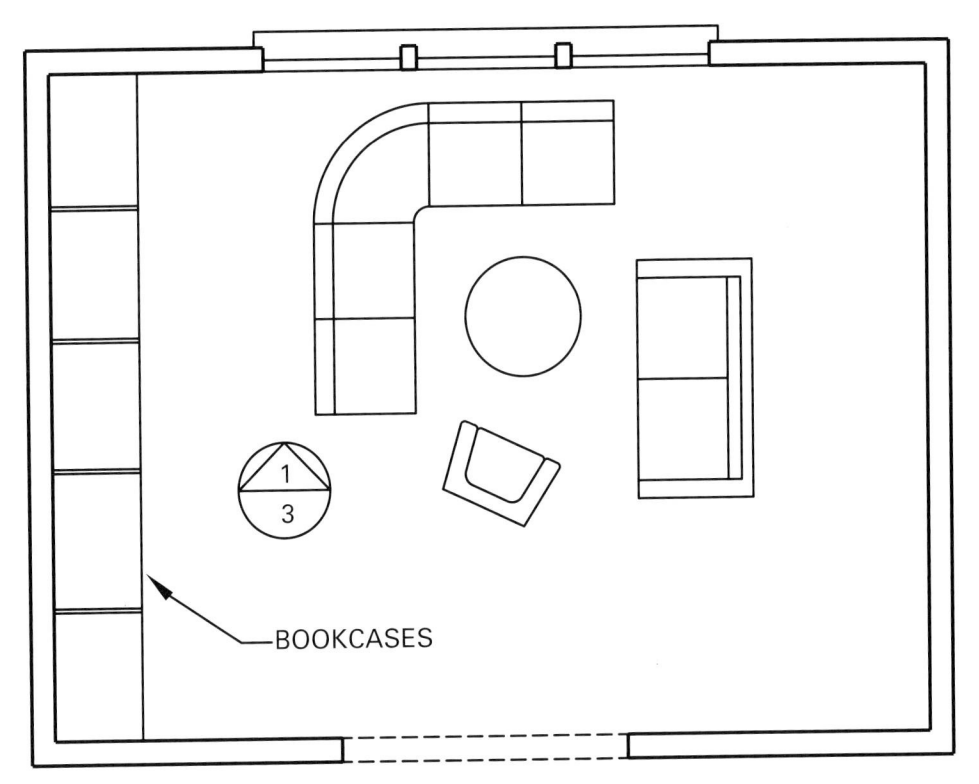

BOOKCASES

1/4" = 1'-0"

NAME	PROJECT	
CLASS	SECTION	

APPENDIX II **441**

Draw an interior elevation where indicated. 9'-0" ceiling.

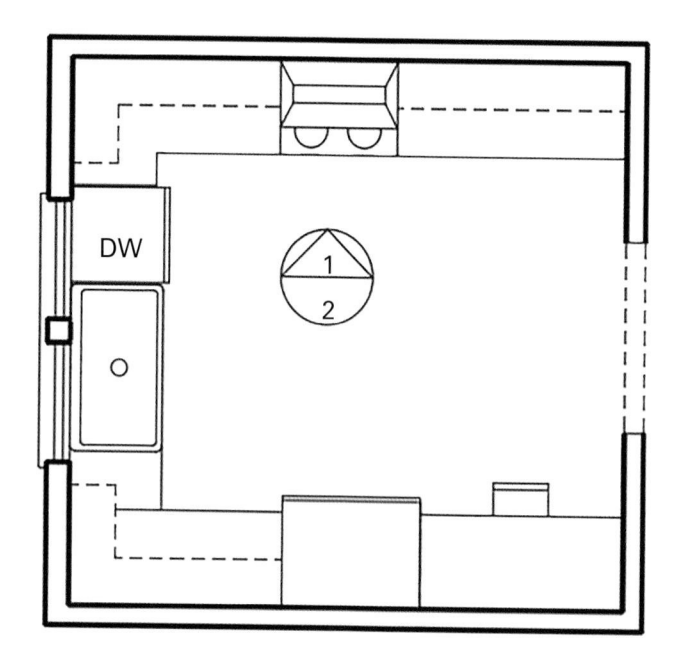

1/4"-1'0"

NAME	PROJECT
CLASS	SECTION

Draw the elevation and section pochés.

Elevation

Ashlar stone	Finish board	Wood siding
Brick	Glass	Concrete block

Section

Brick	Dimension	Glass block
Batting insulation	Finish board	Cast concrete

NAME	PROJECT	
CLASS	SECTION	

Dimension this wood frame plan to ANSI standards.

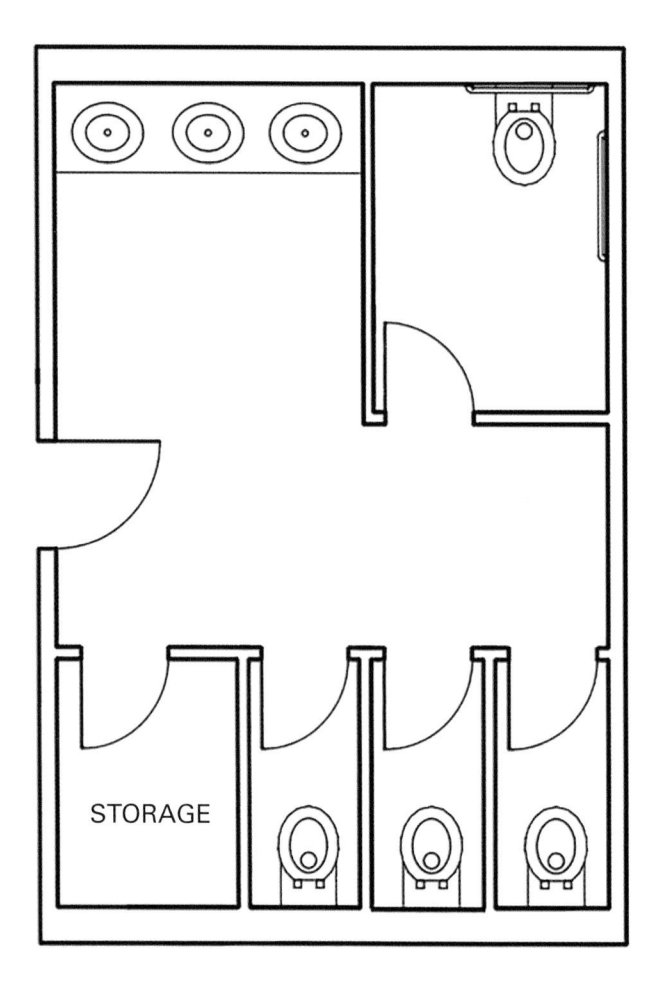

1/4"=1'0"

NAME	PROJECT
CLASS	SECTION

Dimension this wood frame plan to AIA standards.

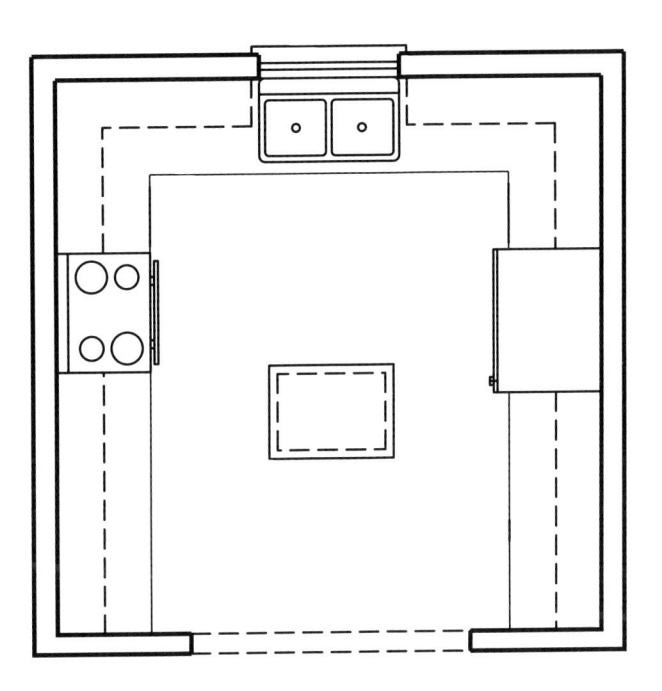

1/4"= 1'0"

NAME	PROJECT
CLASS	SECTION

Dimension to ANSI standards.

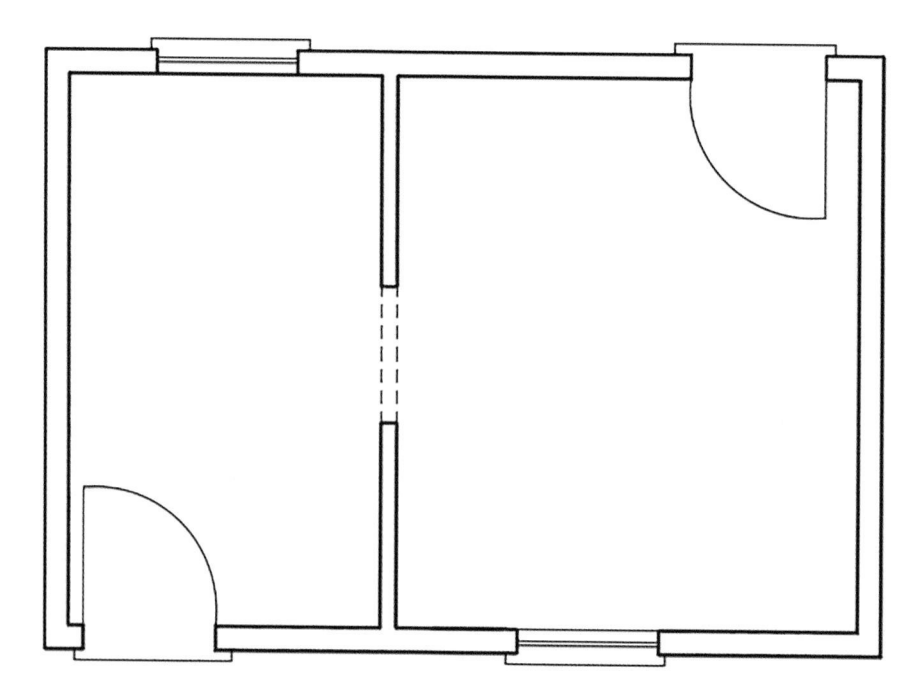

Wood frame

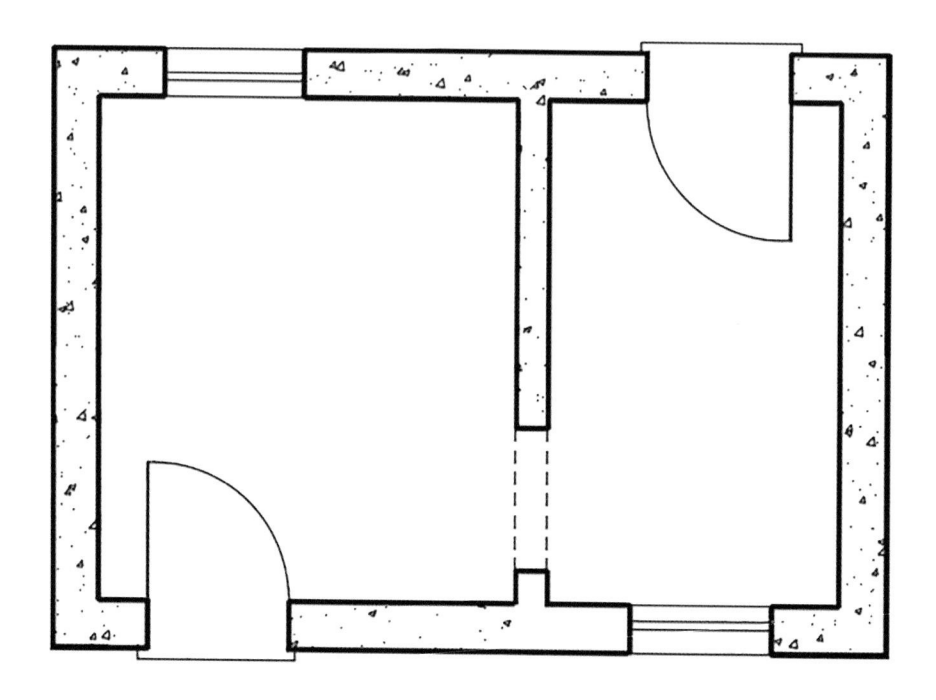

Masonry

1/4"=1'0"

NAME	PROJECT
CLASS	SECTION

Dimension to NKBA standards.

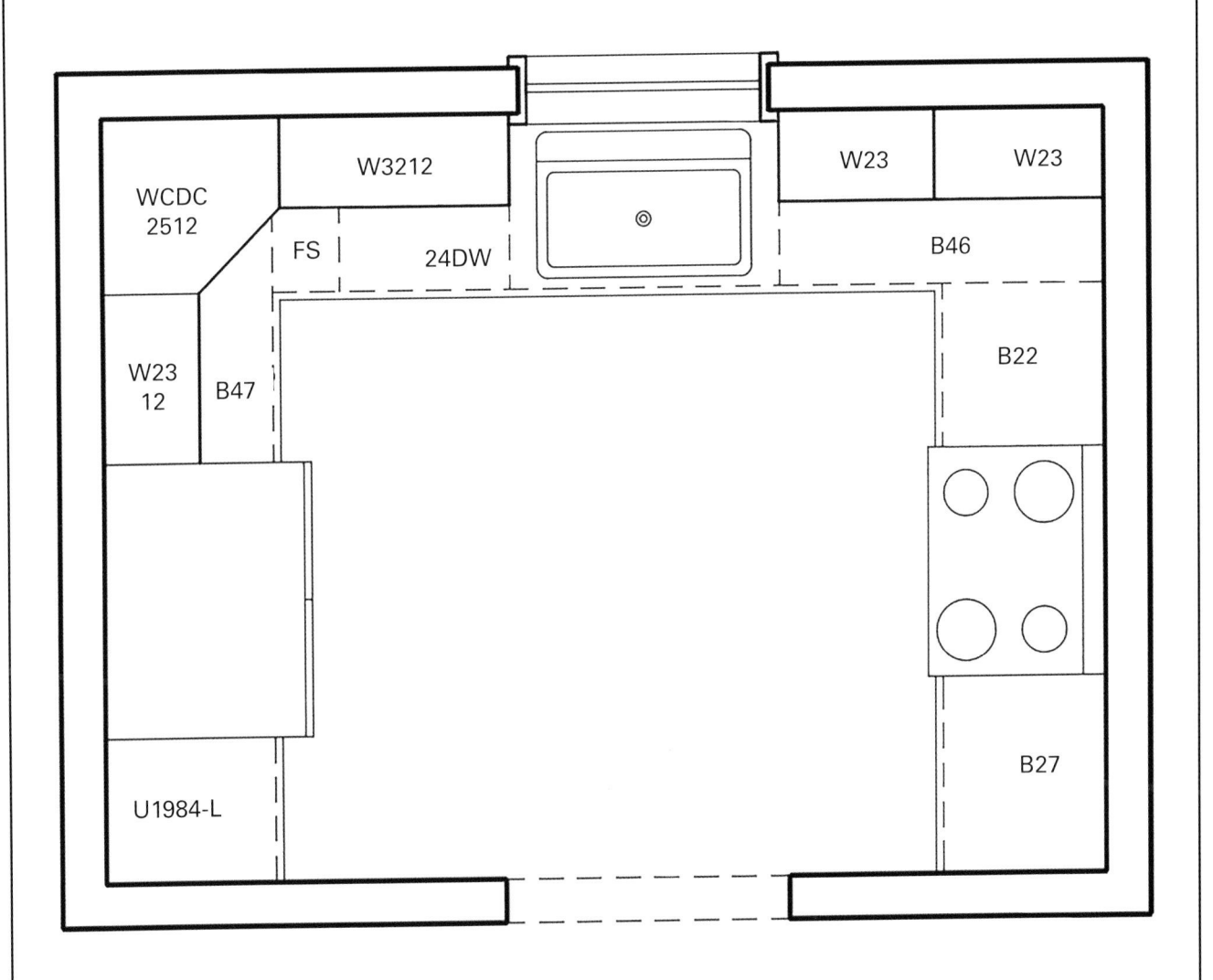

- WCDC 2512
- W3212
- W23
- W23
- FS
- 24DW
- B46
- W23 12
- B47
- B22
- U1984-L
- B27

1/2"=1'0"

NAME	PROJECT	
CLASS	SECTION	

APPENDIX II

Dimension to NKBA standards.

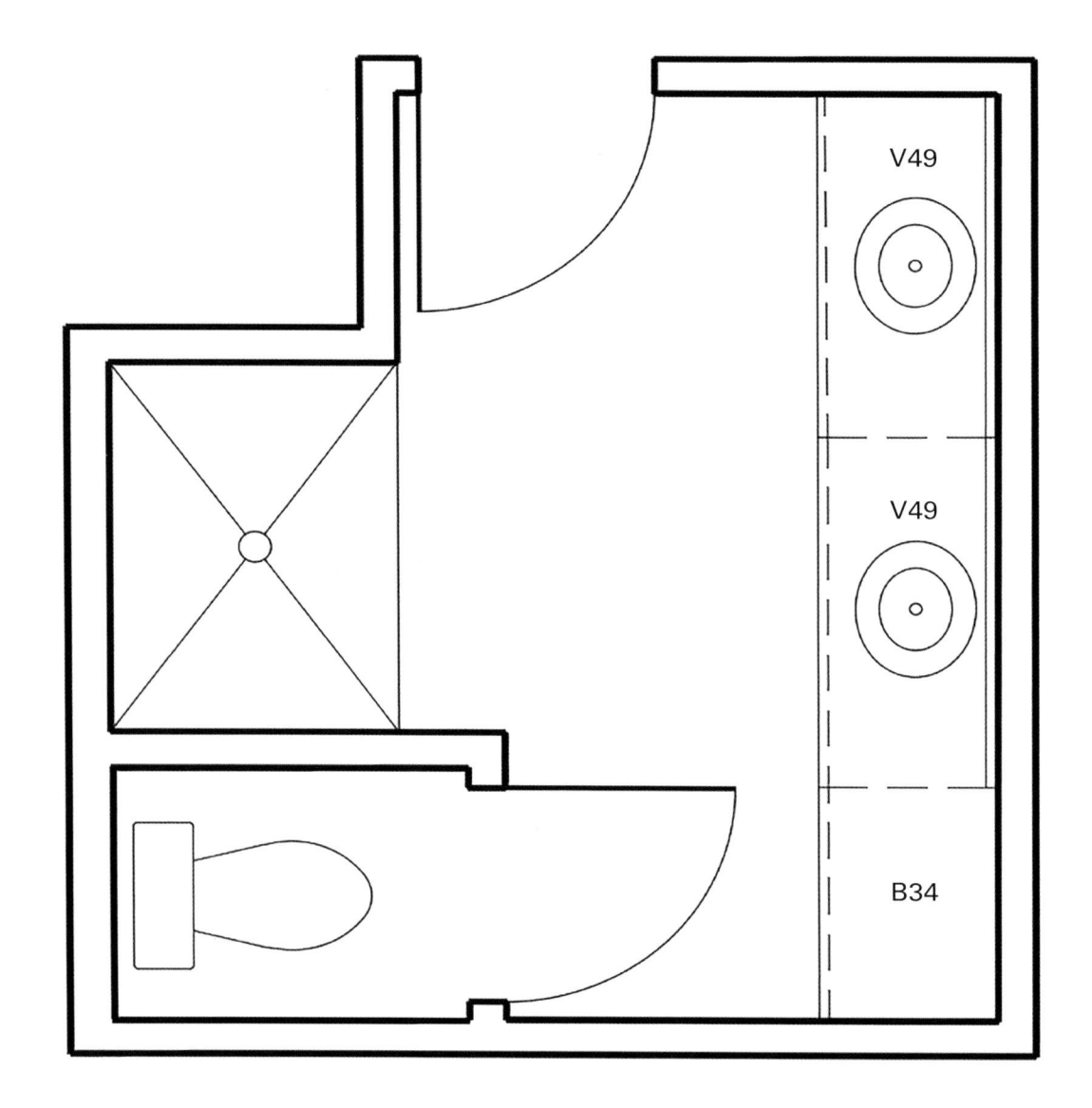

V49

V49

B34

1/2" = 1'0"

NAME	PROJECT
CLASS	SECTION

Calculate the missing dimensions.

18'-4"
5'-10"   6'-9 1/2"   2.

10"
8'-9 1/2"   1.
10"

5'-8 1/2"
6'-9 1/2"
3.
4'-5"

4.   9'-7 1/2"
12'-1 1/2"

1. _____
2. _____
3. _____
4. _____

Not to scale

NAME	PROJECT
CLASS	SECTION

APPENDIX II **449**

Calculate the area of rooms 1, 2, and 3.

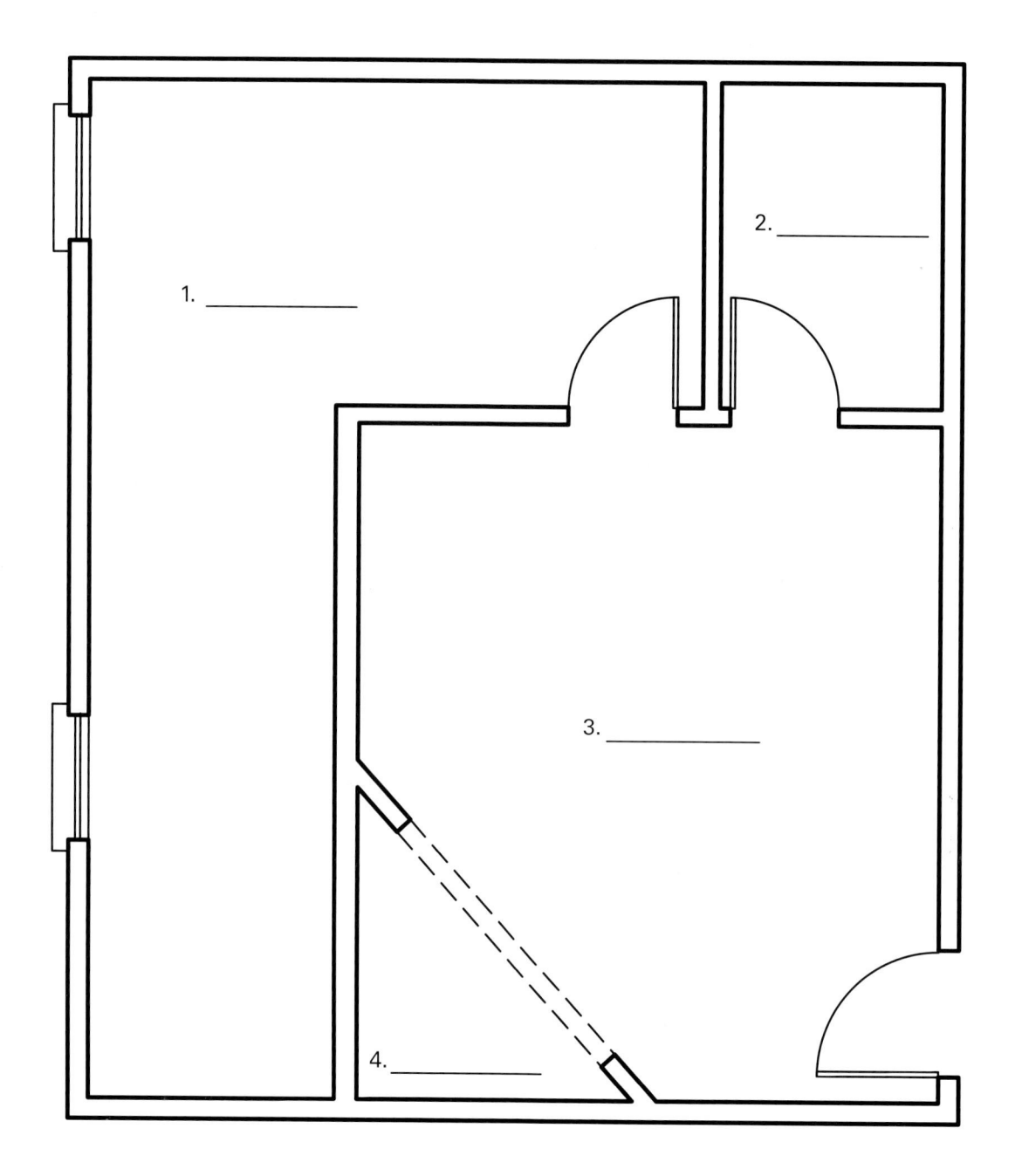

1. _____

2. _____

3. _____

4. _____

1/4" = 1'0"

NAME	PROJECT
CLASS	SECTION

450 APPENDIX II

Draw the door plan symbols.

1. LH Swing

2. RH Dutch

Hall Side

3. Pocket

4. Double Action

5. RH Reverse Swing

Hall Side

6. Double Bifold

7. Accordion

8. Wood Sliding

9. LH RH French

Hall Side

10. Glass Sliding

1/4"=1'-0"

NAME	PROJECT	
CLASS	SECTION	

APPENDIX II **451**

Draw the door and window symbols.

Double hung • Hopper • Awning • 2-sash Casement • Double bifold • RH Swing • LH Swing • Cased opening • Accordion • Sliding • RH Swing • LH Swing • Bay • Revolving • Pocket • LH Swing • RH Swing • RH Swing Door • Glass sliding

1/4" = 1'0"

NAME	PROJECT
CLASS	SECTION

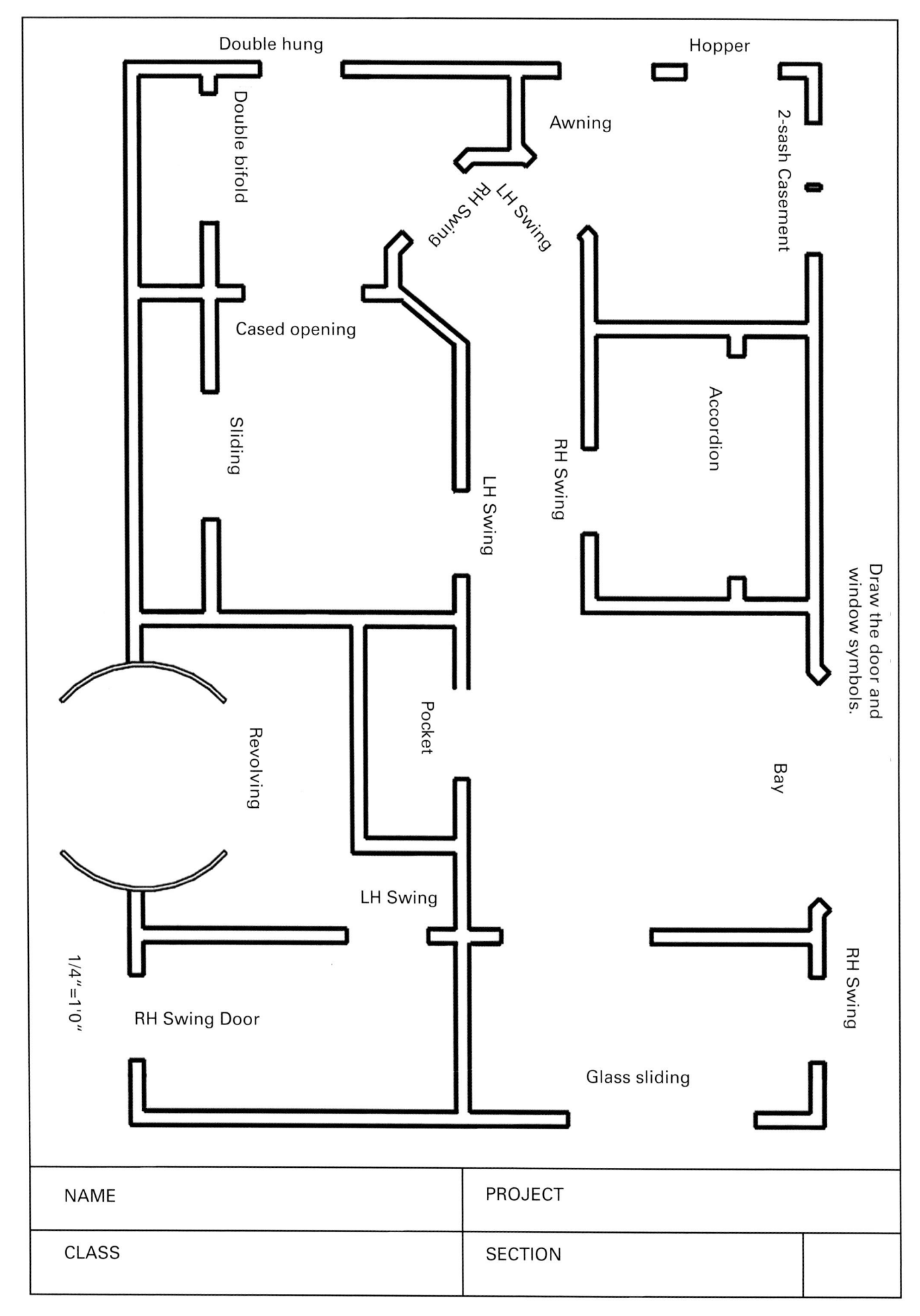

Draw the window plan symbols. 1/4" = 1'-0".

1. Double-hung

4. Awning

2. 2-sash casement

5. Fixed

3. Glass sliding

5. Hopper

7. 45° 2'-0" deep bay°

8. 4-sash bow

NAME	PROJECT	
CLASS	SECTION	

APPENDIX II

Identify the parts.

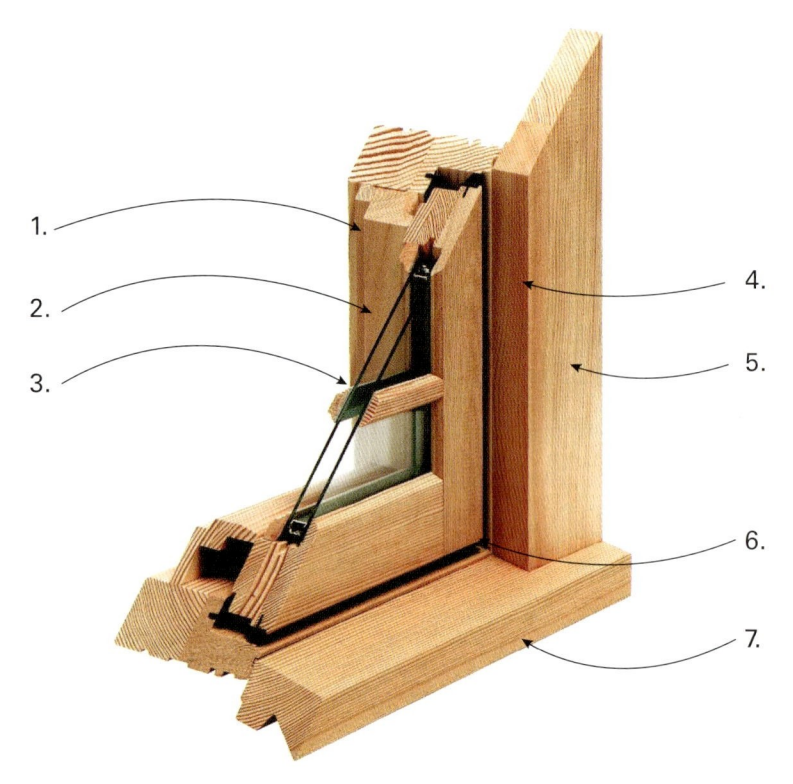

Photo courtesy Loewen, Manitoba, CA.

1. _____

2. _____

3. _____

4. _____

5. _____

6. _____

7. _____

NAME	PROJECT	
CLASS	SECTION	

APPENDIX II

Draw these windows' elevation views.
Place the head on the red line.

1. 5'-0" Double Hung

4. 4'-6" Louvered

2. 4'-0" 2-sash Casement

5. 3'-6" Fixed

3. 4'-0" Sliding

6. 3'-6" Hopper

1/4"-1'0"

NAME	PROJECT
CLASS	SECTION

APPENDIX II

Head

Head

Head

Draw the elevation symbols. 1/4" = 1'-0".

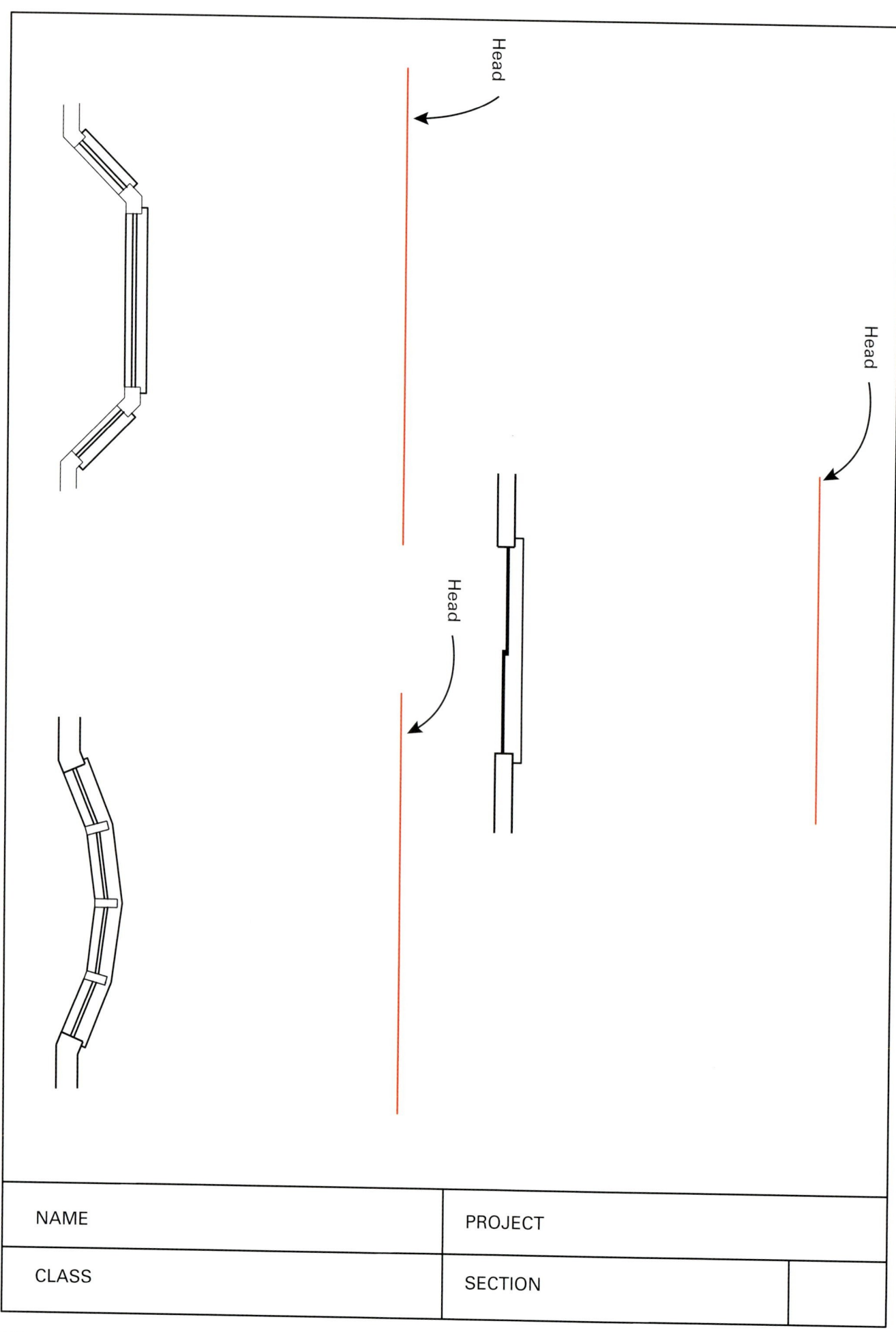

NAME	PROJECT
CLASS	SECTION

Sketch a wall detail.

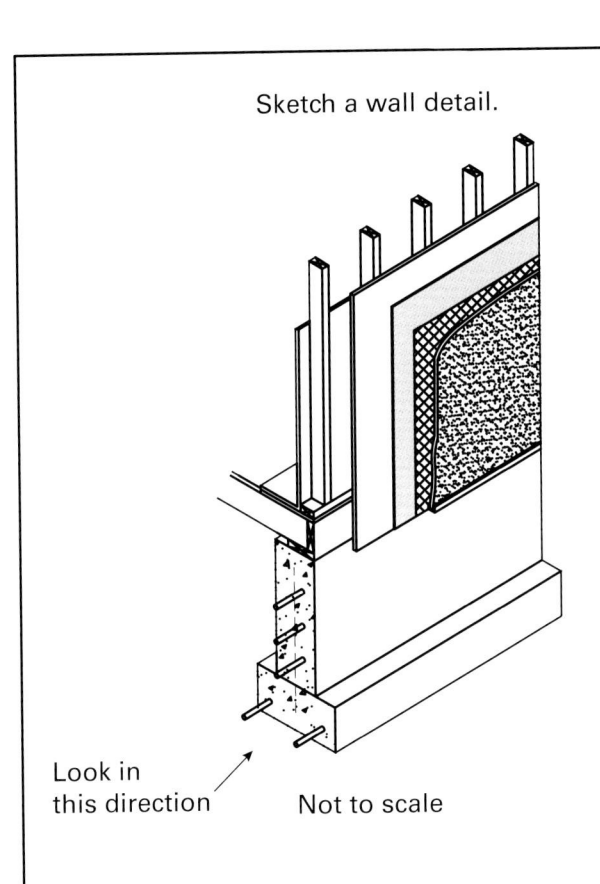

Look in this direction

Not to scale

NAME	PROJECT	
CLASS	SECTION	

APPENDIX II

Sketch a wall detail.

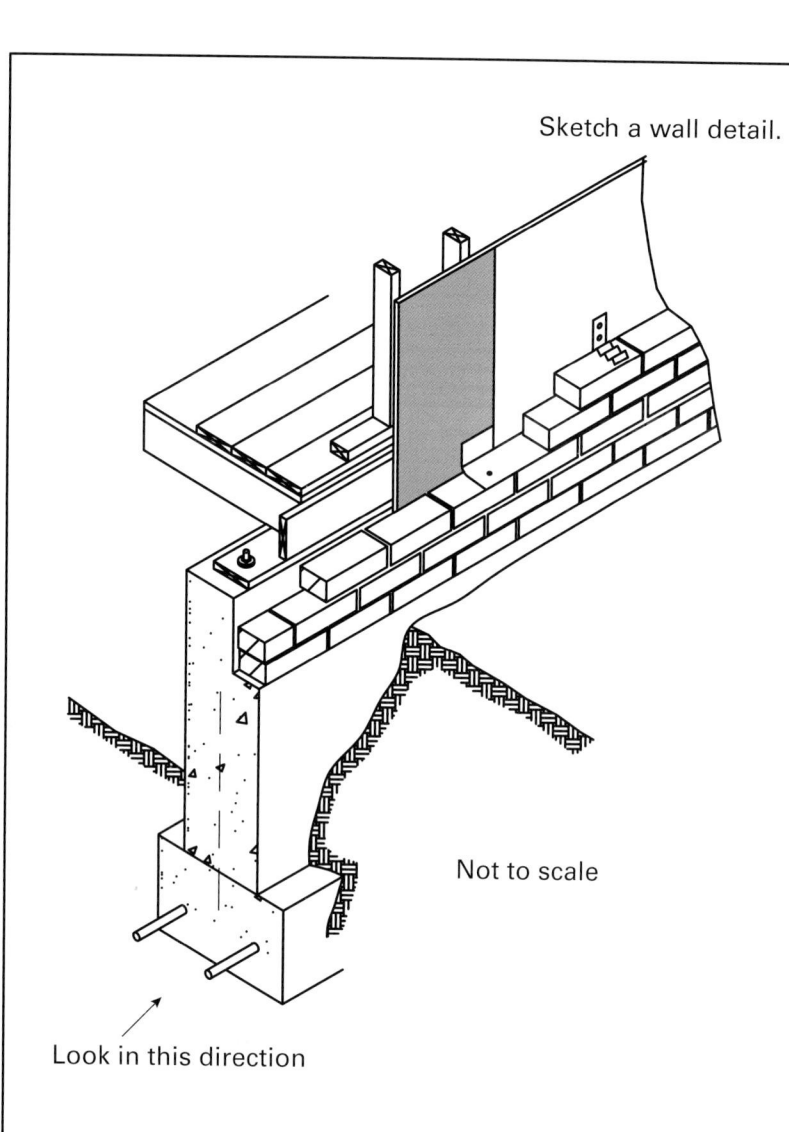

Not to scale

Look in this direction

NAME	PROJECT	
CLASS	SECTION	

APPENDIX II

Draw plan and elevation views of a one-face fireplace. 1/4" = 1'-0".

Plan

Elevation

NAME	PROJECT	
CLASS	SECTION	

Draw the floor pochés.

BATT/LOOSE FILL INSULATION	RIGID INSULATION	METAL LATH AND PLASTER	PLASTER
SAND/MORTAR	GRAVEL	EARTH/COMPACTFILL	TERRAZZO
CARPET ON PAD	ACOUSTICAL TILE	CONCRETE BLOCK	CAST CONCRETE
DIMENSION LUMBER	FINISH BOARD	ASHLAR STONE	RUBBLE STONE
FACE BRICK	FIRE BRICK	METAL	STRUCTURAL CLAY TILE

NAME	PROJECT	
CLASS	SECTION	

Draw a reflected ceiling plan. Center a 2' X 4' acoustic tile ceiling and lights in the bed, bath and game rooms. Put a gypsum board ceiling in the closet and stairwell. Put pull-chain lights in the closet and unfinished areas, a surface-mounted light in the stairwell, and a ventilation fan in the bathroom.

BED

BED

BATH

GAME

CLOSET

UP

UNFINISHED

3/16" = 1'-0"

NAME	PROJECT	
CLASS	SECTION	

Draw an electrical plan.

CEILING SURFACE LIGHT	⊖
CEILING RECESSED LIGHT	⊡
WALL WASH	◐
FLUORESCENT LIGHT	

CHANDELIER	✳
CHIME	CH
CAMERA DOORBELL	CD

CEILING FAN	⚙
PENDANT	⦿
VENTILATION FAN	⊗

SWITCHES	S, S3
DUPLEX OUTLETS	⊖ 220
WEATHERPROOF SWITCH	S_WP

Not to scale

NAME	PROJECT	
CLASS	SECTION	

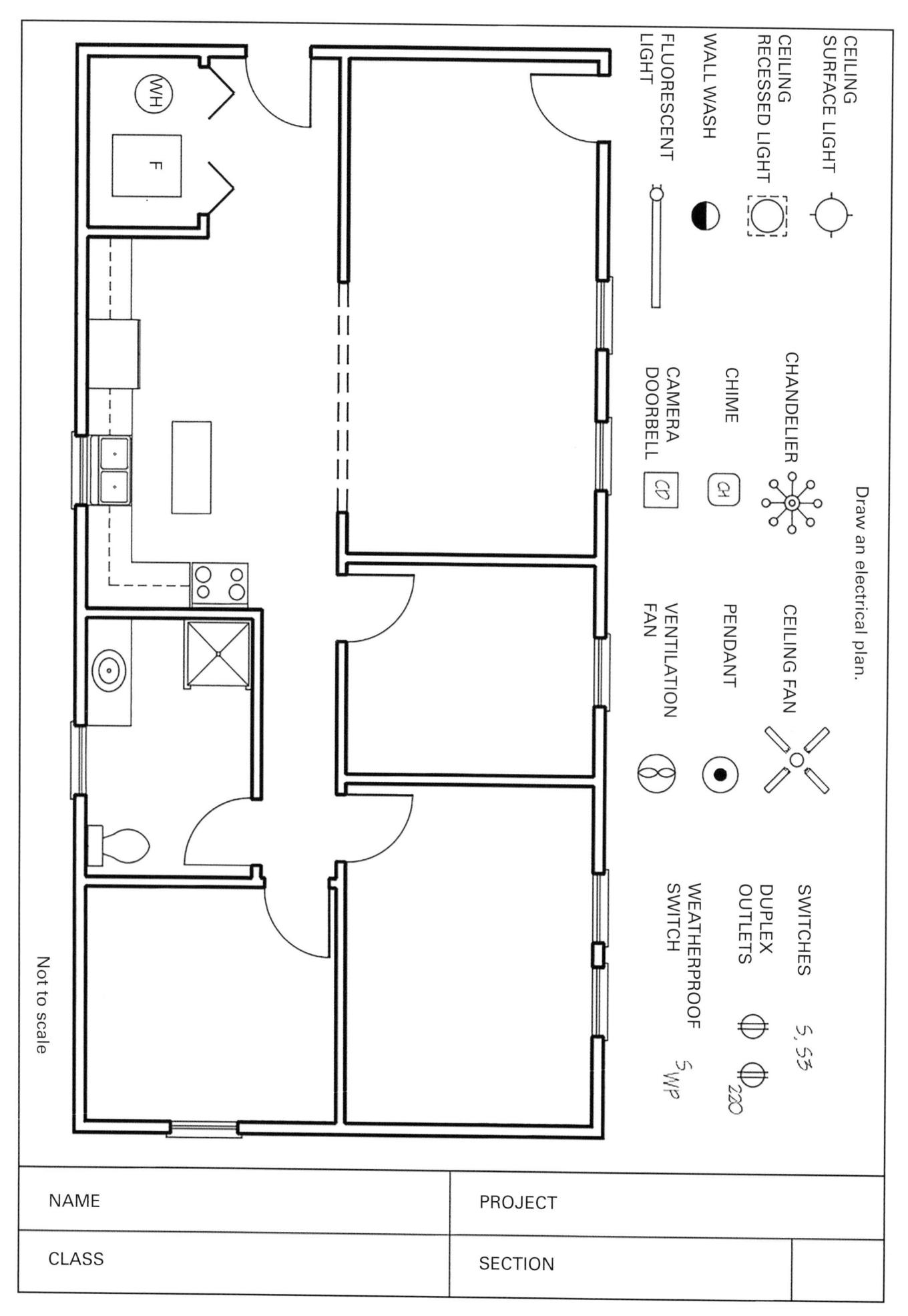

462 APPENDIX II

Draw an electrical plan for this garage using the symbols shown.

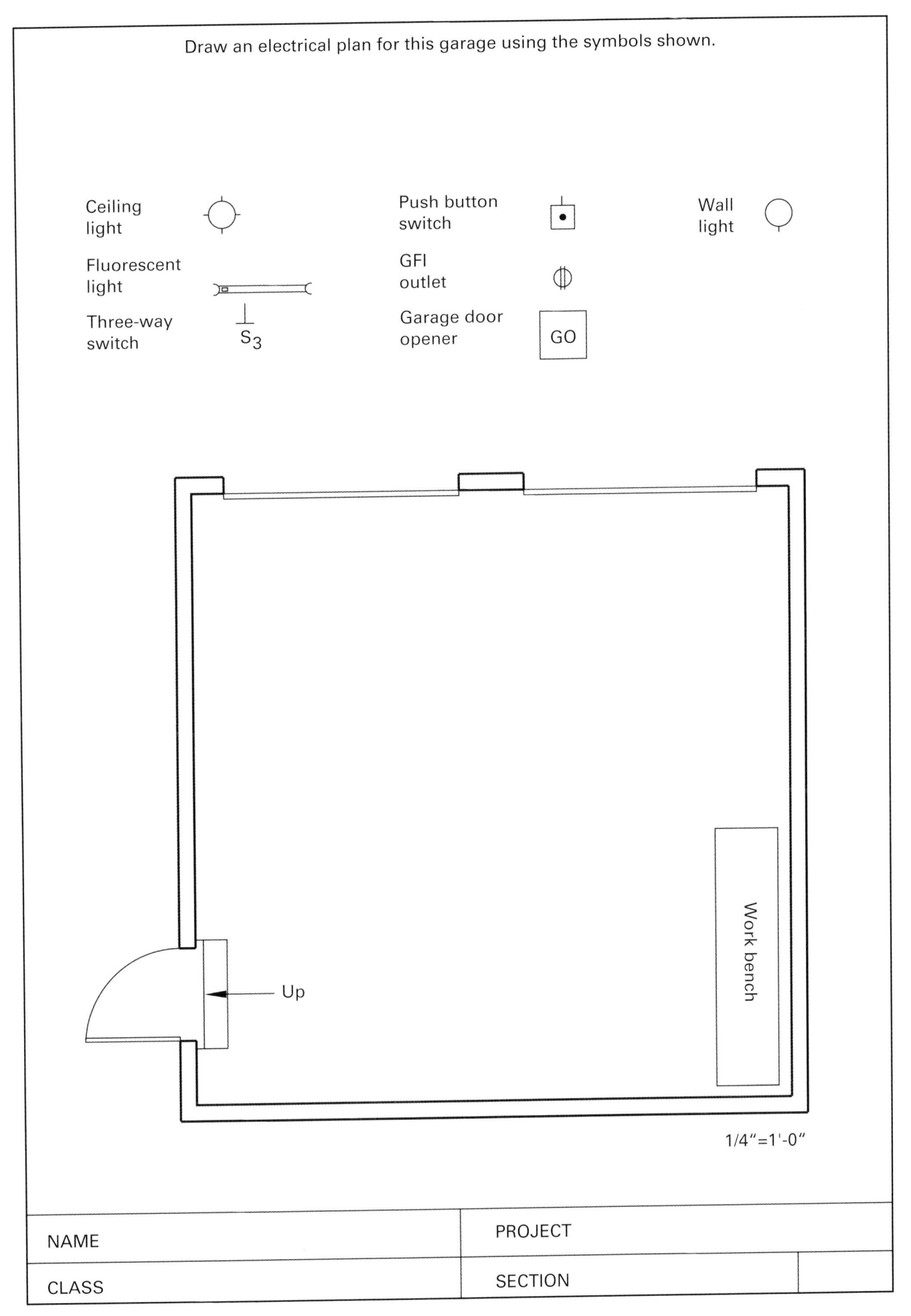

Symbol name		Symbol name		Symbol name	
Ceiling light	○	Push button switch	▪	Wall light	○
Fluorescent light	▭	GFI outlet	⌽	Garage door opener	GO
Three-way switch	S₃				

Up

Work bench

1/4" = 1'-0"

NAME	PROJECT	
CLASS	SECTION	

APPENDIX II   **463**

Draw plan and elevation views of a straight run staircase with 11 treads 11" deep and 3'-0" wide each, and a 3'-0" landing at the top. Risers are 7" tall each.

Plan

start

Elevation

start

1/2" = 1'-0"

NAME	PROJECT
CLASS	SECTION

Center a 7'-6" X 7'-6" sunken area in this living room. Draw two 10 1/2" deep steps on all sides leading into it. Draw one 10 1/2" deep step at the room entrance.

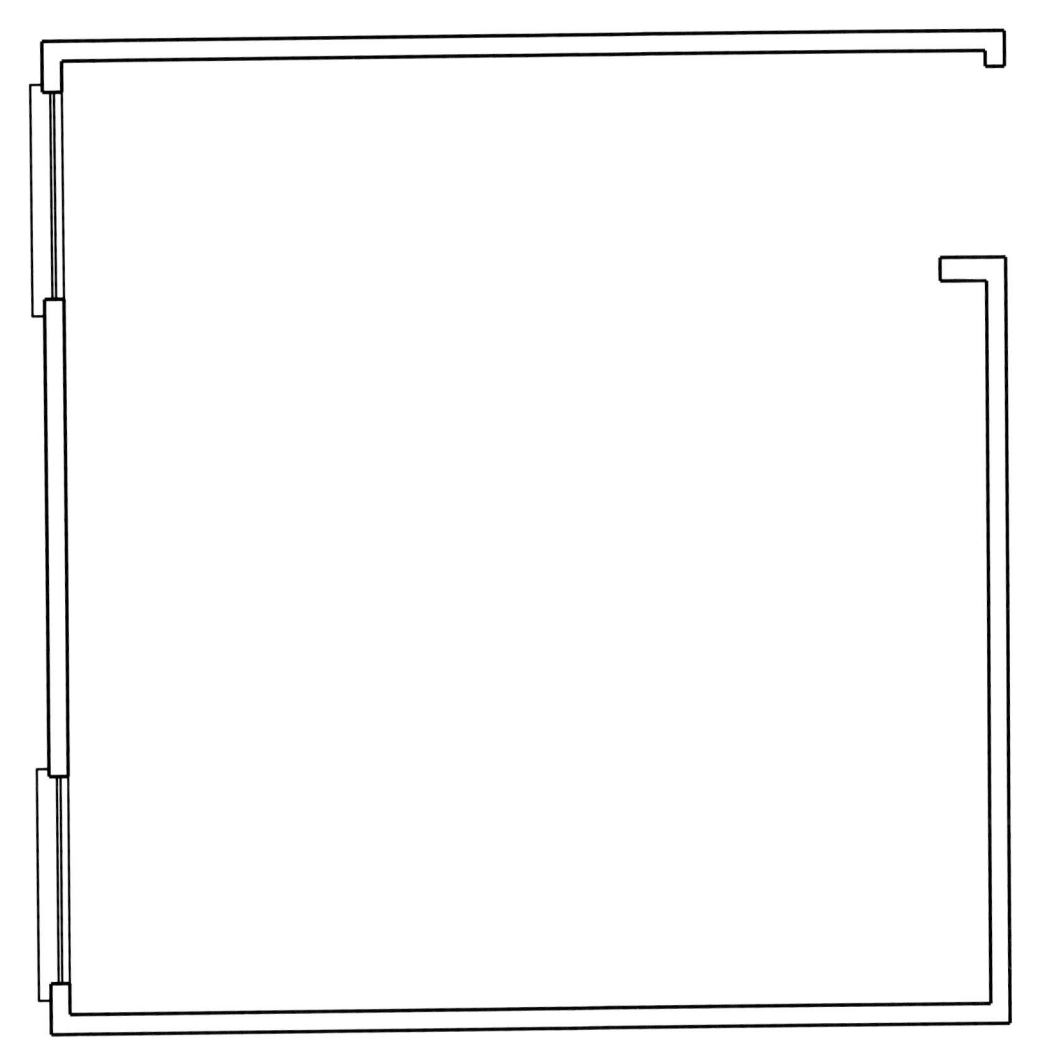

1/4" = 1'-0"

NAME	PROJECT
CLASS	SECTION

Place finish schedule callouts in each room.

Not to scale

NAME	PROJECT
CLASS	SECTION

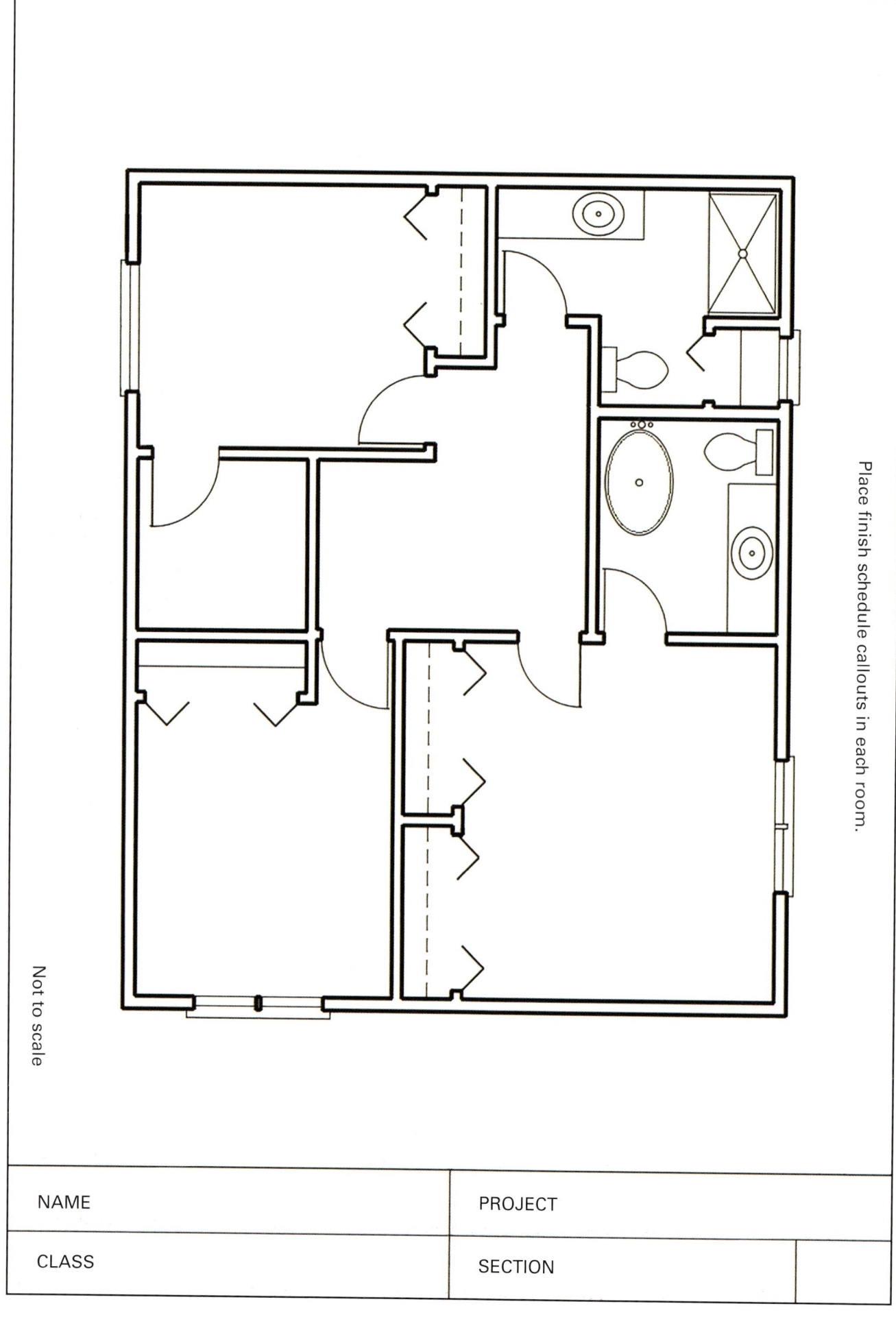

466　APPENDIX II

FINISH SCHEDULE

No.	Room	Floor	Trim	Counters	Walls	Ceiling	Other

Complete the finish schedule.

NAME	PROJECT
CLASS	SECTION

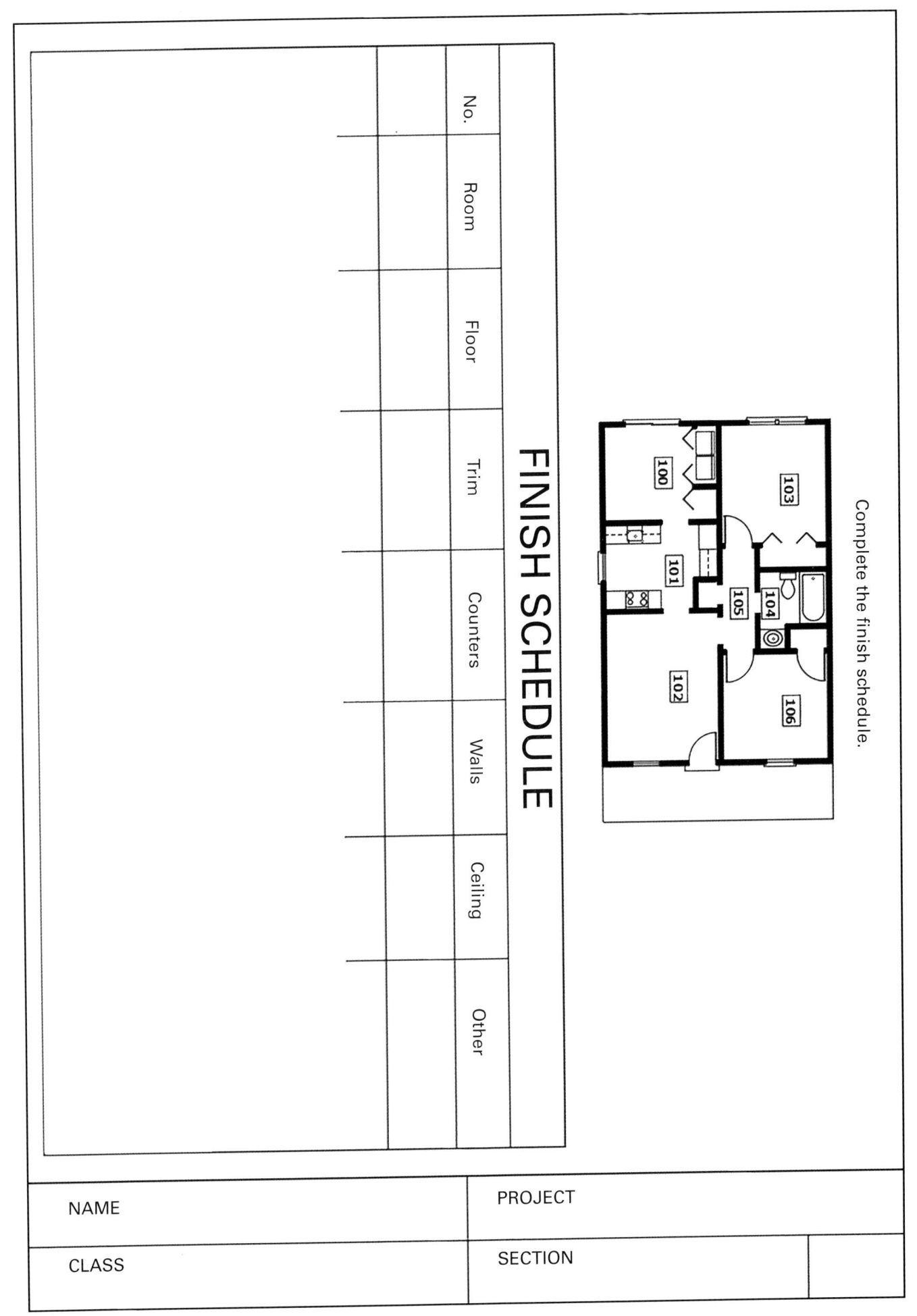

APPENDIX II

FURNITURE KEY

Complete the furniture key for a space of your choice.

Key	Item	Mfgr	Finish	Model #	Qty	Fabric

NAME	PROJECT
CLASS	SECTION

ROOM FINISH SCHEDULE

Complete the schedule for a floor plan of your choice.

Room No.	Room Name	Floor	Walls N	Walls S	Walls E	Walls W	Mat'l	Ht	Notes

NAME

CLASS

PROJECT

SECTION

APPENDIX II

Draw an isometric pictorial.

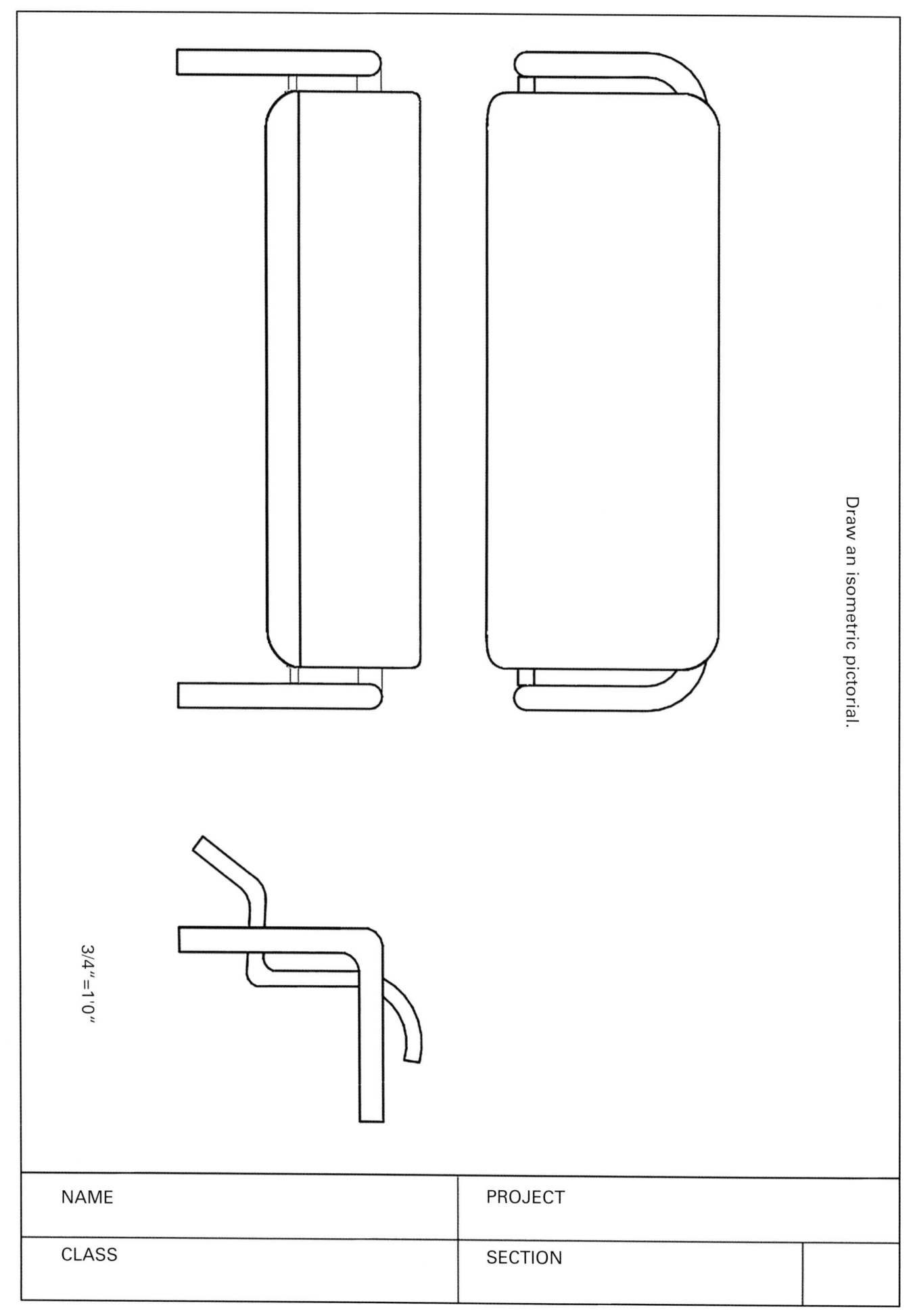

3/4"=1'0"

NAME	PROJECT
CLASS	SECTION

470 APPENDIX II

Draw a cutaway isometric of this plan, twice its size.

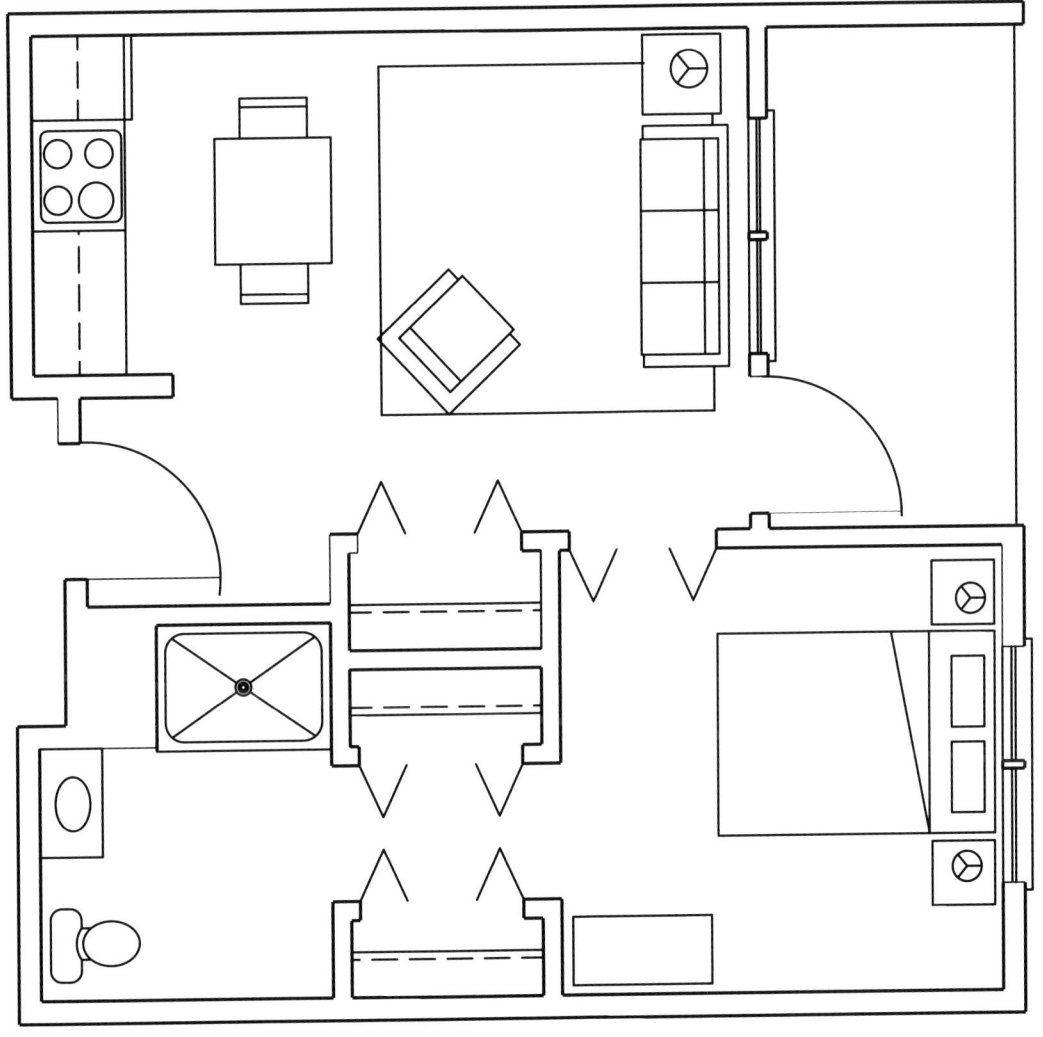

1/4" = 1'-0"

NAME	PROJECT	
CLASS	SECTION	

APPENDIX II **471**

Draw a one-point perspective.

1/2" = 1'-0"

NAME	PROJECT
CLASS	SECTION

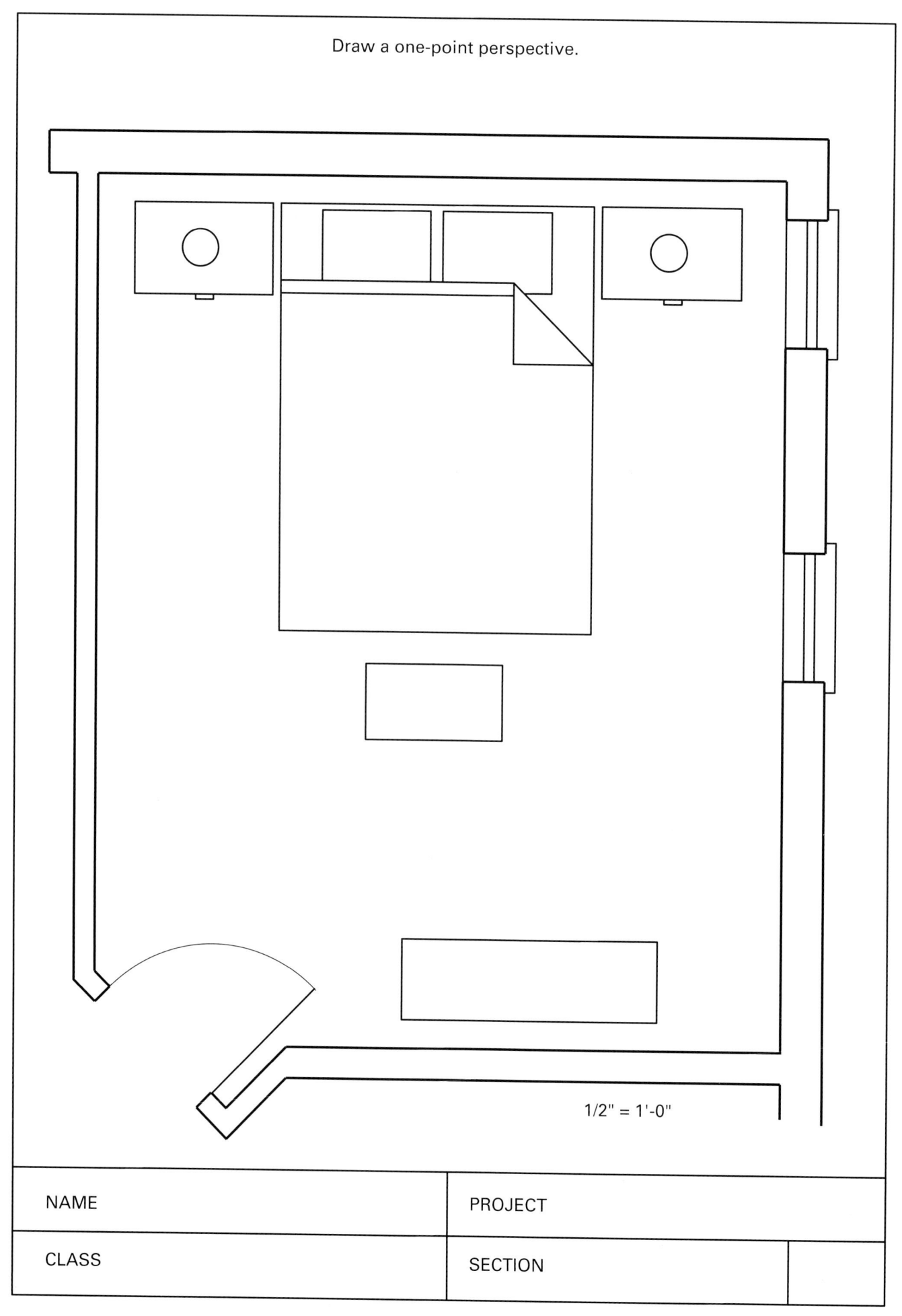

Draw a two-point perspective.

1/4" = 1'-0"

NAME	PROJECT
CLASS	SECTION

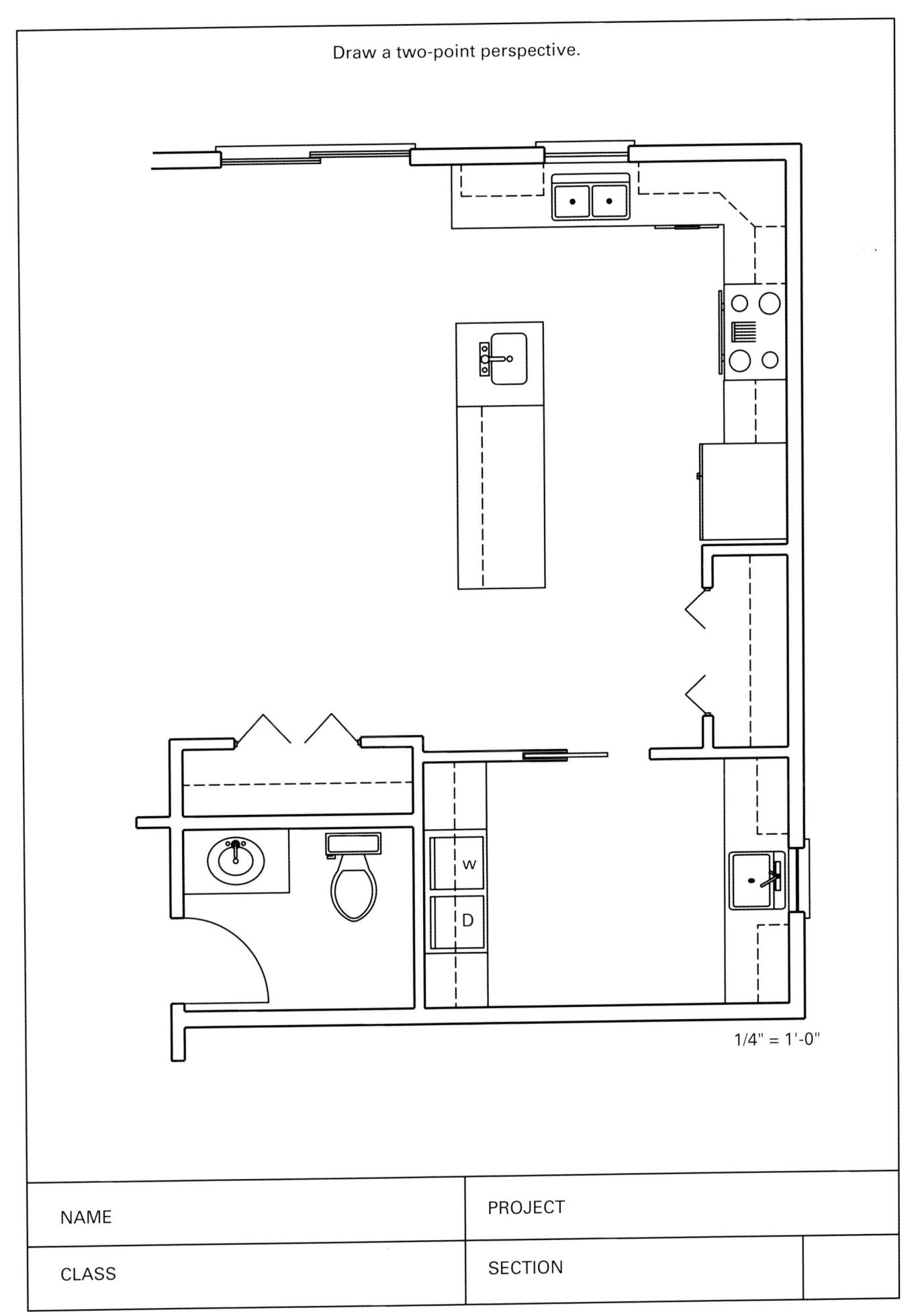

APPENDIX II **473**

Draw a one-point perspective.

1/4"= 1'-0"

NAME	PROJECT
CLASS	SECTION

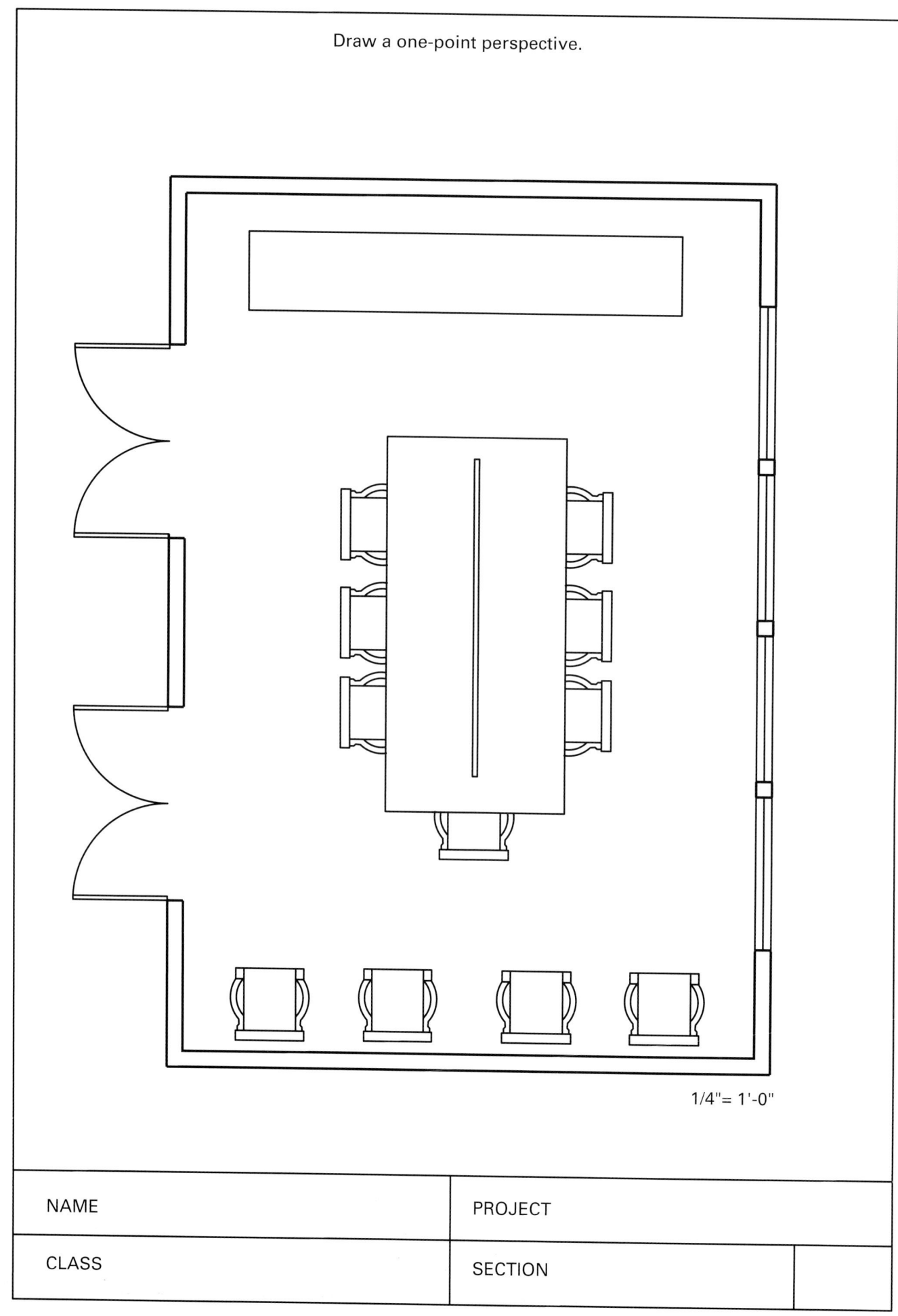

Draw a two-point perspective.

1/4" = 1'-0"

NAME	PROJECT	
CLASS	SECTION	

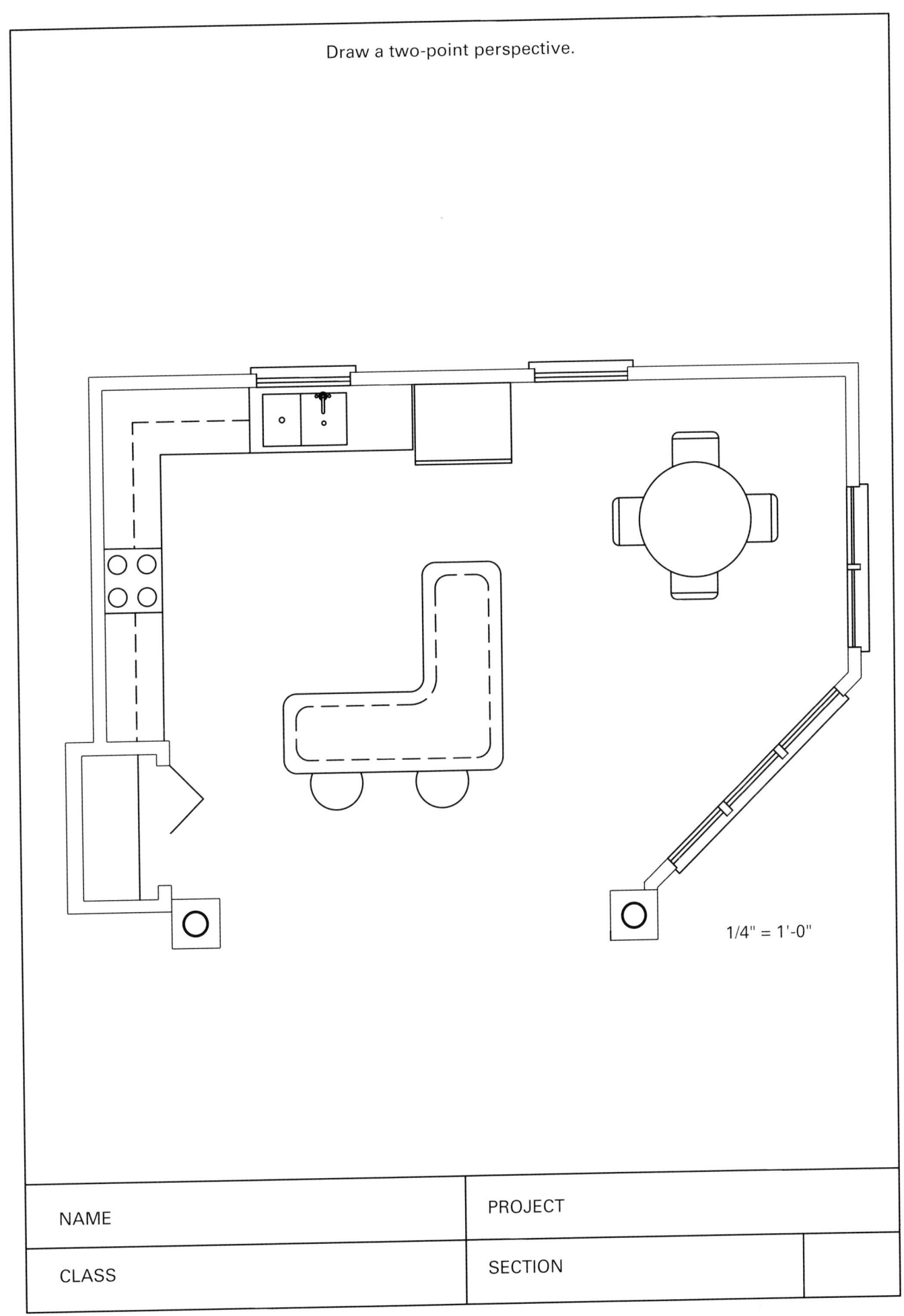

Draw a one- or two-point perspective.

1/2" = 1'-0"

NAME	PROJECT
CLASS	SECTION

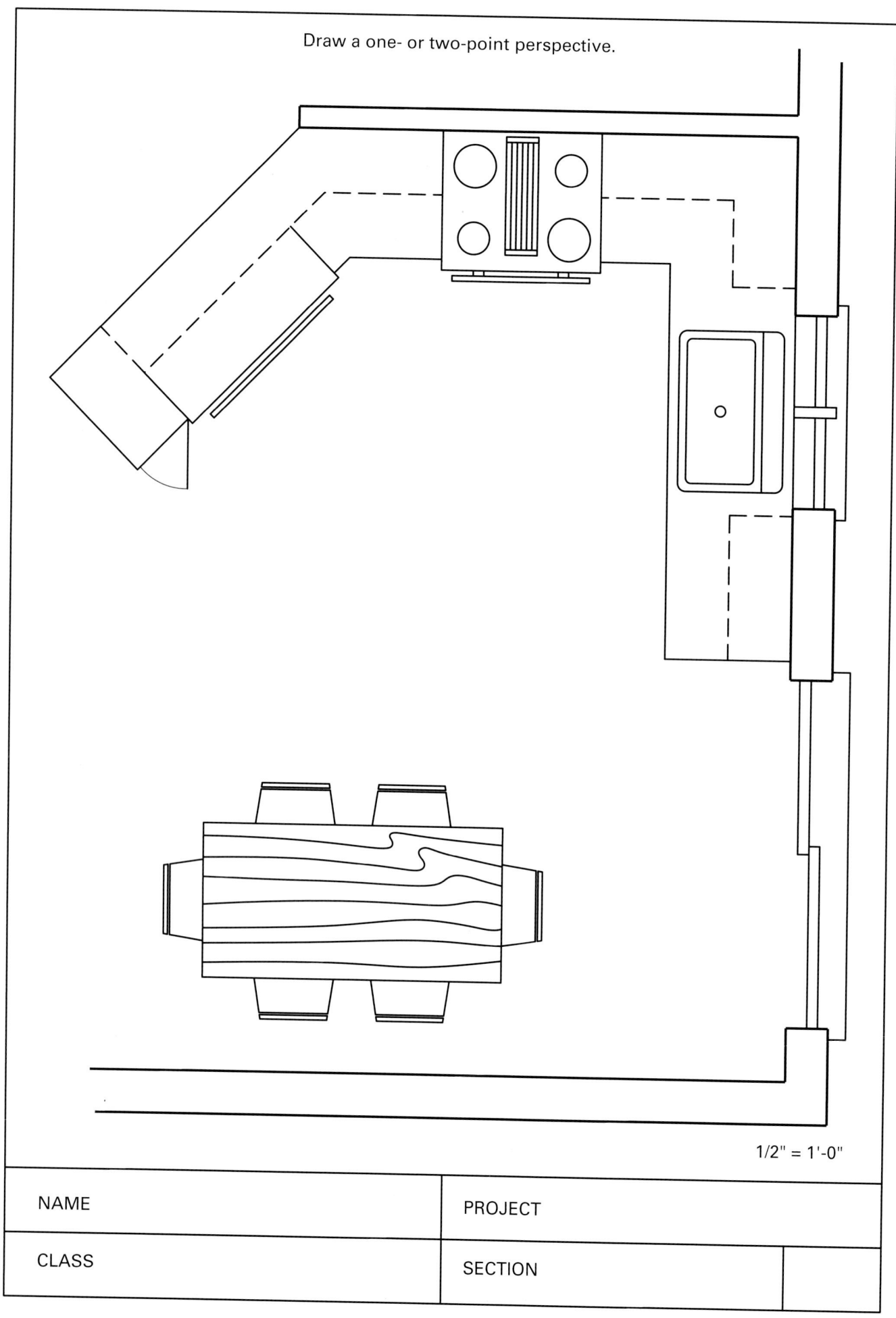

Draw a one- or two-point perspective.

NAME	PROJECT
CLASS	SECTION

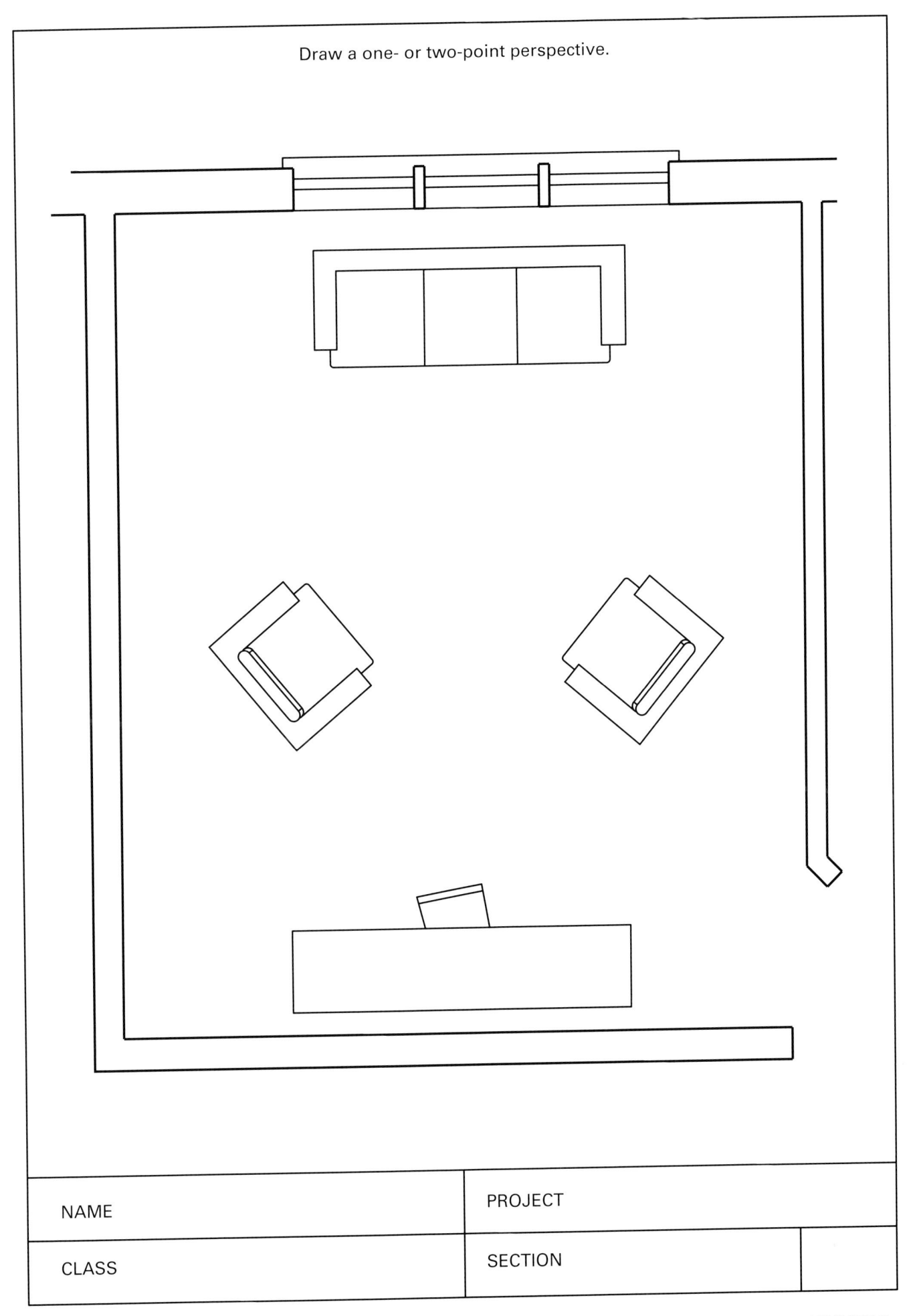

APPENDIX II

GLOSSARY

20/20 a modeling program for the kitchen and bath industry
2D sketch a drawing that shows two of an objects three physical dimensions
3D camera an imaging device that creates depth in photos
3D mouse a device that navigates the 3D environment with dedicated keys
3D print a physical object made from a digital model
3D printing a manufacturing process that converts a digital model into a physical one
3D sketch a drawing that shows an object's three physical dimensions
6/6 bed king-size bed
accessibility accommodation of people with disabilities
accessible and universal design the accommodation of people with disabilities in buildings
accordion door folding door with multiple leaves
active panel the hinged, or operating, portion of a door or window
active solar a heating system that collects and stores energy in solar panels
actual size true size of building components; *see* nominal size
adjacency matrix graphic that shows space relationships
Adobe Acrobat DC a PDF reader and editor
Adobe Illustrator a graphic design program
Adobe Photoshop digital imaging software
AFF abbreviation for Above Finished Floor
air conditioner appliance that cools, filters, and dehumidifies air
air supply vent the opening through which treated air enters a room
alcove corner that houses built-in tubs
aligned dimensioning style in which dimension notes are oriented parallel to dimension lines
American Institute of Architects (AIA) national organization that represents licensed architects and architectural interns
American National Standards Institute (ANSI) source for information on national, regional, and international conformities and conventions
American Society for Interior Designers (ASID) national organization that represents interior designers and interior design students
Americans with Disabilities Act (ADA) civil rights law that prohibits discrimination against individuals with disabilities

Americans with Disabilities Act Accessibility Guidelines (ADAAG) government written code that affects accessibility and access to buildings
Ames lettering guide tool for drawing multiple rows of guidelines that are separated by small spaces
amperage electric current strength
annotation note
ANSI-A117.1 the Accessible & Usable Buildings & Facilities code
appliance device that is attached to a gas or electric line
arch curved structural component
Architect Registration Examination (ARE) test administered by NCARB
architect's scale tool that enables accurate measuring in feet and inch units
architectural floor plan 2D, top-down view of a room or building made by inserting a cutting plane between 4' and 5' above the floor and drawing what is below it
architectural program written description of client needs and wants
area of refuge place where building occupants can safely wait for emergency assistance
area square footage
art knife thin-bladed pencil-like cutting tool
Associate Kitchen and Bath Designer (AKBD) professional association for the kitchen and bath industry
Augmented Reality an app that inserts models of a proposed design onto an existing background
Autodesk AutoCAD computer drafting software
Autodesk ReCap, ReCap Pro photogrammetry software
Autodesk Revit a BIM modeling program
Autodesk Sketchbook a program for digital sketching and coloring
Autodesk University yearly conference for Autodesk software users
automated systems plan a plan that shows electronic devices and the smart home components they operate
awning window a window that is hinged at the top and swings out
axonometric a paraline drawing that is skewed along both horizontal axes
AXP Architecture Experience Program; a licensing program for architects run by NCARB
balance distributing text and graphics evenly on a sheet

balloon framing skeleton frame construction with studs that run continuously from the foundation to the roof

baluster vertical post under a handrail

balustrade assembly containing balusters, newel posts, and handrail

bar joist horizontal load-bearing component made of multiple steel pieces

bar scale a line with distance marks along its length, allowing visual estimates

barn door a surface hung door that slides along a wall-mounted track

base cabinet storage cupboard mounted on the floor

base molding running trim that covers the joint between wall and floor

basement plan top-down view of foundation walls, footings, grade beams, pilasters, and interior spaces

bay window a window with angled sides and projects out from the wall

beam horizontal structural member

bent composite structural member consisting of two columns and a beam or truss spanning between them

bevel sharp point

bifold door a hinged door that slides

bifurcated stair a wide-flight stair that splits into two opposing, narrower flights

BIM abbreviation for Building Information Modeling or Building Information Management

bitmap *see* raster

blind cabinet a corner cupboard with inaccessible or hard-to-access space

block diagram graphic that resembles a floor plan

blocking 1. material that helps slow down the spread of fires; 2. material that strengthens walls and ceilings to support heavy objects

blueprint copies of architectural drawings made via blueprinting technology

bolt a lock that does not require a key

bond arrangement of brick or blocks in a wall

bond beam row of U-shaped blocks that helps hold a wall together

border line a line around the perimeter of the sheet

bow window a curved window that projects from the wall

box bay window a window whose sides are 90° to the wall and projects from the wall

box cutter *see* utility knife

branch duct metal tube that runs from the main duct to individual rooms; part of the HVAC system

branch pipe horizontal tube that runs from a stack pipe to a fixture; part of the plumbing system

break line a line that terminates a feature on a drawing after a clear definition of the feature has been shown

breaker box a wall-mounted steel case with electricity shut-off switches

bubble diagram concept sketch that organizes and articulates ideas and helps the designer visualize how spaces relate to each other

building code rule that governs the design and construction of residential and commercial structures

bumwad *see* tracing paper

cabinet oblique drawing whose depth is drawn half the scale of the height

CAD abbreviation for Computer Aided Drafting

CAD block a digital template

CADD abbreviation for Computer Aided Drafting and Design

California king bed a bed that is not size-standardized

callout a symbol that links two drawings

carpenter's bubble level tool that establishes a horizontal reference line

carpenter's square triangular tool that checks for corners squareness

case goods non-upholstered furniture that provides storage

cased opening a doorway opening without a door

casement window a window that is hinged on the side and swings out

casing a separate frame around a window that hides the joint between it and the wall; *see* trim

cast concrete liquid material poured into forms

cavalier oblique drawing type whose depth is drawn the same scale as its height

ceiling height door a door that is the full height of the wall, or higher than a standard door

center line a line type drawn through the center of a component to indicate its center location or that the component is round

central air *see* forced air

Certified Master Kitchen and Bath Designer (CMKBD) professional certification obtained through NKBA

chair rail running trim that is 36" above the floor to protect walls from chair backs

chandelier lighting fixture suspended from the ceiling that directs light upward

chase wall a wall built behind plumbing fixtures to contain the pipes

Chief Architect software program used in the architecture and interior design fields

chimney vertical, freestanding structure that carries smoke and gas out of the firebox

circle template a stencil with different size circle cutouts

circuit path of an electric current

circuit breakers switches that shut down overloaded circuits

circular stair a stair that rises along a continuous "C," oval, or elliptical shape

circumscribe draw outside, and tangent to, a circle; see inscribe

clean-out a covered hole in a pipe or trap that provides access for unclogging

cleanout valve a plugged hole in the house's main sewer line that provides access for unclogging

clear span interior space unobstructed by columns or load-bearing walls

clearance distance between architectural features or furniture

clerestory a window placed high in the wall

climate control system heating, ventilation, and air-conditioning (HVAC) system

closer hardware that automatically closes a door after it has been manually opened

clutch pencil a mechanical pencil that holds a 2 mm lead

CMU abbreviation for concrete masonry unit

color board a visual that shows textiles and paint chips used in a design

column vertical load-bearing member outside a wall; runs continuously from foundation to roof

commercial furniture plan a non-residential plan that shows furniture layout

compact fluorescent lamp (CFL) an energy-saving light and compact fluorescent tube designed to replace an incandescent light bulb

compass a tool used for drawing circles and arcs

compass north a directional orientation, usually shown straight up

concept sketch drawing that shows a design's main idea for function or appearance

concrete block rectangular unit used in building construction

cone of vision an angle that defines the limits of a person's view

construction calculator device that converts between imperial and metric units; linear, square, and cubic formats; and provides other construction functions

construction documentation drawings that serve as legal and binding instructions for building design and construction

construction line light line for layout work and is not meant to be reproduced on a copy

contract administration design phase in which the job is awarded to a contractor

contrast different-appearing items on a drawing sheet

corner cabinet cupboard configured to maximize usable space

Council for Interior Design Accreditation (CIDA) organization that accredits college interior design programs

Council for Interior Design Qualification (CIDQ) certifying body that develops and administers the NCIDQ

course horizontal row of bricks or concrete blocks

cove light a lighting fixture placed in a recess high on the wall; directs light to the ceiling

crawl space hollow area between the ground and first floor, usually 1' to 3' high

cripple stud a short stud placed above or below a window

criteria matrix graphic that visually shows space requirements

cross bracing structural component that strengthens walls

cross bridging structural component that strengthens floors

crown molding running trim that covers the joint between the top of the wall and the ceiling

curtain wall non-load-bearing outer wall of a building

cutaway isometric a drawing that omits or partially removes walls and roof to show what is behind them

cutting mat vinyl surface for cutting paper and board

cutting plane a symbol on the floor plan that indicates where a section drawing slice is made

data plan a floor plan that shows telecom systems, data communications and security systems

deadbolt a lock that requires a key

dead-end hallway corridor with that allows exit in just one direction

decking horizontal substrate on a roof or floor

demolition plan a top-down view of all items to be removed or relocated

design development design stage when the project goes from concept to workable design

design thinking the process of creative problem solving

detail drawing a drawing made via a vertical or horizontal slice through a small portion of the building to show how it is assembled

diameter distance from one point on the circle through the center to an opposite point on the circle

digital imaging the manipulation of files via software

digital model a 3D drawing created with a modeling program

dimension line a line that spans a feature described by a dimension note

dimension note a number that describes the size of an architectural feature

dimensional lumber wood cut to a specific size such as 2" × 4"

dimensioning the process of indicating the size of architectural features and their location from other features

dimetric a type of axonometric drawing

dimmer switch a switch that adjusts light brightness via a rotating knob

direct modeling software that creates drawings with lines and curves

direct vent a fireplace with a sealed-combustion system

disability a broad term that includes physical and psychological issues

disposable pen a non-refillable ink pen

distribution panel *see* breaker box
divider a tool that that is used for marking distances
door hand a door's hinge position
door schedule a chart with information about a building's doors
double door two swing doors hung from opposite jambs in one opening
double-action door a door that swings in both directions
double-hung window a window with two vertical sliding sashes
double-pane window two sheets of glass separated by an air space
double top plate two plates at ceilings and above doors
double-wythe a masonry wall composed of two walls with a space between them
downdraft vent an opening built into, or near, cooktops that allows air, gas, and liquid to pass through
drafting the art of putting ideas to paper in picture form to explain ideas and create instructions
drafting board a large, smooth surface on which drawing is done
drafting brush a bristled tool that sweeps away eraser crumbs
drafting chair a cushioned seat with rolling castors
drafting tape sticky-backed paper strip that attaches media to a drafting board
drawing number a number on an ID label that tells which drawing to look for; *see* sheet number
drop-in ceiling *see* suspended ceiling
duct an air passageway in a forced-air heating and cooling system
duplex outlet electrical input device with two receptacles
Dutch door a swing door that is cut into top and bottom halves that operate independently
DWG a file format created in AutoCAD
easement a legal right to use another person's property for a specific, limited purpose
egress an exit that allows building occupants to evacuate safely during an emergency
egress door a door that facilitates escape from a building
egress window a window that meets specific size requirements to serve as an emergency exit
electrical plan a top-down view that shows lights, switches, circuits, and outlets
elevation a height drawing of a wall
elevation callout a symbol on the floor plan that indicates where, and of which wall, an elevation drawing is made
elevation view a drawing made by looking straight ahead at the wall and outlining it from corner to corner
elevator a vertical transport vehicle
ellipse a foreshortened circle
enclosed staircase a staircase that is surrounded by walls

EnergyStar a US Environmental Protection Agency voluntary certification program for sustainable design
engineer's scale a tool used for measuring lengths using decimals
engineered wood product building component made of laminated veneers
English units *see* Imperial units
enlargement box dashed line encircling an area to indicate there is a larger-scale drawing elsewhere in the drawing set
entourage people, furniture, décor, and accessories in an architectural drawing that provide visual scale
equipment tangible property used in the operation of a business or system
equipment plan *see* HVAC plan
eraser a piece of soft rubber or plastic used to remove pencil lines
eraser bag mesh fabric that contains eraser flakes, used to remove dirt and smudges
eraser shield a metal template with openings that facilitate fine erasing
evaporative cooler a type of air conditioner
Excel a spreadsheet program used for creating reports and schedules
exit the fully enclosed, fire-protected space between the exit access and the exit discharge
exit access the portion of the building that leads to an exit, including halls and stairs
exit device a panic (push) bar and latch used on out-swinging doors
exit discharge the area between the termination of an exit and a public way
exit stairs an assembly of stairs, fire wall enclosure, and if applicable, doors and an area of refuge
extension line a line that runs perpendicular to the dimension line and defines the endpoints of the feature being dimensioned
exterior elevation height drawing of a building's outside wall
exterior perspective a 3D drawing that shows the outside of a building
Federal Housing Authority government agency that provides mortgage insurance on single and multi-family properties
fenestration arrangement of windows and doors in a building
FFE abbreviation for Furniture Fixture and Equipment
FFE Schedule a chart with information about a plan's moveable equipment
finial cap decorative knob
fire door a door rated to protect against the passage of fire
fire door assembly a door and frame that protects against the passage of fire
firebox the combustion chamber in a fireplace where a fire is contained

fireplace a framed opening in a chimney that holds an open fire

fit sketch a drawing that shows how furniture and equipment fit in a room

fixed features non-moving items on an architectural floor plan

fixed panel non-moving

fixed window a non-moving window

flashlight portable light or phone app

flight *see* staircase

floor plan *see* architectural floor plan

flue vertical duct inside a chimney that goes to the roof

fluorescent a type of lamp

flush door a door with a smooth, even face

fold a crease in sections

footing widened base of a foundation wall, pier, or column that supports and distributes the weight of the loads above it

footprint a building's shape, size, and orientation on the site

forced air system an HVAC system that uses air to transfer heat and cold

foreshortened shorter than true (actual) length

framed cabinet a cupboard with *rails* (horizontal framing pieces) and *stiles* (vertical framing pieces) on the inside of their doors

frameless cabinet a cabinet without rails and stiles (framing pieces) on the insides of their doors

French curve a flat, irregularly shaped piece of plastic used to draw arcs

French door two swing doors that operate opposite each other and have glass inserts

front view height drawing of an object; *see* side view

front-loaded appliance a machine, such as a washer or dryer, with doors on the front

full bath a bathroom containing a lavatory, toilet, and tub or tub/shower combination

full height door *see* ceiling height door

full overlay a cabinet door that has a maximum 1/8" exposed portion around each door and drawer and between the doors; *see* standard overlay

furnace an appliance that produces heat

furniture key a list of furniture

furniture schedule a chart with information about furniture type, quantity, size, finish, fabric, and manufacturer

furniture template a flat piece of plastic with furniture cutouts for tracing

furring strip a wood strip attached to a masonry wall for applying finish materials

Fused Deposition Modeling (FDM) a 3D printing process

Fused Filament Fabrication (FFF) *see* fused deposition modeling

gable ceiling a ceiling that is sloped on two sides

galley kitchen layout where cabinets, appliances, and counters run along two parallel walls

gang multiple switches on one plate

garden window *see* box bay

gate valve a valve that stops the water supply

general note informational note that applies to everything on a project

general oblique drawing type whose depth is drawn between 1/3 and 3/4 scale of the front view

generative design a design exploration process done with cloud computing

GFCI outlet Ground Fault Interrupter outlet; an electrical receptacle with a safety feature that turns off the circuit when sensing water

GIF an image file

GIMP free digital imaging software

girder horizontal load-bearing beam that supports smaller beams

Glulam glue laminated lumber

grab bar a safety device that helps a person maintain balance

grade ground

grade beam a thickened portion of a foundation that supports load-bearing walls

grade line a line that represents the top of the ground

graph paper *see* grid paper

graphic communication the description of ideas and instructions in picture form; *see* drafting

graphite drafting leads for mechanical or clutch pencils

grid paper a paper printed with horizontal and vertical lines at regular increments

grille a cover plate over a return vent

Ground Fault Circuit Interrupter (GFCCI) outlet a breaker that protects an entire circuit electrical receptacle with a safety feature that turns off the circuit when sensing water

grounded outlet an outlet with a third wire that directs power surges to the earth, reducing the risk of appliance damage and personal injury

G-shape kitchen *see* peninsula kitchen

guard assembly of balusters under a handrail to prevent falling

guides hardware installed at the bottom of a door to keep the door from swinging back and forth

gypsum board an interior wall finish of gypsum powder pressed between two sheets of building paper

habitable category of space fit to live in, suitable for habitation; other words for this include occupiable, dwelling, sleeping, and living

habitable space code descriptor for any space in which general living takes place

half bath bathroom with a toilet and lavatory

half-diagonal rule requirement that the minimum distance between two exits be at least one-half of the longest diagonal distance within the building
hallway long passage with doors into rooms on both sides
halogen type of lamp
handle and knob hardware that operates a door latch
handrail continuous horizontal piece on top of a row of balusters
hardware schedule chart with information about hinges, handles, locks and knobs on doors and windows
hatch lines 45° angled lines that indicate an object has been sliced
head top of a door or window
header wood lintel
headroom clear vertical distance between the tread and the ceiling, measured vertically along a sloped plane
hearth fireplace floor
heat exchanger a device that passes heat from its fluid directly to the air inside the house
height line the line on a perspective drawing on which measurements are made
helical *see* circular
hidden line a line that defines an item not visible in the view
high-resolution a file that has at least 300 dpi (dots per inch)
hinges hardware that holds a door to a wall
hip rafter a roof component that extends from the wall plate to the ridge and forms the angle of a hip roof
holder hardware attached to a door and frame to keep the door in an open position
holistic all parts of the project are coordinated and integrated
hopper window a window that is hinged at the bottom and swings into the house
horizontal run the total floor space a flight of stairs takes up, including landings
hose bib outdoor faucet
humidifier appliance that adds moisture to a building
HVAC abbreviation for heating, ventilation, and air-conditioning
HVAC plan a horizontal view that shows heating, ventilation, and air-conditioning systems
ID label an identification symbol under a drawing
imbibed eraser an eraser formulated for use on plastic film
imperial rule full or true scale; 1" = 1"
imperial units English or U.S. Customary units of feet and inches
incandescent a type of lamp
ink pen a writing instrument used for manually created presentation drawings
inset door a door that sits within the cabinet frame, flush with the cabinet box's front
inscribe draw inside, and tangent to, a circle; *see* circumscribe

Insulated Concrete Forms (ICF) hollow foam blocks into which concrete foundations are poured
interior elevation height drawing of a wall inside the building or of the interior side of an exterior wall
interior perspective a 3D drawing that shows the inside of a building
International Building Code (IBC) a model code that governs commercial construction
International Code Council (ICC) a standards organization that writes model codes
International Interior Design Association (IIDA) a global organization that supports commercial interior designers, industry affiliates, educators, students, and companies
International Residential Code (IRC) a model code that governs construction of one- and two-family homes of three stories or less
Irfanview a free digital imaging program
irregular curve *see* French curve
isometric pictorial a drawing type that has two horizontal axes skewed at a 30° angle
jalousie window a window with rows of narrow, horizontal glass slats
jamb side of a door or window
joist horizontal load-bearing beam in ceilings and floors
JPG a photo image file format
key chart of furniture information
key letter callout symbol that links a graphic on the floor plan to a key
key number *see* key letter
keynote geometric shape with a number or letter inside that references a component to a legend or schedule
king bed *see* 6/6 bed
knee space the space needed under a countertop for accessibility
knee wall *see* partial wall
L stair stairway with a landing and turn that form an L shape
lamp 1. bulb in a light fixture; 2. light-providing appliance consisting of an electric bulb, holder, and cover
landing rest area on a stairway
laser tape measure electronic device that records distance with a laser and displays it on a screen
lavatory bathroom
lavatory sink bathroom sink
layout line *see* construction line
lead grade level of hardness or softness
lead holder *see* clutch pencil
lead pointer rotary blade sharpener
leader line a line with an arrow at one end and a local note at the other
leaf a single door panel
leaves plural of leaf

LED abbreviation for light emitting diode; a type of electric light

LEED abbreviation for Leadership in Energy and Environmental Design; a green building rating system

left-hand door a door that is hinged on your left and swings away from you

left-hand reverse a door with the hinge on your left and swings toward you

legend a chart that clarifies symbols on a plan

lettering note making on an architectural drawing

LiDAR a method for measuring distances with a laser light and sensor

lighting fixture schedule a chart with information about permanently attached lighting fixtures

lighting plan a plan that shows all lighting fixtures, switches, and their connecting wires

line 1. a thin (relative to its length) geometric object that is the fundamental symbol of graphic communication; 2. a pipe or electrical conduit

line quality appearance of a line

line type a line pattern that represents a concept

line weight thickness that varies with item importance

lintel a load-bearing beam over a window, door, or fireplace opening

lipped door a door that has a rabbet cut all the way around it on the back edge

lite framed glass insert

load-bearing a component that carries weight besides its own, such as an upper story or roof

local note a short description of a specific feature

location a particular place or position

lock a security device that contains a latch bolt, a lock strike, and a cylinder

longitudinal section a drawing made by slicing parallel to the roof ridge

louvered window see jalousie

low-resolution a blurry or pixelated image

LSC abbreviation for Life Safety Code; covers hazards to human life in buildings

L-shape kitchen a layout where cabinets, appliances, and counters make an L-shape

machine learning the application of artificial intelligence that lets software learn and improve from experience without being specifically programmed

main supply duct a sheet metal air passageway that is attached to the furnace

mantel 1. a frame surrounding a fireplace; 2. the shelf over the fireplace

mark a symbol in a floor plan that links to an elevation or schedule

masonry opening a rough opening (hole in the wall) in a concrete or brick wall

masonry units of brick, concrete block, stone, glass block, structural clay tile, or terra-cotta

MasterFormat a 49-division system of organization for building materials, developed by the Construction Specifications Institute

match line a line type that shows where to piece together a drawing that spans two sheets of paper

matte dull side of the film for drawing

means of egress building occupants' route to exits, especially in an emergency

measuring app a phone program that measures and calculates

mechanical pencil a multipiece, refillable tool available in various tip sizes

mechanical plan a horizontal view that shows heating, ventilation, and air-conditioning systems

medicine cabinet a small, wall-mounted cupboard in a bathroom

mesh the digital surface of polygons that make up a model

metal tape measure manual measuring device

metric scale tool used for measuring architectural drawings in metric units

microlam see Laminated Veneer Lumber

milled lumber wood cut flat, square, and to specific sizes

millwork architectural wood products manufactured in a lumber mill

mini-split ductless heat pump that heats and cools a room

mixed occupancy a building that contains two or more different land uses

model a 3D drawing made with a software modeling program

model code a building code written by a standards organization that is independent of the jurisdiction responsible for adopting and enforcing it

modeling program 3D drawing software

modular construction a building technique that only uses standard-sized, mass-produced components

molding decorative lengths of wood that hide the joints at floor/wall and floor/ceiling intersections

mood board a visual that describes the design essence without being specific

mortar joint a layer of mortar in between courses of masonry

movement the path a viewer's eye takes through a sheet of drawings

mullion a structural horizontal or vertical bar between individual windows

muntin a nonstructural bar that divides a large piece of glass into multiple panes

NAHB abbreviation for National Association of Home Builders, a large network of craftsmen in the building industry

narrow U stair a U stair with a small space between flights

National CAD Standard a set of drawing rules created by the American Institute of Architects, the Construction

Specification Institute, and the National Institute of Building Sciences

NCARB abbreviation for the National Council of Architectural Registration Boards, an organization that recommends model laws, regulations, and other guidelines

NCIDQ abbreviation for the National Council for Interior Design Qualification, an organization that oversees the eligibility, examination, and certification of interior designers

NEC abbreviation for the National Electric Code; a standard for the safe installation of electrical wiring and equipment

newel post the first pole on a balustrade

NFPA abbreviation for the National Fire Protection Association; an organization that writes codes and standards

NKBA abbreviation for the National Kitchen and Bath Association; a national organization that provides education and support to the kitchen and bath industry

NKBA standard rules and requirements written by the NKBA

nominal size *see* actual size

non-grounded outlet an outlet with only two wires, hot and neutral; no grounding wire

non-load bearing wall *see* partition wall

nosing the portion of the tread that overhangs the riser

object line a line that defines the item being drafted

oblique drawing technique where the front of the object is drawn true size and shape and the depth is drawn at an angle between 0° and 90°

occupancy a category that describes what a building is used for

occupant evacuation elevator an elevator for general public use

occupant load the number of people a building is designed to safely hold

one-point perspective a drawing technique that uses one convergence point

open riser a stair step without a riser board

open staircase a staircase with no surrounding walls

orbiting move moving around in a circle in a modeling program

ordinance regulation for aesthetics, traffic, and activities

orientation compass location and direction

ortho a prefix that means straight, perpendicular, or vertical

orthographic drawing a drawing that shows two of an object's three dimensions

orthographic projection a drawing technique that deconstructs a 3D object into multiple 2D views

OSHA abbreviation for Occupational Safety and Health Administration; an organization that writes workplace safety and health codes

outlet 1. connection device in a circuit that allows electricity; 2. hole in a forced-air system through which treated air is returned to a room

overhead sectional door a door that is assembled in sections and moves on a vertical track

painter's tape a sticky-backed strip paper alternative to drafting tape

pane framed sheet of glass within a window

panel door a door with rectangular raised or recessed areas framed by rails and stiles

panic bar door hardware that allows quick egress in an emergency

paraline a drawing technique where all parallel lines remain parallel from start to end

parallel bar a long, straight tool installed on a drafting board used for drawing horizontal lines

parametric modeling a complex computer drawing process that uses features and constraints

parti sketch a drawing that shows the organizing thoughts of a design and the relationship of parts to the whole

partition wall an interior, nonstructural wall that carries only its own weight

passenger elevator a vertical vehicle for general public use

passive panel the portion of a door or window that may only be opened once the active panel is opened

passive solar a heating system that uses a house's design to collect the sun's energy

paver a brick or block installed in the floor

PDF a file format that allows reading and editing text

pendant a lighting fixture suspended from the ceiling that directs light down

permanently attached light fixture an electric unit with a lamp and power source

perspective a drawing technique that represents spaces and objects as we see them

perspective grid a 3D chart that enables quick perspective sketching

photogrammetry a process that creates a digital model from photographs

photo-textured model a model covered with the photos taken during a photogrammetry process

pictorial drawing 3D presentation of an item

picture plane 2D surface onto which the side of an object is projected

picture window a large fixed window

pilaster thickened portion of a wall

pile foundation component consisting of a giant concrete shaft without footings

pivot door a single-panel door that rotates on a spindle

pivot window a window that rotates around a horizontal or vertical axis

pixelation jagged appearance

plan an orthographic view projected onto the horizontal plane
plan north the convention of orienting all plans straight up and calling the top north
plan oblique a drawing that angles both horizontal axes 45°, resulting in the top view being true size and shape
planometric *see* plan oblique
plastic film polyester manual drafting media
plate horizontal board that evenly distribute loads placed on it and ties studs together
platform (western) framing skeleton frame construction in which the studs run from level to level
plenum the space between the ceiling and roof
plotter a large, freestanding inkjet printer that can print large drawings
plumbing fixture a device that draws freshwater and discharges wastewater
plumbing plan a horizontal view that shows the plumbing system and fixtures
plumbing system a network of pipes and controllers that delivers a building's fresh water and removes dirty water
PNG a graphic file format
poché a textural pattern that describes material
pocket door a door that slides in and out of a wall compartment
pop-up space a short-term, temporary retail store
portable drafting board a large drawing surface without a base
post-and-beam framing a building technique that utilizes large structural members spaced far apart
post-on-pad a foundation type consisting of a square pole that rests on a concrete cube
powder room *see* half bath
power plan *see* data plan
prefabricated *see* modular construction
profile 2D shape
programming the research stage of the design process
project statement a description of client needs and wants
project bring forward in orthographic drawing technique
Pronto an early CAD program
proportional scale a tool that calculates percentages to enlarge or reduce an image
prototype a preliminary model of the design solution that tests user reaction to it
protractor a tool used for drawing angles
P-trap a curved piece of pipe that prevents sewer gases from bubbling back into the bowl
public way the area outside a building between the exit discharge and a public street, usually the means of egress's final destination
push-button switch a press- and-release button
quadrant markers printed lines at the circle's four quarters

queen bed 5'0" bed
rabbet groove
radial an arrangement of graphics like rays around a center point
radiant heat an HVAC system that supplies heat directly to the floor instead of to air
radius distance from the center to the circumference of a circle; half of diameter
raft a foundation with no footings; the whole basement floor is the foundation
rafter an inclined beam at the roof
rail 1. horizontal framing component on a window or door; 2. vertical component on top of stair balusters
rake view foreshortened view of an angled object
ramp sloped surface that facilitates building accessibility
range hood appliance that ventilates cooktops and stoves
raster a file format made of pixels
rebar *see* reinforcing bar
recessed lighting a lighting fixture installed above the ceiling
reclaimed wastewater recycled and cleaned wastewater
reflected ceiling plan a view of the ceiling as if it were reflected onto a mirror that is flat on the floor
Regreen a green building certification
render 1. the process of adding shade, shadow, and color to a drawing; 2. a drawing with fine detail and/or shade, shadow, and color
rendering the process of adding shade, shadow, and color to a drawing
repetition using the same items throughout a sheet of drawings
reveal the portion of a cabinet not concealed by a door or drawer
revolving door a multi-panel door that rotates around a vertical axis
Rhinoceros 3D a modeling program
ridge board a board to which roof rafters are attached
right-hand door a door that is hinged on your right and swings away from you
right-hand reverse door a door that is hinged on your right and swings toward you
rigid frame *see* bent
riser a stair step's vertical board
riser diagram an elevation drawing of plumbing pipes
roof framing plan a top-down view of the rafter layout
roof plan a bird's-eye view of the roof that shows hips, valleys, skylights, and anything mounted on it
roof ridge the highest, longest beam on a sloped roof
rotated plan a drawing technique that rotates the floor plan on the drawing board during a perspective drawing setup
rough opening the hole in a framed wall for a door or window
RVT the file format for Autodesk Revit

sandpaper block light grit sandpaper for sharpening 2 mm leads
sash window glass plus the surrounding frame
sash size the dimensions of the glass plus the frame that grips it
scale size of one item in relation to another
schedule 1. chart of information about building components; 2. workflow of a project
schematic an abstract drawing that uses simple lines and symbols to represent physical items
scissors stairs two interlocking U-stairs with two separate means of egress
section detail a drawing that shows a small portion of a building
section view a detail drawing made with a horizontal or vertical slice
septic tank an underground tank used for processing solid waste
service elevator an elevator that accesses equipment and storage rooms
service panel *see* breaker box
service stairs staircase that accesses roof and equipment rooms
sheathing vertical substrate under a wall's finish material
sheet number a number that tells what sheet to find a drawing number on; *see* drawing number.
shop drawing detailed construction drawing derived from the design drawing
SI abbreviation for System International; metric units
side view the part of an object seen from a vertical plane
sight ray a construction line in a perspective drawing
sill the bottom of a window or a door
single plug outlet an outlet with one plug for dedicated circuits
single-hung window a window that has one sliding and one fixed sash
single wall kitchen all appliances and cabinets are arranged along one wall
site visit an information-gathering trip to a construction area
size an object's overall dimensions
skeleton framing wood framing with small structural components spaced closely together
sketch 1. to draw freehand; 2. a freehand drawing
sketching the process of making quick, freehand drawings
SketchUp a modeling program
skylight window in the roof
slab door cabinet door that appears to be made of one solid piece with no raised or recessed profile
slab-on-grade a foundation type that consists of a slab poured over gravel directly on the ground
slide glide parallel to a wall
sliding window a window that slides horizontally in a wall

soffit the underside of an architectural structure
soffit light a lighting fixture attached to the wall that radiates light downward
softener a plumbing system appliance that removes hard minerals from the water
solar heating an HVAC system that collect the sun's energy
sole plate a plate at ground level
space plan a floor plan that shows furniture arrangements
space planning the process of designing an interior to make it beautiful and functional
spatial tension perceived line or link through space between drawings on a sheet
specifications written descriptions of the quality of materials and workmanship standards
spin revolve around a center point
spiral stair flight of stairs that rises around, and is connected to, a pole
split wire outlet an outlet with one plug that is always hot and another that is activated with a switch
spread footing a foundation system that consists of a wall on top of a footing
stack pipe a vertical plumbing pipe that vents and carries away wastewater
staircase a set of stairs and its surrounding walls or structure
stairwell vertical shaft in which the staircase is built
standard a universally recognized conformity
standard overlay a cabinet door that has a minimum of 1" around the door and drawer perimeters; *see* full overlay
station point the viewer location in a perspective drawing
stationary panel the portion of a door or window that doesn't move
steel building material made of iron and carbon
step chair portable stairs
stepped footing a foundation that stair-steps into a sloped site
stick framing *see* skeleton framing
stile vertical framing component on a window or door
stop hardware that protects walls and equipment from the door as it swings open
straight run stair a linear flight of stairs
stringer the structural support of a staircase
strip footing long, horizontal beams that support load-bearing masonry walls
structural *see* load-bearing
structural grid lines lines that annotate a plan's system of columns
stud a vertical load-bearing member inside a wall
subdivided scale the open-divided portion of the architect's scale
substructure the portion of the building below ground
superstructure the portion of the building above ground
surface fixture a light fixture mounted directly to the ceiling

surround 1. the immediate border around a firebox opening; 2. a masonry decorative frame around the whole fireplace

suspended ceiling a grid hung from the underside of the roof into which lighting fixtures and acoustic tiles are placed

swamp cooler *see* evaporative cooler

swing door a door that opens on being pushed or pulled from either side

switch an electrical device that opens and closes an electrical circuit

symmetrical an arrangement of graphics that is identical on both sides of an axis

T stair *see* bifurcated stair

tambour door cabinet door made of multiple separate pieces attached to a flexible backing sheet

tangent geometric objects such as lines and arcs that touch each other

technical compass a tool for drawing circles

technical pen non-disposable ink pen

thermostat a device that automatically regulates temperature in a furnace or air

three-quarter bath a bathroom that has a water closet, lavatory, and shower

three-point perspective a perspective drawing made with three vanishing points

threshold the bottom of the door frame

tick mark an angular line at the intersection of the dimension and extension lines

tilt-up construction a building and a construction technique that lifts pre-made concrete panels into place and ties them together

timber framing *see* post-and-beam framing

tiny home a residence under 600 square feet

title block a box on a drawing sheet that contains identifying information about the project and the drawings on a sheet

toe kick the indented space at the bottom of a cabinet

toe space *see* toe kick

toggle switch an angled, two-position lever used for light switches

top view the part of an object seen from a horizontal plane

total rise distance from finished floor to finished floor

total run horizontal distance that a staircase takes up

tracing paper thin, cheap, and semitransparent paper used for sketching and problem solving

track lighting multiple bulbs on a rail

transom window fixed window above a door

transverse section a section drawing cut perpendicular to the roof ridge

trapezoidal a four-sided shape with at least one set of parallel sides

tread a stair step's horizontal board

triangle a tool used for drawing vertical and angled lines

trim 1. general term for finish millwork; 2. a separate frame installed around a window or door

trimetric drawing type that has horizontal axes skewed at any angle between 0° and 90°

true length actual length

trussed rafter also called a truss, a structural load-bearing member at the roof

T-square a tool used on a drafting board to draw horizontal lines

turning stair a staircase that uses winder treads

twin bed 3'-3" wide, called a 3/3

two-point perspective (1) a drawing technique that utilizes two convergence points (2) a perspective drawing made with two vanishing points

U stair a stairway with two parallel flights

under-cabinet lighting task lights mounted under wall cupboards

uni-directional dimensioning style in which all notes are oriented straight up

Uniform Mechanical Code governing rules for planning and installation of equipment and appliances

universal compass a tool for drawing circles in which most pencils or pens can fit

universal design the philosophy of making buildings and products usable to as many people as possible, regardless of age, height, gender, ability, or disability

U.S. Customary units *see* Imperial units

USGBC abbreviation for U.S. Green Building Council; organization that provides a framework for identifying and implementing sustainable building design

U-shape kitchen a layout of cabinets, appliances, and counters that resembles a U in plan

utility knife a thick-bladed tool for cutting vinyl and heavyweight board

valley the fold in a roof where it changes direction

valley rafter a roof component that runs from a wall plate to the ridge along the valley

vanishing point the location where sets of parallel lines converge in a perspective drawing

vanity cabinet a bathroom cupboard that houses a sink

vector a file format made of mathematically defined lines and curves

vellum semi-opaque, high-quality cotton paper used for ink work

vent an opening that allows combustion by-products, gases, and pressure to escape and fresh air to enter

vent stack a wastewater pipe that rises above the roof, allowing sewer gases to escape

vessel sink a bowl that is raised above the countertop

vinyl cover a semi-soft surface placed on top of a drafting board

vinyl substrate a resilient drafting surface placed atop a board

virtual reality headset a head-mounted device that creates an immersive reality experience for the wearer

visual communication the art of putting ideas to paper in picture form

visual inventory written, sketched, measured, and photographic documentation of existing conditions

volute a curved fitting on top of a handrail

wainscot a protective surface on the lower 4 feet of an interior wall

wall cabinet cupboard attached to a wall

wall rail a handrail mounted to the wall

wall sconce a lighting fixture that directs light up or down

wall section a vertical slice from footing to roof that shows details of construction

wall stack a vertical duct that fits between the wall studs

wall wash a lighting fixture that spreads light uniformly over a wall from top to bottom

wastewater used dirty water

water closet toilet

water heater an appliance that heats and stores water

well 1. water source obtained by digging into an aquifer; 2. hole in an upper floor for stairway placement

wet wall *see* chase wall

wheelchair-accessible allows passage of a wheelchair

wide U stair a staircase with a well hole or large space

wide-format photocopier a copier that supports paper roll widths between 18" and 100"

winder stair *see* turning stair

winder tread *see* trapezoidal

windowsill the bottom of a window

wood frame a structural system made of dimensional lumber

work triangle the area in a kitchen between the stove, sink, and refrigerator

workflow a sequence of processes through which a piece of work passes from initiation to completion

wythe a vertical section of masonry that is one unit thick

zoning regulation for aesthetics, traffic, and activities

Index

2D Drawing, 46–53
    orthographic theory, 47–49
3D camera, 93
3D Drawing, 55–64
    defined, 55
    paraline, 56–59, 328
    perspective, 60–64
3D Modeling, 364, 365, 367
3D mouse, 367
3D printing, 6, 239, 370–372
3ds Max, 364
20-20 Modeling software, 365

A

A-01, 77
Accessibility, 120
    ramps, 311
Accessible and universal design accessibility, 129–131
Accordion door, 187
Accreditation organizations, 5–6
Acoustic tile ceilings, 279, 280
Acrobat DC, 365
Active, 180
Active solar, 285
Actual sizes, masonry, 224
Adjacency matrixes, 7
Adobe Illustrator, 364
Adobe InDesign, 365
Adobe Photoshop, 365
    AutoCAD drawings enhanced with, 364, 368
    entourage adding with, 365, 368, 369
Air conditioners, 283
Air supply vents, 279
Alcove, 126
Aligned, 163
American Institute of Architects (AIA), 5
American National Standards Institute (ANSI) conventions, 162–164
American Society for Interior Designers (ASID), 5
Americans with Disability Act (ADA), 15
Ames lettering guide, 35–36
Amperage, 267
Ancient architectural drafting, 3
Angle drawing, 21–23
Angled wall, swing door symbol in, 184
Annotation, 72, 162
Appliance, 266
Appliances, kitchen symbols and legend, 316, 322
Arches, 239

Architect Experience Program (AXP), 5
Architect Registration Examination (ARE), 5
Architect's scales, 24–25
Architectural dimensioning, 162, 163
Architectural drafting history, 3
Architectural floor plan, 84
    accessible and universal design accessibility, 129–131
    bathroom fixture clearances, 126–127
    bathroom fixture sizes, 124–125
    bathroom space planning, 124
    bedroom furniture, 133–134
    bedrooms, 133
    CAD standards, 113
    closets, 131–132
    dining room furniture, 123
    drafting floor plan, 106–110
    drawing scale, 84–85
    egress and exits, 135–137
    existing space, sketch plan of, 85–88
    fixtures, placement of, 125–126
    floor pochés, 96–97
    habitable space, 137
    hallways, 137
    inking, 110
    islands, 118
    kitchen appliance sizes and clearances, 119–120
    kitchen clearances, 120–122
    kitchen space planning, 115–118
    laundry rooms, 133
    layouts, 115
    line hierarchy, 104–106
    line quality, 104
    living room, 134
    living room furniture, 134–135
    photogrammetry, 93–94
    presentation plan, drafting, 111–112
    rendering, 92
    room measurement, 90–91
    room sizes, 114–115
    second-floor plan, 110
    sketching tips, 91–92
    space planning, 113–114
    symbols, 97–104
    vanity, 125
    visual inventory, 88–90
    wall thicknesses and pochés, 94–95
Architectural notation writing, 28–29
Architectural program, 7
Arcs, 36
Area, 175
Area of refuge, 136, 308
Art knives, 34, 38
Ashlar, 227

Associate Kitchen & Bath Designer (AKBD), 6
Asymmetrical graphics, 78
Augmented reality, 370
AutoCAD, 4, 364, 368
    Photoshop enhancing drawings from, 365–366, 368–369
    sketching and modeling in, 364, 368
Autodesk ReCap Photo, 93
Autodesk Revit, 365
AutoDesk software
    3ds Max, 364–365
    AutoCAD, 364, 368
    Revit, 365, 366, 368
    TrueView, 364
Autodesk University, 372
Automated systems, 266
Automated systems plan, 266
Awning
    wide window sizes, 208
    window, 196–197
Axonometric drawing, 56, 328
A-XX, 76
A-X-XX, 76

B

Balance, 78
Balloon framing, 234
Balusters, 297–298
Balustrades, 297, 298
Bar joists, 239, 241
Barn, 186
Bar scale, 109
Base moldings, 255
Base, stringers, 169
Bathroom
    accessory schedule (chart), 323
    waste removal piping, 291
Bathroom fixture clearances, 126–127
Bathroom fixture sizes, 124–125
Bathroom space planning, 124
Bay, 243
Bay window, 200
    window sizes, 212
Beam, 232
Bearing angle, 261
Bedroom furniture, 133–134
Bedrooms, 133
Bifold, 186
Bifold door, sizing charts, 214
Bifurcated stair, 307
BIM modeling, 365
BIM software, 4, 365
Bitmap, 366
Blind cabinets, 256
Block diagrams, 7, 10, 12
Blocking, 236
Blueprinting, 41
Board drafting, 3

Bolt, 193
Bond, 225, 227
Bond beam, 226
Border, 75
Bow window, 201–202
    window sizes, 212
Box bay window, 201
Box beam, 230
Box cutter, 38
Boxes, tool storage, 40
Branch duct, 283
Branch pipe, 288, 289, 291
Breaker box, 266
Brushes, 34
Bubble diagram, 7
Building codes, 15
    stairs, 296, 312
Building construction
    3D printing, 239
    benefit of, 218
    construction plans, 237–238
    detail drawings, 244–250
    finish materials, 244
    fireplace, 251–255
    foundation, 218
    foundation types, 219–223
    framing items, 236
    importance of understanding, 218
    masonry, 224–228
    MasterFormat, 259–260
    millwork, 255–258
    modular construction, 238
    site plan, 260–262
    steel frame components, 239–242
    steel framing, 242–243
    structural vs. partition wall, 218–219
    tilt-up construction, 228
    wood frame types, 234–236
    wood framing components, 228–231
    wood framing definitions, 231–233
Building Information Modeling/Management (BIM), 4
Building information modeling (BIM) software, 365
Bull nose, 224
Bumwad, 40

C

Cabinet oblique drawing, 59
Cabinets, 256–258
    cabinet door styles, 256
    cabinet door types, 256
    framed vs. frameless, 256
    section and elevation views of, 175, 176
    stringers, 169

CAD standards, 113
California king bed, 133
Callouts, 72, 141, 316–317
Camera, 89
Carpenter's bubble level, 89
Carpenter's square, 89, 90
Cased opening, 190
Case goods, 255
Casement
    wide window sizes, 207
    window, 196, 250
Casing, 184, 248
Cast concrete, 164
Cavalier oblique drawing, 59
Cavity, 225
Cavity walls, 226
Ceiling height door, 189
Ceiling light fixtures, 268
Ceiling plans, reflected, 278–280
Cells, 226
Center lines, 70
Central air, 282–284
Certification organization, 5–6
Certified Bath Designer (CBD), 6
Certified Kitchen Designer (CKD), 6
Certified Master Kitchen & Bath Designer (CMKBD), 6
Chair rails, 255
Chandelier, 268
Charts (schedules), 16
Chase wall, 94
Chest of drawers, 52–54
Chief architect, 365
Chimney, 251
Circles, 330–332, 353
    two-point perspective drafting, 353
Circuit, 266
    drafting, 274
Circuit breakers, 266
Circular stairs, 306–307
Circumscribed, 72
CL (center line), 169
Cleaning of tools, 40
Clean-out valves, 289, 291
Clearances, 169
Clear spans, 230
Clerestory, 203
Client needs, 7
Climate control system, 282–287
Closer, 193
Closet, 131–132
Clothing pattern, 2
Clutch pencils, 33
Codes, 15
    building, stairs, 296, 312
    for common cabinetry items, 170–172
Color boards, 13
Commercial furniture plan, 84
Communication, graphic, 2

Communications systems, 308
Compact fluorescent (CFL), 268
Compass, 34
Computer Aided Drafting (CAD), 4, 364
Computer Aided Drafting and Design (CADD) programs, 4, 364. *See also* Software
Concept, 7
Concrete, 218
Concrete blocks, 224
Concrete masonry units (CMU), 224
Cone of vision (COV), 339, 341–347, 354, 355, 359
Construction, 75
    calculator, 89
    documentation, 7, 16
    lines, 33
Contour lines, 261
Contract administration, 7
Contract administration phase, 16
Contrast, 78
Conversions, decimal scale, 29
Copy enlargement or reduction, 31–32
Copying process, 40–42
Corner cabinets, 256
Council for Interior Design Accreditation (CIDA), 6
Council for Interior Design Qualification (CIDQ), 6
Courses, 227
Cove light, 268
Cover sheet, 77
Crawl space foundation, 223
Cripple studs, 231
Criteria matrixes, 7
Cross bracing, 236
Cross bridging, 236
Crown moldings, 255
Current, electrical, 268
Curtain walls, 234
Curved stairs, 297
Curves, French or irregular, 36
Cutaway drawings, 333–337
Cutaway isometric, 333–337
Cutting mat, 38
Cutting plane, 69–71
Cylindrical break, 70

D

Dark lead pencil, 89
Data, 266
Data plan, 266
Data ports, 266
Deadbolt, 193
Dead-end hallways, 137
Decimal conversions, 29
Decking, 232, 242
Definition of drafting, 2

Demolition plan, 237
Designator, 76
Design development, 7
Design development phase, 14–15
Design layout example, 13
Design process, 7–16
Design thinking, 6
Detail drawings, 244–250
Diagrams, bubble and block, 7, 10
Diameter, 164
Diametric drawing, 58
Digital imaging, 365
Digital model, 93, 364
Digital photocopiers, 41
Digital scans, 41
Dimensional lumber, 228
Dimension drawings, 162
Dimensioning floor plans and elevations, 161
    American National Standards Institute conventions, 162–164
    cabinets, section and elevation views of, 175, 176
    dimension drawings, 162
    interior elevation, 166
    National Kitchen and Bath Association drawings, 166–174
    wood frame vs. masonry dimensions, 164–166
Dimension lines, 74, 162
Dimension notes, 28–29, 74, 162
Dimensions, 261
    2D and 3D, 46
    and bearing angle, 308
Dimmer switch, 268
Dining room, electrical plan, 270
Dining room furniture, 123
Direct modeling, 365
Direct vent, 251
Disabled persons, design for, 311–312
Disposable pens, 36
Divider, 34
Door, 180–190
    accordion, 187
    building construction, 248–250
    callout symbols, 316–317, 325
    cased opening, 190
    door hardware, 192–193
    door types, 180–190
    egress requirements and egress doors, 190–192
    flush and panel doors, 184
    full height, 189
    overhead sectional, 189
    pivot, 188
    revolving, 188
    and rough opening sizes, 190
    schedules (charts), 318, 320, 323

    sliding, 185–187
    swing, 180–184
Door hand, 181
Door hardware, 192–193
Door schedule, 318
Double-action, 182
Double corner, 224
Double-hung window, 194
    fixed window sizes, 209, 211
    wide window sizes, 210
Double-pane glass, 205
Double top, 231
Double-wythe, 226
Downdraft vents, 258
Drafting
    defined, 2
    history, 3–5
    presentation plan, 111–112
Drafting boards, 20
    accessories, 20–30
Drafting brush, 34
Drafting electrical plan, 274–275
Drafting media, 39–40
Drafting orthographically, 51–55
Drafting tape, 34
Drawers, chest of, 52–54
Drawing. *See also* sketching
    2D and orthographic, 46–53
    3D paraline, 56–59, 328
    3D perspective, 60–64, 328–360
    angles, 21–23
    ceiling plans, reflected, 280
    cutaway 3D plans, 333–337
    furniture, orthographic, 49
    guidelines, 35–36
    horizontal lines, 21
    one-point perspective, 59
    to scale, 24–28
    shop, 16
    stairs, 309, 310, 313
    two-point perspective, 59
Drawing number, 141
Drawing scale, 84–85
Drop-in ceilings, 279–281
Duct, 258, 282
Duplex outlet, 266
Dutch, 182
Dwelling, 9
DWG file, 366

E

Easements, 261, 481
Egress, 190
Egress door, 136–137, 191
Egress window, 133, 136
Electrical drawings, 266
Electrical plans, 266, 268, 275–276
    design, 270–274
    drafting, 274–275
    symbols, 268, 271–274

INDEX   **491**

Electricity, 266, 267
  distribution, 266
Electric radiant heat, 284–285
Elevation, 48
  window placement in, 203–204
Elevation callout, 72
Elevation callout symbol, 141–142
Elevation view, 46, 140–146
  draw caddy-corner items in, 145
  drawing, steps for, 142–144
  elevation callout symbol, 141–142
  oblique drawing, 59
  sloped ceiling in, 144, 145
Elevator, 312
Ellipses, 330, 332, 353
Enclosed staircase, 296
EnergyStar, 6
Engineered wood products (EVP), 228
Engineer scale, 260, 398, 481
Enlargement box, 75
Enlargement, copy, 31–32
Entourage, 84, 152–158, 365, 368, 369
Equipment, 266
Equipment plan, 286
Eraser, 33–34
Eraser bag, 104
Eraser shield, 34
Estimating proportions, 63
Evaporative cooler, 285–286
Excel, 365
Excel spreadsheets, 320
Exit, 135, 191
Exit access, 135, 191
Exit device, 193
Exit discharge, 135, 191
Exit stairs, 302, 308
Extension lines, 74, 162
Exterior elevation, 141
Exterior perspective, 340
Exterior stairs, 308–309
Eyeball apparent-size relationships, 63

F

Federal Housing Authority, 114
Fenestration, 110, 180
File cabinet, 51–54
Film, plastic, 39
Finial cap, 297
Finishes, schedule (chart), 316, 320, 322
Finish materials, 244
Fire, 182
Firebox, 251
Fire door assembly, 182
Fireplace, 251–255
  perspective drawing, 346–348
Fit sketch, 86

Fixed (non-moving) panel, 185
Fixed window, 193
Fixed window sizes, double/single hung, 209, 211
Fixtures, plumbing, 289, 291, 293
Flashlight, 89
Flight, 296
Floor plan
  block diagrams becoming, 10
  callouts on, 317, 318
  cutaway isometric drawings, 333–337
  drafting, 106–110
  electrical plans, 270, 274
  power systems, 278
  rough, 10
Floor pochés, 96–97
Flues, 251, 482
Fluorescent, 268
Flush doors, 184
Fold, 180
Footing, 220
Footprint, 261, 349
  building, 313
Forced air, 282–284
Foreshortened lines, 47
Foreshortening, 145
Foundation, 218
  plan, 218
  wall elements, 219–220
Framed opening, 205
French curves, 36
French door, 182
Freshwater supply, 289, 290
Front views, 140
Frost line, 220
Full bath, 124
Full bed, 133
Full height door, 189
Full overlay, 256
Full scale, 28
Furnace, 282, 283
Furniture
  bedroom, 133–134
  dining room, 123
  legends or keys, 318, 321
  living room, 134–135
  perspective drawing, 348–350, 360
Furniture fixtures and equipment (FFE), 318
Furniture and fixtures plans
  commercial, 318, 320
  HVAC, 282–287
  legends for, 318–320
  plumbing, 288–289, 292–293
  power, 275–277
  reflected ceiling, 278–280
  site, 316
Furniture key, 318
Furniture schedule, 318
Furniture template, 38, 86
Furring strips, 244

Fused Deposition Modeling (FDM), 370
Fused Filament Fabrication (FFF), 370

G

Gable ceiling, 359
Gabled roof
  isometric drawing, 328–332
  perspective ceiling, 359
Galley (corridor) kitchens, 116, 117, 270
Gang, 267
Garden window, 201
Gate valve, 289
General oblique drawing, 59
Generative design, 372–373
GFCI outlet, 266, 267
GIF file, 366
GIMP, 365
Girder, 232
Glass size, 205
Glazing, 204
Glue laminated timber (Glulam), 230
Grade (ground) line, 141
Graphic communication, 2
Graphics tablets, 366
Graphite, 33
Graph paper, 49
Grid, 49
Grid, perspective, 59–62
Grilles, 284
Grounded outlet, 267
G-shape (peninsula) kitchens, 117
Guards, stair, 297–298
Guidelines, drawing, 35–36
Guides, 185
Gypsum board, 244

H

Habitable, 9
Habitable space, 137
Half bath, 124
Half-diagonal rule, 136, 191
Hallways, 137, 270, 274
Halogen, 268
Handle, 193
Handrails, 297–298
Hard surfaces, 261
Hardware, 366
Hardware schedule, 318
Hatch lines, 73, 152
Head, 150, 180, 249
Header, 232
Headroom, stair, 300, 305
Heat exchanger, 285
Heating, radiant, 284–285

Heating, ventilation, and air-conditioning (HVAC) plans, 282–287
Height line, 61, 343–345, 348, 350, 352, 353
Height measurements, 91
Helical, 306
Hidden line, 50
Hidden object lines, 69
High-resolution, 366
Hinges, 193
Hip, 236
History of drafting, 3–5
Holder, 193
Holistically, 266
Hopper window, 197–198
Horizon line (HL), 61, 343
Horizontal line drawing, 21
Horizontal run, 301
Hose bib, 288
Hot water heaters, 288
House main and sewer, 291
Human figures, added with Photoshop, 369
Humidifier, 283

I

Identification (ID) labels, 72, 73, 317
Identifiers, 72–73
Imbibed eraser, 33–34
Incandescent, 268
InDesign, 366, 375
Inking, 110
Ink pens, 36–37
Inscribed, 72
Inset, 256
Instructions, legal and binding, 16
Insulated Concrete Forms (ICF), 223
Interior elevation, 141
  dimensioning, 166
  elevation view, 140–146
  section drawing, 146–153
  sheet layout, 153
Interior perspective, 340
International Building Code (IBC), 15, 190
International Code Council (ICC), 15
International Interior Design Association (IIDA), 6
International Residential Code (IRC), 15, 114, 190, 286
Irfanview, 365
Irregular curves, 36
Islands, architectural floor plan, 118
Isometric drafting, 328–337
  circular table, 330–332
  cutaway drawings, 333–337
  gabled roof house, 328–332

Isometric drawing, 56, 58
Isometric pictorial, 328–332

J

Jalousie window, 198
Jamb, 180, 224, 249
Joist, 232
JPG file, 366
Junction boxes, 272

K

Key, 72, 84, 318, 321
Key letter, 316
Keynote, 76, 84
Key number, 316
Kitchen
 electrical plans, 270
 software modeling, 365
Kitchen appliance, sizes and clearances, 119–120
Kitchen clearances, 120–122
Kitchen space planning, 115–118
Knee wall, 94, 218
Knob, 193

L

Laminated veneer lumber (LVL), 230
Lamps, 20, 268
Landings, stair, 300–303
Laser tape measure, 89
Laundry rooms, 133
Lavatory, 125, 288
Law office, mood board example, 13
Leader lines, 74
Leadership in Energy and Environmental Design (LEED), 6
Lead grades, 33
Lead holder, 33
Leads, 33
Left-hand door, 181
Left-hand reverse, 181
Left vanishing point (LVP), 343
Legend, 76, 316–325
Lettering, 80–81
LiDAR, 370
Life Safety Code (LSC), 15
Light emitting diode (LED), 268
Light fixtures, 266–269
Lighting, 266
Lighting fixture schedule, 318
Lighting plan, 266
Lighting symbols, 272
Line hierarchy, 104–106
Line quality, 104

Lines, 68
 border, 75
 dimension, 74
 extension, 74
 foreshortened, 47
 hatch, 73
 horizon, 343
 leader, 74
 match, 75
 quality, 68
 type, 69–72
 weight, 68–69
 weight and type, 68
Line weights, 33, 68–69, 104
Lintel, 232
Lipped, 256
Lites, 184
Living room furniture, 134–135
Living rooms, 134
 electrical plan, 269
 one-point perspective drafting, 354–358
Load-bearing, 90, 218–219
Loads, 218
Local codes, 15
Location, dimension notes, 162
Lock, 193
Long break lines, 70
Longitudinal section, 146, 148–151
Louvered window, 196
Low resolution, 42, 366
L-shape kitchens, 117
L stairs, 302, 303

M

Machine learning, 372
Main, house water supply, 291
Main supply duct, 283
Mantel, 251
Manual drafting, 3–4
Marks, 72, 141
Marks, callout, 316
Masonry, 224–228
Masonry dimensions, 164–166
Masonry opening, 190
Masonry veneer, 166
MasterFormat, 68, 259–260
Mat, 223
Match lines, 75
Mats, cutting, 38
Matte, 40
Means of egress, 133, 135–136, 190–191
Measurements, in visual inventory, 90–91
Measuring app, 89
Mechanical pencils, 32–33
Mechanical plan, 286
Media, drafting, 39–40

Medicine cabinets, 124
Medium density fiberboard (MDF), 230
Metal tape measure, 88
Metric scales, 30
Milled lumber, 164
Millwork, 166, 255–258
Mini-split, 284, 285
Mission style chest, 54–55
Mixed occupancy, 7
Modeling, 4, 364–367, 370, 373
Modeling programs, 4
Modeling software, 370, 375
Models, scale, 14
Modular construction, 238
Molding, 255
Mood boards, 13
Mortar joint, 224
Movement, 78
Multiple drawings, sheet layout for, 78
Mullions, 204
Muntins, 204

N

Narrow U stair, 304
National Association of Home Builders (NAHB), 6
National CAD Standard (NCS), 68
National Council for Interior Design Qualification (NCIDQ), 5
National Council of Architectural Registration Boards (NCARB), 5
National Electric Code (NEC), 15
National Fire Protection Association (NFPA), 15
National Kitchen and Bath Association (NKBA), 6, 106, 166–174
 for elevations, 173–174
 for floor plans, 166–173
Newel posts, 297, 298
Nominal sizes, masonry, 224
Non-grounded outlet, 267
Non-load-bearing, 90, 218
Non-traditional fireplace, 254
North arrow, 77, 260
Nosing, 296
Notation writing, 28–30

O

Object lines, 33, 68
Oblique drawing, 59
Observational skills, 63
Occupancy, 7
 and occupant load, 7

Occupant evacuation elevator, 312
Occupational Safety and Health Administration (OSHA), 15
One-face fireplace, 253
One-point perspective, 59, 60, 354–358
Open riser, 308
Open staircase, 296
Orbiting, 364
Ordinances, 15
Organizations
 professional, 5–6
 standards, 15
Orientation, drawing and compass, 77, 114
Oriented strand board (OSB), 230
Orthographic drawings, 47
Orthographic projection, 47–49
Orthographic sketching, 49–51
Orthographic theory, 47–49
Outlets, 266–267, 270–272
Overhead sectional doors, 189

P

Painter's tape, 34
Pane, 205
Panel, 256
Panel doors, 184
Panic (push) bar, 193
Paper copies, 41–42
Paper, grid or graph, 49–51
Paraline drawing, 56, 328
 types of, 56–59
Parallel bars, 21
Parametric modeling, 365
Partial walls, 218
Parti sketches, 7
Partition wall, 90, 218–219
Passenger elevator, 312
Passive, 180
Patio door, 2 wide door sizes, 213
Pavers, 227
PDF file, 366
Pencils
 clutch, 33
 leads, 33
 mechanical, 32–33
Pencil trick, 63
Pendant, 268
Pens, 36–38
Percentage enlarged or reduced, 31–32
Permanently attached light fixtures, 268
Perspective drawing, 59, 328–360
 fireplaces, 346–348
 furniture, 348–350
 mistakes in, 359–360
 one-point perspective, 354–358
 two-point perspective, 341–352
 windows, 345, 346

INDEX **493**

Perspective grid, 60–62
   one-point perspective grid, 60
   two-point perspective grid, 61–62
Perspective pictorial, 328–360
Phone apps, 42
Photocopying, 41
Photogrammetry, 93–94, 373
Photoshop, 365–366, 368, 369
   AutoCAD drawings enhanced with, 364, 368
Photo-textured (photo-covered) mesh model, 93
Physical conditions, 7
Pictorial drawing, 343, 346, 348, 349
Pictorials, 55
Picture plane, 47
Pilasters, 218
Pile, 223
Pipes, water, 288, 289
Pivot, 188
Pivot door, 188
Pivot window, 199
Pixelation, 366
Plan north, 77
Plan oblique drawing, 59
Planometric drawing, 59
Plans, 266
   defined, 48
   electrical, 268–276
   furniture and fixtures. *See* furniture and fixtures plans
Plastic film, 39, 40
Plastic tape, 38
Plate, 231
Platform, 234
Plenum, 279
Plotter, 367
Plugin software, 366
Plumbing, 288–293
   fixtures and terms, 289, 291, 293
   plans, 291–293
Plumbing fixture, 289
Plumbing plan, 288–291, 293
Plumbing system, 288–291
PNG file, 366
Pochés, 73, 94–95
   floor, 96–97
   legends for, 317
Poché symbols, in section and elevation, 150
Pocket, 185
Point-to-point drawing method, 63
Pop-up space, 6
Post-and-Beam, 234
Post-on-pad foundations, 222–223
Powder room, 124
Power, 266
Power plans, 275–277
Prefabricated, 238

Presentation plan, drafting, 111–112
Prints, 41
Process, design, 7–16
Production drawings, 16
Profiles, 94
Program, architectural, example, 7
Programming, 7
Projection, orthographic, 47
Project statement, 7
Property characteristics, 10
Property lines, 261
Proportional scales, 31–32, 313
Proportions, estimating, 63
Prototypes, 6
Protractors, 24
P-trap, 289
Public way, 190
Push-button switch, 268

Q

Quadrant markers, 182

R

Rabbet, 256
Radial graphics, 78
Radiant heat systems, 282, 284–285
Radius, 164
Raft, 223
Rafter, 232
Rails, 184
Rake, 144
Ramps, 311–312
Random rubble, 227
Range hoods, 258
Raster, 42, 366
Rebar, 218
Recessed lighting, 268
Reclaimed wastewater, 288
Reductions
   architect's scale, 29
   copy, 31–32
Reflected ceiling plans, 203, 266, 278–280
Regreen, 6
Rendering, 38, 92
Repetition, 78
Resolution, file, 365, 366
Return or supply vents, HVAC, 279, 282–284
Reveal, 256
Revit, 4, 365, 366
Revolving door, 188
Rhinoceros 3D, 365
Ridge board, 236
Right-hand door, 181
Right-hand reverse, 181

Right vanishing point (RVP), 343
Rigid frames, 239
Riser diagram, 293
Risers, stair, 296, 305, 308–310
   sizes, 309, 310
Roof framing plan, 237
Roof plan, 237
Roof ridge, 144
Room sizes, architectural floor plan, 114–115
Rotated plan drafting, 341–352
Rough opening/masonry opening, 190, 205
Round tables
   isometric drafting, 330–332
   sketching, 50
RVT file, 366

S

Sandpaper block, 33
Sash, 193
   size, 205
Scale conversions, 29
Scale drawing, 24
Scale models, 14
Scales, 78
   architect's, 24–25
   commonly used, 31
   drawing, 84–85
   metric, 30
   proportional, 31
Scanners, 363, 373
Schedules (charts), 16, 76, 316–318, 320, 325
   bathroom accessories, 323
   doors, 318, 320, 323
   finishes, 315, 320, 322
   furniture, 318–321
   pictorial, 318
   windows, 316, 317, 322, 324
Schematic, 287
Schematic design, 7
Schematic design phase, 7
Scissors stairs, 304–305
Second-floor plan, 110
Section detail, 148
Section drawing, 146
   entourage, 152–158
   hatch lines, 152
   longitudinal section, 148–151
   poché symbols, in section and elevation, 150
   section detail, 148
   section location symbol, 148
   section scale, 146
Section location symbol, 148
Section scale, 146
Section view, 140
Septic tank, 289

Service elevator, 312
Service panels, 266
Service stairs, 296
Setbacks, 261
Sewers, house, 291
Sheathing, 232
Sheet layout, 153
   for multiple drawings, 78
   for single drawing, 78–80
Sheet number, 141
Sheet sequence, 77
Sheet size, 77
Shop drawings, 16, 166
Short break lines, 70
Shutoff valves, 291, 293
Side views, 140
Sight ray, 344–346, 348, 349, 351, 356, 357, 359
Sill, 180, 231, 249
Single drawing, sheet layout for, 78–80
Single-hung window, 194
   fixed window sizes, 209, 211
Single plug outlet, 266
Single-pole, 267
Single wall kitchens, 116, 117
Site plan, 289
   building construction, 260–262
Site visit, 88
Sizes
   dimension notes, 162
   window, 206–214
Skeleton, 234
Sketches
   colored, 14
   fit, 86
   parti and concept, 7
Sketching
   2D and orthographic, 46–53
   3D, 55–64
   defined, 46
   floor plans, existing, 85–88
   observation in, 63
   orthographic, 49–51
   photograph underlay in, 63–64
   proportion estimation, 63–64
   scaling picture to paper, 64
   tips, 63–64, 91
SketchUp Warehouse, 97, 365, 368
Skylights, 203
Slab, 256
Slab-on-grade foundations, 219–220
Slide, 180
Sliding, 185–187
Sliding window, 196
Soffit light, 268
Softener, 288

Software, 370, 375
    BIM, 4, 365
    CAD/CADD, 4, 364
    imaging and graphics, 365, 366
    modeling, 364–367, 370, 373
Soil stacks, 291, 293
Solar heating, 282, 285
Sole, 231
Solid, 225
Space plan, 84
Space planning, 113–114
Spatial tension, 78
Specialty window, 202
Specifications, 16
Spin, 180
Spiral stair, 305–306
Split wire outlet, 266
Spread footing, 220–222
    foundations, 220
Square table, 49–50
Stack pipe, 125, 288
Stack walls, 289, 291
Stages, design process, 7–16
Staircase, 296
Stairs
    components of, 296, 313
    drafting staircases, 300–301, 310–311
    drawing, elevation view, 307, 310–311
    electrical plan, 269, 274
    exit, 302, 308
    exterior, 308–309
    types of, 301–307
Stairwell, 296
Standard overlay, 256
Standards, 68
Standards organizations, 15
Stationary, 180
Station points (SP), 338–339, 341
Steel framing, 242–243
    components, 239–242
Stepped footing, 223
Steps, stairway, 296, 299
Stick framing, 234
Stiles, 184
Stop, 193
Storage of tools, 40
Straight run staircases, 301–302
Stretcher, 224
Stringers, 162
Stringer, stair, 296
Strip footing, 223
Structural grid, 75
Structural wall, 218–219
Stud, 231
Subcontractors, 16
Subdivided scales, 25–28
Substructure, 218
Superstructure, 218

Surround, 251
Suspended ceilings, 279–281
Swamp cooler, 285
Swing, 180–184
    door hand, 181
    door symbol in angled wall, 184
    door symbol in plan, 182–184
    door variations, 182
Switch, 266
Switches, electrical, 266–268, 274
    symbols, 268, 271–273
Symbols, 97–104
    dimension, 74
    electrical, 268, 271–274
    HVAC plans, 287
    in legends and keys, 316, 322
    site plans, 289
    switches, electrical, 268, 271–273
Symmetrical graphics, 78

T

T-1 lines, 318
Table, round
    isometric drafting, 330–332
    sketching, 50
Table, square, sketching, 49–50
Tablet, graphics, 366
Tambour, 256
Tangent, 72
Tankless water heaters, 288
Tank water heaters, 288
Tape, drafting and plastic, 38
Technical compasses, 35
Technical pens, 37
Telephony, 273
Thermostats, 266, 287
Three-point perspective, 340
Tick marks, 74, 162
Tilt-up construction, 228
Timber framing, 234
Tiny homes, 114
Title block, 76, 77
Title sheet, 77
Toe kicks, 170
Toggle switches, 267
Toilets, valves, 289, 291
Tools, drafting, 20
    Ames lettering guide, 35–36
    care and storage of, 40–41
    drafting board accessories, 34–39
    historical, 3–4
    other miscellaneous, 34–39
    scales, 24–28
Total rise, 300
Total run, 310
Tracing paper, 39
    layer sheets of, 91

Tracing templates, 38
Track lighting, 268
Traditional fireplace, 252–253
Transom, 2 wide door sizes, 213
Transom window, 188
Transverse section, 146, 147
Trapezoidal, 303
Traps, plumbing, 291, 293
Trash paper, 40
Treads, stair, 296–301, 303, 309–310
    sizes, 309
Triangle tools, 21–23
Trim (casing), 184, 204
Trimble SketchUp, 365
Trimetric drawing, 56
True length, 47
True line (height line), 343
True scale, 28
Trussed rafter (truss), 232
T squares, 21
T stair, 307
Turning stair, 303
Twin bed, 133
Two-point perspective, 59, 61–62, 341–352
    of a circle, 353
    rotated plan drafting, 341–352

U

Under-cabinet lighting, 268
Unidirectional, 169
Uniform Mechanical Code, 286
Unit size, 205
Universal compasses, 35
Universal design accessibility, 129–131
US Green Building Council (USGBC), 6
U-shape kitchens, 117
U stairs, 304–305
Utility knife, 38
Utility lines, 261, 310
Utility systems
    benefit of, 266
    drop-in ceiling, 280–281
    electrical components, 266–268
    electrical drawings, 266
    electrical plans, 268–275
    HVAC systems, 282–287
    plumbing plan, 288–293
    power plans, 275–277
    reflected ceiling plan, 278–280

V

Valley, 236
Valley roof rafters, 236

Valves, clean-out, 291
Valves, shutoff, 291, 293
Vanishing point, 61, 339
Vanity, 125
Vector, 42
Vector/raster, 366
Vegetation, 261
Vellum, 39, 40
Veneer walls, 226
Vents, 251
    forced-air HVAC, 283
Vent stack, 289
Views, 49–50
Vinyl substrate, 38
Virtual reality headset, 370
Visible object lines, 69
Visual communication, 2
VoIP (Voice over Internet Protocol), 266
Voltage, 272
Volutes, 297

W

Wall rail, stair, 299
Walls
    legends, 317, 318
    stack or plumbing, 289, 291
    stringers, 169
    thicknesses, 94–95
Wall sconce, 268
Wall sections, 220, 244
Wall stack, 283
Wall wash, 268
Waste stacks, 291
Wastewater, 288
    removal, 289, 290
Water closet, 288
Water heaters, 288, 291
Water pipes, 288, 289
Well, 289, 296
    stair, 296
Well hole, 304
Western framing, 234
Wet wall, 94
Wheelchair-accessible doors, 190
White-printing, 41
Wide door sizes
    patio door, 213
    transom, 213
Wide-format photocopiers, 41
Wide U stair, 304
Wide window sizes
    awning, 208
    casement, 207
    double/single hung, 210
Winder stairs, 303
Winder tread, 303

Windows
- awning, 196–197
- bay, 200, 212
- bow, 201–202, 212
- box bay, 201
- building construction, 248–250
- callout symbols, 316, 317
- casement, 196
- clerestory and skylights, 203
- definitions, 204–205
- double-hung, 194
- fixed, 195
- hopper, 197–198
- jalousie, 198
- perspective drawing, 345, 346
- pivot, 199
- schedule (chart), 324
- sizes, 206–214
- sliding, 196
- specialty, 202

Windowsill, 125

Window symbol, 193–194
- elevation, window placement in, 203–204
- window types, 194–203

Wood framing
- components, 228–231
- definitions, 231–233
- dimensions, 164–166
- types, 234–236

Workflow, 364
Work triangle, 116
Woven wire mesh (WWM), 218
Wythe, 226

Z

Zoning, 15